Sustainable Development Goals Series

The **Sustainable Development Goals Series** is Springer Nature's inaugural cross-imprint book series that addresses and supports the United Nations' seventeen Sustainable Development Goals. The series fosters comprehensive research focused on these global targets and endeavours to address some of society's greatest grand challenges. The SDGs are inherently multidisciplinary, and they bring people working across different fields together and working towards a common goal. In this spirit, the Sustainable Development Goals series is the first at Springer Nature to publish books under both the Springer and Palgrave Macmillan imprints, bringing the strengths of our imprints together.

The Sustainable Development Goals Series is organized into eighteen subseries: one subseries based around each of the seventeen respective Sustainable Development Goals, and an eighteenth subseries, "Connecting the Goals", which serves as a home for volumes addressing multiple goals or studying the SDGs as a whole. Each subseries is guided by an expert Subseries Advisor with years or decades of experience studying and addressing core components of their respective Goal.

The SDG Series has a remit as broad as the SDGs themselves, and contributions are welcome from scientists, academics, policymakers, and researchers working in fields related to any of the seventeen goals. If you are interested in contributing a monograph or curated volume to the series, please contact the Publishers: Zachary Romano [Springer; zachary.romano@springer.com] and Rachael Ballard [Palgrave Macmillan; rachael.ballard@palgrave.com].

Uday Chatterjee • Rajib Shaw
Suresh Kumar • Anu David Raj
Sandipan Das
Editors

Climate Crisis: Adaptive Approaches and Sustainability

Springer

Editors
Uday Chatterjee
Department of Geography
Bhatter College, Dantan (Affiliated to Vidyasagar University)
Paschim Medinipore, West Bengal, India

Rajib Shaw
Graduate School of Media and Governance
Keio University
Tokyo, Japan

Suresh Kumar
Agriculture, Forestry and Ecology Group
Indian Institute of Remote Sensing, Indian Space Research Organization (ISRO), Department of Space, Government of India
Dehradun, Uttarakhand, India

Anu David Raj
Agriculture and Soils Department
Indian Institute of Remote Sensing, Indian Space Research Organization (ISRO), Department of Space, Government of India
Dehradun, Uttarakhand, India

Sandipan Das
Symbiosis Institute of Geoinformatics
Symbiosis International (Deemed University)
Pune, Maharashtra, India

ISSN 2523-3084 ISSN 2523-3092 (electronic)
Sustainable Development Goals Series
ISBN 978-3-031-44396-1 ISBN 978-3-031-44397-8 (eBook)
https://doi.org/10.1007/978-3-031-44397-8

This Springer imprint is published by the registered company Springer Nature Switzerland AG
The registered company address is: Gewerbestrasse 11, 6330 Cham, Switzerland

Paper in this product is recyclable.

Dedicated to Young Scholars in the Fields of Atmospheric Science, Climatology, Environmental Science, Hazard and Disaster Management, Geography, Social Science and Sustainability Science as well as Policy Makers

Foreword

It is a great honour to write a foreword for the book entitled *Climate Crisis: Adaptive Approaches and Sustainability* edited by Uday Chatterjee, Rajib Shaw, Suresh Kumar, Anu David Raj, and Sandipan Das and published by Springer.

Climate change and global warming together are having a stupendous impact on the environment, which will ultimately lead to a climate crisis. According to the IPPC report, climate change is real and human beings are responsible for it. The effects of climate change are being witnessed by humanity in the form of global temperature rise and heat waves. Rising temperatures are fueling environmental degradation, natural disasters, weather extremes, food and water insecurity, economic disruption, conflict, and terrorism. Sea levels are rising, the Arctic is melting, coral reefs are dying, oceans are acidifying, and forests are burning. The long-term consequences of this on human activities like agriculture and related activities will be immense. So global warming needs to be limited to overcome this climate emergency and crisis and sustainable solutions should be utilized for this process. It is necessary to reduce our carbon footprint while also supporting vital ecosystem services such as promoting biodiversity, ensuring access to fresh water, improving livelihoods, encouraging healthy diets, and enhancing food security. For this, we can employ nature-based solutions that include improved agricultural practices, land restoration, conservation, and the greening of food supply chains. This volume is a timely publication given the idea of such practices in different parts of the world.

The present book gives a deep insight into the different aspects of climate crisis and its management strategies with experiences from different parts of the world. This edited volume has been scientifically organized into 29 chapters under 5 parts. The parts include Introduction (Part I), Climate Crisis: Geophysical Hazard and Risk Reduction and Mitigation (Part II), Climate Crisis and Smart Agriculture and Food Security (Part III), Climate Crisis and Urban Health (Part IV), and Climate Crisis and Land Water and Forest Sustainability (Part V).

I would like to appreciate and congratulate the dedication and collaborative efforts of all the editors and authors of the book. The book will be a valuable addition to this field and will provide insights and probable measures to overcome this climate crisis. Furthermore, I believe that the book will inspire new actions and advancements in this domain and it will be widely acclaimed by researchers, practitioners, managers, and decision makers, as this concern is going to affect every living being on this earth.

Professor and Ex Head
Department of Geography
Jadavpur University
Kolkata, West Bengal, India

Lakshmi Sivaramakrishnan

Preface

Over the last three decades, climate change has been an important area of observations at all levels, from policymakers to scholars and practitioners. Our changing climate has become a pressing concern of the Anthropocene. Extreme weather, water pollution, food and water shortages, economic instability, violence, and other related problems have affected billions of people throughout the world. Humanity has been cautioned in recent years about the long-term and immediate effects of changes in weather patterns, as anthropogenic factors are the primary cause of such alterations. Rising frequency of severe weather events throughout the world as well as the ramifications of these occurrences have significant impact on national economies and lives of the people. The United Nations (UN) reports that climate change, humanity's biggest tragedy, is occurring far more rapidly than anybody has predicted. Despite this worldwide menace, we are in no way unable to respond. "The climate concern is a race we are losing, but it is a race we can win," quoted by UN Secretary-General Antonio Guterres. Coastal communities throughout the globe are feeling the effects of global warming as ice caps melt and sea levels rise. Extreme weather is resulting in food, water, and environmental crises. Climate crisis repercussions and people's capacity to respond are unevenly distributed. Low-income people without adequate social safety nets and access to limited physical infrastructures are among the most vulnerable groups. Combating climate crisis challenges calls for collective action and adaptive approaches fostering sustainability at the grassroots.

Climate change exacerbates a wide range of problems, including environmental degradation, natural catastrophes, weather extremes, food and water shortages, economic instability, violence, and terrorism. At the onset of severe climatic crisis, mankind will be compelled to take decisive action on a global scale. The threat that global warming poses to our access to safe food and water affects each and every one of us. Carbon sequestration capacity is getting affected due to the degradation of the soil, which is triggered by climate change. Up to 30 percent of the world's food production potential is lost due to soil erosion. On the other hand, climate change reduces the amount and quality of water available for domestic, agricultural, and industrial purposes.

Greenhouse gases (GHGs) level in the atmosphere is rising continuously due to various anthropogenic activities. An overview of UNEP *Emission Gap Assessments* from the past decade indicates that we are on pace for business as usual. The Paris Agreement on climate change asks for keeping temperature "far below" 2 °C and initiate practices to keep it even lower, at 1.5 °C. Temperatures could rise by more than 3 °C by 2100 if no action is taken to curtail global emissions. Polar and mountainous regions are witnessing a rise in sea level faster than ever due to the melting of glaciers and glacial ice sheets. Heat waves, droughts, floods, typhoons, and hurricanes, among other severe weather phenomena, are causing devastation on every continent due to global warming. Millions of peoples are facing the severity of climate-related calamities resulting in huge economic loss in every continent. Climate change also poses a significant threat to international peace and security. Socio-economic tensions and mass displacement are becoming more common as a result of climate change's effects on resources like food and water. Due to the localized effects of climate disturbances, the primary responsibility for addressing and adjusting to climate change falls upon the local urban and rural authorities, as they are at the forefront of climate action and adaptation efforts. Across time, individuals have adapted and managed the impacts of climate fluctuations and extreme events with differing levels of achievement. At the grassroots level, communities have actively sought local remedies for their challenges. The involvement of community members at the grassroots level is crucial for lessening the unpredictability of climate-related issues. In addition, merging different knowledge systems offers a viable path ahead. Indigenous knowledge and practices can serve as a bridge between climate change and the resulting disasters. By identifying and amassing information regarding the workings and attributes of climate change, preparedness against climate-induced disasters can be enhanced. Nature itself is a tool for mitigating the climate crisis occurring all over the globe at a devastating scale. Hence, a bottom-up approach of identifying proper adaptation, mitigation, and risk reduction measures for all ecosystems (natural and artificial) is a necessity of the world for a sustainable and minimum risk future.

This book aims to contribute to the discourse on the climate crisis by bringing together high-quality research on adaptive approaches and sustainability case studies across the world. The book is divided into five parts. The introductory part discusses the framing of climate crisis and its different approaches. It also situates the book within the global discourse. The second part has five chapters engaging with the major geophysical hazards due to the climate crisis. The third part focusses on food security and methods to combat its antagonistic effect due to the climate change. These chapters address the global need to acquire food security through the novel technologies, traditional knowledge, and changes in cropping patterns. Most of the world population is now concentrated in cities and towns for better facilities and a smooth way of life. The pressure increases on urban area have led to diversified concerns such as water availability and air, water, and soil pollution. The fourth part discusses sustainable and climate-resilient urban cities. The chapters in this part cover topics such as sustainable usage of urban habitat and ecosystems, air quality monitoring, combating air pollution, studying urban heat island effect, and implementing nature-based solutions and mitigation measures to reduce its adverse impact. In addition, megacities face significant challenges stemming from the climate crisis, water scarcity, and the need to enhance groundwater recharge in urban areas. To address these issues in an environmentally sustainable manner, it is essential to implement well-designed drainage systems and promote the concept of urban forests. The earth/nature itself is a tool for combating the adverse effects of climate change. Hence, the fifth part focuses on land/soil, forest, and water ecosystems and their sustainable management. The importance of forests in the maintenance of ecological balance is unavoidable. Thus, these chapters discuss the application of mitigation techniques that might strengthen resilience to climate crisis phenomena.

The book explores the recent advances in various tools and techniques for improving the quality of the research theme. Our primary concern is providing readers with impactful research on the current theme of climate change. The critical areas of the proposed book focuses on the emergence of efficient and sustainable adaption and mitigation measures that can reduce or alleviate the adverse consequences that emerge from changing climate. Hence, we have oriented this book as a guide that can begin to bridge the gap between new climate science and identifying the major extreme events due to climate change (global warming). Next, we delve into each event, starting with a detailed examination of the impact of climate change on food security and human habitats along with effective measures to address these threats. We then explore a range of ecosystem-based sustainability approaches aimed at mitigating climate change. This resource will serve as a comprehensive reference for individuals working in the climate sector, providing systematic guidance and information that can support their endeavours. It will also provide an opportunity to link the various ecosystems for adaptation and mitigation measures through a multidimensional integrated approach against climate change.

The book offers adaptation, mitigation, and disaster risk reduction measures that can help to address the challenges of climate-induced disasters and food security in the face of exponential population growth and climate crisis. It is a timely synthesis of the rapidly expanding field of climate science by presenting fresh and thought-provoking ideas that will shape a unified and productive vision for future research. It is intended to be a valuable resource for undergraduate and postgraduate students, Ph.D. scholars, and researchers in the fields of environmental sciences, humanities, social sciences, and geography. Additionally, the book will be highly valuable for policy and decision makers, environmentalists, NGOs, corporate sectors, social scientists, and government organizations. Also, we strongly deemed that it will provide some support for various levels of organizations and administrations for developing and achieving the UN Sustainable Development Goals (SDGs) by 2030 in view of the climate change. These measures provide valuable insights to alleviate these concerns. Hence, it will deliver high-quality research content to the prescribed audience. It advocates for significant shifts in the way we produce food, use land, transport commodities, and power our economies at all levels of society. With the help of SDG 2030 and nature-based solutions, we can make rapid progress toward a cleaner and more resilient future. We are confident that this collective effort will contribute to the creation of a bright and sustainable future.

West Bengal, India Uday Chatterjee

Tokyo, Japan Rajib Shaw

Dehradun, India Suresh Kumar

Dehradun, India Anu David Raj

Pune, India Sandipan Das

Acknowledgments

We express our deepest appreciation to the contributors, colleagues, publishers, researchers, and readers who were instrumental in the creation and publication of *Climate Crisis: Adaptive Approaches and Sustainability*. This exhaustive resource addressing the challenges of climate change was shaped by their expertise, support, guidance, and dedication. We appreciate the opportunity to build upon pioneering research, share applicable examples, and inspire positive change. The unwavering support of our families and the readers' dedication to sustainability are deeply appreciated. Through this book, we can make a difference in addressing the climate crisis and create a more sustainable future. We are also appreciative of the publishing team at Springer for their guidance, expertise, and unwavering dedication to excellence. Their diligent editing, design, and production efforts have transformed the book into a thorough and accessible reader resource.

West Bengal, India — Uday Chatterjee

Tokyo, Japan — Rajib Shaw

Dehradun, India — Suresh Kumar

Dehradun, India — Anu David Raj

Pune, India — Sandipan Das

Disclaimer

The authors of individual chapters are solely responsible for the ideas, views, data, figures, and geographical boundaries presented in the respective chapters of this book and these have not been endorsed in any form, by the publisher, the editor, and the authors of foreword, preambles, or other chapters.

Contents

Part I Introduction

1 **Global Warming and Climate Crisis/Extreme Events** 3
Suresh Kumar, Uday Chatterjee, Anu David Raj, and K. R. Sooryamol

2 **Ecosystem Degradation to Restoration: A Challenge** 19
Shaista Khan, T. H. Masoodi, M. A. Islam, Tahera Arjumand, Azeem Raja, Aafaq Ahmad Parrey, Anushka Pallavi, and Javaid H. Bhat

3 **Exploring the Dynamics of Antarctic Sea Ice over Four Decades Using Geospatial Technology** 35
Niladri Saha, Babula Jena, C. C. Bajish, Sandipan Das, Binaya Kumar Pattnaik, and Uday Chatterjee

4 **Climate Change and Himalayan Glaciers: A Socio-Environmental Concern in Anthropocene Epoch** 53
Aju David Raj, Anu David Raj, and K. R. Sooryamol

5 **Indigenous Strategies and Adaptive Approaches to Scrabble Recent Climate Crisis in Two Districts (Bankura and Purulia) of West Bengal, India** 75
Mainak Sarkar, Partha Gorai, and Biplob Kumar Modak

Part II Climate Crisis: Geophysical Hazard and Risk Reduction and Mitigation

6 **Addressing Climate Crisis Through Coastal Risk Management: The Social Protection Alternative** 105
Sayanti Sengupta

7 **Land Degradation and its Relation to Climate Change and Sustainability** 121
Anu David Raj, Suresh Kumar, Justin George Kalambukattu, and Uday Chatterjee

8 **Social Resilience of Local Communities Due to Tidal Flooding on the North Coast of Semarang City, Indonesia** ... 137
Hari Harjanto Setiawan

9 **Effects of Climatic Risks on Soil Erosion/Desertification in Southern and Northern Nigeria Using GIS/Remote Sensing Analysis** ... 151
Angela Oyilieze Akanwa, Idakwo V. Iko-ojo, I. C. Ezeomedo, F. I. Ikegbunam, P. U. Igwe, L. N. Muoghalu, S. O. Okeke, A. U. Okonkwo, Chinwe Ngozi Odimegwu, K. F. Nkwocha, V. C. Arah, E. I. Madukasi, C. Anukwonke, Joel Mari Bwala, and M. Obidiegwu

10 **Strategies for Compound Urban and Climate Hazards: Linking Climate Adaptation and Sustainability to Address Risk in Environmental Justice Communities** ... 171
Dalia Munenzon and Maria Noguera

Part III Climate Crisis and Smart Agriculture and Food Security

11 **The Role of Indigenous Climate Forecasting Systems in Building Farmers' Resilience in Nkayi District, Zimbabwe** ... 195
Joram Ndlovu, Mduduzi Ndlovu, and Douglas Nyathi

12 **Agroforestry Practices: A Sustainable Way to Combat the Climate Crisis and Increase Productivity** ... 211
Sushil Kumar, Badre Alam, Sukumar Taria, Priyanka Singh, Ashok Yadav, R. P. Dwivedi, and A. Arunachalam

13 **Climate Crisis and Adoption of Climate-Smart Agriculture Technologies** ... 229
Trisha Roy, Justin George Kalambukattu, Abhijit Sarkar, I. Rashmi, Rama Pal, Vibha Singhal, Deepak Singh, and Suresh Kumar

14 **Farming Technologies and Carbon Sequestration Alternatives to Combat Climate Change Through Mitigation of Greenhouse Gas Emissions** ... 253
Knight Nthebere, M. R. Apoorva, Mandapelli Sharath Chandra, M. Bhargava Narasimha Yadav, and T. Ram Prakash

15 **Nature-based Solutions (NbS) for Dryland Agriculture in Semi-Arid Regions of Maharashtra, India: A Short Review with Possible Approaches for Building Climate Resilience** ... 277
Wasim Ayub Bagwan

16 **Smart Farming and Carbon Sequestration to Combat the Climate Crisis** ... 293
K. R. Sooryamol, Suresh Kumar, Anu David Raj, and M. Sankar

17 Alleviation of Climate Catastrophe in Agriculture Through Adoption of Climate-Smart Technologies 307
D. T. Santosh, Subhankar Debnath, Sagar Maitra, Masina Sairam, La Lichetti Sagar, Akbar Hossain, and Debojyoti Moulick

18 Climate Crisis and Adoption of Climate-smart Agriculture Technologies and Models 333
Khadijeh Bazrafkan, Ali Karami, Naser Valizadeh, Samira Esfandyari Bayat, Hajar Zareie, and Dariush Hayati

Part IV Climate Crisis and Urban Health

19 Mainstreaming Biodiversity in Urban Habitats for Enhancing Ecosystem Services: A Conceptual Framework 349
Jayeeta Sen and Meenakshi Dhote

20 Climate-Resilient Agropolitan Approach Towards Sustainable Regional Development of Barddhaman District of West Bengal 369
Tanmoy Basu and Biraj Kanti Mondal

21 Analysing Sustainable Approaches in MGNREGA Works for Climate Change Adaptation: Case Study of Debra Block, West Bengal, India 405
Piu Dutta, Debashish Das, and Debasree Bhadra

22 Urban Heat: UHI and Heat Stress Threat to Megacities 425
Jagabandhu Panda, Asmita Mukherjee, Animesh Choudhury, and Sreyasi Biswas

23 Assessment of LULC Changes and Its Impact on Surface Temperature and Urban Heat Island Conditions in Kolkata During SARS COVID-19 Period 447
Sohini Sen, Debashish Das, Pankaj Chakroborty, and Raja Ghosh

24 Addressing Climate Change Challenges in South Africa: A Study in KwaZulu Natal Province 475
Ifeanyi Michael Smarte Anekwe, Helper Zhou, Mphathesithe Mzwandile Mkhize, and Stephen Okiemute Akpasi

Part V Climate Crisis and Land Water and Forest Sustainability

25 Evaluating the Potential Impact of Climate Change on Glacier Dynamics in Western Himalayas, India 499
Suhail A. Lone and Gh. Jeelani

26 A Tale of Crab Collectors and Fatteners: Negotiating Climate Change in Indian Sundarbans 511
Anindya Basu and Adrija Bhattacharjee

27 Climate Crisis and Wetland Ecosystem Sustainability 529
Suchetana Karmakar, Sk Saruk Islam, Krishnendu Sen, Sayani Ghosh, and Sujoy Midya

28 Land Suitability Assessment for Mulberry-Based Agroforestry Using AHP and GIS Technique in the Northwestern Himalayan Region of Kashmir Valley, India to Achieve Sustainable Agriculture 551
Ruyida Mushtaq, Rajesh Kumar Yadav, Abida Fayaz, Pervez Ahmed, Harmeet Singh, and Jaipreet Singh

29 Climate Crisis and Coastal Risk Management 571
N. P. P. S. Nugawela, A. S. Mahaliyana, and G. Abhiram

About the Editors

Uday Chatterjee is an Assistant Professor at the Department of Geography, Bhatter College, Dantan, Paschim Medinipur, West Bengal, India, and an Applied Geographer with a Doctoral Degree in Applied Geography at Ravenshaw University, Cuttack, Odisha, India. His areas of research interest cover urban planning, social and human geography, applied geomorphology, hazards and disasters, environmental issues, disaster governance, community-based disaster risk management, climate change adaptation, urban risk management, and disaster. He has delivered 10 invited lectures in University Grants Commissions (UGC)-sponsored national seminars and various academic departments of different colleges in India. In addition, he presented 20 papers in national and international seminars/conferences held in India as well as chaired and co-chaired more than 5 technical sessions. He has successfully guided project dissertations to undergraduate students. He has also conducted (Convener) one Faculty Development Programme on "Modern methods of teaching and advanced research methods" sponsored by the Indian Council of Social Science Research (ICSSR), Govt. of India. Dr. Uday Chatterjee has served as the lead editor of Special Issue (S.I) on Urbanism, Smart Cities and Modelling, *Geojournal*, Springer. He is also a book series editor of Developments in Environmental Science, Elsevier (https://www.elsevier.com/books-and-journals/book-series/developments-in-environmental-science). His research work has been funded by the West Bengal Pollution Control Board (WBPCB), Govt. of West Bengal, India. He has served as a reviewer for many international journals. Currently, Dr. Chatterjee is doing an international project, in collaboration with Indonesia, Malaysia, and Japan, funded by APN (Asia Pacific-Global Change Research), and he is the guest editor of the special issue Social Ecology, Human-Well-being and Sustainability, *Global Social Welfare* journal, Springer. He has published 30 research papers, 10 edited books, 2 authored books, 15 book chapters, and 2 conference proceedings.

Rajib Shaw is a Professor at the Graduate School of Media and Governance in Keio University, Japan. He is also the Senior Fellow of Institute of Global Environmental Strategies (IGES) Japan, and the Chairperson of SEEDS Asia and CWS Japan, two Japanese NGOs. Earlier, he was the Executive Director of the Integrated Research on Disaster Risk (IRDR) and was a Professor at Kyoto University. His expertise includes disaster governance, community-based disaster risk management, climate change adaptation, urban risk management, and disaster and environmental education. Professor Shaw is the Chair of the United Nations Science Technology Advisory Group (STAG) for disaster risk reduction and also the Co-chair of the Asia Science Technology Academic Advisory Group (ASTAAG). He is also the CLA (Coordinating Lead Author) for Asia chapter of IPCC's 6th Assessment Report. He is the editor-in-chief of the Elsevier's journal *Progress in Disaster Science*, and series editor of a Springer book series Disaster Risk Reduction. Prof. Shaw has published more than 45 books and over 300 academic papers and book chapters.

Suresh Kumar is a Scientist–G & Group Head, Agriculture, Forestry and Ecology Group at Indian Institute of Remote Sensing (IIRS), Indian Space Research Organization (ISRO), Govt. of India. He has 30 years of vast experience in applications of Geospatial Technologies in Natural Resource Management with specialization in soil resource management, land degradation, and watershed management. He is graduated in Agriculture and has Doctorate in Soil Science from G. B. Pant University of Agriculture and Technology (G.B. P.U. & T., Pantnagar) in 1993, and since then serving at the Indian Institute of Remote Sensing (IIRS) at various Scientist positions. He has done commendable research and has published 22 research papers in international journals, 30 research papers in national journals, and 7 book chapters with CRC/Springer Nature press. He has attended several international conferences and national seminar/symposium and delivered invited lead talk in the seminar/symposium. As faculty member, he is involved in training and education of P.G. diploma and M.Tech. courses in the department. He has supervised 20 M.Sc./M.Tech. students. He has carried out several research projects as PI/Co-I of the projects such as FAO-AEZ-based Agricultural Land Use Planning; National Soil Carbon Pool Assessment, Soil Carbon Dynamic (SCD) Studies, Mountain Ecosystem Processes and Services: Studying impact of Soil Erosion and Nutrient Loss and its impact on Soil Quality; Digital soil mapping using environmental covariates for mountainous region, etc. He was involved in operational projects of National Land Degradation Mapping, National Wasteland Mapping, and Integrated Mission for Sustainable Development (IMSD) projects. He is a Life Member of various professional societies such as Indian Society of Remote Sensing, Dehradun; Indian Society of Soil Survey and Land Use Planning, NBSS&LUP (National Bureau of Soil Survey and Land Use Planning), Nagpur; Indian Association of Soil and Water Conservation, ICAR-IISWC, Dehradun; Farming Systems Research and development Association, Modipuram, PDCSR, Modipuram, Meerut; Association of Agrometeorologist, GAU, Anand, India.

Anu David Raj is currently a Junior Research Fellow at Agriculture and Soils Department, Indian Institute of Remote Sensing, Indian Space Research Organization. He obtained his postgraduate (Integrated) degree in Climate Change Adaptation with a specialization in Agriculture from Kerala Agricultural University. Recently he has published research articles on climate change impact on soil erosion. His areas of research interest cover climate modelling, soil erosion modelling, climate change adaptation, agriculture, and sustainability. He is also pursuing his Ph.D. in forest influence and climate change discipline at Forest Research Institute, Dehradun, India, on "Climate change impact on soil erosion using ^{137}Cs radioisotope tracer technique."

Sandipan Das is an Assistant Professor at Symbiosis Institute of Geoinformatics, Symbiosis International (Deemed University), Pune, India. Dr. Das pursued his Master's degree in Geoinformatics from Pune University and Ph.D. in Geoinformatics from Symbiosis International (Deemed University). During his Ph.D., he was qualified for Geography NET conducted by UGC (June 2013). He has 8 years of teaching and 10 years of research experience. His areas of research interest include forest biomass and carbon assessment, water resources management, groundwater mapping, drought monitoring, and Geospatial Modelling of Environments. He has worked as a Principal Investigator, Co-Principal Investigator, and team member on several research projects funded by the Indian Space Research Organisation (ISRO), Department of Science and Technology (DST), and Ministry of Environment and Forest (MoEF). He has published 20 research papers, 1 book, 7 book chapters, and 4 conference proceedings. He has organized four capacity building geospatial training programs sponsored by DST and ISRO. He is a reviewer for the several scientific journals of international reputes. He is currently supervising two Ph.D. research scholars.

Contributors

G. Abhiram Faculty of Animal Science and Export Agriculture, Department of Animal Science, Uva Wellassa University of Sri Lanka, Badulla, Sri Lanka

Pervez Ahmed Department of Geography, University of Kashmir, Srinagar, India

Angela Oyilieze Akanwa Climate Change Impacts, Sustainability and Adaptation, Department of Environmental Management, Chukwuemeka Odumegwu Ojukwu University (COOU), Ihiala, Anambra State, Nigeria

Stephen Okiemute Akpasi Durban University of Technology, Durban, South Africa

Badre Alam ICAR-Central Agroforestry Research Institute, Jhansi, Uttar Pradesh, India

Ifeanyi Michael Smarte Anekwe University of the Witwatersrand, Johannesburg, South Africa

C. Anukwonke Climate Change and EIA Studies, Department of Environmental Management, Chukwuemeka Odumegwu Ojukwu University (COOU), Ihiala, Anambra State, Nigeria

M. R. Apoorva Department of Soil Science and Agricultural Chemistry, Jayashankar Telangana Agricultural University, Hyderabad, India

V. C. Arah Land Management Studies, Department of Environmental Management, Chukwuemeka Odumegwu Ojukwu University (COOU), Ihiala, Anambra State, Nigeria

Tahera Arjumand Division of NRM, Faculty of Forestry, Sher-e-Kashmir University of Agricultural Sciences and Technology of Kashmir, Ganderbal, Jammu & Kashmir, India

A. Arunachalam ICAR-Central Agroforestry Research Institute, Jhansi, Uttar Pradesh, India

Wasim Ayub Bagwan School of Rural Development, Tata Institute of Social Sciences, Tuljapur, Maharashtra, India

C. C. Bajish National Center for Polar and Ocean Research, Vasco Da Gama, Goa, India

Anindya Basu Department of Geography, Diamond Harbour Women's University, Sarisha, West Bengal, India

Tanmoy Basu Faculty (SACT-1), Department of Geography, Katwa College, Purba Bardhaman, West Bengal, India

Samira Esfandyari Bayat Department of Agricultural Extension and Education, College of Agriculture, Tarbiat Modares University (TMU), Tehran, Iran

Khadijeh Bazrafkan Department of Agricultural Extension and Education, School of Agriculture, Shiraz University, Shiraz, Iran

Debasree Bhadra Departments of Geography, Lady Brabourne College, Kolkata, West Bengal, India

Javaid H. Bhat Central Library, Sher-e-Kashmir University of Agricultural Sciences and Technology of Kashmir, Srinagar, Jammu & Kashmir, India

Adrija Bhattacharjee Department of Geography, Diamond Harbour Women's University, Sarisha, West Bengal, India

Sreyasi Biswas Department of Earth and Atmospheric Sciences, National Institute of Technology Rourkela, Rourkela, Odisha, India

Joel Mari Bwala Land Management Studies, Department of Geography, Faculty of Social Sciences, University of Maiduguri, Maiduguri, Borno State, Nigeria

Pankaj Chakroborty Department of Geoinformatics, NK College (Mumbai University), Mumbai, Maharashtra, India

Mandapelli Sharath Chandra AICRP on Integrated Farming System, Jayashankar Telangana Agricultural University, Hyderabad, Telangana, India

Uday Chatterjee Department of Geography, Bhatter College, Dantan (Affiliated to Vidyasagar University), Paschim Medinipur, West Bengal, India

Animesh Choudhury Department of Earth and Atmospheric Sciences, National Institute of Technology Rourkela, Rourkela, Odisha, India

Debashish Das Department of Architecture, Jadavpur University, Kolkata, West Bengal, India

Sandipan Das Symbiosis Institute of Geoinformatics, Symbiosis International (Deemed University), Pune, Maharashtra, India

Aju David Raj Department of Environmental Science, School of Earth Sciences, Central University of Rajasthan, Rajasthan, India

Anu David Raj Agriculture and Soils Department, Indian Institute of Remote Sensing, Indian Space Research Organization (ISRO), Dehradun, Uttarakhand, India
Forest Research Institute, Dehradun, Uttarakhand, India

Subhankar Debnath School of Agriculture and Bioengineering, Centurion University of Technology and Management, Paralakhemundi, Odisha, India

Meenakshi Dhote Department of Environmental Planning, School of Planning and Architecture, New Delhi, India

Piu Dutta Department of Architecture, Jadavpur University, Kolkata, West Bengal, India

R. P. Dwivedi ICAR-Central Agroforestry Research Institute, Jhansi, Uttar Pradesh, India

I. C. Ezeomedo Climate Change Impacts, Sustainability and Adaptation, Department of Environmental Management, Chukwuemeka Odumegwu Ojukwu University (COOU), Ihiala, Anambra State, Nigeria

Abida Fayaz Department of Geography and Disaster Management, University of Kashmir, Srinagar, India

Raja Ghosh Synprox Solutions, Kolkata, West Bengal, India

Sayani Ghosh Department of Sociology, The University of Burdwan, Burdwan, West Bengal, India

Partha Gorai Department of Zoology, Sidho-Kanho-Birsha University, Purulia, West Bengal, India

Dariush Hayati Department of Agricultural Extension and Education, School of Agriculture, Shiraz University, Shiraz, Iran

Akbar Hossain Division of Soil Science, Bangladesh Wheat and Maize Research Institute, Dinajpur, Bangladesh

P. U. Igwe Ecological Studies, Department of Environmental Management, Chukwuemeka Odumegwu Ojukwu University (COOU), Ihiala, Anambra State, Nigeria

F. I. Ikegbunam Ecological Disaster and Management, Department of Environmental Management, Chukwuemeka Odumegwu Ojukwu University (COOU), Ihiala, Anambra State, Nigeria

Idakwo V. Iko-ojo Climate Change, Department of Urban and Regional Planning, Faculty Environmental Sciences, University of Maiduguri, Maiduguri, Borno State, Nigeria

M. A. Islam Division of NRM, Faculty of Forestry, Sher-e-Kashmir University of Agricultural Sciences and Technology of Kashmir, Ganderbal, Jammu & Kashmir, India

Sk Saruk Islam Department of Zoology, Raja Narendra Lal Khan Women's College, Midnapore, West Bengal, India

Gh. Jeelani Department of Earth Sciences, University of Kashmir, Srinagar, India

Babula Jena National Center for Polar and Ocean Research, Vasco Da Gama, Goa, India

Justin George Kalambukattu Agriculture and Soils Department, Indian Institute of Remote Sensing, Indian Space Research Organization (ISRO), Dehradun, Uttarakhand, India

Ali Karami Department of Soil Science, School of Agriculture, Shiraz University, Shiraz, Iran

Suchetana Karmakar Department of Zoology, Raja Narendra Lal Khan Women's College, Midnapore, West Bengal, India

Shaista Khan Division of NRM, Faculty of Forestry, Sher-e-Kashmir University of Agricultural Sciences and Technology of Kashmir, Ganderbal, Jammu & Kashmir, India

Suresh Kumar Agriculture, Forestry and Ecology Group, Indian Institute of Remote Sensing, Indian Space Research Organization (ISRO), Dehradun, Uttarakhand, India

Sushil Kumar ICAR-Central Agroforestry Research Institute, Jhansi, Uttar Pradesh, India

Suhail A. Lone Department of Earth Sciences, University of Kashmir, Srinagar, India

E. I. Madukasi Pollutants Studies, Department of Environmental Management, Chukwuemeka Odumegwu Ojukwu University (COOU), Ihiala, Anambra State, Nigeria

A. S. Mahaliyana Faculty of Animal Science and Export Agriculture, Department of Animal Science, Uva Wellassa University of Sri Lanka, Badulla, Sri Lanka

Sagar Maitra M.S. Swaminathan School of Agriculture, Centurion University of Technology and Management, Paralakhemundi, Odisha, India

T. H. Masoodi Division of NRM, Faculty of Forestry, Sher-e-Kashmir University of Agricultural Sciences and Technology of Kashmir, Ganderbal, Jammu & Kashmir, India

Sujoy Midya Department of Zoology, Raja Narendra Lal Khan Women's College, Midnapore, West Bengal, India

Mphathesithe Mzwandile Mkhize Mangosuthu University of Technology, Umlazi, South Africa

Biplob Kumar Modak Department of Zoology, Sidho-Kanho-Birsha University, Purulia, West Bengal, India

Biraj Kanti Mondal Department of Geography, Netaji Subhas Open University, Kolkata, West Bengal, India

Debojyoti Moulick Department of Environmental Science, University of Kalyani, Nadia, West Bengal, India

Asmita Mukherjee Department of Earth and Atmospheric Sciences, National Institute of Technology Rourkela, Rourkela, Odisha, India

Dalia Munenzon Hines College of Architecture and Design, University of Houston, Houston, TX, USA

L. N. Muoghalu Policy and Administration, Department of Environmental Management, Chukwuemeka Odumegwu Ojukwu University (COOU), Ihiala, Anambra State, Nigeria

Ruyida Mushtaq Department of Geography and Disaster Management, University of Kashmir, Srinagar, India

Joram Ndlovu School of Social Sciences, Howard College, University of KwaZulu-Natal, Durban, South Africa

Mduduzi Ndlovu Faculty of Humanities and Social Sciences, Lupane State University, Bulawayo, Zimbabwe

K. F. Nkwocha Climate Change and Sustainability, Department of Geography, Faculty of Social Sciences, University of Maiduguri, Maiduguri, Borno State, Nigeria

Maria Noguera Hines College of Architecture and Design, University of Houston, Houston, TX, USA

Knight Nthebere Department of Soil Science and Agricultural Chemistry, Jayashankar Telangana Agricultural University, Hyderabad, India

N. P. P. S. Nugawela Faculty of Animal Science and Export Agriculture, Department of Animal Science, Uva Wellassa University of Sri Lanka, Badulla, Sri Lanka

Douglas Nyathi School of Social Sciences, Howard College, University of KwaZulu-Natal, Durban, South Africa

M. Obidiegwu Climate Change and Sustainability, Department of Environmental Management, Chukwuemeka Odumegwu Ojukwu University (COOU), Ihiala, Anambra State, Nigeria

Chinwe Ngozi Odimegwu Land and Property Policy, Taxation and Administration, Department of Estate Management, Faculty Environmental Sciences, Chukwuemeka Odumegwu Ojukwu University (COOU), Ihiala, Anambra State, Nigeria

S. O. Okeke Landscape Design and Conservation, Department of Environmental Management, Chukwuemeka Odumegwu Ojukwu University (COOU), Ihiala, Anambra State, Nigeria

A. U. Okonkwo Waste and Land Management, Department of Environmental Management, Chukwuemeka Odumegwu Ojukwu, University, (COOU), Ihiala, Anambra State, Nigeria

Rama Pal ICAR-Indian Institute of Soil and Water Conservation, Dehradun, Uttarakhand, India

Anushka Pallavi School of Agriculture (SoA), Himalaya University, Dehradun, Uttarakhand, India

Jagabandhu Panda Department of Earth and Atmospheric Sciences, National Institute of Technology Rourkela, Rourkela, Odisha, India

Aafaq Ahmad Parrey Division of NRM, Faculty of Forestry, Sher-e-Kashmir University of Agricultural Sciences and Technology of Kashmir, Ganderbal, Jammu & Kashmir, India

Binaya Kumar Pattnaik Symbiosis Institute of Geoinformatics, Symbiosis International (Deemed University), Pune, Maharashtra, India

T. Ram Prakash AICRP on Weed Management, Jayashankar Telangana Agricultural University, Hyderabad, Telangana, India

Azeem Raja Division of NRM, Faculty of Forestry, Sher-e-Kashmir University of Agricultural Sciences and Technology of Kashmir, Ganderbal, Jammu & Kashmir, India

I. Rashmi ICAR-Indian Institute of Soil and Water Conservation, Research Centre, Kota, Rajasthan, India

Trisha Roy ICAR-Indian Institute of Soil and Water Conservation, Dehradun, Uttarakhand, India

La Lichetti Sagar M.S. Swaminathan School of Agriculture, Centurion University of Technology and Management, Paralakhemundi, Odisha, India

Niladri Saha Symbiosis Institute of Geoinformatics, Symbiosis International (Deemed University), Pune, Maharashtra, India

Masina Sairam M.S. Swaminathan School of Agriculture, Centurion University of Technology and Management, Paralakhemundi, Odisha, India

M. Sankar Indian Institute of Soil & Water Conservation (ICAR-IISWC), Dehradun, Uttarakhand, India

D. T. Santosh School of Agriculture and Bioengineering, Centurion University of Technology and Management, Paralakhemundi, Odisha, India

Abhijit Sarkar ICAR-Indian Institute of Soil Science, Bhopal, Madhya Pradesh, India

Mainak Sarkar Department of Zoology, Bankura Christian College, Bankura, West Bengal, India
Department of Zoology, Sidho-Kanho-Birsha University, Purulia, West Bengal, India

Jayeeta Sen Department of Environmental Planning, School of Planning and Architecture, New Delhi, India

Krishnendu Sen Department of Plant Pathology, Bidhan Chandra Krishi Viswavidyalaya, Mohanpur, West Bengal, India
Department of Microbiology, Vidyasagar University, Midnapore, West Bengal, India

Sohini Sen Department of Architecture, Jadavpur University, Kolkata, West Bengal, India

Sayanti Sengupta Social Protection and Climate, Red Cross Red Crescent Climate Centre, The Hague, The Netherlands

Hari Harjanto Setiawan National Research and Innovation Agency (BRIN), Jakarta, Indonesia

Deepak Singh ICAR-Indian Institute of Soil and Water Conservation, Dehradun, Uttarakhand, India

Harmeet Singh Department of Geography, University of Kashmir, Srinagar, India

Jaipreet Singh Department of Geography, Cluster University of Kashmir, Srinagar, Jammu and Kashmir, India

Priyanka Singh ICAR-Central Agroforestry Research Institute, Jhansi, Uttar Pradesh, India

Vibha Singhal ICAR-Indian Institute of Soil and Water Conservation, Dehradun, Uttarakhand, India

K. R. Sooryamol Indian Institute of Soil & Water Conservation (ICAR-IISWC), Dehradun, Uttarakhand, India

Sukumar Taria ICAR-Central Agroforestry Research Institute, Jhansi, Uttar Pradesh, India

Naser Valizadeh Department of Agricultural Extension and Education, School of Agriculture, Shiraz University, Shiraz, Iran

Ashok Yadav ICAR-Central Agroforestry Research Institute, Jhansi, Uttar Pradesh, India

M. Bhargava Narasimha Yadav Department of Soil Science and Agricultural Chemistry, University of Agricultural Sciences, Dharwad, Karnataka, India

Rajesh Kumar Yadav Mohanlal Sukhadia University, Udaipur, Rajasthan, India

Hajar Zareie Department of Agricultural Extension and Education, College of Agriculture, University of Tehran, Tehran, Iran

Helper Zhou University of KwaZulu Natal, Durban, South Africa

Part I

Introduction

1 Global Warming and Climate Crisis/Extreme Events

Suresh Kumar, Uday Chatterjee, Anu David Raj, and K. R. Sooryamol

Abstract

Global warming has emerged as a major global threat after industrialization and has become more prominent in recent years. The increase in temperature is a result of enhanced greenhouse gas emissions, which are generated from natural and anthropogenic sources. This can trigger a vigorous hydrological cycle, enhancing extreme weather events and causing severe losses to human beings and the environment. Therefore, this chapter describes the evidence, causes, and consequences of climate change while also seeking suitable adaptation and mitigation strategies from a sustainable socio-economic and environmental perspective. Soil has been identified as one of the major realms that can ameliorate the adverse effects of climate change and help fulfill global food security. Site-specific adaptation and mitigation measures at the watershed level would be more suitable for achieving climate resilience with the aid of the global SDG framework.

Keywords

Greenhouse gases · Global warming · Climate change adaptation · Extreme weather events · Food security · Sustainability

S. Kumar (✉)
Agriculture, Forestry and Ecology Group, Indian Institute of Remote Sensing, Indian Space Research Organization (ISRO), Dehradun, Uttarakhand, India

U. Chatterjee
Department of Geography, Bhatter College, Dantan (Affiliated to Vidyasagar University), Paschim Medinipore, West Bengal, India

A. David Raj
Agriculture and Soils Department, Indian Institute of Remote Sensing, Indian Space Research Organization (ISRO), Dehradun, Uttarakhand, India

Forest Research Institute, Dehradun, Uttarakhand, India

K. R. Sooryamol
Indian Institute of Soil & Water Conservation (ICAR-IISWC), Dehradun, Uttarakhand, India

1.1 Introduction to Global Warming, Climate Change, and Extreme Events

In the face of continuing climate change, poverty, inequality, and environmental degradation, understanding the connections between climate and international development is a matter of urgency, emphasized WMO Secretary-General Prof. Petteri Taalas. While the terms "climate change" and "global warming" are often used interchangeably, they have distinct meanings. There is a significant difference between the two. Generally, global warming refers to the long-term heating of the Earth's atmosphere that has been observed since the pre-industrial era,

U. Chatterjee et al. (eds.), *Climate Crisis: Adaptive Approaches and Sustainability*, Sustainable Development Goals Series, https://doi.org/10.1007/978-3-031-44397-8_1

primarily due to anthropogenic activities such as the burning of fossil fuels. This has led to an increase in heat-trapping greenhouse gases in the atmosphere. As a result, the Earth's local, regional, and global climates have been undergoing long-term changes, which correspond to climate change.

These temperature changes in the Earth's climate system also affect the hydrological cycle, causing it to become more vigorous. It has been reported that atmospheric moisture levels have increased by 5–20% since the 1970s (IPCC 2021). A warmer atmosphere can absorb more moisture from the surface, potentially leading to more frequent extreme weather events due to the intensified hydrological conditions. This raises concerns about the possibility of severe, pervasive, and irreversible impacts resulting from global warming. There are numerous observed consequences of these variations that align with the aforementioned statement (NASA n.d.-a).

The average global temperature has risen by about 1 °C since the pre-industrial period, and it is currently increasing at a rate of more than 0.2 °C per decade (IPCC 2018). Human activities since the 1950s have undoubtedly contributed to this unprecedented warming trend, which has been occurring at a faster rate than in previous centuries. One of the most noticeable impacts of global warming is the increased occurrence of extreme weather events. Droughts, heatwaves, heavy rainfall, hurricanes, tornadoes, and wildfires are among the most common extreme weather events, and they have been observed to be more frequent and intense in recent years (Eckstein et al. 2021). Additionally, flooding and drought, as consequences of climate change, have resulted in millions of people being displaced, leading to increased poverty, hunger, limited access to essential services, widened inequality, and hindered economic development. Moreover, these extreme weather events have the potential to escalate civil conflicts (UN-SDG n.d.). Throughout recent year, various extreme weather disasters have occurred worldwide, ranging from recent floods in Libya and Pakistan to wildfires in Europe.

Increasing numbers of billion-dollar disasters serve as an indicator of the impact of extreme weather on the global economy. According to an Ipsos survey conducted for the World Economic Forum, it is expected that up to 35% of individuals will be forced to relocate due to climate change within the next 25 years (WEF 2022). Extreme weather events are projected to become more frequent, intense, and impactful because of climate change. Experts emphasize the urgent need to rapidly reduce pollution to prevent a climate catastrophe, as time is running out. A total of 504 extreme weather patterns and events were assessed, and, in 71% of the cases, they were found to be more susceptible or more severe due to human-induced climate change. Climate change was found to decrease the likelihood or severity of 9% of these events or trends, indicating that 81% of all events had some anthropogenic influence. The remaining 20% of occurrences and trends were either uncertain or showed no signs of human influence. When examining data from the past 20 years, studies on excessive heat (30%), rainfall or flooding (25%), and drought (16%) dominate the literature. Together, these account for two-thirds of all published research (CarbonBrief n.d.). This highlights the frequency of extreme events and the need for the development of adaptation and mitigation measures to alleviate the adverse impacts of climate change.

1.2 Climate Change: Evidences, Causes, and Consequences

During the 1970s, systematic scientific assessments of the impact of human activity on climate change began. It is now widely recognized that humans significantly influence the climate system. There is abundant evidence of climate change derived from various sources, including ice cores, rock samples, tree rings, and satellite observations. These sources provide evidence of both natural and anthropogenic climate changes. Multiple lines of evidence support this perspective, such as increasing temperatures, declining ice sheets and snow cover, glacier retreat, rising sea levels, diminishing Arctic sea ice, a growing frequency of extreme weather events, and increasing ocean acidification (WMO 2021).

The greenhouse effect is essential for supporting life on Earth, as it allows for the transfer and regulation of heat. However, human-generated emissions in the atmosphere are trapping and impeding the release of heat from the Earth into space. The five most significant greenhouse gases are carbon dioxide (CO_2), nitrous oxide (N_2O), methane (CH_4), chlorofluorocarbons (CFCs), and water vapor (H_2O). While there is historical evidence indicating that the Sun has played a role in past climate variations, the current warming of the climate cannot be solely attributed to solar activity. Anthropogenic activities are the primary drivers of the current climate change. Scientists have foreseen several anticipated outcomes because of climate change, and these manifestations are already becoming evident. These include the melting of ice sheets, rising sea levels, and more frequent and intense heatwaves and extreme weather events. It is evident that human-generated greenhouse gases are projected to continue increasing global temperatures. As the extent of warming intensifies, there is an increasing likelihood of severe, widespread, and irreversible impacts occurring (IPCC 2021).

The concentration of CO_2 in the atmosphere is a significant factor that enhances the greenhouse effect, contributing to global warming. Apart from its role as a heat-trapping gas, CO_2 is emitted by natural sources like forest fires and volcanic eruptions, as well as human activities such as the burning of fossil fuels. The graph from Mauna Loa Observatory in Hawaii depicts atmospheric CO_2 levels in recent years, excluding natural seasonal fluctuations (Fig. 1.1). Additionally, CO_2 levels over the Earth's last three glacial cycles can be determined based on air bubbles trapped in ice sheets and glaciers. Human activities have greatly contributed to an increase in atmospheric CO_2, with levels rising by 50% since the beginning of industrialization. This means that the amount of CO_2 has increased by 150% since 1750. Notably, there has been a clear change in the concentration of CO_2, with levels increasing from 365 parts per million (ppm) in 2002 to over 420 ppm at present (NASA n.d.-b). Figure 1.1 provides a time series representation of the concentration of various greenhouse gases in the atmosphere.

1.2.1 Evidences

One common approach to demonstrate the evolution of our climate is by comparing mean temperatures of the Earth across different time periods. Generally, land surfaces have experienced a greater temperature rise compared to the oceans. As a result, the Arctic snow cover has been significantly impacted by the drastic temperature changes in the northern latitudes, as well as the increasing shallowness of ocean temperatures. These changes have also led to the rapid disappearance of the Arctic and Antarctic ice caps. For instance, in West Antarctica, Milillo et al. (2022) observed swift rates of glacier retreat. To project the implications of Antarctica on future sea-level rise, it is essential to understand the recent history of Thwaites Glacier and the mechanisms driving its current retreat. Results indicate that Thwaites Glacier has experienced prolonged periods of fast retreat over the past 200 years (Graham et al. 2022). Increasing temperatures are causing many glaciers to retreat, which has negative effects on downstream water resources in the Himalayan region (Drenkhan et al. 2019; IPCC 2019). Thakur et al. (2023) reported that the Gangotri Glacier needs to be closely monitored in the coming years using spatiotemporal remote sensing data and field-based surveys, as it is particularly prone to rapid retreat and significant loss of glacial area soon.

As mentioned earlier, increased solar radiation is causing both air and water temperatures to rise, which is negatively affecting ice caps. Over the twentieth century, global sea levels have risen due to the melting of these ice caps. Consequently, it is expected that as global warming continues to spread, sea levels will rise at an accelerated rate (Rahmstorf 2010). The continuous and exacerbated rise in sea levels will intensify the risk of flooding, storm surges, tropical cyclones, and heavy rainfall. These events will not only pose a

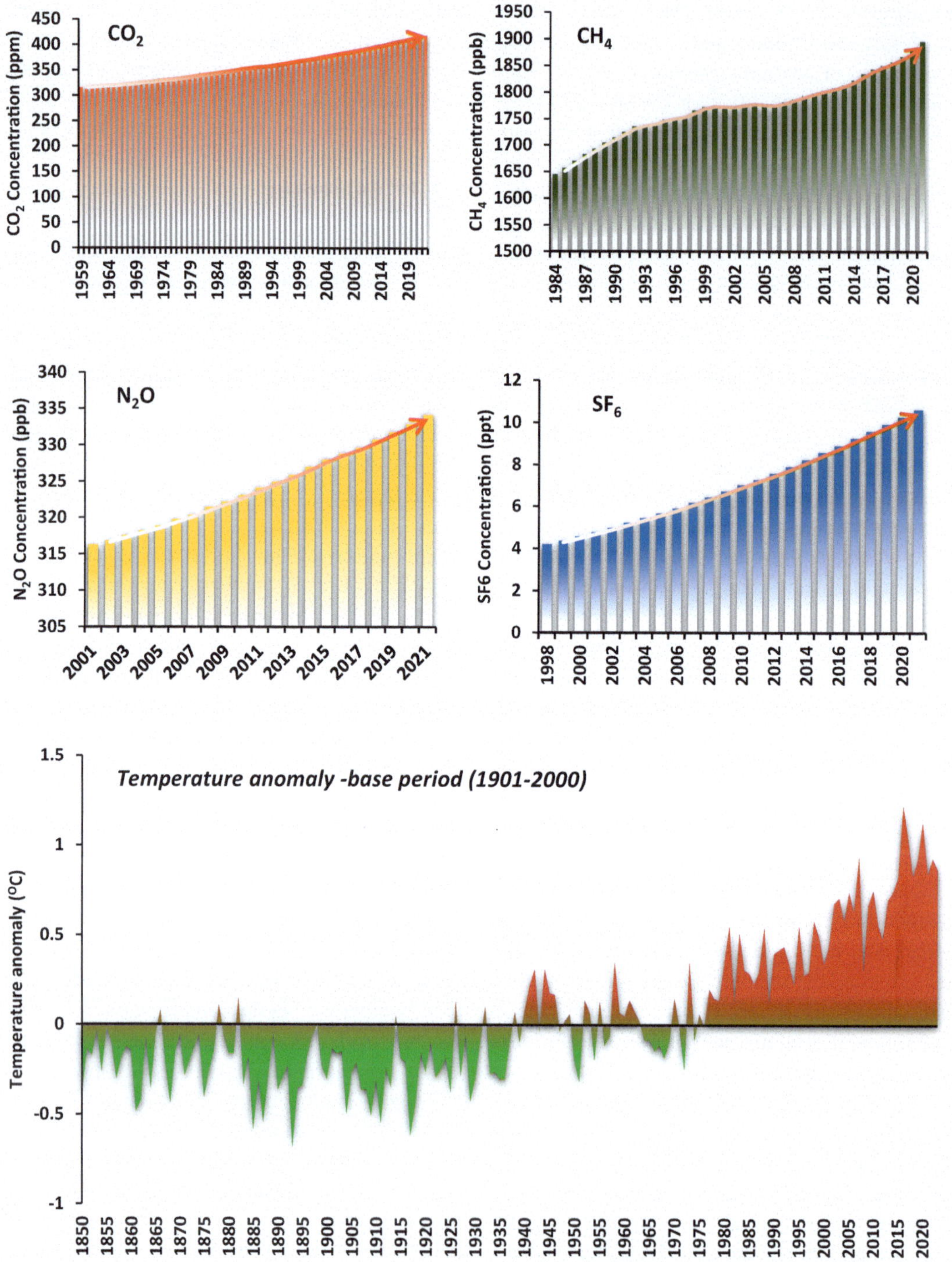

Fig. 1.1 Temporal change in the atmospheric concentration of various greenhouse gases and temperature anomaly (GHG concentrations data obtained from https://gml.noaa.gov/ccgg/trends/; temperature anomaly data obtained from https://www.ncei.noaa.gov/access/monitoring/climate-at-a-glance/global/time-series)

threat to coastal infrastructure and habitats but also lead to the submergence of low-lying coastal ecosystems. The consequences include jeopardized health, well-being, food, and water security, and ultimately a complete collapse of the system (IPCC 2022).

Extensive research and analyses have been conducted to identify the ways in which accelerated climate change has adversely affected biodiversity. There is growing concern that the current rate of climate change may lead to the extinction of numerous animal and plant species. Various studies indicate that climate change is a primary factor contributing to the ongoing loss of biodiversity (MEA 2005; Guo et al. 2017). According to Tol (2009), climate change has far-reaching consequences beyond just changes in precipitation and temperature. It also impacts biodiversity through potential changes in nutrient availability, ocean acidification, and the spread of invasive species into new habitats. Habibullah et al. (2022) further emphasize that climate change significantly affects the frequency and severity of natural disasters, as well as temperature and precipitation patterns, all of which have a substantial impact on biodiversity loss.

In certain oceanic regions, the negative impacts of ocean acidification and warming are evident in aquaculture and fishing. Coastal inundation and ocean acidity contribute to increased corrosion of shipwrecks and submerged ruins, and pose a greater threat to sacred sites, including burial grounds (IPCC 2022). Sea ice melt exposes seawater to the environment, facilitating rapid absorption of atmospheric CO_2, which leads to a decrease in alkalinity and a decline in pH and aragonite saturation state. Predictions suggest a further drop in pH, particularly in higher latitudes where sea ice retreat is occurring. However, Arctic warming could potentially offset future increases in aragonite saturation (Qi et al. 2022).

These impacts collectively provide evidence of global warming, which has been driven by countries' pursuit of greater economic growth. Rosales' (2008) inferred that "economic expansion is the principal driver of climate change-related biodiversity loss" remains accurate in this context.

1.2.2 Causes

Human activities and natural processes are the main drivers of climate change, impacting the planet's energy balance. Human actions have significantly altered the Earth's surface and atmosphere, leading to ongoing changes (IPCC 2007a, b). While human activities are the primary cause of climate change, there are also significant natural factors that influence the climate system (C32/IPCC 2013a). Throughout Earth's history, natural phenomena such as volcanic eruptions, variations in solar radiation, tectonic shifts, and even slight changes in Earth's orbit have contributed to certain climate changes. The findings of Stern and Kaufmann (2014) support the notion that human activity is partially responsible for the observed global temperature increase, which in turn affects the global carbon cycle. Thus, climate change is influenced by a combination of natural and human-induced factors. However, numerous studies have demonstrated that human activities are the primary drivers of climate change in the modern era (Fakana 2020). While it is true that some changes are gradual and beyond human control, human-induced climate change far surpasses them in terms of speed and magnitude.

Fossil fuels, including coal and crude oil, account for over 75% of greenhouse gas emissions and nearly 90% of CO_2 emissions, making them the largest contributors to climate change (UN, n.d.-a). In addition, activities such as deforestation, livestock grazing, biodiversity loss, and land degradation also have significant impacts on the climate and temperature of the planet. A 2 °C temperature increase from pre-industrial levels has severe implications for ecosystems, human health, and well-being. Changes in rainfall patterns due to human activity have resulted in more frequent extreme weather events worldwide. Climate change poses additional pressures on land, exacerbating existing threats to livelihoods, biodiversity, human settlements, infrastructure, and food systems. Warmer temperatures contribute to a more active hydrological cycle, increasing the likelihood of severe droughts and/or floods in certain regions (Fakana 2020). Moreover, there is a significantly higher risk of major environmental changes with potentially disastrous consequences.

Deforestation has detrimental effects as trees play a crucial role in regulating temperature by absorbing CO_2 from the atmosphere. Forests are vital to the Earth's climate system in numerous ways. One of their significant contributions in combating climate change is through photosynthesis, where they convert atmospheric CO_2 into living biomass (UN-FAO 2010). Forests act as natural sinks for CO_2 absorption from the atmosphere. In addition to the greenhouse gases, other pollutants like soot, which are not greenhouse gases, have both warming and cooling effects and are associated with issues such as poor air quality. N_2O, like CO_2, is a long-lasting greenhouse gas that accumulates in the atmosphere over years to millennia. Natural factors such as solar radiation fluctuations and volcanic activity are estimated to have contributed less than plus or minus 0.1 °C to global warming between 1890 and 2010 (EU n.d.). The burning of coal, oil, and gas releases CO_2 and N_2O as by-products. CH_4, a significant greenhouse gas, is produced by ruminant animals during digestion, while N_2O emissions result from the use of nitrogen-based fertilizers. These emissions can have warming potentials up to 23,000 times greater than CO_2 emissions (EU n.d.). Despite efforts to improve agricultural practices, such as the use of nitrogen-based fertilizers, N_2O concentrations in the atmosphere continue to increase. Consequently, human activities such as livestock farming, rice cultivation, natural gas usage, and landfill operations contribute to a substantial increase in CH_4 and other greenhouse gases, exacerbating climate change.

1.2.3 Consequences/Effects

The term "extreme weather events" is defined by the IPCC as follows: "The occurrence of a value of a weather variable above (or below) a threshold value near the upper (or lower) ends of the range of observed values of the variable" (IPCC 2012). Extreme weather and climatic conditions can lead to disasters in vulnerable areas, negatively impacting both natural and societal systems. In the past, such extreme events were often rare in specific locations, allowing time for natural and human systems to recover. However, due to climate change, several extreme weather and climate events are becoming more frequent, intense, and prolonged, reducing the time for recovery (IPCC 2012, 2013a, b).

Urban heat islands (UHI) and climate change are increasingly affecting public health, the economy, urban infrastructure, and the urban environment, making extreme heat a significant concern for cities (Keith et al. 2019). It is highly likely that the frequency, severity, and duration of heat events will increase over time due to climate change, and there is evidence that the hottest temperatures may warm up faster than the seasonal average in many locations (Horton et al. 2016). Extreme heat events pose a serious threat to global crop production systems, with implications for prices and food security (Lesk et al. 2016). The impact of excessive heat on livestock has been extensively studied, with historical decreases in dairy and meat output and increased livestock mortality during heat events (Dunn et al. 2014; Vitali et al. 2015). Extreme heat events can directly harm terrestrial ecosystems by causing death or injury to organisms and indirectly make them more vulnerable to subsequent disturbances such as disease, pests, fire, and drought (Teskey et al. 2014). Rising temperatures in many regions have contributed to desertification and land degradation. Irregular precipitation patterns and unpredictable wet seasons can lead to crop failures and uncertainty regarding planting times. Deforestation and urbanization, two human activities that significantly alter land conditions, can amplify these negative effects on a local scale. Global temperature rise will affect processes related to soil erosion, vegetation loss, and food security through crop losses and changes in the food supply. If current trends continue, we are on track to exceed critical thresholds, leading to increased risks of food supply instability, including frequent food shocks and rising food prices across regions (Gaupp 2020).

The occurrence of drought is presented as an example of how it can directly and indirectly impact future vulnerability, in addition to being a current risk. Drought is defined as a period of unusually dry weather that leads to a significant hydrological imbalance (IPCC 2012). Areas that are becoming desertified and arid can be seen as experiencing varying levels of permanent drought. Drought can occur in humid and semi-arid regions during specific seasons or prolonged periods of time. Drought can be categorized as follows: meteorological drought is characterized based on the intensity and duration of dryness. Hydrological drought is caused by rainfall deficits that affect surface or groundwater supplies. Agricultural drought occurs when the moisture needs of a particular crop are not met due to hydrological or meteorological drought. Socioeconomic drought refers to the impact on a community when water or agricultural shortages occur, or when demand for a specific economic commodity exceeds supply due to weather-related water supply deficits (Ebi and Bowen 2016).

Global warming has amplified the hydrological cycle, leading to increased cloud cover, latent heat fluxes, and frequency of climate extremes (Huntington 2006; IPCC 2007a, b). As temperatures continue to rise, it is expected that the hydrological cycle will intensify (Wu et al. 2013). Although it has been suggested that changes in the distribution of rainfall may accentuate inequalities between dry and wet regions, this has been refuted for land-based variations (Donat et al. 2016). A key finding from the research is that ecological responses to more extreme precipitation patterns will depend on the ambient precipitation levels, meaning that xeric, mesic, and hydric ecosystems may respond differently to this aspect of climate change (Knapp et al. 2008). Extreme precipitation can lead to significant runoff and soil erosion when the soil surface becomes saturated. This runoff can carry large amounts of sediment, leading to the clogging of downstream river channels and loss of soil nutrients, thereby decreasing soil productivity (Jia et al. 2022). Therefore, it is crucial to enhance the resilience of the global food system, as it is predicted that extreme events like drought will continue to increase throughout the twenty-first century.

A tropical cyclone, commonly known as a hurricane, is a rapidly rotating storm that forms over tropical oceans. It features an "eye" at the center, where weather conditions are calm and cloud-free. Surrounding the eye is the eyewall, which consists of intense thunderstorms, and the system has a low-pressure center with swirling clouds. Tropical cyclones can have a diameter of up to 1000 km, but typically range from 200 to 500 km. They are known for their strong winds, heavy rainfall, and high waves, and can occasionally cause devastating storm surges and coastal flooding (Alimonti et al. 2022). Predictions based on theory and high-resolution dynamical models indicate that greenhouse warming will lead to changes in the strength of tropical cyclones. It is projected that by 2100, there will be an increase in the intensity of 2–11%. However, current modeling studies suggest a global average decrease of 6–34% in the occurrence of tropical cyclones (Knutson et al. 2010). Over the past 50 years, tropical cyclones have been responsible for nearly 2000 disasters, resulting in the loss of thousands of lives and causing economic losses exceeding US$ 1400 billion (WMO n.d.). A tornado is a powerful vortex that extends from the surface to the cloud base and is capable of causing damage at the surface. Predicting how climate change will impact tornadoes in the future can be challenging due to the complex interplay of risk factors that may increase or decrease (Alimonti et al. 2022).

Forest fires play a significant role in shaping the structure and composition of vegetation in an area. They contribute to the mosaic landscape and impact biogeochemical cycles, such as the carbon cycle. The fire regime, consisting of fire frequency, size, intensity, seasonality, type, and severity, influences ecosystems by interrupting or terminating life cycles. The complex relationships between these fire regime components and the structure and function of forest ecosystems are heavily influenced by weather and climate (Flannigan and Harrington 1988; Johnson 1992). When considering the impact of global warming on vegetation, it is important to note that fire may have a greater effect on species distribution,

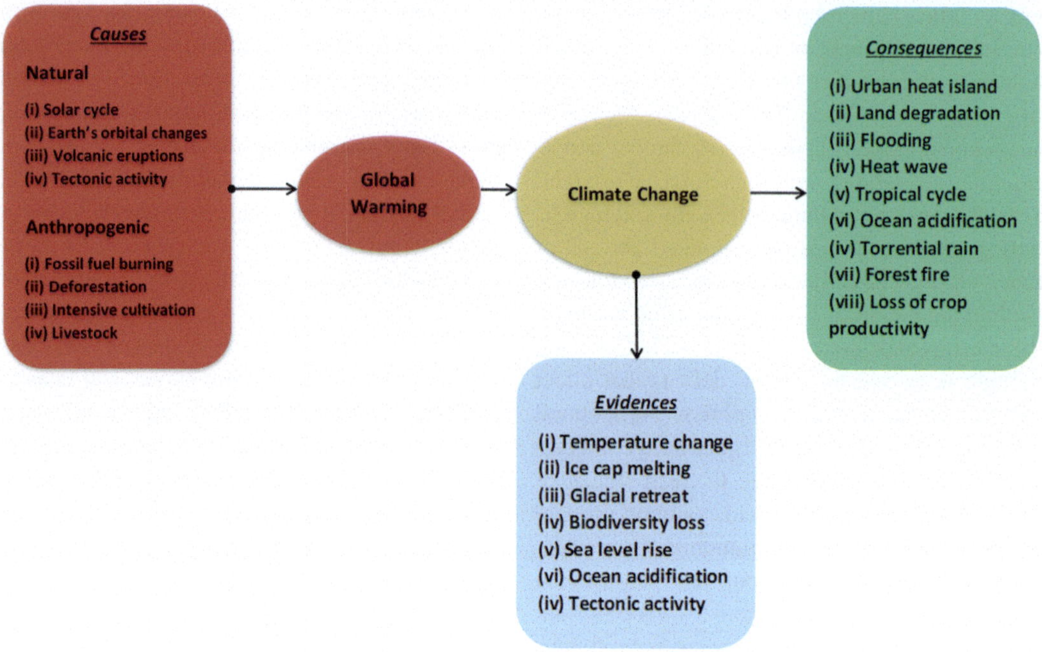

Fig. 1.2 General characterization of climate change

migration, substitution, and extinction than the direct consequences of temperature change (Weber and Flannigan 1997). Figure 1.2 provides an overall mechanism of global warming and how it has become a major global concern as climate change.

1.3 CO_2 as a Major Indicator for Global Warming

The Earth's surface temperature is regulated by naturally occurring greenhouse gases (GHGs) such as CO_2, CH_4, N_2O, and chlorinated hydrocarbons. However, reports have shown that the significant temperature variations are primarily due to increased GHG emissions caused by human activities. Approximately 60% of global warming is attributed to the effects of H_2O content, while 20% is attributed to CO_2. The focus on CO_2 as a major indicator of global warming, as opposed to H_2O, may be due to its longer residence time (50–200 years) in the atmosphere and its influence on H_2O levels. Even a slight change in CO_2 concentration can lead to increased evaporation and H_2O content (Letcher 2019). The concentration of CO_2 has been steadily increasing over the years, with NASA's latest measurements indicating a current concentration of 419 ppm as of January 2023.

Most adaptation and mitigation measures suggest the use of CO_2 removal technologies to combat global warming. Forest management practices, biomass energy conversion, direct air carbon capture methods, improved weathering of rocks, soil carbon sequestration, and negative emission technologies are among the techniques aimed at reducing carbon emissions and maintaining carbon pools worldwide. Agriculture systems, particularly due to the significant carbon storage capacity of soils, play a significant role as a source of carbon emissions. Factors such as land use conversion, reduced biomass, soil degradation from erosion and overgrazing, and fuel consumption contribute to CO_2 emissions in agriculture systems. Disturbed soil can release long-buried and conserved carbon. Therefore, understanding the response of soils to global warming is crucial (FAO 2016; Terlouw et al. 2021).

1.3.1 Relationship Between Soils and Climate Change

One meter of soil contains approximately three times the amount of carbon found in vegetation and roughly twice as much as that in the atmosphere (Abdullahi et al. 2018). Understanding the response of soils to global warming is crucial for evaluating carbon cycle feedbacks, as even small losses from this large pool can have a significant impact on future atmospheric CO_2 concentrations (Smith 2012). Historically, a portion of this carbon (40–90 Pg C) has been lost from soils in managed ecosystems, with some persisting in the atmosphere. Climate change predictions suggest that future changes in soil carbon will range from slight deficits to modest gains, but these patterns vary significantly across regions. The future response of soil carbon will depend on the complex interplay of differences in carbon losses from decomposition and gains from enhanced productivity, as well as the effects of increased temperature and decreased soil moisture on decomposition rates. Globally, soil carbon sequestration offers a significant and cost-effective mitigation potential for reducing climate change (Smith 2012).

In comparison to the 620 Pg of carbon in terrestrial biota and detritus and the 780 Pg of carbon in the atmosphere, soil represents the largest active terrestrial carbon reservoir, estimated at around 2500 Pg of carbon (Lal 2010). Additionally, the Earth's crust contains 90,000,000 Pg of carbon in its geological structures, 38,000 Pg of carbon as dissolved carbonates in the ocean, 10,000 Pg as gas hydrates, and 4000 Pg as fossil fuels (Rustad et al. 2000). While any soil has the capacity to store carbon, highly managed systems such as arable and agroforestry soils offer the greatest potential for human control over carbon sequestration. Soil management practices like no-till systems, compared to intensive tillage, often result in reduced CO_2 emissions from the soil and increased carbon uptake. However, it is important to note that management choices can also release carbon from the soil, turning it into a net emitter of greenhouse gases. Examples of management practices that can increase CO_2 emissions from soils include ploughing perennial soils for agricultural purposes, adopting more destructive tillage methods, and draining wetlands (Brevik 2012). Approximately 47% of annual global human CH_4 emissions come from agriculture (Smith et al. 2007). The production of CH_4 in soil is associated with anaerobic decomposition of organic matter. Rice (*Oryza sativa* L.) cultivation is the primary anthropogenic source of soil-derived CH_4, while natural soil-derived CH_4 primarily originates from wetlands (Stępniewski et al. 2011). About 58% of anthropogenic N_2O emissions also come from agriculture (Smith et al. 2007). In summary, human management practices can significantly influence soil processes that produce or deplete CO_2, CH_4, and N_2O.

Climate change is expected to have significant consequences on the soil system. Changes in CO_2 levels, temperature, and precipitation patterns will lead to variations in soil interactions, decomposition rates, and soil organic carbon levels. Soil organic carbon plays a crucial role in soil structure, fertility, microbial populations, and other important soil qualities. Rising temperatures are expected to have a detrimental effect on carbon allocation to the soil, resulting in a decrease in soil organic carbon and creating a positive feedback loop in the global carbon cycle (Wan et al. 2011). The impact of climate change on soil carbon dynamics, considering the interactions between vegetation, landscape, soil quality, and parent material, has been examined by Anderson (1992). Soil degradation resulting from climate change restricts the planet's capacity to store carbon. Climate change also affects the quantity and quality of drinking water and agricultural water, leading to increasing uncertainty regarding food security in many parts of the world where traditional crops have thrived for centuries.

Climate change will not only affect soil processes and qualities but soils themselves will also have an impact on climate. Although research on the specific effects of climate change on soil processes and properties is still in its early stages, it is already known that soil organic matter dynamics, including soil organisms and properties associated with organic matter, soil water, and soil erosion, will be affected. While there is still much to learn about how climate change will impact soils, previous research and our understanding of soil qualities and processes can provide some insight into the expected effects (Brevik 2012). Studies have shown that soil has the potential to mitigate the adverse effects of climate change-induced consequences, although further research is needed to fully understand and harness this potential.

1.4 Climate Crisis and Food Security

According to the Food and Agriculture Organization (FAO 2012), food security relies on four pillars: availability, accessibility, utilization, and stability. With the global population expected to increase, there is a projected 50% rise in the demand for agricultural products by 2030, emphasizing the need for sustainable reinforcement of food systems. Climate change can directly impact agricultural production through factors such as rainfall variations leading to droughts or floods, as well as temperature fluctuations that affect the length of the growing season. Indirectly, climate change can also influence market dynamics, food prices, and distribution networks (Gregory et al. 2005). These impacts may hinder progress towards a hunger-free society and pose risks to food availability and system consistency. To address these challenges, substantial funding is required for mitigation and adaptation efforts in developing a "climate-smart food system" that can withstand the impacts of climate change on food security (Wheeler and Von Braun 2013).

The growth of the global population and the increase in malnourished individuals suggest a resurgence of world hunger, reversing the previous downward trend. One contributing factor to this is land use changes combined with population growth, which has led to a decrease in per capita agricultural land use for crops and livestock over the past decade. The need to increase production within limited arable land puts pressure on productive areas, pushing their capacity to the limit (FAO 2022). Climate change exacerbates this situation by exacerbating land degradation processes. Land degradation is primarily caused by soil erosion, the continued removal of natural vegetation, decline in soil quality, and the expansion of agriculture. Unfortunately, around three-quarters of global soil erosion is attributed to the agriculture sector (FAO and ITPS 2015).

Since 1950, global food supply has more than doubled, but often at the expense of farming practices that contribute to high rates of soil erosion. Altered and intensified rainfall and temperature patterns directly affect land degradation. Projected changes in rainfall patterns pose future challenges in erosion-induced land degradation. The combined effects of climate change and land degradation lead to changes in soil erosion rates, vegetation cover, and decomposition rates. Soil is being lost from at least 20% and possibly up to 33% of the world's farmland, threatening its long-term productivity. If topsoil continues to be lost at the current rate, global food production per person will eventually decline. It is crucial to urgently realign national priorities worldwide to slow down population growth and invest in farming technologies that preserve the farmland base (Brown 1981).

Global warming has implications for food and water security for everyone. Climate change, a direct driver of land and soil degradation, limits the Earth's capacity to retain carbon in its soil. Currently, up to 30% of food is lost or wasted due to erosion, affecting areas inhabited by 500 million people (IPCC 2018). Climate change also reduces the quantity and quality of available water for agriculture and drinking purposes. Food security has become increasingly precarious as many regions struggle to sustain crops that have thrived for centuries. These impacts disproportionately affect vulnerable and impoverished populations. Global warming is expected to widen the economic disparity between the wealthiest and poorest countries.

Climate change exacerbates the risk of land degradation and food insecurity, leading to reduced production, income, and increased poverty. Changes in land use and land cover patterns due to climate change further contribute to the threat of land degradation. Land degradation, in turn, is a significant source of greenhouse gas emissions. Cultivation practices are responsible for a considerable loss of organic carbon, ranging from 20% to 60%, from agricultural lands. Soil organic carbon and soil organic matter, crucial components for soil quality and productivity, are highly vulnerable to climate change and associated land degradation processes. Conventional agricultural management practices often result in lower soil organic carbon content and higher carbon emissions. The loss of nutrient-rich topsoil, along with organic carbon depletion, reduces crop yields (FAO 2022). In addition to carbon emissions, agriculture is a significant source of non-carbon greenhouse gases. The conversion of peatlands for various purposes, including cultivation, has resulted in approximately 11–15% of peatlands becoming a source of greenhouse gas emissions (FAO 2020). To meet dietary needs, marginal lands have been converted into cultivation, making them susceptible to erosion and carbon depletion. In summary, climate change acts as a risk multiplier, exacerbating both land degradation and food security challenges.

Approximately 52% of global agricultural land is currently experiencing moderate to severe degradation (FAO 2018). Expanding cultivation and restoring degraded lands present significant challenges in terms of time, effort, and cost. It is crucial to transition agricultural production systems towards sustainable land management practices to mitigate land degradation. This can be achieved through the implementation of site-specific soil and water management techniques. Adopting integrated approaches that promote land quality, carbon sequestration, and climate change mitigation provides sustainable solutions to address land degradation and its impacts on food security. In addressing these challenges, it is essential for science, policy, and geopolitical security imperatives to collaborate and offer a comprehensive perspective on the issue. By proactively tackling the root causes of land degradation, rather than solely dealing with its consequences in the future, we can effectively respond to the increasingly complex climate crisis. Cooperation among stakeholders along the human–nature nexus is vital to develop and implement systemic initiatives that reshape food systems and revitalize rural livelihoods. Understanding the escalation of conflicts due to climate change through research on food systems is crucial. Such research helps policymakers, stakeholders, and international organizations make informed decisions based on robust scientific knowledge (Läderach et al. 2021).

1.5 A Way Forward? Sustainability

Even though research has established the reality of climate change, it has also demonstrated that we still could reverse its effects. This will require significant changes in all aspects of human civilization, including food production, land use, transportation, and energy generation. While technology has contributed to climate change, it can also be leveraged to reduce net emissions and improve the health of our planet. Existing technology alone can potentially reduce over 70% of today's emissions (UN n.d.-a, b). Electric vehicles are emerging as a prominent solution, and renewable energy has already become the most cost-effective energy source in many regions. Nature-based solutions provide us with breathing space as we work towards decarbonizing our economy. By adopting these solutions, we can reduce our carbon footprint while simultaneously promoting critical ecosystem functions such as species diversity, freshwater availability, livelihood improvement, and food security. Examples of nature-based solutions include sustainable farming practices, land restoration, conservation efforts, and environmentally conscious food supply chains. With the help of accessible new technologies and nature-based solutions, we can collectively move towards a more sustainable and resilient future. Governments, corporations, civil society, youth, and academia must come together to create a green future where disease is minimized, justice is upheld, and people and the environment coexist harmoniously. By joining forces, we can pave the way for a better world.

1.5.1 Role of SDG in Mitigating Climate Crisis

By 2030, a significant portion of the global population is at risk of displacement due to the consequences of climate change. Immediate action is crucial to save lives and livelihoods and prevent the devastating impacts of climate change. Achieving the 17 Sustainable Development Goals outlined in the 2030 Agenda for Sustainable Development is essential for creating a better future for all. One key commitment is the implementation of the United Nations Framework Convention on Climate Change, which involves raising $100 billion annually by 2020 from various sources to meet the needs of developing countries in their efforts to mitigate climate change. The Green Climate Fund should also be fully operationalized through capitalization to support climate action (SDG Tracker n.d.). SDG 13 emphasizes the integration of climate change policies into national frameworks, capacity building for economic impacts and mitigation, and raising public awareness.

To address the climate crisis and ensure a safe future below 1.5 °C, global greenhouse gas emissions must be reduced by 50% within the next 10 years. While such high goals may seem daunting, experts suggest that individuals and decision-makers can take meaningful action to combat climate change. Learning about sustainable living practices and exploring the actions we can take as individuals and as a society can contribute to a healthier planet. With the current wealth of knowledge and solutions available, we could prevent the catastrophe and provide people with improved habitats and livelihoods. It is important to recognize that climate change is a long-term issue that is already unfolding, which creates uncertainties for decision-makers as they shape the future. However, by prioritizing climate action and working collectively, we can navigate these challenges and create a more sustainable and resilient world for generations to come.

Nature-based approaches, such as better farming methods, land rehabilitation, conservation, and sustainable green food distribution networks, offer effective ways to address climate change and mitigate its disastrous consequences. By drawing inspiration from nature and harnessing technological innovations, we can pave the way towards a future that is both reliable and scalable. Collaboration among governments, businesses, civic society, young people, and academia is crucial in creating a green future where suffering is minimized, justice is upheld, and harmony between people and the environment is restored. To achieve this, it is imperative to work towards the 17 goals outlined in the 2030 Agenda for Sustainable Development (Fig. 1.3). SDG 13 emphasizes the need to prioritize climate change mitigation measures in national policies, enhance capacity building for climate-related advisories, influence reduction, mitigation efforts, and raise awareness through campaigns (UNEP n.d.). Recognizing that climate change poses the most significant obstacle to sustainable development, we must take action to prevent millions of people from falling into severe poverty. With access to a wealth of information and solutions, we could avert this crisis and improve the lives of people worldwide. Understanding the interconnectedness between climate and global relations is crucial in addressing the challenges posed by climate change, unemployment, injustice, and environmental degradation. By embracing sustainable practices and pursuing collaborative efforts, we can overcome these challenges and build a more resilient and equitable future for all.

1.6 A Framework of Climate Resilience for Sustainability: Watershed Level Measures

To achieve sustainability in the face of climate change, an effective framework is needed to foster climate-resilient communities. Given the significant variations in climate, topography, and vegetation across the globe, site-specific research and frameworks are highly suitable for implementing policies at the local scale. Watershed-based approaches prove to be more efficient, as watersheds serve as natural hydrological units where crucial natural processes take place. A watershed refers to an area where rainfall drains into a common outlet point, with ridges serving as the boundaries based on topographical charac-

Fig. 1.3 Sustainable Development Goals illustrated as SDG tree and details of SDG-13 (climate action)

teristics that direct water flow. Therefore, adopting a watershed-based approach for natural resource management is highly suitable. Researchers have already recognized watershed management as the ideal approach for integrated natural resources management in rainfed areas (Wani et al. 2003).

Addressing climate-induced issues at their root through adaptation and mitigation measures proves more efficient than blindly applying remedial measures. This is why soil and water conservation measures often employ a watershed-based method. The primary requirement is to characterize the basic natural resources (soil and vegetation) of the watershed. Monitoring and analyzing weather and hydrological parameters are crucial for understanding the processes at play. Continuous monitoring helps build an understanding of the dominant physical processes within the watershed, providing insights for identifying watershed characteristics. This knowledge enables preparedness against various hazards such as soil erosion, disasters like flooding and drought, and associated adverse impacts like crop and livelihood loss. Recent advancements in remote sensing technologies, sensors, and Internet of Things (IoT) approaches efficiently aid in assessing problems, thereby proposing suitable socio-economic and environmental sustainability measures for the watershed. A sustainable watershed should be self-reliant in fulfilling the major needs of the people residing within it. International treaties, nature-based solutions, and the SDG framework further enhance this approach, enabling the achievement of sustainability and climate resilience at the local level, particularly at the village level.

1.7 Conclusions

This chapter provides an overview of global warming, climate change, and their consequences, with a particular focus on extreme weather events. Human activities have led to alterations in the Earth's atmosphere, resulting in global warming and triggering climate change. Scientific bodies have already established that climate change will have widespread impacts on the planet, making it increasingly challenging to survive in these conditions. Extreme weather events, characterized by sudden and severe consequences, are responsible for numerous deaths. Achieving sustainability is crucial for building resilience to climate change, reducing emissions, and capturing excess carbon from human sources. Researchers have recognized the significant potential of soil in mitigating global warming. The SDG framework offers substantial potential for implementing measures to address the adverse impacts of climate change. We suggest a site-specific adaptation and mitigation measures to enhance climate resilience and sustainability.

References

Abdullahi AC, Siwar C, Shaharudin MI, Anizan I (2018) Carbon sequestration in soils: the opportunities and challenges. In: Carbon capture, utilization and sequestration. https://doi.org/10.5772/intechopen.79347

Alimonti G, Mariani L, Prodi F et al (2022) A critical assessment of extreme events trends in times of global warming. Eur Phys J Plus 137:112. https://doi.org/10.1140/epjp/s13360-021-02243-9

Anderson JM (1992) Responses of soils to climate change. In: Advances in ecological research, vol 22. Academic Press, London, pp 163–210

Brevik EC (2012) Soils and climate change: gas fluxes and soil processes. Soil Horiz 53(4):12–23

Brown LR (1981) World population growth, soil erosion, and food security. Science 214(4524):995–1002

CarbonBrief (n.d.) Extreme weather, attributing extreme weather to climate change. Available at https://www.carbonbrief.org/mapped-how-climate-change-affects-extreme-weather-around-the-world/. Accessed on 4 Feb 2023

Donat MG, Lowry AL, Alexander LV, O'Gorman PA, Maher N (2016) More extreme precipitation in the world's dry and wet regions. Nat Clim Chang 6(5):508–513

Drenkhan F, Huggel C, Guardamino L, Haeberli W (2019) Managing risks and future options from new lakes in the deglaciating Andes of Peru: the example of the Vilcanota-Urubamba basin. Sci Total Environ 665:465–483

Dunn RJH, Mead NE, Willett KM, Parker DE (2014) Analysis of heat stress in UK dairy cattle and impact on milk yields. Environ Res Lett 9:64006. https://doi.org/10.1088/1748-9326/9/6/064006

Ebi KL, Bowen K (2016) Extreme events as sources of health vulnerability: drought as an example. Weather Clim Extremes 11:95–102

Eckstein D, Künzel V, Schäfer L (2021) The global climate risk index 2021. Germanwatch, Bonn, 50p

European Union (n.d.) Climate action. Accessed from https://climate.ec.europa.eu/climate-change/causes-climate-change_en. Accessed on 27 Feb 2023

Fakana ST (2020) Causes of climate change. Glob J Sci Front Res H Environ Earth Sci 20(2):7–12

FAO (2012) Technical note: FAO methodology to estimate the prevalence of undernourishment. FAO, Rome. www.fao.org/fileadmin/templates/es/SOFI_2012/sofi_technical_note.pdf

FAO (2016) The state of food and agriculture 2016: climate change, agriculture and food security. FAO, Rome

FAO (2018) Land and water. Food and Agriculture Organization. Retrieved from: http://www.fao.org/land-water/land/httpwwwfaoorgsoils-portalen/en/

FAO (2020) Peatland mapping and monitoring: recommendations and technical overview. FAO, Rome. https://doi.org/10.4060/ca8200en

FAO (2022) The state of the world's land and water resources for food and agriculture – systems at breaking point. Main report. FAO, Rome. https://doi.org/10.4060/cb9910en

FAO & ITPS (Intergovernmental Technical Panel on Soils) (2015) Status of the world's soil resources: main report. FAO, Rome. www.fao.org/3/i5199e/i5199e.pdf

Flannigan MD, Harrington JB (1988) A study of the relation of meteorological variables to monthly provincial area burned by wildfire in Canada 1953–80. J Appl Meteorol 27:441–452

Gaupp F (2020) Extreme events in a globalized food system. One Earth 2(6):518–521

Graham AG, Wåhlin A, Hogan KA, Nitsche FO, Heywood KJ, Totten RL, Smith JA, Hillenbrand C-D, Simkins LM, Anderson JB, Wellner JS, Larter RD (2022) Rapid retreat of Thwaites Glacier in the pre-satellite era. Nat Geosci 15(9):706–713

Gregory PJ, Ingram JS, Brklacich M (2005) Climate change and food security. Philosophical Transactions of the Royal Society B: Biological Sciences 360(1463):2139–2148

Guo D, Desmet PG, Powrie LW (2017) Impact of the future changing climate on the Southern Africa biomes and the importance of geology. J Geosci Environ Prot 5:1–9. http://www.scirp.org/journal/gep

Habibullah S, Din BH, Tan SH et al (2022) Impact of climate change on biodiversity loss: global evidence. Environ Sci Pollut Res 29:1073–1086. https://doi.org/10.1007/s11356-021-15702-8

Horton RM, Mankin JS, Lesk C, Coffel E, Raymond C (2016) A review of recent advances in research on extreme heat events. Curr Clim Change Rep 2:242–259

Huntington TG (2006) Evidence for intensification of the global water cycle: review and synthesis. J Hydrol 319:83–95

Intergovernmental Panel on Climate Change (IPCC) (2007a) Climate change 2007 – the physical science basis. Cambridge University Press, Cambridge

Intergovernmental Panel on Climate Change (IPCC) (2007b) Climate change 2007: the physical science basis. Summary for policymakers. Cambridge University Press, New York. (19 August 2008; www.ipcc.ch/ipccreports/ar4-wg1.htm)

Intergovernmental Panel on Climate Change (IPCC) (2012) In: Field CB, Barros V, Stocker TF, Qin D, Dokken DJ, Ebi KL, Mastrandrea MD, Mach KJ, Plattner G-K, Allen SK, Tignor M, Midgley PM (eds) Managing the risks of extreme events and disasters to advance climate change adaptation. A special report of Working Groups I and II of the Intergovernmental Panel on Climate Change. Cambridge University Press, Cambridge/New York, 582 pp

Intergovernmental Panel on Climate Change (IPCC) (2013a) In: Stocker TF, Qin D, Plattner G-K, Tignor M, Allen SK, Boschung J, Nauels A, Xia Y, Bex V, Midgley PM (eds) Climate change 2013: the physical science basis. Contribution of Working Group I to the fifth assessment report of the Intergovernmental Panel on Climate Change. Cambridge University Press, Cambridge/New York

Intergovernmental Panel on Climate Change (IPCC) (2013b) Summary for policymakers. In: Stocker TF, Qin D, Plattner GK, Tignor M, Allen SK, Boschung J, Nauels A, Xia Y, Bex V, Midgley PM (eds) Climate change 2013: the physical science basis. Contribution of Working Group I to the Fifth Assessment Report of the Intergovernmental Panel on Climate Change. Cambridge University Press, Cambridge/New York, p 28

Intergovernmental Panel on Climate Change (IPCC) (2018) Summary for policymakers. In: Masson-Delmotte V, Zhai P, Pörtner H-O, Roberts D, Skea J, Shukla PR, Pirani A, Moufouma-Okia W, Péan C, Pidcock R, Connors S, Matthews JBR, Chen Y, Zhou X, Gomis MI, Lonnoy E, Maycock T, Tignor M, Waterfield T (eds) Global warming of 1.5°C. An IPCC special report on the impacts of global warming of 1.5°C above pre-industrial levels and related global greenhouse gas emission pathways, in the context of strengthening the global response to the threat of climate change, sustainable development, and efforts to eradicate poverty. Cambridge University Press, Cambridge/New York, pp 3–24. https://doi.org/10.1017/9781009157940.001

Intergovernmental Panel on Climate Change (IPCC) (2019) Climate change and land: an IPCC special report on climate change, desertification, land degradation, sustainable land management, food security, and greenhouse gas fluxes in terrestrial ecosystems

Intergovernmental Panel on Climate Change (IPCC) (2021) Arias, P.A., N. Bellouin, E. et al., 2021: Technical summary. In: Masson-Delmotte V, Zhai P, Pirani A, Connors SL, Péan C, Berger S, Caud N, Chen Y, Goldfarb L, Gomis MI, Huang M, Leitzell K, Lonnoy E, Matthews JBR, Maycock TK, Waterfield T, Yelekçi O, Yu R, Zhou B (eds) Climate change 2021: the physical science basis. Contribution of Working Group I to the sixth assessment report of the Intergovernmental Panel on Climate Change. Cambridge University Press, Cambridge/New York, pp 33–144. https://doi.org/10.1017/9781009157896.002

Intergovernmental Panel on Climate Change (IPCC) (2022) In: Pörtner H-O, Roberts DC, Tignor M, Poloczanska ES, Mintenbeck K, Alegría A, Craig M, Langsdorf S, Löschke S, Möller V, Okem A, Rama B (eds) Climate change 2022: impacts, adaptation and vulnerability. Contribution of Working Group II to the sixth assessment report of the Intergovernmental Panel on Climate Change. Cambridge University Press, Cambridge/New York, 3056 pp. https://doi.org/10.1017/9781009325844

Jia H, Wang X, Sun W, Mu X, Gao P, Zhao G, Li Z (2022) Estimation of soil erosion and evaluation of soil and water conservation benefit in terraces under extreme precipitation. Water 14(11):1675

Johnson EA (1992) Fire and vegetation dynamics: studies from the North American boreal forest. Cambridge University Press, Cambridge, p 129

Keith L, Meerow S, Wagner T (2019) Planning for extreme heat: a review. J Extreme Events 6(03n04):2050003

Knapp AK, Beier C, Briske DD, Classen AT, Luo Y, Reichstein M, Smith MD, Smith S, Bell JE, Fay PA, Heisler JL, Leavitt S, Sherry RA, Smith B, Weng E (2008) Consequences of more extreme precipitation regimes for terrestrial ecosystems. Bioscience 58(9):811–821

Knutson T, McBride J, Chan J et al (2010) Tropical cyclones and climate change. Nat Geosci 3:157–163. https://doi.org/10.1038/ngeo779

Läderach P, Pacillo G, Thornton P, Osorio D, Smith D (2021) Food systems for peace and security in a climate crisis. Lancet Planet Health 5(5):e249–e250

Lal R (2010) Managing soils and ecosystems for mitigating anthropogenic carbon emissions and advancing global food security. Bioscience 60:708–721. https://doi.org/10.1525/bio.201

Lesk C, Rowhani P, Ramankutty N (2016) Influence of extreme weather disasters on global crop production. Nature 529:84–87. https://doi.org/10.1038/nature16467

Letcher TM (2019) Why do we have global warming? In: Managing global warming. Academic Press, Amsterdam, pp 3–15

Milillo P, Rignot E, Rizzoli P, Scheuchl B, Mouginot J, Bueso-Bello JL, Prats-Iraola P, Dini L (2022) Rapid glacier retreat rates observed in West Antarctica. Nat Geosci 15(1):48–53

Millennium Ecosystem Assessment (MEA) (2005) Ecosystems and human well-being: synthesis. Island Press, Washington, DC. https://www.cifor.org/knowledge/publication/1888/

NASA (n.d.-a) Global climate change, global warming vs. climate change. Accessed from https://climate.nasa.gov/global-warming-vs-climate-change/. Accessed on 26 Jan 2023

NASA (n.d.-b) Vital signs. Accessed from https://climate.nasa.gov/vital-signs/carbon-dioxide/. Accessed on 26 Feb 2023

Qi D, Ouyang Z, Chen L, Wu Y, Lei R, Chen B, Feely RA, Anderson LG, Zhong W, Lin H, Polukhin A, Zhang Y, Zhang Y, Bi H, Lin X, Luo Y, Zhuang Y, He J, Chen J, Cai WJ (2022) Climate change drives rapid decadal acidification in the Arctic Ocean from 1994 to 2020. Science 377(6614):1544–1550

Rahmstorf S (2010) A new view on sea level rise. Nat Clim Change 1:44–45. https://doi.org/10.1038/climate.2010.29

Rosales J (2008) Economic growth, climate change, biodiversity loss: distributive justice for the global north and south. Conserv Biol 22(6):1409–1417. https://doi.org/10.1111/j.1523-1739.2008.01091.x

Rustad LE, Huntington TG, Boone RD (2000) Controls on soil respiration: implications for climate change. Biogeochemistry 48:1–6. https://doi.org/10.1023/A:1006255431298

SDG Tracker (n.d.) Sustainable development goals/climate action, sustainable development goal 13-take urgent action to combat climate change and its impacts. Accessed from https://sdg-tracker.org/climate-change. Accessed on 4 Mar 2023

Smith P (2012) Soils and climate change. Curr Opin Environ Sustain 4(5):539–544

Smith P, Martino D, Cai Z, Gwary D, Janzen H, Kumar P, McCarl B, Ogle S, O'Mara F, Rice C, Scholes B, Sirotenko O (2007) Agriculture. In: Metz B et al (eds) Climate change 2007: mitigation. Contribution of Working Group III to the fourth assessment report of the Intergovernmental Panel on Climate Change. Cambridge University Press, Cambridge, pp 497–540

Stępniewski W, Stępniewski Z, Rożej A (2011) Gas exchange in soils. In: Hatfield JL, Sauer TJ (eds) Soil management: building a stable base for agriculture. SSSA, Madison, pp 117–144

Stern DI, Kaufmann RK (2014) Anthropogenic and natural causes of climate change. Clim Chang 122:257–269. https://doi.org/10.1007/s10584-013-1007-x

Terlouw T, Bauer C, Rosa L, Mazzotti M (2021) Life cycle assessment of carbon dioxide removal technologies: a critical review. Energy Environ Sci 14(4):1701–1721

Teskey R, Wertin T, Bauweraerts I, Ameye M, McGuire MA, Steppe K (2014) Responses of tree species to heat waves and extreme heat events. Plant Cell Environ 38:1699–1712. https://doi.org/10.1111/pce.12417

Thakur PK, Pandey P, Chauhan P (2023) Assessment of dynamics of frontal part of Gangotri glacier, India, from 2017 to 2021 using remote sensing data. J Indian Soc Remote Sens. https://doi.org/10.1007/s12524-022-01655-z

Tol RSJ (2009) The economic effects of climate change. J Econ Perspect 23(2):29–51. https://www.jstor.org/stable/27740523

UNEP (n.d.) GOAL 13: climate action, take urgent action to combat climate change and its impacts. Available at https://www.unep.org/explore-topics/sustainable-development-goals/why-do-sustainable-development-goals-matter/goal-13. Accessed on 7 Feb 2023

United Nations (n.d.-a) Climate action, causes and effects of climate change. Available at https://www.un.org/en/climatechange/science/causes-effects-climate-change. Accessed on 6 Feb 2023

United Nations (n.d.-b) Shaping our future together UN 75 2020 and beyond, the climate crisis – a race we can win. Available at https://www.un.org/en/un75/climate-crisis-race-we-can-win. Accessed on 6 Feb 2023

United Nations Food and Agricultural Organization (FAO) (2010) Global forest resources assessment

Unitend Nations-Sustainable Development Goals (n.d.) Goal 13: take urgent action to combat climate change and its impacts. Available at https://www.un.org/sustainabledevelopment/climate-change/. Accessed on 7 Mar 2023

Vitali A, Felici A, Esposito S, Bernabucci U, Bertocchi L, Maresca C et al (2015) The effect of heat waves on dairy cow mortality. J Dairy Sci 98:4572–4579. https://doi.org/10.3168/jds.2015-9331

Wan Y, Lin E, Xiong W, Li Y, Guo L (2011) Modeling the impact of climate change on soil organic carbon stock in upland soils in the 21st century in China. Agric Ecosyst Environ 141:23–31. https://doi.org/10.1016/j.agee.2011.02.004

Wani SP, Singh HP, Sreedevi TK, Pathak P, Rego TJ, Shiferaw B, Iyer SR (2003) Farmer-participatory integrated watershed management: Adarsha watershed, Kothapally India. Case 7

Weber MG, Flannigan MD (1997) Canadian boreal forest ecosystem structure and function in a changing climate: impact on fire regimes. Environ Rev 5:145–166

Wheeler T, Von Braun J (2013) Climate change impacts on global food security. Science 341(6145):508–513

WMO (2021) WMO: climate change threatens sustainable development. Accessed from https://public.wmo.int/en/media/press-release/wmo-climate-change-threatens-sustainable-development. Press Release Number: 22092021. Accessed on 26 Jan 2023

WMO (n.d.) Tropical cyclones. https://public.wmo.int/en/our-mandate/focus-areas/natural-hazards-and-disaster-risk-reduction/tropical-cyclones

World Economic Forum, Calvo HG, Markovitz G (ed) (2022) Global public braces for 'severe' effects of climate change by 2032, New survey finds. Published on 15 Sept 2022. Available at https://www.weforum.org/press/2022/09/world-bracing-for-severe-effects-of-climate-change-new-survey-finds/

Wu P, Christidis N, Stott P (2013) Anthropogenic impact on Earth's hydrological cycle. Nat Clim Change 3:807–810

2 Ecosystem Degradation to Restoration: A Challenge

Shaista Khan, T. H. Masoodi, M. A. Islam, Tahera Arjumand, Azeem Raja, Aafaq Ahmad Parrey, Anushka Pallavi, and Javaid H. Bhat

Abstract

Ecosystem degradation is one of the big challenges of our time. It is disrupting the nature's balance due to the overexploitation of natural resources such as water, soil, and air. Particularly the degradation of farmland and the forest ecosystems will indeed harm the entire biological systems if not taken under consideration. To sustain the life of living beings on the earth and to address the impacts of the climate crisis, the significance of restoring two important ecosystems, namely, farmland and forests, must be recognized. Restoration through land distribution by means of environment friendly farming practices has the capacity to enhance agricultural productivity, as well by providing other ecosystem benefits at both the arable farmland and the landscape scale. Ecosystem restoration is a simple natural process with effective results. Meaning that we have to reduce or eliminate pressure on our ecosystem and at times nature can itself restore or employ corrective actions. In this chapter, we have highlighted the importance of farmland and forest ecosystems; their current degradation status, causes of degradation, and possible alternatives to restore these ecosystems by employing different rehabilitation approaches. The chapter also represents a basic concept and the need for restoration methods as well as describes how ecosystem degradation can be converted into restoration by the application of various techniques. Furthermore, proper measures should be taken by the state and central government agencies together with local people to bring out good results from restored ecosystems sustainably through the restoration programs.

Keywords

Ecosystem · Environment · Farmland · Forest · Restoration

2.1 Introduction

In an ecosystem, the living and non-living components interact with each other to continue proper functioning of the surrounding environ-

S. Khan (✉) · T. H. Masoodi · M. A. Islam
T. Arjumand · A. Raja · A. A. Parrey
Division of NRM, Faculty of Forestry, Sher-e-Kashmir University of Agricultural Sciences and Technology of Kashmir, Ganderbal, Jammu & Kashmir, India
e-mail: drthmasoodi@skuastkashmir.ac.in

A. Pallavi
School of Agriculture (SoA), Himalaya University, Dehradun, Uttarakhand, India

J. H. Bhat
Central Library, Sher-e-Kashmir University of Agricultural Sciences and Technology of Kashmir, Srinagar, Jammu & Kashmir, India

U. Chatterjee et al. (eds.), *Climate Crisis: Adaptive Approaches and Sustainability*, Sustainable Development Goals Series, https://doi.org/10.1007/978-3-031-44397-8_2

ment. This interaction provides us with multiple services like food, water resources, raw materials, and recreational spaces. Natural forces such as flood, drought, typhoon, global warming, pollution, and forest fires as well as, anthropogenic activities including depletion of natural resources (Maurya et al. 2020), greenhouse gas emissions leading to both climate change and ocean acidification (Malhi et al. 2020), which causes destruction of wildlife habitat, and loss of biodiversity (Kideghesho et al. 2006), thereby, increasingly threaten the viability and resilience of natural ecosystems. Moreover, the intricate interaction among different ecosystems also implies that the transformation of one ecosystem can direct towards the damage of another ecosystem type (Keith et al. 2022). Hence, there is less assurance that ecosystems will sustainably produce many of the services ranging from raw materials, fresh air, and water, to food and a stable environment. According to the World Economic Forum (2020), nearly 1.9 million square kilometers of undisturbed ecosystems, that is roughly the size of Mexico, have been lost between 2000 and 2013. The losses are generally observed locally as ecosystems lose their potential to regulate the effects of climate change and, therefore, consequences will reverberate around the world (Keith et al. 2022). The increased recognition of ecosystem services needs to shift the world's focus towards ecosystem restoration. Now the question arises of how recovery will happen leading to restoration and then how it can assist ecosystems to retrieve their potential to generate diverse benefits. Hence, we require nature-based alternatives such as restoration techniques that assist in the recovery of degraded, damaged, and destroyed ecosystems (Griscom et al. 2019). Meaning that we have to reduce or eliminate pressure on our ecosystem and, at times, nature can itself restore or employ corrective steps. Moreover, ecological restoration is an effective strategy for climate crisis and disaster risk reduction. We must take the deliberate measures to shift from crisis to remedial, while in doing so, we must be acquainted that the remediation of nature is crucial to the continued existence of our planet the earth and human race. Among various ecosystems, farmland is one of the main ecosystems covering about one-third of the terrestrial earth's surface (Bengtsson et al. 2019), while forest ecosystem contains a larger range of biodiversity than any other ecosystems on earth (Brockerhoff et al. 2017). Farmland and forest degradation are probably the most widely known forms of ecosystem degradation as they quickly and radically threaten the sustainable livelihood security and alter the structure of the habitat. Therefore, this chapter discusses the importance of farmland and forest ecosystems, their current scenario, different forms of degradation, and possible restoration methods.

2.2 Farmland Ecosystem

Farmlands are the most vital ecosystems to sustain the life of human beings on the planet earth (Fig. 2.1). Globally, agriculture is a predominant form of land use management system covering about 40% of the terrestrial surface of the earth (FAO 2009) (Table 2.1). Arable lands and rangelands provide numerous benefits for instance food, fiber, fodder, and other requisites like biodiversity habitat for bats and birds to beetles and worms, and substantial trees from farmlands offer spiritual, cultural, and economic advantages as well (UNCCD 2017). According to the estimates, agriculture sustains the daily livelihoods of approximately 2 billion people (Searchinger et al. 2019) and contributes around 90% of food calories and 80% of protein and fats for livestock production (Viana et al. 2021). Moreover, agriculture contributes to local economies directly through job creation, sales, support services, and businesses, and also by delivering lucrative subsidiary markets like food processing (FAO 2017). But the way we are utilizing land resources leads to the loss of their potential and vitality. In developing countries like India, an abundant and large agricultural sector yields around 16% of GDP and 10% of export income. The country ranks the second largest in the world after the United States

Fig. 2.1 Farmlands (**a**. Paddy fields **b**. Maize fields)

Table 2.1 Current status of farmland ecosystem

Region	Total area under farmlands	Area presenting reduced productivity (%)
World	4.8 billion ha	20
India	159.7 million ha	30

Source: FAO (2021) and UNEP (2021)

(Table 2.1), by supporting a huge human population (18%) and livestock population (15%) across the globe (FAO 2011). Furthermore, it possesses large gross irrigated cropland of 82.6 million hectares at the global level and, hence, is counted among the top three international producers of several crops, involving rice, wheat, pulses, peanuts, cotton, vegetables, and fruits.

2.3 Degradation of Farmlands

Farmland is degrading owing to the depletion of crop yields and livestock productivity (Table 2.1). Farmland and its surrounding are prone to an unintentional or deliberate degradation in their physical, chemical, and biological states by a range of farming activities like intensive cultivation and ploughing practices, overgrazing, the removal of hedges and trees, and large monocultures are causing wind and rain wear away precious soil surface. Over dose of fertilizer is lowering the soil quality and contaminating streams which chiefly results in changes to the biological diversity, air quality, climate, the quality and quantity of water, and soil conditions. Globally, almost 80% of arable land is imposed by the following different forms of degradation (UNEP 2021):

2.3.1 *Vegetation loss*: As per the report of the United Nations FAO, approximately 10 million ha of forests are removed each year. Vegetation degradation decreases the potential of nature to provisional ecosystem services. It causes climate change, flooding, desertification, and enhanced greenhouse gases in the air.

2.3.2 *Soil degradation*: Physical degradation, chemical deterioration, and soil erosion are the different types of soil degradation. It increases sedimentation and pollutants in rivers and streams. Vegetation loss due to grazing, fires, and cultivation causes soil depletion. It is reported that about 2 billion ha of soil surface have been degraded across the world.

2.3.3 *Soil salinization*: Soil salinization generally occurs in arid areas and causes decline or death of plants. Moreover, removal of trees, irrigation, drainage, and farming causes alteration in fertility and structure, with subsequent acidification and salinization of soils (Wade et al. 2008).

2.3.4 *Aridity*: It promotes the vulnerability of soil wash and deterioration of nutrient cycle. The arid zone covers 14.6%, the semiarid 12.2%, and hyper-arid region 4.2% of the total surface area of the earth.

2.3.5 *Depletion of soil carbon*: Deforestation leads to the carbon loss including soil erosion. At the worldwide level, carbon loss accounts for about 21,616 million tons annually.

Globally, nearly one-fifth part of farmland is influenced by soil erosion alone. According to Pravalie et al. (2021), about 2.5% have been amplified in soil erosion during 2001 to 2012, mainly because of agriculture expansion and deforestation. About half of the global arable land is already under mechanical- and chemical-intensive agriculture, owing to high inputs of nutrients, water requirements, and energy (Alhameid et al. 2017). It is estimated that degraded lands could decline food yields by 12%, thereby triggering food prices to get higher by 30% by 2040 (Kopittke et al. 2019). In the European Union, about 12 million hectares of heavily eroded crop fields tend to an annual loss of 1.25 billion EUR in agricultural productivity as per Panagos et al.'s (2018) investigation. It is reported by Jang et al. (2020) that the depletion in fertility of soil containing maize fields in the USA charges farmers an extra fertilizer of more than half a billion dollars in a year, while as in China, over the half of total farmland has suffered from degradation and only 14% of the surface area remains fit for crop generation (Deng and Li 2016). Local inhabitants continue to go through unemployment, food insecurity, and health problems (Dasgupta 2021). In India, the green revolution improved food production but the same revolution had unintended consequences like soil degradation, water consumption, and run-off. The major common cause for the degradation of approximately 80% of non-irrigated farmland is water erosion and wind erosion which is 17%, followed by alkalinity/salinity in 2% of land area, and 1% of waterlogged area. As per the Pocket Book of Agricultural Statistics (2020), only 66% of the farmland ecosystem remains rainfed in the country. Moreover, according to Tilman et al. (2011), agriculture contributes to a number of environmental issues like:

Climate change: The climate crisis is happening at an unprecedented rate and is the major issue of our time. However, agriculture is a primary source of emitted methane and nitrous oxide, thereby contributing to global warming.

Deforestation and biodiversity loss: Globally, agriculture is the chief cause of deforestation resulting in biodiversity and habitat loss.

Risks of genetic engineering: Several genetically modified crops risk the unrelenting vulnerability of insects to the most important pesticides of nature.

Irrigation problems: Irrigation results in enhanced water evaporation rate, affecting air temperature and surface pressure in addition to atmospheric moisture states.

Toxic pollutants: Toxic substances from farming practices include pesticides, sediments, pathogens, nutrients, salts, and metals.

Soil degradation: Agricultural practices account for around 28% of soil degradation such as tilling which breaks up and destroys soil structure, and kills many beneficial fungi and bacteria that exist there.

Wastes: Agricultural waste is an unwanted product generated as a result of farming activities, i.e., fertilizer, manure, herbicides, and pesticides.

Repeated extortion and exploitation of land and disturbance of the ecosystem in India have retorted back in the appearance of calamities, for example, the recent cyclones Yass, Uttarakhand, and Tauktae tragedy. The environmental scientists have considered ruthless anthropogenic actions accountable for such kinds of natural consequences taking a toll on the survival of humans, biodiversity, ecosystem, and infrastructure of a location. The possible primary causes for raised vulnerabilities are the following: construction without proper planning, susceptible socio-economic structure, over and unmanaged utilization of natural resources, urban sprawl, conflicts between human–wildlife, inadequate institutional potency, and climate change variability.

2.4 Forest Ecosystems

Forests make the earth fit for the biological systems by supplying fresh air, water, and shelter. They absorb carbon from the atmosphere (Harris et al. 2021) and moderate the climate (FAO and UNEP 2020), hence, are a critical defense against global warming. Forests are dwelling places for nearly all of the planet's wonderful biodiversity (Table 2.2). Forest ecosystems offer recreation, a feeling of well-being, and shade and sustain the livelihood security of billions of inhabitants across the globe (Fig. 2.2). They made a contribution to precipitation, foster groundwater recharge, regulate stream flow (FAO 2019), and supply water needs to one-third of largest cities of the world for drinking purposes (HLPE 2017). Forests provide food, medicines, shelter, energy resources, and generate employment opportunities (FAO 2018). A vigorous forest canopy is important for a long-term implementation of the hydroelectric projects. They also protect habitats from floods, landslides, and avalanches in mountainous areas. India is one of the 17 mega-diverse countries and hosts about 8% of the globally recognized fauna and flora. The country is home to a variety of forest types: temperate and subtropical montane forests, dry and moist tropical forests, scrub, and alpine forests. Forests have significantly contributed to the human lives by generating a fresh environment and a diversity of valuable forest resources for food and medicine (Khan et al. 2022). As well, Indian forests sustain the livelihoods of almost 275 million people by providing fuel wood, food, fodder, and other minor forest resources. Furthermore, forest provides many ecosystem services for instance amelioration of climate, carbon sequestration, in situ conservation of biodiversity, and safeguarding ecological species that have been replaced by uneatable species and pollution abatement (Deva 2021).

2.5 Degradation of Forests

After un-irrigated farmland, forest ecosystems remain the most vulnerable to degradation. Because forests are under unrelenting pressure due to increasing populace growth and demands for more land surface and resources to fulfill the basic needs of life. Over 80% of the earth's forest ecosystem has been fragmented, degraded, or cleared (Table 2.2), causing a serious threat to the world's climate, biodiversity, and the livelihood security of hundreds of millions of population. As per recent reports, even though deforestation has decreased in these years but during the years 2015–2020, the globe still lost nearly 10 million hectares of forests in a year (FAO and UNEP 2020). By 2050, the global vegetation cover could be reduced by about 223 million hectares if these rates prolong unabated (Bastin et al. 2019). Fires, diseases, pests, drought, invasive species, and adverse climatic events are affecting around 122 million hectares of forest area on an average per annum (IUFRO 2018). Moreover, deforestation could affect nearly 1.75 billion forest fringe communities involving native local people as well as those who do jobs in official or private forest-related enterprises. It adds to the hazard of water logging besides the partition of various ecosystems, enhancing human–wildlife conflicts (Gibb et al. 2020). Many forest ecosystems are also degraded due to firewood cutting, logging, and by pollution. Though trees outside the forest are declining in number because of intensive agriculture practices and creating ways to houses, roads, and dams. As per the report of the UNEP and ILRI (2020), it has been found to be related to the pandemic of animal-borne infections as well, for instance, COVID-19. The degradation and extensive loss of indigenous forests are now identified as the chief environmental problem. Logging operations, deforestation, and other disturbances between 2001 and 2019 have resulted

Table 2.2 Current status of forest ecosystem

Region	Total forest area	Depletion in productivity (%)	References
World	4.06 billion ha	10.34	UNEP (2021)
India	71.3 million ha	18	ISFR (2021)

Fig. 2.2 Role of forest ecosystem

in huge emissions of 8.1 ± 2.5 $GtCO2e$ per annum (Harris et al. 2021). Despite immense ongoing losses, the forests still occupy greater than 30% of the earth's surface. According to the India State of Forest Report (2021), India's forest and tree cover increased by more than 1500 km^2 between 2019 and 2021, bringing the total forest area to 80.9 million hectares of the country. Even if India has revealed an increase in the total area of forest cover, still there are several locations in the country that have shown a decline in the natural forest cover. Both the quality and the quantity of forests are reducing primarily due to anthropogenic pressures (Fig. 2.3). Clearing forests for many developmental purposes is stopping the cover of native forest lands. Moreover, drivers of degradation lead to the deterioration of forest growing stocks in India (Fig. 2.4).

Thus, these factors lead to the problems of food shortage, exposing soil to direct rain and heat, climate change, loss of biodiversity, flooding, displacement of native communities, health issues, and consequently economic loss. The forests are under tremendous pressure due to the rapid increase in human and livestock population to meet the increasing demands for fodder, grazing, small timber, and other forest produce. Change in vegetation cover, rise in temperatures, shortage of drinking water and rapid deforestation, corridor fragmentation, and habitat destruction may create a big threat to the extinction of wild fauna and flora.

2.6 Why Restoration?

We need to recreate a stable network with different ecosystems that sustain us. After the degradation of an ecosystem, we have three options, namely, rehabilitation (partial displacement of the original ecosystem), enhancement (alternative ecosystem), and restoration (meaning bring back to pre-disturbance condition) (Fig. 2.5) or allow further degradation. In broad terms, rehabilitation of an ecosystem means to repair and replace the basic structures and functions of the environment, which have been disrupted or eliminated by means of disturbance (Cook 2005), whereas the ecological enhancement connotes the manipulation of natural landscapes to enhance and improve its ecological functions while at the same time safeguarding the health of humans and the environment (ITRC 2004). Restoration means the occurrence of indigenous vegetation, the creation of tree plantations, or the implementation of regenerative farming practices. It can help in the recovery of degraded, damaged, or destroyed ecosystems and is necessary to ensure food provision for a rapidly growing populace, mitigating the effects of the climate crisis as well as halting biodiversity loss. Therefore, ecosystem restoration is the method of reversing back and halting degradation, consequently producing improved ecosystem benefits, and restores biological diversity (UNEP 2021). Active restoration is labor intensive, however, essential for cultiva-

Fig. 2.3 Causes of forest degradation (**a**. deforestation, **b**. overgrazing, and **c**. urbanization)

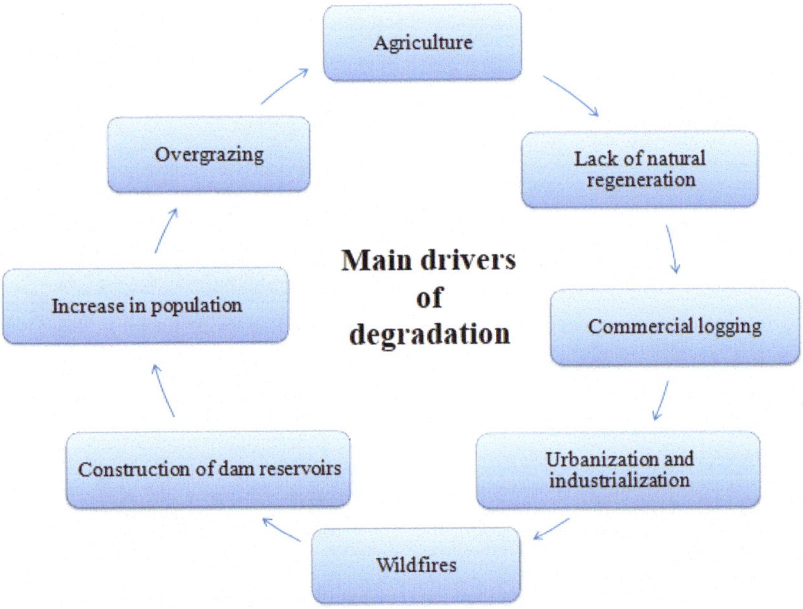

Fig. 2.4 Major drivers of forest degradation

tion to restore. More importantly, ecosystem restoration provides an effective solution to address the effects of the climate crisis (SER 2021) by enhancing resilience and reducing susceptibility to extreme events.

Hence, the amelioration of degraded environments has remarkable capacities to progress the Sustainable Development Goals (Fig. 2.6) by providing the following benefits:

2.6.1 *Health & welfare*: Air pollution leads to nine million premature deaths each year. Therefore, re-establishing blue and green spaces can contribute a significant part to the human welfare.

2.6.2 *Climate change mitigation & adaptation*: It can significantly contribute to more than one-third of the overall climate change mitigation required by 2030 to maintain

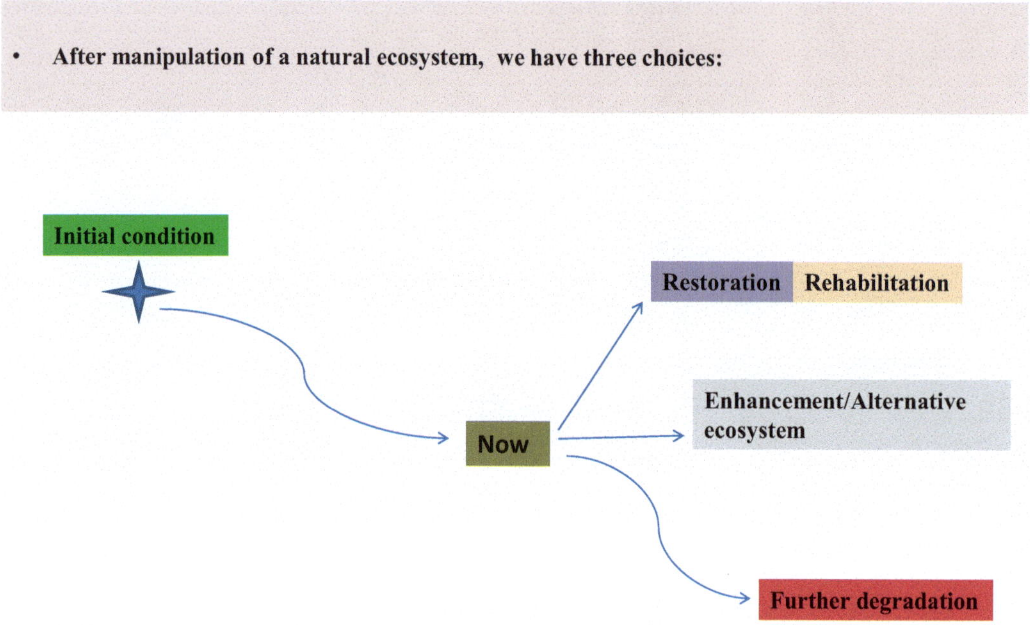

Fig. 2.5 Basic concept of restoration

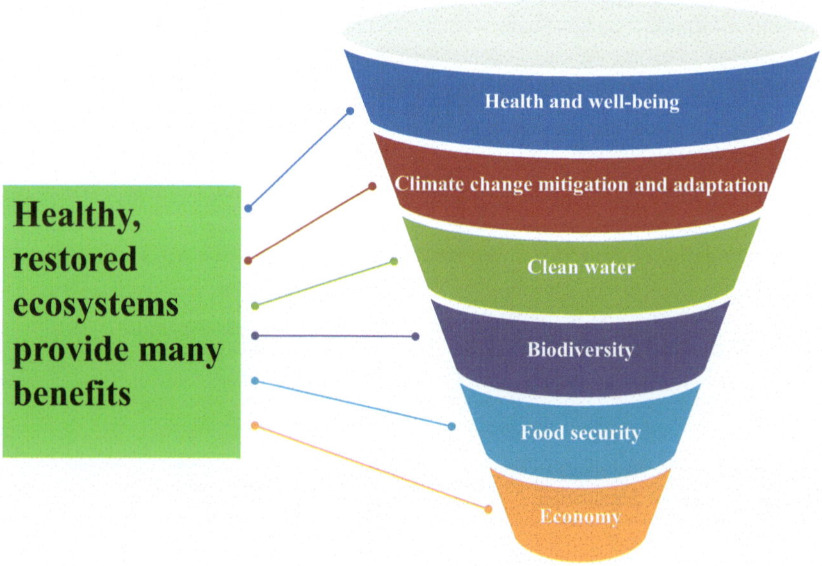

Fig. 2.6 Benefits to be obtained from restored ecosystems

global warming to just lower than 2 °C. It also plays a vital role in individual's adaptation to climate change and reduces exposure to extreme events.

2.6.3 *Clean water*: Internationally nearly 81% of cities can reduce contamination in water by integrating restoration in agriculture and practices of forest protection.

2.6.4 *Biodiversity*: The restoration of biodiversity centers on rehabilitating (reintroducing) species to the ecosystems from where they have been found extinct.

2.6.5 *Food security*: Food security for nearly 1.3billion populace can be significantly enhanced by employing solely agroforestry practices. Restoration facilitates the alleviation of poverty and provides opportunities for the young by means of improved resource access and through employment generation.

2.6.6 *The economy*: Ecosystem restoration creates livelihood opportunities and about half of the international GDP is found reliant on the nature.

2.7 Restoring Farmland Ecosystem

Restoration of farmland is fundamental for the sustainability of agriculture practices and the surrounding environment. Ways to revive farmlands include reduced tillage, growing a range of crops, including trees, and use of more natural pest control and fertilizer. These steps can re-establish carbon provisions in soils, making them more fertile so that nations can nourish their growing populations without utilizing even more land. On croplands, a range of operations can make contribution to ecosystem re-establishment, involving climate-smart agriculture, organic farming, conservation farming techniques, agro-ecology, intensification and sustainable land management, bio-innovations, regenerative agriculture, and integrated production systems (Saturday 2018). The restoration of farmland also creates habitat for wildlife. Restoring degraded cropland and rangeland is the topmost priority to feed a rising population and protect forests from deforestation for fresh farmland ecosystems. Bringing back degraded sites into fruitful productive utilization can also transform them into carbon sinks. Abandoned farmland ecosystems can be a source of emissions if soils are left to wash away. In an attempt to protect soil from erosion, several biological efforts are utilized such as the use of crop residues as manure, buffers, and conditioner application in direct contact with the soil surface. Even agroforestry practice has the capacity to enhance soil carbon by 21%, phosphorus by 11%, and inorganic nitrogen by 46% and subsequently diminish soil erosion by 50% (Muchane et al. 2020), in addition to food security. By means of reconstruction of fencing and recharge stabilization, retaining remnant vegetation, runoff interception earthworks, revegetation, and water table lowering, saline agricultural lands can be restored. Creative ideas/solutions are also applied like the improvement of natural bioremediation by using microbial inoculants and biostimulation processes. The bioremediation technique involves the use of living beings, such as bacteria and microbes, in the degradation of pollutants and encounter soil pollution. Such natural techniques can be improved by biostimulation, which means supplying inputs to speed up the activities of present bioaugmentation and microbiomes that is adding up more microbes to remove particular contaminants (Goswami et al. 2018). Amelioration by land distribution through environment-friendly arable farming operations has the significant capacity to enhance agricultural yield and provides other ecosystem benefits at both the arable farmed field and landscape scale. However, restoration by land separation would provide these triple benefits only at the landscape scale as this restoration type is at the expense of field-level agricultural production. However, recovery can be time-consuming, requiring decades to centuries to reach natural or pre-cultivation states and, in a few cases, soils continue to degrade with no active restoration.

For example, to appraise the unsustainable agriculture in one of India's states namely Andhra Pradesh, the state is focusing on the utilization of Zero-Budget Natural Farming, and the method relies on the most recent technical investigation in ecology and entrenched in Indian culture. It serves as a substitute to exorbitant-cost synthetic inputs-based agriculture and the objective of the program is to make aware that all growers in the state use either less or free cost commonly resourced inputs, for instance, cow urine and cow dung to attain 100% chemical-free farming by 2024. The current study revealed that natural farming enhanced carbon sequestration, soil fertility, biodiversity and raised farmer's income, lessen pesticide uses, water utilized for electricity consumption and irrigation, and improved more than 40% of crop resilience (Galab et al. 2019). The technique engages *waaphasa* (aeration of soil) which was created to store about 1400–3500 cubic meters of water per acre in a paddy cropping time period and thereby, less water utilization in paddy farming and also reducing farmers' dependence on groundwater reservoirs for both irrigation and the electricity generation (CSTEP 2020).

2.8 Restoring Forest Ecosystem

Restoring the forest ecosystem is a decisive strategy for better climate regulation, improving erosion and flood control, diversification of food and non-food resources, and providing livelihood opportunities for indigenous communities. There are billions of individuals who depend on forest resources for their food, water, and livelihoods. Unfortunately, the major driver of forest loss is the food system. Reviewing the ways people raise and utilize food sources can aid in decreasing the strain on forest ecosystems. Moreover, the degraded and abandoned farmland ecosystems can be perfect for forest remediation, which connotes raising stands of forest and plants in landscapes that involve active farming. Therefore, the process of forest restoration involves re-establishing trees on former forest lands and rehabilitating the condition of degraded forests, while conserving wild plants and animals and also preserving the water and soil resources. Restoration of forests also signifies developing patches of trees and woods in landscapes including active farmlands and villages. Restoring and rehabilitating natural forests is the solely scientific way to tackle the climate crisis. Managing, protecting, and rehabilitating forests offer nearly two-thirds of the total mitigation capacities of all nature-based alternatives. Forests help in the reduction of greenhouse gas emissions by storing and sequestering carbon dioxide from the atmosphere. According to the UN Decade on Ecosystem Restoration, around 26 gigatons of greenhouse gas emissions can be removed from the atmosphere by restoring and rehabilitating the ecosystem. Reversing the forest cover loss by means of restoration and preventing the further degradation of forests both contribute greatly to the global efforts of mitigating the climate crisis. Restoring the forest ecosystem represents regrowing and reducing the stress on forests so that trees are re-established in nature. In current forests, indigenous species can be planted to reproduce the tree cover. Though, in several cases, forest plants will re-grow naturally by natural regeneration. Successful restoration of forests must involve setting away some land for fast-growing native species feasible for charcoal, like eucalyptus, to replace wood from natural native forests in some countries. Apparently, forest restoration is a powerful process in conserving and protecting both the natural and the manmade landscape, but cautious consideration must be attempted when evaluating the best approach to fulfill the restoration objectives. A number of the operations employed when pursuing the restoration of forests include afforestation, invasive species management, and restoring historic groundwater levels.

2.9 Restoration Methods

In general, there are various ways by which degradation can be converted into restoration. Some of the approaches are listed below.

2.9.1 Regenerative Agriculture

The technique is an alternative option for generating food having less or constant net positive environmental as well as social effects (Newton et al. 2020).

It is a rehabilitation and conservation approach where focus is given to topsoil regeneration. The approach helps in recreating soil organic matter and restoring abandoned soil biological diversity. It affects both carbon sequestration and improves the water-cycle processes. The techniques like using cover crops, decreased crop rotation, tillage, use of composts as well as avoiding synthetic fertilizers help in upgrading soil health, soil carbon, and restoring biodiversity on-farm. Restorative agriculture has five key principles such as (i) elimination of chemical, physical, and mechanical field treatment; (ii) biodiversity enhancement; (iii) utilization of all-year cover crop plants; (iv) conservation of live roots of perennial plants; (v) integration of livestock into crop yield. Regenerative farmers promote and protect associations between people, water bodies, land, livestock, wildlife, and also microbial existence in soil. Therefore, it enhances the environmental, economic, and social dimensions of sustainable food production. For example, General Mills has assured to sustain regenerative agriculture over a million acres of farmland ecosystem by 2030.

2.9.2 Climate Smart Agriculture

Climate-smart agriculture (CSA) is a process that assists in changing agricultural food systems direct towards the green as well as climate resilient performances. CSA sustains approaching globally agreed objectives for instance the Paris Agreement and the Sustainable Development Goals. It intends to grab three main goals: adapting and establishing resilience to climate variability; sustainably enhancing agricultural productivity and livelihoods; and if possible lowering and eliminating greenhouse gas emissions. The CSA also plays a potent task in counteracting climate change and variability effects and reducing pressure on natural forest ecosystems (Nkumulwa and Pauline 2021). For example, agroforestry tree components can enhance the potential of seasonal crop plants to tolerate dry periods and, therefore, shun overall crop failure in the farmland while ensuring food security.

2.9.3 Forest and Landscape Restoration

Forest and landscape restoration (FLR) presents a framework for employing restorative interventions that together address main environmental challenges like land and soil degradation, loss of biodiversity, lack of sustainable livelihoods, water scarcity, and climate change mitigation and adaptation (Guariguata et al. 2021). By altering the abandoned soils, forests, agricultural lands, and watersheds consequently results in retrieving their ecological functionality. Recreating various social, economic, and ecological functions on a landscape provides different kinds of ecosystem provisions and benefits that promote multiple groups of stakeholders. The Bonn Challenge as well as regional programs have initiated political support to start the restoration of forest landscapes on 150 and 350 million ha by 2020 and 2030, respectively.

2.9.4 Phytoremediation

Phytoremediation is basically the usage of green plants and accompanying soil microbial population to deplete the concentration levels or toxic impacts of pollutants in the surrounding environment. Phytoremediation technology is generally recognized as a cost-efficient ecological restoration biotechnology (Greipsson 2011). It has many types based on the objectives and type of contaminants present in the growing soil and water media such as phytoextraction, phytodegradation, rhizofiltration, phytovolatilization, and phytostabilization. Plants have the capacity to accumulate ionic compounds even at low concentrations present in the soil by means of root systems. By extending the root system into the soil mixture and by developing the rhizosphere ecosystem, plants extract pollutants and modulate their biological availability resulting in the rehabilitation of polluted ecosystems and also improving the fertility of the soil.

2.9.5 Phytocapping

Phytocapping is a sustainable, an eco-friendly and cost-effective practice to cap old landfill/ dump sites and to cover new landfill areas, with an aim to reduce leachate production by lowering water intrusion into waste products and to moderate greenhouse emissions and resulting odor. In phytocapping technology, trees are exploited as "bio-pumps" and "rainfall interceptors" and soil surface as "storage" of water resources (Venkatraman and Ashwath 2007). Phytocapping decreases percolation by means of the following three mechanisms: (a) sequestration of moisture in the soil layers, (b) canopy interception of rainfall, and (c) evapotranspiration of stored water sources. This technology provides the following benefits like geotechnical, ecological, environmental, and social characteristics. It restores the space already occupied by landfills into an abundant vegetation cover possessing inherent significances for the ecology and environment of the surroundings.

2.9.6 Nucleation Techniques

In this technique, a small nucleus of vegetation is utilized for the regeneration of vegetation. The vegetation is restored through natural succession processes (Reis et al. 2010). The vegetation nuclei serve the purpose of attracting plants and animals, thereby allowing them to colonize the area. There are six different nucleation techniques that are generally used in restoration programs as:

I. Insertion of artificial perches or small trees or shrubs in degraded areas to accelerate plant succession.
II. Soil with a higher amount of seeds and microorganisms are transported from non-degraded areas to the degraded sites.
III. Key-tree species that naturally occur in those areas known as Anderson group plants are planted to increase the genetic variation.
IV. A dense group of different plant species is planted at equal distances on the island.
V. Natural perches are planted to boost regeneration.
VI. The area is left to naturally regenerate without human intervention.

2.9.7 Miyawaki Forest: An Eco-sustainable Afforestation Technique

The Miyawaki forest introduced by Japanese ecologist Akira Miyawaki is an opportunity to take part in ecosystem rehabilitation. The method is considered to guarantee that the growth of the plant is 10 times quicker and consequently the established plantation is 30 times denser than the normal plantation. The process comprises planting indigenous plant species in the same zone and growing without any maintenance afterwards the establishment period of 3 years. Moreover, Miyawaki-developed forests grow quickly; about 1 m in height per year and an area without vegetation can be converted into an indigenous forest rapidly in a decade with plants 10 m in height. Then, it will be entirely developed in another decade, with native vegetation covering various layers. The approach works very well in urban environments. It is different from traditional forests in the way that Miyawaki forest areas found to possess abundant biodiversity and are confined to capture much carbon emissions from the atmosphere.

2.9.8 Sustainable Agriculture

Sustainable agriculture can be defined as the use and management of the agricultural ecosystems in a way that sustains its biodiversity, regeneration capacity, yields, potential, and vitality to do work, to fulfill economic, ecological, and social purposes at the local, national, and worldwide levels and that without harm to other ecosystems. This approach is intended to preserve the surrounding environment, expand the natural resource base of the earth, and sustain and improve the fertility of the soil. It seeks to enhance the quality of life of farmers, advance environmental stewardship, increase productivity of fiber and food needs, and more importantly increase farmer's annual income. Therefore, it is a method to meet human's food and fiber needs at present and to ensure that future generations will meet their own needs without any compromise as present generations are obtaining.

2.10 Conclusion

Ecosystems are the base of our life. Various anthropogenic actions and some natural forces are damaging these ecosystems and endangering the lives of biological systems on the earth. The degradation of the ecosystem is occurring at an alarming rate. The degradation has caused a severe impact on food and livelihood security. Globally the detrimental impact has been well-recognized but the efforts to restore the ecosystems are rather slow or unrecognizable. The loss of agriculture and the forest ecosystem will indeed harm the entire humanity. However, it is time to work together and take suitable approaches to combat the degradation of both the farmlands and forests. In such circumstances, the restoration approach is an opportunity for all the stakeholders to save the planet before it is doomed. Ecosystem restoration is a natural and easy process with effective results. The increased recognition of ecosystem services needs to shift the world's focus towards ecosystem restoration. But, the implementation should be done with full dedication and integrity. Furthermore, ecosystem degradation one of the challenges at present can be converted into restoration in various ways like regenerative agriculture, phytoremediation, phytocapping, CSA, forest and landscape restoration, sustainable agriculture, Miyawaki forest, and nucleation techniques. Deliberate attempts by the state and central governments including the local people can bring good results while focusing on native species through the organization of various eco-restoration programs.

References

Alhameid A, Tobin C, Maiga A, Kumar S, Osborne S, Schumacher T (2017) Intensified agroecosystems and changes in soil carbon dynamics. In: Al-Kaisi MM, Lowery B (eds) Soil health and intensification of agro-ecosytems. Academic Press, Amsterdam, pp 195–214

Bastin JF, Finegold Y, Garcia C, Mollicone D, Rezende M, Routh D, Zohner C, Crowther TW (2019) The global tree restoration potential. Science 365(6448):76–79. https://doi.org/10.1126/science.aax0848

Bengtsson J, Bullock JM, Egoh B, Everson C, Everson T, O'Connor T, O'Farrell PJSHG, Lindbor R (2019) Grasslands—more important for ecosystem services than you might think. Ecosphere 10. https://doi.org/10.1002/ecs2.2582

Brockerhoff EG, Barbaro L, Castagneyrol B et al (2017) Forest biodiversity, ecosystem functioning and the provision of ecosystem services. Biodivers Conserv 26:3005–3035. https://doi.org/10.1007/s10531-017-1453-2

Centre the Study of Science, Technology and Policy (CSTEP) (2020) Life cycle assessment of ZBNF and non-ZBNF: a preliminary study in Andhra Pradesh. https://cstep.in/drupal/node/932

Cook GD (2005) Ecosystem rehabilitation. Lake Reservoir Manage 21(2):218–221. https://doi.org/10.1080/07438140509354431

Dasgupta P (2021) The economics of biodiversity: the Dasgupta review. HM Treasury, London

Deng X, Li Z (2016) Economics of land degradation in China. In: Nkonya E, Mirzabaev A, von Braun J (eds) Economics of land degradation and improvement: a global assessment for sustainable development. Springer, Cham. https://doi.org/10.1007/978-3-319-19168-3_13

Deva MAW (2021) Eco restoration and climate change in J&K. The transition to a low carbon economy is the most significant investment opportunity of our lifetimes. Greater Kashmir Newspaper. https://www.greaterkashmir.com/todays-paper/eco-restoration-and-climate-change-in-jk

FAO (2011) The state of the world's land and water resources for food and agriculture (SOLAW) – managing systems at risk. Food and Agriculture Organization of the United Nations/Earthscan, Rome/London. http://www.fao.org/nr/solaw/en/

FAO (2017) The future of food and agriculture: trends and challenges. FAO, Rome

FAO (2018) The state of the world's forests 2018. FAO, Rome

FAO (2019) Forests: nature-based solutions for water. Unasylva 90. http://www.fao.org/3/ca6842en/CA6842EN.pdf

FAO (2021) Land use statistics and indicators statistics. Global, regional and country trends 1990–2019. FAOSTAT Analytical Brief Series No 28. Rome. https://www.fao.org/3/cb6033en/cb6033en.pdf

FAO and United Nations Environment Program (UNEP) (2020) The state of the world's forests. Forests, biodiversity and people. FAO, Rome. https://doi.org/10.4060/ca8642en

Food and Agriculture Organization of the United Nations (FAO) (2009) Statistics from www.faostat.fao.org. Updated April 2009. Rome: FAO

Galab S, Reddy P, Raju DSR, Ravi C, Rajani A (2019) Impact assessment of zero budget natural farming in Andhra Pradesh-Kharif 2018–19: a comprehensive approach using crop cutting experiments. Centre for Economic and Social Studies, Telangana

Gibb R, Redding DW, Chin KQ, Donnelly CA, Blackburn TM, Newbold T, Jones KE (2020) Zoonotic host diversity increases in human dominated ecosystems. Nature 584:398–402. https://doi.org/10.1038/s41586020-2562-8

Goswami M, Chakraborty P, Mukherjee K, Dey S, Mitra G, Tribedi P (2018) Bioaugmentation and biostimulation: a potential strategy for environmental remediation. J Microbiol Exp 6(5):223–231. https://doi.org/10.15406/jmen.2018.06.00219

Greipsson S (2011) Phytoremediation. Nat Educ Knowl 3(10):7p

Griscom BW, Lomax G, Kroeger T, Fargione JE, Adams J, Almond L, Bossio D, Cook-Patton SC, Ellis PW, Kennedy CM, Kiesecker J (2019) We need both natural and energy solutions to stabilize our climate. Glob Chang Biol 25:1889–1890. https://doi.org/10.1111/gcb.14612

Guariguata MR, Atmadja S, Baral H, Boissiere M, Brady M, Chomba S, Cronkleton P, Djoudi H, Duchelle A, Duguma L, Elias M, Gitz V, Graudal L, Harrison R, Ihalainen M, Jamnadass R, Jalonen R, Kassa H, Kettle C, Lawry S, Li Y, Locatelli B, Martius C, Meybeck A, Minang P, Murdiyarso D, Nawir A, Robiglio V, Sist P, Somarriba E, van Noordwijk M, Wardell A, Vagen T-G, Winowiecki L (2021) Forest and landscape restoration. FTA highlights of a decade 2011–2021 series. Highlight No. 4. Bogor, Indonesia: The CGIAR Research Program on Forests, Trees and Agroforestry (FTA). https://doi.org/10.17528/cifor/008214

Harris NL, Gibbs DA, Baccini A, Birdsey RA, de Bruin S, Farina M, Fatoyinbo L, Hansen MC, Herold M, Houghton RA, Potapov PV, Suarez DR, Roman-Cuesta RM, Saatchi SS, Slay CM, Turubanova SA, Tyukavina A (2021) Global maps of twenty-first century forest carbon fluxes. Nat Clim Chang 11(3):234–240. https://doi.org/10.1038/s41558-020-00976-6

HLPE (High Level Panel of Experts on Food Security and Nutrition) (2017) Sustainable forestry for food security and nutrition. A report by the high level panel of experts on food security and nutrition of the committee on world food security. HPLE, Rome. http://www.fao.org/3/i7395e/i7395e.pdf

India State of Forest Report (ISFR) (2021) Forest survey of India. Ministry of Environment Forest and Climate Change, Dehradun

International Union of Forest Research Organizations (IUFRO) (2018) Global fire challenges in a warming world. In: Robinne FN, Burns J, Kant P, de Groot B, Flannigan MD, Kleine M, Wotton DM (eds) Occasional Paper No. 32. IUFRO, Vienna

Interstate Technology & Regulatory Council (ITRC) (2004) Making the case for ecological enhancement. ECO-1. Interstate Technology & Regulatory Council and the Wildlife Habitat Council, Washington, DC. http://www.itrcweb.org

Jang WS, Neff JC, Im Y, Doro L, Herrick JE (2020) The hidden costs of land degradation in US maize agriculture. Earth's. Future 9(2). https://doi.org/10.1029/2020EF001641

Keith DA, Ferrer-Paris JR, Nicholson E et al (2022) A function-based typology for Earth's ecosystems. Nature 610:513–518. https://doi.org/10.1038/s41586-022-05318-4

Khan S, Masoodi TH, Islam MA, Wani AA, Gatoo AA (2022) Ethnomedicinal study of wild plants used by fringe communities in Temperate Forests of Himalayan Kashmir. India Phytomed Plus 2:100251. https://doi.org/10.1016/j.phyplu.2022.100251

Kideghesho JR, Nyahongo JW, Hassan SN, Tarimo TC, Mbije NE (2006) Factors and ecological impacts of wildlife habitat destruction in the Serengeti ecosystem in northern Tanzania. AJEAM-RAGEE 11:917–932

Kopittke PM, Menzies NW, Wang P, McKenna BA, Lombi E (2019) Soil and the intensification of agriculture for global food security. Environ Int 132:105078. https://doi.org/10.1016/j.envint.2019.105078

Malhi Y, Franklin J, Seddon N, Solan M, Turner MG, Field CB, Knowlton N (2020) Climate change and ecosystems: threats, opportunities and solutions. Phil Trans R Soc B375:20190104. https://doi.org/10.1098/rstb.2019.01014

Maurya PK, Ali SA, Ahmad A, Zhou Q, da Silva CJ, Khane E, Ali H (2020) An introduction to environmental degradation: causes, consequence and mitigation. In: Kumar V, Singh J, Kumar P (eds) Environmental degradation: causes and remediation strategies, pp 1–20. https://doi.org/10.26832/aesa2020-edcrs-01

Muchane MN, Sileshi GW, Gripenberg S, Jonsson M, Pumarino L, Barrios E (2020) Agroforestry boosts soil health in the humid and sub-humid tropics: a meta-analysis. Agric Ecosyst Environ 295:106899. https://doi.org/10.1016/j.agee.2020.106899

Newton P, Civita N, Frankel-Goldwater L, Bartel K, Johns C (2020) What is regenerative agriculture? A review of scholar and practitioner definitions based on processes and outcomes. Front Sustain Food Syst 4:577723. https://doi.org/10.3389/fsufs.2020.577723

Nkumulwa HO, Pauline NM (2021) Role of climate-smart agriculture in enhancing farmers' livelihoods and sustainable forest management: a case of villages around Songe-Bokwa Forest, Kilindi District. Tanzania Front Sustain Food Syst 5:671419. https://doi.org/10.3389/fsufs.2021.671419

Panagos P, Standardi G, Borrelli P, Lugato E, Montanarella L, Bosello F (2018) Cost of agricultural productivity loss due to soil erosion in the European Union: from direct cost evaluation approaches to the use of macroeconomic models. Land Degrad Dev 29(3):471–484. https://doi.org/10.1002/ldr.2879

Pocket Book of Agricultural Statistics (2020) Government of India, Ministry of Agriculture & Farmers Welfare, Department of Agriculture and Farmers Welfare

Pravalie R, Patriche C, Borrelli P, Panagos P, Rosca B, Dumitrascu M, Nita I, Savulescu I, Birsan M, Bandoc G (2021) Arable lands under the pressure of multiple land degradation processes. A global perspective. Environ Res 194:110697. https://doi.org/10.1016/j.envres.2020.110697

Reis A, Bechara FC, Tres DR (2010) Nucleation in tropical ecological restoration. Sci Agric 67:244–250. https://doi.org/10.1016/j.ncon.2014.09.002

Saturday A (2018) Restoration of degraded agricultural land: a review. J Environ Health Sci 4(2):44–51. https://doi.org/10.15436/2378-6841.18.1928

Searchinger T, White R, Hanson C, Ranganathan J (2019) Creating a sustainable food future: a menu of solutions to feed nearly 10 billion people by 2050. WRI, Washington, DC

Society for Ecological Restoration (SER) (2021) Climate change is here: ecological restoration can help us meet this moment. In: Four reasons why we should all be talking about restoration in light of the last UN report. SER, Washington, DC. https://www.ser.org/news/576671/Climate-change-is-here-ecological-restoration-can-help-us-meet-this-moment.htm

Tilman D, Balzer C, Hill J, Befort L (2011) Global food demand and the sustainable intensification of agriculture. PNAS 108(50):20260–20264. https://doi.org/10.1073/pnas.1116437108

UNEP and ILRI (2020) Preventing the next pandemic: zoonotic diseases and how to break the chain of transmission. UNEP and International Livestock Research Institute, Nairobi. https://www.unep.org/resources/report/preventing-future-zoonotic-disease-outbreaks-protecting-environmentanimals-and

United Nations Convention to Combat Desertification (UNCCD) (2017) The global land outlook, 1st edn. UNCCD, Bonn

United Nations Environmental Programme (UNEP) (2021) Making peace with nature: a scientific blueprint to tackle the climate, biodiversity and pollution emergencies. UNEP, Nairobi. https://www.unep.org/resources/making-peace-nature

Venkatraman K, Ashwath N (2007) Phytocapping: an alternative technique to reduce leachate and methane generation from municipal landfills. Environmentalist 27:155–164. https://doi.org/10.1007/s10669-007-9014-y

Viana CM, Freire D, Abrantes P, Rocha J, Pereira P (2021) Agricultural land systems importance for supporting food security and sustainable development goals: a systematic review. Sci Total Environ 806(3). https://doi.org/10.1016/j.scitotenv.2021.150718

Wade MR, Gurr GM, Wratten SD (2008) Ecological restoration of farmland: progress and prospects. Philos Trans R Soc Lond Ser B Biol Sci 27(1492):831–847. https://doi.org/10.1098/rstb.2007.2186

World Economic Forum (2020) Nature risk rising: why the crisis engulfing nature matters for business and the economy. https://www.weforum.org/agenda/2020/09/humansdestroyed-ecosystem-mexico-2000-2013

3 Exploring the Dynamics of Antarctic Sea Ice over Four Decades Using Geospatial Technology

Niladri Saha, Babula Jena, C. C. Bajish, Sandipan Das, Binaya Kumar Pattnaik, and Uday Chatterjee

Abstract

The present climate change phase has had contrasting effects on sea ice in the Arctic and Antarctic. There has been a startling loss of ancient ice in the Arctic, leading to a dramatic reduction in sea-ice area and thickness. No evidence of analogous shifts in the Antarctic has been found till 2014. This work seeks to understand the issue presented in the title by synthesizing relevant research from the literature. This chapter begins by reviewing important recent records and defining major geographical and meteorological aspects and sea-ice characteristics of the two polar areas. More specifically satellite data revealed that the Antarctic sea-ice extent is increasing at a rate of 0.7 ± 0.4 percent decade^{-1} (1979–2018). On the contrary, the Arctic sea ice showed a decreasing trend as an optimal consequence of global warming at a rate of −4.8 ± 0.2 percent decade^{-1} (1979–2018). This study explored 40 years (1979–2018) of Antarctic sea-ice remote sensing data to explore its overall and regional trends. Anomaly has been computed over the previous 3 years (2015, 2016, 2017, and 2018) to determine what form of anomaly has arisen in the Antarctic area. The findings of this study have the potential to understand the drivers and dynamics of Antarctic sea ice and contribute to future climate projections.

Keywords

Antarctic · Arctic · Sea-ice extent · Climate change · Remote sensing

3.1 Introduction

Antarctica, the southernmost continent on Earth, is a unique and remote region with a vast expanse of ice-covered land and ocean. The Southern Ocean that surrounds Antarctica plays a crucial role in regulating the Earth's climate and ocean circulation. One of the defining features of the Southern Ocean is its sea-ice cover, which exhibits a distinct seasonal cycle and interannual variability. Understanding the variability and anomalies of Antarctic sea ice is crucial for predicting future climate change and its impacts on

N. Saha · S. Das (✉) · B. K. Pattnaik
Symbiosis Institute of Geoinformatics, Symbiosis International (Deemed University), Pune, Maharashtra, India

B. Jena · C. C. Bajish
National Center for Polar and Ocean Research, Vasco Da Gama, Goa, India
e-mail: bjena@ncpor.res.in

U. Chatterjee
Department of Geography, Bhatter College, Dantan, (Affiliated to Vidyasagar University), Paschim Medinipur, West Bengal, India

U. Chatterjee et al. (eds.), *Climate Crisis: Adaptive Approaches and Sustainability*, Sustainable Development Goals Series, https://doi.org/10.1007/978-3-031-44397-8_3

global systems, including sea level rise and ocean circulation. In this context, the study of Antarctic sea-ice variability and anomalies has become an increasingly important area of research in the field of climate science.

Sea ice in the remote polar region is found in the Arctic Ocean (AO) and the Southern Ocean (SO). In Antarctica, sea-ice extent expands to the highest 19 million square kilometers in winter (September) (Desilets and Zreda 2014), and it retreats to the lowest 3.2 million square kilometers during summer (February). On the contrary, in the Arctic, sea ice decreased to its lowest at 6.4 million square kilometers in summer (September) and it expanded to its highest at 15.4 million square kilometers in winter (February) (Petersen et al. 2019). It is quite challenging to research and keep an eye on the sea ice because it is located in such a distant part of the polar regions. Several field expeditions have been attempted in the Antarctic area by researchers. The issue with this strategy is that researchers can only collect data in a very narrow location. Nonetheless, remote sensing is beneficial for monitoring sea ice because of its synoptic, repeating, and multispectral properties (Swift and Cavalieri 1985). Airborne instruments and satellites are used by the researchers to collect the data that show the sea-ice zones in the polar region, the motion of the sea ice, the temperature of sea ice, and many other variables. The satellite can easily detect sea ice in the visible, infrared, and microwave regions of the electromagnetic spectrum (Ludwig et al. 2020). The variation in the reflectivity, temperature, and emissivity of the sea-ice surface compared to the open sea water makes it one of the ideal applications in the field of remote sensing.

Sea ice is a key element of the mass transfer system as well as of global energy circulation (Jungclaus et al. 2013). It has a strong insulation capacity which can resist the momentum, mass, and energy transfer between ocean and atmosphere (Murphy et al. 2013). Also, it has a great impact on the ocean water circulation by influencing the density and salinity of the upper and bottom layers of the ocean (Timmermann et al. 2009). The key variables of the sea-ice characterization are extent, area, concentration, thickness, and types. First-year ice (FYI) and multiyear ice (MYI) are the two principal varieties of sea ice depending on their age and stage of development. FYI is the seasonal ice cover, whereas MYI is older, thicker ice that has endured at least one summer melt (Maslanik et al. 2011).

Monitoring and characterization of the sea ice are very challenging because not only sea-ice extent (SIE) changes seasonally but also the sea ice is dynamic and drifts out in several places (White et al. 2004). Current satellite-derived data indicate that global warming has an asymmetrical impact on polar sea ice (Melsheimer et al. 2023). Since 1979, the sea-ice extent in Antarctica has seen a modest expansion (Knorr and Lohmann 2014), while the sea-ice extent in the Arctic area has decreased considerably, as predicted (Vihma 2014). However, since late 2016, the Antarctic sea ice has experienced a decreasing trend that takes the interest of the climatologist.

Sea ice has a bright white surface, which returns the incoming solar radiation (ISR) to space, which is termed as albedo. The amount of albedo of the open seawater is ranging from 10% to 15% whereas sea ice reflects back 80% of ISR (Hanson 1961). If it is covered with fresh snow, then nearly 98% of ISR can be returned back to space. Due to these factors, the polar area absorbs hardly any sunlight; hence, the climate there is consistently cold. Various scientific studies reveal that the sea-ice extent of Antarctica is dramatically retreating so there will be very less ice space to reflect back the ISR which can increase the global temperature rapidly (Alekseev et al. 2022).

The sea ice resists coastal erosion by acting as a barrier to the wind and the ocean waves. Global oceanic thermohaline circulation is also controlled by the sea ice (Guo et al. 2020). In the winter months, when sea ice forms on the ocean, it ejects the salt as brine to the underlying water. Thus, the salinity and the density of the ocean water increased relative to the surrounding seawater, causing it to sink. The cold denser polar ocean water sinks and flows towards the equatorial region whereas the warmer surface water moves towards the polar region to replace it (Hay 2011).

In Antarctica, a powerful ocean current that circles the continent is a major factor in the formation and persistence of Antarctic sea ice than changes in temperature. The Antarctic circumpolar current prevents warmer sea water from the north from entering the Antarctic sea-ice zone (Baker et al. 2007). However, during 2016–2018, it was observed that the sea ice was decreasing in Antarctica (Ludescher et al. 2019). The evolution of remote-sensing technology had a significant impact on the monitoring and observation of the Cryosphere, particularly the polar area, which is one of the most dominant factors in the global climate (Ma et al. 2016). The advancement of satellite-based remote sensing observation widens the application of climatological research (Salleh et al. 2014). The polar sea-ice cover has a profound impact on ocean–atmospheric circulation, environmental cycles, marine ecology, etc. In this study, the sea-ice extent and its anomalies were analyzed to understand the reason for the unprecedented decline in the Antarctic sea ice. This study will provide an overview of the current understanding of Antarctic sea-ice variability and anomalies, highlighting the main drivers and impacts on the Earth's climate system.

3.2 Data and Methodology

Data used for this study is primarily sourced from satellite observations, which provide a comprehensive and consistent record of the Antarctic sea-ice extent over the past few decades (Fig. 3.1). The dataset is based on passive microwave data collected by the special sensor microwave imager (SSMI) and the special sensor microwave imager/sounder (SSMI/S) instruments onboard various polar-orbiting satellites. The dataset from which sea-ice area and extent are calculated are derived from the following microwave radiometer: Nimbus 7 Scanning Multichannel Microwave Radiometer (SMMR), Defence Meteorological Radiometer (DMSP) series of F8, F11, and F13 SSMI, and F17 SSMI/S. Nimbus 7 was launched in October 1978 and the SMMR captured the data every other day from 26th October 1978 through 20th August 1987. The first SSMI was launched on the DMSP F8 satellite in June 1987, and the sequence of F8, F11, and F13 SSMIs collected data on a daily basis for most of the periods from 9 July 1987 to the end of 2007, after which the F13 SSMI began to degrade. The F17 SSMIS was the second SSMIS in orbit and was launched in November 2006, with a daily data record beginning in mid-December 2006. For analyses, the trend of sea-ice data is collected from the National Snow and Ice Data Center (NSIDC) in Boulder, Colorado.

The SMMR, SSMI, and SSMIS data are used to calculate ice concentrations (percentage areal coverages of ice) with the NASA Team algorithm (Cavalieri et al. 1997), and the ice concentrations are mapped at a grid cell size of 25 × 25 km. Figure 3.2 shows the methodological framework followed in this study.

Further, the sea ice was derived by computing the areas of those pixels with more than 15% sea-ice concentration. The sea-ice concentration and extent values were analyzed to determine the spatial and the temporal variability of Antarctic sea ice. The analysis involved computing trend estimates, seasonal patterns, and inter-annual variability of sea-ice extent and concentration. The results of the analysis were interpreted in the context of existing literature on Antarctic sea-ice variability and its drivers. The findings were also compared with other studies to validate the results and provide new insights into the dynamics of Antarctic sea ice. The methodology adopted for this study involved processing and analyzing satellite data to determine the variability and anomalies of Antarctic sea ice.

3.3 Results and Discussion

The sea ice covers a large area in the polar region ranging from 18 million square kilometers to 26 million square kilometers for both the Arctic and the Antarctic region. However, global climate change has a contradicting effect in both polar regions. The Antarctic sea-ice extent (SIE) trend is increasing at a rate of 0.7 ± 0.4 percent decade^{-1}, whereas the Arctic sea-ice extent exhibits a declining trend at a rate of −4.8 ± 0.2 percent

Fig. 3.1 Antarctic sea-ice extent, April 2019

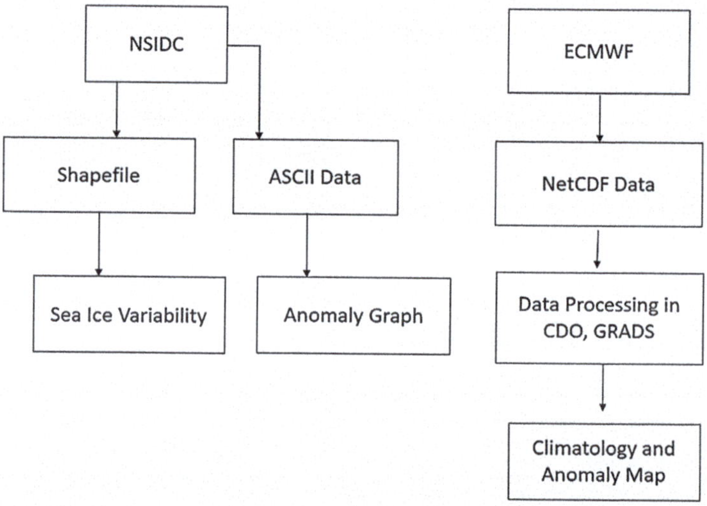

Fig. 3.2 Flow chart of methodology

decade^{-1} for the period of 1979–2018. The maximum reduction in sea-ice extent in the Arctic region was observed in the winter of July–August–September, i.e., −10.1 ± 0.7 percent decade^{-1}. Overall, the global (Arctic and Antarctic) SIE shows a negative trend at a rate of −1.9 ± 0.3 percent decade^{-1} (Table 3.1). The Antarctic region experienced a drastic decline in SIE after 2014 (Fig. 3.3a, b) which involved high variations in summer and autumn (Fig. 3.3b). The observed decline in sea ice since 2016 was attributed to polar cyclones (Turner et al. 2020; Jena et al. 2022a), SAM and ENSO (Turner et al. 2020), strengthening of Amundsen Sea Low (ASL) (Turner et al. 2022) and atmospheric planetary waves, anomalous warming of ocean mixed layer, and atmospheric temperature (Jena et al. 2022b; Turner et al. 2020). SIE of the Arctic region showed a significant relationship with a change of year (R^2 = 0.89) which was not observed in the case of Antarctica (R^2 = 0.06) (Figs. 3.3a and 3.4a).

The sea-ice area is denoted by the actual area covered by the sea ice which is usually less than SIE. The sea ice area for a grid cell is calculated by summing the concentrations of sea ice within

Table 3.1 Yearly and seasonal Antarctic, Arctic, and polar sea-ice extent trends for the period 1979–2018 with estimated standard deviations

	Years		Winter (JAS)		Spring (OND)		Summer (JFM)		Autumn (AMJ)	
Sector	$10^3 km^2 yr^{-1}$ (*R*)	Percent $decade^{-1}$	$10^3 km^2 yr^{-1}$ (*R*)	Percent $decade^{-1}$	$10^3 km^2 yr^{-1}$ (*R*)	Percent $decade^{-1}$	$10^3 km^2 yr^{-1}$ (*R*)	Percent $decade^{-1}$	$10^3 km^2 yr^{-1}$ (*R*)	Percent $decade^{-1}$
Antarctic	9.2 ± 5.7 (1.5)	0.7 ± 0.4	9.1 ± 5.1 (1.7)	0.5 ± 0.2	3.6 ± 7.4 (0.4)	0.2 ± 0.4	10.2 ± 7.8 (1.3)	2.4 ± 1.8	13.8 ± 8.3 (1.6)	1.2 ± 0.7
Arctic	−55.3 ± 3.0 (−18.3)	−4.8 ± 0.2	−75.3 ± 5.7 (−13)	−10.1 ± 0.7	−60.0 ± 4.2 (−14)	−5.7 ± 0.4	−45.7 ± 3.0 (−15)	−3.0 ± 0.2	−40.1 ± 2.7 (−14.3)	−3.0 ± 0.2
Polar	−46.1 ± 7.1 (−6.4)	−1.9 ± 0.3	−66.2 ± 7.6 (−8.6)	−2.6 ± 0.3	−56.4 ± 8.9 (−6.2)	−2.1 ± 0.3	−35.4 ± 9.5 (−3.7)	−1.8 ± 0.4	−26.3 ± 9.5 (−2.7)	−1.1 ± 0.4

Note: *JAS* July–August–September, *OND* October–November–December, *JFM* January–February–March, *AMJ* April–May–June

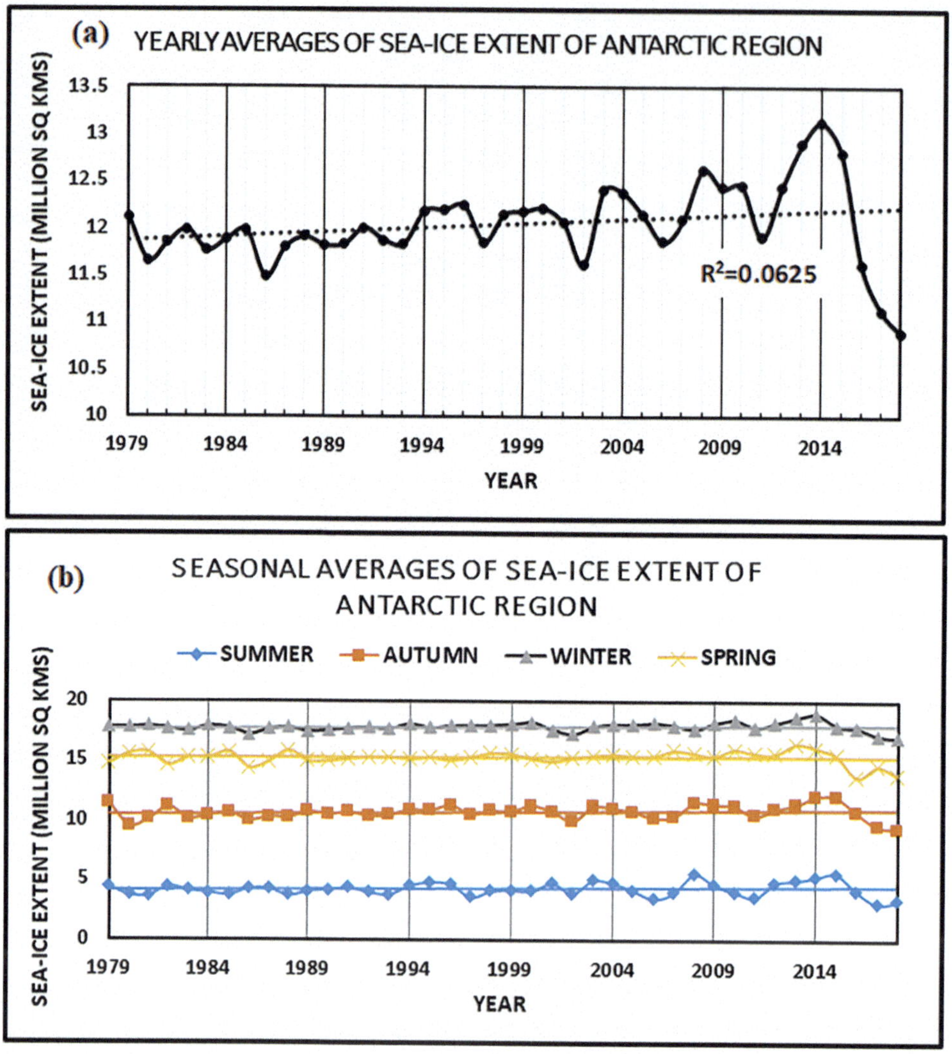

Fig. 3.3 (**a**) Yearly and (**b**) seasonal average of the Southern Ocean sea-ice extents, 1979–2018

each cell. In Antarctica, the sea-ice area is expanding at a rate of 1.1 ± 0.5 percent decade^{-1} and the Arctic is showing a negative trend at a rate of −2.4 ± 0.5 percent decade^{-1} and the entire polar sea-ice area shows a retreating trend around −0.6 ± 0.4 percent decade^{-1} (Table 3.2). Similar to SIE, the sea-ice area of the Arctic region experienced a maximum reduction in the winter of July–August–September. Figure 3.5 shows the gradual declining trend of SIE in the polar region with a rapid decrease after 2014, which might be the cumulative impact of the Antarctic region SIE at that time period (Fig. 3.3b).

The Southern Ocean contains five different sectors: Weddell Sea (WS), Indian Ocean (IO), Western Pacific Ocean (WPO), Ross Sea (RS), and Bellingshausen/Amundsen Sea (BAS) (Fig. 3.6). Analysis of the sea-ice extent data for the Antarctic region as a whole and for each five sectors for the period of 1979–2017 exhibits an expanding trend, except the BAS sector. The seasonal and yearly averaged SIE data show a positive trend for the Southern Ocean with an annual increase of 1.2 ± 0.4 percent decade^{-1} (Table 3.3). In terms of the different seasons, autumn displays the largest expanding trend

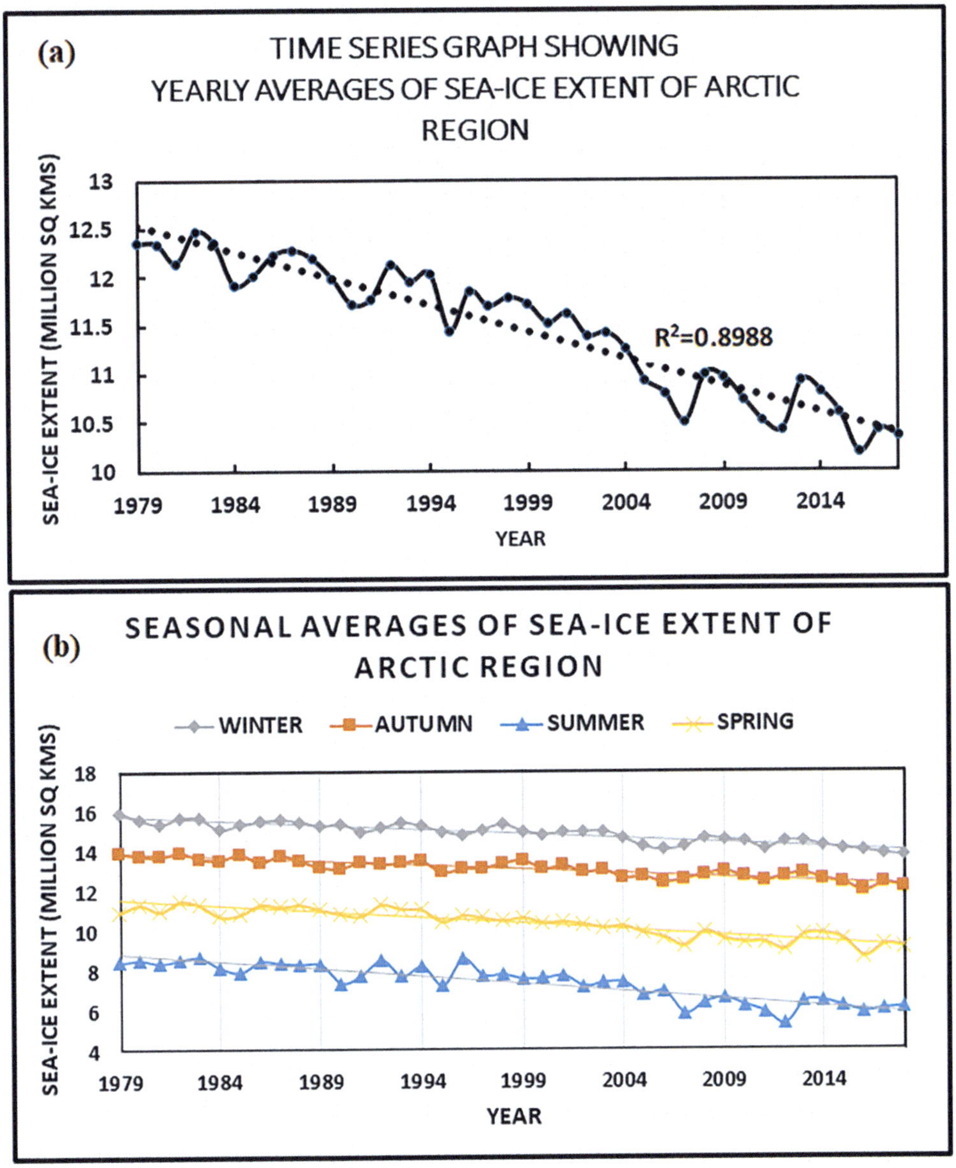

Fig. 3.4 (**a**) Yearly and (**b**) seasonal average of Northern Ocean sea-ice extents, 1979–2018, with the corresponding lines of least squares fit

(20.4 ± 7.7 $10^3 km^2 yr^{-1}$) followed by summer and winter in the Southern Hemisphere.

The WS is the largest sector among the five sectors (Longitudinal extent: 60°W- 20°E). Therefore, in comparison to other sectors, this region showed the highest increasing trend SIE which is 10.1 ± 3.0 percent decade^{-1} during the summer period of Jan–Feb–Mar. Decreasing trends in the SIE were observed in spring (−0.5 ± 0.9 percent decade^{-1}) and winter (−0.29 ± 0.7 percent decade^{-1}). Overall, the WS sector showed an annual increase of 1.2 ± 0.7 percent decade^{-1}. Seasonally, variability in trends could be observed between different sectors. In winter, the largest increasing trend is observed in the RS sector (1.7 ± 0.8 percent decade^{-1}) followed by BAS (1.5 ± 1.4 percent decade^{-1}) with a negative trend in the WS. In spring, the positive

Table 3.2 Yearly and seasonal Antarctic, Arctic, and polar sea-ice area trends for the period 1979–2018 with estimated standard deviation

Sector	Yearly		Winter (JAS)		Spring (OND)		Summer (JFM)		Autumn (AMJ)	
	$10^3 km^2 yr^{-1}$ (*R*)	Percent $decade^{-1}$	$10^3 km^2 yr^{-1}$ (*R*)	Percent $decade^{-1}$	$10^3 km^2 yr^{-1}$ (*R*)	Percent $decade^{-1}$	$10^3 km^2 yr^{-1}$ (*R*)	Percent $decade^{-1}$	$10^3 km^2 yr^{-1}$ (*R*)	Percent $decade^{-1}$
Antarctic	10.2 ± 4.6 (2.1)	1.1 ± 0.5	10.0 ± 4.6 (2.1)	0.7 ± 0.3	3.9 ± 6.7 (0.5)	0.3 ± 0.6	8.3 ± 5.0 (1.6)	3.1 ± 1.9	18.5 ± 7.3 (2.5)	2.3 ± 0.9
Arctic	−22.6 ± 4.8 (−4.6)	−2.4 ± 0.5	−42.1 ± 6.0 (−6.9)	−8.4 ± 1.2	−30.1 ± 6.9 (−4.3)	−3.5 ± 0.8	−7.3 ± 5.2 (−1.4)	−0.5 ± 0.4	−10.8 ± 3.9 (−2.7)	− 1.0 ± 0.3
Polar	−12.4 ± 7.4 (−1.6)	−0.6 ± 0.4	−32.0 ± 7.4 (−4.3)	−1.6 ± 0.3	−26.0 ± 10.1 (−2.5)	−1.3 ± 0.5	0.9 ± 8.4 (0.1)	0.06 ± 0.5	7.6 ± 9.4 (0.8)	0.4 ± 0.5

Note: *JAS* July–August–September, *OND* October–November–December, *JFM* January–February–March, *AMJ* April–May–June

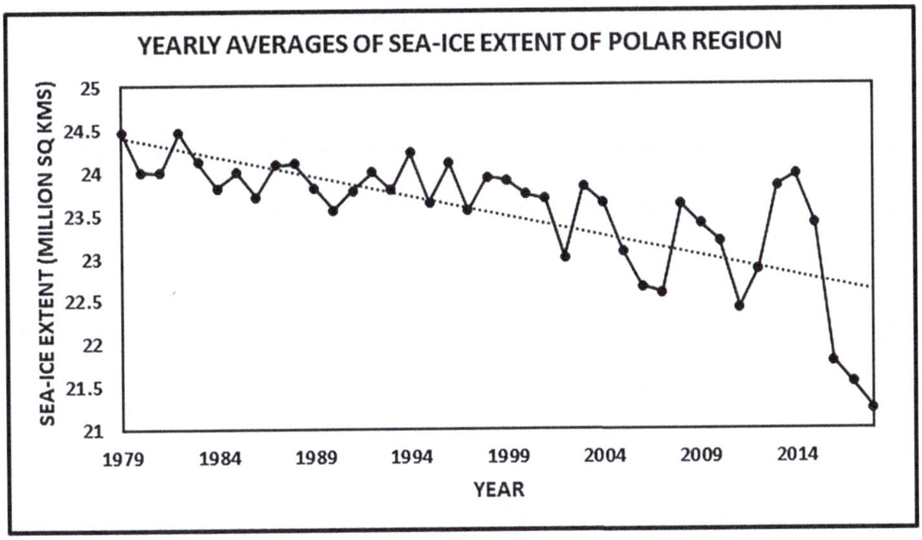

Fig. 3.5 Yearly average of entire polar region sea-ice extents, 1979–2018, with the corresponding lines of least squares fit

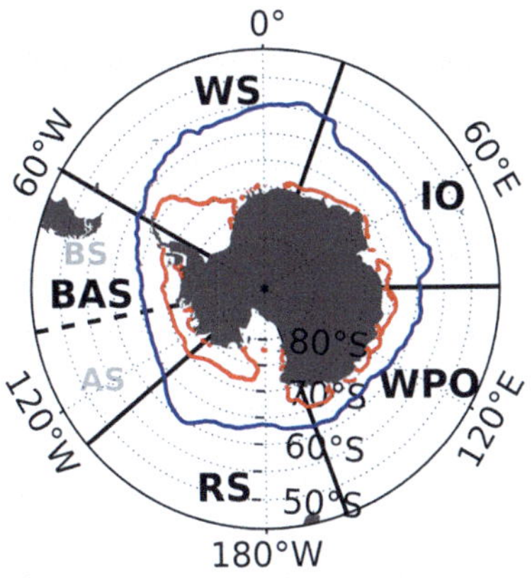

Fig. 3.6 Geographical sectors of Antarctica. Weddell Sea sector (WS, 60oW-20oE), Indian Ocean sector (IO, 20o-90oE), Western Pacific Ocean sector (WPO, 90o-160oE), Ross Sea sector (RS, 160oE-130oW), Bellingshausen and Amundsen Seas sector (BAS, 60o-130oW), and Southern Ocean (SO, 0o-360o). The colored lines represent the sea-ice edge (SIE) of the climatology of sea-ice concentration (%) from satellite data for the months of February (red) and September (blue). (Source: Bajish et al. 2021)

trend is observed in RS > WPO > IO, while negative trends prevail in BAS and WS. In the summer season, the trends show the highest values with increasing SIE in WS (10.1 ± 3.0 percent decade^{-1}) and decreasing SIE in BAS (−17 ± 2.9 percent decade^{-1}). In autumn, the maximum trend was observed in WPO (3.6 ± 1.3 percent decade^{-1}) and the minimum in the BAS sector (−6.4 ± 2.0 percent decade^{-1}). Annually, WS, IO, WPO, and RS show an expanding trend like the entire Southern Ocean but the BAS sector exhibits a retarding trend at a rate of −2.8 ± 1.2 percent decade^{-1} (Table 3.3). Only in the winter season, the BAS sector shows a positive trend around 1.5 ± 1.4 percent decade^{-1} whereas during the summer season, it shows the highest negative trend at a rate of −17 ± 2.9 percent decade^{-1} (Table 3.3). BAS sectors showed the highest reduction of SIE in both summer and autumn compared to other sectors.

The SIE in all the sectors showed strong interannual variability for different seasons during 1979–2017 (Fig. 3.7), which are linked to large-scale atmospheric climate models such as the southern annular mode (SAM), El Niño—Southern Oscillation (ENSO) and zonal wave 3 (Kohyama and Hartmann 2016; Simpkins et al. 2012; Raphel 2004). Further, SIE exhibits strong intra-annual variations with a maximum extent in winter (Fig. 3.8b) and a minimum extent in summer (Fig. 3.8a), with this dynamic and strong variability contributing towards modulating the regional and global climate. The largest contributor to the total SIE variability in the

Table 3.3 Yearly and seasonal sea-ice extent trend of the Southern Ocean as a whole and for the five sectors for the time period of 1979–2017

Sector	Yearly		Winter (JAS)		Spring (OND)		Summer (JFM)		Autumn (AMJ)	
	$10^3km^2y_r{-1}$ (R)	Percent decade^{-1}	$10^3km^2yr^{-1}$ (R)	Percent decade^{-1}	$10^3km^2yr^{-1}$ (R)	Percent decade^{-1}	$10^3km^2yr^{-1}$ (R)	Percent decade^{-1}	$10^3km^2yr^{-1}$ (R)	Percent decade^{-1}
SH	14.6 ± 5.1 (2.8)	1.2 ± 0.4	13.4 ± 4.7 (2.8)	0.7 ± 0.2	9.6 ± 6.9 (1.3)	0.6 ± 0.4	14.9 ± 7.7 (1.9)	3.4 ± 1.8	20.4 ± 7.7 (2.6)	1.8 ± 0.7
WS	5.7 ± 3.5 (1.6)	1.2 ± 0.7	−1.8 ± 4.5 (−0.4)	−0.29 ± 0.7	−2.9 ± 5.1 (−0.5)	−0.5 ± 0.9	16.9 ± 5.14 (3.2)	10.1 ± 3.0	10 ± 4.5 (2.3)	2.7 ± 1.1
IO	2.8 ± 1.8 (1.5)	1.4 ± 0.9	3.0 ± 3.0 (0.9)	0.9 ± 0.9	1.5 ± 3.4 (0.4)	0.5 ± 1.2	2.6 ± 1.4 (1.8)	7.5 ± 4.0	4.3 ± 2.1 (2)	2.9 ± 1.4
WPO	2.8 ± 1.3 (2)	2.2 ± 1.1	1.9 ± 2.4 (0.7)	1.0 ± 1.2	1.3 ± 2.1 (0.6)	0.9 ± 1.4	3.7 ± 1.6 (2.3)	7.5 ± 3.2	4.2 ± 1.5 (2.7)	3.6 ± 1.3
RS	7.3 ± 2.9 (2.4)	2.5 ± 1.0	7.0 ± 3.4 (2)	1.7 ± 0.8	10.2 ± 3.8 (2.6)	2.8 ± 1.0	2.7 ± 4.1 (0.6)	2.4 ± 3.7	9.4 ± 3.9 (2.4)	3.2 ± 1.3
BAS	−4.2 ± 1.8 (−2.2)	−2.8 ± 1.2	3.3 ± 3.1 (1.4)	1.5 ± 1.4	−0.7 ± 3.5 (−0.2)	−0.3 ± 1.9	−11.1 ± 1.8 (−5.9)	−17 ± 2.9	−8.3 ± 2.7 (3.1)	−6.4 ± 2.0

Note: *SH* Southern Hemisphere, *WS* Weddell Sea, *IO* Indian Ocean, *WPO* Western Pacific Ocean, *RS* Ross Sea, *BAS* Bellingshausen and Amundsen Sea

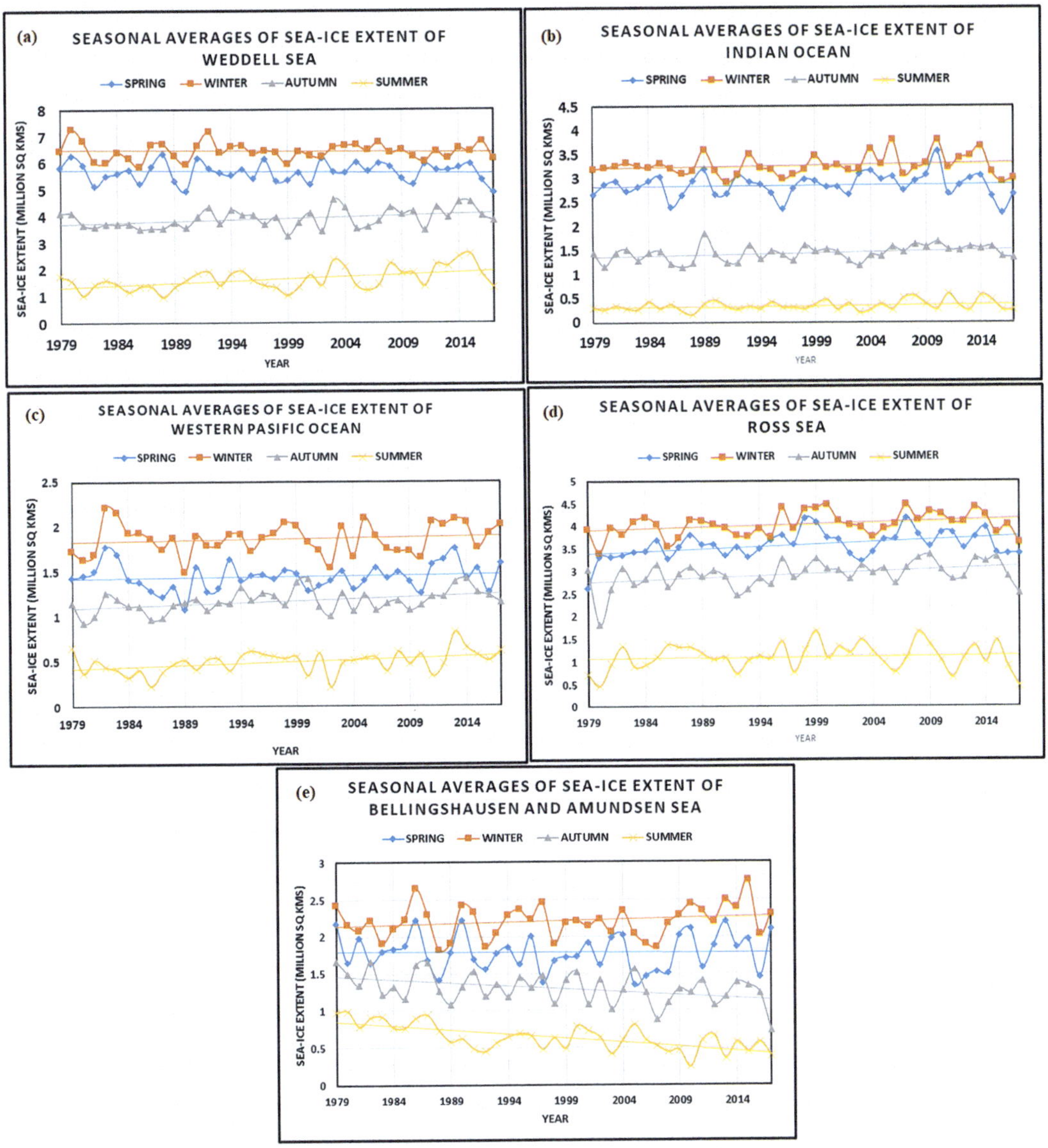

Fig. 3.7 Seasonal average of sea-ice extent from November 1978 to December 2017 of (**a**) Weddell Sea, (**b**) Indian Ocean, (**c**) Western Pacific Ocean, (**d**) Ross Sea, and (**e**) Bellingshausen and Amundsen Sea

Southern Ocean is the WS sector, followed by RS, IO, BAS, and WPS (Fig. 3.7). During winter, the season of maximum SIE, the interannual variability is less for all sectors. Moreover, WS has the largest extent (>6 million sq. km), and WPO has the least SIE, with BAS occasionally being the lowest contributor in some years. However, in summer, the season of lowest SIE, large interannual variability is evident in all sectors (Fig. 3.8a). The drastic decline in SIE after 2015 is prominent in WS and RS sectors, whereas IO, WPS, and BAS, which has the lowest SIE, do not show much change in the extent.

In the context of sea-ice area, all sectors of Antarctica except BAS had shown increasing trends in 39 years (1979–2017) (Table 3.4). Based on the data available during these periods, the BAS sector shows a retarding trend at a rate of -3.2 ± 1.4 percent decade^{-1}. Weddell Sea experienced the lowest negative trend for the sea-ice area (-0.3 ± 0.6 percent decade^{-1}) in winter, which is not the same for the other sectors.

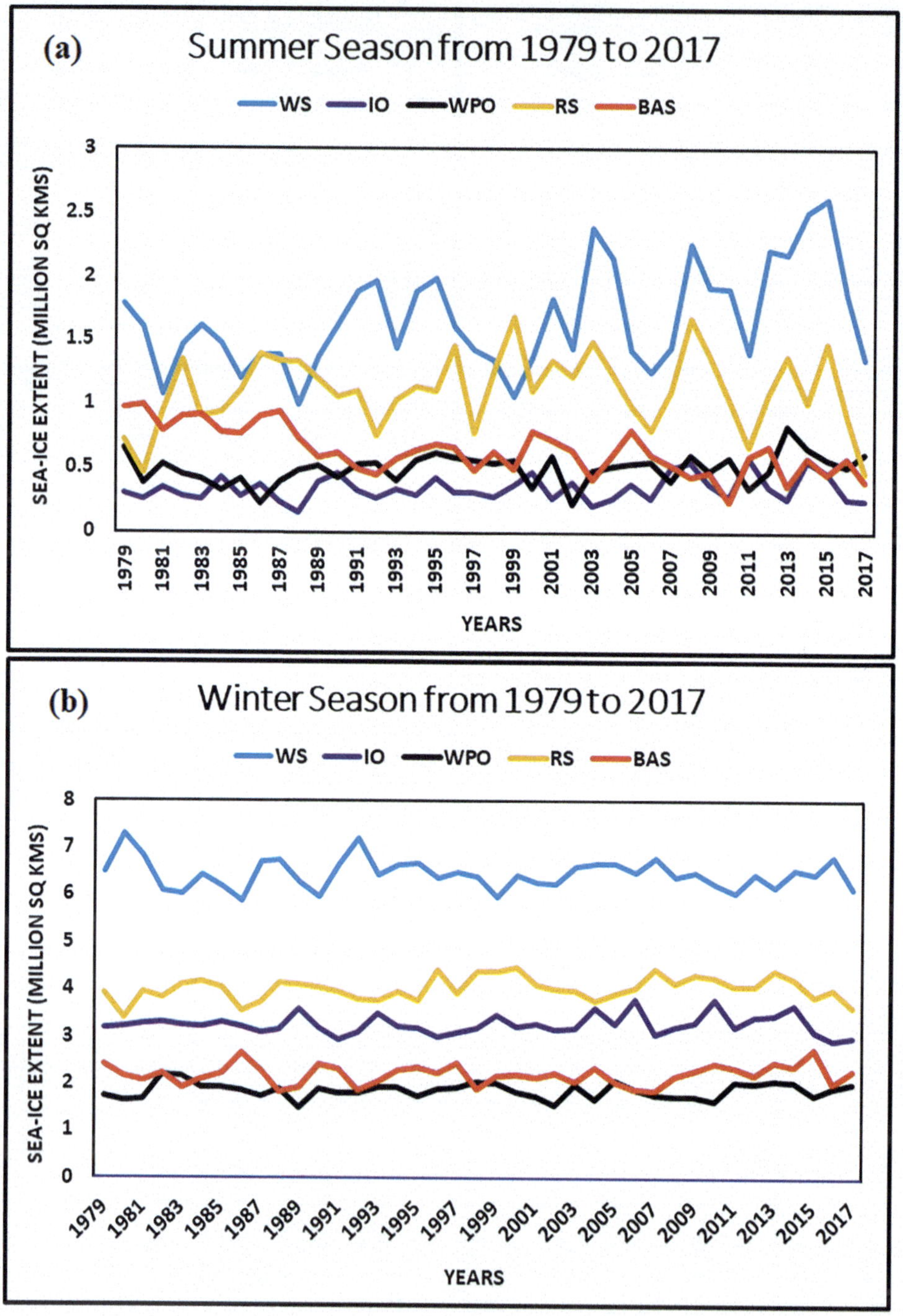

Fig. 3.8 Average SIE of Antarctica from 1979 to 2017 in (**a**) summer and (**b**) winter

BAS sector had a maximum retarding trend in spring, summer, and autumn which is −1.2 ± 2.0, −21.9 ± 3.8, and −6.4 ± 2.3 percent decade^{-1} respectively. The Ross Sea had a maximum increasing trend in winter (1.9 ± 0.9 percent decade^{-1}) and spring (3.1 ± 1.2 percent decade^{-1}). During summer and autumn, the Western Pacific Ocean showed the highest increasing trend of 12.0 ± 4 and 5.9 ± 1.6 percent decade^{-1}, respectively (Table 3.4). This explains that different

Table 3.4 Yearly and seasonal sea-ice area trend of the Southern Ocean as a whole and for the five sectors for the time period of 1979–2017

Sector	Yearly		Winter (JAS)		Spring (OND)		Summer (JFM)		Autumn (AMJ)	
	$10^3 km^2 yr^{-1}$ (R)	Percent decade^{-1}	$10^3 km^2 yr^{-1}$ (R)	Percent decade^{-1}	$10^3 km^2 yr^{-1}$ (R)	Percent decade^{-1}	$10^3 km^2 yr^{-1}$ (R)	Percent decade^{-1}	$10^3 km^2 yr^{-1}$ (R)	Percent decade^{-1}
SH	13.7 ± 4.5 (3)	1.5 ± 0.5	12.4 ± 4.7 (2.6)	0.9 ± 0.3	7.7 ± 6.8 (1.1)	0.7 ± 0.6	11.6 ± 5.1 (2.2)	4.4 ± 1.9	23.0 ± 7.2 (3.1)	2.9 ± 0.9
WS	5.2 ± 2.9 (1.7)	1.5 ± 0.8	−1.8 ± 3.6 (−0.5)	−0.3 ± 0.6	−2.0 ± 5.0 (−0.4)	−0.5 ± 1.2	14.0 ± 3.6 (3.8)	11.9 ± 3.8	10.8 ± 4.0 (2.6)	3.6 ± 1.3
IO	2.1 ± 1.5 (1.4)	1.5 ± 1.1	−2.4 ± 2.7 (0.8)	0.9 ± 1.1	0.8 ± 2.7 (0.3)	0.4 ± 1.4	1.6 ± 0.9 (1.8)	9.3 ± 5.1	3.7 ± 1.7 (2.1)	3.9 ± 1.8
WPO	3.2 ± 1.0 (3.1)	2.8 ± 1.1	2.5 ± 1.8 (1.3)	1.9 ± 1.4	2.3 ± 1.4 (1.6)	2.5 ± 1.5	3.6 ± 1.2 (2.9)	12.0 ± 4.0	4.5 ± 1.2 (3.5)	5.9 ± 1.6
RS	6.0 ± 1.1 (2.5)	4.0 ± 1.2	6.1 ± 3.0 (2)	1.9 ± 0.9	8.2 ± 3.1 (2.6)	3.1 ± 1.2	0.5 ± 2.7 (0.2)	1.0 ± 4.6	9.3 ± 3.5 (2.5)	4.1 ± 1.6
BAS	−3.2 ± 1.4. (−2.2)	−3.2 ± 1.4	2.0 ± 2.6 (0.7)	1.3 ± 1.7	−1.4 ± 2.5 (−0.5)	−1.2 ± 2.0	−8.2 ± 1.4 (0.003)	−21.9 ± 3.8	−5.4 ± 1.9 (−2.7)	−6.4 ± 2.3

Note: *SH* Southern Hemisphere, *WS* Weddell Sea, *IO* Indian Ocean, *WPO* Western Pacific Ocean, *RS* Ross Sea, *BAS* Bellingshausen and Amundsen Sea sectors of Antarctic regions exhibit distinctive and dynamic trends in the context of the sea-ice area. Although certain regions, like the BAS sector of Antarctica, had a drop in sea-ice cover between 1979 and 2014; generally, the Antarctic sea ice exhibited a minor increase in its extent. Changes in the Antarctic short-term trend can be explained by the inherent unpredictability of the atmosphere, ocean, and sea ice. Ocean variables, such as the inflow of cold fresh water from melting ice shelves, may have had a part in the Antarctic pattern of sea-ice change, in addition to changes in surface wind patterns across the continent.

3.4 Sea-Ice Anomaly

Sea-ice anomaly explains the difference between the sea-ice value (area, extent, or concentration) at a given time and the long-term average. Therefore, an anomaly tells us how close to or away from the average are the area, extent, or concentration in a given month. In this study, 37 years (1979–2015) have been considered as a standard climatological baseline. As an expected result of global warming, the Arctic sea ice is retreating at a rapid rate (Fig. 3.4a), but in the case of the Antarctic sea ice, it shows the opposite. From satellite-derived data (1979–2018), it has been observed that the Antarctic sea ice has been expanding till 2014 (Fig. 3.3a) and overall sea ice is retreating with regional heterogeneity. The Antarctic sea-ice anomaly of 2015 (Fig. 3.9a) clearly shows a positive anomaly in summer, and, later, it overlaps with the climatological line. Predominantly, negative anomalies of the Antarctic sea ice were observed after 2015 in comparison with 37-year climatological data (Fig. 3.9b–d). In 2016, a negative anomaly in SIE was observed in the winter and the spring season (Fig. 3.9b). Gradually the negative anomaly was observed throughout all the seasons of 2017 (Fig. 3.9c) and 2018 (Fig. 3.9d).

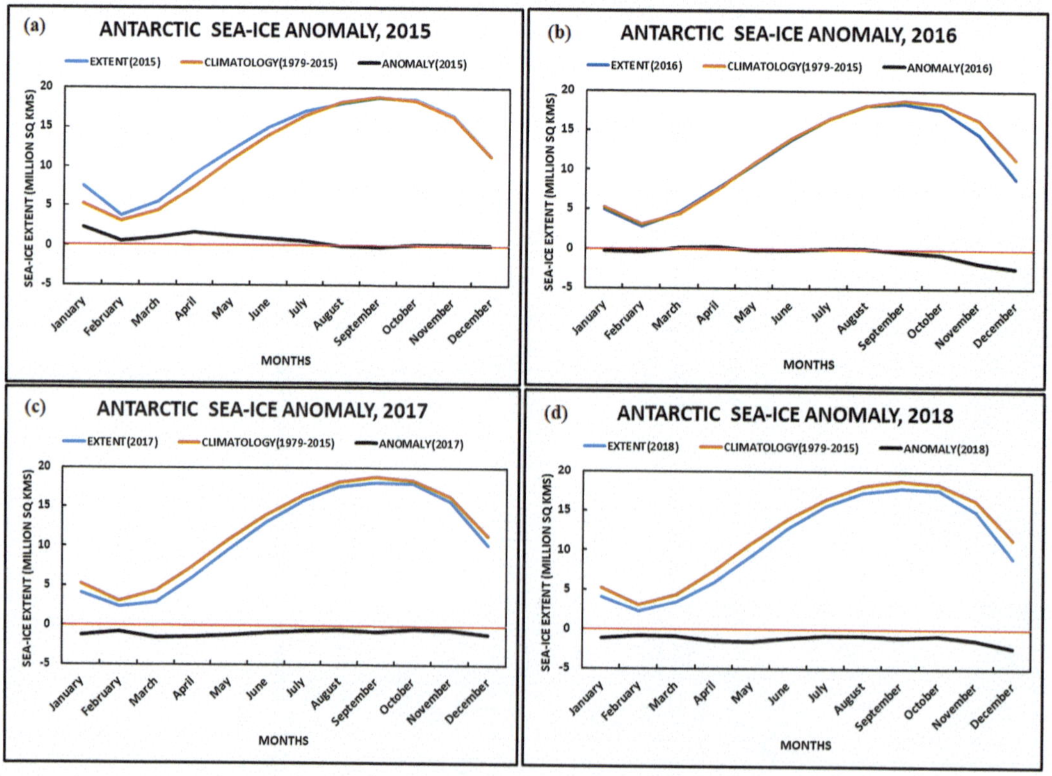

Fig. 3.9 Antarctic sea-ice anomaly of (**a**) 2015, (**b**) 2016, (**c**) 2017, and (**d**) 2018

The investigation of the monthly time series of the Antarctic SIE anomalies clearly shows a drastic decrease from spring 2016, which continued throughout the study period (Fig. 3.9). To further understand the spatial variability of this decrease within a year, we examine the monthly spatial anomaly maps of 2018. Moreover, the monthly anomalies of SIE are negative for all the months in 2018. Analysis of spatial sea-ice maps in terms of climatology indicated maximum sea-ice concentration and extent in September and minimum in February (Fig. 3.10). The seasonal growth of sea-ice cover from March to September and the retreat from October to February is evident. The sea-ice anomaly maps showed the predominance of large negative anomalies throughout all seasons except for some regional heterogeneities consisting of both positive and negative anomalies (Fig. 3.11). The monthly anomaly maps clearly highlight the importance of the regional

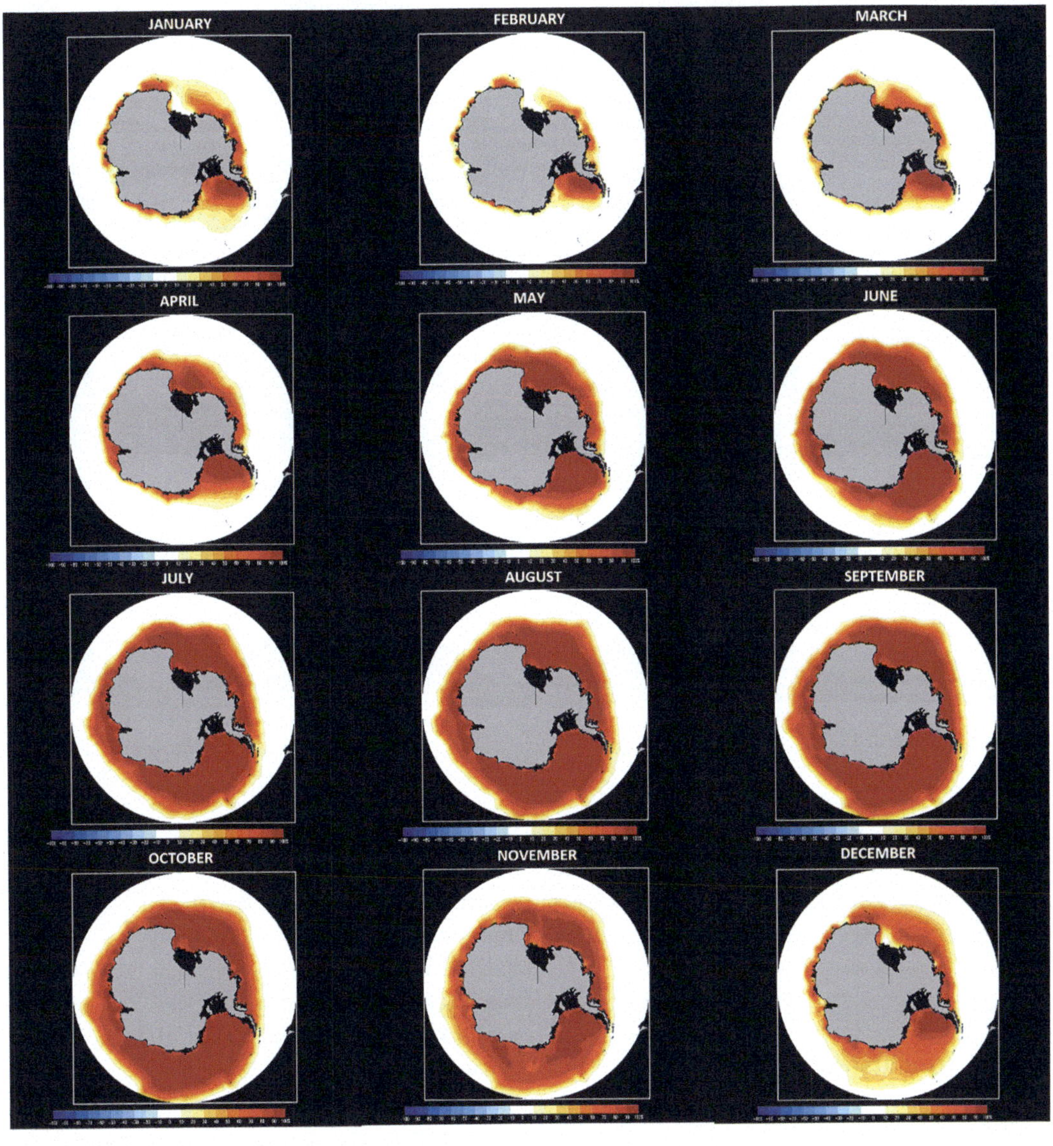

Fig. 3.10 Mean of Antarctic sea-ice concentration from 1979 to 2015

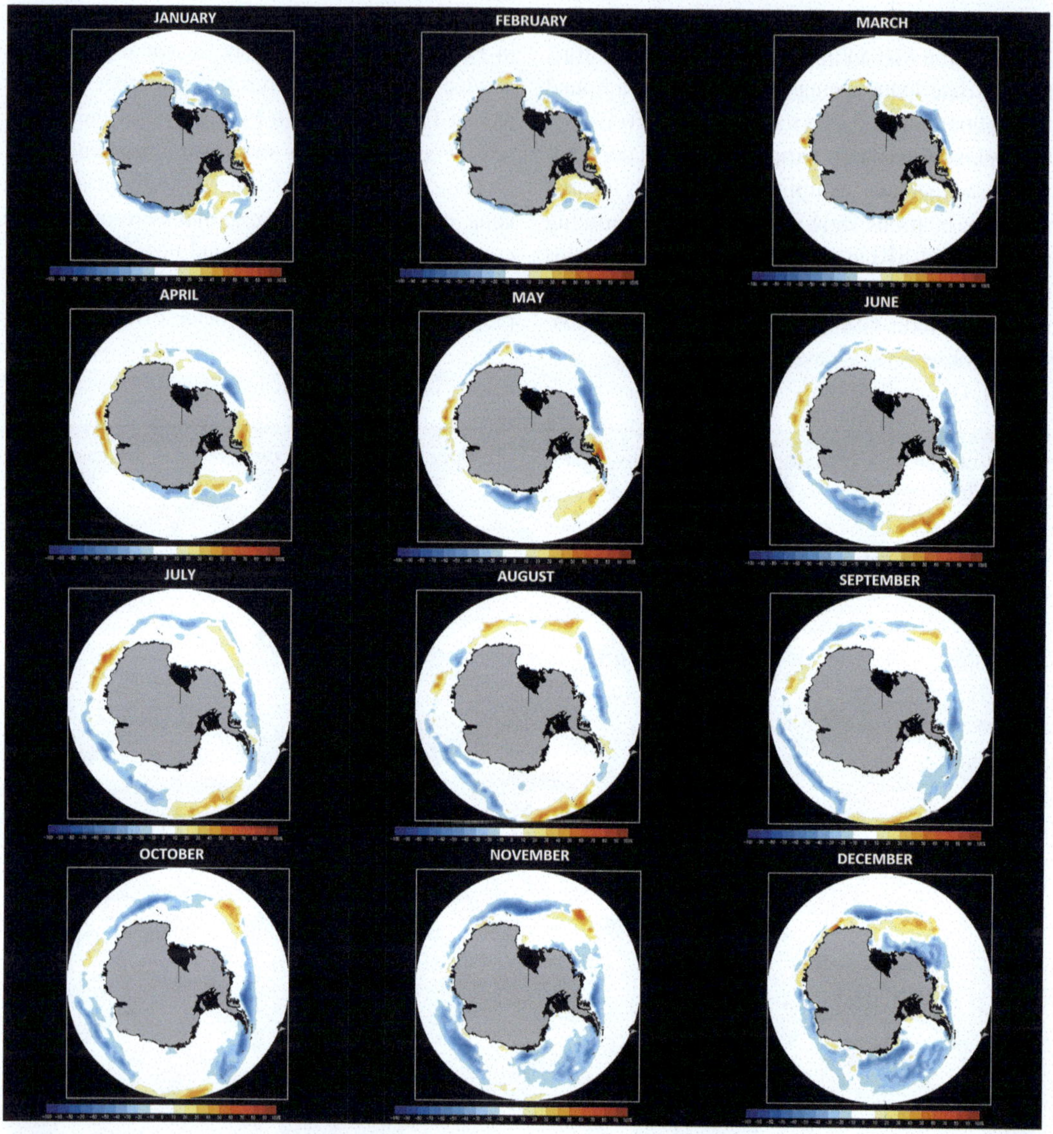

Fig. 3.11 Anomaly of sea-ice concentration in 2018

variability of the sea ice around Antarctica, which is mainly affected by a combined effect of anomalous synoptic events such as polar cyclones along with large-scale atmospheric circulations and oceanic processes (Jena et al., 2022a).

3.5 Conclusion

In this study, 40 years (1979–2018) of climatological data were analyzed to understand the regional, seasonal, and yearly sea-ice trend that highlights the complex nature of Antarctic climatic conditions. As an influence of global warming, Arctic sea ice has shown a declining trend, but Antarctica shows the exception. Satellite equipped with passive microwave sensor has been monitoring the sea ice since 1979, which revealed that the sea ice of Antarctica is expanding at a trend of 2.1% per decade up to 2015. From 2016 to 2019, February sea-ice data, it was found that the SIE is retreating in the Southern Ocean with regional increasing–decreasing heterogeneity. The anomaly was calculated for the last 4 years (2015, 2016, 2017, and 2018) to understand the closeness of the sea-ice data to its mean climatological value (1979–2015). Up to 2015, SIE shows a positive anomaly which refers to the more amount of SIE than the expected mean value. However, for the last 3 years, it has shown a negative anomaly in the Antarctic sea-ice region.

Sea-ice extent and area were well documented in this study for both hemispheres, although the satellite observations have not been fully comprehended. Due to the complexity of the earth's system, climate change is a continuous process involving a variety of responses from various sectors. Polar ice can be considered as a perfectly preserved database, explaining the cumulative impact of climate change on the planet. Satellite monitoring of Antarctic sea ice will experience more advancement in the upcoming years, including the development of various models and analysis tools. Such studies will add further research to understand the relationship of Antarctic sea-ice dynamics with global climate change.

Acknowledgments The authors would like to thank the Director of NCPOR for their encouragement and support. The authors also acknowledge various organizations such as the National Snow and Ice Data Center (NSIDC), National Oceanic and Atmospheric Administration (NOAA), the European Centre for Medium Range Weather Forecast (ECMWF), the National Centre for Atmospheric Research (NCAR), Technische Universita¨t Dresden, and the Australian Antarctic Data Centre for making various datasets available in their portals.

References

Alekseev G, Vyazilova A, Smirnov A (2022) Influence of sea surface temperature in the tropics on the Antarctic sea ice during global warming. J Mar Sci Eng. https://doi.org/10.3390/jmse10121859

Bajish CC, Jena B, Anilkumar N (2021) Is the Indian monsoon rainfall linked to the Southern Ocean sea ice conditions? Weather Clim Extrem 34:100377. https://doi.org/10.1016/j.wace.2021.100377

Barker PF, Filippelli GM, Florindo F, Martin EE, Scher HD (2007) Onset and role of the antarctic circumpolar current (PDF). Deep-Sea Res II 54(21):2388–2398

Cavalieri DJ, Parkinson CL, Gloersen P, Zwally HJ (1997) Arctic and Antarctic sea ice concentrations from multichannel passive-microwave satellite data sets: October 1978–September 1995 – user's guide – NASA TM 104647, 17. Goddard Space Flight Center, Greenbelt, MD

Desilets D, Zreda M (2014) Cosmic-ray hydrometeorology BT – encyclopedia of remote sensing. In: Njoku EG (ed) Encyclopedia of earth sciences series. Springer, New York, pp 83–86. https://doi.org/10.1007/978-0-387-36699-9_208

Guo Y, Yu Y, Lin P, Liu H, He B, Bao Q, An B, Zhao S, Hua L (2020) Simulation and improvements of oceanic circulation and sea ice by the coupled climate system model FGOALS-f3-L. Adv Atmos Sci 37:1133–1148. https://doi.org/10.1007/s00376-020-0006-x

Hanson KJ (1961) The albedo of sea-ice and ice islands in the Arctic ocean basin. Arctic 14:188–196. http://www.jstor.org/stable/40506917

Hay WW (2011) Can humans force a return to a 'Cretaceous' climate? Sediment Geol 235:5–26. https://doi.org/10.1016/j.sedgeo.2010.04.015

Jena B, Bajish CC, Turner J, Ravichandran M, Anilkumar N, Kshitija S (2022a) Record low sea ice extent in the Weddell Sea, Antarctica in April/May 2019 driven by intense and explosive polar cyclones. Nat Portfolio J Clim Atmos Sci 5(19). https://doi.org/10.1038/s41612-022-00243-9

Jena B, Bajish CC, Turner J, Ravichandran M, Kshitija S, Anilkumar N et al (2022b) Mechanisms associated with the rapid decline in sea ice cover around a stranded ship in the Lazarev Sea, Antarctica. Sci Total Environ 821:153379. https://doi.org/10.1016/j.scitotenv.2022.153379

Jungclaus JH, Fischer N, Haak H, Lohmann K, Marotzke J, Matei D, Mikolajewicz U, Notz D, von Storch JS (2013) Characteristics of the ocean simulations in the Max Planck Institute Ocean Model (MPIOM) the ocean component of the MPI-Earth system model. J Adv Model Earth Syst 5:422–446. https://doi.org/10.1002/jame.20023

Knorr G, Lohmann G (2014) Climate warming during Antarctic ice sheet expansion at the Middle Miocene transition. Nat Geosci 7:376–381. https://doi.org/10.1038/ngeo2119

Kohyama T, Hartmann DL (2016) Antarctic sea ice response to weather and climate modes of variability. J Clim 29:721–741. https://doi.org/10.1175/JCLI-D-150301.1

Ludescher J, Yuan N, Bunde A (2019) Detecting the statistical significance of the trends in the Antarctic sea ice extent: an indication for a turning point. Clim Dyn 53:237–244. https://doi.org/10.1007/s00382-018-4579-3

Ludwig V, Spreen G, Pedersen LT (2020) Evaluation of a new merged sea-ice concentration dataset at 1 km resolution from thermal infrared and passive microwave satellite data in the Arctic. Remote Sens. https://doi.org/10.3390/rs12193183

Ma J, Hung H, Macdonald RW (2016) The influence of global climate change on the environmental fate of persistent organic pollutants: a review with emphasis on the Northern Hemisphere and the Arctic as a receptor. Glob Planet Change 146:89–108. https://doi.org/10.1016/j.gloplacha.2016.09.011

Maslanik J, Stroeve J, Fowler C, Emery W (2011) Distribution and trends in Arctic sea ice age through spring 2011. Geophys Res Lett 38. https://doi.org/10.1029/2011GL047735

Melsheimer C, Spreen G, Ye Y, Shokr M (2023) First results of Antarctic sea ice type retrieval from active and passive microwave remote sensing data. Cryosphere 17:105–126. https://doi.org/10.5194/tc-17-105-2023

Murphy EJ, Hofmann EE, Watkins JL, Johnston NM, Piñones A, Ballerini T, Hill SL, Trathan PN, Tarling GA, Cavanagh RA, Young EF, Thorpe SE, Fretwell P (2013) Comparison of the structure and function of Southern Ocean regional ecosystems: the Antarctic Peninsula and South Georgia. J Mar Syst 109–110:22–42. https://doi.org/10.1016/j.jmarsys.2012.03.011

Petersen MR, Asay-Davis XS, Berres AS, Chen Q, Feige N, Hoffman MJ, Jacobsen DW, Jones PW, Maltrud ME, Price SF, Ringler TD, Streletz GJ, Turner AK, Van Roekel LP, Veneziani M, Wolfe JD, Wolfram PJ, Woodring JL (2019) An evaluation of the ocean and sea ice climate of E3SM using MPAS and Interannual CORE-II Forcing. J Adv Model Earth Syst 11:1438–1458. https://doi.org/10.1029/2018MS001373

Raphael MN (2004) A zonal wave 3 index for the Southern Hemisphere. Geophys Res Lett 31:1–4. https://doi.org/10.1029/2004GL020365

Salleh SA, Latif ZA, Pradhan B, Wan Mohd WMN, Chan A (2014) Functional relation of land surface albedo with climatological variables: a review on remote sensing techniques and recent research developments. Geocarto Int 29:147–163. https://doi.org/10.1080/10106049.2012.748831

Simpkins GR, Ciasto LM, Thompson DWJ, England MH (2012) Seasonal relationships between large-scale climate variability and Antarctic sea ice concentration. J Clim 25:5451–5469. https://doi.org/10.1175/JCLI-D-1100367.1

Swift CT, Cavalieri DJ (1985) Passive microwave remote sensing for sea ice research. EOS Trans Am Geophys Union 66:1210–1212. https://doi.org/10.1029/EO066i049p01210

Timmermann R, Danilov S, Schröter J, Böning C, Sidorenko D, Rollenhagen K (2009) Ocean circulation and sea ice distribution in a finite element global sea ice–ocean model. Ocean Model 27:114–129. https://doi.org/10.1016/j.ocemod.2008.10.009

Turner J, Guarino MV, Arnatt J, Jena B, Marshall GJ, Phillips T, Bajish CC, Clem K, Wang Z, Andersson T, Murphy EJ, Cavanagh R (2020) Recent decrease of summer sea ice in the Weddell Sea, Antarctica. Geophys Res Lett 47:e2020GL087127. https://doi.org/10.1029/2020GL087127

Turner J, Holmes C, Harrison TC, Phillips T, Jena B, Francois TR et al (2022) Record low Antarctic sea ice cover in February 2022. Geophys Res Lett 49:e2022GL098904. https://doi.org/10.1029/2022GL098904

Vihma T (2014) Effects of Arctic sea ice decline on weather and climate: a review. Surv Geophys 35:1175–1214. https://doi.org/10.1007/s10712-014-9284-0

White WB, Gloersen P, Simmonds I (2004) Tropospheric response in the Antarctic circumpolar wave along the sea ice edge around Antarctica. J Clim 17:2765–2779. https://doi.org/10.1175/1520-0442(2004)017<2765:TRITAC>2.0

4 Climate Change and Himalayan Glaciers: A Socio-Environmental Concern in Anthropocene Epoch

Aju David Raj, Anu David Raj, and K. R. Sooryamol

Abstract

Glaciers are playing a prime role in the regulation of the global climate system and cover around 10% of the total land area of the planet. For almost a century, glaciers have been explored as sensitive climatic indicators. Some of the world's largest and most gorgeous glaciers can be found in the Himalayan range, spanning across eight countries in Asia. The adverse effects due to climate change may cause serious socio-economic and environmental concerns for the community of Himalayas. Hence, this chapter discusses the climate change impacts on Himalayan glaciers in socio-environmental aspect with an extreme significance in regard to the Anthropocene epoch. It delivers various methods for assessment, climate change impacts, influence of livelihoods, and mitigation and adaptation strategies. The in-depth synthesis of previous studies will aid in getting a better understanding of existing knowledge and gaps areas in Himalayan glaciology. It will also help decision-makers to devise critical measures for mitigating the consequences of probable threats.

Keywords

Snow and Glaciers · Glacial Retreat · Soil Erosion · GLOF · Climate Change Adaptation · Mitigation

A. David Raj (✉)
Department of Environmental Science, School of Earth Sciences, Central University of Rajasthan, Rajasthan, India
e-mail: ajudraj77@gmail.com

A. David Raj
Agriculture and Soils Department, Indian Institute of Remote Sensing, Indian Space Research Organization (ISRO), Dehradun, Uttarakhand, India

Forest Research Institute, Dehradun, Uttarakhand, India

K. R. Sooryamol
Indian Institute of Soil & Water Conservation (ICAR-IISWC), Dehradun, Uttarakhand, India

4.1 Introduction

The global temperature has risen by 0.8 °C to 1.2 °C because of anthropogenic activity, and it is anticipated to climb to 1.5 °C by 2030 (IPCC 2018). According to UNFCCC, climate change is defined as a change in climate that is induced by human activity that modifies the structure (composition of gases) of the global atmosphere, in supplement to natural climate variability recognized over identical time periods (UNFCC 1992). Although there are some variations relying on IPCC, it implies a change in the condition of the climate that could be defined by fluctuations in the average and/or variability of its features and that lasts for decades or more. While it can

U. Chatterjee et al. (eds.), *Climate Crisis: Adaptive Approaches and Sustainability*, Sustainable Development Goals Series, https://doi.org/10.1007/978-3-031-44397-8_4

simply be defined as any change in climate over time, particularly caused by natural variability or anthropogenic activities. Regional climate changes, notably temperature rises, have an impact on several ecological systems across all geographies and oceans. Glaciers are one among the system which has a role in regulating the climate change. Any significant mass of perpetual ice existing on land due to the recrystallization of snow or other kinds of solid precipitation exhibits signs of previous or current flow which is known as a glacier. The mass and thermal fluctuations of glaciers and permafrost are vividly reflected in changing climatic patterns (Chaujar 2009). Therefore, variations in snow, ice, and frozen ground have consequences on hydrological systems, terrestrial marine, and freshwater biological systems.

The Himalayas are one of the world's most vital mountain ranges, with a unique ecology that makes it particularly sensitive, as well as crucial in the face of climate change scenarios. The Great Himalayas geographical area is the world's biggest cryosphere beyond the polar regions, and it is home to several rivers that provide water to over 800 million inhabitants (Li et al. 2016; Hegdahl et al. 2016). It is the most snow-capped region following the two polar areas in terms of ice reserves, thus "third pole" became a well-known nickname for the Himalayas (Schild 2008). Figure 4.1 shows one such snow–/ice-capped mountains of the Himalaya near Harsil, Uttarkashi. The Himalayas are also known as Asia's "water tower" (Xu et al. 2009). There are over 15,000 glaciers in the area, holding over 12,000 km^3 of freshwater. The Indus Basin and the Ganga-Brahmaputra Basin are two of the world's largest and most widespread river systems, with snow and glacier melt accounting for a significant amount of runoff water (IPCC 2007). Himalayan glaciers do not appear to be adapting to climate change swiftly enough. The Himalayan region's fragile environments are particularly vulnerable to natural disasters, raising concerns about the region's existing and future climate change implications (Cruz et al. 2007). The Himalayan region is confronted with various environmental challenges such as floods, droughts, and landslides, which are attributed to the impact of climate change (Barnett et al. 2005). Temperature data from the Himalayas consistently reveal a temperature increase, however, at varying speeds according to the locations and seasons (Bhutiyani et al. 2010; Dimri and Dash 2011; Khattak et al. 2011). Comparable trends are observed in the western Indian Himalayas, where Dimri and Dash (2011) indicate a rise in the number of warm days and a reduced number of cold days between 1975 and 2006. Glacial lakes are natural dams that confine volumes of water behind loose glacier moraine walls. The lake will collapse beyond a certain point, causing catastrophic occurrences like those seen in the glacial lake outburst floods (GLOF). Meltwater-induced soil erosion is a serious problem that has a negative impact on people's livelihoods, agriculture, and the environment. Hence, the Himalayan ecosystem serves as a source of freshwater, a habitat of several organisms and human beings, the livelihood of millions of people, and needs to be sustainably managed to keep the ecosystem without the threat of climate change.

In this context, this chapter examines the impact of climate change on Himalayan glaciers, including glacial retreat, glacial lake formation, glacial lake outburst floods (GLOF), soil erosion, and precipitation. Also, this chapter discusses its impact on the local community, ecosystem, and water productivity, as well as possible mitigation strategies in this Anthropocene epoch. Thus, the study emphasizes the importance of long-term monitoring of glacier mass balance, growth, and melting. Soil erosion caused by snowmelt runoff water is a minimally explored topic in the Himalayan region. This chapter also attempts to undertake a case study in the Himalayan area on glacier melt and land surface temperature.

4.1.1 Anthropocene Epoch

The "Anthropocene" is a postulated new geological era stemming from substantial human-driven disruptions to the Earth system's structure and functionality, particularly the climate system.

Fig. 4.1 Snow/ice-capped mountains of the Himalayas

Humankind is causing changes to the Earth's life sustenance system because of the rapid expansion of the economies. Carbon dioxide emissions are increasing at a rate that has never been seen before in the history of mankind. The notion of Anthropocene was first explicitly stated in a concept article published by Paul J. Crutzen in *Nature* journal in 2002. He stated "For the past three centuries, the effects of humans on the global environment have escalated. Because of these anthropogenic emissions of carbon dioxide, global climate may depart significantly from natural behavior for many millennia to come. Thus, it seems appropriate to assign the term "Anthropocene" to the present, in many ways human-dominated, geological epoch, supplementing the Holocene—the warm period of the past 10—12 millennia" (Crutzen 2002). He indicated that the escalation of population pressure has led to a rise in the consumption and exploitation of natural resources, resulting in the occurrence of this event. This consumption leads to an increase in the concentration of greenhouse gases, especially the carbon dioxide (CO_2). This phenomenon serves as the primary radiative forcing mechanism that contributes to the escalation of global temperatures, ultimately resulting in a shift in the Earth's climate.

Since 2000, global CO_2 concentrations have risen at a rate of roughly 20 parts per million per decade, which is up to ten times higher than any other persistent CO_2 rise in the preceding 800,000 years (Bereiter et al. 2015; Lüthi et al. 2008). After 1970, the worldwide average temperature has risen at a pace of 1.7 °C per century, relative to a long-term drop of 0.01 °C per century for the previous 7000 years (Marcott et al. 2013). This rise in temperature is considered as the major driving force for glacial retreat and snowmelt. The change in the volume of snow/glacier from the Earth has serious consequences as it is a major component of ice feedback mechanisms. Hence, the energy and mass balance at the glacier surface is directly influenced by climatic change. Variations in mass balance induce quantity and thickness differences across longer durations, which influence ice flow via intrinsic displacement and basal sliding (Cuffey and Paterson 2010). Apart from this, it also has some indirect consequences to the human being. A common example is fluctuations in agricultural water availability fluctuation in glacier volume upstream. This is extremely important in the case of Himalayan rivers and cultivation in its downstream areas. Hence, it is explicit that the climate change, glacial melt, and Anthropocene are interlinked which is characterized by more exploitation of natural resources and leads to the change in the composition of Earth's atmosphere. Thus, sustainable management of natural resources is very much necessitated in the Anthropocene epoch instead of exponential exploitation.

4.2 Himalayan Glaciers

The Himalayan range is one of the world's most recently formed mountain ranges, consisting primarily of sedimentary and metamorphic rock. It was formed by a continental collision or orogeny at the convergent plate boundary (Major Himalayan Thrust) between the Indo-Australian Plate and the Eurasian Plate, according to the modern plate tectonic theory. Following Antarctica and the Arctic, the Himalayan mountainous region has the world's third-largest glacier ice. Mount Everest, K2 as well as other notable peaks are all found in this region, thus, which is known as the world's "Third Pole" due to its massive ice storage (Schild 2008). The Indian Himalayan glaciers are separated into three geographical sections, classified as the Western Himalaya, Central Himalaya, and Eastern Himalaya. The Himalayas are 2400-km-long mountain range fueled by two distinct climatic systems. Each summer, the meltwater from these glaciers replenishes the region's rivers and streams, including some of Asia's largest river systems such as the Indus, Ganges, and Brahmaputra. During the next few decades, melting glaciers are unlikely to create substantial variation in water supply in lowland areas (National Research Council 2012). The cryosphere serves as a critical supply of freshwater in the Hindu Kush Himalayan area, which includes the Pamirs, Tien Shan, and Tibetan Plateau Mountain ranges. More than 240 million people live in mountainous areas, relying on the rivers and streams from these mountainous regions. Freshwater coming from the cryosphere is critical for agricultural production, hydroelectric, inland navigation, and traditional and religious usage in the region, in addition to offering a supply of water for humans, cattle, and wildlife. While a variety of variables influence water flow, such as rainfall, groundwater, and springs, the existence of the cryosphere has resulted in consistent water flows over the Hindu Kush Himalayan area. However, as glaciers, snow, and permafrost melt at alarming rates, these patterns are shifting (Omair Ahmad 2021). The change in melting pattern and pace will severely affect the livelihoods of people's lives in the downstream of the Himalayas.

Glaciers in the Hindu Kush Himalayan area are anticipated to lose 10–30 percent of their mass by 2030. This number is predicted to rise to 25–35 percent by 2050 (Dasgupta and Shyamoli 2021). If governments do not implement ambitious emissions-reduction measures, glacier mass loss is expected to reach 35 percent in the Karakoram by 2080–2100, 45 percent in the Pamir mountains by 2080–2100, and 60–95 percent in the eastern Himalayas by 2080–2100 (Ahmad 2021). Pandey et al. (2011) examined the satellite imagery of 26 glaciers in the western Indian Himalayas from 1975 to 1989, 1992, 2001, and 2007, and discovered that all of them were retreating. The Gangotri Glacier, a well-studied Indian glacier, revealed no retreat between 2006 and 2010 (Kargel et al. 2011), despite a significant retreat rate in previous decades. The remoteness of the location, height, geography, and debris cover on the glaciers, as well as the restricted availability of large-scale maps, optimal aerial photography, and high-resolution satellite images, all hampered the study of glaciers in the Himalayas. International Centre for Integrated Mountain Development (ICIMOD) provided the land use/land cover map of the Hindukush Himalayan Karakoram (HKH) region. Figure 4.2 illustrates the land use/land cover map of the Himalayas of the year 2021. The extent of snow and glaciers across the Himalayas is clearly visible.

4.3 Assessment of Glacial Mass Loss

In glaciology, the mass change is a critical metric for assessing the glacial mass loss. It governs snow/ice dynamics and glacier activity as a direct response to climatic changes. The basic aim of glaciological study is to measure mass balance, which is of considerable scientific and practical

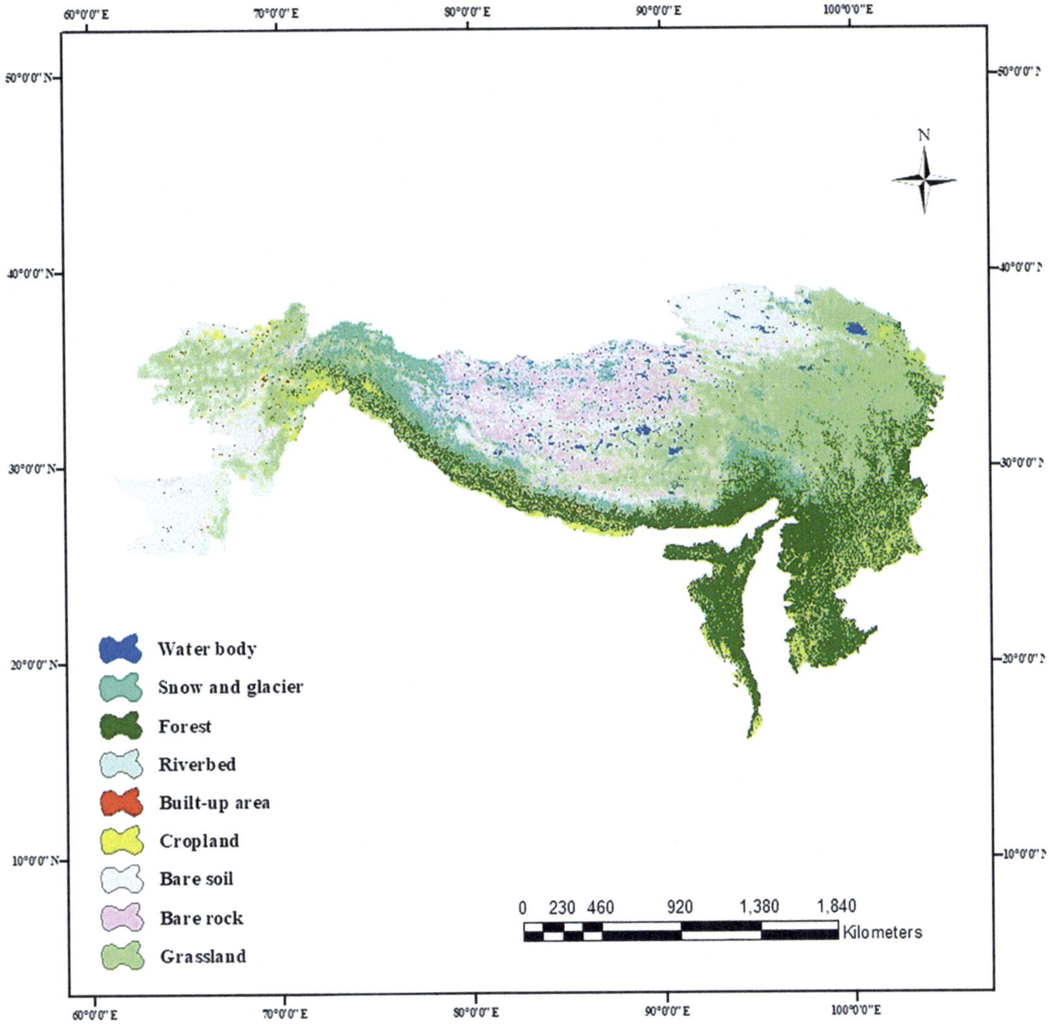

Fig. 4.2 Land use/land cover map of Hindukush Karakoram Himalaya. (Source: ICIMOD)

importance (Hagg et al. 2004). The glaciological, geodetic, and hydrological are the major approaches used to determine the mass balance (Hoinkes 1970). The most common approach for establishing the mass budget is to quantify the source and dissemination on the glacier top in relation to the previous year's melt horizon. The thickness and density of snow and ice are usually measured in cores or pits in the accumulation region. Stakes are used to determine how much ice has melted (Hagg et al. 2004). While the geodetic method is based on comparing precise topographic maps and calculating the volume change between photogrammetric surveys. The water equivalent may be estimated using the snow and ice mean densities. Because altitude variations in the glacier surface are governed by not only accumulation and ablation, as well as by ice movement, this approach only enables deductions for the entire glacier (Hoinkes 1970). Glacier mass balance is computed as a storage component in the water balance using the hydrological approach. The glacier mass budget for the entire watershed is calculated by subtracting runoff and evapotranspiration from precipitation. Because the outcome is based on a difference in huge figures, substantial relative errors can occur. Due to the well-known inaccuracies in point measurements, basin precipitation is particularly imprecise (Sevruk 1982).

Glaciologists have modified or created a range of devices throughout the last 50 years. They have quickly embraced and adapted new technologies to better understand ice's size, spread, inner structure, movement, deformation, accretion rate, boundary conditions, physical qualities, and chemical contents. Glaciology will benefit greatly from modern satellite remote-sensing technologies. For continuous monitoring of glacier temperatures, strain rates, and other parameters, low-power, automated data collectors with microprocessor control and continuous solid-state storage will be widely deployed. For data gathering at the ice-bedrock interface, new approaches might be devised. Advances in ice drilling may be down-bore-hole monitoring techniques, and certainly developments in laboratory ice analysis techniques could be expected (Zwally 1987). This advancement and novel technologies help to aid more reliable estimation of glacial mass loss. Although the lack of hydro-meteorological and glaciological data in the HKH region is a significant constraint for reliable studies (Cogley 2011; Kargel et al. 2011). Since its beginning, remote sensing has proven to be an effective tool for collecting data on glaciers.

The recent introduction of Geographic Information Systems (GIS) and Global Positioning Systems (GPS) has established a powerful tool for analyzing and monitoring glacier temporal dynamics. Remote sensing elements are an unprecedentedly advanced and useful medium for studying glaciers that are frequently located in remote, inaccessible, and harsh locations because of their synoptic view, repeating coverage, and temporal coverage. Sensors installed on airborne or spaceborne platforms can collect glaciers in a variety of situations. The introduction of the very first earth observation technology satellite in the early 1970s sparked a huge interest in using remote sensing to study glaciology's different fields. The introduction of GIS has made it much easier to discover glaciers and track their geographical extent over time using multi-temporal pictures (Allen 1998; Binaghi et al. 1993; Garelik et al. 1996; Li et al. 1998; Sohn et al. 1998). Data from various sources may now be added to snowmelt runoff modeling. Recent advancements in GPS have brought new options for monitoring glacier dynamic motion and predicting glacier mass balance on a more regular and frequent basis. GPS methods are especially well suited to repetitive glacier mapping because of their speed and precision (Gao and Liu 2001). The assessment method of glacial mass loss has improved a lot but in the Himalayas significant measurement and modeling gaps persist. The inaccessibility and ruggedness of terrain and extreme weather events are the some of the major constraints in here. Local support, institutional support, and proper administrative policies can overcome data scarcity to a certain extent.

4.4 Climate Change Impacts

Climate change is the prime threat to Himalayan glaciers and each year huge mass of snow and ice is melting which contributes to high-surface runoff to rivers. Glacial retreat, glacial lake formation, glacial lake outburst, change in precipitation, and glacial soil erosion are the major concerns fueled by climate change. This section briefly exemplifies the major climate change impacts on Himalayan glaciers.

4.4.1 Glacial Retreat

Glacier retreat is slow and cannot be considered as a direct consequence of climate change, but glacier volume loss is an immediate response to climatic fluctuations and may be used as an excellent indication (Haeberli and Hoelzle 1995; Kargel et al. 2011). Since 1970, the glaciers of Mount Everest have been diminishing (Bolch et al. 2011). Since 1950, most Nepalese and Chinese Himalayan glaciers have halted accumulating mass (Kehrwald et al. 2008). Except for the Karakoram area, the Himalayan glaciers have all receded significantly during the last century (Hewitt 2005). The exception is owing to the region's unique characteristics compared to the rest of the Himalayas, such as orographic circumstances, an all-year accumulation regime, avalanche concentration, and ablation buffering due to dense debris cover (Hudaa et al. 2021). Glacier mass balance measurements and estimations based on field, ratio of accumulation area, equilibrium line altitude (ELR), and geodetic data indicate a large increase in mass waste of Himalayan glaciers during the previous three to four decades. The total loss of glacial ice during the previous four decades has been calculated to be 19 ± 7 meters. This translates to a loss of 443 ± 136 Gt of glacier mass in the Indian Himalayas, out of a total of 3600 to 4400 gigatons of glacial stored water (Kulkarni and Karyakarte 2014).

Dobhal et al. (2008) found that since 1992–1993, the average glacier decline has been −0.32. m w.e.a^{-1}. Over the course of the investigation, the vertical mass balance gradient has grown, demonstrating a decrease in snow accumulation well above ELA and an increase in ablation below it. The Himalayan glaciers, like every other glacier on the planet, are actively degrading. According to the mass balance pattern in Dokriani Glacier, the annual mean ablation rate ranged between −2.5 and −3.5 m w.e. a^{-1}, whereas the average annual accumulation rate was 0.45 mw.e. a^{-1} during the study period in the 1990s (Dobhal et al. 2004). Medium-sized glaciers like Naradu and Shaune Garang have receded by 550–923 meters. The region's altitude distribution has an impact on the mass balance of such glaciers. Between 1987 and 1995, the Naradu glacier had an ELA of 5030 m and an AAR of 0.71, but the lower altitude Shaune Garang glacier had an ELA of 4740 m and an AAR of 0.42, suggesting a faster retreat rate on the lower glacier. Massive glaciers such as Jorya Garang and Baspa Bamak retreated even less (425 and 380 meters, respectively) than medium-sized ones (Kulkarni and Bahuguna 2002).

According to the current research, Gangotri Glacier has melted 0.41 ± 0.03 sq. (~0.01 sq. km year–1) km between 1965 and 2006 from the front, between 1968 and 2001, the Gangotri glacier withdrew around 764 ± 19 m (23.2 ± 0.6 m/year) (Bhambri, 2012), whereas a field-based GSI study revealed that it lost about 720 meters length (28.8 m/year) (Srivastava 2004). Kulkarni et al. (2007) stated that between 1988 and 2003, field investigations were conducted on the Chhota Shigri glacier and data show a retreat of 800 meters. By the year 2003, field images of the glacier terminal region showed changes in glacial morphology, with the whole region coated with debris, indicating glacial retreat and a loss in the glacier's debris-carrying capacity. This glacier will become a rock glacier if the current trend continues (Kulkarni et al. 2007). The central part of certain bigger glaciers, such as Zemu, has an ice thickness of more than 200 meters. Between 1909 and 2005, the Zemu glacier receded 863 meters. Between 1988 and 2000, however, the retreat was interrupted by a 92-meter advance (7.67 per year). During this time, the glacier's area of coverage grew. In concisely, Zemu withdrew from 1976 to 1988, advanced for 12 years (from 1988 to 2000), and then retreated again. As a result, the influence of climate change on the glacier cannot be determined only by these short-term fluctuations in Sikkim (Luitel et al. 2012). The various glacial retreat rate estimated by different researchers in Himalayan glacier is tabulated in Table 4.1. Their estimations indicated that the glaciers are retreating primarily as a result of climate change.

4.4.2 Glacial Lake Formation

A glacial lake is a pond of water that is formed because of glacial dynamics. They arise when a glacier erodes terrain and then retreats, filling the resulting depression. Earth has lost more than half of its glaciers since the Ice Age. This, along with the present trend of retreating glaciers due to climate change, has resulted in a transition from solid to liquid water, causing a rise in the size and number of glacial lakes throughout the world. The formation of glacial lakes is increasing year by year. For example, based on Landsat imagery from 2014, a total of 3235 glacial lakes with a total area of 255.8 31.6 km^2 were found in the Hengduan Shan (Wang et al. 2017). Salerno et al. (2012) show that supraglacial lake production is influenced by the inclination of the glacier where the lakes are located. Because of the modest tran-

Table 4.1 List of some major glaciers and their retreat rate

Serial no.	Glacier	Observed period	No. of years	Retreat rate (m/decade)	Reference
1	Gangotri	1935–1996 1962–1999 1965–2006 2017–2021	61 37 41 4	188 338 200 68(m/y)	Sangewar and Kulkarni (2011) Naithani et al. (2001) Bhambri et al. (2012) Thakur et al. (2023)
2	Beas Kund	1963–2003	40	190	Sangewar and Kulkarni (2011)
3	Samudra Tapu	1962–2000	38	200	Kulkarni et al. (2006)
4	ChotaShigri	1962–1995	33	70	Sangewar and Kulkarni (2011)
5	Dokriani	1962–2007	45	170	Sangewar and Kulkarni (2011)
6	Zemu	1976–2005	29	140	Raina (2009)
7	Adikailash	1962–2002	40	130	Sangewar and Kulkarni (2011)
8	Bara Shigri	1906–1995	89	300	Sangewar and Kulkarni (2011)
9	Satopanth	1962–2006	44	220	Nainwal et al. (2008)
10	Hamtah	1961–2005	44	80	Sangewar and Kulkarni (2011)

sit of debris, which raises the ablation rate, and minimal transport of fresh snow and ice, which reduces the fluid velocity of the glacier termini, the slope of the glacier upstream encourages the creation of supraglacial lakes.

Because of the build-up of meltwater behind loosely cemented end-moraine dams, glacier shrinkage and retreat in the Sikkim Himalayas have led to the development of new glacial lakes and the expansion of old ones. These lakes are naturally unstable and prone to catastrophic draining, posing a risk to society and property in the lowlands beneath (ICIMOD 2011). On debris-covered glaciers, supraglacial lakes and ponds can cause hotspots of mass loss. While a lot of study has gone into understanding lateral lake expansion, little is understood about the rates and processes that determine lake deepening. Due to the scarcity of observations of lake bottoms, this information gap remains to a significant extent (Mertes et al. 2017). Glacier type is one of the most key regulators on the commencement of lake formation. Negative mass balances can cause long valley glaciers to thin while retaining a reasonably steady terminus location. In the first case, the supraglacial melt may concentrate on the glacier surface and produce tiny melt ponds. Individual ponds can combine into enormous lakes if glacier structure obstructs drainage. This has resulted in the formation of some of Nepal's, Tibet's, and Bhutan's greatest glacial lakes. Lakes such as Tsho Rolpa (3.2 km), Thulagi (2 km), and Imja (1.3 km) in Nepal, and Lugge Tsho (1.8 km) in Bhutan are examples. Smaller lakes are usually connected with glaciers that are shorter and steeper. The drainage of ice water from the glacier top to proglacial sites, where it gathers behind terminal moraine ridges, is aided by a high surface gradient. Because the Neoglacial maxima's terminal moraines can surpass 100 m in height, the associated lakes may still hold 1–20 million cubic meters of water, posing a major risk (Hanisch et al. 1999; Pant and Reynolds 1999).

4.4.3 Glacial Lake Outburst Flood (GLOF)

Glacier floods, sometimes known as GLOFs, are created by the draining of natural dammed lakes in, on, or near glaciers. GLOFs are not a modern phenomenon, but as glaciers throughout the world retreat and temperatures rise, the likelihood of their occurrence in several mountain ranges has increased. Glacier floods are the most severe and far-reaching glacial threat, posing the greatest risk of disaster and devastation (Richard and Gay 2003). Earthquakes may potentially play a role in causing lake eruptions by acting as external triggers (Sakai et al. 2000). Lake water overflows and erodes the moraine dam, resulting in catastrophic breakdown once the hydrostatic pressure surpasses the restraining lithostatic pressure, causing outbursts from moraine-dammed lakes. In most cases, a trigger mechanism is required, such as a displacement wave from an ice or rock avalanche, or a dissolving ice core within the dam Lake outbursts caused by moraine damming are referred to by several names. The term "glacial lake outburst flood" (GLOF) is used in the Himalayas to characterize catastrophic lake outbursts caused by proglacial moraine damming (Richardson and Reynolds 2000). Unusual weather precipitation, lake outbreaks, increasing temperatures, flash flooding, and rock avalanches from destabilized slopes causing road closures are all effects of climate change that the Himalayas are now experiencing. So far, few researches on the effects of climate change on GLOFs have been carried out in the Himalayas (Chalise et al. 2006; Shrestha and Aryal 2011; Bajracharya et al. 2008; Fujita et al. 2013). It is crucial that the government creates a climate change knowledge database so that it can plan for mitigating the effects and adjusting to anticipated changes. GLOFs are largely unexplored, and more study is needed to forecast and mitigate their consequences (Kumar and Murugesh Prabhu 2012).

Because of the combined impacts of climate change and deforestation, the frequency of GLOF occurrences in the Hindu Kush Himalayan (HKH) area has increased during the second half of the twentieth century. Satellite observations of the region's mountaintop lakes have indicated a steady expansion in the size and volume of many of these glacial lakes at high altitudes, increasing the risk of a devastating outburst of flood affecting large populations and wreaking havoc on the Himalayan belt's valuable socioeconomic infrastructure and development assets (UNDP 2010; Kumar and Murugesh Prabhu 2012). It may cause serious a threat to downstream infrastructure, people, hydroelectric projects, and capacity of reservoirs. Conflicts with glacial dangers are becoming increasingly obvious as human activities expand deeper into the world's high alpine regions. In the Himalayas, scores of people and cattle have perished in the previous 50 years because of catastrophic flows from high-altitude lakes. The expenses of destroying a commercial hydroelectric power project in Nepal, for example, have been estimated to reach above \$500 million. The effects of a loss of generated power might endure a generation or longer, and a severe and very devastating occurrence could jeopardize Nepal's economic progress. Even the possibility of glacial risks may be sufficient to limit national funding for rural development in severe circumstances. As a result, the hazard is currently being considered at the national planning level within the Himalayan region's host governments (Chhetri 1999).

The area most vulnerable to glacial outburst floods is in the eastern Himalayas, offering life-threatening scenarios downstream. The eastern Himalayas are warmer than the western Himalayas since they are located at a lower latitude. According to scientists, this area is presently twice as prone as neighboring regions to experience glacial outburst floods. However, if global warming continues, dangers might quadruple by 2050, with new hotspots forming further west (Zheng et al. 2021). Outburst floods are threatening hydroelectric power facilities in the Himalayas, as well as the Karakoram highway, a multibillion-dollar commerce route linking China and Pakistan, and the Trans-Alaska pipeline, which passes through mountains with several glacial lakes. Oil-filled pipelines might be ruptured by an outburst flood, causing an environmental calamity (National Snow and Ice Data Center 2022).

4.4.4 Precipitation

The native population of the Himalayas relies on river systems for drinkable water, sanitation, trade, and fishing, as well as energy production and cultivation, which are among the key economic sectors in the area (Ménégoz et al. 2013). As a result, precipitation is critical to the local society and people's well-being. Climate change has a substantial influence on water security, and low socioeconomic status, avoidance, and adaptation to climate change are more difficult in this region. Bhutiyani et al. (2010) found a significant decrease pattern in monsoon and mean yearly precipitation in the northwest Indian Himalayas from 1866 to 2006. Between 1901 and 2003, there was a surge in pre-monsoon rainfall over the western Indian Himalayas (Guhathakurta and Rajeevan 2008).

Rainfall estimates by GCMs would be less reliable than temperature projections, although they indicate the larger unpredictability involved with rainfall. Western Nepal, Uttarakhand, Himachal Pradesh, and Bhutan will get more monsoon rainfall in 2071–2100 than base-period rainfall, according to the PRECIS regional simulation study conducted by Rupa Kumar et al. (2006). Likewise, according to SRES and RCP scenarios, Gupta and Kumar (2017) and Sooryamol et al. (2022) found a rise in rainfall over the Indian Mid-Himalayas and lesser Himalayas. David Raj et al. (2022) also stated that there may be an increasing rainfall over Shivalik Himalayas based on SRES and RCP scenarios. Sabin et al. (2020) highlighted the anticipated changes based on CMIP5 models, a box-whisker analysis is provided. In both RCP4.5 and RCP8.5, future anticipated variations in rainfall extremes exhibit considerable rises in R95p, reflecting an increased risk of extreme precipitation over the HKH. The central Himalayas, specifically, are expected to see a significant surge in R95p in the twenty-first century. The maximum consecutive 5-day rainfall (RX5day) also represents a clear climb, indicating that rainfall extremes will become more intense in the future. In the later part of the twenty-first century, the Coupled Model Intercomparison Project Phase 5 (CMIP5) models predict that total rainfall would rise faster in the western Himalaya and Karakoram than it does in the central and eastern Himalayas (Kapnick et al. 2014). Employing statistically downscaled data from eight CMIP5 GCMs, a study indicated that the climate change in a Himalayan River basin found a 14 percent increase in Monsoon rainfall by 2050 (Rajbhandari et al. 2016). Wu et al. (2017) evaluated the changes anticipated in average and extreme climates across the Hindu Kush Himalayan region using 21 CMIP5 models and found that rainfall extremes are expected to exaggerate in the future over the region. CMIP5 models reliably anticipate a significant rise in seasonal average rainfall across the central Himalayas in the middle and the late twenty-first century under various emission scenarios (Kadel et al. 2018).

4.4.5 Glacial Erosion

Glacial erosion involves both processes that take place explicitly in conjunction with the mobility of glacial ice over its bed, such as abrasion, quarrying/plucking, and physical and chemical erosion by subglacial meltwater, as well as mechanisms that are amplified or reconfigured by glaciation, including fluvial and mass wasting. Glacial erosion is well-known for being extremely effective and having a significant impact on terrain and tectonic processes (Hallet et al. 1996; Berger et al. 2008). The estimation of water and sediment yields from glacierized basins, as well as the shape and size of suspended matter, is critical for hydropower program development, construction, and implementation. Glacial erosion and other factors that make sediment obtainable for transit result in massive amounts of sediment being conveyed in glacier melt streams. A glacial river in Iceland and a neighboring non-glacier-fed river show a five-fold variation in sediment production, implying that glacial abrasion produces additional debris for stream conveyance than non-glacial weathering and erosion mechanisms (Embleton and King 1975). Harbor and Warburton (1993) also discovered that glacial processes cause faster erosion rates than nonglacial processes, leading to higher proglacial river sediment outputs. It is difficult to measure or simulate glacier erosion values, but findings from a range of techniques prove that glaciers and ice sheets erode at rates ranging from 0.001 mm $year^{-1}$ under cold-based ice to 100 mm $year^{-1}$ under temperate ice, with subsequent subglacial bedrock erosion values in the majority of the circumstances far larger than protracted averages (Koppes 2022). Furthermore, hydrological and glacial alterations are projected to exacerbate many H-K rivers that are already high and worldwide above-average sediment concentrations (Jonell et al. 2017). This material enhances hydro-abrasive erosion, leading to frequent reduced productivity and lower element lifetimes, as well as swift reservoir sedimentation, lowering HPP efficacy (Padhy and Saini 2008).

Meanwhile, in the Himalayan region, the quantification of sediment values began in the nineteenth century on the Gangotri Glacier melt river (Everest 1832), but these investigations were not pursued for this glacier or explored for other glaciers in the vicinity. However, combined sediment production calculations for the Ganga and Brahmaputra rivers are around 1.0×10^9 t $yr.^{-1}$ (Subramanian 1993), relative to worldwide yearly sediment flux calculations of roughly 15×10^9 t $yr.^{-1}$ (Milliman and Meade 1983). Most Himalayan rivers deliver a lot of suspended sediment, although the Bhagirathi River has an elevated specific sediment output. The tender age of the mountains, the sharp elevation gradient and feeble physical features of soils present on these slopes, huge and vigorous glaciers, extreme intensity monsoonal precipitation, natural weathering processes, and tectonic instability of the area are all natural features that contribute to high rates of sediment conveyance from the Himalayas. The existence of glaciers in a drainage basin adds a significant amount of sediment to the proglacial stream. Glaciers erode the mountain slopes and transfer rock and stones to lower valleys in high alpine locations. In the Himalayan region, estimating sediment transport in streams is critical for the preparation, developing, and functioning hydropower facilities (Singh et al. 2003). Singh et al. (2003) investigated suspended sediment transport from the Dokriani Glacier in the Garhwal Himalayas of the Bhagirathi basin and discovered a large sediment production from the glacial basin. Debris and silt are brought to the glacier top by erosion, debris flows, rock falls, and snow avalanches, partly masking the glaciers in the Indian Himalayas (Singh 2006). The Gangotri, Dokriani, and Rakiot glacier catchments in the Himalayas have erosion rates of 1.8, 1, and 1.5 mm, respectively (Gardner and Jones 1984; Singh et al. 2003; Haritashya et al. 2006). The Chaturangi glacier in the Garhwal Himalaya has a significant sediment production, according to estimates of suspended sediment concentration and meltwater flow (Bisht et al. 2020). The suspended sediments and eroded soil will become a huge threat to the storage of dams/reservoirs in the Himalayas. It will drastically diminish the storage capacity of the reservoir.

4.4.6 Impact on People's Livelihood and Ecosystem

The ice resource is expected to decrease, and it will have a significant impact on future dry season discharge. Low flows in Nepalese as well as other Himalayan regions are often 10–20 times lower than high flows (Myint and Hofer 1998). The dry season streams are likely to reduce as the resource base shrinks. This will have a significant impact on not only hydropower development but also industrial and agricultural development, as well as a severe shortage of water for human consumption (Chalise et al. 2006). Most of the precipitation is currently received in the form of rainfall rather than snow, according to 70% of respondents. Because there is less snowfall and it lasts for a shorter period, the possibilities of it compacting and turning to ice are lessened. This is thought to be the reason for the glaciers' observed negative mass balance and rapid retreat. Winter is said to be the most severe in the study region during November and February–March. This time frame has apparently shrunk dramatically in the last ten to fifteen years. A majority of 89% of participants have reported a reduction in the duration of the winter season, leading to an early melting of snow. This phenomenon has had an adverse impact on agro-horticultural activities. The greatest impact of the changes, according to the respondents, is on the receding of glaciers across valleys (Rautela and Karki 2015) (Fig. 4.3).

Glacier and snowmelt, as well as water flow into wetlands, sustain these plant species. Although the availability of water is important, overgrazing and climate change are other factors (Zhou et al. 2005). The climatic zones of the region support a diverse range of animals and habitats that occur along a clear humidity gradient. In the northwest, the vegetation shifts from subtropical semidesert and thorn steppe development to tropical evergreen forests in the southeastern Himalayas (Schickhoff 2005). Climate change will very certainly have further severe consequences for these mountains in the next decades, including major cascade effects on streamflow, groundwater recharge, natural calamities, and biodiversity; ecosystem composition, organization, and functioning; and human livelihoods (Nijssen et al. 2001; Parmesan 2006; Bates et al. 2008; Ma et al. 2009). Thus, the Himalayan people are suffering a wide range of natural and manmade calamities for the past several years. Valleys of the Himalayas are most prone to disasters like flash flooding and other climate change-related events. Climate change badly affects the people's livelihood and sources of income. Studies show that GLOF-like potential hazards will wipe out the region leading to a catastrophic disaster. So, there is a need for early warning systems and other mitigation strategies which will reduce the aftereffects of the disaster.

4.5 Mitigation and Adaptation

Five key methods that Himalayan peoples manage their natural resources using TEK (Traditional Ecological Knowledge) have been discovered from an interview and a survey of Himalayan communities: (1) biodiversity protection, (2) afforestation, (3) agroforestry, (4) soil management, and (5) a carbon-negative way of life. People in the Himalayas are not often aware that their traditional knowledge is helping to mitigate climate change, but it is and it should be noticed and may be replicated elsewhere (Salick et al. 2014). Although they are well aware of the climate change in general particularly the rise in temperature, melting glaciers, and irregular rain patterns, they are also less aware of how it is affecting them (Byg and Salick 2009). People in the Himalayas are actively mitigating climate change by implementing TEK. Holy site protection, afforestation, tree crops, soil carbon absorp-

Fig. 4.3 A view of the Himalayan village (Bagori) in Harsil Valley (Apple orchard)

tion through mulch and manure integration, and carbon-negative lifestyles all had mitigating impacts. Traditional ecological knowledge does not identify climate change or its mitigation in and of itself, but it does give useful climate change mitigation approaches. These TEK climate change mitigation practices are beneficial not just to Himalayan peoples but also to the rest of the globe (Salick et al. 2014). Preventive actions, such as reducing the level of the glacial lake, or interim measures, such as constructing an early warning system or shifting susceptible infrastructure to higher elevations of threatened valleys, can all be used to control glacial hazards (Hanisch et al. 1998).

Adaptation refers to real modifications performed in response to changes in the climate system that have been seen or expected. According to the case study conducted by Carey et al. (2012), the ideal climatic change glacier hazard management strategy involves: identifying and predicting hazards associated with environmental change; reducing the chances of those events occurring through appropriate engineering strategies; and removing all vulnerable populations, property, and infrastructure from the hazard zone. This is also part of an ideal disaster risk reduction and climate change adaptation plan, which includes several socio-environmental components. While achieving all of these is unrealistic or impracticable, the combination of these aspects provides a benchmark against which to analyze the acts that were executed or neglected. The next stage is to develop new adaptation practices based on these factors—plans to identify new and current dangers and, over time, to minimize human susceptibility to glacial hazards. As a result, the examination of possible GLOF dangers must be expanded from the glacier's lower border and glacial lake dam to avalanche-initiating zones near mountain summits above lakes. People should be kept out of flood and avalanche routes through long-term development goals and community planning. Existing communities may not be able to be relocated, but land use planning can assist in minimizing danger. Given the variety of perceived dangers in the region, it is equally critical to disseminate risk information to local residents via education and access to up-to-date information. Early warning systems might help mitigate the effects of an upcoming GLOF or avalanche (Carey et al. 2012).

4.5.1 Sustainability

The impoverished, frequently rapidly growing societies near rivers and lakes, which rely on climate-sensitive assets and are vulnerable to extreme weather, are perhaps the most likely to be impacted. As weather extremes grow increasingly severe and/or recurring, their socioeconomic costs are expected to rise. Thus, the sustainability of these communities is much necessitated for mitigating the adverse impact of climate change. Various sustainable development goal especially Goal-13 climate action states the way for achieving the sustainability in the context of climate change. It exemplifies that all countries must improve their resilience and adaptation capabilities in the face of climate-related threats and natural disasters. Incorporate climate change policies, approaches, and preparation into policy initiatives, schemes, and planning. Increase climate change mitigation, adaptation, impact reduction, and early warning ability through education, sensitization, and enhance the institutional and human capacity.

The relationship between the melting of the Himalayan glaciers and planetary well-being is intricate. Since the Himalayan region's planetary health is now in grave danger due to the Himalayan glaciers' accelerated melting, hundreds of millions of people are now at risk (Talukder et al. 2021). The existing glacier change models are overparameterized and unable to limit important features of the internal dynamics and interactions of glaciers with the climate, such as the impacts of avalanching and debris cover (Rounce et al. 2020). Regional-scale understanding of the dynamics and mass change of High-Mountain Asia glaciers has improved as a result of recent remote-sensing observations (Kääb et al. 2015). Due to negative consequences on water resources, ecosystems, agriculture, hydroelectric, livelihood, communities, leisure, heritage, natural hazards, and other associated activities, the retreat of glaciers is a huge concern that poses significant challenges (Talukder et al. 2021). Pakistan's economic underpinning, irrigated agriculture is highly dependent on water supplies originating from the upper Indus' mountain sources. Hence, any change in the number of accessible resources due to socioeconomic or climatic change might have a severe impact on both the ecosystem and the food security. The sustainability of water supplies does not appear to be most threatened by climatic trends. Nevertheless, inquiry and understanding of the link between climate, glaciology, and runoff are far from complete; previous past climatic experiences may not be a fair prediction of the future (Archer et al. 2010).

Responses can be analyzed in terms of data accessibility, capability of scientific analysis, and management's reaction to changing conditions (Archer et al. 2010). Despite having a wealth of natural resources, most of its population is marginalized and yet subsists. Unscientific resource use is accelerating environmental degradation and worsening the effects of natural disasters. To re-establish equilibrium between economic interest and ecological imperatives while considering socio-cultural norms, a new paradigm must be developed (Singh 2006). Like other mountain ecosystems, the IHR's inhabitants rely significantly on local natural resources and primary production from sectors like agriculture, forestry, cattle, and others to survive. The dependence of the continuously expanding population on limited resources, the absence of technologies that can address the unique challenges posed by mountains, and increased production to meet the demand are all contributing to resource depletion and farmer marginalization, which ultimately encourages poverty (Samal et al. 2003). Thus, sustainable management of the natural resources is the key solution for adapting and mitigating against the climate change. It will provide a sustainable livelihood for the mountain community and reduce the consequences of climate change and associated concerns.

4.6 Case Study

Gangotri Glacier is one of the most vital and studied glaciers located in the Bhagirathi River catchment. Various studies indicate that the glacier is retreating very rapidly. Bhagirathi River is a perennial river with the glacial origin. The water flowing through the Bhagirathi River has a major role in downstream water availability and power generation. It is considered as the prime source of water for the Tehri dam. Keeping this in view, we attempted to map snow cover and land surface temperature mapping of the upper part of the Bhagirathi River basin during last the 20 years.

4.6.1 Materials and Methods

This study provides a glimpse of the past 30-year condition of the upper Bhagirathi River basin in terms of snow cover and land surface temperature. We have used different series of Landsat satellite data for the generation of results. The Normalized Difference Snow Index (NDSI) was used to identify the snow-covered area. The bands required for generating NDSI were downloaded from USGS Earth Explorer. Snow cover can be detected using the NDSI ratio of the difference in VIS (Green) and SWIR reflectance:

$$\text{NDSI} = \left(\left(\text{band } 4 - \text{band } 6\right) / \left(\text{band } 4 + \text{band } 6\right)\right)$$

A pixel with NDSI >0.0 is considered to have some snow present. A pixel with NDSI <= 0.0 is a snow-free land surface (Riggs et al. 2016). We have generated three NDSI maps for 2001, 2011, and 2021 with Landsat 4/5 used for 2001 and 2011, while Landsat 8 used for 2021, respectively. Based on the data availability and cloud cover, we selected 20 January 2001, 25 January 2011, and 20 January 2021 for the generation of maps. ERDAS 2018 version software was used for generating the NDSI map of the study area. Similarly, the MODIS 11 A1/A2 Land Surface Temperature (LST) was obtained from Level-1 and Atmosphere Archive & Distribution System

Distributed Active Archive Center (LAADS DAAC). Based on the cloud cover and data availability, we selected January 19, 2001, 2011, and 2021 LST over the study area. ArcGIS 10.1 version was used for processing and generating the LST map of the study area.

4.6.2 NDSI and LST

The NDSI illustrates the area covered with snow/glacier and its value ranging from −1 to +1. Figure 4.4 illustrates the NDSI of the year 2001 and it is explicit that except the valley region, most of the regions are covered with snow/glacier. The negative value indicates bare soil or without snow-covered surface. During the year 2011, it is visible that a huge portion of snow are disappeared from the surface. It indicates a drastic decrease in snow cover during the year January 2011 compared to January 2001. Similarly, the latest January 2021 showed a decrease in snow cover during the month of January. For estimating the areal change in snow-covered area during the month of January, we estimated the snow-covered area (NDSI>0.0) using raster reclassification. It was found that the total snow-covered area (based on Fig. 4.4) in 2001 was 756.4 km^2, 631.4 km^2 during 2011, and 514.7 km^2 during 2021. It indicates the approximate extent of snow/glacial area loss/retreat in the meantime. Because of climate change, the temperature increases globally, thus the highly sensitive snow/glacial area is reducing as a consequence of this. This promotes higher snowmelt runoff, glacial erosion, downstream water availability, GLOF, glacial lake formation, etc.

As mentioned earlier, the snow cover decreases and, at the same time, it leads to an increase in the land surface temperature. Figure 4.4 illustrates the land surface temperature during 2001, 2011, and 2021 and the extent of hotter surface is increasing. Apart from this, low-value temperature (region) is increasing, and high-value temperature (region) is increasing. Thus, increased LST further decreases the snow-covered region and eventually affects the ecosystem. The higher temperature value (°C) was 17.8 during January 2021, 19.1 during 2011, and 22.9 during the year 2021. Similarly, low-temperature value increased from −22.2 during 2001 to −15.5 during the year 2021. All the observations indicate that the snow cover is decreasing from the Bhagirathi catchment.

Figure 4.5 shows the reduction/change of the areal extent of the snow-covered region from 1984 to 2020. The TCC explicitly delivers the variation in snow cover and the retreat of glaciers. Apart from this, due to global warming, the thickness of snow/glaciers is reducing. The Intergovernmental Panel on Climate Change (IPCC) predicts that worldwide glaciers, excluding the Greenland and Antarctic ice sheets, will decline by 15% to 55% by 2100, irrespective of whether global temperatures rise by 2 °Celsius. If the climate continues to warm significantly faster, glaciers all over the globe might drop up to 85% of their volume. Glacier retreat frequently follows atmospheric carbon dioxide levels, and as glaciers progress to melt, the depletion of fresh water for drinking and a variety of other anthropogenic activities will be felt.

4.7 Summary and Conclusion

The melting of glaciers acts as an indicator of global warming. For almost a century, glaciers have been researched as sensitive climatic indicators. The most prevalent emphasis of alpine glacier observations is on changes in terminal behavior to determine glacier response to weather change. The most reliable indication of short-term glacier sensitivity to climate change is annual mass balance measurements. Glacier withdrawal is hastening, posing significant difficulties to humans and the environment. During protracted droughts, especially late in the summer, after the seasonal snowpack has thawed away, glaciers supply populations and habitats with a consistent source of river flow and potable water in many locations.

Most Himalayan glaciers are shrinking at similar rates as glaciers elsewhere. Therefore, the Himalayan glacier's response to fast climate

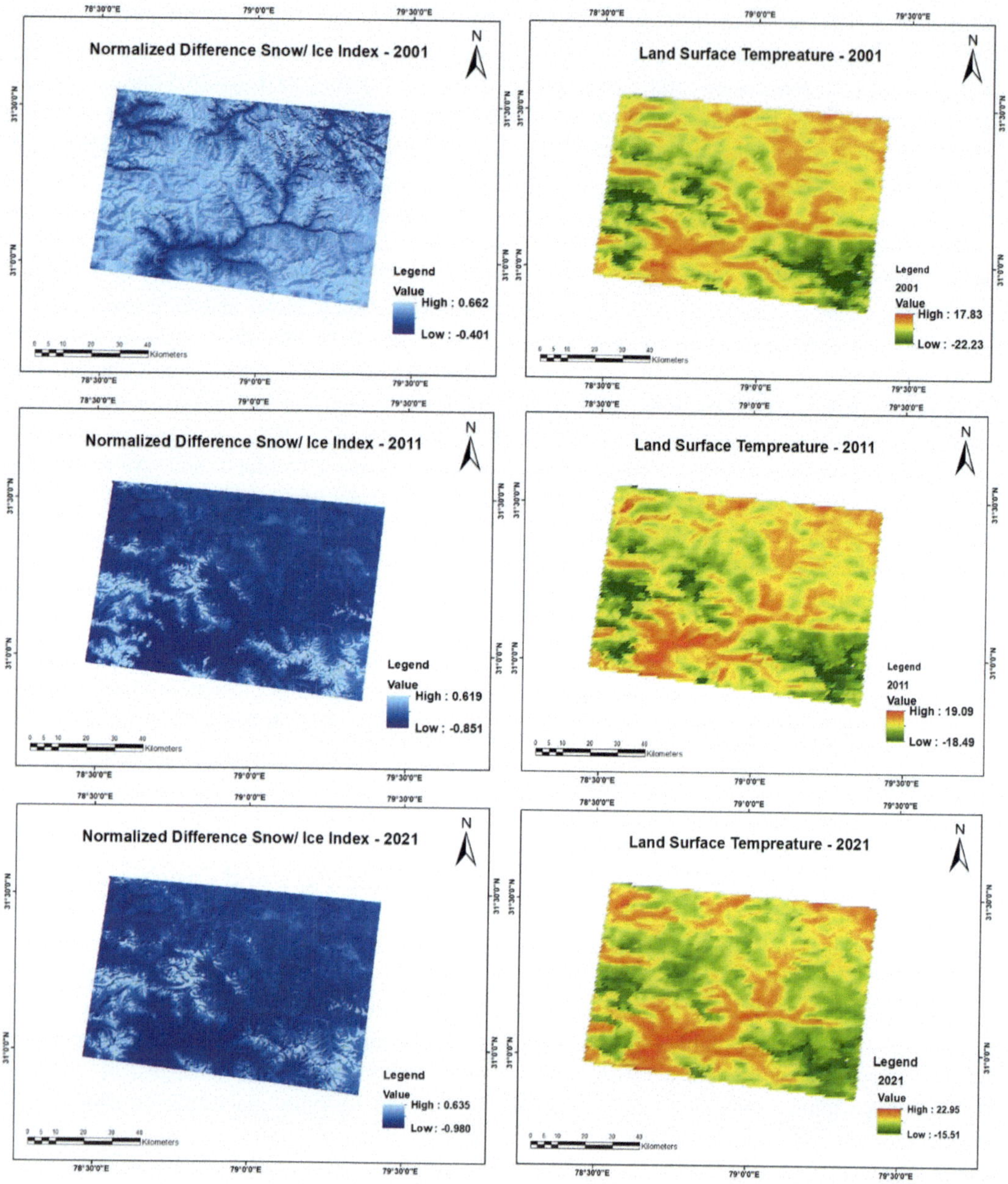

Fig. 4.4 Time series NDSI and LST map of the upper Bhagirathi River basin

change is a significant challenge. The purpose of this study is to emphasize the impact of climate change on Himalayan glaciers such as (i) glacial retreat, (ii) glacial lake formation, (iii) glacial lake outburst floods (GLOF), (iii) soil erosion, (iv) precipitation, etc. and its influence on the local community, ecosystem, water productivity, and finding possible mitigation strategies; also analyzing methods and data used by various researchers. Compared to traditional approaches, which are usually time-consuming, laborious, and sometimes spatially impractical in isolated places, remote sensing has shown to be an excellent alternative for evaluating glaciers in inaccessible mountains and simultaneously monitoring large numbers of glaciers.

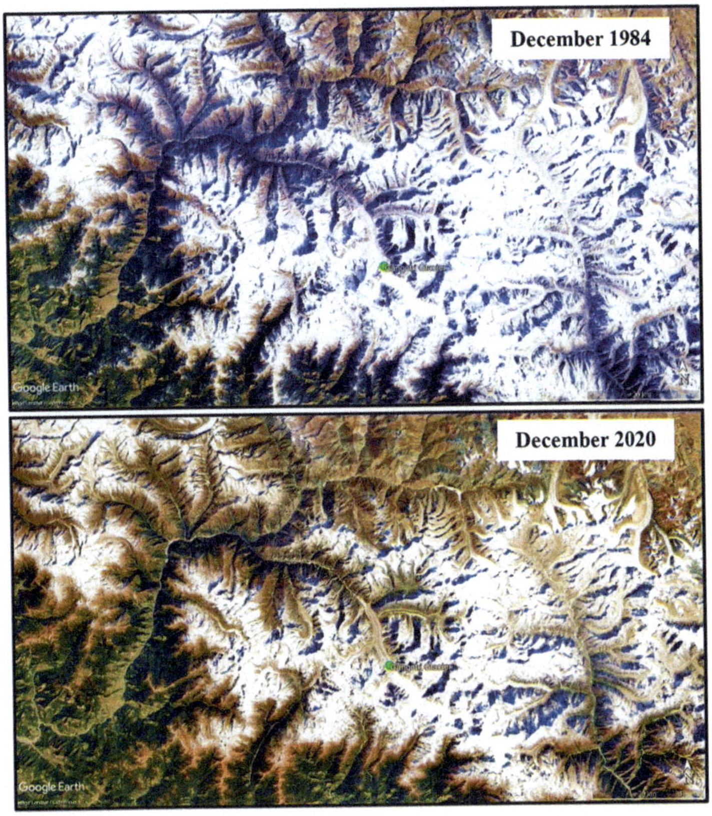

Fig. 4.5 True color composite (TCC) of the snow-covered area of upper Bhagirathi River basin; Gangotri glacier marked green in color. (Source: Google Earth Pro)

There is a lot of research gap in Himalayan glaciology due to the orography of the Himalayas. Most of the knowledge is based on satellite data. This study highlights the necessity of tracking glacier mass balance, growth, and melting for an extended period. Soil erosion caused by snowmelt runoff water is a minimally explored topic in the Himalayan region. This detailed analysis of past studies will help researchers in the field to conduct more studies and to get an idea about previous works and gaps in Himalayan glaciology; also help decision-makers to formulate vital strategies mitigate the aftereffects of potential hazards. The lack of a comprehensive broad study of Himalayan glaciology will emphasize the importance of this chapter and contribute to the development of the proposed book since this chapter discusses the extreme weather events impact and their mitigation strategies.

References

Ahmad O (2021) Explainer: The disappearing glaciers of the Himalayas. The third pole. https://www.thethirdpole.net/en/climate/glaciers-himalayas-melting/

Allen TR (1998) Topographic context of glaciers and perennial snowfields, glacier National Park. Montana Geomorphology 21(3–4):207–216

Archer DR, Forsythe N, Fowler HJ, Shah SM (2010) Sustainability of water resources management in the Indus Basin under changing climatic and socio-economic conditions. Hydrol Earth Syst Sci 14(8):1669–1680

Bajracharya SR, Mool PK, Shrestha BR (2008) Global climate change and melting of Himalayan glaciers. Melting glaciers and rising sea levels: Impacts Implications, pp 28–46

Barnett TP, Adam JC, Lettenmaier DP (2005) Potential impacts of a warming climate on water availability in snow-dominated regions. Nature 438:303–309

Bates B, Kundzewicz Z, Wu S (2008) Climate change and water. Intergovernmental Panel on Climate Change Secretariat

Bereiter B, Eggleston S, Schmitt J, Nehrbass-Ahles C, Stocker TF, Fischer H et al (2015) Revision of the EPICA dome C CO2 record from 800 to 600 kyr before present. Geophys Res Lett 42(2):542–549

Berger AL, Gulick SP, Spotila JA, Upton P, Jaeger JM, Chapman JB et al (2008) Quaternary tectonic response to intensified glacial erosion in an orogenic wedge. Nat Geosci 1(11):793–799

Bhambri R, Bolch T, Chaujar RK (2012) Frontal recession of Gangotri glacier, Garhwal Himalayas, from 1965 to 2006, measured through high-resolution remote sensing data. Curr Sci 102(3):489–494

Bhutiyani MR, Kale VS, Pawar NJ (2010) Climate change and the precipitation variations in the Northwestern Himalaya: 1866–2006. Int J Climatol 30:535–548

Binaghi E, Madella A, Madella P, Rampini A (1993) Integration of remote sensing images in a GIS for the study of alpine glaciers. In: Winkler P (ed) Remote sensing for monitoring the changing environment of Europe. Proceedings 12th EARSEL symposium, Hungary. Balkema, Rotterdam, pp 173–178

Bisht H, Kotlia BS, Kumar K, Arya PC, Sah SK, Kukreti M, Chand P (2020) Estimation of suspended sediment concentration and meltwater discharge draining from the Chaturangi glacier. Garhwal Himalaya Arabian Journal of Geosciences 13(6):1–12

Bolch T, Pieczonka T, Benn DI (2011) Multi-decadal mass loss of glaciers in the Everest area (Nepal Himalaya) derived from stereo imagery. Cryosphere 5:349–358

Byg A, Salick J (2009) Local perspectives on a global phenomenon—climate change in eastern Tibetan villages. Glob Environ Chang 19:156–166

Carey M, Huggel C, Bury J, Portocarrero C, Haeberli W (2012) An integrated socio-environmental framework for glacier hazard management and climate change adaptation: lessons from Lake 513, cordillera Blanca. Peru Climatic Change 112(3):733–767

Chalise SR, Shrestha ML, Bajracharya OR, Shrestha AB (2006) Climate change impacts on glacial lakes and glacierized basins in Nepal and implications for water resources. IAHS Publ 308:460

Chaujar RK (2009) Climate change and its impact on the Himalayan glaciers–a case study on the Chorabari glacier, Garhwal Himalaya. India Curr Sci:703–708

Chhetri MP (1999) Disaster management in Nepal: problems and solutions. In: Ingleton J (ed) Natural Disaster Management. U.N. Secretariat for the IDNDR, Tudor Rose Holdings Ltd, Leicester, pp 223–224

Cogley G (2011) Present and future states of Himalaya and Karakoram glaciers. Ann Glaciol 52(59):69–73

Crutzen PJ (2002) The "anthropocene". *Journal de Physique IV (Proceedings)*. 12(10):1–5

Cruz RV, Harasawa H, Lal M, Wu S, Anokhin Y, Punsalmaa B, Honda Y, Safari M, Li C, HuuNinh N (2007) Asia. Climate change 2007: impacts, adaptation and vulnerability. In: Parry ML, Canziani OF, Palutikof JP, Van Der Linden PJ, Hanson CE (eds) Contribution of working group II to the fourth assessment report of the intergovernmental panel on climate change. Cambridge University Press, Cambridge, UK, pp 469–506

Cuffey K, Paterson WSB (2010) The physics of glaciers, 5th edn. Massachusetts Academic Press, Burlington, VT

Dasgupta A, Shyamoli S (2021) Himalayan glacial retreat—an interpretation of the glacial mass balance model. Int J Adv Eng Manag (IJAEM):1054–1063

Dimri AP, Dash SK (2011) Wintertime climatic trends in the Western Himalayas. Clim Chang. https://doi.org/10.1007/s10584-011-0201-y

Dobhal DP, Gergan JT, Thayyen RJ (2004) Recession and morphogeometrical changes of Dokriani glacier (1962-1995) Garhwal Himalaya. India Curr Sci Bangalore 86(5):692–696

Dobhal DP, Gergan JT, Thayyen RJ (2008) Mass balance studies of the Dokriani glacier from to, Garhwal Himalaya, India. Bull Glaciol Res 25:9–17

Embleton C, King CAM (1975) Glacial and periglacial geomorphology

Everest R (1832) Some observations on the quantity of earthy matter brought down by the Ganga River. Royal Asiatic Society, Bangal 1:238–242

Fujita K, Sakai A, Takenaka S, Nuimura T, Surazakov AB, Sawagaki T, Yamanokuchi T (2013) Potential flood volume of Himalayan glacial lakes. Nat Hazards Earth Syst Sci 13(7):1827–1839

Gao J, Liu Y (2001) Applications of remote sensing, GIS and GPS in glaciology: a review. Prog Phys Geogr 25(4):520–540

Gardner JS, Jones NK (1984) Sediment transport and yield at the Rakiot glacier, Nanga Parbat, Punjab Himalaya. In: Miller KJ (ed) The international Karakoram project 1. Cambridge University Press, Cambridge, pp 184–197

Garelik IS, Kotlyakov VM, Osipova GB, Tsvetkov DG (1996) Computer analysis of the dynamics of pulsating glaciers. Mapp Sci Remote Sens 33(3):207–216

Guhathakurta P, Rajeevan M (2008) Trends in the rainfall pattern over India. Int J Climatol 28(11):1453–1469

Gupta S, Kumar S (2017) Simulating climate change impact on soil erosion using RUSLE model– a case study in a watershed of mid-Himalayan landscape. Journal of Earth System Science 126(3):1–20

Haeberli W, Hoelzle M (1995) Application of inventory data for estimating characteristics of and regional climate-change effects on mountain glaciers: a pilot study with the European Alps. Ann Glaciol 21:206–212

Hagg WJ, Braun LN, Uvarov VN, Makarevich KG (2004) A comparison of three methods of mass-balance determination in the Tuyuksu glacier region, Tien Shan, Central Asia. J Glaciol 50(171):505–510

Hallet B, Hunter L, Bogen J (1996) Rates of erosion and sediment evacuation by glaciers: a review of field data and their implications. Glob Planet Chang 12(1–4):213–235

Hanisch J, Delisle G, Pokhrel AP, Dixit AM, Reynolds JM, Grabs WE (1998) The Thulagi glacier lake, ManasluHimal, Nepal * Hazard assessment of a potential outburst. In: Proceedings eighth international congress, international association for engineering geology and the environment. Vancouver, Canada, pp 2209–2215

Hanisch J, Pokhrel AP, Dixit AM, Grabs WE, Reynolds JM (1999) GLOF mitigation strategies * lessons learnt from studying the Thulagi glacier lake. Nepal [abstract] Journal of Nepal Geological Society 20:163

Harbor J, Warburton J (1993) Relative rates of glacial and nonglacial erosion in alpine environments. Arct Alp Res 25(1):1–7

Haritashya UK, Singh P, Kumar N, Gupta RP (2006) Suspended sediment from the Gangotri glacier: quantification, variability and associations with discharge and air temperature. J Hydrol 321:116–130. https://doi.org/10.1016/j.jhydrol.2005.07.037

Hegdahl TJ, Tallaksen LM, Engeland K, Burkhart JF, Xu C-Y (2016) Discharge sensitivity to snowmelt parameterization: a case study for upper Beas basin in Himachal Pradesh, India. Hydrol Res. https://doi.org/10.2166/nh.2016.047

Hewitt K (2005) The Karakoram Anomaly? Glacier Expansion and the 'Elevation Effect,' Karakoram Himalaya. Mt Res Dev 25:332–340

Hoinkes H (1970) Methoden und Mo¨glichkeiten von Massenhaushaltsstudien auf Gletschern: Ergebnisse der MessreiheHintereisferner (O¨ tztaler Alpen) 1953–1968 [Ways and means of studies of glacier mass balance]. Z Gletscherk Glazialgeol 6(1–2):37–90

Hudaa MB, Lone MA, Kumar R, Khoshoueid SR (2021) A review on the implications of changing climate on the water productivity of Himalayan glaciers

ICIMOD (2011) Glacial Lakes and glacial Lake outburst floods in Nepal

IPCC (2007) Climate change 2007: impacts, adaptation and vulnerability: summary for policymakers. In: Working group II contribution to the intergovernmental panel on climate change fourth assessment report. IPCC, Geneva

IPCC (2018) Global warming of 1.5oC. An IPCC special report on the impacts of global warming of 1.5oC above pre-industrial levels and related global greenhouse gas emission pathways, in the context of strengthening the global response to the threat of climate change, sustainable development, and efforts to eradicate poverty

Jonell TN, Carter A, Böning P, Pahnke K, Clift PD (2017) Climatic and glacial impact on erosion patterns and sediment provenance in the Himalayan rain shadow, Zanskar River. NW India Geol Soc Am Bull 129:820–836

Kääb A, Treichler D, Nuth C, Berthier E (2015) Brief Communication: Contending estimates of 2003–2008 glacier mass balance over the Pamir–Karakoram–Himalaya. Cryosphere 9(2):557–564

Kadel I, Yamazaki T, Iwasaki T, Abdillah MR (2018) Projection of future monsoon precipitation over the Central Himalayas by CMIP5 models under warming scenarios. Clim Res 75(1):1–21

Kapnick SB, Delworth TL, Ashfaq M, Malyshev S, Milly PC (2014) Snowfall less sensitive to warming in Karakoram than in Himalayas due to a unique seasonal cycle. Nat Geosci 7(11):834–840

Kargel JS, Cogley JG, Leonard GJ, Haritashya U, Byers A (2011) Himalayan glaciers: the big picture is a montage. Proc Natl Acad Sci 108:14709–14710

Kehrwald NM, Thompson L, Tandong Y, Mosley-Thompson E, Schotterer U, Alfimov V, Beer J, Eikenberg J, Davis M (2008) Mass loss on Himalayan glacier endangers water resources. Geophys Res Lett 35:L22503. https://doi.org/10.1029/2008GL035556

Khattak MS, Babel MS, Sharif M (2011) Hydro-meteorological trends in the upper Indus River Basin in Pakistan. Clim Res 46:103–119

Koppes MN (2022) 4.09 - Rates and Processes of Glacial Erosion. In: Jack J, Shroder F (eds) Treatise on Geomorphology, 2nd edn. Academic Press, pp 169–181. https://doi.org/10.1016/B978-0-12-818234-5.00032-8

Kulkarni AV, Bahuguna IM (2002) Glacial retreat in the Baspa basin, Himalaya, monitored with satellite stereo data. J Glaciol 48(160):171–172

Kulkarni AV, Bahuguna IM, Rathore BP, Singh SK, Randhawa SS, Sood RK, Dhar S (2007) Glacial retreat in Himalaya using Indian remote sensing satellite data. Curr Sci:69–74

Kulkarni AV, Dhar S, Rathore BP, Raj KBG, Kalia R (2006) Recession of SamundraTapu glacier, Chadra river basin, Himachal Pradesh. J Indian Soc Remote Sensing 34(1):39–46

Kulkarni AV, Karyakarte Y (2014) Observed changes in Himalayan glaciers. Curr Sci:237–244

Kumar B, Murugesh Prabhu TS (2012) Impacts of climate change: glacial lake outburst floods (GLOFs). In: *Climate change in Sikkim patterns, impacts and initiatives*. Information and Public Relations Department, Government of Sikkim, Gangtok

Kumar KR, Sahai AK, Kumar KK, Patwardhan SK, Mishra PK, Revadekar JV, Kamala K, Pant GB (2006) High-resolution climate change scenarios for India for the 21st century. Curr Sci:334–345

Li Z, Sun W, Zeng Q (1998) Measurements of glacier variation in the Tibetan plateau using Landsat data. Remote Sens Environ 63(3):258–264

Li H, Xu C-Y, Beldring S, Tallaksen LM, Jain SK (2016) Water resources under climate change in Himalayan basins. Water Resour Manag 30:843–859

Luitel KK, Shrestha DG, Sharma NP, Sharma RK (2012) Impact of climate change on east-Rathong glacier in Rangit Basin, West Sikkim. In: Climate change in Sikkim patterns, impacts and initiatives. Information and Public Relations Department, Government of Sikkim, Gangtok

Lüthi D, Le Floch M, Bereiter B, Blunier T, Barnola JM, Siegenthaler U, Stocker TF (2008) High-resolution carbon dioxide concentration record 650,000–800,000 years before present. Nature 453(7193):379–382

Ma X, Xu J, Luo Y, Prasad Aggarwal S, Li J (2009) Response of hydrological processes to land-cover and climate changes in Kejie watershed, south-West China. Hydrological Processes: An International Journal 23(8):1179–1191
Marcott SA, Shakun JD, Clark PU, Mix AC (2013) A reconstruction of regional and global temperature for the past 11,300 years. Science 339(6124):1198–1201. https://doi.org/10.1126/science.1228026
Ménégoz M, Gallée H, Jacobi HW (2013) Precipitation and snow cover in the Himalaya: from reanalysis to regional climate simulations. Hydrol Earth Syst Sci 17:3921–3936. https://doi.org/10.5194/hess-17-3921-2013
Mertes JR, Thompson SS, Booth AD, Gulley JD, Benn DI (2017) A conceptual model of supra-glacial lake formation on debris-covered glaciers based on GPR facies analysis. Earth Surf Process Landf 42(6):903–914
Milliman JD, Meade RH (1983) World-wide delivery of river sediment to the oceans. J Geol 91(1):1–21
Myint AK, Hofer T (1998) Forestry and key Asian watersheds: a paper prepared as background document for the Asia-Pacific forestry outlook study of the food and agriculture organization with financial support from the Canadian international development agency (CIDA). ICIMOD
Nainwal HC, Negi BDS, Chaudhary M, Sajwan KS, Gaurav A (2008) Temporal changes in rate of recession: evidences from Satopanth and Bhagirath Kharak glaciers, Uttarakhand, using Total Station survey. Curr Sci 94(5):653–660
Naithani AK, Nainwal HC, Sati KK, Prasad C (2001) Geomorphological evidences of retreat of the Gangotri glacier and its characteristics. Curr Sci 80(1):87–94
National Research Council (2012) Himalayan glaciers: climate change, water resources, and water security. National Academies Press
National Snow and Ice Data Center (2022) | What are glacial lakes? https://nsidc.org/cryosphere/icelights/2013/05/ebb-and-flow-glacial-lakes#:~:text=How%20glacial%20lakes%20form,fills%20the%20hole%20left%20behind
Nijssen B, O'Donnell GM, Hamlet AF, Lettenmaier DP (2001) Hydrologic sensitivity of global rivers to climate change. Clim Chang 50(1):143–175
Padhy MK, Saini RP (2008) A review on silt erosion in hydro turbines. Renew Sust Energ Rev 12:1974–1987
Pandey AC, Ghosh S, Nathawat MS (2011) Evaluating patterns of temporal glacier changes in greater Himalayan range, Jammu & Kashmir, India. Geocarto Int 26:321–338
Pant SR, Reynolds JM (1999) Subsurface electrical imaging techniques for the investigation of Thulagi glacier Lake dam [abstract]. J Nepal Geolog Soc 20:80
Parmesan C (2006) Ecological and evolutionary responses to recent climate change. Annu Rev Ecol Evol Syst 37:637–669
Raina VK (2009) Himalayan glaciers: a state-of-art review of glacial studies, glacial retreat and climate change. MoEF Discussion Paper 56
Raj AD, Kumar S, Sooryamol KR (2022) Modelling climate change impact on soil loss and erosion vulnerability in a watershed of Shiwalik Himalayas. Catena 214:106279
Rajbhandari R, Shrestha AB, Nepal S, Wahid S (2016) Projection of future climate over the Koshi River basin based on CMIP5 GCMs. Atmos Clim Sci 6:190–204
Rautela P, Karki B (2015) Impact of climate change on life and livelihood of indigenous people of higher Himalaya in Uttarakhand, India. Am J Environ Protect 3(4):112–124
Richard D, Gay M (2003) Guidelines for scientific studies about glacial hazards. Survey and prevention of extreme glaciological hazards in European mountainous regions. Glaciorisk Project, Deliverables. Website: http://glaciorisk.grenoble.cemagref.fr
Richardson SD, Reynolds JM (2000) An overview of glacial hazards in the Himalayas. Quat Int 65:31–47
Riggs, G. A., Hall, D. K., & Román, M. O. (2016). NASA S-NPP VIIRS snow products collection 1 user guide
Rounce DR, Hock R, Shean DE (2020) Glacier mass change in High Mountain Asia through 2100 using the open-source python glacier evolution model (PyGEM). Front Earth Sci 7:331
Sabin TP, Krishnan R, Vellore R, Priya P, Borgaonkar HP, Singh BB, Sagar A (2020) Climate change over the Himalayas. In: Assessment of climate change over the Indian region. Springer, Singapore, pp 207–222
Sakai A, Chikita K, Yamada T (2000) Expansion of a moraine-dammed glacial lake, TshoRolpa, in RolwalingHimal. Nepal Himalaya Limnology Oceanography 45(6):1401–1408
Salerno F, Thakuri S, D'Agata C, Smiraglia C, Manfredi EC, Viviano G, Tartari G (2012) Glacial lake distribution in the Mount Everest region: uncertainty of measurement and conditions of formation. Glob Planet Chang 92:30–39
Salick J, Ghimire SK, Fang Z, Dema S, Konchar KM (2014) Himalayan alpine vegetation, climate change and mitigation. J Ethnobiol 34(3):276–293
Samal PK, Palni LMS, Agrawal DK (2003) Ecology, ecological poverty and sustainable development in central Himalayan region of India. Internat J Sustain Develop World Ecol 10(2):157–168
Sangewar CV, Kulkarni AV (2011) Observational studies of the recent past. In: Report of the Study Group on Himalayan glaciers prepared for the Office of the Principal Scientific Adviser to the Government of India. PSA
Schickhoff U (2005) The upper timberline in the Himalayas, Hindu Kush and Karakorum: a review of geographical and ecological aspects. Mountain Ecosyst:275–354
Schild A (2008) ICIMOD's position on climate change and mountain systems. Mt Res Dev 28:328–331
Sevruk B (1982) Methods for correction for systematic error in point precipitation measurements for operational use. WMO Rep. 589. Geneva, World Meteorological Organization. (Operational hydrology rep. No. 21)

Shrestha AB, Aryal R (2011) Climate change in Nepal and its impact on Himalayan glaciers. Reg Environ Chang 11:65–77

Singh JS (2006) Sustainable development of the Indian Himalayan region: linking ecological and economic concerns. Curr Sci:784–788

Singh P, Ramasatri KS, Kumar N, Bhatnagar NK (2003) Suspended sediment transport from the Dokriani glacier in the Garhwal Himalayas. Hydrol Res 34(3):221–244

Sohn HG, Jezek KC, van der Veen CJ (1998) Jakobshavn glacier, West Greenland: 30 years of spaceborne observations. Geophys Res Lett 25(14):2699–2702

Sooryamol KR, Kumar S, Regina M, David Raj A (2022) Modelling climate change impact on soil erosion in a watershed of north-western lesser Himalayan region. J Sedimentary Environ:1–22

Srivastava D (2004) Recession of Gangotri glacier. In: Srivastava D, Gupta KR, Mukerji S (eds) Proceedings of Workshop on Gangotri Glacier, Lucknow. Geological Survey of India, Special Publication Number 80, pp 21–32

Subramanian V (1993) Sediment load of Indian rivers. Curr Sci:64-928-930

Talukder B, Matthew R, Bunch MJ, Hipel KW, Orbinski J (2021) Melting of Himalayan glaciers and planetary health. Curr Opin Environ Sustain 50:98–108

Thakur PK, Pandey P, Chauhan P (2023) Assessment of dynamics of frontal part of Gangotri glacier, India, from 2017 to 2021 using remote sensing data. J Indian Soc Remote Sens:1–8

UNDP (2010) Project facts – HIMALAYAS - glacial Lake outburst flood (GLOF) risk reduction in the Himalayas. Regional Glacial Lake Outburst Flood (GLOF) Risk Reduction Project in the Himalayas

United Nations Framework Convention On Climate Change (1992) pp. 33. Available at https://unfccc.int/files/essential_background/background_publications_htmlpdf/application/pdf/conveng.pdf. Accessed 13 March 2001

Wang XIN, Chai K, Liu S, Wei J, Jiang Z, Liu Q (2017) Changes of glaciers and glacial lakes implying corridor-barrier effects and climate change in the Hengduan Shan, southeastern Tibetan plateau. J Glaciol 63(239):535–542

Wu J, Xu Y, Gao XJ (2017) Projected changes in mean and extreme climates over Hindu Kush Himalayan region by 21 CMIP5 models. Adv Clim Chang Res 8:176–184

Xu J, Grumbine RE, Shrestha A, Eriksson M, Yang X, Wang Y, Wilkes A (2009) The melting Himalayas: cascading effects of climate change on water, biodiversity, and livelihoods. Conserv Biol 23:520–530

Zheng G, Allen SK, Bao A et al (2021) Increasing risk of glacial lake outburst floods from future third pole deglaciation. Nat Clim Chang 11:411–417. https://doi.org/10.1038/s41558-021-01028-3

Zhou H, Zhao X, Tang Y, Gu S, Zhou L (2005) Alpine grassland degradation and its control in the source region of the Yangtze and yellow Rivers, China. Grassland Sci 51(3):191–203

Zwally HJ (1987) Technology in the advancement of glaciology. J Glaciol Cambridge University Press 33(S1):66–77. https://doi.org/10.3189/S0022143000215840

5 Indigenous Strategies and Adaptive Approaches to Scrabble Recent Climate Crisis in Two Districts (Bankura and Purulia) of West Bengal, India

Mainak Sarkar, Partha Gorai, and Biplob Kumar Modak

Abstract

A population's susceptibility to climatic change varies spatially as well as temporally. Indigenous and rural people have evolved the sensitivity necessary to recognize any changes that may occur in the environment and to build adaptive strategies in response to climate change due to their intimate relationship with natural ecosystems. This kind of traditional knowledge can be used to design solutions for disaster risk reduction that are cost-effective, inclusive, and sustainable. Traditional communities have spent a lot of time decoding environmental changes and devising intricate plans to adapt. Long-term survivors of climate change and other environmental pressures may provide valuable insights about climate adaptation. In the two districts of West Bengal with the most indigenous people, numerous adaptation strategies and approaches were deciphered during the study. For example, during drought, indigenous people in these places try to use every possible way to procure available nutrients, coupled with unorthodox diets like clotted animal blood, to meet intense nutrient needs. Likewise, the pitcher watering system is very popularly used by the indigenous people of these regions to overcome high salinity and extreme aridity. This study depicts the native techniques and adaptation strategies of these regions and researchers could acquire a few signs to fight the current situations of environmental emergency.

Keywords

Climate · Environment · Indigenous · Strategies · West Bengal

5.1 Introduction

Climate change alters weather patterns and raises the likelihood of weather-related events (Houghton 2009). Rural communities around the world are experiencing significant disruptions to their ways of life as a direct result of climate change (Kihila 2018). Among the climate-related occurrences that have been reported are droughts, flooding, and violent ocean currents (Malley et al. 2008; Wang et al. 2013). As a result, these

M. Sarkar
Department of Zoology, Bankura Christian College, Bankura, West Bengal, India

Department of Zoology, Sidho-Kanho-Birsha University, Purulia, West Bengal, India

P. Gorai · B. K. Modak (✉)
Department of Zoology, Sidho-Kanho-Birsha University, Purulia, West Bengal, India

U. Chatterjee et al. (eds.), *Climate Crisis: Adaptive Approaches and Sustainability*, Sustainable Development Goals Series, https://doi.org/10.1007/978-3-031-44397-8_5

events have a negative impact on people's ability to go about their daily lives as normal, limiting their access to basic necessities like food, shelter, water, and sanitation (Sommer et al. 2013). Two West Bengal districts, Bankura and Purulia, have experienced severe drought for decades. Despite its prevalence, there is no universal agreement on how severe this danger actually is (Bhunia et al. 2020). When drought conditions prevail, farmers and pastoralists, who rely on water and pasture for the production of crops and the maintenance of animals, can suffer a significant blow to their ability to provide for themselves and their families (Kihila 2018). Communities have created their own coping and adaptation mechanisms over time using indigenous knowledge (IK) in response to the aforementioned challenges (Gorjestani 2000). As a result of these methods, communities have been better able to adapt and keep their lives going. This provides support for the idea that indigenous methods are worthwhile and significant. IK has been shown to play a role in climate change management through its contribution to the creation of local coping and adaptation techniques that can complement the global knowledge system (Nyong et al. 2007). Improved natural resource management and more robust coping and adaptation techniques can both benefit from IK, which can provide a framework for understanding people's relationships with the natural world (Nyong et al. 2007). While coping and adaptation techniques refer to reactive and proactive changes or adaptations in response to climate change, IK in this context refers to the knowledge that is assigned to a particular group (Fabricius et al. 2007). Indigenous coping or adaptation strategies use IK to adjust to a changing environment (Fabricius et al. 2007). In this scenario, IK is crucial because community members can gather climate knowledge through time and understand the changes and possible tactics (Leonard et al. 2013). Community members can, therefore, contribute extensive information about possible strategies, and adaptation and mitigation plans can be made using the climate change data.

"Indigenous knowledge" (IK) is information that has been passed down orally from one generation to the next, as well as ideas, experiences, practices, and information that may have started locally or elsewhere but have been changed by natives and incorporated into their way of life through legends, folklore, and rituals (Tarafder and Debnath 2021). Native people in a community pass on their knowledge from one generation to the next so that they can adapt to their social, economic, and agroecological environments (Senanayake 2006). Kearney et al., in 2007, say that the valuable practices and knowledge of indigenous people can make a big difference in ecological sustainability. Many studies have shown that indigenous or traditional knowledge can be passed down from one generation to the next through cultural rituals and oral transmission. This information has served as the foundation for several humanitarian endeavors all around the world, including medical treatment, farming, preservation, food preparation, and education (Senanayake 2006). Technical and non-technical domains both contribute significantly to IK. Beliefs and practices, music, flora, social ecology, religious taboos, conversational norms, and weather patterns are just a few examples (Anandaraja et al. 2008). Ecosystem restoration relies heavily on IK, which is essential for a number of reasons (Lakhani 2019). Traditional knowledge helps local communities make educated decisions for sustainable development. Indigenous peoples' knowledge of the interrelationships of natural resource management, aquaculture, agriculture, ecological zones, forest management, and game management is more than commonly recognized (Sillitoe 1998). According to Ulluwishewa (1993), sustainable development requires environmental conservation and resource availability. He adds that indigenous wisdom that has helped native people manage their environment for generations may be applied to sustainable development.

Due to their dependence on natural ecosystems, indigenous and rural people have learned to adapt to climate change (Timilsena and Devkota 2022). Traditional wisdom can illuminate fair, cost-effective disaster mitigation (Timilsena and Devkota 2022). Despite growing interest in IK, others remain skeptical. IK is not used more in research and quality management because of this (Ifejika Speranza et al. 2010). IK systems, which include techniques to preserve a population's natural resources, were formed because their users believed that they were important for survival. It guides indigenous communities' agricultural, non-agricultural, health, resource exploitation, service, and other decisions. Climate change is humanity's biggest challenge (Cunningham 1992). Indigenous people are especially vulnerable to the effects of a global temperature shift on their biological systems (Grieves 2009). Indigenous people face civil rights and inequality issues when responding to climate change due to institutional and legal limits. Strengthening and sustaining indigenous communities require linking indigenous peoples' adaptive talents to catastrophe preparedness, long-term economic plans, and environmental conservation (Abioye et al. 2014; Altman et al. 2018; Costanza et al. 2014). Indigenous communities utilize short-term adaptation techniques because they lack the money and technology to adapt to the changing world (Florin and Wandersman 1990).

India has the second-largest population and one of the highest natural disaster risk scores. Drought is a nationwide natural disaster. Bankura is the state's most drought-prone rainfed district. Drought is a regular occurrence throughout the northwest part of the district and requires urgent attention (Goswami 2017). Averaging 1400 mm annually, this district's agriculture and agricultural production are highly dependent on a small window of time with highly variable rainfall (Goswami 2017). Purulia district is widely recognized as the area within West Bengal most at risk from the consequences of drought. The area has earned the nickname "the region with a stone heart" as a result of this. Bankura and Purulia's sloping, lateritic, and permeable soil lacks subsoil moisture, threatening crops. Seasonal precipitation patterns increase crop yield variability (Kar et al. 2012). The Standardized Precipitation Index can depict actual rainfall as a standard deviation from the likelihood distribution of rainfall (SPI). Bankura and Purulia experience seasonal droughts (Table 5.1) with a frequency and severity of 9 and 10, respectively, before the monsoon, 3 and 4, respectively, during the monsoon, and 4 and 5, respectively, after the monsoon. These values are based on the standard precipitation index (SPI) (Kar et al. 2012). Agricultural drought typically causes crop failure, human and animal famine, land degradation, disruption of numerous economic operations, disease spread, and population and livestock displacement (Hazra et al. 2017). Droughts increase floods and wildfires. Droughts impact the national economy and food supplies beyond individual farmers (Hazra et al. 2017).

India's tribal communities have distinct identities and a shared cultural history. Most Indian tribes live in forests and highlands, following their traditional lifestyles (Sarkar and Modak 2022). In the state of West Bengal, the Western Rarh region encompasses the entire Purulia and Bankura districts in their entirety. There are 40 distinct cultural groups that are recognized as scheduled tribes (STs) in the Indian state of West Bengal (Article 342 of the Constitution of India). According to the Census of India from 2011, people of scheduled tribes (ST) make up 18.27% of the population in the Purulia district and 10.36% of the population in the Bankura district. These two districts also have a richly diverse cultural heritage (Basu 2020). The purpose of this study is to evaluate the indigenous local knowledge, capabilities, and traditional practices of resource management and adaptation skills in these areas in relation to the fight against drought.

Table 5.1 Number of drought prone blocks, areas, monsoon rainfall, and non-monsoon rainfall of Bankura and Purulia district of West Bengal

	Bankura	Purulia
Number of drought prone blocks	7	20
Drought prone area (km^2)	2185	6305
Monsoon rainfall (mm)	1159.70	1163.00
Non-monsoon rainfall (mm)	387.50	344.30

Modified from Halder and Sadhukhan (2012)

5.2 Materials and Methods

5.2.1 Study Area

West Bengal's Western Rarh region includes Purulia and Bankura (Fig. 5.1). The current study is focused on Bankura and Purulia (23°13′ 52.68″ N, 87°4′ 42.24″ E). The semi-arid Western Rarh region of West Bengal is dominated by these two districts. Bankura and Purulia have hot, humid summers, and nice winters. In winter, the average temperature is 10–20 °C, whereas in summer, it is 32–40 °C. The monsoon season lasts from June through September and averages 1100 mm. Laterite dominates Bankura and Purulia. Laterite soil has high iron and aluminum but low nitrogen and phosphorus, making crop growth difficult. Deep forests and lush greenery characterize Bankura and Purulia. Sal, Teak, Mahua, and Palash dominate the deciduous woodlands. Therapeutic herbs and spices include amla, haritaki, neem, and turmeric (De and De 2021). Indigenous people live in rough and wooded villages. These rural and indigenous people have the lowest wages in the nation. *Santhals*, *Bhumijs*, *Kherias*, *Lodhas*, *Mundas*, *Oraons*, *Paharias*, and *Birhores* are popular.

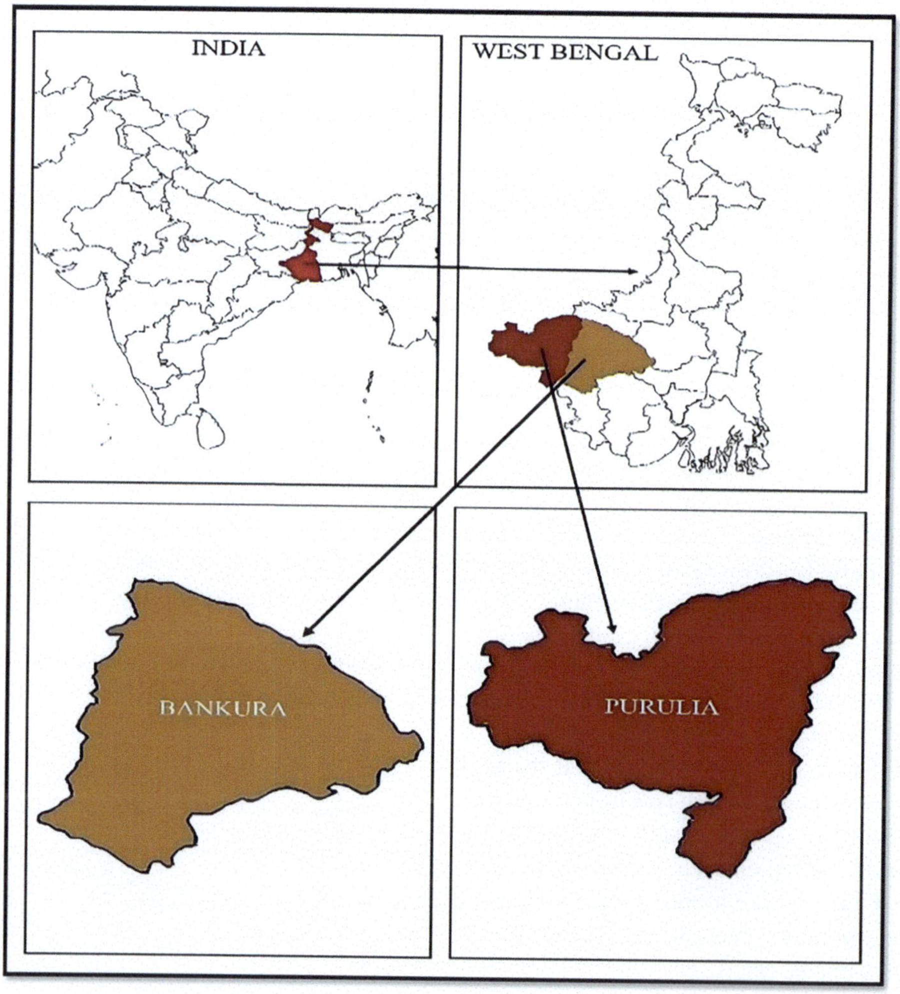

Fig. 5.1 Study area (Bankura and Purulia). (Source: Author)

5.2.2 Data Collection

Primary data were gathered through interviews, which included both in-depth, one-on-one probes and broader, more generalized conversations. A routine question-and-answer session survey of 42 informants (34 males and 8 females) from principally three tribes (Santhal, Bhumij, and Lodha) was done in a number of outlying villages of the study region between April 2021 and June 2022. Data on local IK and local practices for disaster risk reduction were gathered from a variety of stakeholders in the preliminary data collection. Proposals were made for key informant interviews (KIIs) and target group discussions (TGD). Evaluation of natural hazards, management of local values and resources, risk reduction, and risk aversion were all discussed in detail in a TGD. To learn more about the local population's exposure to natural catastrophes, their potential causes, and the unique risks they face, a series of key informant interviews (KIIs) was undertaken. The interview was conducted in a methodical fashion, with a schedule of predefined questions and efficient recording procedures. To get even more specifics, a targeted interview was also done. Several homes were marked as "essential" for interviews due to the presence of enlightened tribal members living there. Important people in the village, such as the village chief or a respected educator, contributed the data. One male (the expert) from each household was interviewed, and the elderly female residents of those homes also provided insightful commentary. While interacting with natives is essential to the research process, it is often difficult due to their restrictive nature regarding communication. Thus, a local guide—who may or may not be a tribe member—was used to translate between interviewee and interviewer. The study examines tribal life's food, water, and coping habits. Respondents' unwillingness to share personal information out of worry for the safety and integrity of their revered, cloaked, and unharmed usual coping strategy is the survey's biggest obstacle. The second interviewee confirmed that this study's data is specific and convincing.

5.3 Result and Discussions

The results of our study, which encapsulated IK relevant to the fight against drought and included solutions to conserve food, water, and design their houses to reduce temperature, are shown below.

5.3.1 Navigating Food Choices, Eating Habits, and Food Preservation Techniques During Times of Scarcity

Humans have a symbolic relationship with food, which means that it serves as more than just a source of fuel. Food serves a crucial symbolic role in maintaining cultural identity. And indigenous people follow these customs religiously since they are essential to their sense of community. Indigenous people's food systems are understood to be made up of regional, natural, and socially acceptable ingredients. Traditional food systems of indigenous peoples are defined as consisting of ingredients that are locally obtained, occur naturally, and have been approved by the community's culture (Kuhnlein et al. 2013). Primary dietary data were collected using standardized questionnaires and focus groups. Recipe collection and documentation are complete. Thirty-four tribal households' 24-h dietary recalls were recorded. This cross-sectional study examined indigenous Bankura and Purulia residents' famine-related dietary patterns. This diet chart (Tables 5.2 and 5.3) includes several common and uncommon foods (Fig. 5.2). Indigenous people do not always like year-round cuisine. These foods will be eaten when the drought deepens. Thus, favored foods will be unavailable. Thus, food changes even without a drought.

5.3.1.1 Rice and Its Other Derivatives

The research shows that rice is the staple food for the vast majority of the indigenous people (Bisai and Dutta 2021). They eat at least twice daily and often even three times. Roasted potatoes and tomatoes cooked in a touch of mustard oil, coupled with

Table 5.2 Survival diet (plant): the traditional foods eaten by the indigenous people of Bankura and Purulia in times of drought

S. no.	Scientific name	Local name	Part(s) use	Time of collection	Collection method	Preparation	Importance given by local people
1.	*Oryza sativa*	Dhalamar/ Marjhola	Grains	Throughout the year	Hand picking	Boiled with water	High
2.	*Alangium salvifolium*	Ankra/Dhela	Fruits	May–June	Hand picking	The juicy part of the ripe fruit is eaten	Little
3.	*Allium cepa*	Piyaj shak	Leaves	January–March	Cut with the sickle	Collected leaves are cut into small pieces and fried with potato slice, salt, turmeric, and oil. Sometimes curry is made with spices	High
4.	*Amaranthus spinosus*	Kanta bhazi shak/Kanta note	Leaves	May–August	Hand picking	Collected leaves are cut into small pieces and fried with garlic and oil	Moderate
5.	*Averrhoa carmbola*	Kamranga	Fruits	June–July	Hand picking	Collected fruits are cut into small pieces and the chutney is made with oil mustard and sugar	Moderate
6.	*Azadirachta indica*	Neem	Leaves	March–May	By using pole with hooks	Collected leaves are cut into small pieces and fried with potato/brinjal/flat-beans and oil Collected young leaves are cut and pasted and mixed with chickpea flour or rice husk, onion, salt, and chilli to make cutlet-like Boda	High
			Flower			Collected flowers are mixed with chickpea flour or rice husk, onion, salt, and chilli to make cutlet-like Boda	Moderate
7.	*Bambusa* sp.	Bans	Sprouting shoots	July–August	A cover of the clay pot is given during germination, after a few days, it cuts with a sharp cutter	Collected materials are cut into small pieces and fried with onion, chilli, salt, turmeric, and oil	High
8.	*Bauhinia racemosa*	Kanchan	Flower	February–May	By using pole with hooks	–	Little
9.	*Bauhinia vahlii*	Chihore	Seeds	May–August	Hand picking	–	Little
10.	*Borassus flabellifer*	Tal	Fruits, kernel, shells	April–August	With sharp cutter or by using pole with hooks	–	Moderate
			Sap	March–May			High
11.	*Brassica nigra*	Sorisha shak	Leaves	March–April	Hand picking	Collected leaves are cut into small pieces and fried with garlic and oil	Moderate

12.	*Buchanania latifolia*	Pial	Fruits, seeds	April–May	Hand picking	–	High
13.	*Butea monosperma*	Palash	Roots	May–July	Uprooted with hand	–	Little
14.	*Capparis zeylanica*	Rohini	Fruits	June	Hand picking	–	High
15.	*Centella asiatica*	Thankuni shak	Leaves	June–August	Hand picking	Collected leaves are cut into small pieces and fried with garlic and oil	Moderate
16.	*Chenopodium album*	Note shak/ Gandhari shak	Leaves, shoot	May–August	Hand picking	Collected leaves are cut into small pieces and fried with garlic and oil	Moderate
17.	*Chorchorus* sp.	Lal kudrum/ pat	Leaves	May–July	Hand picking	Collected leaves are cut into small pieces and fried with garlic and oil	Little
18.	*Ciecar arietinum*	Buot/chola shak	Leaves	February–March	Hand picking	Collected leaves are cut into small pieces and fried with garlic and oil	Moderate
19.	*Colocasia esculenta*	Kochu shak	Leaves	May–August	Hand picking	Collected leaves are cut into small pieces and fried with garlic and oil	High
20.	*Colocasia nymphaeifolia*	Alti ahak	Leaves	May–August	Hand picking	Collected leaves are cut into small pieces and fried with garlic and oil	High
21.	*Comellina sinensis*	Kana Shak	Leaves	May–August	Hand picking	Collected leaves are cut into small pieces and fried with garlic and oil	Little
22.	*Cordia nyxa*	Buch	Fruits	July–August	Hand picking or by using pole with hooks	–	Little
23.	*Dioscorea* sp.	Kham Alu/ Genthi/ Shushni Alu	Tuber	January–March	Excavated with the pickaxe	Collected tuber are cut into small pieces and fried with mustard seeds, salt, turmeric powder and oil Sometime a type of palatable dish has makes with pigeon meat	High
24.	*Diospyros melanoxylon*	Kend	Fruits	April–June	Hand picking or by using pole with hooks	The softer parts of the ripe fruit are eaten	High
			Leaves			Collected leaves are left to semi-dry under the shed and cut into diamond shape to make the cigar	Moderate
25.	*Enhydra fluctuans*	Hemcha Shak	Leaves	Throughout the year	Hand picking	Collected leaves are cut into small pieces and fried with garlic and oil	Moderate
26.	*Feronia indica*	Kot bel	Fruits	May–July	By using pole with hooks	Softer parts of fruits, salt, chilli, sugar, and oil are crushed on flat stone mortar and pastel to make the chutney	Moderate

(continued)

Table 5.2 (continued)

S. no.	Scientific name	Local name	Part(s) use	Time of collection	Collection method	Preparation	Importance given by local people
27.	*Ficus benghalensis*	Bot	Fruits	May–June	Hand picking or by using pole with hooks	–	Little
28.	*Ficus religiosa*	Pipor shak/ Joird shag	Leaves	April–May	Hand picking	Collected leaves are cut into small pieces and fried with garlic and oil	Moderate
29.	*Flacourtia indica*	Bainchi	Fruits	March–July	Hand picking	–	Moderate
30.	*Lagenaria vulgaris*	Lau shak	Leaves	Throughout the year	Hand picking	Collected leaves are cut into small pieces and fried with garlic and oil	High
31.	*Lathyras* sp	Khesari shak	Leaves	July–August	Hand picking	Collected leaves are cut into small pieces and fried with garlic and oil	Little
32.	*Luffa aegyptiaca*	Porolta	Fruits	June–August	Hand picking	–	Moderate
33.	*Lycoperdon* sp	Pootka chati	Fruitbody	June–August	Hand picking	–	High
34.	*Madhuca indica*	Mahua	Flower and fruits	March–May	Hand picking or by using pole with hooks	–	Moderate
35.	*Marsilea minuta*	Sooshni shak	Leaves	Throughout the year	Hand picking	Collected leaves are cut into small pieces and fried with garlic and oil	High
36.	*Mentha longifolia*	Pudina	Leaves	Throughout the year	Hand picking	–	High
37.	Mollugo oppositifolia	Gima shak	Leaves	April–July	Hand picking	Collected leaves are cut into small pieces and fried with garlic and oil Collected leaves are mixed with chickpea flour or rice husk, onion, salt and chilli to make cutlet-like Boda	High
38.	*Moringa oleifera*	Soinla shak/ sogina/ Moonga shak	Leaves	Throughout the year	Hand picking	Collected leaves are cut into small pieces and fried with garlic and oil	High
39.	*Morus alba*	Tunt	Fruits	May–August	Hand picking		Moderate
40.	*Nymphaea speciosu*	Pandma phhol	Rhizomes	March–May	Hand picking or by using pole with hooks	Collected rhizomes are cut into small pieces and makes curry	High
41.	*Physalis minima*	Fotka fol	Fruits	June–July	Hand picking	Taken simply or makes chutney	Moderate
42.	*Rhaphinus sativus*	Mula shak	Leaves	Throughout the year	Hand picking	Collected leaves are cut into small pieces and fried with garlic and oil	High

43.	*Rivea hypocratiformis*	Bon pui	Leaves	May–June	Hand picking	Collected leaves are boiled with lentils	Moderate
44.	*Schleichera trijuga/Schleichera oleosa*	Kusum	Fruits, seeds, oil	June–July	Hand picking or by using pole with hooks	Taken simply or makes chutney	Moderate
45.	*Semecarpus anacardium*	Bhela	Accessory of fruit, seeds	April–May	By using pole with hooks	Semi-dry in sun light	High
46.	*Sesbania grandiflora*	Bok Phul	Flower	June–July	By using pole with hooks	Collected flowers are cut and pasted and mixed with chickpea flour or rice husk, onion, salt and chilli to make cutlet-like Boda	Moderate
47.	*Shorea robusta*	Sal	Leaves	Throughout the year	Hand picking or by using pole with hooks	Collected leaves are left to semi-dry under the shed and cut into diamond shape to make the cigar	Moderate
48.	*Sterculia foetida*	Kath badam/ Baxa badam	Seeds	April–May	By using pole with hooks	Seeds are hammered with stone and shells are removed	Moderate
49.	*Syzygium cumini*	Jam	Fruits	June–July	Hand picking or by using pole with hooks	–	High
50.	*Tamarindus indica*	Tentul	Leaves	Throughout the year	By using pole with hooks	The leaves are sprinkled in hot oil and mixed with starch of rice	Moderate
			Fruits	March–April		The fruit's pulp extracted in the starch of rice, with salt, chilli and oil	High
			Seeds			Soaked seed is boiled and mixed with flower of mahua or shal to make mush or "latha"	Moderate
51.	*Terminalia belerica*	Bohera	Seeds	March–April	By using pole with hooks	Seeds are hammered with stone and shells are removed	Little
52.	*Ventilago calyculata*	Rairui	Seeds	March–April	Hand picking	–	Little
53.	*Vicia faba*	Batla	Seeds	January–March	Hand picking	Seeds are fried on hot clay or iron pot	Little
54.	*Volvariella* sp	Pual chati	Fruitbody	June–August	Hand picking	Collected fungi are cut into small pieces and fried with onion, salt, turmeric, and oil. Most of the times soups are made with spices	High

(continued)

Table 5.2 (continued)

S. no.	Scientific name	Local name	Part(s) use	Time of collection	Collection method	Preparation	Importance given by local people
55.	*Zizyphus* sp.	Kul	Fruits	January–March	Hand picking or by using pole with hooks	Ripe fruits are eaten. Pickles and chutney are made with ripe fruit. Fully ripe fruits are dried in the sun and collected for future use. The dried fruits are crush to make powder by the husking pedal. Fruit powders are sprinkled in hot oil and mixed with starch of rice	High
			Seeds			The goats are fed ripe and semi-ripe fruits under the trees. The seeds are collected from the dung pellets of the goat. Then the seeds are smashed. The soft parts inside the seed are fried to eat	Little
56.	*Zizyphus oxyphylla*	Shiya kul	Fruits	January–March	Hand picking	Ripe fruits are eaten.	Moderate

Source: Present Study

Table 5.3 Survival diet (animal): the traditional foods eaten by the indigenous people of Bankura and Purulia in times of drought

S. no.	Scientific name	Local name	Part(s) use	Time of collection	Collection method	Preparation	Importance given by local people
1.	*Oecophylla smaragdina*	Kurkut (weaver ant)	Whole insect	April–June	Hand picking	Cooked in mustered oil, consumed with rice, also available in the form of "chutney"	High
2.	*Bombyx mori*	Lumang Tiju (silk worm)	Whole insect	April–June	Hand picking	Cooked in mustered oil, consumed solely	High
3.	*Bellayma bengalensis*	Gugli	Flesh	Throughout the year	Collected from shallow water	Consumed with rice after frying it with mustered oil	High
4.	*Glossogobius giuris*	Velsa	Meat	April–June	Collected from shallow water	Consumed with rice after frying with garlic and oil	High
5.	*Heteropneustes fossilis*	Magur	Meat	April–June	Collected from shallow water	Consumed with rice after frying with garlic and oil	High
6.	*Channa puctatus*	Chang	Meat	April–June	Collected from shallow water	Consumed with rice after frying with garlic and oil	High
7.	*Channa stratus*	Gorui	Meat	April–June	Collected from shallow water	Consumed with rice after frying with garlic and oil	High
8.	*Puntius ticto*	Punti	Meat	Throughout the year	Collected from shallow water	Consumed with rice after frying with garlic and oil	High
9.	*Esomus danricus*	Darka	Meat	April–June	Collected from shallow water	Consumed with rice after frying with garlic and oil	High
10.	*Macrognathus* sp.	Genti	Meat	April–June	Collected from shallow water	Consumed with rice after frying with garlic and oil	High
11.	*Varanus* sp.	Gosaap/Satna (monitor lizard)	Flesh	Throughout the year (collected especially in summer)	Collected from nearby forest	Consumed with rice after making curry	High
12.	*Ptyas* sp.	Snake	Flesh	Throughout the year (collected especially in summer)	Collected from nearby forest	Consumed with rice after making curry	High
13.	*Gallus gallus*	Bon Morog/ Khukri (red jungle fowl)	Meat	Throughout the year (collected especially in summer)	Collected from nearby forest	Consumed with rice after making curry	High

(continued)

Table 5.3 (continued)

S. no.	Scientific name	Local name	Part(s) use	Time of collection	Collection method	Preparation	Importance given by local people
14.	*Ortygormis pondicerianus*	Titir (gray francolin)	Meat	Throughout the year (collected especially in summer)	Collected from nearby forest	Consumed with rice after making curry	High
15.	*Streptopelia decaocto*	Ghughu (Indian collared dove)	Meat	Throughout the year (collected especially in summer)	Collected from nearby forest	Consumed with rice after making curry	High
16.	*Treron phoenicoptera*	Harial (yellow footed green pigeon)	Meat	Throughout the year (collected especially in summer)	Collected from nearby forest	Consumed with rice after making curry	High
17.	*Rattus norvegicus*	Indur (wild rat)	Meat	Throughout the year (collected especially in summer)	Collected from nearby forest	Consumed with rice after making curry	High
18.	*Hystrix indica*	Sahi (porcupine)	Meat	Throughout the year (collected especially in summer)	Collected from nearby forest	Consumed with rice after making curry	High
19.	*Maccaca rhesus*	Banor (monkey)	Meat	Throughout the year (collected especially in summer)	Collected from nearby forest	Consumed with rice after making curry	High
20.	*Funambulus* sp.	Kathberali (squirrel)	Meat	Throughout the year (collected especially in summer)	Collected from nearby forest	Consumed with rice after making curry	High
21.	*Lepus nigricolis*	Khera (rabbit)	Meat	Throughout the year (collected especially in summer)	Collected from nearby forest	Consumed with rice after making curry	High
22.	*Pteropus* sp.	Badur (bat)	Meat	Throughout the year (collected especially in summer)	Collected from nearby forest	Consumed with rice after making curry	High
23.	*Felis lybica*	Bon biral (wild cat)	Meat	Throughout the year (collected especially in summer)	Collected from nearby forest	Consumed with rice after making curry	High

Source: Present study

Fig. 5.2 (continued) sources to compensate water loss during summer; (**k**) gugli (*Bellamya bengalensis*) is consumed in high amount during summer; (**l**) pupa of tasar math (*Antheraea mylitta*) is an excellent fodder for poultry; (**m**) hunting down the wild cat (*Felis lybica*); (**n**) capturing of Rabbit/Khera (*Lepus nigricolis*); (**o**) Ghughu/Indian Collared Dove (*Streptopelia decaocto*)—meat of it is very popular among many tribal in this area (*Photo credit*: Avisek Patra); (**p**) meat of snake (*Ptyas* sp.) is very much popular during summer among the tribal (*Photo credit*: Avisek Patra). (Source: Author)

Fig. 5.2 Some of the traditional food items of this region that are consumed during summer: (**a**) bamboo shoot (*Phyllostachys edulis*); (**b**) genthi (*Dioscorea alata*) is grown in the field by some enthusiastic tribal farmers; (**c**) Hencha shaak (*Enhydra Fluctuans*) is used in the daily diet during summer; (**d**) kaju fruit (*Anacardium occidentale*); (**e**). Kend fruit (*Diospyros melanoxylon*) is a daily diet for the tribal people during summer; (**f**). kochra/mahul fruit (*Madhuca indica*); (**g**) piyal fruit (*Buchanania lanzan*) is one of the popular fruits during the hot season; (**h**) preparation of Hariya; (**i**) Hariya at its initial stage of making (paste form); (**j**) rice gruel is one of the main

roasted drumstick leaves, and rice are a staple side dish. They had boiled rice and a variety of veggies for lunch. A delicacy called Marjhola, which comprises of sticky rice cooked with a significant amount of rice gruel and sojne (*Moringa oleifera*), ranks as one of the oldest traditional cuisines that are eaten during times of drought. It is also one of the most commonly consumed foods. The rice and veggies were cooked in a clay oven that was brought to a boil. The final step before serving is to salt the food. In the evening, individuals have the identical food, that was previously consumed at lunch. When there is a severe drought, they are forced to eat rice gruel (called "Dhala mar" in the local language), which is formed from broken-up leftover rice. This "dhala maar" is crucial in supplying the necessary water during this dry season. The residents of these areas have a strong belief that a certain preparation of rice gruel has the ability to bring down one's core temperature, which can be beneficial during the blistering heat. They dehydrated the leaves of the *Tamarindus indica* plant, then ground the dry leaves into a powder and combined it with the rice gruel. They guzzle down the entirety of this liquid before heading for their daily activity. This food will serve as a cooling agent for one's body.

Bankura and Purulia's indigenous people consume Hariya, a traditional liquor (Fig. 5.2h and i). This region includes Santals, Mundas, Oraons, and Lodhas. Rice beer is made from Oryza sativa. Bakhar and raw rice ferment this rice product. Some regions call it Renuboti. Starting culture is essential for "Hariya" cooking. Rice dust fermented with plant, root, and leaf extracts creates "Bakhar" or "Renuboti," a starter culture used to inoculate microorganisms (Dhal et al. 2010). Keep Bakhar cool and dry. About 2–3 g per 200 g of cooked rice is typical. "Hariya" requires a clean earthenware jar that must be cleaned in the sun. Soaking, boiling, and mat-drying the rice follows. Next, mix the dehydrated rice with the "Bakhar" starting culture. Lead lids keep the mixture fresh in clay jars. The container is then darkened for 3–4 days. After three to 4 days, the mixture is diluted one to six with drinking water. Next, filter the liquid using a little, clean cloth. Drink "Hariya" now. After fermentation, "Hariya" has 11–12% alcohol (Ghosh et al. 2014). Hariya is crucial to the local indigenous inhabitants during the harsh summer heat.

5.3.1.2 Various Cereals, Leafy Green Vegetables, and Others

The indigenous people of this region depend on grain for nourishment during this time of extreme instability. The names of these plants are Marua (*Eleusine coracana*), Bajra (*Pennisetum typhoideum*), and Junar (*Sorghum bicolor*). They prepare "Lapsi Ghata," a local grain dish, during severe droughts. This dish requires patience and basic cooking abilities. They pounded the crops using a "Dheki," a threshing implement. Second, the crushed grains are divided into three size groups and sieved to remove the tiniest grains. In a clay pot, they put the largest grains in water and boil them, then add the medium grains, and then the smallest grains. The final presentation uses a "sal" (*Shorea robusta*) leaf plate. They occasionally made delicious "Pithas," which are pancakes, with these cereals. Indigenous Nigerians also eat cereals. Akamu, made from millet cereals, is a popular Nigerian dish (Nwachukwu et al. 2010).

During droughts, local tribal people depend on a variety of "saag" (leafy green vegetables). In particular, they like Kana saag (*Comellina sinensis*), Hemcha saag (*Enydra fluctuans*), Kural saag (*Bauhinia* sp.), Marchi saag (*Catharanthus pusillus*), Hurhuria saag (*Cleome monophylla*), Kolmi saag (*Ipomoea aquatica*), Kulekhara saag (*Leucas mollissima*), Kedo saag (*Limnophila indica*), Gara saag (*Polygonum barbatum*), Barial saag (*Sida cordata*), Satgithia saag (*Spermacoce hispida*), and Notey saag (*Amarantus viridis*). They collected those green leafy vegetables from agricultural land with little water, mixed them with leftover broken rice, and boiled the mixture with a little water. Salt and roasted green chilies topped the oatmeal. Piper saag (*Ficus* sp.), a drought-resistant leafy green vegetable, has antipox properties (Yarmolinsky et al. 2012). During this crucial time, this porridge will be one of the key sources of nutrients. These leafy greens contain calcium, iron, and beta-carotene (Ghosh-Jerath et al. 2016). They also like Mulhan (*Nelumbo nucifera*) shoots, Koril (*Bambusa* sp.) new shoot tips, and lotus plant rhizomes.

5.3.1.3 Fruits, Tubers, and Leftover Food

Indigenous people of these areas will need to rely on edible wild fruit and tuber resources as an auxiliary or quasi-source of food against the conventional counterparts in the near term to cope with the increased demand for food during climate-related hazards and to satisfy nutritional requirements. This will allow them to overcome the shortage that occurs during periods of drought. Hence, plant resources solve several food issues. Also, many of these wild foods are culturally significant to locals and related to their indigenous customs (Bhujel et al. 2018). Traditional methods include boiling Genthi (*Dioscorea* sp.) tubers and other varieties like Khesari (*Actinoscirpus grossus*), Simchiru ara (*Cyanotis axillaris*), Kana Kanda (*Nymphoides indica*), and fruits like Piyal (*Buchanania lanzan*), Vela (*Semecarpus anacardium*), Kusum (*Schleichera oleosa*), Amla (*Emblica officinalis*), Madal (*Annona squamosa*), Amra (*Spondias pinnata*), and Kaju (*Anacardium occidentale*). Piyal (*Buchanania lanzan*) fruit and seed are edible. "Chironji" is the seed's edible nut. Its green and ripe fruits create various prickles. Kend (*Diospyros melanoxylon*) is delicious. Locals eat this fruit throughout the summer because they believe that it is refreshing and astringent (Kumari and Kumar 2021). The native inhabitants have discovered several methods to remove the bitterness, making the cereal pleasant as a snack and in roasted, ground, and some other versions (Kumar et al. 2017). The local indigenous people trimmed the plant into a spherical form, rinsed it twice in a nearby creek (if water was available), and dried it in the sun. This caused its resentment to go. They then take it in this form after boiling it with tamarind seed the following day, so baula's whole renovation takes 2 days.

Many indigenous people in the far regions have lost their ability to provide for themselves due to the prolonged drought. They live almost entirely off of a local specialty cuisine known as Potom. The aforementioned items are simply leftovers and scavenged foodstuffs from a nearby ceremonial event. Sal leaves (*Shorea robusta*) are utilized to produce a wrapper for the meal after it has been harvested, sun-dried to the maximum extent feasible, and then cooled. During this period of tremendous poverty, this potom will provide an enormous service. You would not find this potom in the supermarket, and it is much more unlikely to appear in a community that distributes rations on a regular basis.

During this time, "Mohul latha" was a popular meal among the locals. Boiled tamarind seed and dried Mohul fruit (*Madhuca indica*) are used to create this meal. After that, it is molded into a semi-solid sphere, which is then eaten. The seed of tamarind was allowed to dry out in the sun, followed by being pulverized into a powder, and finally being ingested by mixing it with sticky rice. All of these items are liquids, and together they can make up for the massive loss of water that has occurred during this disastrous time.

5.3.1.4 Wild Animals as a Nutritious Source

Protein is an essential part of a balanced diet and vital to human survival. Yet, indigenous inhabitants in these areas met their protein needs by eating a wide variety of foods, even throughout this devastating period. The "Kurkut" food item is a popular choice because it is made entirely of weaver ant eggs (*Oecophylla smaragdina*). The Santhal people call eggs like this "Hau," and they have their own word for them. Once the eggs have been gathered from the trees, they are cooked in mustard oil adding salt, chilli spices, as well as other seasonings. The locals in the Himalayan region to the east also enjoy eating red ant eggs (Chowdhury et al. 2015). They also eat the "Lumang tiju," or mulberry silkworm (Bombyx mori). The locals here regularly consume "gugli" (snail, *Bellayma bengalensis*) from the murkiest parts of the water body. Although the rivers, lakes, and ponds have all but dried up due to the extreme heat, they have been catching fish. They commonly catch Velsa (*Glossogobius giuris*), Chang (*Channa puctatus*), Gorui (*Channa stratus*), Puti (*Puntius ticto*), Darka (*Esomus danricus*), and Genti (*Macrognathus* sp.). The majority of the seafood they consumed was grilled with minimal oil. Meat from monitor lizards (*Varanus* sp.), wild brown rats (*Rattus norvegicus*), and other native animals are common fare for some locals, especially the Santhals. The Bon Morog/Khukri (*Gallus gallus*), Titir (*Ortygormis pondicerianus*), Harial (*Treron phoenicoptera*), and Ghughu (*Streptopelia decaocto*) are only some of the birds they bring back from the neighboring forest (Chanda and Mukherjee 2012). Meat cooked in a curry with spices like turmeric, cumin, coriander, chilies, and black pepper will become a staple in their diet (Gorai et al. 2022). More than 25 non-conventional animals, such as the monkey (*Maccaca rhesus*), Wild cat (*Felis lybica*), bat (*Pteropus* sp.), porcupine (*Hystrix indica*), Kathberali (*Funambulus* sp.), Khera (*Lepus nigricolis*), rat (*Rattus norvegicus*), snake (*Ptyas* sp.), and Indian Monitor Lizard (*Varanus* sp.), are pursued their flesh by the less fortunate residents of this area. Due to financial hardship, some peasants have taken to using dried blood in their curries. In addition, certain native communities are rumored to enjoy a dish of raw snake blood and steaming rice (Modak 2010). Meat from monitor lizards (*Varanus* sp.) is highly prized by the indigenous populaces of Africa, Australia, and many regions of Asia (Hoffman and Cawthorn 2012).

5.3.1.5 Feeding Material for Pets and Livestock

When temperatures rise to unsafe levels, people must ensure the safety of both themselves and their domestic animals. The indigenous people who lived here before used the various fruits and leaves to provide for their livestock. These communities' understanding of local flora is crucial during food shortages, as it allows them to harvest enough veggies to survive. They travel to surrounding forests in search of young plants and different sections of those plants (Dangol and Maharjan 2012). Some examples of plants and their parts that are used as feed are listed below. Leaves from both the Assatha tree (*Ficus religiosa*) and the Arjun tree (*Terminalia arjuna*) are utilized as feed. Tender branches of the vela tree (Semecarpus anacardium) are also utilized as fodder. The green leaves of the ankra plant (*Alangium salvifolium*) are used as livestock feed. To feed cattle, fresh Buch (*Cordia nyxa*) leaf is quite beneficial. Leaves of the kural tree (*Bauhinia racemose*) are used as livestock feed. Bans, or bamboo (*Dendrocalamus strictus*), has fodder-worthy leaves. Green leaves from the Gamhar plant (*Gomelina arborea*) are used as feed. Leaves of the Indian rosewood (Sisoo) tree (*Dalbergia sissoo*) are eaten by livestock. Leaves from the sirish tree (*Albezia lebbeck*) are used as animal feed. Cattle feed is made from the seeds as well as the leaves of the tun (*Toona ciliate*) plant. Leaves from the Amlaki or Indian gooseberry (*Phyllanthus emblica*) are used as bovine fodder, as are those from the Bohera or *Terminalia belerica*, both of which are regarded as high-quality forage.

5.3.1.6 Role of Wild Edible Mushrooms in Indigenous Communities' Food Security

Throughout the beginning of history, indigenous cultures have collected wild mushrooms using ethnomycological expertise, prepared them with food items, and consumed them for their many daily uses, both edible and medicinal, as secondary food resources. To achieve this goal, one must first acquire the mushrooms before using them in any culinary capacity (Debnath et al. 2019). The natives of this area include a wide variety of seasonal wild mushrooms in their diet at this time. Many of these are particularly noteworthy; these include the Bali (*Agaricus* sp.), Puyal (*Volvariella* sp.), Putka (*Marasmius* sp.), Parab (*Agaricus campestris*), Patra (*Agaricus pracimosus*), and Mura (*Psalliota* sp.) mushrooms. It has been discovered that the flavor, palatability, and nutritional profile of edible wild species of mushrooms are generally superior to that of their produced counterparts (Lampman 2010).

5.3.1.7 Food Preservation Techniques for Long-Term Sustainability

Indians have always preserved food. Preventing rotting and prolonging food's shelf life made it safe to eat (Varghese et al. 2022). Anti-microbial chemicals and insect repellents, drying, brewing, salting, pickling, freezing, heating, smoking, and chilling in natural freezers like streams and underground pits are some of the first ways to preserve goods (Mobolade et al. 2019). Food-storage traditions vary by region. These indigenous people sun-dry food to prevent spoiling during hunger. Sun-drying grain is standard. High-moisture grains require longer to sun-dry. Grain and seed storage are best around 10–12% moisture. This reduces grain moisture, lengthens dormancy, and eliminates insects and pests (Kumar and Singh 2013). Sun-dried Indian palm seed and fruit pulp drinks were popular in the hot climate. They ground sun-dried tamarind leaves with other veggies to eat. Indigenous food preservation systems are cheaper, more efficient, and more accessible than high-tech ones. It is crucial to use time-tested strategies to sustain agriculture (Sundaramari et al. 2011).

5.3.2 Strategies for Effective Water Management During Times of Scarcity

Using and conserving water is critical. Smallholder farmers can boost crop yields by managing agricultural water resources (Barron and Noel 2008). Climate change-induced droughts harm farmers' productivity and revenue, underlining the need for better water conservation and utilization. Farmers have utilized many water-saving methods for millennia. Rainwater harvesting (RWH) increases the amount of water available for farming and livelihoods and reduces peak flows after heavy rains (Ravishankar et al. 2020). Every watershed management plan should prioritize water supply protection and expansion. In drought-prone areas, micro-watershed-based development may be the greatest way to improve the region (Deshpande and Narayanmoorthy 2000). Table 5.4 depicts many methods that the local indigenous population employ to keep their water supply sustainable (Fig. 5.3).

Table 5.4 Strategies used by indigenous people of this region to manage water resources

S. no.	Indigenous process	Brief description	Benefit
1.	Pitcher watering system	The process involves burying "Kalsi" water-filled jars adjacent to the plants. These jars feature narrow bottom openings. A small piece of fabric is placed through the opening to allow water to escape and hydrate the soil. This aids plant watering	Prevent plants from drying out
2.	Cultivation of sugarcane	Farmers cultivate seedlings by burying their lower sections in water from riverbank canals. Soil is tilled in the afternoon before planting so that the night-time dew can help the seeds take root. They harrow the land at dawn to soak up the moisture and prevent it from evaporating	Cultivation
3.	Embankment on river	This area's local people embank a lower river segment. In summer, once the river slows, these embanked portions of the stream have ample water for farming and household usage	Growth of plants, grasses, and other forage crops

(continued)

Table 5.4 (continued)

S. no.	Indigenous process	Brief description	Benefit
4.	Roof rainwater harvesting system	They dig a 50 ft^2 pit near their residence. High precipitation around Kal Baisakhi (heavy storm with rain) or rainy season directs roof top water to the pit via earthen open outflow for at least 3 months of home use	Continuous water supply
5.	Water storages beside the river	Invaluable to the locals, these water reserves emerged as a result of soil erosion. To access these natural water depots for drinking, bathing, and other uses, people will travel vast distances from their homes	Naturally available water
6.	Aquifer	A natural water supply is one that forms when water seeps out of the ground and collects at a certain depth. Water from this source does not evaporate throughout the summer because it is a natural resource. Typically, this source's runoff will create a stream, which may serve as the supply of water for a bigger water body or as a useful agricultural water source	Natural source of water during dry period
7.	Hapa or Doba	Hapa, the big hole in the center of the pond. Water collects here after a heavy downpour or seeps through the ground and becomes trapped between the rocks and soil	Availability of water during dry period
8.	Natural waterfalls	Tribal people in the hills Purulia and Bankura regions rely heavily on the natural waterfalls that mark the landscape. During the wet season, various types of waterfalls carry vast quantities of water. The adjacent bodies of water benefit from this runoff water. Although these waterfalls dry up over time, especially in the summer, the low water slows down and runs like a tiny stream, providing the sole water for the surrounding people and animals	Availability of water during dry period
9.	Contour bunds	Bunds are placed along the contour to distribute the weight of the water from the runoff. Bunds are constructed to lessen the amount of water that flows off a site, increase the amount that is absorbed, and redirect any excess to a central drainage system	Reduce runoffs

(continued)

Table 5.4 (continued)

S. no.	Indigenous process	Brief description	Benefit
10.	Naturally created water storage beside the river	Several rivers have their beginnings in the little hills and undulating regions of Purulia and Bankura, despite the region's status as a plateau. Most of these rivers dry up in the summer, but they are brimming with water during the monsoon. Nonetheless, the soil erosion that has developed water storage along the river bank has proven to be an invaluable resource for the locals. To access these natural water reserves for drinking, bathing, and other uses, people travel vast distances away from their homes located along the river	Availability of water during dry period

Adapted from Bauri et al. (2020)

5.3.3 Strategies of Indigenous People to Beat the Heat

5.3.3.1 Architectural Design of House to Reduce Indoor Temperature in Summer

Indigenous communities vary in size. Their cottages are scattered across the land's altitudes. These homes usually have no windows or small ones (Fig. 5.4a, b). A typical indigenous dwelling is a sal wood, mud-covered rectangle structure with numerous uses. The hut's roof is grass or straw over a double-sloped wooden structure. Affluent residences have khappar tile roofs. The house has a rectangular base. A partition wall divides the single room into a small kitchenette and a larger bedroom. Their houses have rammed walls. "Rammed earth" is a building technique that uses numerous layers of soil to produce a sturdy wall. Walls can be thick and heavy because thermal mass helps control temperature and humidity (Patadia 2020). The walls absorb heat during the day and gradually release it at night, maintaining the building's temperature and humidity. To keep animals cool, they built a goat enclosure, pigsty, and cattle shed near the main house and covered the ground and roof with straw. However, the chickens and other birds are caged off the living space. They built temporary leaf dwellings near collecting sites and water sources far from their main campsite in the wild forest. They keep a palm-leaf hand fan at home to cool down. Domestic furniture includes date palm leaf mats, string cots, clay pots, gourd jars, brass and silver cutlery, and bamboo baskets.

5.3.3.2 Ancient Practices and Usage of Cotton Towels

Indigenous people in these areas are very good at keeping their age-old traditions alive, such as offering jaggery and water to anyone who comes to their house and also taking both of these things before going to their respective places of work. This is done despite the sweltering heat that prevails during the summer months in these areas. The indigenous population relies heavily on jaggery-water drink to combat the heat. The sugarcane-based sweetener jaggery has been used for centuries and is renowned for its calming effects (Kumar et al. 2008). It aids in maintaining a healthy internal temperature and alleviates discomfort caused by excessive heat. A common beverage among indigenous people is a mixture of jaggery and water. The electrolytes in this beverage can assist in making up for what you have lost via perspiration (Ahmad Mir et al. 2019).

Indigenous folks also use cotton towels (Fig. 5.4c) to combat the scorching temperatures. Cotton allows air to move throughout the body

Fig. 5.3 Some traditional water conservation methods: (**a**) pitcher watering system; (**b**) naturally occurring water body. Indigenous people of these regions create bunds around it to store the water; (**c**) stream water collection by rural people; (**d**) naturally occurring waterbody, provide water throughout the year. These waterbodies did not dry out during heat; (**e**) collection of water by tribal people; (**f**) digging a pit in the middle of the pond. Look carefully that the people here provide boundaries around it to impede the pollution and trespass; (**g**) digging a pit at agricultural field. These pits provide essential water for the vegetables in the field; (**h**) cultivation of sugarcane beside a human-made waterbody at one side of the river. The river may dry out in the heat, but this water body provides sufficient water for the sugarcane. (Source: Author)

Fig. 5.4 (**a**) A typical tribal house with architectural design with no window; (**b**) the roof of the house with khappar tiles. Look they are soaking Indian palm (*Ziziphus mauritiana*) on the roof; (**c**) local people performing the rituals to their communal deity for rain. Look carefully at the right-sided man carrying a cotton towel in his shoulder to calm the temperature. (Source: Author)

and thus helps to keep the wearer cool (Hota and Behera 2014). Its high absorption rate means it may effectively remove sweat from the skin, relieving any associated pain. A common way for indigenous people to beat the heat is to wet cotton cloths with cool water and wipe their cheeks and necks.

As a whole, indigenous people's approach to dealing with the heat is simple and effective: they use jaggery, water, and cotton towels. These cures have been utilized for years, and they are all-natural and simple to obtain. Native peoples can beat the heat in even the most oppressive environments by adopting some simple lifestyle changes.

5.3.4 Role of Indigenous Women During Famine

During the famine period, the Indigenous women in this region assumed a significant role and collaborated with their male counterparts. In certain instances, their contributions may surpass those of their male counterparts. In the context of food preservation, the sun-drying technique is predominantly executed by female members of the household. The individual consistently tends to perform their household responsibilities while their male partner is occupied with their daily professional obligations. In times of scorching heat, it is customary for women to provide jaggery and water to visitors and household members when they return from work, which helps to restore their electrolytic balance along with a calming effect. The majority of culinary activities are typically carried out by female members of the household. Women are primarily responsible for carrying the watered clay pot from the site to the house.

Research findings of Sengupta and Ghosh's (2022) indicate that a considerable proportion of women (60%, 58.95%, and 54.12% in Arsha, Barabazar, and Manbazar II blocks, respectively, of Purulia district) have been involved in the collection of water, as per the survey conducted. A significant proportion of women actively participate in the decision-making process concerning water utilization. Moreover, the study highlights that women's involvement in water resource management is particularly effective in the study region during the dry season. The limited technical skills of women in matters related to water and food resource management restrict the application of their knowledge to household activities, thereby limiting their impact beyond the domestic sphere. It is recommended that a systematic approach be taken to prioritize the involvement of women in the management of water and food resources in drought-prone regions such as Bankura and Purulia districts to promote sustainable resource management (Sengupta and Ghosh 2022).

Everyone requires water and food. This catastrophe may threaten food and water. The water and food crises are caused by a growing population and anthropogenic activities that contribute to global warming and climate change (Aberoumand and Deokule 2009). Wild foods increase consumer choice. Ethnobotanical investigations have shown that about 7000 wild plant species are cultivated for human consumption (Grivetti and Ogle 2000). Some indigenous communities eat more than 200 plant species (Kuhnlein et al. 2009); Rathore (2009) estimates that 600 plant species in India have nutritional value; and DeFoliart (1992) estimates that 1000 edible insect species are eaten worldwide. Around 1069 wild fungus species provide protein (Boa 2004). Bushmeat and fish provide 20% of protein in 60 emerging nations (Bennett and Robinson 2000).

Here, people know how to use food in diverse ways and preserve it for climate-change-related shortages. These meals may also boost the community's nutrition due to nutritional fortification. To solve indigenous people's specific problems, it must raise awareness of and utilize indigenous food systems' atypical feedstocks. Locals eat several native plants and animals. Foods with low nutritional value may cause insufficient dietary intake (Ghosh-Jerath et al. 2016). Health campaigns and education must encourage native food nutrition. Thus, eating native and wild foods sustainably may benefit natives. This may satisfy their long-term nutritional and health needs. Indigenous food preparation, preservation, and storage help achieve these goals. Rural households without resources must apply traditional food handling, conservation, and preservation practices. Modernizing indigenous practices reduces food insecurity (Kuyu and Bereka 2020).

Overuse and waste of water have generated the worldwide water crisis. Bankura and Purulia are typical. Bankura and Purulia are drought-prone West Bengal districts. Women in rural villages often walk far to gather water in clay pots, especially on hot days. Due to the water conflict, they value water conservation more. These rural residents have developed many water management practices over generations to assist them in weather adversity. For ages, Bankura and Purulia residents have suffered from terrible weather. They learned disaster-related local knowledge to reduce natural calamities. IK helps communities adapt over time. Indigenous people in their native habitat gained this knowledge. It is wonderful for learning about the local ecosystem and might be used to reduce risks in other surrounding neighborhoods. Drawing on indigenous wisdom helps empower local people and spread information about natural disaster mitigation (Shaw et al. 2008). A summary of drought-related challenges and their adaptive mechanisms are listed in Table 5.5.

Table 5.5 Summary of drought-related challenges and drought-related adaptation mechanisms

Drought-related challenges	Drought-related adaptation mechanisms
Drying up of rivers, springs, and other waterbodies	Conventional techniques for water conservation, such as the utilization of pitcher watering systems and the establishment of aquifers, have been employed
Lack of fodder and less productive land	The gathering of diverse categories of untamed comestible flora from the adjacent forest region is solely predicated on their traditional knowledge
Lack of nutrients due to agricultural loss	There exist diverse alternative means of procuring nutrients from both flora and fauna, which are primarily rooted in IK systems
Scorching heat and rotting of food	The utilization of traditional indigenous methods for food preservation, such as sun drying, serves as a means to inhibit food spoilage
Unbearable temperature	The implementation of a particular architectural design for residential buildings aimed at reducing temperature, coupled with the utilization of cotton towels to regulate body temperature, is a potential approach to achieve thermal comfort The consumption of jaggery water derived from sugarcane has been found to regulate the body's internal temperature and mitigate the discomfort associated with excessive heat
Loss of biodiversity	No such adaptive mechanisms
Loss of forest	Sacred groves represent the sole remaining domain of forested areas, as no other comparable adaptive mechanisms are currently in place

Source: Present study

5.4 Recommendations

Traditional indigenous methods of preserving food and water supplies are significant and should be recognized for their contributions (Fig. 5.5). Sustainable resource management is something indigenous people have been doing for generations (Holt 2005); their expertise can shed light on and offer answers to contemporary challenges like water and food scarcity. So, in the context of food and water conservation, it is crucial for policymakers, scholars, and the general public to recognize and learn from IK. The following actions can be taken to promote and integrate IK into a sustainable system:

Engage with indigenous communities: The indigenous communities and their leaders must be consulted to learn about their food and water conservation methods and customs. To do this, it is important to speak with indigenous groups, community leaders, and elders. By doing so, we can increase cooperation, appreciation, and knowledge sharing throughout diverse communities (Hunt 2013).

Incorporate IK into research and policy: Conservation practices for both food and water sources should be studied and codified, and IK should be integrated whenever possible. For this purpose, it is important to include members of indigenous communities in studies and discussions, and to incorporate their insights and knowledge into policy documents and suggestions. The success, longevity, and cultural appropriateness of conservation activities can be ensured by the incorporation of IK into research and policy (Mauro and Hardison 2000).

Support indigenous-led conservation initiatives: When it comes to conserving and managing their resources, indigenous communities frequently function as stewards of the land and water systems that they inhabit. Assisting indigenous-led conservation programs can help preserve and expand indigenous peoples' traditional knowledge and traditions, as well as give them the means to continue their conservation work. Money, help with technology, and other attempts to boost capabilities would all help (Artelle et al. 2019).

Foster cross-cultural education and knowledge exchange. To help people appreciate and understand each other, it is important for the general public to know about the expertise of indigenous peoples in saving food and water. This can be accomplished by facilitating educational opportunities for indigenous and non-indigenous people through workshops, seminars, and cultural exchange programs. By

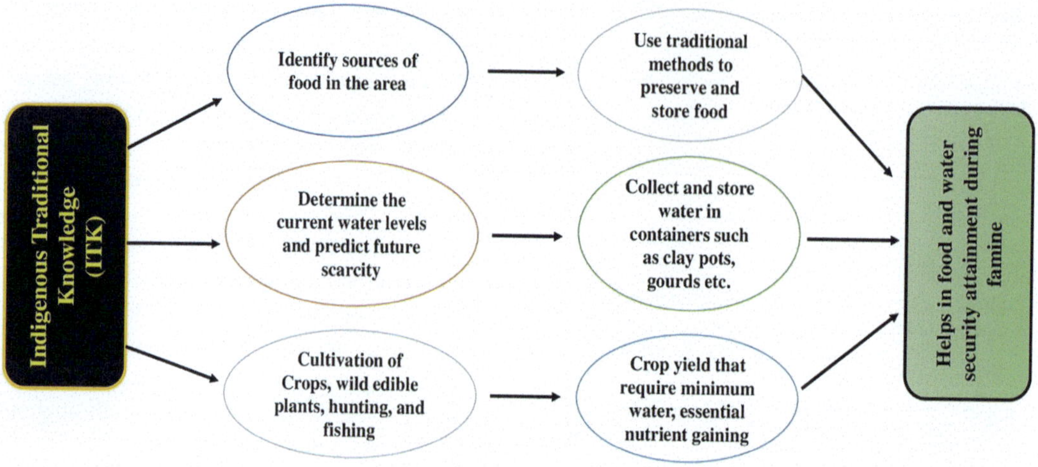

Fig. 5.5 Indigenous traditional knowledge (ITK) flowchart for food and water security attainment during scarcity. (Source: Author)

encouraging people from different backgrounds to learn from one another and share their experiences and perspectives, we may strengthen relationships between these groups and encourage a more comprehensive and equitable approach to conservation (Caligiuri 2014).

5.5 Conclusions

To deal with modern problems like climate change, it is important to bring back and improve IK systems. In areas where the climate is unstable or likely to change, traditional knowledge is increasingly recognized as a vital resource that needs to be transmitted to protect livelihoods and assure food security. This chapter seeks to document the many approaches taken by communities and indigenous peoples to adapt to the effects of climate change.

IK originates from generations of locals constantly watching and engaging with their environments (Dekens 2007). They prioritize quality and location. It differs from scientific understanding, which is based on synchronic observations and is more accurate and broader (Dekens 2007). Modern scientific study sometimes ignores other issues. Indigenous perspectives are often ignored. Indigenous groups were convinced by government and private efforts. Bankura and Purulia residents are encouraged to enter the twenty-first century. Youths adjust rapidly. They are behind other indigenous people due to their natural reserves. Middle-aged and older people still avoid development projects. Today's indigenous youngsters desire modern amenities. Young people prefer day work, unlike their century-old forest economy. These activities improve youth quality of life but lose a generation's forest resource management skills (Gorai et al. 2022). Their beliefs and practices are diluted. When a nation develops, it must enhance living circumstances and economic prospects. The ideal would be improving living without compromising culture. Indigenous people naturally use a variety of natural resources, but many are being turned into low-skilled employees, and new tribe members have the same understanding of the forest and forest products as their ancestors. Traditional knowledge loss threatens biocultural frameworks, biodiversity, and ecological functioning. Current research shows indigenous communities have several natural catastrophe mitigation measures. One is reducing operations to reduce disaster damage. They employ natural resources, most of which they grow, to simplify their diet and get all the nutrients they need. They ingeniously manage water supplies. These tactics might provide you with a lot for little effort. As climate change damages our ecosystem and escalates natural calamities, these indigenous answers may be ideal.

Acknowledgments The authors owe a great debt of gratitude to Dr Jaladhar Karmakar, whose research on traditional culture has been groundbreaking. For their assistance and cooperation during our study, we are indebted to the local indigenous communities.

References

Aberoumand A, Deokule SS (2009) Determination of elements profile of some wild edible plants. Food Anal Methods 2:116–119

Abioye AA, Zaid YA, Egberongbe HS (2014) Documenting and disseminating agricultural indigenous knowledge for sustainable food security: the efforts of agricultural research libraries in Nigeria. Libri 64(1):75–84. https://doi.org/10.1515/libri-2014-0007

Ahmad Mir N, Ashutosh A, Ahmad Shergojry S, Ahmed Wani S, Ahmad Sheikh F (2019) Effect of induced transportation stress in goats supplemented with vitamin C and jaggery during hot dry season. Biol Rhythm Res 50(3):389–399. https://doi.org/10.1080/09291016.2018.1452591

Altman J, Kerins S, Hunt J, Ens E, May K, Russell S, Fogarty W (2018) Indigenous cultural and natural resource management futures. Review of caring for our country: Australia's natural resource management investment initiative. CAEPR topical issue no. 9/2011, pp 1–28

Anandaraja N, Rathakrishnan T, Ramasubramanian M, Saravanan P, Suganthi NS (2008) Indigenous weather and forecast practices of Coimbatore district farmers of Tamil Nadu. Indian J Tradit Knowl 7(4):630–633

Artelle KA, Zurba M, Bhattacharyya J, Chan DE, Brown K, Housty J, Moola F (2019) Supporting resurgent indigenous-led governance: a nascent mechanism for just and effective conservation. Biol Conserv 240:108284

Barron J, Noel S (2008) Agricultural water management in smallholder farming systems: the value of soft components in meso-scale interventions. Stockholm Environment Institute Project Report. Agricultural water management in smallholder farming system

Basu A (2020) Cultural identity and sustainability in santal indigenous community of Birbhum district, India. Archaeologies 16(3):492–504. https://doi.org/10.1007/s11759-020-09413-8

Bauri K, Gorai P, Modak BK (2020) Indigenous knowledge and practices on water conservation and management in Purulia District, West Bengal. J Environ Sociobiol 17(1):89–98

Bennett EL, Robinson JG (2000) Hunting of wildlife in tropical forests: implications for biodiversity and forest peoples. WCS, New York

Bhujel D, Chhetri G, Rai YK (2018) Wild edible plants used by ethnic communities in Kalimpong district of West Bengal, India. NeBIO 9(4):314–326

Bhunia P, Das P, Maiti R (2020) Meteorological drought study through SPI in three drought prone districts of West Bengal, India. Earth Syst Environ 4(1):43–55. https://doi.org/10.1007/s41748-019-00137-6

Bisai S, Dutta S (2021) Traditional food practices of LODHA: a gathering-hunting indigenous community of WEST BENGAL, India. Researchgate. https://www.researchgate.net/publication/355436368_TRADITIONAL_FOOD_PRACTICES_OF_LODHA_A_GATHERINGHUNTING_INDIGENOUS_COMMUNITY_OF_WEST_BENGAL_INDIA#fullTextFileContent

Boa ER (2004) Wild edible fungi: a global overview of their use and importance to people. FAO, Rome

Caligiuri P (2014) Many moving parts: factors influencing the effectiveness of HRM practices designed to improve knowledge transfer within MNCs. J Int Bus Stud 45:63–72

Census of India (2011) West Bengal, series 20, part XII-B, district census handbook, Kolkata, village and town wise primary census abstract. Directorate of Census Operations, West Bengal

Chanda S, Mukherjee A (2012) Animal resources linked with the life of Birhor community settled in Ayodhya hills, Purulia District, West Bengal. Indian J Appl Pure Biol 27(1):31–36

Chowdhury S, Sontakke PP, Bhattacharjee J, Bhattacharjee D, Debnath A, Kumar A, Datta J (2015) An overview of edible insects in eastern Himalayas: indigenous traditional food of Tribal's. Ecol Environ Conserv 21:1941–1946

Costanza R, Kubiszewski I, Giovannini E, Lovins H, McGlade J, Pickett KE, Ragnarsdóttir KV, Roberts D, De Vogli R, Wilkinson R (2014) Development: time to leave GDP behind. Nature 505(7483):283–285. https://doi.org/10.1038/505283a

Cunningham AB (1992) The role of ethnobotany and indigenous knowledge in conservation of plant genetic resources. Google Scholar. https://journals.co.za/doi/pdf/10.10520/AJA00123013_74

Dangol DR, Maharjan KL (2012) Spatial and temporal dynamics of flora in forest, grassland and common land ecosystems of Western Chitwan, Nepal. J Int Dev Cooper 18(4):77–92

De LC, De T (2021) Nutrient rich foods in human diet as immunity boosters. Journal of Pharmacognosy and Phytochemistry. https://www.phytojournal.com/archives/2021/vol10issue3/PartC/10-2-342-706.pdf

Debnath S, Debnath B, Das P, Saha AK (2019) Review on an ethnomedicinal practices of wild mushrooms by the local tribes of India. J Appl Pharm Sci 9(8):144–156. https://doi.org/10.7324/JAPS.2019.90818

DeFoliart GR (1992) Insects as human food: gene DeFoliart discusses some nutritional and economic aspects. Crop Prot 11(5):395–399

Dekens J (2007) Local knowledge for disaster pepared-ness: a literature review, *https://core.ac.uk*, https://core.ac.uk/download/pdf/48027198.pdf

Deshpande RS, Narayanmoorthy A (2000) Appraisal of watershed development programme across regions in India. Gokhale Institute of Politics and Economics (GIPE), Pune. https://doi.org/10.21648/arthavij/1999/v41/i4/115909

Dhal NK, Pattanaik C, Reddy S (2010) Bakhar starch fermentation—a common tribal practice in Orissa. Indian J Tradit Knowl 9:279–281

Fabricius C, Folke C, Cundill G, Schultz L (2007) Powerless spectators, coping actors, and adaptive co-managers: a synthesis of the role of communities in ecosystem management. Ecol Soc 12(1):1–16. Retrieved from http://globaltigerinitiative.org/download2/Fabricus_Powerless_Spectators.pdf

Florin P, Wandersman A (1990) An introduction to citizen participation, voluntary organizations, and community development: insights for empowerment through research. Am J Community Psychol 18(1):41–54. https://doi.org/10.1007/BF00922688

Ghosh K, Maity C, Adak A, Halder SK, Jana A, Das A, Parua S, Kumar P, Mohapatra D, Pati BR, Mondal KC (2014) Ethnic preparation of Haria, a rice-based fermented beverage, in the province of lateritic West Bengal, India. Ethnobot Res Appl 12(1):39–49

Ghosh-Jerath S, Singh A, Magsumbol MS, Kamboj P, Goldberg G (2016) Exploring the potential of indigenous foods to address hidden hunger: nutritive value of indigenous foods of Santhal tribal community of Jharkhand, India. J Hunger Environ Nutr 11(4):548–568. https://doi.org/10.1080/19320248.2016.1157545

Gorai P, Dey A, Modak BK (2022) Birhors of Purulia: marching toward mainstream of the society. In: Indigenous people and nature. Elsevier, pp 29–60. https://doi.org/10.1016/B978-0-323-91603-5.00009-9

Gorjestani N (2000) Indigenous knowledge for development, opportunities and challenges. Paper presented at the Traditional Knowledge, Geneva. http://www.worldbank.org/afr/ik/ikpaper_0102.pdf

Goswami A (2017) Analysis of meteorological drought and its trend in Bankura District of West Bengal, India. AFHABEC, pp 1–6

Grieves V (2009) Aboriginal spirituality: aboriginal philosophy, the basis of aboriginal social and emotional wellbeing, vol 9. Cooperative Research Centre for Aboriginal Health, Darwin

Grivetti LE, Ogle BM (2000) Value of traditional foods in meeting macro-and micronutrient needs: the wild plant connection. Nutr Res Rev 13(1):31–46

Halder S, Sadhukhan D (2012) Adaptability of water management strategies in the context of climate change for drought prone areas of West Bengal. In: International conference on impact of climate change on water resources, Karunya University, Coimbatore, India

Hazra S, Roy S, Mitra S (2017) Enhancing adaptive capacity and increasing resilience of small and marginal farmers of Purulia and Bankura districts, West Bengal to climate change. DRCSC Report, pp 1–114

Hoffman LC, Cawthorn DM (2012) What is the role and contribution of meat from wildlife in providing high quality protein for consumption? Anim Front 2(4):40–53. https://doi.org/10.2527/af.2012-0061

Holt FL (2005) The catch-22 of conservation: indigenous peoples, biologists, and cultural change. Hum Ecol 33:199–215

Hota P, Behera B (2014) Understanding the household perceptions of and adaptations to heat wave: a case study. J Ecol Econ Stat 33(2):98–117

Houghton J (2009) The complete briefing. Cambridge University Press, New York

Hunt J (2013) Engagement with indigenous communities in key sectors. ACT, Canberra

Ifejika Speranza C, Kiteme B, Ambenje P, Wiesmann U, Makali S (2010) Indigenous knowledge related to climate variability and change: insights from droughts in semi-arid areas of former Makueni District, Kenya. Clim Chang 100(2):295–315. https://doi.org/10.1007/s10584-009-9713-0

Kar B, Saha J, Saha JD (2012) Analysis of meteorological drought: the scenario of West Bengal. Indian J Spatial Sci 3(2):1–11

Kearney J, Berkes F, Charles A, Pinkerton E, Wiber M (2007) The role of participatory governance and community-based management in integrated coastal and ocean management in Canada. Coast Manag 35(1):79–104

Kihila JM (2018) Indigenous coping and adaptation strategies to climate change of local communities in Tanzania: a review. Clim Dev 10(5):406–416

Kuhnlein H, Erasmus B, Spigelski D (2009) Indigenous peoples' food systems. FAO. Centre for Indigenous People's Nutrition and Environment, Rome

Kuhnlein HV et al (2013) Indigenous peoples' food systems and Well-being: interventions and policies for healthy communities. CABI. https://www.cabdirect.org/cabdirect/abstract/20133239133

Kumar S, Singh AK (2013) Traditional agricultural knowledge followed by tribal farmers of Uttar Pradesh, India. Search Res 4(3):43–47

Kumar S, Parveen F, Goyal S, Chauhan A (2008) Indigenous herbal coolants for combating heat stress in the hot Indian Arid Zone. Indian J Tradit Knowl 7(4):679–682

Kumar S, Das G, Shin HS, Patra JK (2017) *Dioscorea* sp. (a wild edible tuber): a study on its ethnopharmacological potential and traditional use by the local people of Similipal Biosphere Reserve, India. Front Pharmacol 8:52

Kumari A, Kumar A (2021) Assessment of genetic diversity of Diospyros melanoxylon Roxb. (kendu) growing in Jharkhand, India using RAPD markers. J Plant Sci Res 37(1):23–31

Kuyu CG, Bereka TY (2020) Review on contribution of indigenous food preparation and preservation techniques to attainment of food security in Ethiopian. Food Sci Nutr 8(1):3–15
Lakhani, S. (2019). *How Indigenous knowledge systems can play a crucial role in environment protection and sustainable development (In Indian context).* https://countercurrents.org/2019/09/how-indigenous-knowledge-systems-can-play-a-crucial-role-in-environment-protection-and-sustainable-development-in-indian-context/
Lampman MA (2010) How folk classification interacts with ethnoecological knowledge, a case study from Chiapas Mexico. J Ecol Anthropol 14(1):39–51. https://doi.org/10.5038/2162-4593.14.1.3
Leonard S, Parsons M, Olawsky K, Kofod F (2013) The role of culture and traditional knowledge in climate change adaptation: insights from East Kimberley, Australia. Glob Environ Chang 23(3):623–632. https://doi.org/10.1016/j.gloenvcha.2013.02.012
Malley ZJU, Taeb M, Matsumoto T, Takeya H (2008) Linking perceived land and water resources degradation, scarcity and livelihood conflicts in southwestern Tanzania: implications for sustainable rural livelihood. Environ Dev Sustain 10(3):349–372. https://doi.org/10.1007/s10668-006-9069-9
Mauro F, Hardison PD (2000) Traditional knowledge of indigenous and local communities: international debate and policy initiatives. Ecol Appl 10(5):1263–1269
Mobolade AJ, Bunindro N, Sahoo D, Rajashekar Y (2019) Traditional methods of food grains preservation and storage in Nigeria and India. Ann Agric Sci 64(2):196–205
Modak B (2010) Animal resource dependence of the under-privileged people of Purulia District: a report. Biodivers Envis Newsl 2(1) West Bengal Biodiversity Board, Govt of West Bengal 2(1):5–6
Nwachukwu E, Achi OK, Ijeoma IO (2010) Lactic acid bacteria in fermentation of cereals for the production of indigenous Nigerian foods. Int Scholars J 2(1):1–6
Nyong A, Adesina F, Osman Elasha B (2007) The value of indigenous knowledge in climate change mitigation and adaptation strategies in the African Sahel. Mitig Adapt Strateg Glob Chang 12(5):787–797. https://doi.org/10.1007/s11027-007-9099-0
Patadia NA (2020) Role of circular economy in the indigenous built environment: an assessment of design and construction potential of circular building materials in an American Indian community, Doctoral dissertation. Arizona State University, Tempe
Rathore M (2009) Nutrient content of important fruit trees from arid zone of Rajasthan. J Hortic For 1(7):103–108
Ravishankar K, Nagasree K, Nirmala G, Rama Rao CA, Raju BMK, Beevi A, Rohit J, Pankaj PK, Ramana DBV, Srinivas I, Vijaya Kumar S, Sindhu K, Srinivasarao C (2020) Farmers perceptions, attitude and adaptations towards climate change in selected districts of India: implications from an adaptation view. Curr J Appl Sci Technol 39(48):379–395
Sarkar M, Modak BK (2022) Rituals and festivals of indigenous people of Chota Nagpur plateau of West Bengal: a positive correlation with the environment. In: Indigenous people and nature. Elsevier, pp 465–491. https://doi.org/10.1016/B978-0-323-91603-5.00020-8
Senanayake SGJN (2006) Indigenous knowledge as a key to sustainable development. J Agric Sci:87–94. https://doi.org/10.4038/jas.v2i1.8117
Sengupta N, Ghosh K (2022) Women's role in water resource management: a case study on upper catchment area of Kangsabati river basin under Purulia district, West Bengal. Sustain Water Resour Manage 8(4):107. https://doi.org/10.1007/s40899-022-00685-2
Shaw R, Takeuchi Y, Uy N, Sharma A (2008) Indigenous knowledge: disaster risk reduction, policy note. United Nations International Strategy for Disaster Risk Reduction, Kyoto
Sillitoe P (1998) The development of indigenous knowledge: a new applied anthropology. Curr Anthropol 39(2):223–252. https://doi.org/10.1086/204722
Sommer R, Glazirina M, Yuldashev T, Otarov A, Ibraeva M, Martynova L, Bekenov M, Kholov B, Ibragimov N, Kobilov R, Karaev S, Sultonov M, Khasanova F, Esanbekov M, Mavlyanov D, Isaev S, Abdurahimov S, Ikramov R, Shezdyukova L, de Pauw E (2013) Impact of climate change on wheat productivity in Central Asia. Agric Ecosyst Environ 178:78–99. https://doi.org/10.1016/j.agee.2013.06.011
Sundaramari M, Ganesh S, Kannan GS, Seethalakshmi M, Gopalsamy K (2011) Indigenous grain storage structures of South Tamil Nadu. Indian J Tradit Knowl 10(2):380–383
Tarafder S, Debnath D (2021) Use of indigenous disaster adaptive practices to reduce risk in the southern part of West Bengal. Res Rev Int J Multidiscip 6(1):256–270. https://doi.org/10.31305/rrijm.2021.v06.i01.050
Timilsena NP, Devkota KM (2022) Indigenous knowledge and local practices for disaster risk reduction: a study of Kailali district. Geogr J Nepal 15:119–130. https://doi.org/10.3126/gjn.v15i01.42890
Ulluwishewa R (1993) Indigenous knowledge systems for sustainable development: the case of pest control by traditional paddy farmers in Sri Lanka. J Sustain Agric 3(1):51–63. https://doi.org/10.1300/J064v03n01_05
Varghese SM, Parisi S, Singla RK, Begum ASA (2022) Trends in food chemistry nutrition and Technology in Indian sub-Continent Food Preservation. Springer International Publishing Cham, pp 45–53
Wang H-W, Kuo P-H, Shiau J-T (2013) Assessment of climate change impacts on flooding vulnerability for lowland management in southwestern Taiwan. Nat Hazards 68(2):1001–1019. https://doi.org/10.1007/s11069-013-0670-6
Yarmolinsky L, Huleihel M, Zaccai M, Ben-Shabat S (2012) Potent antiviral flavone glycosides from *Ficus benjamina* leaves. Fitoterapia 83(2):362–367

Part II

Climate Crisis: Geophysical Hazard and Risk Reduction and Mitigation

6 Addressing Climate Crisis Through Coastal Risk Management: The Social Protection Alternative

Sayanti Sengupta

Abstract

Climate change is one of the most concerning challenges faced by countries today, with far-reaching impacts on lives and livelihoods. Worsening climate extremes disturb ecosystems, disrupt economic activities, and endanger the ones most vulnerable. Natural disasters like cyclones or floods are being exacerbated by climate change and heavily impacting lives, livelihoods, property, and assets, especially in the coastal areas. To minimize the socio-economic impacts of climate hazards and reduce poverty and inequality among coastal communities, comprehensive coastal risk management has become an integral part of the policy focus in both national and international agendas.

This chapter explores the specific risks that Sundarbans coastal communities in India face due to climate change. Based on primary data and secondary literature, the chapter presents techniques adopted at the individual and community levels for coping with climate hazards in Sundarbans. Recognizing the weaknesses of such coping techniques, the chapter provides an in-depth understanding of social protection programs, which could be used as viable tools against poverty and climate-related impacts. The chapter establishes social protection as a relevant policy option for effective coastal risk management by providing practical examples of how social protection schemes could build the climate resilience of coastal communities.

Keywords

Climate change · Coastal areas · Coastal communities · Social protection · Sundarbans

6.1 Introduction

Coastal areas, acting as boundaries between the earth's landmasses and oceans/seas, are unique features of the planetary landscape. They harbor a teeming array of natural vegetation and animal species, on the one hand, and extensive manmade infrastructure and human population, on the other hand. Close to 2.4 billion people inhabit the 620,000 kilometers of coastline on this planet, living within 100 kilometers of the coast (NASA n.d.; United Nations 2017). Approximately 10 percent of the global population, which amounts to over 600 million individuals, reside in coastal regions with an elevation of less than 10 meters from the sea level (United Nations 2017) and are, thus, exposed to unprecedented climate change-induced sea level rise.

S. Sengupta (✉)
Social Protection and Climate, Red Cross Red Crescent Climate Centre, The Hague, The Netherlands
e-mail: sengupta@climatecentre.org

U. Chatterjee et al. (eds.), *Climate Crisis: Adaptive Approaches and Sustainability*, Sustainable Development Goals Series, https://doi.org/10.1007/978-3-031-44397-8_6

Human inhabitation along the coastal areas implies that a number of economic activities prevail there, including fishing, aquaculture, tidal plantations, shore drilling, and other commercial activities through ports, tourism, and recreation. Coastal ecosystems play a determining role in the sustainability of these activities, and thereby have an impact on the well-being of the coastal populations and their livelihoods. Beyond the economic benefits, coastal areas play a protective role by buffering storms and reducing wave impacts on inland areas. On the one hand, these benefits have resulted in attracting human settlements to the coasts at a rate higher than inland areas. On the other hand, human activities have resulted in coastal areas undergoing tremendous land use changes, exploitation and extraction, and higher risks from sea level rise and extreme weather events due to climate change (Nature Communications 2020).

Coastal areas are at risk of loss and degradation over the past two to three decades—including marshes (50% either lost or degraded), mangroves (35%), and reefs (30%) (Barbier et al. 2008)—and they continue to decline worldwide. In the last decades, there has been widespread shrinking of the cryosphere, with mass loss of ice sheets and glaciers; the rate of ocean warming has more than doubled and is affecting coastal ecosystems (IPCC 2019). Coastal ecosystems, and thereby coastal areas, are rapidly being altered by natural as well as anthropogenic causes (Roy et al. 2021). According to IPCC (2022), climate change has resulted in substantial damage and increasingly irreversible losses to coastal and open ocean marine systems.

Coastal areas, being highly populated regions of the world, are highly exposed to extreme weather events and other climate change-related impacts. About 89% of all the disasters that occurred between 1970 and 2012 were due to storms and flooding (Goldenberg 2014). Coastal storms and floods have become highly common over the years and are increasing in intensity and frequency due to climate change (IPCC 2022). Proximity to the ocean water increases the exposure of these areas to tsunamis, storm surges, cyclones, etc., and results in serious threats to the population living there.

South Asian coasts are highly populated, and close to 27% of the Indian population lives in coastal areas (Roy et al. 2021). India or rather the coast of the Bay of Bengal is expected to witness a significant increase in coastal flooding and inundation by 2100 (Kirezci et al. 2020). The Sundarban forests, spanning across 10,000 square kilometers across India and Bangladesh, are the largest mangrove forests in the world, located at the super confluence of the river Hooghly (distributary of the river Ganga), Padma, Brahmaputra, and Meghna, with the Bay of Bengal (Pani et al. 2013). Unlike a linear coastline, the Ganga–Brahmaputra delta is highly inundated and comprises of a complex system of mangrove forests, numerous coastal islands, and tidal waterways.

The Sundarbans coastal area lying within the Indian sub-continent is occupied by close to 4.6 million people (Rosencranz et al. 2020) who are highly dependent on the ecosystem services provided by the coast and the forest. This unique landscape is at the forefront of facing climate change impacts, with studies indicating that sea level rise, storm surge intensification, and water salinization will adversely affect the health, nutrition, and livelihoods of the inhabitants (Dasgupta et al. 2020). Extreme weather events and low coping capacity of the coastal communities result in a range of challenges, like loss of income opportunities, reduced access to education and health care, out-of-pocket expenses to prevent and recover from disasters, and overall reduction in human capital.

Coastal communities in Sundarbans adopt various strategies to cope with these challenges, often relying on their traditional knowledge and expertise. While these strategies help in managing the hazard impacts to some extent, it is not adequate for building long-term climate resilience. Policy instruments like social protection, that are traditionally designed to protect vulnerable people from life-cycle risks (e.g., old age, unemployment, work accidents), are increasingly being considered as useful tools that can address the socio-economic impacts arising due to climate change (Costella et al. 2021).

This chapter takes a closer look at the various climate-related risks faced by the Indian

Sundarbans and highlights the social, economic, and health impacts caused as a result. Based on primary data, the chapter describes a few short- and long-term coping techniques adopted by communities in the Sundarbans coastal area against climate risks. In addition, the chapter presents social protection as an innovative policy option for coastal risk management and explores how different social protection schemes can be introduced, replicated, and adopted to protect some of the poorest and most vulnerable coastal populations in the world.

While there is adequate literature on different physical and infrastructural interventions that can be undertaken to reduce impacts from climate hazards, there is hardly any literature or research done on how social protection policies can become part of coastal climate risk management strategies. To address this gap in knowledge, the objective of the chapter has been to provide an overview of climate hazards and impacts faced by Sundarbans coastal communities, and present potential social protection program options that can promote alternative livelihoods, compensate for loss of income, improve access to social and healthcare services, and enable coping and adaptive capacities among coastal communities. These options can be further explored by the state and national government, as well as humanitarian actors, to either modify existing social policies or introduce new ones to offer better protection to the vulnerable coastal populations in. Re-emphasizing the need to protect this ecologically fragile and resource-rich ecosystem lends rationale to the research conducted for developing the chapter. To the author's knowledge, this is the first academic work that explores social protection options for managing climate risks that affect the Indian Sundarbans and its coastal population.

6.2 Methodology

This chapter has been developed on the basis of primary data collected during fieldwork and secondary literature reviewed. As part of the authors' previous field study, a number of research tools were used to collect information. This includes key informant interviews, household surveys, focused group discussions, a resource mapping exercise, and a few participatory rural appraisal (PRA) techniques. PRA techniques used for data collection included: (a) developing seasonal crop calendars to document climate-induced changes in cultivation patterns, (b) resource mapping to identify the crucial community resources and their locations that are currently exposed to climate hazards, (c) developing a timeline chart to enable a historical mapping of large-scale climate hazards that have affected the coastal communities, (d) problem ranking exercise to identify main challenges faced by communities due to climate hazards, and (e) service/amenities mapping exercise to identify what women and men rely on for assistance during/after disasters. Annex 1 presents a schematic overview of the findings from these five PRA techniques.

The sampling technique used for selecting the location for data collection was multi-staged: (a) purposive sampling technique was selected to identify three habitations in the Mousuni island of Sundarbans, within the Namkhana block, South 24 Parganas district of West Bengal, on the basis of their high vulnerability scores in the 2009 District Human Development Report (GoWB 2009); (b) a random sampling technique was used to identify households for the survey; and (c) a convenient sampling technique was used to identify the key informant interviews and focused group discussion participants. The primary data collection was completed in 2016–2017, while the secondary literature review has been completed between 2022 and 2023.

Survey respondents during the primary data collection phase were asked both open-ended and closed-ended questions. Information relating to hazard impacts and coping strategies were mostly answered through open-ended questions during the survey, key informant interviews, or focused group discussions, to allow respondents to be subjective and descriptive, thus enabling an in-depth understanding of the ground realities.

For the desk research component, an exhaustive search of academic papers, peer-reviewed articles, and reports was conducted. The literature search was completed using academic search engines including Google Scholar, Science

Direct, Web of Science, Open Thesis, and Scopus, and then identifying additional grey literature through a Google search.

The central research question that the chapter seeks to answer is: *How can innovative options like social protection support vulnerable communities in the Sundarbans coastal areas to increase their capacity to cope, adapt, and address climate risks?*

The primary data collected during the field research on climate hazards and impacts in Sundarbans is described in Sect. 6.3, as well as Sect. 6.4, and highlights the coping techniques used by vulnerable populations in these coastal areas. In Sect. 6.5, social protection's potential role in addressing the impacts of climate hazards is explored. The chapter relies on primary and secondary literature and proposes a few practical solutions that can be piloted.

6.3 Climate Hazards and Impacts in the Sundarbans Coastal Area

Coastal areas, like any other natural ecosystem, provide a range of services and resource opportunities. Climate change and anthropogenic actions have put these resources under threat by degrading ecosystem services over space and time (Nicholls et al. 2007; Nandy and Ahammad 2012). Torresan et al. (2008) have shown that sea level rise, increased level of inundation and storm flooding, coastal erosion, seawater ingress, and temperature increase are some of the common climate-related hazards that make coastal areas vulnerable. Communities closely linked to coastal environments including small islands (like the Sundarbans) face heightened vulnerability to changes in sea level and weather events. If current patterns continue, the growing exposure and vulnerability of coastal communities are anticipated to result in a substantial rise in risks from erosion and land loss, flooding, salinization, as well as the compounding effects of mean sea level rise and extreme events (IPCC 2019). The geographical location of the Sundarbans coastal area makes it unique and prone to several natural disaster-related risks. Beyond natural risks, climate change exacerbates the intensity and severity of climate events and disrupts lives and livelihoods in many ways. The three major climate-related risks affecting the Sundarbans coastal area is discussed below:

1. Sea level rise: The Sundarbans coastal area also faces acute threat from sea level rise (Dasgupta et al. 2020), especially by virtue of being an archipelago of islands with high levels of exposure. The sea level has risen by 3 centimeters per year on average in the last two decades (Augustin 2019). Raha et al. (2012), through a time series analysis using satellite imageries between 1924 and 2008, has shown the continued erosion of several Sundarbans islands. Recent reports have suggested that the sea level may rise by 1 m or more in this century, resulting in an increase of 1 billion vulnerable people (Dasgupta et al. 2020). Hazra et al. (2002) in their paper make a claim of visible changes in the shape, size, and geomorphology of the islands. Topographic maps, trend analysis, and satellite data show degraded mangrove swamps and mudflats, increased salinization and formation of saline banks, and most importantly, the overall reduction in the total land area due to erosion and submergence. Data from NASA's Landsat satellite imagery indicate that the sea level in the Sundarbans has experienced an annual average increase of 3 centimeters (1.2 inches) over the past 20 years. Furthermore, within the last 40 years, the region has witnessed a loss of nearly 12 percent of its shoreline (Muller 2020). In the last two decades, Sundarbans reported one of the fastest rates of coastal erosion in the world, well above the global average (Augustin 2019). For some islands, the rate of coastal erosion is as high as 40 m per year, and, at this rate, these islands will cease to exist in the next 50–100 years (Dasgupta et al. 2020).
2. Cyclones and storm surges: The Sundarbans coastal area is regularly affected by cyclones and storm surges, which have become more frequent and severe due to climate change.

Dasgupta et al. (2020) suggest that cyclones and storm surges cause significant damages to Sundarbans, including loss of life, damage to infrastructure, and destruction of crops and fisheries. The Indian Sundarbans currently lie in the highest impact zone from tropical cyclones in the Bay of Bengal (Dasgupta et al. 2020). Two highly devastating cyclones in 2007 and 2009 resulted in drastic changes in the mangroves and coastal communities in Sundarbans, and the heightened frequency of storms in the Bay of Bengal could potentially lead to greater destruction of the coastal mangroves in the coming years (Paul et al. 2017), thus degrading the natural barrier these forests provided to the coastal communities.

3. Floods: Flooding, during and after cyclones, storm surges, or regular riverine/flash floods is common in Sundarbans and is expected to be influenced by increasingly erratic monsoons (Dasgupta et al. 2020), which can result in stagnation of water, which in turn leads to water-borne diseases (Mahadevia Ghimire and Vikas 2012). This results in health issues and can lead to heightened risks to the physiological adaptive capacity of individuals in the long run.

While not a climatic event, increasing salinity is a major cause of concern in this region. A natural neo-tectonic shift has resulted in land subsidence in the Eastern Bengal basin causing more of the Ganges water to flow towards Bangladesh (Ahuja 2022). Several dams, weirs, and barrages built on the Ganga–Brahmaputra–Meghna river system divert fresh water upstream, resulting in a lack of freshwater downstream to dilute the saline water brought in regularly by the tides (Ahmad 2020). The eastward meandering of the Ganga River, combined with climate-induced changes in sea level, temperature, and rainfall, can be expected to decrease freshwater influx, and intensify river-water salinization by 2050, reaching near-ocean salinity in many areas (Dasgupta et al. 2020). This also increases the salinity level of groundwater on which coastal populations in Sundarbans are heavily dependent. As sea levels rise, saltwater intrudes further inland, leading to saltwater intrusion into freshwater sources and having significant impacts on agriculture and drinking water supplies in the Sundarbans.

The hazards discussed above result in a number of impacts. Physical losses are rampant along the island coasts of the Sundarbans. Recurrent flooding, regularly occurring cyclones, and coastal erosion result in the loss of land and infrastructure, such as roads, buildings, and health centers. Embankments, which are crucial against cyclones, storm surges, high waves, and flooding, frequently collapse (Tenhunen et al. 2023). Cyclone Aila, the most devastating storm recorded in the Indian Sundarbans, ravaged numerous sea and riverine embankments, breaching them and uprooting mangroves in 2009 (Das and Das 2022). According to the same study, during Amphan, another super cyclone that hit Sundarbans in 2020, embankments were damaged in several ways in 500 sites with 200 villages inundated.

The unique and diverse ecosystem of the Sundarbans coastal area is also under threat due to climate hazards. Rising sea levels and increased salinity are causing significant biodiversity loss in the Sundarbans, including the loss of mangrove species and the displacement of wildlife. The altered hydrological and salinity pattern would result in changes in not only plant habitats but also fish and animals. This would disrupt the overall biological equilibrium that is necessary for any ecosystem to function (Uddin et al. 2013). The mere existence of many plant and animal species in the low-elevation areas of the Sundarban forests is threatened by cyclones (TWC India 2020).

Beyond the threat to the natural or geophysical environment, the population living on 106 islands in Sundarbans are recurrently impacted by floods, cyclones, sea level rise, and bank erosion, as well as an increasing number of hot days, drought conditions, and lack of rainfall in the recent years. This causes food insecurity and results in loss of livelihood opportunities (Masum 2012), resulting in a push factor that causes outmigration of the people from Sundarbans, especially the men. Cyclone Aila, for instance, caused saltwater intrusion due to embankment breaching

(Das and Das 2022) and raised the ground salt to a record level, pushing many to make a decision to migrate over the next years (Tenhunen et al. 2023). In search of better job opportunities or income, several villagers migrate to the cities, and the women and families are often left behind (Das and Das 2022).

The effects on traditional livelihoods as a result of climate change depend on the nature and severity of the physical impacts on agriculture, water availability and quality, disaster-proneness, hospitability of the physical environment caused by increasing temperature, and changing water regimes to pathogenic activity and coastal inundation (Afjal Hossain et al. 2012). Cyclones like Aila have short- as well as long-term impacts on food security, livelihoods, education, and security of people. The lack of industry and major dependence on agriculture in the Sundarbans make the livelihoods of people increasingly uncertain (Das and Das 2022). A survey conducted by Hazra et al. (2002) reveals that 7000 people in Sundarbans have been displaced from their original habitats due to either sea level rise, coastal erosion, cyclone, or coastal flooding. Salinity and waterlogging emerge as severe challenges for practicing agriculture in Sundarbans. Paddy being the main crop is brutally affected as the soil is rendered infertile after floods leave behind saline residues. Considering the threat posed by rising sea levels and coastal flooding, it is projected that approximately one million individuals will require relocation from highly vulnerable areas in the Indian Sundarbans by 2050. Furthermore, it is likely that managed relocation efforts will need to be undertaken on a larger scale in the future (Panda 2020).

Another major problem that the residents here face is the significant land loss that occurs every year due to riverbank erosion and surges or cyclones. The capacity of people to rehabilitate themselves is limited and thus they suffer from chronic poverty. The increase in climate refugees due to repetitive cyclonic activities causes problems related to accommodation, employment, water, and food (Bhattacharjee et al. 2011). Resettlement activities from inhabited islands like Lohachara and Ghoramara to the nearby Sagar Island had been initiated in the late 1970s, to families who had lost their lands due to unprecedented land erosion. The Sagar Island itself is currently facing rapid erosion and is increasingly becoming vulnerable and unfeasible as a relocation option for newly displaced populations (Panda 2020; Sinha 2022).

6.4 Addressing Climate Change Impacts in the Sundarbans Coastal Area

The effects of climate-induced changes on coastal areas and communities are posing growing challenges to existing governance initiatives aimed at devising and implementing adaptation strategies across various scales, from local to global. In certain instances, these challenges are pushing these initiatives to their utmost capabilities. At a micro level, human vulnerability to natural hazards and poverty are interdependent and play a major role in determining the capacity of populations to address the risks and impacts of climate hazards. The primary data collected for the research has shown that coastal populations facing the adversities of a changing climate and associated disadvantages for several years have developed their own way of minimizing the impacts. Some of these techniques have been listed here. There has been a distinction made between short-term coping, which refers to measures taken immediately after or at the time of the disaster event, while long-term coping measures are those that try to address salinity or loss of embankment due to riverbank erosion.

6.4.1 Short-Term Coping Techniques

Bamboo huts raised on plinths or stilts are quickly constructed when floods occur due to high tides or excessive rainfall. During cyclonic storms or coastal flooding, embankments are breached causing saline water to advance rapidly inland towards the villages. Breached embankments often prevent the outflow of water, resulting in waterlogged areas. The standing water rises higher level due to tides, and often floods households located in the surveyed villages of Mousuni island. Families are often forced to abandon their dwellings as long as the stagnant water does not recede back. In such cases, people take refuge in the shelters built by NGOs or local governments. Data collected during interviews suggest that inhabitants build raised platforms on plinths only after a breach in the nearby embankments becomes apparent, thus resulting in a very short response time and a limited window of opportunity to take additional coping measures. While this is an example of a community-based early warning system where the community helps in spreading news of an embankment breach, this presents an opportunity to improve awareness and train communities to act even earlier based on forecasts.

When there is water logging or flooding, freshwater ponds located along the coastlines of Sundarbans get affected. The quality of water from the tube wells also deteriorates and gets contaminated due to increased salt levels left behind by retreating saline water. To address this, the villagers often drill deeper wells to tap into deeper aquifers. While this acts as a short-term coping capacity, over-exploitation of aquifers can potentially become a maladaptive practice.

The freshwater ponds, once they become saline, are unsuitable for aquaculture and may take several years to return to their natural state. Households that own bigger ponds and have the economic means use pumps to drain out the salt water from their ponds and manually dredge the upper layers of the soil to remove the salty mud layers from the surface. During the next monsoons, freshwater fills up these ponds, rendering them usable again. During the period in between, when the pond waters are saline, coastal communities largely depend on already contested public sources of water like tube wells, which can sometimes result in tension and conflict.

Many of the households surveyed in Mousuni owned their own dinghies (small boats), which are made use of when the roads and huts are submerged during floods. They are often forced to temporarily shift all their valuables onto these boats to other islands, to protect them from getting inundated or damaged when the flood water retreats.

Survey findings also suggest that during floods and cyclones, food and freshwater shortages affect the coastal communities. To cope with that, the Gram Panchayat (the lowest tier of the Indian decentralized governance structure) at times in the past has provided every affected household with a tarpaulin sheet, locally referred to by the villagers as "paper," to cover the roofs of the makeshift houses. Local NGOs additionally support with dry food, water, and medicine supplies. Community kitchens are also formed on an ad-hoc basis for flooded neighborhoods. Arrangements are also made by the Gram Panchayat for spraying mosquito repellents, carbolic acid, and such other chemical repellents to prevent the spread of diseases from the stagnant water.

The Sundarbans coastal area is characterized mainly by rural households who own cattle, poultry, and other domesticated animals. Recurrent saline water ingression poses a serious threat to households owning these animals as every year, there is an acute shortage of fresh water to feed them. When the easily accessible freshwater ponds become saline, carrying additional water from tube wells to feed the livestock burdens families. As a result, some families resort to negative coping techniques like selling of cattle. Others who can afford it were found to send their cattle across the rivers via boats to other islands, to seek temporary shelter in their relatives' sheds/farms. However, the back-and-forth transport of livestock results in additional financial costs for the families and is a major reason for foregoing livestock-rearing practice in the Indian Sundarbans.

Through government initiatives, flood houses or shelters have been constructed in high elevated areas, inland from the coastline, to provide shelter to families that are unable to construct their own makeshift huts. However, a flooded house has limited capacity and cannot accommodate more than 350–400 people, and this can result in a lack of adequate shelter during hazards. Some families interviewed have reported to take shelter in local schools until the floodwater recedes.

6.4.2 Long-Term Coping Techniques

The stagnant water in the aftermath of a flood or cyclone has far-reaching impacts that continue beyond the post-monsoon months. Even after the flood water has retreated or subsided, the salts brought inland from the sea contaminate the soil and the water in ponds. This makes it difficult to practice agriculture or fishing for years, like in the case of cyclone Aila. Farmers and fishermen, the two most predominant livelihoods in Sundarbans, suffer a steep decline in income opportunities after cyclones and floods. However, over the years, the households have adopted alternative crops which can withstand high salinity. Newer cropping patterns are being explored and local NGOs are playing a key role in supporting farmers in making this shift.

Erosion of agricultural land, loss of property, and infrastructure have pushed many people into poverty and forced others to migrate to different cities in search of work. While some permanently migrate and settle on another island or in nearby cities like Kolkata, most of them move to different Indian states like Kerala and Tamil Nadu in search of work. A consistent increase in the number of people migrating out of Sundarbans shows that this has become a common coping technique of populations living here.

While not recorded as part of the data collection, the literature suggests that employment in eco-tourism is also an alternative livelihood option that the people in Sundarbans are increasingly adopting. The coping techniques used by populations in the Sundarbans coastal areas are quite minimal, and the focus is mainly on temporary coping measures instead of long-term solutions against future risks and increasing weather events. The sustainability of such coping techniques is subject to resource availability, at the household as well as government level.

6.5 Social Protection as an Innovative Option for Coastal Risk Management

"Structurally weak" groups, like those interviewed for this chapter, are highly vulnerable and require support in addressing the underlying drivers of vulnerability in the long term (UNDP and Malik 2014). Ahammad et al. (2013) state that vulnerability to sea level rise and climate change is strongly influenced by the characteristics of diverse socio-economic development and ecological capacity to adapt to a new situation. While the short- and long-term coping strategies discussed above play a role in helping these populations to survive in the coastal areas, they are not adequate in building climate resilience, especially in a changing climate.

Climate-related risks are expected to worsen the impacts on the Sundarbans ecosystem from cyclones, floods, storm surges, erosion, and salinization. Increased risks to health and well-being, reduced income opportunities, declining land, and labor productivity will impoverish families further and make it difficult to move out of poverty (Dasgupta et al. 2020). Literature suggests that poverty eradication interventions need to address multiple and compounding vulnerability drivers like chronic poverty, food insecurity, inequality, violence, instability, weak governance, and impacts of climate change (Olsson et al. 2014). Islam and Winkel (2017) identify that pre-existing social inequality in societies results in climate change disproportionately impacting disadvantaged groups primarily in three ways: (a) increase in the exposure of the disadvantaged groups to the adverse effects of climate change; (b) increase in their susceptibility to damage caused by climate change; and (c) decrease in their ability to cope and recover from the damage suffered.

Acknowledging existing social inequalities as the baseline, this chapter now discusses how "social protection" can be a potential option for addressing Islam and Winkel's (2017) three ways that enable climate change to affect disadvantaged groups disproportionately.

Social protection is a series of policies and programs aimed at protecting individuals from shocks across the life cycle to mitigate poverty and inequality (ILO 2017). FAO (2017) expands on the definition to state that social protection is a set of policies and programs that address economic, environmental, and social vulnerabilities to food insecurity and poverty by protecting and promoting livelihoods.

Social protection systems comprise a range of schemes that are adopted and implemented by countries and actors based on their capacities. Table 6.1 shows the different social protection schemes that can commonly be used.

Non-contributory schemes are those that do not generally require contributions from the recipients of support and can be of two forms: social assistance schemes and social care policies/interventions. Social assistance schemes can be of various types: (a) social transfers that provide cash or in-kind support to vulnerable groups; (b) public works' programs where manual labor is provided by recipients in exchange for cash or food; (c) fee waivers for education or health services; and (d) subsidies provided for purchasing fuel and/or food.

Social care policies and interventions can include providing basic health-care services, psycho-social support, pre and post-natal care for young mothers, as well as various kinds of training and outreach programs for improving living conditions and access to basic services for the recipients of social care.

Contributory schemes, on the other hand, are funded by regular payments made by participants to protect them from life-cycle risks. This includes insurance schemes that protect contributing members from unemployment or disability as well as pension schemes for old age. Another form of contributory scheme is the crop/livestock insurance that offers income protection to insurance takers from crop failure or livestock-related diseases.

Active labor market policies are aimed at the working-age population and help in facilitating employment, entry into the labor market, skills training, etc.

National governments have a number of social protection options to choose from, and some of these options can be viable for supporting populations in the Sundarbans coastal area. External actors like international NGOs, humanitarian agencies, as well as development partners may be keen on supporting such interventions, especially when they are nationally owned (Table 6.2).

1. Social assistance
   (a) Cash transfers: Coastal communities in the Indian Sundarbans would benefit from systematic and reliable cash transfers. These cash transfers, depending on the fiscal capacity, could be targeted to vulnerable groups like widowed women, who often lose their husbands due to attacks from the maneater tigers in the Sundarbans. Other typically vulnerable groups like orphan children, persons with

Table 6.1 Typology of social protection

Non-contributory					Contributory	
Social assistance						
Social transfers	Public works	Fee waivers	Subsidies	Social care	Social insurance	Active labor market policies
Cash transfers Vouchers In-kind transfers School feeding programs	Cash for work Food for work	Health Education	Fuel Food	Psychological therapy Mentoring Rehabilitation support	Insurance for Unemployment Maternity Disability Work accidents Old-age pensions Crop/livestock insurance	Work sharing Training Job search services

Source: OPM (2017), Loewe and Schüring (2021)

Table 6.2 A summary table showing potential options for using social protection to support the Sundarbans coastal population

Non-contributory					Contributory	
Social assistance						
Social transfers	Public works	Fee waivers	Subsidies	Social care	Social insurance	Active labor market policies
Cash transfers are given to at-risk livelihoods like farmers, fishermen, honey collectors, or traditionally vulnerable groups like widowed women, persons with disabilities, orphans, or the elderly after disasters. Vouchers and in-kind support can also be given right after disasters to the most vulnerable.	Public works' programs can support building community rainwater harvesting systems, community toilets, tanks, tube wells, and flood houses on raised platforms, cleaning and dredging drainage channels prior to monsoons. The activities can be tailor made and revised based on the needs of coastal communities.	Fee waivers can be offered to households with children affected during a disaster. Fee waivers can be coupled with school meals or take-home food packets to enable children to get nutrition in the aftermath of a disaster.	Fuel, food, and transport subsidies can be given to support coastal communities in the months following a disaster. Subsidies can be given to support the re-construction of buildings and infrastructure at the household or community level.	Social care services can be provided to support communities' access to utilities. Health services can include training to improve awareness related to water, sanitation, and hygiene (WASH) services, with a focus on menstrual hygiene management.	Crop and livestock insurance can be extended to farmers and livestock owners, with premiums subsidized by the government or external actors, to protect them against crop failure and losses due to rainfall uncertainties, flooding, or cyclones. If there are some formally employed people living and working in Sundarbans, a social insurance scheme for such people could be created which they and their employers regularly contribute. This can act as a contingency fund during disasters.	At risk, livelihoods can be identified and given training to learn skills for alternative employment opportunities. Farmers can be trained in cultivating salinity-resistant crop varieties. Job support can be provided for those able and willing to migrate.

disabilities, and the elderly could also be targeted. Regular cash transfers can enable households to purchase productive assets or make investments that reduce their overall poverty and improve their capacity to address climate hazards. In case of financial constraints, a one-time lump sum cash transfer can be made to vulnerable groups, prior to the flood or cyclone season, to enable them to safeguard their assets, protect their property, and minimize damages. Cash or in-kind transfers can be conditional in nature, where recipients receive benefits only after fulfilling certain conditions. Conditions can be used to promote behavioral change among coastal communities; for example, refraining from deforestation and logging of mangroves, which is a common negative coping technique observed in the Sundarbans coastal areas.

(b) Vouchers or in-kind support: During and after cyclones or floods, distribution of cash may be difficult. Humanitarian agencies, donors, and NGOs sometimes may have a preference to provide non-cash support. In these cases, providing vouchers that coastal communities can use to purchase essentials is a useful way of reducing distress. During previous disaster events, several organizations have provided in-kind supplies like packaged drinking water, dry food, medicines, tarpaulin sheets, and such other amenities for the re-construction of damaged infrastructure and property. Providing vouchers and/or in-kind support as part of a national social protection scheme could enable larger coverage, quality control as well as building public trust towards governance systems. In-kind transfers in the form of food packets during crises can also be provided to students through school feeding programs.

(c) Public works: Public works' schemes are popular social protection interventions and can be used to support coastal communities in many ways. Before the cyclone/flooding season, public works' activities can be designed and deployed in a way that prevents water stagnation and allows for the swift outflow of flood waters into the sea. Strengthening of embankments, digging freshwater retention ponds, constructing infiltration channels, etc. are some of the other activities that can be included under the public works' scheme. The Mahatma Gandhi national rural employment guarantee scheme (MGNREGS) is the largest public works' intervention in the world and is also implemented in Sundarbans where activities like growing mangrove nurseries and construction of cement roads have been undertaken (WBPRD 2021). The range of activities under the scheme can be further extended to promote the construction of climate smart infrastructure.

(d) Fee waivers: Fee waivers for school-going children in the aftermath of a hazard can be an option to support families indirectly. Relieving the financial burden of paying school fees can help reduce dropout rates, which is a common negative coping strategy among families strapped for cash.

(e) Subsidies: Subsidies or lowering/maintaining prices at an affordable level in the aftermath of a climate hazard is an option to reduce the financial difficulties of affected households. For example, transport subsidies could be provided on routes between Sundarbans and nearby cities during/after a hazard, to enable people to migrate affordably if necessary. Fuel or food subsidies are also useful for these communities.

2. Social care: Social care services could include a comprehensive assessment of the needs of vulnerable groups during climate disasters, and customize support based on these needs. This support can be in the form of assisting affected families to find support groups, establishing contacts with relatives, providing counseling services to deal with post-traumatic stress, etc. Public health centers and the Accredited Social Health Activists (ASHA workers) are functional in most Indian states and can be leveraged in the Sundarbans coastal area to deliver social care services.

3. Social insurance: Sundarbans coastal area is dominated by a rural population, with high levels of informality. As a result, most people are employed in non-contractual jobs and are unable to participate in formal social insurance schemes like unemployment or disability insurance and have no pensions for old age. In the long term, the government can explore the feasibility of increasing formal employment

relations by simplified contractual setups among individual farmers, fishermen, small traders, etc. with state-run cooperatives. In the short to medium term, it may be more feasible to develop a crop or livestock insurance scheme for coastal households, with subsidized premium payments from the government side.

4. Active labor market policies: The working age group in Sundarbans coastal area can be supported by active labor market policies that support in accessing gainful employment. This could be in the form of training that helps learn relevant skills for the job market or improve upon current farming, fishing, or aquaculture techniques that can withstand increasing salinity challenges. Job-matching services can also be provided so that workers are informed on skill requirements in neighboring states and can effectively migrate.

While there are several policy options available for using social protection to protect the coastal communities, the choices are often dependent on the political will, fiscal envelope, technical capacity, and the willingness of people to adopt such interventions. South Asian countries including India are using different measures for managing climate risks, and social protection can be one of the instruments in the mix that can help people in Sundarbans cope and adapt to the climate change impacts. Being chronically poor and living in a fragile ecosystem, coastal populations of Sundarbans are predominantly vulnerable, and according to Islam and Winkel (2017), they are also at risk of being disproportionately affected in three ways. Figure 6.1 presents the conceptual framework of how social protection can prevent this.

There are three ways in which climate change impacts get manifested among the disadvantaged coastal populations in Sundarbans.

First, it increases the exposure of people to the adverse impacts of climate change. People living in these areas are poor and vulnerable, exposed to risks from floods, cyclones, and lately to alarming levels of coastal erosion, resulting in loss of land and property. Labor market policies, as a form of social protection intervention, can support people to move out of the islands and integrate into labor markets in neighboring cities and states. This change in location reduces exposure of the people who move out of the high hazard prone coastal islands.

Second, the susceptibility to damages caused by climate change increases among disadvantaged groups as they have little capacity or agency to take preventive or protective measures. Social protection interventions like public works schemes can commission the construction of climate smart infrastructure like concrete embankments and streets, raised tube wells, freshwater collection tanks, and even flood houses that can provide shelter during floods. Infrastructure that can withstand cyclonic storms and recurrent flooding can greatly reduce the susceptibility of coastal populations to climate change-related damages.

Third, disadvantaged people have reduced the ability to cope and recover from the damages caused due to climate change. Cash transfers, regular or one-time, can be used as a safety net to protect people from falling into poverty during and after disasters. It can help and act as an income supplement for those livelihoods that get severely affected due to different coastal hazards. Cash transfers can also enable productive investments that enable coastal populations to improve their adaptive capacity, adopt new coping strategies, and minimize the losses incurred during disasters.

Theoretically, these options are viable, relevant, and can prove to be effective in reducing the costs of disasters in Sundarbans coastal areas. Small-scale pilot studies introduced by the national and state governments by adopting a few of the options presented here can be a way forward in safeguarding coastal populations against climate change impacts.

6.6 Conclusion

Climate change disproportionately affects marginalized and disadvantaged groups, including low-income communities, women, and indigenous people living in coastal areas like the Sundarbans. Sundarbans are highly vulnerable to

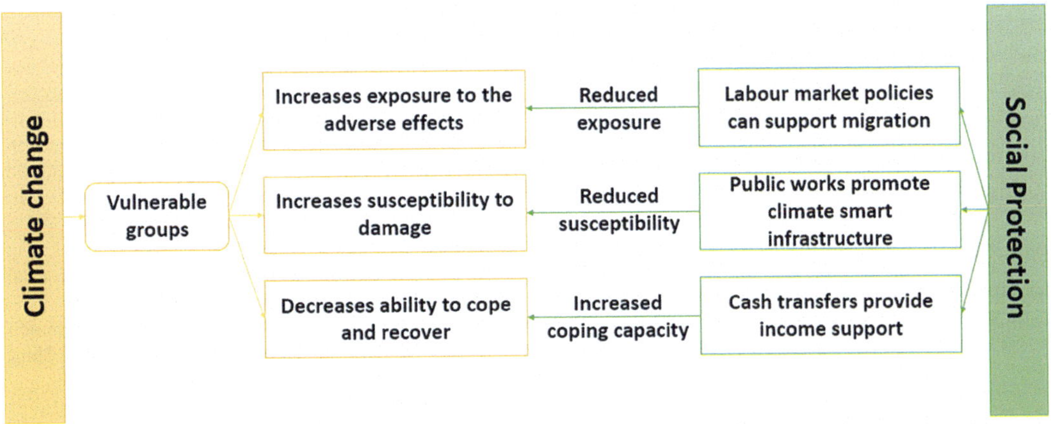

Fig. 6.1 Conceptual framework showing social protection options for climate change. (Source: Author's own)

extreme weather events like cyclones, storm surges, sea level rise, and these can be expected to worsen in intensity and frequency due to climate change. Coastal communities have developed traditional knowledge over generations, which helps them to adapt to the changing environment. They have knowledge about the behavior of tides, monsoon, and cyclones and how to prepare for them. Sundarbans communities are highly dependent on fishing, agriculture, and forestry. To reduce their dependence on a single livelihood, they have diversified their income-generating options by engaging in other activities such as tourism, honey harvesting, and crab farming.

Populations living in the Sundarbans coastal areas adopt short- and long-term coping techniques like raising huts on plinths, using boats for commute, transferring assets to safer places, etc. The Sundarbans mangrove forests are vital to the ecosystem, providing a habitat for many species and helping to protect the coast from storm surges. With support from the state and local NGOs, community-based adaptation initiatives such as planting mangrove forests and growing nurseries have also been initiated as an investment towards the long-term vision of protection against climate change.

While these initiatives improve the coping capacity of coastal communities, building climate resilience will require interventions that address underlying structural challenges, like poverty and socio-economic vulnerability. At the same time, policies will need to contribute to enhancing human capital, which is a reliable pathway out of poverty. Social protection is a critical tool for reducing poverty through its income support objectives, while building resilience of coastal populations against climate change through its capacity-building objectives. As sea level rise, extreme weather events become more frequent, and ocean acidification increases, and coastal communities are facing unprecedented challenges that threaten their livelihoods, health, and wellbeing. Social protection measures can help these communities cope with the impacts of climate change by providing assistance during times of need, building resilience, and supporting long-term adaptation efforts.

For example, cash transfers can help people recover from losses due to a natural disaster or offset the costs of adapting to a changing climate. Food assistance can help to ensure that vulnerable populations have access to adequate nutrition in the face of crop failures or food price spikes. Health care and education can also play a critical role in building resilience to the impacts of climate change by promoting overall well-being and knowledge about how to adapt to changing conditions.

By investing in disaster risk reduction measures through public works and cash transfers, social protection can help communities prepare for and respond to climate-related hazards. This can include measures such as building sea walls, planting mangroves, and strengthening community-based organizations that can respond to emergencies and coordinate relief efforts.

Social protection can support long-term adaptation efforts by providing resources and support to communities as they transition to more sustainable and climate-resilient livelihoods. This may include investing in sustainable agriculture, promoting eco-tourism, or supporting alternative livelihoods that are less vulnerable to the impacts of climate change.

Overall, social protection can be a critical tool for coastal areas facing climate change, like the Sundarbans. By increasing livelihood options, promoting food security and improving access to healthcare, social protection programs can support India's long-term adaptation efforts, and enable coastal communities to cope with the impacts of climate change, thereby building a more sustainable and resilient future.

Annex 1: Schematic Representing Findings from PRA Methods Applied for Data Collection

References

Afjal Hossain M, Imran Reza M, Rahman S, Kayes I (2012) Climate change and its impacts on the livelihoods of the vulnerable people in the southwestern coastal zone in Bangladesh. In: Climate change and the sustainable use of water resources, pp 237–259. https://doi.org/10.1007/978-3-642-22266-5_15

Ahammad R, Nandy P, Husnain P (2013) Unlocking ecosystem based adaptation opportunities in coastal Bangladesh. J Coast Conserv 17:833–840

Ahmad O (2020, December 9) Reporting common issues in transboundary Sundarbans. The Third Pole. https://www.thethirdpole.net/en/climate/reporting-common-issues-in-transboundary-sundarbans/#:~:text=Multiple%20dams%2C%20weirs%20and%20barrages%20divert%20much%20of,adds%20to%20the%20effect%20of%20sea%20level%20rise

Ahuja A (2022, September 26) Climate change in Sundarbans: loss of Mangroves, adaptation and mitigation | News. NDTV-Dettol Banega Swasth Swachh India. https://swachhindia.ndtv.com/climate-change-in-sundarbans-loss-of-mangroves-adaptation-and-mitigation-69004/

Augustin J (2019) On the front line of climate change in India's Sundarbans. Mongabay Environmental News. https://news.mongabay.com/2019/10/sundarbans-climate-change-tigers-india/

Barbier EB, Koch EW, Silliman BR, Hacker SD, Wolanski E, Primavera J, Granek EF, Polasky S, Aswani S, Cramer LA, Stoms DM, Kennedy CJ, Bael D, Kappel CV, Perillo GME, Reed DJ (2008) Coastal ecosystem-based management with nonlinear ecological functions and values. Science 319(5861):321–323. https://doi.org/10.1126/science.1150349

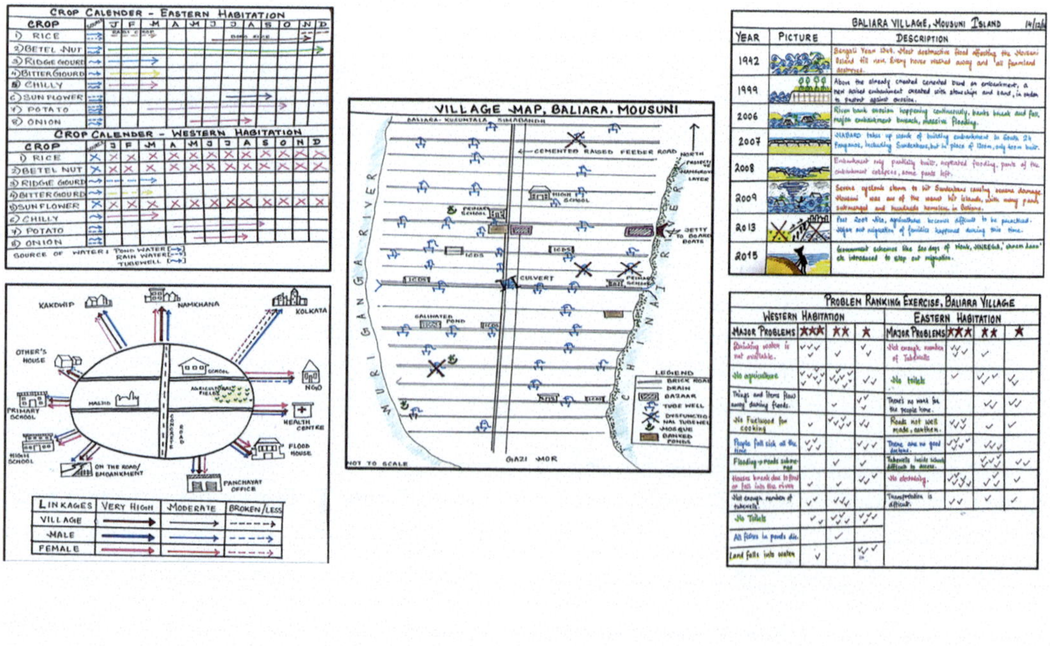

Bhattacharjee T, Chatterjee S, Verya B, Dutta D (2011) The sinking "beauty" – Sundarbans. Geospatial World Forum. Conference paper. Hyderabad. https://geospatialworldforum.org/2011/proceeding/pdf/tapaleena.pdf

Costella C, Mccord A, Aalst M, Holmes R, Ammoun J, Barca V (2021) Social protection and climate change: scaling up ambition. https://doi.org/10.13140/RG.2.2.15187.91680

Das K, Das K (2022) Aila to Yaas – legacies of destruction: case studies from selected sites of Indian Sundarban. Indian J Spat Sci Spring Issue 13(1):2249–3921. https://www.academia.edu/85185127/Aila_to_Yaas_Legacies_of_Destruction_Case_Studies_from_selected_Sites_of_Indian_Sundarban

Dasgupta S, Wheeler D, Sobhan MI, Bandyopadhyay S, Paul T (2020) Coping with climate change in the Sundarbans: lessons from multidisciplinary studies. World Bank Publications

Food and Agriculture Organization (FAO) (2017) FAO social protection framework: promoting rural development for all (ISBN 978-92-5-109703-8). Retrieved from http://www.fao.org/3/a-i7016e.pdf

Goldenberg S (2014) Eight ways climate change is making the world more dangerous. The Guardian

Government of West Bengal (GoWB) (2009) District human development report: South 24 Parganas. Development and Planning Department

Hazra S, Ghosh T, DasGupta R, Sen G (2002) Sea level and associated changes in the Sundarbans. Sci Cult 68(9/12):309–321

http://wbprd.gov.in/writereaddata/Sucess_Stories_mgnrega/South%2024%20Parganas.pdf

https://www.im4change.org/docs/963TITEL_SOUTH%2024.pdf

ILO (2017) World social protection report 2017–19: Universal social protection to achieve the Sustainable Development Goals. International Labour Organization, Geneva. https://www.ilo.org/global/publications/books/WCMS_604882/lang%2D%2Den/index.htm

IPCC (2019) Summary for policymakers. In: Pörtner H-O, Roberts DC, Masson-Delmotte V, Zhai P, Tignor M, Poloczanska E, Mintenbeck K, Alegría A, Nicolai M, Okem A, Petzold J, Rama B, Weyer NM (eds) IPCC special report on the ocean and cryosphere in a changing climate. In press

IPCC (2022) Climate change 2022: impacts, adaptation, and vulnerability. In: Pörtner H-O, Roberts DC, Tignor M, Poloczanska ES, Mintenbeck K, Alegría A, Craig M, Langsdorf S, Löschke S, Möller V, Okem A, Rama B (eds) Contribution of Working Group II to the sixth assessment report of the Intergovernmental Panel on Climate Change. Cambridge University Press. In Press

Islam N, Winkel J (2017) Climate change and social inequality. United Nations, Department of Economic & Social Affairs (DESA) Working Paper No. 152 ST/ESA/2017/DWP/152;wp152_2017.pdf (un.org)

Kirezci E, Young IR, Ranasinghe R, Muis S, Nicholls RJ, Lincke D, Hinkel J (2020) Projections of global-scale extreme sea levels and resulting episodic coastal flooding over the 21st century. Sci Rep 10(1):11629

Loewe M, Schüring E (2021) Introduction to the handbook on social protection systems. In: Schüring E, Loewe M (eds) Handbook on social protection systems. Edward Elgar Publishing, pp 1–35

Mahadevia Ghimire K, Vikas M (2012) Climate change–impact on the Sundarbans, a case study. Int Sci J Environ Sci 2(1):7–15

Masum SJH (2012) Climate change impact on the poor people of the Sundarbans community in Bangladesh. Coastal Development Partnership (CDP). Retrieved 20 Mar 2019. https://www.researchgate.net/publication/308468453_Climate_Change_impact_on_the_Poor_People_of_the_Sundarbans_Community_in_Bangladesh

Muller N (2020, May 1) In the Indian Sundarbans, the sea is coming. The Diplomat. https://thediplomat.com/2020/05/in-the-indian-sundarbans-the-sea-is-coming/

Nandy P, Ahammad R (2012) Navigating mangrove resilience through the ecosystem based adaptation approach: lessons from Bangladesh. Sharing Lessons on Mangrove Restoration

NASA (n.d.) Living ocean | Science Mission Directorate. https://science.nasa.gov/earth-science/oceanography/living-ocean/

Nature Communications (2020) Sea change in coastal science. Nat Commun 11:4601. https://doi.org/10.1038/s41467-020-18333-8

Nicholls RJ, Hanson S, Herweijer C, Patmore N, Hallegatte S, Corfee-Morlot J, Chateau J, Muir-Wood R (2007). Ranking of the world's cities most exposed to coastal flooding today and in the future'. No. 1. OECD Environment Working Paper

Olsson L, Opondo M, Tschakert P, Agrawal A, Eriksen SE (2014) Livelihoods and poverty. In: Field CB, Barros VR, Dokken DJ, Mach KJ, Mastrandrea MD, Bilir TE, Chatterjee M, Ebi KL, Estrada YO, Genova RC, Girma B, Kissel ES, Levy AN, MacCracken S, Mastrandrea PR, White LL (eds) Climate change 2014: impacts, adaptation, and vulnerability. Part A: Global and sectoral aspects. Contribution of working group II to the fifth assessment report of the intergovernmental panel on climate change. Cambridge University Press, Cambridge, United Kingdom and New York, NY, USA, pp 793–832

Oxford Policy Management (OPM) (2017) Shock-responsive social protection systems research: literature review (2nd edn.). Oxford Policy Management, Oxford, UK

Panda A (2020, December 17). Climate change, displacement, and managed retreat in coastal India. migrationpolicy.org. https://www.migrationpolicy.org/article/climate-change-displacement-managed-retreat-india

Pani DR, Sarangi SK, Subudhi HN, Misra RC, Bhandari DC (2013) Exploration, evaluation and conservation of salt tolerant rice genetic resources from Sundarbans region of West Bengal (PDF). J Indian Soc Coast Agric Res 30(1):45–53

Paul AK, Ray R, Kamila A, Jana S (2017) Mangrove degradation in the Sundarbans. In: Coastal wetlands: alteration and remediation. Springer, Cham, pp 357–392

Raha A, Das S, Banerjee K, Mitra A (2012) Climate change impacts on Indian Sunderbans: a time series analysis (1924–2008). Biodivers Conserv 21:1289–1307

Rosencranz A, Nath R, Chowdhury A (2020, August 27) Protecting the Indian Sundarbans in a global health crisis – policy forum. Policy Forum

Roy J, Datta S, Kapuria P, Guha I, Banerji R, Miah MG et al (2021) Coastal ecosystems and changing economic activities: challenges for sustainability transition. In: Disaster resilience and sustainability. Elsevier, pp 397–424

Sinha D (2022, September 12) Sundarbans' climate refugees face an uncertain future. dw.com. https://www.dw.com/en/forgotten-people-sundarbans-climate-refugees-forced-to-move-again/a-61162969

Tenhunen S, Uddin MJ, Roy D (2023) After cyclone Aila: politics of climate change in Sundarbans. Contemp South Asia 31(2):1–14

The Weather Channel (TWC), India (2020, February 24) Climate change poses multiple threats to delicate Sundarban Ecosystem: CMS COP13. The Weather Channel. https://weather.com/en-IN/india/news/news/2020-02-24-sundarbans-ecosystem-facing-threat-due-to-climate-change

Torresan S, Critto A, Dalla Valle M, Harvey N, Marcomini A (2008) Assessing coastal vulnerability to climate change: comparing segmentation at global and regional scales. Sustain Sci 3:45–65

Uddin MS, Shah MAR, Khanom S, Nesha MK (2013) Climate change impacts on the Sundarbans mangrove ecosystem services and dependent livelihoods in Bangladesh. Asian J Conserv Biol 2(2):152–156

United Nations (2017) Factsheet: people and oceans. In The ocean conference. https://www.un.org/sustainabledevelopment/wp-content/uploads/2017/05/Ocean-fact-sheet-package.pdf

United Nations Development Programme (UNDP), Malik K (2014) Human development report 2014: sustaining human progress-reducing vulnerabilities and building resilience (PDF). UN

West Bengal Panchayat & Rural Development Department (WBPRD) (2021) Success stories related to Raising of Nursery (Mangrove) 2020–2021

Land Degradation and its Relation to Climate Change and Sustainability

7

Anu David Raj, Suresh Kumar, Justin George Kalambukattu, and Uday Chatterjee

Abstract

Climate change is increasingly threatening the globe, and, currently, we are facing various weather extremities frequently. It may affect ecosystem health, extinction of species, and risk to people's lives and livelihoods. Although there are numerous factors that contribute to land degradation, distinguishing between them can be challenging. The mutual impacts of these processes on one another, such as how land degradation affects climate change and vice versa, add an additional complex dimension to the system. The degradation of land is caused by a wide variety of interactions involving physical, chemical, biological, and anthropogenic processes. While elements of the land systems, such as soils, water, and biota, are frequently subjected to pressures that promote land degradation, further supplementary elements are simultaneously influenced by complex interactions. In the context of this complexity, conducting a practical assessment of the primary land degradation processes can help to shed light on and categorize the different pathways through which climate change may exacerbate the land degradation processes. Soil organic carbon serves as a common link between land degradation and climate change, presenting an opportunity for its mitigation. To achieve sustainability under these circumstances, this chapter focuses on the scientific understanding of how land degradation impacts the climate change and vice versa, as well as the relationship between soils and climate change.

Keywords

Global warming · Land degradation · Soil erosion · Carbon loss · Food security · Sustainability

A. David Raj (✉)
Agriculture and Soils Department, Indian Institute of Remote Sensing, Indian Space Research Organization (ISRO), Dehradun, Uttarakhand, India

Forest Research Institute, Dehradun, Uttarakhand, India
e-mail: anudraj@iirs.gov.in

S. Kumar
Agriculture, Forestry and Ecology Group, Indian Institute of Remote Sensing, Indian Space Research Organization (ISRO), Dehradun, Uttarakhand, India

J. G. Kalambukattu
Agriculture and Soils Department, Indian Institute of Remote Sensing, Indian Space Research Organization (ISRO), Dehradun, Uttarakhand, India

U. Chatterjee
Department of Geography, Bhatter College, Dantan, (Affiliated to Vidyasagar University), Paschim Medinipore, West Bengal, India

U. Chatterjee et al. (eds.), *Climate Crisis: Adaptive Approaches and Sustainability*, Sustainable Development Goals Series, https://doi.org/10.1007/978-3-031-44397-8_7

7.1 Introduction to Climate Change and Land Degradation

The term "climate crisis" describes the consequences of climate change and global warming. This expression is used to emphasize the importance of implementing significant climate change mitigation measures and to highlight the danger that global warming poses to the entire world. Climate change is the extended-period alteration of temperature and weather patterns of a region. These variations might be brought on by natural processes, like variations in the solar cycle, change in Earth's rotation axis, etc. Even so, since the eighteenth century, climate change has been predominantly caused by anthropogenic activity, particularly by the burning of fossil fuels like coal, oil, and gas. Carbon dioxide (CO_2), methane (CH_4), nitrous oxide (NO_2), and water vapor are some examples of greenhouse gases that are contributing to climate change. The major emitters are energy industry, transportation, buildings, agriculture, and land use change. According to the IPCC, recent changes are unparalleled across centuries to thousands of years in terms of their pace, intensity, and duration (IPCC 2021). The past few decades indicate that significant global climate changes were attributed to enhanced anthropogenic sources that altered the elemental concentration of the earth's atmosphere (IPCC 2007). There is no doubt that human activity, for the most part, has contributed to the warming of our climate. The assessment of non-climatic global trends, such as biodiversity loss, unsustainable management of natural resources, land degradation, urbanization, human population, and socio-economic disparities, are contrasted with the impacts and risks of climate change as well as adaptation strategies.

The biosphere encompasses the terrestrial component, which includes the ecological processes, landscapes, habitats, and structures that support the plants, near-surface air, water, and other lifeforms within that system. One of the unavoidable effects of the climate crisis is land degradation, which is mostly the result of the combined effects of soil, water, and forest deterioration. The likely effects of the climate crisis on land include shifts in soil physico-chemical and biological properties, losses in soil nutrients due to higher runoff rates, and soil erosion. The probability of a climate crisis, however, is further exacerbated by the growing use of forest wood, fossil fuel, biomass, and rising greenhouse gas emissions. Therefore, land degradation mechanisms that are visibly responsive to climate change pressures that encompass a wide range of processes, including erosion, declines in soil organic carbon (SOC), salinization, sodification, waterlogging in dry ecosystems, etc. These mechanisms involve a diverse array of biologically mediated processes, such as forest invasion, species migration, pest outbreaks, and the combined effects of biological soil crusting and increased forest fires (Olsson et al. 2019).

Forests are a critical resource for regulating the planet's climate, sequestering a significant amount of CO_2 from the atmosphere, and maintaining ecological stability. Climate change is projected to affect the frequency of forest disturbances, as well as the availability of commodities and services, based on their specific locations and ecosystems. Climate change has had a severe impact on flora and fauna, particularly at regional and local scales, and has reduced the capacity of forest ecosystems to regulate themselves. Furthermore, according to Sajjad et al. (2022), climate change has a significant impact on agricultural productivity, making it a complex challenge to provide for the world's population in the face of such changes (World Population Review 2023). It is anticipated that global grain production of maize and wheat will decline by 3.8% and 5.5%, respectively, due to the definite negative impact of climate change on agricultural production (Lobell et al. 2011).

One significant adverse factor that affects the health and quality of watersheds is the accelerated soil erosion (Hazbavi et al. 2019). Given the spatial variability of land degradation challenges, geophysical approaches related to soil systems are crucial for enhancing system resilience and developing appropriate restoration and rehabilitation strategies (Rodrigo-Comino et al. 2018). Although much less attention is given to the issues such as soil erosion, salinization, alkalinization, acidification, and soil compaction. Contemporary scientific research primarily focuses on the risks of climate change, often neglecting the growing exposure to natural hazards. Therefore, research efforts on soil erosion, soil quality, and food security should prioritize addressing land degradation and climate change. Quantitative soil conservation research, focusing on areas with high rates of soil erosion, provides solid scientific support for ecosystem management, and reducing local soil erosion. Furthermore, there is a significant likelihood that land degradation and climate change will result in global socioeconomic and security instability. Achieving land degradation neutrality requires accurate monitoring of soil erosion processes and measuring soil and water yields. At present, the most practical and suitable mitigation strategy that can effectively address the climate problem is sustainable land management.

7.2 Land Degradation

Land degradation is the short-term or long-term reduction in the land's or soil's capacity for production. In comparison to soil degradation, land degradation has a broader scope as it encompasses all detrimental changes to the ecosystem's ability to deliver goods and services, including those linked to biodiversity, water, environment, and economy. The term "soil degradation" refers to a change in the ecosystem's ability to produce products and services for (Fig. 7.1).

its beneficiaries because of a change in the soil health condition (FAO 1979). Irrelevant policies and exercises, such as the transformation of productive arable lands into other uses including urbanization, inefficient management of government subsidies, land segmentation, over herding, illegal logging, and ineffective irrigation and cultivation techniques, are some of the reasons that have caused agricultural land to degrade over time. This degradation has led to a loss of productivity,

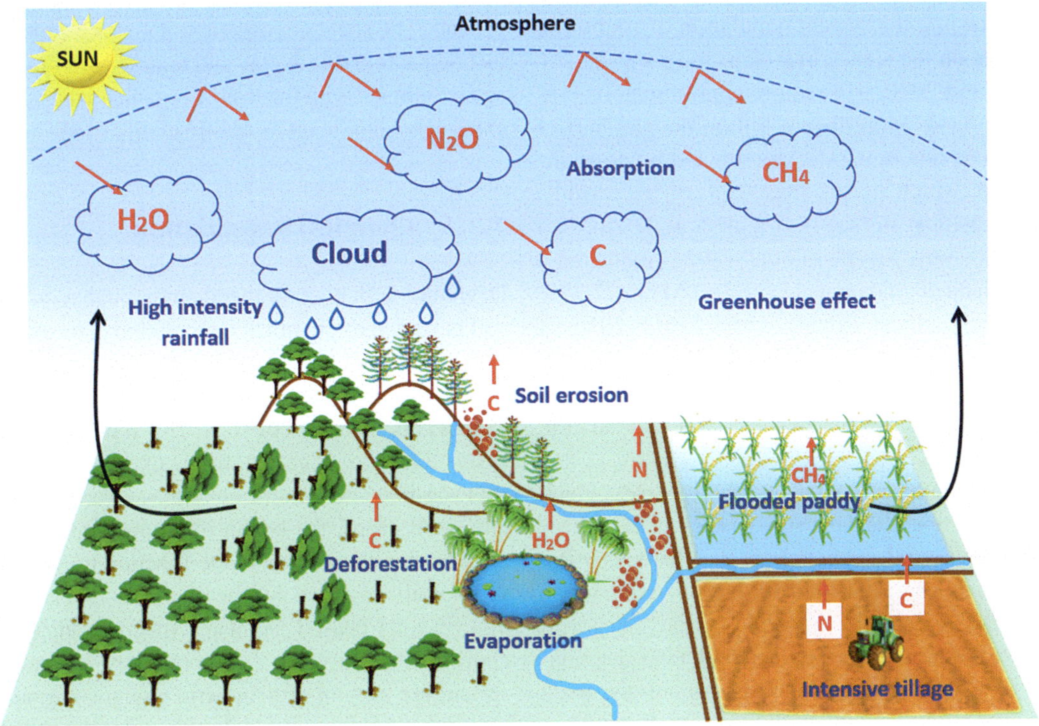

Fig. 7.1 Land degradation, climate change, and their interlinkages

despite the significance of ensuring high agricultural productivity (Ziadat et al. 2022). Hence, it is crucial to prioritize the management of natural resources in order to preserve the global environmental benefits provided by ecosystem services and to safeguard agricultural productivity, ultimately ensuring food security and livelihoods.

Due to soil erosion and desertification, the production of numerous lands has declined by 30 to 40%. Africa could experience yield reductions of up to 40% because of past soil erosion, with an average loss of 8.2 percent. In South Asia, a total annual production loss of 36 million tonnes is attributed to water erosion and wind erosion, valued at $5400 million and $1800 million respectively. Globally, the loss of 75 billion tonnes of soil annually costs approximately $400 billion, which translates to about US$70 per person each year (Eswaran et al. 2019). The FAO estimates that 33 percent or more of the world's land is degraded (FAO, n.d., www.fao.org). Roughly 3% of the world's land area falls into the category of prime or Class I land, and this does not include any tropical areas. The remaining 8% of land is classified as Classes II and III. Feeding the current population of six billion people on the planet, as well as the projected 7.6 billion in 2020, will require approximately 11% of the land. More than a billion people, half of whom reside in Africa, are affected by desertification, which occurs over 33% of the world's geographical surface (Eswaran et al. 2019). In a recent study, Sreenivas et al. (2021) indicated a total land degradation area of 91.2 M ha (27.77% of geographical extent) during 2015–2016 of India. Global statistics demonstrate the severity of land degradation, highlighting the need for assessing the intricacies of land degradation processes.

There are various kinds of processes which are affecting the land/soil based on their geographical and climatological characteristics. Physical, chemical, and biological degradation are the three most frequently used categories used to classify the processes of land degradation. Most land degradation issues are discussed in these three categories (Keesstra et al. 2018).

7.2.1 Physical Degradation

Physical degradation deals with deteriorating the physical quality of land or soil. Soil structure, infiltration, and water-holding capacity all have a significant impact on erosion and compaction. These characteristics could, therefore, be used to evaluate the physical degradation of a soil system. The result of a complex interaction of processes, known as erosion, causes the loss of soil fertility, organic matter, and topsoil. This loss of topsoil is concerning as it is the layer of soil with the highest capacity to retain both water and nutrients (Conway et al. 2018). However, there are other factors that are important besides these indigenous effects. The off-site consequences of erosion can also cause significant harm. Sediment and related materials have the potential to pollute and suffocate wetlands and marine habitats that are located downhill from eroded areas. In agricultural soils worked by large machinery, the type of compaction plays a significant role in (Gürsoy 2021) both shallow and deeper soil layers. Shallow compaction can occasionally be brought on by people or animals trampling the ground. The societal costs of soil sinking caused by compaction due to clay and nutrient soil drainage, combined with the loss of organic matter, are extremely significant (Keesstra et al. 2018).

7.2.2 Chemical Degradation

Degradation that alters the chemical behavior of soil is known as chemical degradation. The overuse of farmyard manure and synthetic fertilizers, such as nitrogen and phosphorus, can result in the chemical degradation of soils. This includes processes like eutrophication, where excessive nutrients wash out and runoff into surface waters. The quality of crops, the health of animals and people, and the biotic and abiotic functioning of the soil are all impacted by the chemical deterioration of soils caused by heavy metals, radionuclides, and organic compounds. Irrigated agriculture in semi-arid and arid climatic regions faces the development of salt-affected soils where salts available in sub-surface soil move to the surface through evaporation and accumulated on the surface known as the salinization process. Salinization threatens the extended viability of irrigated farming, which is crucial for food security. Salinization process may increase due to a rise in temperature and reduction in rainfall in climate change scenarios in some areas in arid and semi-arid climatic regions. The hydromorphic soils will become more saline due to the effects of global warming and increased aridity on climate conditions (Okur and Örçen 2020). Farming activities enhance this process, which can influence the surface and subsurface soil. Nutrient availability in soils is strongly influenced by soil pH, and excessively acidic soils can have a significant negative impact on crop and pasture yields. The accumulation of soluble organic and inorganic acids in the soil often outpaces their neutralization, leading to a process of ionization. This ionization results in an increased concentration of free H^+ ions in the soil solution (Rengel 2011).

7.2.3 Biological Degradation

Soil organic matter (SOM) is known to enhance virtually every aspect of soil functioning, including agricultural productivity, soil structure, water retention, infiltration capability, nutrient management, and the prerequisites for a thriving soil ecosystem and a globally significant carbon pool. SOM is also beneficial in supporting soil microorganisms, which play a crucial role in various soil quality parameters. The conversion of natural grasslands and forests into agricultural land has had the most significant human-made impact on SOM (Eswaran et al. 1993; Jackson et al. 2017; Keesstra et al. 2018). Between 2015 and 2030, intensive farming is projected to result in the loss of 32 Gigatons of carbon from the landscape (UNCCD, n.d.). In this context, biological degradation of soil is defined as the reduction or eradication of one or more important populations of soil microorganisms, often accompanied by corresponding changes in biogeochemical functioning within the associated ecosystem.

7.3 Interlinkages Between Land Degradation, Carbon Loss, Climate Change and Sustainability

The soil, water, and biotic components of the land, as well as the interactions between them, can all be impacted by land degradation processes. The mechanisms that influence the soil have drawn greater attention than other types of land degradation. Water and wind erosion, which have been present since the beginning of agriculture and now dominate, are the most widespread and extensively studied processes of land degradation that impacts soils. The effects of erosion processes on the environment also extend to areas where transportation and sediment deposition take place.

Soil erosion plays an important role in distributing soil carbon, i.e., SOM in the landscape. Soil carbon is exposed to the surface and mineralized to CO_2 to release in the atmosphere. The loss of organic matter in soils is among the most important chemical degradation processes in the circumstance of climate change. SOM pools have been depleted in agricultural soils because of the rise in respiration rates caused by tillage and the decline in below-ground plant biomass responses. These reductions have been expanded by the immediate consequences of warming, which affect both arable and non-arable lands (Lamberty et al. 2018). The question of whether the simultaneous stimulation of decomposition and productivity in more humid and carbon-rich ecosystems could prevent effects on soil carbon is still up for debate (Crowther et al. 2016; van Gestel et al. 2018). In forests, harvesting can also lead to reductions in organic matter, particularly when it is extensive, such as when using forest residues for energy generation (Achat et al. 2015). Soil is considered as the largest carbon sink followed by biotic and atmosphere pools (Fig. 7.2). Carbon in soils comprises of root materials, decomposing plants, and SOC, bacteria, fungi, and invertebrates. Through proper land treatment, it can enhance soil carbon reserves and gain several advantages in addition to reducing greenhouse gas emissions. Hence, it is explicit that SOM

Fig. 7.2 Carbon stocks in various biosphere components. (Source: Modified from IUCN; https://www.iucn.org/resources/issues-brief/land-degradation-and-climate-change)

serves as both a crucial link with the climatic system and a "hub" for degradative processes.

The latest estimates place the annual cost of the loss of ecosystem services caused by desertification and land degradation around US$6.3 trillion and US$10.6 trillion (ELD 2015). Due to the challenge of accurately assessing the ripple effects and externalities of land degradation, these substantial costs have not received sufficient attention. Countries frequently consider how it will affect food production while ignoring ecosystem services like water supply, regulation, and reduced carbon sequestration. Sustainable land management entails increasing carbon reserves while reducing carbon emissions. SOC plays a crucial role in determining the fertility and water-holding capacity of soils. It significantly influences the soil's ability to support biodiversity and ensure food availability. Regions with higher soil productivity and carbon stores tend to exhibit greater resilience in both socio-economy and ecosystems.

To enhance carbon in the soil, a wide range of agro-ecological techniques can be applied, such as agroforestry, fallowing; that is, soil left fallow for at least a year, and regulated herd mobility for sustainable pasture management. With the correct assistance, these are well-known techniques that are even indigenous to many regions. Recent research indicates that managing soil carbon is the most affordable approach to combat climate change. For example, rangelands contain over a third of all terrestrial above- and below-ground carbon storage. Through improved rangeland management, they might be able to store an additional 1300–2000 million Mt. of CO_2 by 2030 (Tennigkeit and Wilkes 2008). Small increases in SOC can have a significant impact on the global carbon cycle and CO_2 concentration. Even a 1% increase in carbon stocks in the top one meter of soil would surpass the annual anthropogenic CO_2 emissions from burning fossil fuels (IUCN 2015). Overturning land degradation and increasing SOC levels present one of the most certain and cost-effective strategies for achieving multiple successes.

Countries might have a look at several strategies to raise SOC stocks. Using policy and financial tools, it is possible to speed up sustainable land management to enhance SOC in a manner that concurrently combats desertification, avoids the extinction of biodiversity, and helps to mitigate and adapt to climate change. Other measures usually involve treating land-based strategies for mitigating climate change as integral parts of both international and national strategies, spreading knowledge of and insight into the numerous advantages of sustainable land management, and making sure that SOC is completely described throughout all spheres as an indicator of those numerous advantages. Countries can make progress and improve by monitoring and reporting trends in land cover and carbon stock characteristics above and below ground, in adherence to the three key land-based progress indicators set forth by the United Nations Convention to Combat Desertification (UNCCD) (IUCN 2015).

7.4 Soil Erosion—A Socio-Economic and Environmental Concern Worldwide

The natural environment and the human social system are two interconnected and complicated that are affected by land degradation (Barrow 1994). One of the biggest difficulties for land managers is land degradation, which includes soil erosion, as it affects about 60% of the world's land surface (Pimentel 2006). Particularly one of the most problematic and raising serious issues globally is soil erosion (Guerra et al. 2017). When one recognizes that only about 11% of the world's land surface can be deemed as best form or Class I land and that this should indeed feed the 6.3 billion people alive today and the 8.2 billion anticipated in the year 2020, one can see how important the issue of land degradation is for global food security and the quality of the environment (Reich et al. 2001). Land degradation is mostly caused by water and wind erosion, which together account for 84% of the world's degraded land, rendering severe erosion one of the most important environmental issues on the planet (Blanco and Lal 2008; Rhodes 2017).

Resource management initiatives either succeed or fail based on the existent of the two systems. In addition to socioeconomic factors like land occupancy, marketing, institutional support, profit, and public health, political factors like political stability and incentive programs also contribute to land degradation (Sivakumar and Stefanski 2007). The interaction between socioeconomic contexts and biophysical elements makes it difficult to determine how erosion affects soil quality and vice versa. The minimal soil depth required to secure productivity and environmental regulatory capability determines how erosion affects soil quality (Lal et al. 2018). Through their effects on soil quality, accelerating soil erosion and other degrading processes affect agronomic productivity and the ecosystem. All three dimensions of soil quality, physical, chemical, and biological are negatively impacted by soil erosion, but soil physical quality is particularly affected because of deteriorating soil structure and an unbalanced soil–water regime. The solution to the issues of food insecurity and the deterioration of the environment's quality will be greatly aided by an understanding of the complex and dynamic connection between soil erosion and soil quality. A major concern to the world's food, energy, and environmental supplies and habitats is soil degradation, specifically regarding water quality and the greenhouse effect (Lal 2018). A rather more ideal tactic than focusing on cross-slope soil conservation methods is the application of biological and mechanical strategies to enhance soil quality through soil fortification, the integration of organic matter, and the usage of soil organisms (Hellin 2003). Since 99% of all food is generated in the soil, it is the foundation of agricultural production (Pimentel 2006). Soil production directly affects food security. According to Blanco-Canqui et al. (2008), one of the main reasons for the decline in soil productivity and the rise in the danger of global food insecurity is intensified soil erosion. This suggests the necessity of checking soil erosion for maintaining a sustainable food security worldwide. Otherwise, it will grow as a whole sector concern which may bring the globe to severe chaos in terms of food security.

7.4.1 Climate Change Impact on Soil Degradation

Based on the currently available knowledge, soil degradation processes have significantly increased in the past few decades, and there are indications that these processes will continue to increase, if necessary, mitigation measures are not taken. Human activities promote or exacerbate soil deterioration processes. Discrete, recurrent extreme weather events and climate change also contribute to adverse effects on soil (Montanarella 2007). Climate change is specifically recognized as one of the primary causes of soil degradation in the United Nations Convention to Combat Desertification (UNCCD) definition. Understanding a region's climate resources and the likelihood of climate-related or induced natural disasters is essential for effectively assessing sustainable land management techniques (Sivakumar and Stefanski 2007). Temperature and precipitation play major roles in soil formation, development, and the distribution of terrestrial plants. The relationships between soil degradation, precipitation, and climate change are closely intertwined. Climate change can exacerbate land degradation by altering the geographical and temporal patterns of temperature, rainfall, solar radiation, and winds. The increased occurrence of droughts, severe floods, and related environmental issues, such as recurrent dust storms observed in China in recent years, are largely attributed to the linkages between climate change and soil degradation (Qian and Zhu 2001). These issues continue to impact biological productivity and environmental sustainability.

7.4.2 Adaptation and Mitigation Measures to Address Climate Change

The adaptation and mitigation measures are the only available way to fight against the adverse impact of climate change. These strategies encourage people to take actions that protect soils from deterioration or move toward a more intensive use, perhaps raising the livelihood of the people (Salinas and Mendieta 2013). Combating land degradation needs to integrate social and economic policies related to poverty, population growth, rural–urban migration, and uncontrolled exploitation of natural resources. Policy always does not result in direct mitigation of land degradation but they favor both explicit and implicit effects in the reduction of land degradation and desertification. Agronomic, economic, and social policies were listed as three types of policies for reducing land degradation by Halbac-Cotoara-Zamfir et al. (2020).

The world's limited natural resources are under extreme stress because of the planet's rising population and changing climate. The foundational natural resources for the agricultural production system are soil and water. The main causes of the degradation of natural resources are anthropogenic and unfavorable natural activity. Soil erosion is one of the major causes of the deterioration of soil and depletion of water resources among the different processes of degradation. In India, water erosion has deteriorated to around 68.4% of the total land area. Mechanical/engineering/structural and agronomical/biological approaches are the two categories of soil and water conservation measures. Terracing, bunding, trenching, check dams, gabion structures, stone boulders, crib walls, etc. are examples of mechanical measures. Biological measures are vegetative measures that include forestry, agroforestry, horticulture, and agricultural/agronomic operations (Sarvade et al. 2019). Kumawat et al. (2020) provided a detailed description of various soil and water conservation measures.

7.5 Climate Change as a Factor of Land Degradation

Temperature and precipitation play crucial roles in the formation and evolution of soil, as well as in determining the potential distribution of terrestrial plants. Rainfall is the primary climatic factor in identifying areas at risk of desertification and possible land degradation. While rainfall is essential for plant growth and dispersion, its variability and extremes can contribute to soil erosion and other forms of land degradation (Sivakumar and Stefanski 2007). It is well-documented that human-induced climate change can intensify the hydrological cycle (Pendergrass and Knutti 2018). Changes in rainfall patterns are expected to result in alterations in vegetation cover and composition, which can both contribute to and be a consequence of land degradation. The type, biomass, and diversity of plants in arid regions are strongly influenced by climate. However, climate change often acts as a secondary driver compared to other significant human stressors. For example, vegetation cover plays a critical role in determining soil loss from water (Nearing et al. 2005). The occurrence of different combinations of these variations can lead to fluctuations in the processes that drive increases and decreases in soil erosion.

The aridity will increase, temperatures will change due to global warming, and hydromorphic soils will become more salinized. Arid regions are particularly vulnerable to soil salinization and desertification. Salinization refers to the accumulation of salts soluble in water in the soil. Salinity affects the metabolism of soil organisms in the early stages and reduces soil productivity. However, as it progresses, it eliminates all plant life and other species that depend on the soil, transforming once fertile and productive land into arid and decertified regions. With rising temperatures, water and soil salinity are projected to increase, leading to a significant increase in the need for irrigation (Okur and Örçen 2020). Increased alkalinity in harvested products could potentially result in higher soil acidity in managed ecosystems with increased biomass output due to higher temperatures and/or increased rainfall. Increased rainfall-induced leaching of basic cations will transfer alkalinity from the soil exchange complex into surface waters and groundwater, leading to acidified soils (Rengel 2011).

In conclusion, climate change raises the risk of land degradation, impacting both its likelihood and severity. However, it is challenging to attribute the exact cause of land degradation solely to climate change due to various complicating factors. Nevertheless, considering that climate change accelerates most degradation processes, it is evident that without improved land management, climate change will contribute to worsening land degradation. Erosion reduces land resilience, diminishing its ability to support plants that can absorb CO_2, thereby contributing to global warming. Soils alone have the potential to absorb a significant amount of greenhouse gases each year, equivalent to approximately 5% of total annual human-produced GHG emissions. Effective land management can help preserve soils, enabling them to support the growth of more plants that absorb carbon (WRI, n.d.).

7.6 Land Degradation Impact on Climate Change

The impact of land degradation processes on climate, whether or not linked to changes in land use/land cover, differs from the previous section's emphasis on climate change and its effects on land degradation. The impacts of land degradation on CO_2 and other greenhouse gases (GHGs), surface albedo, and other physical factors that affect the global radiative balance are significant. Due to the size and movement of these reservoirs in the global carbon cycle, land degradation activities that affect soil and terrestrial biota have a significant influence on CO_2 exchange with the atmosphere. Erosion, the most common form of land degradation, leads to the separation of topsoil material, which typically contains high levels of organic carbon stocks, facilitating its mineralization and release of CO_2. Areas experiencing depletion have a high potential for carbon emissions, as both natural and human-induced erosion can result in net carbon loss from relatively stable subsurface pools near accumulation areas. In addition to the initial pulse of CO_2 emissions associated with the onset of agriculture and crop harvesting, agricultural management strategies can either enhance or reduce carbon losses to the atmosphere (Olsson et al. 2019).

Among the physical impacts of land degradation, changes in surface albedo have the most noticeable effect on the net global radiative balance and overall climate warming/cooling. Soil carbon leakage has been degrading the land for thousands of years, but its magnitude has significantly increased in the last two centuries (Sanderman et al. 2017). Intensive agricultural practices and soil erosion associated with agriculture contribute to CO_2 and N_2O emissions. The expansion of agriculture and deforestation for agricultural purposes further contribute to CO_2 emissions. Implementing agroforestry, adopting no-till farming, cultivating perennial crops, increasing soil organic matter, preventing erosion, and promoting deforestation and afforestation can help mitigate the impacts of land degradation on climate change.

While the causes and impacts of both land degradation and climate change are well-documented, their interconnections are not fully understood. Further research is needed to explore how these processes interact in different social–ecological systems worldwide and how societies can adapt to these dual challenges (Reed and Stringer 2016). Agroecological systems are particularly vulnerable to climate change due to land degradation, which also hampers the effectiveness of available adaptation strategies (Webb et al. 2017). Therefore, an integrated approach that addresses both land degradation and climate change is crucial for creating a sustainable and resilient pathway for both human well-being and the environment.

7.7 Sustainability of Natural Resources

Sustainability is the key solution to address both climate change and land degradation. In 1992, the UN proposed "Agenda 21," which largely embodies this concept by emphasizing economic growth, social development, environmental preservation, and promoting fairness to foster human development. Sustainable development, as defined by the IPCC (2001), means meeting the present needs without compromising the ability of future generations to meet their own needs. The factors that affect a nation's ability to adapt to land degradation and climate change, such as land and climate variability and extreme weather events, are also the ones that define its capacity for sustainable development. Ecological vulnerability, along with social and economic vulnerabilities, extends beyond the potential for disasters and the effectiveness of management.

The most valuable natural resources include land, soil, forests, and water. Sustainable conservation of these three ecosystem components can enhance our habitat and environment. Key challenges to sustainability include declining soil fertility and an increase in uncultivable land (Vasisht et al. 2003). While the United Nations Agreement to Prevent Desertification has promoted better information exchange between scientists and policymakers, there is still room for researchers to contribute concrete evidence to international policy (Stringer and Dougill 2013). Policies need to provide a framework that considers the complex cause-and-effect relationship between agricultural practices and climate change. The agricultural sector contributes significant greenhouse gas emissions, and the increasing levels of emissions, rising temperatures, and shifts in precipitation patterns impact agricultural production and the natural habitat in which it occurs (Agovino et al. 2019). Hence, transition to a sustainable eco-friendly farming practices is much required for the future. In developing countries, there are already some connections between climate change policies and the sustainable development agenda, such as energy efficiency, renewable energy, transportation, and sustainable land-use policies. Climate change policies can benefit local ecosystems, although they have received limited attention from policymakers thus far. Conversely, regional and national initiatives addressing traffic, air quality, access to energy services, and energy diversification can also help reduce greenhouse gas emissions (Beg et al. 2002).

These varying social, economic, and environmental characteristics worldwide highlight the need for site-specific sustainable measures for ecosystem components (land/soil, water, and forest) rather than generalized approaches.

Table 7.1 Soil-landscape unit-wise SOC, soil erosion-induced carbon loss, and sustainability index

Soil unit	Topography	Land use/land cover	SOC in soil (%)	Soil loss (t ha^{-1} yr.$^{-1}$)	Carbon loss (t ha^{-1} yr.$^{-1}$)	Soil sustainability index
HS12	Very steep	Scrubland	0.67	42.8	0.7	Low
HS14		Maize	2.44	30.2	0.5	Medium
HS21	Steep to very steep	Forest	2.24	20.1	0.4	High
HS23	Moderately steep	Paddy	1.23	24.0	0.4	Medium
HS24		Maize	0.96	30.3	0.5	Low

However, generalized approaches are also necessary to conceptualize the overall idea of sustainable ecosystem management. Therefore, local self-government, institutional support at the local level, and engagement of local communities are essential for this approach. An induction-based approach, rather than deduction, is suggested to better achieve the construction of sustainable ecosystems.

7.8 A Case Study: Soil Erosion and Soil Quality in Relation to Sustainability of Soils

SOC serves as a key indicator for assessing the status of land degradation and the impact of climate change. Furthermore, soil erosion plays a significant role as the primary physical process in regulating the spatial redistribution of SOC through erosion, transport, and deposition. Therefore, this study focuses on examining SOC, soil erosion, and the associated carbon loss for assessing sustainability.

7.8.1 Study Area

The study was conducted in the north-western region of the Himalayas in India, specifically in the Lesser Himalayas. The average annual rainfall in the study area is 2200 mm, and temperatures typically range from 15.8 to 33.3 °C. The elevation in the study area varies from 835 to 1286 m. The soils in the study area are primarily derived from alluvium parent material. The predominant soil types in the Lesser Himalayas are sandy loam to loam textures. The land use/land cover in the study area is mainly comprised of cropland (rice, maize, and mustard), forest (Sal-*Shorea Robusta*), and scrubland, as shown in Table 7.1. Based on topography, slope, and land use, the watershed is divided into five soil-landscape units.

7.8.2 Materials and Methods

The watershed is divided into five unique soil-landscape units for comparison of various soil-landscape units. The SOC data were obtained from the Indian Institute of Remote Sensing (IIRS) Soil Database (Sooryamol 2020). SWAT model predicted soil loss values were obtained from a previous publication (Sooryamol et al. 2022). The carbon loss was estimated using the carbon enrichment ratio published elsewhere (Kumar et al. 2023 *in press*). SOC is a crucial intrinsic characteristic of soil that plays a significant role in promoting soil health and is widely recognized as a key indicator. The loss of soil and the associated carbon loss are important factors in determining the resilience of soil against external pressures, including climate change impacts such as water erosion, as well as anthropogenic pressures such as grazing and intensive cultivation. SOC, soil loss, and associated carbon loss were considered to assess the sustainability of the landscape and land use/cover units. A ranking scheme based on expert opinions was adopted to identify the sustainable soil-landscape units. A higher rank was assigned to units with low SOC values, while lower ranks were given to units with higher soil loss and associated carbon loss. The resulting parameters are presented in Table 7.1:

$$LSF(Z) = \left(\frac{a-l}{u-l}\right)$$

The linear score of soil is Z, the soil quality value is "a," and the lower and upper values are "l" and "u." Weights (Wi) were assigned based on expert opinions, with the highest weight given to SOC content, followed by soil loss and soil erosion-induced carbon loss, respectively:

$$SSI = \sum_{i=1}^{n} W_i \times S_i$$

Hence, the soil sustainability index can be described as the:

$$Soil\,Sustainability\,Index(SSI) = f\,(Soil\,organic\,carbon\,content, Soil\,loss, Carbon\,loss)$$

The SSI scores were subsequently categorized into three classes: low, medium, and high. Higher scores were associated with higher sustainability, while lower scores indicated lower sustainability.

The highest average soil loss was observed from scrubland (42.8 t ha^{-1} $yr.^{-1}$), followed by maize fields (30.2 t ha^{-1} $yr.^{-1}$) and paddy fields (24.0 t ha^{-1} $yr.^{-1}$). The lowest soil loss was observed from the Sal (*Shorea robusta*) Forest with a value of 20.1 t ha^{-1} $yr.^{-1}$. Maize fields showed the highest SOC content, as farmers applied farmyard manure and cow dung for the crops. Sparsely vegetated scrubland exhibited lower SOC content due to continuous soil erosion, resulting in a soil loss of 42.8 t ha^{-1} $yr.^{-1}$. On the other hand, the moderately dense Sal Forest had the lowest erosion. Consequently, the forest exhibited low SOC loss, while scrubland showed the opposite trend. The paddy fields also exhibited low soil erosion compared to maize and scrubland. The standing water in the paddy fields helps reduce further removal of soil particles from the surface, although overflow from the terraced paddy fields contributes to most of the soil erosion. Based on the values of the aforementioned attributes, a sustainable soil-landscape unit was identified and illustrated in Fig. 7.3. Steep slope maize fields also fell under the low sustainable class. Paddy fields and moderately steep slope maize fields were categorized as medium sustainability. The results indicate that forest land cover is highly sustainable against the adverse impacts of climate change. In cultivated areas, there is a possibility for climate change to increase the rate of soil loss unless conservation strategies or proper land use plans are implemented (David Raj et al. 2022). This study emphasizes the significance of forest cover in reducing soil erosion and carbon loss and highlights the importance of conservation through reforestation and afforestation approaches. Additionally, the cropping system should change to climate-resilient smart agriculture to achieve a sustainable socio-economic and environmental state.

7.9 Conclusion

This chapter explores the scientific understanding of the interconnectedness between land degradation and climate change, with a focus on attaining sustainability in the face of the climate crisis. The intricate nature of this relationship poses challenges in identifying land degradation hotspots and implementing precautionary measures. Soil erosion, which directly impacts food security, is a major global concern in land degradation. It leads to the removal of nutrient-rich topsoil, transporting it into reservoirs and oceans, and causing significant on-site and off-site impacts. Climate change exacerbates this issue unless appropriate measures are adopted. In tropical hilly and mountainous terrains, high-intensity rainfall and unsustainable land management practices contribute to soil degradation. Furthermore, acidification, compaction, and decline in organic matter content are the adverse impacts of land degradation. SOC serves as a crucial indicator for evaluating soil quality and overall sustainability. A study conducted in the lesser Himalayan landscape supports the finding that landscapes with higher organic carbon content

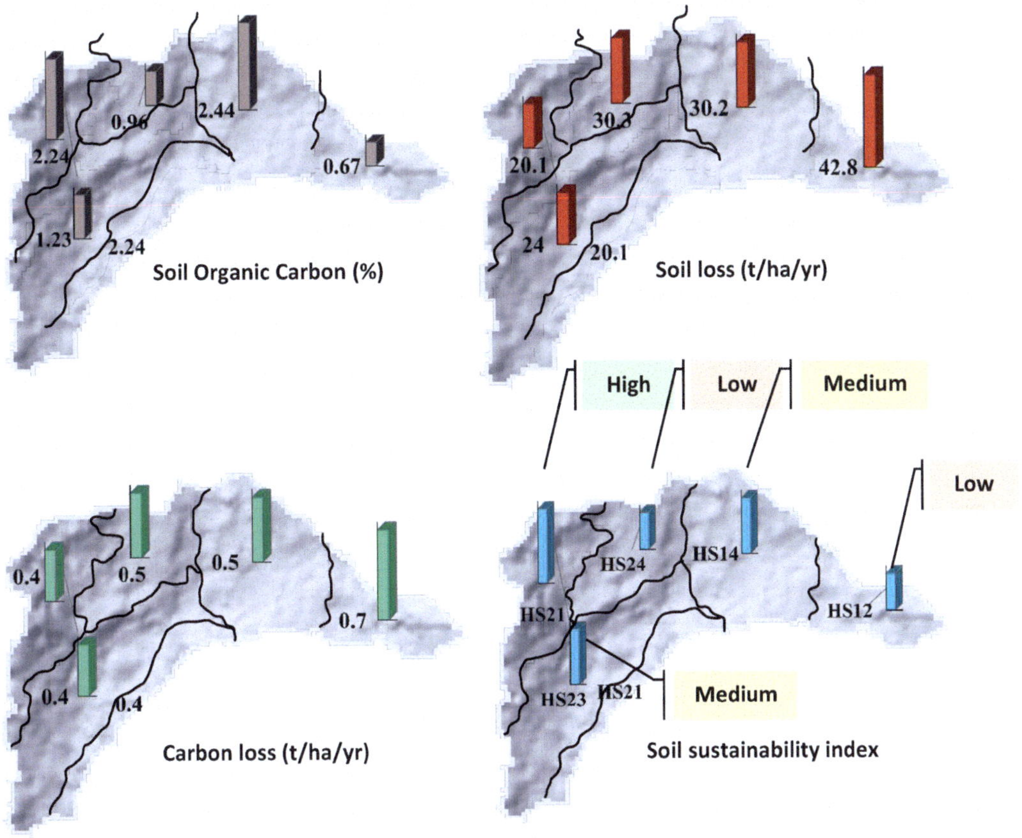

Fig. 7.3 Soil organic carbon, soil loss, carbon loss, and sustainability index of the Pasta micro-watershed

and lower soil erosion are highly sustainable in the face of climate change and its effects. Implementing site-specific sustainable management practices can help us achieve a sustainable environment for the well-being of both humanity and the natural ecosystem.

References

Achat DL, Fortin M, Landmann G, Ringeval B, Augusto L (2015) Forest soil carbon is threatened by intensive biomass harvesting. Sci Rep 5(1):15991

Agovino M, Casaccia M, Ciommi M, Ferrara M, Marchesano K (2019) Agriculture, climate change and sustainability: the case of EU-28. Ecol Indic 105:525–543

Barrow CJ (1994) Land degradation. Development and breakdown of terrestrial environments. Cambridge University Press, New York, NY, USA

Beg N, Morlot JC, Davidson O, Afrane-Okesse Y, Tyani L, Denton F, Rahman AA (2002) Linkages between climate change and sustainable development. Clim Pol 2(2–3):129–144

Blanco H, Lal R (2008) Principles of soil conservation and management, vol 167169. Springer, New York

Blanco-Canqui H, Lal R, Blanco-Canqui H, Lal R (2008) Soil erosion and food security. Princip Soil Conservat Manag:493–512

Conway LS, Yost MA, Kitchen NR, Sudduth KA, Veum KS (2018) Cropping system, landscape position, and topsoil depth affect soil fertility and nutrient buffering. Soil Sci Soc Am J 82(2):382–391

Crowther TW, Toddbrown KEO, Rowe CW, Wieder WR, Garey JC, Machmuller MB, Bradford MA (2016) Quantifying soil carbon losses in response to warming. Nature 540:104–108. https://doi.org/10.1038/nature20150

David Raj A, Kumar S, Sooryamol KR (2022) Modelling climate change impact on soil loss and erosion vulnerability in a watershed of Shiwalik Himalayas. Catena 214:106279

ELD Initiative (2015) The value of land: Prosperous lands and positive rewards through sustainable land management. Available at: www.eld-initiative.org

Eswaran H, Lal R, Reich PF (2019) Land degradation: an overview. Resp Land Degradat:20–35

Eswaran H, Van den Berg E, Reich P (1993) Organic carbon in soils of the world. Soil Sci Soc Am J 57:192–194

FAO (1979) A provisional methodology for soil degradation assessment. FAO, Rome, p 84

Guerra AJT, Fullen MA, Jorge MDCO, Bezerra JFR, Shokr MS (2017) Slope processes, mass movement and soil erosion: a review. Pedosphere 27:27–41

Gürsoy S (2021) Soil compaction due to increased machinery intensity in agricultural production: its main causes, effects and management. Technol Agri:1–18

Halbac-Cotoara-Zamfir R, Smiraglia D, Quaranta G, Salvia R, Salvati L, Giménez-Morera A (2020) Land degradation and mitigation policies in the Mediterranean region: a brief commentary. Sustainability 12(20):8313

Hazbavi Z, Sadeghi SHR, Gholamalifard M (2019) Dynamic analysis of soil erosion-based watershed health. Geogr Environ Sustain 3:43–59

Hellin J (2003) From soil erosion to soil quality. Leisa-Leusden 19:10–11

Intergovernmental Panel on Climate Change (IPCC) (2001) Climate change 2001 – impacts, adaptation, and vulnerability. Cambridge University Press, Cambridge, UK, p 1032

IPCC (2007) Climate change (2007): impacts, adaptation and vulnerability. In working group II contribution to the fourth assessment report of the intergovernmental panel on climate change. Cambridge University Press, Cambridge, UK

IPCC, Masson-Delmotte V, Zhai P, Pirani A, Connors SL, Péan C, Berger S, Zhou B (2021) Climate change 2021: the physical science basis. Contribution of working group I to the sixth assessment report of the intergovernmental panel on climate change 2

IUCN (2015) Land Degradation and Climate Change The multiple benefits of sustainable land management in the drylands, Accessed from https://www.iucn.org/sites/default/files/2022-07/land_degradation_issues_brief_cop21_031215.pdf

Jackson RB, Lajtha K, Crow SE, Hugelius G, Kramer MG, Piñeiro G (2017) The ecology of soil carbon: pools, vulnerabilities, and biotic and abiotic controls. Annu Rev Ecol Evol Syst 48(1)

Keesstra S, Mol G, De Leeuw J, Okx J, De Cleen M, Visser S (2018) Soil-related sustainable development goals: four concepts to make land degradation neutrality and restoration work. Land 7(4):133

Keesstra S, Nunes J, Novara A, Finger D, Avelar D, Kalantari Z, Cerdà A (2018) The superior effect of nature based solutions in land management for enhancing ecosystem services. Sci Total Environ 610:997–1009

Kumar S, Sooryamol KR, David Raj A, Kalambukattu JG, Mariappan S (2023) Use of advanced techniques to estimate soil erosion and associated carbon loss in the Himalayan region. In: Mishra G, Giri K, Nath AJ, Francaviglia R (eds) Soil carbon dynamics in Indian Himalayan region. Springer

Kumawat A, Yadav D, Samadharmam K, Rashmi I (2020) Soil and water conservation measures for agricultural sustainability. Soil Moist Importance 23

Lal R (2018) Soil quality and food security: the global perspective. In: Soil quality and soil erosion. CRC press, pp 3–16

Lal R, Mokma D, Lowery B (2018) Relation between soil quality and erosion. In: Soil quality and soil erosion. CRC press, pp 237–258

Lobell DB, Schlenker W, Costa-Roberts J (2011) Climate trends and global crop production since 1980. Science 333:616–620

Montanarella L (2007) Trends in land degradation in Europe. Climate Land Degradat:83–104

Nearing MA et al (2005) Modeling response of soil erosion and runoff to changes in precipitation and cover. Catena 61:131–154. https://doi.org/10.1016/j.catena.2005.03.007

Okur B, Örçen N (2020) Soil salinization and climate change. In: Climate change and soil interactions. Elsevier, pp 331–350

Olsson L, Barbosa H, Bhadwal S, Cowie A, Delusca K, Flores-Renteria D, Hermans K, Jobbagy E, Kurz W, Li D, Sonwa DJ, Stringer L (2019) Land Degradation. In: Shukla PR, Skea J, Buendia EC, Masson-Delmotte V, Pörtner H-O, Roberts DC, Zhai P, Slade R, Connors S, van Diemen R, Ferrat M, Haughey E, Luz S, Neogi S, Pathak M, Petzold J, Pereira JP, Vyas P, Huntley E, Kissick K, Belkacemi M, Malley J (eds) Climate Change and Land: an IPCC special report on climate change, desertification, land degradation, sustainable land management, food security, and greenhouse gas fluxes in terrestrial ecosystems

Pendergrass AG, Knutti R (2018) The uneven nature of daily precipitation and its change. Geophys Res Lett 45:11,980–11,988. https://doi.org/10.1029/2018GL080298

Pimentel D (2006) Soil erosion: a food and environmental threat. Environ Dev Sustain 8:119–137

Qian W, Zhu Y (2001) Climate change in China from 1880 to 1998 and its impact on the environmental condition. Clim Change 50:419–444

Reed MS, Stringer LC (2016) Land degradation, desertification and climate change: anticipating, assessing and adapting to future change, 1st edn. Routledge. https://doi.org/10.4324/9780203071151

Reich PF, Numben ST, Almaraz RA, Eswaran H (2001) Land resource stresses and desertification in Africa. In: Bridges EM, Hannam ID, Oldeman LR, de Vries FWT P, Scherr SJ, Sompatpanit (eds) Responses to Land Degradation. Proc. 2nd. International Conference on Land Degradation and Desertification, Khon Kaen, Thailand, Oxford Press, New Delhi, India

Rengel Z (2011) Soil pH, soil health and climate change. Soil Health Climate Change:69–85
Rhodes CJ (2017) The imperative for regenerative agriculture. Sci Prog 100(1):80–129
Rodrigo-Comino J, Senciales JM, Cerdà A, Brevik EC (2018) The multidisciplinary origin of soil geography: a review. Earth-Sci Rev 177:114–123
Sajjad H, Kumar P, Masroor M, Rahaman MH, Rehman S, Ahmed R, Sahana M (2022) Forest vulnerability to climate change: a review for future research framework. Forests 13(6):917
Salinas CX, Mendieta J (2013) Mitigation and adaptation investments for desertification and climate change: an assessment of the socioeconomic return. Mitig Adapt Strateg Glob Change 18:659–672
Sanderman J, Hengl T, Fiske GJ (2017) Soil carbon debt of 12,000 years of human land use. PNAS USA 114:9575–9580
Sarvade S, Upadhyay VB, Kumar M, Khan MI (2019) Soil and water conservation techniques for sustainable agriculture. In: Sustainable agriculture, Forest and environmental management. Springer, Singapore, pp 133–188
Sivakumar MV, Stefanski R (2007) Climate and land degradation—an overview. Springer, Berlin Heidelberg, pp 105–135
Sooryamol KR (2020) Potential impact of climate change on surface runoff and sediment yield in a watershed of lesser Himalayas. M.Sc. Thesis (Climate Change Adaptation), Kerala Agricultural University, Thrissur, 115
Sooryamol KR, Kumar S, Regina M, David Raj A (2022) Modelling climate change impact on soil erosion in a watershed of north-western lesser Himalayan region. J Sedimentary Environ 7(2):125–146
Sreenivas K, Sujatha G, Mitran T, Janaki Rama Suresh KG, Ravisankar T, Rao PVN (2021) Decadal changes in land degradation status of India. Curr Sci 121(4):539–550
Stringer LC, Dougill AJ (2013) Channelling science into policy: enabling best practices from research on land degradation and sustainable land management in dryland Africa. J Environ Manag 114:328–335
Tennigkeit T, Wilkes A (2008) An assessment of the potential for carbon finance in rangelands. Working paper no. 68
van Gestel N, Shi Z, van Groenigen KJ, Osenberg CW, Andresen LC, Dukes JS, Hovenden MJ, Luo Y, Michelsen A, Pendall E, Reich PB, Schuur EAG, Hungate BA (2018) Predicting soil carbon loss with warming. Nature 554(7693):E4–E5. https://doi.org/10.1038/nature25745
Vasisht AK, Singh RP, Mathur VC (2003) Economic implications of land degradation on sustainability and food security in India. Agropedology 13(2):19–27
Webb NP, Marshall NA, Stringer LC, Reed MS, Chappell A, Herrick JE (2017) Land degradation and climate change: building climate resilience in agriculture. Front Ecol Environ 15(8):450–459
World Population Review. Walnut, United States. Available online: www.worldpopulationreview.com. Accessed on 16 February 2023
WRI, The causes and effects of soil erosion, and how to prevent it. February 7, 2020 By Dede Sulaeman and Thomas Westhoff. https://www.wri.org/insights/causes-and-effects-soil-erosion-and-how-prevent-it. Accessed on 20 February 2023
Ziadat FM, Zdruli P, Christiansen S, Caon L, Monem MA, Fetsi T (2022) An overview of land degradation and sustainable land Management in the near East and North Africa. Sustain Agric Res 11:11–24

Social Resilience of Local Communities Due to Tidal Flooding on the North Coast of Semarang City, Indonesia

8

Hari Harjanto Setiawan

Abstract

This chapter describes the social resilience of the northern coastal communities of Semarang City in facing tidal floods. In the last few years, tidal floods in this region have become more frequent in intensity and higher in volume. The global climate crisis impacts the lives of the northern coastal communities of Semarang City. The causes of the climate crisis are human actions such as using fossil fuel vehicles, deforestation, and greenhouse gases. This global warming causes the ice at the poles to melt so that the volume of seawater increases. This incident resulted in tidal flooding in the lowlands around the north coast of Semarang City. Tidal floods can disrupt community work activities. Schoolchildren sometimes cannot go to school. The community spends a lot of money to raise the ground level in their home area. The community experiences two vulnerabilities due to tidal floods, namely social vulnerability and economic vulnerability. Communities must adapt to the two vulnerabilities experienced. In particular, this chapter will explain community resilience as a form of adaptation to these two vulnerabilities caused by the climate crisis. In particular, this chapter will describe the condition of the people affected by the tidal flood in the city of Semarang. Second, the vulnerability of the community affected by the tidal flood is described. Third, the resilience of the people of Semarang City in dealing with tidal floods.

Keywords

Climate Crisis · Tidal Flood · Vulnerability · Adaptation · Social Resilience

8.1 Introduction

The city of Semarang is one of the areas that has a coast in the north of the island of Java, Indonesia. The climate crisis that has occurred in recent years has caused several areas in Semarang City to experience tidal floods with more frequent intensity and higher tidal volumes. Tidal floods are rising sea levels caused by the gravity of celestial bodies. The sun's and moon's gravity on the mass of seawater on earth causes tidal flooding. This rise in sea level is what causes tidal flooding. Tidal floods in Semarang every year continue to increase due to the melting of polar ice triggered by global warming. The tidal flood harmed the people living on the north coast of Semarang City. The impact felt by the community is the disruption of activities and the decline in residential infrastructure (Hilmi et al. 2022).

H. H. Setiawan (✉)
National Research and Innovation Agency (BRIN), Jakarta, Indonesia
e-mail: hari.harjanto.setiawan@brin.go.id

U. Chatterjee et al. (eds.), *Climate Crisis: Adaptive Approaches and Sustainability*, Sustainable Development Goals Series, https://doi.org/10.1007/978-3-031-44397-8_8

The climate crisis is a matter of concern to the residents of the north coast of Semarang City. Climate change can result in tidal flooding caused by land subsidence and sea level rise. Tidal flooding is a severe problem in Semarang City, Central Java. The impact of tidal floods will affect people's physical, economic, environmental, and health aspects (Shymbin and Nongbri 2022). According to data from the Regional Disaster Management Agency, there have been seven tidal floods in Semarang City in 2022. The worst floods occurred in May. Extreme weather caused by global warming resulted in tidal floods. The dikes at the "Tanjung Emas" harbor could no longer withstand this high tide, so the embankment broke, severely impacting three sub-districts on Semarang's north coast.

The climate crisis is a natural event and a human contribution. Excessive groundwater extraction is a human activity that can cause tidal flooding. Human exploitation of coastal lands causes a decrease in the groundwater table. Environmental changes that continue to decline in quality will impact the social life of residents on the north coast of Semarang City. Tidal floods will disrupt human activities from a physical and socio-economic aspect (Laignel et al. 2023). Many people have moved because they have lost their homes and livelihoods. The tidal flood also disrupted children's educational activities because they could not attend school.

Tidal floods can affect the economic conditions of local communities (Hino et al. 2019). The impact of flooding is known to disrupt household income. People were forced to take time off work because their homes were submerged. This economic impact is felt by the people who get daily income. They risk not earning on that day. Disrupted income will cause social problems, namely poverty. Tidal floods and poverty are two problems that often occur in coastal areas. Tidal floods cause poverty because they damage people's assets and sources of income (Adnan et al. 2020). The tidal flood resulted in job changes accompanied by a decrease in revenue. This social and economic vulnerability is the focus of this chapter.

The community faces two choices: to stay or move from the location. The desire of every citizen to survive in an environment alert for tidal floods is closely related to one's ability to cope with the disaster. The vulnerability has three crucial elements: the lack of protection during a crisis, people living under pressure, and the slow response to disasters (Hoq et al. 2021). Communities living on the north coast of Semarang City experience vulnerability from economic and social aspects. On the economic aspect, people experience a decrease in household income, so their needs are unmet. In the social aspect, people are prone to becoming poor and committing crimes in the form of theft.

The vulnerability of the northern coastal community of Semarang requires strengthening adaptation to tidal flooding. Community resilience is not just the ability to deal with disasters. However, people can survive in a disastrous environment (Setiawan 2023). Resilience has several components: reaction to threats, planning, prevention, avoidance, and response. Changes influence the resilience of coastal communities in dealing with disasters in the socio-economic community and the community's quality of life. Economic capacity will impact the community to adapt to tidal floods. The economy is essential to resilience because it relates to economic activity (Disse et al. 2020).

This chapter will describe the resilience of the northern coastal community of Semarang City in adapting to tidal flooding. There are two resilience that are of concern, namely economic resilience and social resilience. Financial resilience is the ability of the economy to return due to economic difficulties. At the same time, resilience is the ability of community groups to deal with pressure caused by disturbances that occur due to changes in social conditions. In particular, this tire will answer three research questions. First, what is the condition of the northern coastal community in Semarang City facing tidal flooding? Second, what are the vulnerabilities experienced by the people in the tidal flood environment? Third, how is the community's resilience in dealing with tidal floods?

8.2 The Rationale of the Study

8.2.1 Climate Crisis Global

The pollution and growth of greenhouses worldwide are causing the global climate crisis. The tidal flood disaster in Semarang City impacts the worldwide climate crisis, which cannot be separated from other regions or countries (Dahal et al. 2019). The global crisis incident greatly affected the human ecosystem in the world. The global crisis can disrupt human life and community resilience (Guo et al. 2020). Reducing the risk of tidal floods is done in collaboration between various parties and the local community. This is a form of adaptation to overcome the problem of tidal flooding. Communities on the north coast of Semarang City who are affected need to be involved in the adaptation process to make changes. The involvement of all parties involved is necessary to reduce the impact.

The climate crisis causes ice to melt at the north and south poles, so sea levels rise. In certain places, such as on the north coast of Semarang, it can cause tidal flooding, thereby disrupting residents' activities. Farms and fish ponds were flooded. Factory workers are also disturbed because they cannot work. This condition will affect their resilience and global resilience. The climate crisis has caused erratic seasons in various parts of Indonesia, causing disturbances to several aspects of human life (Filho et al. 2022). Adaptation to tidal floods is crucial to reduce the risks posed. This is necessary to ensure the mitigation is successful. Communities on the north coast of Semarang City can prepare themselves to make changes. The global community is contributing to addressing the impact of the climate crisis, and we are reducing it significantly (Winsemius et al. 2018).

8.2.2 Impact of Tidal Flood in Semarang

Adaptation to the climate crisis must be the same as the disasters society faces (Raza et al. 2019). If not comparable, then there will be extinction. The climate crisis is causing sea levels to rise. This condition will impact the people of Semarang City, located on the coast. Tidal floods impact these conditions where sea water will inundate the land. Coastal abrasion will continue to occur and result in the sinking of settlements on the coast (Setyowati et al. 2021). Small islands are starting to sink, impacting Indonesia as an archipelagic country. The climate crisis also affects people's lifestyles and the species that live in the sea. Ecosystem-based adaptation is a priority agenda for Indonesia, a member of the G-20.

Human actions that caused the climate crisis must be held responsible for the tidal flood in Semarang. Responsibility starts with the individual not taking activities that could cause the climate crisis. Because it will impact many people and the ecology globally, this human action will cause global warming and human life (Abbass et al. 2022). The climate crisis is a threat to society, especially in coastal settlements. Settlement mitigation and modification are needed to reduce the risk of climate crisis in Semarang (Bilotto and Karen 2023).

The climate crisis causes social and economic vulnerability globally. Stagnant water due to tidal floods causes food security to decrease (Purakayastha et al. 2019). The production of agricultural land and fish ponds has been reduced, which has resulted in decreased food security. This requires adaptation measures to the environment so that economic growth is not hampered (Baarsch et al. 2019). It has been proven that tidal floods frequently occur on the northern coast of Semarang City, which causes damage to agricultural land and fish ponds. Industrial secretarial workers were also affected because their access to work was disrupted. This causes vulnerability to poverty and other social problems (Koomson et al. 2020). The vulnerability of the northern coastal communities of Semarang City is grouped into two, namely economic vulnerability and social vulnerability.

8.2.3 Social Resilience of Local Communities as an Adaptation Strategy

The global climate crisis has increased the intensity of tidal floods in Semarang City in recent years. Disaster impact reduction must be reduced through public awareness. This is a concern for the community to build resilience to adapt to the climate crisis (Wang et al. 2019). All humans must participate in overcoming and adapting to current environmental changes (Ford et al. 2020). Events in Semarang that the climate crisis increases the vulnerability of indigenous peoples. This local wisdom aims to pay attention to and preserve nature (Jumriani et al. 2021). So the tidal flood handling program in Semarang must respect local knowledge.

The tidal flood disaster management program is carried out in a partnership that involves many stakeholders. The people of Semarang City have the principle of mutual respect and uphold justice. The most important collaboration is with residents because those who feel the effects of the tidal flood are more affected. The tidal flood prevention program must incorporate the local wisdom of the community. Learning from a community in northeastern Ghana found that a program to increase community knowledge about the climate crisis and a program to increase the ability of extension workers to share information on the climate crisis is appropriate adaptation action (Chan and Amling 2019).

A campaign approach through public communication can appropriately influence knowledge, perceptions, attitudes, and behavior regarding adaptation to the climate crisis (Abunyewah et al. 2020). If implemented on the north coast of Semarang City, collaboration with local communities and awareness of the climate crisis will be essential. How the adaptation of the people of Semarang City in dealing with tidal floods must be understood as a form of resilience. Adaptation and resilience are interrelated words (Metcalfe et al. 2020). So the social resilience of local communities is a form of adaptation strategy to environmental changes (Fig. 8.1).

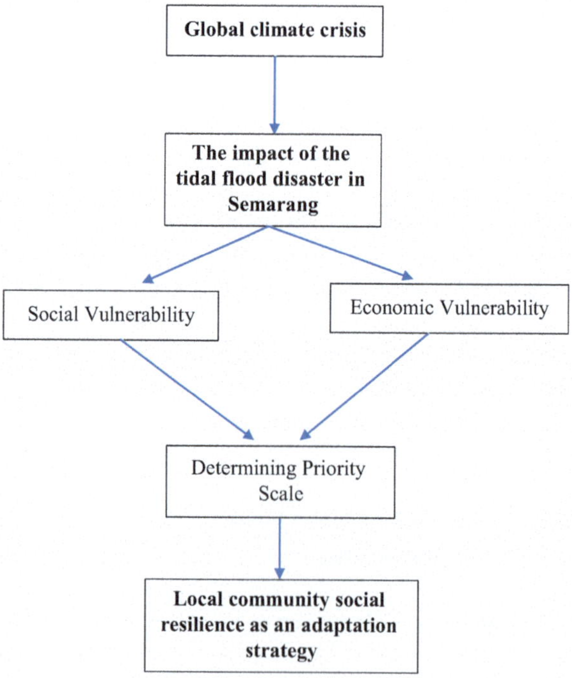

Fig. 8.1 The rationale of the study

8.3 Materials and Methods

This chapter describes the resilience of coastal communities in Semarang City in facing tidal floods due to the global climate crisis. The impact felt by the people who experienced the tidal flood is a social and economic aspect. Next, it describes the adaptation of local communities (Westoby et al. 2020) in dealing with tidal floods. The method used is qualitative because it will explain in detail the experience of the community in coping with tidal surges. Data were collected through interviews (Prentice 2017), observation of the impact of tidal floods (Busetto et al. 2020), and literature studies (Jones and McCoy 2019).

Qualitative data were collected from the beginning of December 2022 until the end of January 2023. Qualitative data were obtained through interviews and observations of the community, community leaders, local government, and experts. The total number of informants is 16 people. The in-depth interviews last about 60 to 90 minutes. The community provides information about the tidal flood experience and the adaptations made. Community leaders and local government provide information about disaster management programs. Information about the tidal flood was also obtained from electronic data analysis and printed news. Literature studies from scientific journals from various countries related to the climate crisis will strengthen the analysis of this chapter.

The collected data are analyzed thematically (Sundler et al. 2019). That is, all data are adjusted to a predetermined theme. The theme chosen was about the community's social resilience due to tidal flooding on the north coast of Semarang City, Indonesia. This chapter has limitations, namely the time to observe disaster-affected areas. Observations were made when there was no disaster. In addition, the conditions described above are limited to situations in Semarang City. So it cannot be generalized for the condition of Indonesia as a whole.

8.4 Results and Discussion

8.4.1 Tidal Flood Conditions in Semarang

Semarang City is the center of government and economy of Central Java Province, Indonesia. Semarang City is the fifth metropolitan city in Indonesia after Jakarta, Bandung, Medan, and Surabaya. The city of Semarang has an area of 363.4 square kilometers. This area has a strategic geographical location as the center of government of Central Java Province. Semarang consists of 16 districts and 177 villages. The sub-districts in Semarang City that have the highest population density compared to other sub-districts are Central Semarang, West Semarang, East Semarang, and North Semarang. The most densely populated 16 districts in Semarang City are North and Central Semarang. Tidal floods often hit Semarang City and its connected rivers.

Semarang City has a coastal area in the north. The city of Semarang has a coast to the north which is prone to tidal flooding. The coastal area of Semarang has a sloping topography, with most of the site being almost at sea level. Meanwhile, Semarang City has a coastal area with a coastline length of ±13.6 kilometers. The coastal area of Semarang City is experiencing tidal flooding caused by rising sea levels and land subsidence. The areas affected by tidal flooding are on the north coast and along the river. Furthermore, tidal waves enter the coastal area of Semarang City through three main rivers, namely Kali Semarang, Kali Baru, and Kali Banger.

The districts most affected by the tidal floods were Bandarharjo, Tanjung Mas, and Kemijen. The tidal flood affected 900 families in Bandarharjo Village. The inundation area reaches 125 hectares. The residents of Kemijen Village affected by the tidal flood were 1245 families. The affected area is 39 hectares.

While the most affected area is Tanjung Mas Village, the area that is inundated by water reaches 300 hectares with a population of 8335 people. To overcome tidal flooding, the Semarang

City Government has the ambition to build a beach belt in Kampung Tambak Lorok Semarang, which is expected to overcome the tidal problem. Currently, work on the coastal belt is underway. The beach belt built in the Tambaklorok area is up to 1.2 kilometers long.

Tidal floods are also caused by land subsidence, a natural phenomenon due to soft soil compression. Land subsidence is also caused by human activities, namely the physical load of buildings and groundwater extraction. This human activity causes soil conditions in Semarang to experience compression. If this continues, the inundation caused by tidal floods will expand yearly because the water level continues to rise. They are building a beach belt to hold back the waves. It is hoped that tidal flooding in Semarang City will decrease. The tidal flood risk map in the city of Semarang can be seen in Fig. 8.2, marked in red.

Every year, the land surface in the city of Semarang always falls. This low ground level causes the city of Semarang to be lower than the sea level. So to overcome tidal floods with sea belts, it is not enough. Efforts must be made so that the seawater does not go down. One of the causes of land subsidence is groundwater extraction. Data from the local government show that the groundwater table is decreasing by 10 centimeters yearly. The city of Semarang must always be prepared for extreme weather when it occurs. The impact of the tidal flood was that the electricity went out, and hundreds of workers at Tanjung Emas Port were sent home. In addition to workers who cannot come to work, schoolchildren must also be escorted by emergency boats. Several school buildings were also flooded and had to be closed, so this condition greatly hampered children's education (Fig. 8.3).

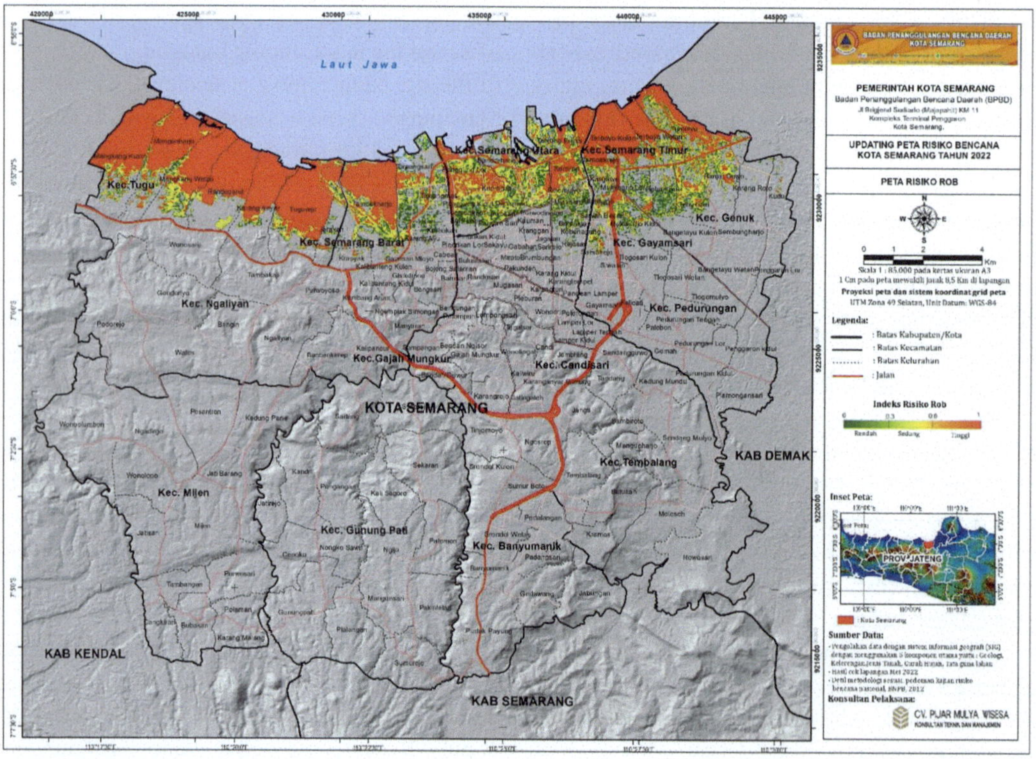

Fig. 8.2 Tidal flood risk index in Semarang City. (Source: Semarang City Regional Disaster Management Agency, 2022)

Fig. 8.3 Tidal flood conditions in the city of Semarang

The sea belt is a coastal protective structure using geotextiles in elongated sacks filled with water and sand. The main obstacle to developing the coastal belt is its cost. So local governments have difficulty without assistance from the central government. The coastal belt will reduce the waves hitting the coast. This beach belt can minimize abrasion. So the coastal belt serves to break up incoming waves, reduce erosion, and hold sediment.

The phenomenon associated with tidal flooding is land subsidence (Putiamini et al. 2022). This condition caused more severe tidal flooding than in previous years. The coast of Semarang City is decreasing by around 10 to 15 centimeters yearly. Several houses and public facilities began to sink. If allowed to continue, the land area will be increasingly narrow. One of the areas close to the beach is the Tanjung Emas sub-district. At first, this area was hit by tidal floods once a month, then once a week, and almost daily. Some buildings are uninhabitable today because they have sunk into the sea. Residents in the area eventually chose to move because their homes were no longer livable (Binder et al., 2015).

8.4.2 Impact of the Tidal Flood Disaster in Semarang

8.4.2.1 Social Vulnerability

Social vulnerability is essential to understanding disaster risk and responding effectively (Tate et al. 2021). Initially, the residents raised their houses so that the tidal water would not enter the house. The longer the residents are, no longer able to face tidal floods. The average resident's house has been raised 50 centimeters from before. Their home therapy is still in the tidal floodwater. Self-help residents have also tried to elevate the road. The cost of upgrading a house or road costs a lot. Buying land to raise the road for one truck costs one million rupiahs. Tidal floods will quickly damage people with motorbikes or cars because salt water easily corrodes them. Usually, tidal water comes every morning until evening or evening until night.

Social vulnerability factors include urban density, low-income households, population changes, and employment (Kirby et al. 2019). Ultimately, residents chose to leave their homes because they could no longer deal with this tidal flood. Residents no longer have their own homes. Most of them rent houses in places that are safe from tidal flooding. People who have lived for 25 years on the north coast of the city of Semarang can tell the difference. Many houses have sunk. The problems experienced by the community are very complex. In addition to natural phenomena in the form of tidal flooding, the people of Semarang City are also socially vulnerable. Conditions like this require residents to survive. The lack of jobs causes people to be forced to violate social norms to stay (Papadopoulos et al. 2021).

The characteristics of the northern coastal community in Semarang City are related to their social life. Social life is related to social relations between community members, such as the length of stay of bonds and social interactions between communities. On the north coast of Semarang City, the people have quite close social ties. The community actively participates in social activities. The community's characteristics also affect the community's resilience in facing disturbances in the settlement environment because there is a sense of kinship among community members. Social organizations that developed include recitations for mothers, recitations for fathers, associations, and sports groups. An organization that cares about the environment is an association of gentlemen who discuss environmental cleanliness and environmental safety but never discuss the tidal flood relief they experienced.

8.4.2.2 Economic Vulnerability

People owe it intending to elevate their houses so they do not sink. Residents have repeatedly raised their homes. But the community's efforts were in vain because the floods were worsening, and the house was eventually no longer habitable. The cost to elevate the place is prohibitive. People are forced into debt so they do not drown. According to the community, he has raised his house six times. In the beginning, there were people whose multi-story houses now only lived on one floor. Residents who have lived for 26 years have terraced houses now only live on one floor. Meanwhile, residents with one-story homes with a height of 6 meters now only have one roof left.

Most people living on Semarang's north coast work as entrepreneurs and laborers. This work is vulnerable to adaptation because community work will affect the income generated. Low income is only enough to meet their daily needs causing the community to be unable to carry out routine home repairs due to tidal floods. A few economically capable people are not vulnerable to adapting because they can set aside their income to repair their houses damaged by the tidal flood. Economic characteristics can also be seen in land ownership. The place you live in is your right. This causes the people to choose to stay because they do not want their houses to be empty and flooded by tidal water. This disaster caused physical damage that affected the economic condition of the community.

Economic vulnerability is the economy's shock exposure (Sahana et al. 2021). This vulnerability can be caused by the tidal disaster that hit the northern coast of Semarang. Such conditions must be considered to strengthen economic resilience. Economic resilience is a family to recover from shocks due to tidal floods. The tidal floods disrupted people's livelihoods. People are forced to look for side jobs or other jobs to meet family needs. Disrupted work causes their income to decrease. When the Tidal floods, the people cannot work, so their income is minimal. The reduced income was due to the tidal floods when they could not work. The community must set aside their income for home improvement in this condition.

8.4.3 Local Community Social Resilience as an Adaptation Strategy

8.4.3.1 Cause Society Survives

Some people on the north coast of Semarang City have moved, but some still live in their homes. Vulnerability is a condition that a person and the environment feel and see from the social and economic aspects (Sugiura et al. 2020). A government program also causes people not to leave their homes. Every individual's desire to survive closely relates to one's ability to deal with disasters. This ability is influenced by the vulnerability factors faced by each individual. The vulnerability is in the form of social and economic vulnerability (Huq et al. 2020).

The work factor causes society to survive. Work is an economic factor that can cause people to determine where to live (Byrne 2022). Communities that are not yet established choose to stay in settlements submerged by tidal floods. Their excuse is that they do not have enough funds to buy a house or move to another location. Meanwhile, people who work as private employees with good incomes also choose to stay because the work location is close to the settlement's location. For the community, the economic factor is more critical than the tidal inundation problem that always occurs daily. Another factor is total revenue. The total income of people who are still vulnerable will affect efforts to deal with tidal floods.

People with middle to lower income choose to stay in settlements prone to tidal floods due to economic problems. In contrast, people with a more stable income and who are not vulnerable in dealing with tidal floods choose to stay because they are close to the location of work. For the people, many changes have been made to the condition of their houses. The people have often piled up land or rebuilt their homes damaged by the tidal flood. Therefore, the people chose to stay afloat because, for them, a house the same height as the road could prevent tidal inundation from entering the house. Vulnerabilities arise due to poor housing, hazardous locations, and unsafe conditions (Mahbubur Rahman et al. 2022). The people affected by tidal floods also experience this condition.

Factors that cause communities to survive are related to economic vulnerability and environmental vulnerability (Sahana et al. 2021). Economic vulnerability includes livelihoods, total income, location of work, and amount of savings. The amount of income affects the ability of each individual to carry out adaptation actions. Identification of economic vulnerability can also be based on the location of community work. People whose work locations are close to residential areas are not prone to moving to other places. Social entrepreneurship programs are needed for communities like this to increase the family's economic resilience. This program aims to reduce poverty (Setiawan et al. 2021).

In contrast, people whose work locations are far from settlements will be vulnerable to moving to other areas. Meanwhile, environmental vulnerability is related to the status of home ownership and the type of house. This semi-permanent house makes the community less vulnerable to tidal floods, while places of the wooden type will be very vulnerable to tidal floods. The type of house ownership is closely related to the economic level of society. Communities with high economic levels will be able to build more permanent homes. Meanwhile, people with a low economic level will make foxes from wood.

8.4.3.2 Form of Community Resilience

Community resilience consists of four: social, economic assets, social capital, community authority, and functional and structural potential (Pasca et al. 2022). Resilience is the ability of every individual and society to deal with disasters. This incident can be a lesson for each individual to be better prepared to deal with disasters if they happen again. Community resilience talks about facing disasters and overcoming vulnerabilities. The lesson learned by the community is the increased ability of the community to survive in an environment affected by a disaster (Papadopoulos et al. 2021). Resilience relates to adaptations made and preparations made before a disaster occurs. The adaptation made by the community is by repairing house buildings, repairing some damaged parts of the house to prevent stagnant water from entering the house, and raising the floor of the house. The people expanded the foundations of their homes by filling in the land, which they bought at a high price. Less able people can adapt by making a barrier in front of the house.

The adaptation form of building a house by renovating is the right way because the time needed to renovate is not as much as to rebuild (Eriksen et al. 2021). The community adapts because of the limited ability to live in other locations, so the community makes a lot of effort to adapt. Economically vulnerable people cannot make any repairs to their houses. Even though they are not economically capable, people still adapt by providing boards in front of their doors and supporting household furniture with bricks. Economic resilience can be seen in how the community deals with tidal inundation following the community's economic conditions. Economic resilience consists of the type of community work, community income, amount of savings for home repairs, and forms of adaptation undertaken by the community (Almutairi et al. 2020). The type of vulnerable community work is related to the amount of people's income sufficient to meet their daily needs, not to repair damage to houses due to tidal floods.

Vulnerable people's income means that only a few people can set aside their income to save for the cost of repairing houses and the damaged environment. To achieve economic resilience, people must have side jobs to meet their daily needs (Formetta and Feyen 2019). People who have side jobs by opening a shop in front of the house. The income can be used to meet daily needs. His income from a side job is not as high as his main job's. The existence of business assistance and loans from financial institutions is a factor in the community's economic resilience. The government or the private sector provides business assistance to increase the community's economic resilience. Economic resilience can be seen in loans from financial institutions (Allam and Jones 2019). Only a few people cooperate by borrowing from financial institutions, while the financial institutions in question are banks. The community is not interested in borrowing from financial institutions because the loan interest is high enough that the community feels unable to repay the loan and the goods.

The community does not have economic resilience because the community's economic conditions are classified as vulnerable to adaptation. Social resilience is an organization that cares about the environment and tidal flood problems (Colombano et al. 2021). This organization also cares about housing and access to education and health. Social organizations that care about the environment are associations. The association of gentlemen discusses social service, cleanliness, environmental safety, and others. This association has never discussed dealing with tidal floods because tidal floods will always occur in the community. Efforts made can only temporarily overcome tidal floods. Access to health facilities is a community health center. Access to educational facilities is classified as complete, namely the existence of a school. There are government policies that help underprivileged communities. Communities are assisted to be able to adapt to facing tidal flood problems. The description above shows that the community already has social resilience (Saja et al. 2019).

8.5 Limitations of the Study

This chapter can not only answer the research questions that have been set. The information presented cannot be generalized because the characteristics of people outside Semarang are different. This research teaches people with similar traits to increase social resilience in adapting to tidal floods. People who have other factors can adapt to their respective characteristics. This study has limited time for observation, so that it has an impact on research results. Researchers did not observe directly when the event occurred. However, the tidal flood incident can be observed from social media recorded by the public. This research cannot be used as a standard of measurement because of the time difference and the ever-changing social conditions of society.

8.6 Recommendations

The risk of tidal flooding due to the climate crisis can be reduced with planned actions. Based on the analysis, the consequences of the climate crisis include (1) the community considers this normal. (2) Society must always be aware of the actual cause. (3) The resilience of the local community must be increased to help each other when tidal floods occur. (4) The government must build global awareness because it protects against tidal floods caused by human activities that cause global warming. (5) The community needs relocation because their houses are no longer livable. (6) Spatial planning is key in adapting to the climate crisis.

To reduce environmental hazards, disaster mitigation is carried out on the north coast of Semarang. This activity aims to increase local community preparedness for disaster response and risk reduction. The northern coastal community of Semarang carried out several activities, such as determining the risk of tidal flooding, planning development by involving affected districts, increasing public awareness of tidal floods, carrying out physical and non-physical disaster management, identifying sources of danger in disasters, managing natural resources by well, and use advanced technology to build early warning systems.

The world community needs to be aware of how climate affects tidal flooding. As a result, residents of the coastal city of Semarang are made aware of the global battle against tidal flooding. Special protection mechanisms must be developed for communities affected by climate change to increase community resilience. The environment will benefit from this event by strengthening resilience. This incident is proof that the impacts of climate change are real. As a result, everyone must defend the earth. This initiative aims to minimize carbon emissions by planting mangrove trees to stop erosion.

8.7 Conclusions

The incident experienced by the northern coastal community of Semarang City, which was affected by the tidal flood, became a lesson for all humans. Climate change causes global warming, so the polar ice caps melt. The increase in the earth's temperature causes various changes in nature and human life. The seawater that rises to the mainland is called tidal flood, which significantly impacts the lives of people on the north coast of Semarang City. Humans must always be aware of protecting the earth to maintain its sustainability. Tidal floods on Semarang City's coast are increasingly occurring, and the intensity is increasing. Initially, the community survived by raising houses and roads, despite the high costs. There are two choices faced by the community, namely, to stay or move. People who are not strong will move by renting houses in areas unaffected by flooding.

The vulnerability experienced by the people is affected by the tidal flood in social and economic aspects. Although this disaster did not cause many casualties, the material losses resulted in the loss of houses, public facilities, and jobs. This condition causes the household economy to become disrupted and vulnerable. This vulnerability creates social problems in the form of poverty and criminal behavior. Disaster mitigation to reduce risk, capacity building, and public awareness must be carried out to deal with this disaster threat. Green technology in planting mangrove trees is an effort that needs to be developed for the community as an effort to mitigate global warming.

Community resilience is an adaptation in dealing with tidal floods. Analysis of community resilience in dealing with tidal floods on the north coast of Semarang City is a form of adaptation to natural conditions. Various events are analyzed by looking for relationships with one another. The sustainability of the lives of the northern coastal communities of Semarang City depends on the awareness of the world's population. Awareness to reduce global warming must be done immediately, starting with individual attention. Even the local government's policy to build a coastal belt is only temporary because the water level will rise due to global warming.

The government must prepare a long-term program through a social campaign about the impact of the climate crisis and create a movement to plant trees. Central Government, Regional Government, and all elements of society must work together to provide social security for people affected by climate change. The involvement of the global community has contributed to minimizing the impact of tidal floods on the north coast of Semarang City. The government must campaign this issue in international meetings to reach a global agreement to protect the earth.

Acknowledgments The authors thank the Research Center for Social Welfare Villages and Connectivity, National Research and Innovation Agency (BRIN), for supporting this research in 2022.

References

Abbass K, Qasim MZ, Song H, Murshed M, Mahmood H, Younis I (2022) A review of the global climate change impacts, adaptation, and sustainable mitigation measures. Environ Sci Pollution Res 29(28):42539–42559. https://doi.org/10.1007/S11356-022-19718-6

Abunyewah M, Gajendran T, Maund K, Okyere SA (2020) Strengthening the information deficit model for disaster preparedness: mediating and moderating effects of community participation. Internat J Disaster Risk Reduct 46:101492. https://doi.org/10.1016/J.IJDRR.2020.101492

Adnan MSG, Abdullah AYM, Dewan A, Hall JW (2020) The effects of changing land use and flood hazard on poverty in coastal Bangladesh. Land Use Policy 99:104868. https://doi.org/10.1016/J.LANDUSEPOL.2020.104868

Allam Z, Jones DS (2019) Climate change and economic resilience through urban and cultural heritage: the case of emerging Small Island developing states economies. Economies 7(2):62. https://doi.org/10.3390/ECONOMIES7020062

Almutairi A, Mourshed M, Ameen RFM (2020) Coastal community resilience frameworks for disaster risk management. Nat Hazards 101(2):595–630. https://doi.org/10.1007/S11069-020-03875-3/TABLES/6

Baarsch F, Granadillos JR, Hare W, Knaus M, Krapp M, Schaeffer M, Lotze-Campen H (2019) The impact of climate change on incomes and convergence in Africa. World Dev 126. https://doi.org/10.1016/j.worlddev.2019.104699

Bilotto F, Karen, Christie-Whitehead M, Malcolm B, Matthew, Harrison T (2023) Carbon, cash, cattle and the climate crisis. Sustain Sci 2023(1):1–17. https://doi.org/10.1007/S11625-023-01323-2

Busetto L, Wick W, Gumbinger C (2020) How to use and assess qualitative research methods. Neurol Res Pract 2(1):1–10. https://doi.org/10.1186/S42466-020-00059-Z/TABLES/1

Byrne D (2022) A worked example of Braun and Clarke's approach to reflexive thematic analysis. Qual Quant 56(3):1391–1412. https://doi.org/10.1007/S11135-021-01182-Y/FIGURES/D

Chan S, Amling W (2019) Does orchestration in the global climate action agenda effectively prioritize and mobilize transnational climate adaptation action? Internat Environ Agreements: Politics Law Economics 19(4–5):429–446. https://doi.org/10.1007/S10784-019-09444-9/FIGURES/6

Colombano DD, Litvin SY, Ziegler SL, Alford SB, Baker R, Barbeau MA, Cebrián J, Connolly RM, Currin CA, Deegan LA, Lesser JS, Martin CW, McDonald AE, McLuckie C, Morrison BH, Pahl JW, Risse LM, Smith JAM, Staver LW et al (2021) Climate change implications for tidal marshes and food web linkages to estuarine and coastal nekton. Estuaries Coasts 44(6):1637–1648. https://doi.org/10.1007/S12237-020-00891-1

Dahal B, Kumar SAP, Li Z (2019) Topic modeling and sentiment analysis of global climate change tweets. Soc Netw Anal Min 9(1):1–20. https://doi.org/10.1007/S13278-019-0568-8/TABLES/4

Disse M, Johnson TG, Leandro J, Hartmann T (2020) Exploring the relation between flood risk management and flood resilience. Water Securit 9:100059. https://doi.org/10.1016/J.WASEC.2020.100059

Eriksen S, Schipper ELF, Scoville-Simonds M, Vincent K, Adam HN, Brooks N, Harding B, Khatri D, Lenaerts L, Liverman D, Mills-Novoa M, Mosberg M, Movik S, Muok B, Nightingale A, Ojha H, Sygna L, Taylor M, Vogel C, West JJ (2021) Adaptation interventions and their effect on vulnerability in developing countries: help, hindrance or irrelevance? World Dev 141:105383. https://doi.org/10.1016/J.WORLDDEV.2020.105383

Filho WL, Ternova L, Fayyaz MM, Abubakar IR, Kovaleva M, Donkor FK, Anuga SW, Matamanda AR, Djekic I, Umar IA, Olooto FM, Meirelles M, Nagy GJ, May J, May M, Ebhuoma E, Begum H (2022) An analysis of climate change and health hazards: results from an international study. Internat J Climate Change Strategies Manag. https://doi.org/10.1108/IJCCSM-08-2021-0090

Ford JD, King N, Galappaththi EK, Pearce T, Mcdowell G, Harper SL (2020) The resilience of indigenous peoples to environmental change. One Earth 2(6):532–543. https://doi.org/10.1016/j.oneear.2020.05.014

Formetta G, Feyen L (2019) Empirical evidence of declining global vulnerability to climate-related hazards. Glob Environ Chang 57:101920. https://doi.org/10.1016/J.GLOENVCHA.2019.05.004

Guo C, Sim T, Ho HC (2020) Impact of information seeking, disaster preparedness and typhoon emergency response on perceived community resilience in Hong Kong. Internat J Disaster Risk Reduct 50:101744. https://doi.org/10.1016/J.IJDRR.2020.101744

Hilmi E, Amron, Christianto D (2022) The potential of high tidal flooding disaster in North Jakarta using mapping and mangrove relationship approach. IOP Conference Series: Earth Environmental Sci 989(1):012001. https://doi.org/10.1088/1755-1315/989/1/012001

Hino M, Belanger ST, Field CB, Davies AR, Mach KJ (2019) High-tide flooding disrupts local economic activity. Sci Advances 5(2). https://doi.org/10.1126/SCIADV.AAU2736/SUPPL_FILE/AAU2736_SM.PDF

Hoq MS, Raha SK, Hossain MI (2021) Livelihood vulnerability to flood Hazard: understanding from the flood-prone Haor ecosystem of Bangladesh. Environ Manag 67(3):532–552. https://doi.org/10.1007/S00267-021-01441-6/FIGURES/3

Huq ME, Shoeb AZM, Hossain MA, Fahad S, Kamruzzaman MM, Javed A, Saleem N, Adnan KMM, Sarker SA, Ali MY, Sarven MS (2020) Measuring vulnerability to environmental hazards: qualitative to quantitative. Environ Climate Plant Vegetat Growth 421–452. https://doi.org/10.1007/978-3-030-49732-3_17/COVER

Jones KML, McCoy C (2019) Reconsidering data in learning analytics: opportunities for critical research using a documentation studies framework. Learn Media Technol 44(1):52–63. https://doi.org/10.1080/17439884.2018.1556216

Jumriani J, Mutiani M, Putra MAH, Syaharuddin S, Abbas EW (2021) The urgency of local wisdom content in social studies learning. Literat Rev Innovat Social Studies J 2(2):103–109. https://doi.org/10.20527/IIS.V2I2.3076

Kirby RH, Reams MA, Lam NSN, Zou L, Dekker GGJ, Fundter DQP (2019) Assessing social vulnerability to flood hazards in the Dutch Province of Zeeland. Internat J Disaster Risk Sci 10(2):233–243. https://doi.org/10.1007/S13753-019-0222-0/FIGURES/2

Koomson I, Villano RA, Hadley D (2020) Effect of financial inclusion on poverty and vulnerability to poverty: evidence using a multidimensional measure of financial inclusion. Soc Indic Res 149(2):613–639. https://doi.org/10.1007/S11205-019-02263-0/TABLES/6

Laignel B, Vignudelli S, Almar R, Becker M, Bentamy A, Benveniste J, Birol F, Frappart F, Idier D, Salameh E, Passaro M, Menende M, Simard M, Turki EI, Verpoorter C (2023) Observation of the coastal areas, estuaries and deltas from space. Surv Geophys 2023:1–48. https://doi.org/10.1007/S10712-022-09757-6

Mahbubur Rahman M, Sadequr Rahman M, Jerin T (2022) Social vulnerability to earthquake disaster: insights from the people of 48th ward of Dhaka South City. Bangladesh. https://doi.org/10.1080/17477891.2022.2085075

Metcalfe SE, Schmook B, Boyd DS, De la Barreda-Bautista B, Endfield GE, Mardero S, Manzón Che M, Medina González R, Munguia Gil MT, Navarro Olmedo S, Perea A (2020) Community perception, adaptation and resilience to extreme weather in the Yucatan peninsula. Mexico Regional Environ Change 20(1):1–15. https://doi.org/10.1007/S10113-020-01586-W/FIGURES/2

Papadopoulos O, Lopez-Andreu M, Jamalian M (2021) Violation and lack of awareness of employment rights in the United Kingdom's hotel industry: isolation, fragmentation and barriers to labour enforcement. Ind Relat J 52(4):315–330. https://doi.org/10.1111/IRJ.12337

Pasca P, De Simone E, Ciavolino E, Rochira A, Mannarini T (2022) A higher-order model of community resilience potential: development and assessment through confirmatory composite analysis based on partial least squares. Qual Quant 1–22. https://doi.org/10.1007/S11135-022-01400-1/TABLES/8

Prentice CM (2017) Interviews, qualitative. International Encyclopedia Intercultural Communicat 1–8. https://doi.org/10.1002/9781118783665.ieicc0108

Purakayastha TJ, Bera T, Bhaduri D, Sarkar B, Mandal S, Wade P, Kumari S, Biswas S, Menon MCHP, Tsang DCW (2019) A review on biochar modulated soil condition improvements and nutrient dynamics concerning crop yields: pathways to climate change mitigation and global food security. Chemosphere 227:345–365. https://doi.org/10.1016/j.chemosphere.2019.03.170

Putiamini S, Mulyani M, Petala MP, Soesilo TEB, Karsidia A (2022) Social vulnerability of coastal fish farming community to tidal (rob) flooding: a case study from Indramayu. Indonesia J Coastal Conservat 26(2):1–15. https://doi.org/10.1007/S11852-022-00854-7/TABLES/4

Raza A, Razzaq A, Mehmood SS, Zou X, Zhang X, Lv Y, Xu J (2019) Impact of climate change on crops adaptation and strategies to tackle its outcome: a review. Plants 8(34). https://doi.org/10.3390/plants8020034

Sahana M, Rehman S, Paul AK, Sajjad H (2021) Assessing socio-economic vulnerability to climate change-induced disasters: evidence from Sundarban biosphere reserve, India. Geology Ecology Landscapes 5(1):40–52. https://doi.org/10.1080/24749508.2019.1700670/SUPPL_FILE/TGEL_A_1700670_SM7300.ZIP

Saja AMA, Goonetilleke A, Teo M, Ziyath AM (2019) A critical review of social resilience assessment frameworks in disaster management. Internat J Disaster Risk Reduct 35:101096. https://doi.org/10.1016/J.IJDRR.2019.101096

Setiawan HH (2023) The resilience of indigenous peoples in facing disaster risks due to extreme climate change on Alor regency. East Nusa Tenggara Indonesia 491–513. https://doi.org/10.1007/978-3-031-26143-5_21

Setiawan HH, Nuryana M, Susantyo B, Purwanto AB, Sulubere MB, Delfirman. (2021) Social entrepreneurship for beneficiaries of the program Keluarga Harapan (PKH) toward sustainable development. IOP Conference Series: Earth Environ Sci 739(1):012053. https://doi.org/10.1088/1755-1315/739/1/012053

Setyowati DL, Hardati P, Benardi AI, Hamid N, Anugrahanto YD (2021) The role of the disaster preparedness Group in Adapting Abrasion to communities affected by abrasion on the north coast of Rembang, Central Java. In: Proceedings of the 6th international conference on Education & Social Sciences (ICESS 2021), 578, pp 61–66. https://doi.org/10.2991/ASSEHR.K.210918.012

Shymbin BW, Nongbri G (2022) Impact of sand mining on the physical health of the river and the livelihood of the people: a case study of Umtyngngar River. Meghalaya:221–242. https://doi.org/10.1007/978-3-030-77572-8_11

Sugiura M, Nouchi R, Honda A, Sato S, Abe T, Imamura F (2020) Survival-oriented personality factors are associated with various types of social support in an emergency disaster situation. PLoS One 15(2):e0228875. https://doi.org/10.1371/JOURNAL.PONE.0228875

Sundler AJ, Lindberg E, Nilsson C, Palmér L (2019) Qualitative thematic analysis based on descriptive phenomenology. Nurs Open 6(3):733–739

Tate E, Rahman MA, Emrich CT, Sampson CC (2021) Flood exposure and social vulnerability in the United States. Nat Hazards 106(1):435–457. https://doi.org/10.1007/S11069-020-04470-2/TABLES/6

Wang Z, Liu J, Xu N, Fan C, Fan Y, He S, Jiao L, Ma N (2019) The role of indigenous knowledge in integrating scientific and indigenous knowledge for community-based disaster risk reduction: a case of Haikou Village in Ningxia, China. Internat J Disaster Risk Reduct 41:101309. https://doi.org/10.1016/J.IJDRR.2019.101309

Westoby R, McNamara KE, Kumar R, Nunn PD (2020) From community-based to locally led adaptation: evidence from Vanuatu. Ambio 49(9):1466–1473. https://doi.org/10.1007/S13280-019-01294-8/METRICS

Winsemius HC, Jongman B, Veldkamp TIE, Hallegatte S, Bangalore M, Ward PJ (2018) Disaster risk, climate change, and poverty: assessing the global exposure of poor people to floods and droughts. Environ Dev Econ 23:328–348. https://doi.org/10.1017/S1355770X17000444

9 Effects of Climatic Risks on Soil Erosion/Desertification in Southern and Northern Nigeria Using GIS/Remote Sensing Analysis

Angela Oyilieze Akanwa, Idakwo V. Iko-ojo, I. C. Ezeomedo, F. I. Ikegbunam, P. U. Igwe, L. N. Muoghalu, S. O. Okeke, A. U. Okonkwo, Chinwe Ngozi Odimegwu, K. F. Nkwocha, V. C. Arah, E. I. Madukasi, C. Anukwonke, Joel Mari Bwala, and M. Obidiegwu

Abstract

Global warming has triggered changes in climate resulting in diverse impacts across the globe. Nigeria is experiencing extreme rainfall and flooding causing massive erosional problems in the south while sparse rainfall has led to desertification in the north. Using the concept of climate action, this chapter employed both primary and secondary sources to examine the effects of climatic risks on soil erosion/desertification in southern and northern Nigeria using GIS/remote-sensing analysis. We gathered data using GPS, remote sensing, field observations, transect walks, discussions, and photography. Remotely sensed data were used for land use/land cover mapping of gullies in the southeast. Similarly, change detection of desert areas in the north was captured for a 20-year period (2000–2020). Our findings showed gullies and bare surfaces were predominant in the southeast region. Also, over a 20-year period in the north, a negative deficit of (−9.79%) in vegetation cover was observed, 8.91% increase in bare surfaces was recorded and built-up areas increased by 0.63%. Notably, there was an

A. O. Akanwa (✉) · I. C. Ezeomedo
Climate Change Impacts, Sustainability and Adaptation, Department of Environmental Management, Chukwuemeka Odumegwu Ojukwu University (COOU), Ihiala, Anambra State, Nigeria
e-mail: angela.akanwa1@gmail.com; ic.ezeomedo@coou.edu.ng

I. V. Iko-ojo
Climate Change, Department of Urban and Regional Planning, Faculty Environmental Sciences, University of Maiduguri, Borno State, Nigeria
e-mail: idawovicks@gmail.com

F. I. Ikegbunam
Ecological Disaster and Management, Department of Environmental Management, Chukwuemeka Odumegwu Ojukwu University (COOU), Ihiala, Anambra State, Nigeria
e-mail: iikebunam@gmail.com

P. U. Igwe
Ecological Studies, Department of Environmental Management, Chukwuemeka Odumegwu Ojukwu University (COOU), Ihiala, Anambra State, Nigeria
e-mail: pu.igwe@coou.edu.ng

U. Chatterjee et al. (eds.), *Climate Crisis: Adaptive Approaches and Sustainability*, Sustainable Development Goals Series, https://doi.org/10.1007/978-3-031-44397-8_9

increase of 0.24% in water bodies which could be attributed to analysis being carried out during the wet season where there is a minimal amount of rainfall to increase stream and river levels. Findings from the study indicate that there is an extreme case of vegetation loss, enlarging bare surfaces which signify extreme levels of desertification. It is expedient to engage all concerned stakeholders, actors, organizations at national and community levels for information exchange, policy advocacy, and climate action, particularly in southern and northern Nigeria.

Keywords

Soil erosion · Desertification · Drought · Climate action · Remote sensing · South/Northern Nigeria

Abbreviations

GIS	Geographical Information System
GPS	Global positioning system
LGA	Local government areas
NGO	Non-governmental Organizations
SED	Soil Erosion and Desertification

9.1 Introduction

Overtime, there have been scientific evidence of a shift in Nigeria's climate observing rising temperature, irregular rainfall patterns, increase in sea level and flooding, and drier lands added to prevalent extreme weather events in different regions of the country (Chatterjee et al. 2022; Akanwa and Joe-Ikechebelu 2019; Elisha 2017; Akanwa et al. 2022a, b).

L. N. Muoghalu
Policy and Administration, Department of Environmental Management, Chukwuemeka Odumegwu Ojukwu University (COOU), Ihiala, Anambra State, Nigeria
e-mail: leomuo2003@yahoo.com

S. O. Okeke
Landscape Design and Conservation, Department of Environmental Management, Chukwuemeka Odumegwu Ojukwu University (COOU), Ihiala, Anambra State, Nigeria
e-mail: so.okeke@coou.edu.ng

A. U. Okonkwo
Waste and Land Management, Department of Environmental Management, Chukwuemeka Odumegwu Ojukwu, University, (COOU), Ihiala, Anambra State, Nigeria
e-mail: auokonkwo@gmail.com

C. N. Odimegwu
Land and Property Policy, Taxation and Administration, Department of Estate Management, Faculty Environmental Sciences, Chukwuemeka Odumegwu Ojukwu University (COOU), Ihiala, Anambra State, Nigeria
e-mail: cn.odimegwu@coou.edu.ng

K. F. Nkwocha
Climate Change and Sustainability, Department of Geography, Faculty of Social Sciences, University of Maiduguri, Borno State, Nigeria
e-mail: kelechnkwocha@unmaid.edu.ng

V. C. Arah
Land Management Studies, Department of Environmental Management, Chukwuemeka Odumegwu Ojukwu University (COOU), Ihiala, Anambra State, Nigeria
e-mail: arahvictor@gmail.com

E. I. Madukasi
Pollutants Studies, Department of Environmental Management, Chukwuemeka Odumegwu Ojukwu University (COOU), Ihiala, Anambra State, Nigeria
e-mail: madify2002@gmail.com

C. Anukwonke
Climate Change and EIA Studies, Department of Environmental Management, Chukwuemeka Odumegwu Ojukwu University (COOU), Ihiala, Anambra State, Nigeria
e-mail: cc.anukwonke@coou.edu.ng

J. M. Bwala
Land Management Studies, Department of Geography, Faculty of Social Sciences, University of Maiduguri, Borno State, Nigeria
e-mail: joelmaribwala@unimaid.edu.ng

M. Obidiegwu
Climate Change and Sustainability, Department of Environmental Management, Chukwuemeka Odumegwu Ojukwu University (COOU), Ihiala, Anambra State, Nigeria
e-mail: michaelobidiegwu@gmail.com

Climate change refers to long-term shifts in temperature and weather patterns which could be attributed to varied drivers though these changes are traceable to unsustainable human actions (Chatterjee et al. 2022). There are different challenges associated with climate change across the country due to the existing tropical climate and dual precipitation regimes. The North experiences low rainfall while high rainfall is observed in the southwest/east. Based on this, climatic risks such as aridity, drought, and desertification are occurring in the north while flooding and erosion have become a glaring nightmare in the south (Akande 2017; Nkechi 2016; Akanwa and Ezeomedo 2018). Interestingly, vulnerability analysis has proven that the northern region of Nigeria has extreme exposure to global warming when compared to the southeast region (Nkwocha et al. 2019). These climatic changes have determined rainfall distribution in both regions resulting in diverse land degrading problems such as soil erosion and desertification.

Soil/land degradation is a severe environmental issue that affects about 1.9 billion ha of land globally. Notably, over 24 billion tons of soil resource is lost annually by land degrading processes such as erosion or desertification (Yousuf et al. 2022; Babu 2016). Soil degradation can occur either as erosion or a desertification problem depending on location and other related factors. There is a relationship between soil erosion and desertification since they are both induced by similar drivers though occurring in different regional areas in Nigeria and around the globe under diverse natural and human factors.

Soil erosion is defined as a process by which grains of sand are dragged by the action of water or wind (FAO 2014). Soil erosion is referred as the process of removing soil and vegetal resources, plant nutrients, from the top surface by the various agents of denudation, and man's action (Akanwa et al. 2017). Soil provides the foundational support system for human, animal, and plant existence particularly agricultural production that feeds over 90% of the world population (Borrelli et al. 2020) indicating its indispensability. When soils and lands are degraded, it escalates issues such as erosion and land fragmentation rendering the soil infertile thereby reducing the quality and production levels of food crops (Muoghalu and Akanwa 2021; Okorafor et al. 2017).

A desert area can be identified with extreme water loss that cannot sustain planetary life or existence (Bellamy (2007), while desertification is an expansion of deserts due to human activity. United Nations (2012) defined a desert as an excessively dry land that severely lacks sufficient rainfall to support life. It is located in regions with arid climates particularly the Sahara desert and Northern parts of Nigeria. The dry climatic conditions added to human actions can hinder the growth and development of vegetation and animal life. Considering the unsustainable actions of man on land, it makes the soil susceptible to intractable issues (Akanwa and Ezeomedo 2018). Unregulated soil exploitation in Nigeria has led to severe soil loss and desertification which are immense environmental hazards faced in different parts of the country (Ofomata 1985).

Generally, Nigerian soil is undergoing dynamic changes occurring in several parts of the nation under different geological, climatic, and soil conditions. Southern Nigerians have experienced gross soil erosional problems for a long time so that farming activities have been severely affected creating low food production due loss of farmlands (Adiaha 2021). Major human drivers of soil degradation vary from deforestation, poor crop management/agronomic practices, and harsh climatic conditions among others (Okorafor et al. 2017). Also, natural factors such as the action of water or wind combine human drivers to sponsor rapid loss of topsoil, nutrient depletion, physical degradation, and reduction of soil biodiversity. These combined actions and more are responsible for soil degradation and erosion whereby the consequences and occurrences differ amounting to a global environmental phenomenon.

Soil erosion has been a serious concern since the 1920s during the inception of the southeast region, exemplified by the creation of a forest reserve and its protection against erosion in Enugu State precisely Udi (Skyes 1940). Although soil erosion occurs in all parts of

Nigeria, it is most prevalent in south-eastern Nigeria which is one-third of southern Nigeria, comprising of Abia, Anambra, Ebonyi, Enugu, and Imo states (Igwe 2012; Yusuf et al. 2019). This is because the clay soils are strongly weathered being formed from an older geologic location sourced from heavily weathered materials that are easily erodible (Oguike and Mbagwu 2009a, b).

There are existing long-aged gullies occurring within the length and breadth of south-eastern Nigeria. Unfortunately, presently new gullies are emerging at an unprecedented rate that threaten the survival of adjoining indigenous communities and their livelihoods (Akanwa and Ezeomedo 2018). Similarly, though with different occurring land degradation attributes, northern Nigeria has experienced extended dry periods over time resulting in drought and desertification. Northern Nigeria has experienced desert encroachment since the 1920s, though the situation was worsened during the food scarcity witnessed in the early 1970s. Desertification affects 15 northernmost states of the country (Jaiyeoba 2002).

Generally, environmental degradation particularly soil erosion and desertification experienced in the southern and northern Nigeria have become an impediment to economic growth in Nigeria (Oriavwote 2019). Interestingly, this study applied remote sensing and global information systems (GIS) to cover and provide multi-spectral and temporal satellite images to detect the affected regions. This will not only improve the accuracy of the existing gullies and arid lands but also initiate proper monitoring. Thereby, informing decision-making related to sustainable land management and climate action.

There is a desperate need for climate action(s) added to sustainable land development and management approaches that should limit unregulated human and natural drivers from degrading our lands. Hence, our chapter focused on the critical examination of these drivers, mapping out the extent of land degradation in the southern and northern parts of Nigeria to provide possible solutions to the present situation. Precisely, this study captured the extent of soil erosion by mapping the gullies in the southeast while a 20-year period of desertification spread in northern Nigeria from 2000 to 2020 was also examined.

9.2 Climate Action as Means to Minimize Soil Erosion/Desertification in Nigeria

Climate action is defined as diverse corporate strategies and efforts embarked on to mitigate the consequences of a changing climate by corresponding acts, policies, and advocacy measures (https://bing.com/search?q=what+is+climate+action). It covers all the means strategized at minimizing the emission of greenhouse gases, adaptation by indigenous groups, action plans, and awareness of climate action. According to the recent IPCC Climate Report (2022), climate change has expedited global temperature rise with resultant severe weather occurrences such as intense rainfall, floods, uncontrolled fires, severe heat, and drought around the globe.

Climate change has affected the African continent and Nigeria precisely creating extreme inequalities that negatively affected human life, economies, and lifestyles. There have been records of immense destruction, migration of indigenous communities, loss of properties, and wellbeing due to extremely unpredictable events (IPCC Report 2014; Akande 2017; Ebele and Emodi 2016; Akanwa et al. 2022a, b). In south-eastern Nigeria, the length of rainfalls and their strength have exacerbated floods that bring about immense erosion (Akanwa et al. 2022a, b).

However, the decline in precipitation levels and high temperatures have triggered droughts and desert spread in the northern Nigeria (Nkwocha et al. 2019). Lake Chad and other lakes in the north have become susceptible to drying up (Dioha and Emodi 2018; Elisha 2017). Interestingly, greenhouse gas emissions are rising globally with the mean surface temperature estimated to increase over the twenty-first century reaching far beyond 3 °C this century.

Additionally, findings from scientific investigations have shown that human activities contribute to climate change risks thereby exacerbating environmental degradation processes such as deforestation, flooding, landslides, erosion menace, and desertification in Nigeria. Proper monitoring of our ecological resources is still a drawback to many developing countries (Ibitoye and Adebayo 2010). As urbanization in Nigeria is becoming fast, a rapid problem embalmed with the consequent challenges that lead to a more severe damage to the fragile natural resources and the ecosystem (Ibrahim 2018).

As cities grow, there is an increased demand for land that sponsors forest loss and environmental devaluation (Sulaiman et al. 2017). There are estimations of the global explosion of the human population by 2050. Where 67% of these population will be domiciled in cities in developing countries. Urban growth will readily promote environmental and economic inequalities (Abalaka and Kebiruumor 2019; Okeke et al. 2020).

Interestingly, urbanization sponsors deforestation that aids climate emergency causing desertification and urban heat island in Maiduguri Metropolis (Jimme et al. 2020; Nkwocha et al. 2019).

West and Wodike (2019) maintained that uncontrollable urbanization in Rivers State has resulted in environmental degradation such as air pollution, deforestation, erosion, and problem of waste management. Sadiq and Surayya (2019) stressed that an increase in population in Yola, South Local Government Area in Adamawa state has led to over-exploitation of vegetative cover leading to deforestation and desert encroachment in the area. The rapid growth of urban centers and urban migration has brought an unprecedented increase in soil erosion and desertification in Nigeria. Forest reserves in many parts of the country are gradually disappearing due to large-scale development and neglect from appropriate authorities. Obviously, soil degradation in Nigeria requires urgent attention from both the state and the federal government to avoid more severe environmental disasters in the near future.

Apart from urbanization, the Nigeria Erosion and Watershed Management Project (NEWMAP 2019) reported that about 90% of damaged lands in the country were due to poor termination of drains during road construction activities. More than 160 buildings were affected by erosion as a result of the construction of the second Niger Bridge in Anambra state (Daily Trust Newspaper 2020). Medugu et al. (2014) opined that the construction of mass housing estates in Abuja, Nigeria, has resulted in some forms of land pressure that led to soil pollution and risks of desertification and drought especially behind Godab and Ipent estates in the area. Onyejekwe et al. (2022) also asserted that soil erosion and gullying in the southeast region have resulted from the combination of poor construction practices.

Notably, the Paris Agreement encouraged all countries and continents to monitor and respond to the global threat of climate change by minimizing the global temperature increase below 2 °C. More so, to take successful steps to limit the temperature rise to as low as 1.5 °C, this was agreed upon on 4 November 2016. This is why, climate action must be considered as a veritable tool in minimizing climate change and its impacts. Observing that climate change is connected to all 16 goals of the 2030 agenda for Sustainable Development. So that, if we fail in taking climate actions, we fail in all measures of achieving all the sustainable development goals.

Nigeria adopted the Paris Agreement at the COP21 in Paris on 12 December 2015 and has taken major strides in providing a national climate action plan. Nigeria has taken action work especially in areas including minimizing carbon emissions, cleaner energy, and a range of other measures that will reduce emissions and increase adaptation efforts (Anukwonke et al. 2021). However, Nigeria still requires international cooperation to aid in achieving a low-carbon economy. The Working Group III (IPCC 2022) report provides a recent series of global observations, carbon emissions, and mitigation prospects for climate change.

9.3 Land/Soil Degradation in South/Northern Nigeria

The southeast gully erosion crisis has been ongoing before the British colonial era. According to Egboka and Okpoko (1984), the Agulu-Nanka gully started forming around 1850. The gully erosion covers an area of about 1100 km and the rate of growth is estimated at 20–50 m per year (see Plate 9.1). The British Colonial masters and the local inhabitants attempted to control the gullying by the construction of small dams and planting trees, but all the measures failed. This gully is one of the largest in Nigeria at 66 meters deep, 2900 m long, and 349 m wide, still guzzling red earth from underneath people's homes and farms, at a very alarming rate (World Bank 2012). Also, Grove (1951) studied land degradation in parts of the former Eastern and Northern provinces of the country indicating their menace. Ofomata (1964, 1965, 1966, 1967, 1973, 1978, 1980, 1981, 1984, 1985) studied the problem extensively in the southeast where population and land distribution are low compared to other parts of Nigeria. Other studies were carried out to indicate the severe consequences of gully erosion in Benue and Delta states as well (Ologe 1988; Jeje 1978; Sada and Omuta 1979).

The southeastern region lies within Awka Orlu uplands and Enugu-Awgu-Okigwe escarpment (Chiemelu et al. 2013), the nature of the soils has been extensively weathered making it easily eroded by wind and water. Notably, soil degradation involves the physical, chemical, and biological deterioration of soils (Celik 2005). Accordingly, studies carried out on gully erosion in the southeast have indicated the porosity and erodibility of the soils making them vulnerable to erosion (Onu 2006; Eboh and Lemchi 1994; Adekalu et al. 2007; Okpala 1990; Chiemelu et al. 2013). Further, common human activities practiced in the southeast fuel soil erosion examples include bush burning, continuous cultivation, and mining on hillside slopes (Akanwa and Ikechebelu 2021; Nwachukwu and Onwuka 2011; George et al. 2008; Okagbue and Uma 1987; Onu 2005; Osadebe and Akpokodje 2007; Eze 2007; Egboka and Orajaka 1990; Ezezika and Adetona 2011; Idowu and Oluwatosin 2008; Oguike and Mbagwu 2009a, b).

Desertification and soil erosion are the most glaring land degradation menace in Nigeria and historically both of them began in the 1920s in the north and southern parts of the country, respectively. Over time, glaring impacts of desert encroachment have been observed in 1968 in the northern region particularly Sokoto, Zamfara, Katsina, Yobe, and Borno states (Olagunju 2015). These five northern states have experienced an extended season of aridity for 15 years from 1981 to 1997 with 1987 being the worse hit for drought.

In recent times, drought has spread from 5 to 15 northern states in Nigeria (Jaiyeoba 2002). Unfortunately, these states supply most of the country's agricultural products like beans, melon, onions, cows, and rams, among others (Audu and

Plate 9.1 The effects of uncontrolled gully erosion over years in the Agulu Nanka community, in Anambra State where houses, lands, economic trees, and lives have been lost. (Source: Authors 2022)

Adie 2018) worse still, the desert conditions are fast spreading southwards (Cullis et al. 2019; Otegbeye 2004). Regrettably, gullies and desert conditions are wicked problems that sabotage the process of sustainable development (West and Wodike 2019). Since 2009, the Nigerian Meteorological Organization has monitored drought occurrences in the north with its monthly updates reported in a Bulletin. However, in spite of the awareness of gullies and desert spread, human activities coupled with the intensified changing climate have continued to accelerate these intractable phenomena in Nigeria (Akanwa and Joe-Ikechebelu 2019).

9.4 Materials and Methods

We employed qualitative, remote sensing, and GIS methods in this chapter to examine the effects of climatic risks on soil erosion/desertification in southern and northern Nigeria using GIS/remote-sensing analysis. The study examined the land degradation issues and monitored the extent of soil erosion and desertification in southern and northern Nigeria, respectively. This was done by capturing the extent of soil erosion and producing a location map of all erosion sites in southeastern states while in the north, a 20-year period of desertification spread was carried out.

We adopted primary and secondary sources of data. The primary data include GIS and satellite images derived from Landsat 8 (2021) for southeast states. Landsat 7 (2000) and Landsat 8 (2020) imageries were used to detect the land use/cover change over a 20 in the north. Remote-sensing data were processed using digital image-processing techniques. The ArcGIS 10.3 software was used to produce the land use map, and Arcmap was employed to derive unique features on the image. A supervised classification technique was used to produce information on land uses. Further, qualitative data include in-depth interviews with place-based researchers, observation, photographs, and discussion with a research team from the study regions. GIS and GPS were adopted to derive the maps and place the geographic positions of the SED. Secondary sources include reviewed literature from the Internet, studies on soil erosion and desertification, government newsletters, and fact sheets on gully erosion and desertification.

9.5 Results and Discussion

The analysis and results of our chapter were based on the data generated from qualitative, remote-sensing, and GIS-adopted methods. The results from remotely sensed data and GIS of southern and northern regions were analyzed and interpreted. Further, the summary of data gathered from interviews and discussions with placed-based researchers which include socio-economic, environmental/soil characteristics, and consequences of gullies and desert spread were presented. Additionally, observations and photographs were presented alongside secondary information. However, the unique experiences of interviewed key informants who are place-based researchers were captured over specific issues or a general experience agreed/accepted by group members.

9.6 Soil Erosion and Its Effects in Southeastern Nigeria

Results from the supervised land use and land cover map of selected southeastern states using Landsat 8 (2021) satellite image is shown in Fig. 9.1. The supervised map was employed to classify the observed land uses indicating five land uses. Black dots indicate the gully sites, bare surfaces were shown in yellow, while built-up areas were reflected in red, vegetation in green shade and water bodies represented in blue. It is notable that gullies and bare surfaces were predominant in the four states in Nigeria.

Results from remotely sensed data and discussions revealed that soil erosion is active in all six regions of Nigeria, namely, south–south, southeast, southwest, and north central (middle belt), northeast, and northwest though in different intensities. Erosion sites were found virtually in all states of the federation, but with a higher mag-

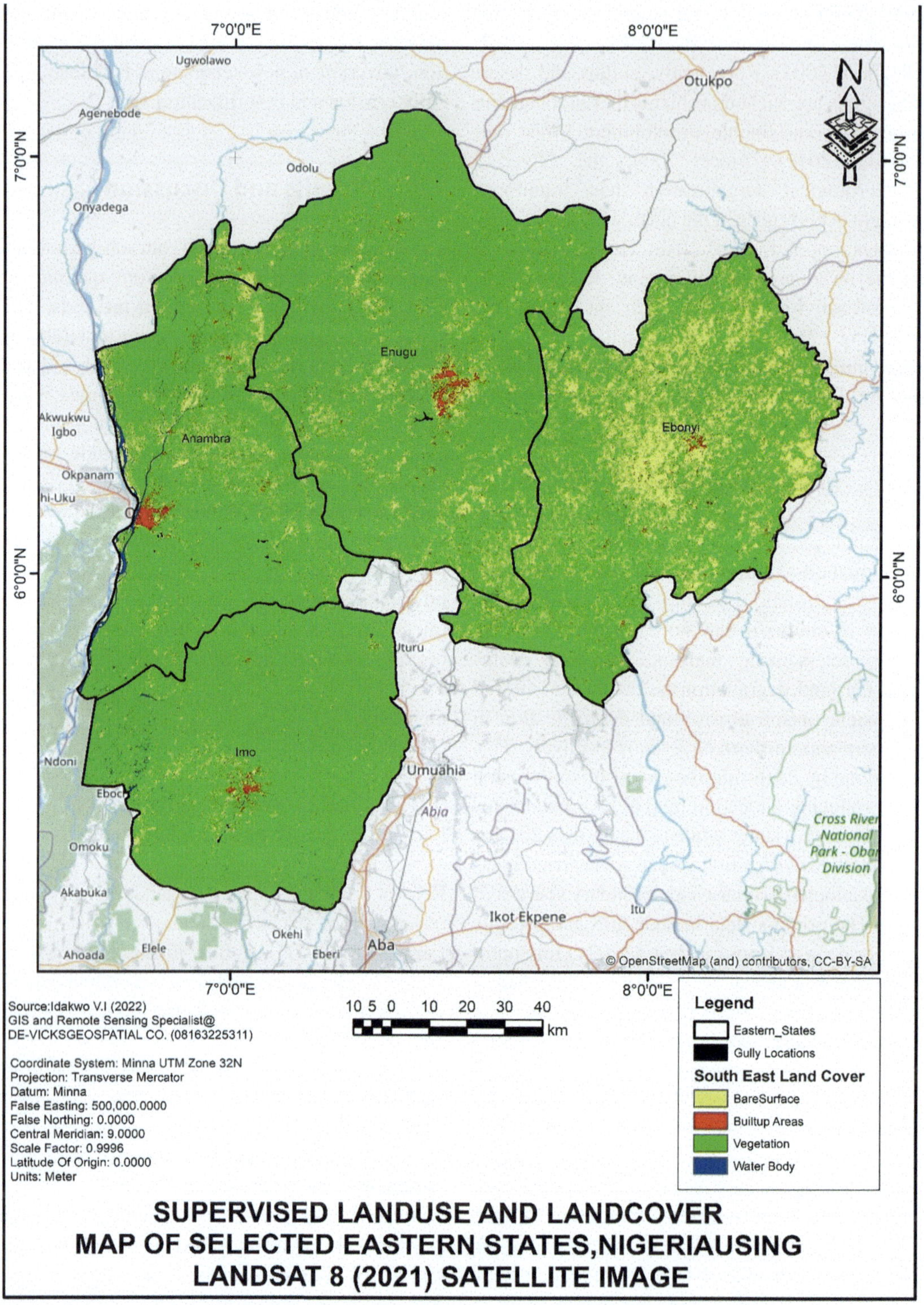

Fig. 9.1 Map of southeast showing four states and distribution of gullies. (Source: Authors 2022)

nitude in the southeastern part of the country. The researchers informed that the distribution of gully sites in the country was in increasing order of southeast, middle belt, southwest, south–south, northwest, and northeast. So, the southeast region is where the most extreme type of soil erosion (gully erosion) occurs which is the highest in Nigeria (Akanwa and Ezeomedo 2018). This region is synonymous with gully erosion found almost in all communities. Table 9.1 shows the spread of varying erosion sites in the five southeastern states of Nigeria. One of the key informants interviewed, a researcher specified that...

> Soil erosion is a factor of climate change because it destroys vegetation which is a sink for carbon dioxide (CO2) and exposes soil biomass which oxidized to form CO2, a potent greenhouse gas. Unarguably, gully erosion is an effect of climate change due to the extreme weather events, manifested in rainfall of high intensities which detaches and transports soil particles. Although gully erosion occurs in all parts of Nigeria, it is predominant in southern Nigeria, particularly Anambra State with many numerous small, medium and large gullies of humongous dimensions. (F.I. Ikegbunam)

Accordingly, Igbokwe et al. (2003) and Egboka (2004) revealed the spread of studied erosion sites in diverse degrading conditions. In Abia, there are 300 sites and Anambra has 700. Ebonyi has 250 gullies, Enugu has 600 gullies, and Imo has 400. New sites keep developing during each rainy season due to flooding and torrential rainfall in the south east (Akanwa et al. 2022a, b).

Studies revealed that about one-tenth of Anambra state is covered with gully sites and it keeps expanding. At least for the past 15 years, about 2000 gullies have been in existence thereby destroying about 36.4 km^2 of land meant for farming purposes (Egboka 2004; Ogbonna 2009; Umeugochukwu et al. (2013). According to the study by Ogbonna (2012) in Orlu, Imo state showed that for the past 24 years, the gully areas have drastically increased from 24.49 km^2 and it is still expanding. In Abia state, this menace of soil erosion occurs in Amucha, Isuikwuato, Ohafia, Abriba, and Arochukwu Local Government Areas. Generally, the adverse environmental impacts of soil erosion in Nigeria (Okuh and Osumgborogwu 2019; Enete 2014; Mezie and Nwajiaku 2020; Akanwa and Ezeomedo 2018) were loss of lives, sedimentation of water bodies, loss of properties, reduced agricultural production, soil loss, and displacement of human populations that are prevalent.

Among the five studied southeastern states of Nigeria, Anambra has the highest number of active gullies that keep spreading. Agulu, Nanka, and Oko communities of the state are the worst hit (see Plates 9.1 and 9.2). The estimated number of gullies in the country is 3000 with the regional soil erosion crisis accelerating alarmingly in recent decades, threatening about 6% of Nigeria's land mass (World Bank 2012), while Anambra state is faced with huge consequences affecting food security and the livelihood of farmers (Akanwa and Ezeomedo 2018). According to a researcher in Anambra state, who specializes in soil erosion informed that...

> Gully erosion is a hazard, creating footprints in the soil with the potential to form irreversible badland such as that of Agulu-Nanka-Oko in Anambra State. It has created numerous environmental and

Table 9.1 Distribution of gully erosion sites in southeastern Nigeria

Site no.	State	No. of gullies	State of the gully site	Control measures
1	Abia	300	Mostly active/some dormant	Not successful
2	Anambra	700	Mostly active	Not successful yet
3	Ebonyi	250	Mostly minor gully sites	No records
4	Enugu	600	Some active/some dormant	None
5	Imo	450	Some active/some dormant	Not successful yet

Source: Egboka et al. (2019)

Plate 9.2 Access into community houses destroyed by gully erosion in Oko, Anambra State (2022). (Source: Authors 2022)

> socio-economic problems that include soil loss, sedimentation of water bodies, loss of biodiversity, damage of properties, destruction of roads and footpaths, closing of 'eyes' of aquifers of rivers, springs, streams and lakes as well as swallowing of arable lands. (P.U. Igwe)

Anambra state government has disclosed that gullies have destroyed about 70% of the land area (Nzeagwu 2022). Worse still, the damage posed by erosion menace is usually comprehensive and total. Everything is eaten up by the deep gullies from buildings, farmlands, infrastructure, and livelihoods to many others (see Plates 9.1, 9.2, and 9.3).

Erosion affects real estate by complete destruction of properties; decrease in property values; exposure of building foundations; destruction of roads, beaches, and harbors (Wokekoro 2020) (see Plate 9.2); damage to access roads or other infrastructure; loss of both rental and capital.

Studies have shown that property and land values tend to rise rapidly in areas with good transportation networks and less in areas without such improvements. There is a very strong relationship between location and property value and accessibility increases property values. Rental values of residential and commercial properties in areas with erosion-devastated roads will be less than what is obtained in other areas. Likewise, the capital value (total worth) of land and buildings in such areas. The extent of the impact on transaction prices of real estate depends on the property distance from the erosion site. The closer the property is to the erosion site, the more the impact on the value. An example is given of two erosion sites in Anambra state where the effect of erosion on land and property values is seen.

The first erosion site is located along Nitel Road, Alatiapom off Onitsha Owerri Road in Ihiala, Anambra state. The erosion was caused by indiscriminate active laterite excavation on the site. It has led to the loss of farmlands and declining land values. A plot of land in the neighborhood sells between five million and seven million Naira (11,643.15–16,334.66 USD) (ngn.currencyrate.today/conversion). But parcels of land close to the erosion site are sold at 1.5 million–2 million Naira (3544.24–4737.09 USD).

The second erosion site is located at Hundred Foot Road, Nnewi, Anambra State. The erosion menace commenced in the 1990s but has been managed by the residents of the area. The site being close to the popular Agbaedo Motorcycle Spare parts was used as a dump for metallic scraps. In later years, when people became aware of the economic values of those scrap metals,

Plate 9.3 Massive destruction of community lands by erosion in Anambra State. (Source: Authors 2022)

scavengers embarked on the excavation of the scraps used in the erosion control and, by the year 2017, the erosion became active again.

Investments in real estate around the area became unattractive and land prices were as low as one million Naira (2.349.53 USD) per plot compared to five million in adjoining neighborhoods. In the year 2019, the state government filled the gully and constructed water channels which brought the erosion to a halt. Today, a piece of land in the area sells between 20 million and 28 million Naira (47,370.91-66,319.28 USD) and rental values have also appreciated considerably.

9.7 Effects of Desertification in Northern Nigeria

Further, results from remotely sensed data and GIS of northern states (2000 and 2020) over a 20-year period were analyzed and interpreted as shown in Figs. 9.2 and 9.3. The supervised classification approach was employed to differentiate the images. This was carried out to connect the digital land cover classification to the spectral features of the image to provide vital data observed on a map (Akanwa et al. 2017) and therefore used to determine the extent of desertification in these states.

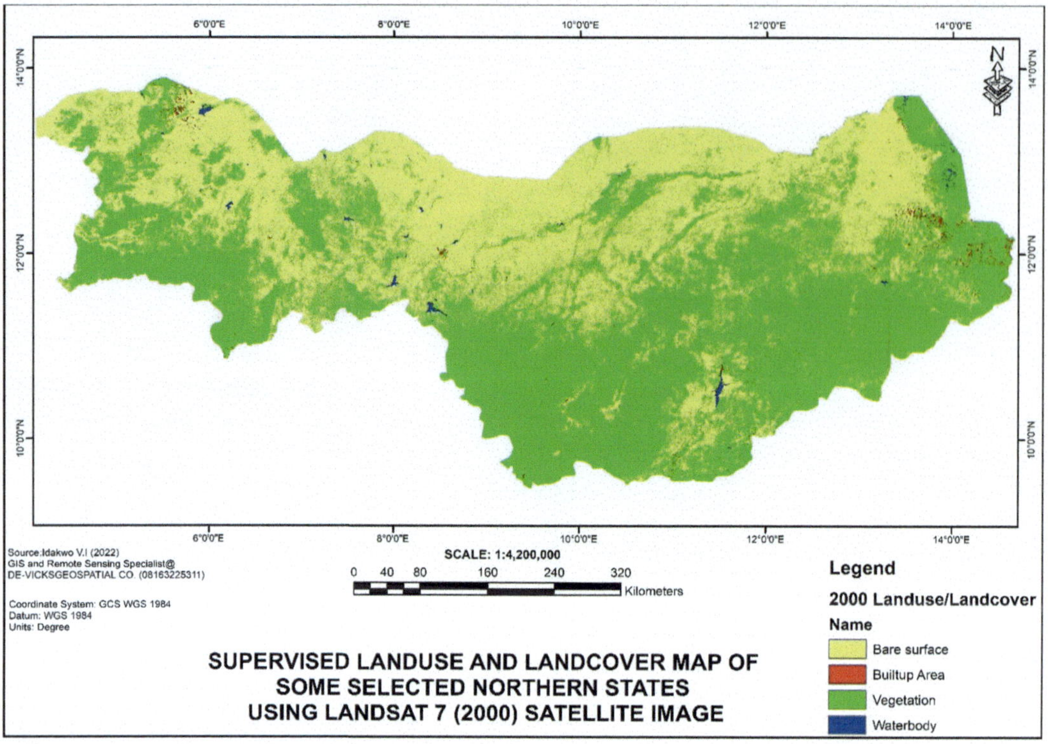

Fig. 9.2 Map of northern Nigeria (2000) showing a vast loss of green area. (Source: Authors 2022)

Table 9.2 Desertification in northern Nigeria States for the year 2000

Name	Area	Percentage
Bare surface	14,860,680.4	46.94
Built-up area	232,694.5	0.73
Vegetation	16,433,632.1	51.90
Waterbody	134,891.0	0.43

Table 9.3 Desertification in northern Nigerian States for the year 2020

Name	Area	Percentage
Bare surface	17,682,931	55.85
Built-up area	431,485.7	1.36
Vegetation	13,334,268	42.11
Waterbody	213,212.8	0.67

However, the supervised classification technique in Fig. 9.2 identified four classifications of land cover in the area. Figure 9.2 and Table 9.2 show that the bare surface occupied was 46.94%, while the built-up area was 0.73%, vegetation 51.90%, and water bodies occupied 0.43%. Interestingly, Fig. 9.3 and Table 9.3 show a 20 shift in land uses of these states where bare surface occupied was 55.85%, while the built-up area was 1.36%, vegetation 42.11%, and water bodies occupied 0.67%. Comparing these two maps Figs. 9.2 and 9.3 and the contents of Tables 9.2 and 9.3, a notable difference and a drastic land use/cover change in the region is seen after 20 years in Table 9.4. Findings showed that the percentage change for bare surfaces is 8.91%, while the built-up area was 0.63%, vegetation −9.79%, and water bodies occupied was 0.24% (see Table 9.4).

Unarguably, findings from the results indicated that there is a negative deficit of −9.79% in vegetation cover in studied northern states over the 20-year period. Also, a notable increase in bare surfaces is observed from 46.94% in the

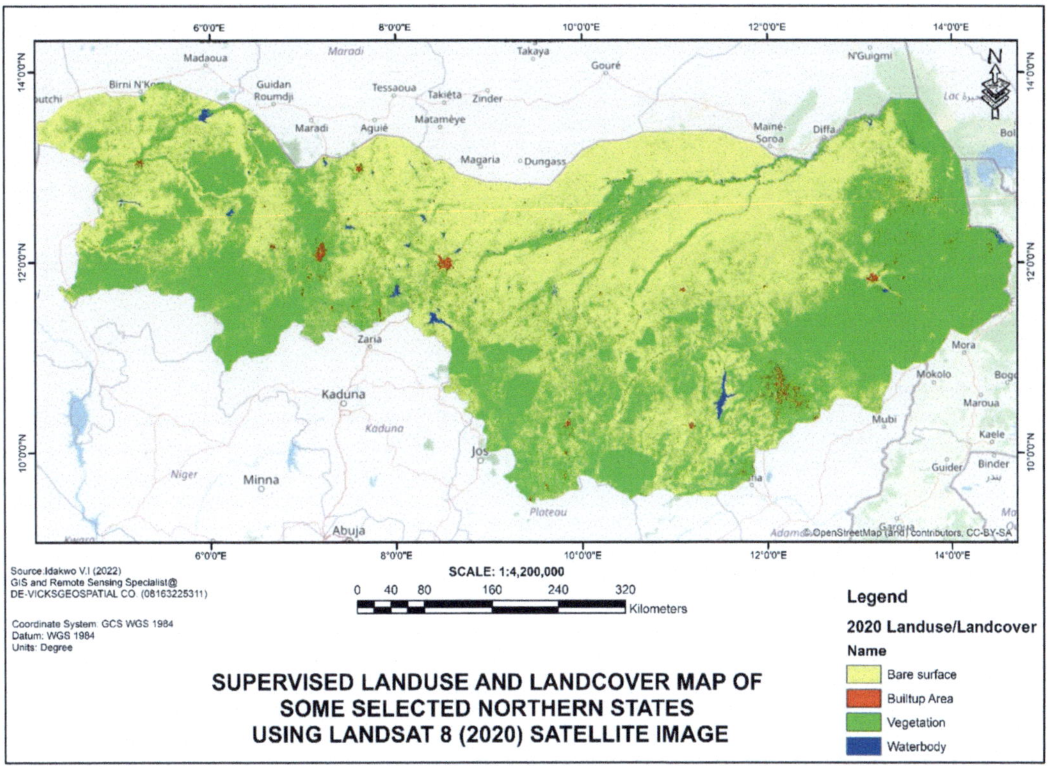

Fig. 9.3 Map of northern Nigeria (2020) showing green cover change within a 20-year period. (Source: Authors' Experimental Analysis 2022)

Table 9.4 Percentage change between the years 2000 and 2020

Land use	2000		2020		Percentage change
Name	Area	Percentage	Area	Percentage	2022–2000
Bare surface	14,860,680.4	46.94	17,682,931.4	55.85	8.91
Built-up area	232,694.5	0.73	431,485.7	1.36	0.63
Vegetation	16,433,632.1	51.90	13,334,268.2	42.11	−9.79
Waterbody	134,891.0	0.43	213,212.8	0.67	0.24

year 2000 to 55.85% in 2020 indicating an 8.91% increase. Similarly, the built-up areas have increased with a 0.63% difference over the years. However, there is an increase of 0.24% in water bodies which could be because this analysis was carried out during the wet season where there is a reasonable amount of rainfall to increase stream and river levels.

Interestingly, the findings from the land use/cover change derived from map analysis are an indicator that there is an extreme case of vegetation loss, enlarging bare surfaces and built areas, and extreme levels of desertification (see Plate 9.4). In accordance with the findings of this chapter, an interviewed researcher based in north, Maiduguri, Borno State informed that…

> The rapid rates of desert spread is alarming and this calls for immediate actions as desert are fast taking over most northern states in Nigeria where vegetal resources such as plants, trees and grasses

Plate 9.4 A large area of dry land facing desert spread in northern Nigeria. (Source: Authors 2022)

> have been lost to bushes. There are huge expanse of dry lands with predominant features of deserts in the north. (F.C. Nkwocha)

Desertification occurs in the three Sahelian northern regions of Nigeria: north central (middle belt), northeast, and northwest where the precipitation is low. According to Jaiyeoba (2002), 15 northern states have been affected, including Katsina, Sokoto, Borno, Kano, Jigawa, Zamfara, Kebbi, Niger, and Adamawa, among others. Like soil erosion, desertification is an environmental hazard in Nigeria. From the perspective of Audu and Adie (2018), desertification has affected 580,841 km^2, out of 927,892 km^2 total land area of Nigeria and its 62 million people. There are predictions that about 50 to 70% of some northern states namely Bauchi, Borno, Gombe, Jigawa, Kano, Katsina, Kebbi, Sokoto, Yobe, and Zamfara States have been besieged by deserts. Unfortunately, these states harbor about 27 million people accounting for 38% of Nigeria's total land area.

Therefore, a lot more national and international interventions are needed to curb the hazards already created by both desertification and soil erosion in the country. Such interventions must endeavor to integrate the perspectives of the affected people who have become acclimatized to the two environmental problems.

In 1998, Babsal opined that it is a common ecological problem of eight northern Nigeria which affects areas above latitude 12°N, namely, Borno, Yobe, Jigawa, Kano, Katsina, Zamfara, Sokoto, and Kebbi with a total land area of about 125,000 km^2 comprising about 13% of the country's land area (Babsal 1998).

Regrettably, another northern researcher with a focus on desertification added that…

> Nigeria's land area is over nine million kilometer square, however, about five million kilometer square covering over sixty per cent of the total land area has been affected by deserts and it is still expanding. (Joel Mari Bwala)

For example, in Damaturu, about 11 local councils have suffered desert encroachment for over 20 years (Olagunju 2015). Worse still, the widespread of deserts in these northern states is estimated to be about 600 meters accounting for the displacement of over 20 million people and their animals into the south (Olagunju 2015).

9.8 Intervention Strategies and Sustainable Pathways

Deserts create dunes that cover a large expanse of agricultural lands, oasis, and ponds, with negative implications on livestock production and food security. Nigeria loses about 350,000 hectares of land every year to desert encroachment (Tercula 2002). Also, Nigeria loses about five billion dollars annually due to the widespread of desert in most northern areas (Odiogor 2010) where the Sahara Desert is gradually advancing southwards at the rate of 6.0% every year. Although the Great Green Wall Programme (GGWP) and the Nigerian Erosion and Watershed Management Programme (NEWMAP) are two intervention strategies in place for desertification and soil erosion in Nigeria, respectively, the two phenomena are fast acquiring more territories.

Soil conservation measures should be adopted to reduce the problem of soil degradation. The most important aspect of it is to control soil erosion and desertification. There is a urgent need to adopt measures such as agroforestry, rainwater harvesting, retaining walls, contour terracing and ridging to conserve nutrients and water run-off, cautionary expansion of cultivated sites mixed farming, strip cropping, and early planting (Akanwa et al. 2019, 2020; Adiaha 2021).

The measures should depend on the indigenous features such as land slope, soil type, water bodies, and farming patterns and lifestyles. Notably, southeast soils are easily eroded and considerations should be made on the type of agricultural products that should be cultivated giving preference to cover crops that enrich and protect the soil surface from erosion (Maitra et al. 2020).

Cover crops play protective roles in managing rainfall intensity and floods that are prevalent in the southeast and also supplying the soil with nutrients rich in nitrogen (Gloria 2022; Anjali 2020).

On the other hand, the practice of afforestation should be encouraged as it reduces the rate of run-off and infiltration, hence, controlling soil erosion. It is also regarded as the most effective measure against desertification. The planting of trees and shrubs establishes the soil; therefore, active cultivation should be encouraged to enable soil protection (Adiaha 2021; Ul Zaman et al. 2018).

Other factors of desertification such as overgrazing and bush burning should be monitored and discouraged by disallowing indiscriminate cattle grazing in lands allotted for agricultural production. Instead of the practice of outdoor grazing which is a long-time tradition in the north, herdsmen can embrace the concept of creating grazing reserves in preferred locations. This will make management easier coupled with improved practices. Additionally, indiscriminate bushfires and burning should be disallowed considering their harmful tendencies on the soil.

An all-inclusive engagement in advocacy, public environmental education, carbon monitoring, mobilization of resources, and grassroot participation by NGOs, CBOs, youths, and community members is expedient for climate actions and also, the implementation of Nigerian climate change policy (Federal Ministry of Environment 2021; Amanchukwu 2015; Daisy 2020) (see Plate 9.5). Unarguably, Nigerian youths account for 70 per cent of the 217 million population, which represents 151 million youths and 42 percent of the 70 percent are under 15 years.

> Nigerian youths can serve as key resource in climate actions for a sustainable enviornment such as awareness programmes, protection of marginal lands, tree planting, sustainable agricultural practices and transition to green energy. (Angela Oyilieze Akanwa).

Also, similar studies centered on the use of GIS, remote sensing, and land use and land mapping of areas suffering erosion and desertification should be encouraged. This will provide an appropriate means for monitoring changes over time. Their continuous application should provide suitable soil management and environmental monitoring (Igwe 1999, 2003; Akanwa et al. 2017).

9.9 Conclusion

The effects of land degradation such as soil erosion and desertification in southeastern and northern Nigeria, respectively, have become

Plate 9.5 Youths can play a role in information on human actions on soil degradation. (Source: Authors fieldwork 2022)

alarming requiring immediate attention. This is a major concern for legislators, government actors, stakeholders, community members, and individuals. Presently, these land degradation challenges have attracted local, national, and international attention on the need for climate action, soil erosion, and desert monitoring in Nigeria though these efforts have proved to be abortive. Formulation of climate policy is a necessity to reduce human-induced greenhouse emissions responsible for large-scale shifts in weather patterns. Also, attracting local and international funding of programs, projects, and extension services while ensuring their implementation, coordination, and monitoring. Researchers should carry out appropriate research in support of erosion and desertification control as well as project implementation assistance. Additionally, the application and interpretation of remote-sensing data, such as aerial photographs and satellite images, remain valuable.

There is a great need for community participation noting that conservation structures at gullies in Nigeria have failed woefully as they have been skewed towards engineering/mechanical measures of concrete works, ignoring the perspectives of the community residents who have developed some coping strategies. The result is that the hitherto stabilized gullies are now expanding and new ones are emerging at an unprecedented rate which calls for further studies.

Also, prioritization of erodibility of soil as a factor linked to the severity of gully erosion in gully erosion studies. Changing the narrative that gullies have side effects only to the understanding that it has some benefits such as sand mining and employment for sand miners, loaders, and food sellers. The management of gullies from the perspectives of the community residents who must be involved in monitoring and reporting on the gullies as a feedback mechanism.

Conflicts of Interest The authors have declared that there are no existing competing interests.

References

Abalaka ID, Kebiruumor (2019) Urban growth assessment and its impact on deforestation in Kwali Area Council, Abuja Nigeria using geospatial technology. Glob Sci J 7(10) Online: ISSN 2320-9186

Adekalu KO, Olorunfemi IA, Osunbitan JA (2007) Grass mulching effect on infiltration, surface runoff and soil loss of three agricultural soils in Nigeria. Bioresour Technol 98(4):912–917

Adiaha M (2021) Soil erosion in Nigeria: 44th conference of Soil Science Society of Nigeria on: climate-smart soil management, soil health/quality and land management: synergies for sustainable development

Akande A (2017) Geospatial analysis of extreme weather events in Nigeria (1985–2015) using self-organizing maps. Adv Meteorol. https://doi.org/10.1155/2017/8576150

Akanwa AO, Ezeomedo (2018) Changing climate and the effect of gully erosion on Akpo community farmers. J Ecol Nat Resour. Medwin Publishers 2(6):2–12

Akanwa AO, Ikechebelu NJ (2021) Clay/sand mining and diminishing greener security and increased climate risks in Nigeria. In: Manjor KJ (ed) Natural resources conservation and advances for sustainability. Elsevier. ISBN: 978-0-12-822976-7

Akanwa AO, Joe-Ikechebelu N (2019) The developing world's contribution to global warming and the resulting consequences of climate change in these regions: a Nigerian case. In: Tiefenbacher J (ed) Global warming and climate change. Intech Open, London., 2020. https://doi.org/10.5772/intechopen.85052

Akanwa AO, Okeke FI, Nnodu VC, Iortyom ET (2017) Quarrying and its effect on vegetation cover for a sustainable development using high resolution satellite image and GIS. Environ Earth Sci 76(14):1. https://doi.org/10.1007/s12665-017-6844-x

Akanwa AO, Mba HC, Jiburum U, Ogboi KC (2019) Strategies for combating climate change. In: Manjor KJ, Ram SM, Arnab B, Dhiraj KY (eds) Sustainable agriculture forest and environmental management. Springer Nature, Singapore. https://doi.org/10.1007/978-981-13-6830-1_12

Akanwa AO, Mba HC, Ogbuene EB, Nwachukwu MU, Anukwonke CC (2020) Potential of agroforestry and environmental greening for climate change minimization. In: Abhishek R, Manjor KJ, Dhiraj KY, Arnab B (eds) Climate change and agroforestry system. CRC-Apple Academic Press/Taylor & Francis, Oxfordshire. https://www.amazon.com/Climate-Change-Agroforestry-Systems

Akanwa AO, Okedo-Alex IN and Joe-Ikechebelu NN (2022a) Persistent climate-driven floods: a voiced case of indigenous women experiencing eco-social and health risks in a southeast Community in Anambra State, Nigeria, Chapter 22. In: Chatterjee U, Everard M, Kashyap A, Mahata D, Panda G.(eds) Indigenous people and nature, pp. 495–523, Elsevier

Akanwa AO, Joe-Ikechebelu NN, Okedo-Alex IN, Okafor KJ, Omoruyi FA, Okeke J, Amobi NS, Enweruzor AC, Obioma CE, Izunobi IP, Nwakacha TO, Oranu CB, Anazodo NI, Okeke CA, Ugwuoke UE, Umeh UM, Ogbuefi EO, Echendu PT (2022b) The causes and consequences of climate -driven temporary displacement of women and children in Anambra State, Nigeria. In: Tiefenbacher JP (ed) Climate issues in Asia and Africa – examining climate, its flux, the consequences, and society's responses. Intech Open, London. http://www.intechopen.com/chapters 82644

Amanchukwu RN (2015) Climate change education in Nigeria: the role of curriculum review. Education 5(3):71–79. https://www.researchgate.net/profile/Nwachukwu_Prince_Ololube/publication/283081778_Climate_Change_Education_in_Nigeria_The_Role_of_Curriculum_Review/links/56292c5508ae518e34 7c967b/Climate-Change-Education-in-Nigeria-The-Role-of-Curriculum-Review.pdf

Anjali D (2020). Analysis and design of retaining wall: a review. International conference on communication and information processing. https://doi.org/10.2139/ssrn.3648731

Anukwonke CC, Tambe EB, Nwafor DC, Malik KT (2021) Climate change and interconnected risks to sustainable development in 'climate change, the social and scientific construct. ISBN: 978-3-030-86290-9. Springer Publishers

Audu IA, Adie LA (2018) Desertification in Northern Nigeria: causes and consequences. Environ Stud 1(2):20–37

Babsal and Co. Ltd (1998) Katsina State environmental action plan, final report. FEPA under World Bank Assisted Programme. Babsal and Co. Ltd, Ikeja, Lagos

Babu S (2016) The history of soil conservation: this is not a new problem! https://eco-intelligent.com/2016/12/17/the-history-of-soil-conservation/

Bellamy P (2007) Academic dictionary of the environment. Academic Press, New Delhi

Borrelli, Robinson DA, Panagos P, Montanarella CB (2020) Land use and climate change impacts on global soil erosion by water. PNAS 117(36):21994–22001

Celik I (2005) Land use effects on organic matter and physical properties of soil in a Southern Mediterranean Highland of Turkey. J Agric Sci Technol 13:727–732

Chatterjee U, Akanwa AO, Kumar S, Singh SK, Roy AD (2022) Ecological footprints of climate change: adaptive approaches and sustainability. Springer Nature, Cham. http://link.springer.com/book/10.1007/978-3-031-15501-1 or https://link.springer.com/

Chiemelu N, Okeke F, Nwosu K, Ibe C, Ndukwu R, Ugwuotu A (2013) The role of surveying and mapping in erosion management and control: a case of Omagba erosion site, Onitsha, Anambra State, Nigeria. J Environ Earth Sci 3(11):11–18

Cullis J, Kunneke MM, Willem H (2019) Urbanization, climate change and its impact on water quality and economic risks in water scarce and rapidly urbanizing catchment: case study of the Berg River catchment. H2 Open J 2(1):146–167

DailyTrust(2020)2ndNigerBridge:Anambracommunities raise alarm over erosion. Retrieved from https://dailytrust.com/2nd-niger-bridge-anambra-communities-

Daisy D (2020) The carbon brief profile: Nigeria. https://www.carbonbrief.org/the-carbon-brief-profile-nigeria

Dioha MO, Emodi NV (2018) Energy-climate dilemma in Nigeria: options for the future. IAEE Energy Forum. https://www.google.com/url?sa=t&rct=j&q=&esrc=s&source=web&cd=108&ved=2ahUKEwihv4iA2-7kAhVoc98KHWNKDtw4MhAWMDl6BAgQEAI&url=https%3A%2F%2Fwww.iaee.org%2Fen%2Fpublications%2Fnewsletterdl.aspx%3Fid%3D465&usg=AOvVaw3qHlxFJnRFCXox9HBg4d-I

Ebele NE, Emodi NV (2016) Climate change and its impact in Nigerian economy. J Sci Res Rep 10(6):1–13. http://www.journaljsrr.com/index.php/JSRR/article/view/21917/40737

Eboh EC, Lemchi JI (1994) Population pressure and indigenous land tenure in Eastern Nigeria: implications for land tilting. J Rural Dev Admin 26(3):77–82

Egboka BCE (2004) Distress call and plea to the senate committee for urgent actions against floods, soil/gully erosion/landslides disasters in the southeastern Nigeria. Paper presented to senate committee on Environmental; roads/Erosion senate delegation

Egboka B, Okpoko E (1984) Gully erosion in the Agulu-Nanka region of Anambra State, Nigeria. Environmental Science IAHS-AISH Publication

Egboka BCE, Orajaka IP (1990) Implications of palaeo- and neotectonics in gully erosion-prone areas of southeastern Nigeria. Nat Hazards 3(3):219–231

Egboka BCE, Orji AE, Nwankwoala HO (2019) Gully erosion and landslides in southeastern Nigeria: causes, consequences and control measures. Glob J Eng Sci 2(4):GJES.MS.ID.000541

Elisha I (2017) Evidence of climate change and adaptation strategies among grain farmers in Sokoto State, Nigeria. IOSR J Environ Sci Toxicol Food Technol (IOSR-JESTFT) 11(3):1–7

Enete IC (2014) Impacts of climate change on agricultural production in Enugu State, Nigeria. J Earth Sci Climat Change 5(9):234. https://www.omicsonline.org/openaccess/impacts-of-climate-change-on-agricultural-production-in-enugu-state-nigeria-2157-7617.1000234.php?aid=32633

Eze HI (2007) Effect of rainfall intensity and energy in gully development in northeastern Enugu State. Niger J Technol 26(1):91–96

Ezezika OC, Adetona O (2011) Resolving the gully erosion problems in southeastern Nigeria: innovation through public awareness and connectivity-based approaches. J Soil Sci Environ Manage 2(10):286–291

Federal Ministry of Environment (2021) National Climate Change Policy for Nigeria, (2021–2030) Department of Climate Change

Food and Agricultural Organization (FAO) of the United Nations (2014) Soil erosion. Accessed at: www.diccionariomedioambiente.org/diccionariomedioambiente-en/en/cpo-erosion-bis.asp, 20 Apr 2022

George NA, Obot I, Akpanetuk N (2008) Geoelectrical investigation of erosion and flooding using the lithologic compositions of erosion and flood-stricken road in Ukanafun Local Government Area, Akwa Ibom State, Southern Nigeria. Disaster Advance 1(4):46–51

Gloria, W (2022) Will building a retaining wall really stop erosion?. https://captainpatio.com/will-a-retaining-wall-stop-erosion

Grove AT (1951) Land use and soil conservation in parts of Onitsha and Owerri Provinces geological survey of Nigeria. Bulletin No 21

Ibitoye M, Adebayo I (2010) Urban growth and development of soil erosion in part of humid region in Southwest Nigeria. Conference of spatial planning and development. https://www.researchgate.net/publication/236861826urbangrowthanddeevelopment_ofsoil_

Ibrahim E (2018) Urbanization and its impacts in Nigeria: a review. https://www.researchgate.net/publication/323755967_urbanization_and_its_impacts_innigeria_

Idowu OJ, Oluwatosin GA (2008) Hydraulic properties in relation to morphology of a tropical soil with hardened plinthite under time land use types. Trop Sub-trop Agro Ecosyst 8(4):145–155

Igbokwe JI, Ojiako JC, Nnodu VC (2003) Monitoring, characterization and controlling of floodwater erosion using remote sensing techniques. Proceedings of the technical session of the 38th annual general meeting and conference of Nigerian Institution of Surveyors, Lokoja, 123–134

Igwe CA (1999) Land use and soil conservation strategies for potentially highly erodible soils of Central-Eastern Nigeria. Land Degrad Dev 10:425–434

Igwe CA (2003) Erodibility of soils of the upper rainforest zone, southeastern Nigeria. Land Degrad Dev 14:323–334

Igwe CA (2012). Gully erosion in southeastern Nigeria: role of soil properties and environmental factors, University of Nigeria, pp 157–171. https://www.intechopen.com. https://doi.org/10.5772/51020

IPCC Climate Change (2014) Synthesis report. Contribution of Working Groups I, II and III to the Fifth Assessment Report of the Intergovernmental Panel on Climate Change [Core Writing Team, Pachauri RK and Meyer LA (Eds)]. IPCC, Geneva, Switzerland, 151 pp

IPCC Climate Change report (2022) Impacts, adaptation, and vulnerability. Contribution of Working Group II to the Sixth Assessment Report of the Intergovernmental Panel on Climate Change [Pörtner H-O, Roberts DC, Tignor M, Poloczanska ES, Mintenbeck K, Alegría A, Craig M, Langsdorf S, Löschke S, Möller V, Okem A, Rama B (Eds)]. Cambridge University Press. Cambridge University Press, Cambridge, UK and New York, NY, USA, 3056 pp https://doi.org/10.1017/9781009325844

Jaiyeoba IA (2002) Environment in Africa atlases: Nigeria. Les Edition JA, Paris, pp 122–123

Jeje LK (1978) A preliminary investigation into soil erosion on cultivated hillsides in parts of Western Nigeria. Ceo-Eco-Trop I:175–187

Jimme MA, Musa AA, Sambo GH (2020) Urbanization and its effects on the environment in Maiduguri Metropolis, Borno State, North East, Nigeria. Retrieved http://oer.tsuniversity.edu.ng/wp-content/uploads/2020/12/jjsms-vol.1.-no.-3

Maitra S, Shankar T, Banerjee P (2020) Potentials and advantages of Maize-Legume intercropping system. In: Maize – production and use. IntechOpen. https://doi.org/10.5772/intechopen.91722

Medugu M, Majid R, Elnfaty H (2014) The environmental effect of mass housing development in Abuja, Nigeria. https://www.researchgate.net/publication/34105123

Mezie EO, Nwajiaku AI (2020) Investigation into the causes of gully erosion in parts of Anambra. J Sci Technol Res 2(1):132–142

Muoghalu LN, Akanwa AO (2021) Ecological intensification for sustainable agriculture: the Nigerian perspective. In: Jhariya MK, Meena RS, Banerjee A (eds) Ecological intensification of natural resources for sustainable agriculture. Springer, Singapore. https://doi.org/10.1007/978-981-33-4203-3_15

NEWMAP (2019) 90% of gully erosion in Nigeria caused by road construction. Retrieved from https://www.google.com/amp/s/www.today.ng/news/nigeria/newmap-90-gully-erosion-nigeria-caused-road-construction-249776/amp

Nkechi O (2016) Mitigating climate change in Nigeria: African traditional religious values in focus. Mediterr J Soc Sci 7(6):299–308. https://www.mcser.org/journal/index.php/mjss/article/view/9612

Nkwocha KF, Akanwa AO, Jimme MA, Iheukwumere SO, Joe-Nkechebelu N (2019) Perception and impacts to climate emergency in Maiduguri Urban, Nigeria: a case for community based adaptation approach (CBPR). Rev Environ Earth Sci 6(2):24–33

Nwachukwu OI, Onwuka MI (2011) Land degradation and food crisis-causes, impact and soil conservation efforts in Nigeria: globalization and rural development in Nigeria. In: Nwachukwu I, Ekwe KC (eds) . Michael Okpara University of Agriculture Umudike, Nigeria, pp 231–232

Nzeagwu U (2022) Gully erosion destroys 70% of Anambra land area: urges NEWMAP, partners to declare state of emergency over menace. The Guardian National Daily. https://guardian.ng.news.gully

Odiogor H (2010) Special report on desertification in Nigeria: the sun eats our land. Vanguard Newspaper, May 3, 2010. Accessed at: https://www.vanguardngr.com/2010/05/special-report-on-desertification-in-nigeria-the-sun-eats-our-land, 25 Apr 2022

Ofomata GEK (1964) Soil erosion in the Enugu region of Nigeria. African Soils IX(2):289–348

Ofomata GEK (1965) Factors of soil erosion in Enugu area of Nigeria. Nigerian Geograph J 8:45–59

Ofomata GEK (1966) Quelques observations sur l'eboulement d'Awgu, Nigeria Oriental. Bull de PIFAN. Serie A XXVm:433–443

Ofomata GEK (1967) Some observations on relief and erosion in eastern Nigeria. Revue de Geomorph Dynamise XVU:21–29

Ofomata GEK (1973) Village erosion at Ozuitem, east central state of Nigeria. Ikenga 11:64–74

Ofomata GEK (1978) Man as a factor of soil erosion in southeastern Nigeria. Geo-Eco-Trop 1:143–154

Ofomata GEK (1980) Perspectives on environmental deterioration in Nigeria. The Tropical Environment 1:6–19

Ofomata GEK (1981) Actual and Potential Erosion in Nigeria and Measures of Control. Paper presented at the 9th Annual Conference of the Soil Science Society of Nigeria, Owere, Dec 6–10, 1981

Ofomata GEK (1984) Erosion in the Forest Zone of Nigeria. Paper presented at the 27th Annual Conference of the Nigerian Geographical Association, Univ. of Nigeria, Nsuka, Mar 1984

Ofomata GEK (1985) Soil erosion in Nigeria, The views of a geomorphologist University of Nigeria inaugural lecture series no. 7. University of Nigeria Press, Nsukka

Ogbonna JU (2009) Macro-analytic study of the impact of gully erosion in Old Imo State of Nigeria. Am J Geograph Inf Syst 1(2):23–32

Ogbonna JU (2012) Examining the vulnerability of gully erosion in the Old Imo State using logidtic regression models and GIS. Am J Geograph Inf Syst 2:35–42

Oguike PC, Mbagwu JSC (2009a) Variations in some physical properties and organic matter content of soils of coastal plain sand under different land use types. World J Agric Sci 5(1):63–69

Oguike PC, Mbagwu JSC (2009b) Variations in some physical properties and organic matter content of soils of coastal plain sand under different land. Int J Dev Econ Sustain 7(2):18–27

Okagbue CO, Uma KO (1987) Performance of gully erosion control measures in southeastern Nigeria. Forest Hydrol Watershed Manag 167:163–172

Okeke AU, Azuh D, Fredrick UO, Tetsuya K (2020) Assessment of land use impact and seepage erosion contributions to seasonal variations in riverbank stability: the Iju River, South West Nigeria. Groundw Sustain Dev. https://doi.org/10.1016/j.gsd.2020.100448

Okorafor O, Akinbile CO, Adeyemo AJ (2017) Soil erosion in south eastern Nigeria: a review. Scientific Research Journal (SCIRJ) 5:8–10. ISSN 2201-2796

Okpala AO (1990) Nigerian population growth and its implications for economic development. Scand J Dev Altern 9(4):63–77

Okuh D, Osumgborogwu JE (2019) Adjustments to the hazards of gully erosion in rural Southeast Nigeria: a case study of Anambra communities. Appl Ecol Environ Sci 7(1):11–20

Olagunju TE (2015) Drought, desertification and the Nigeria environment. A review. J Ecol Nat Environ 7(7):196–209

Ologe OO (1988) Soil erosion characteristics. In: Proc national workshop on ecological disaster on soil erosion. Federal Ministry of Science and Technology, pp 26–28

Onu DO (2005) Determinants of farm – level soil conservation and erosion control adoption and utilization behaviours in the ecologically vulnerable areas of Imo State, Nigeria. J Rural Dev 24(4):521–543

Onu DO (2006) Socioeconomic factors influencing farmers adoption of Alley farming Technologies under intensified agriculture in Imo State Nigeria. Philipp Agric Sci 89(2):521–543

Onyejekwe S, Obi JC, Ismail E (2022) Assessment of the causes of erosion and gullying along the Leru–Nkwoagu, Amuda (Isuochi)–Mbala (Isuochi) Highway, Southeastern Nigeria: a case study. Retrieved from https://www.springerprofessional.de/en/assessment-of-the-causes-of-erosion-

Oriavwote VE (2019) Economic implications of environmental degradation in Nigeria: the environmental Kuznets curve relevant to Nigeria. European Centre for Research Training and Development, UK (www.eajournals.org) 2053–2199

Osadebe CC, Akpokodje EG (2007) Statistical analysis of variability in properties of soils in gully erosion sites of Agulu-Nanka-Oko area, southeastern Nigeria. J Min Geol 43(2):197–202

Otegbeye GO (2004) Drought and desertification: challenges of afforestation in Nigerian and Semi-Arid Region Savanna 19:32–40

Sada PO, Omuta GED (1979) Auchi Urban Area – a preliminary report on Erosion and Environmental Problems and Planning. In P.O. Sada and G.E.D. Omuta Ifds.): spatial perspectives in national development. Proceedings of the 22nd Annual Conference of the Nigerian Geographical Association, Benin, March 28 to April 1, 1979, 130–152

Sadiq A, Surayya A (2019) Assessment of substantive causes of soil degradation on farmlands in Yola South LGA, Adamawa State, Nigeria. Int J Sci Res Publ 9(4):537–540. Shivalikfoot Hills. 327–335

Skyes RA (1940) A history of and erosion work at Udi. Farm Forest 1:3–6

Sulaiman I, Musa M, Mohdnadzri R (2017) Urban growth assessment and its impact on deforestation in Bauchi Metropolis, Nigeria using remote sensing and GIS techniques. J Eng Appl Sci 12(6):345–353. ISSN: 1819-6608

Tercula IA (2002) Desertification: water scarcity hits Sokoto farmers, Sunday trust newspaper of 9th February, 2015. Accessed at: www.sundaytrust.com, 25 Apr 2022

Ul Zaman M, Bhat S, Sharma S, Bhat O (2018) Methods to control soil erosion – a review. Int J Pure App Biosci 6(2):1114–1121. https://doi.org/10.18782/2320-7051.6462

Umeugochukwu PO, Ezeaku PI, Chude VO (2013) Impact of soil erosion on land degradation in UGA southeastern Nigeria. Niger J Soil Sci 22(1):26–36

United Nations (2012) United Nations convention to combat desertification. Accessed at https://www.unccd.int/en/about-the-convention/pages/text-part-1.aspx, 20 Apr 2022

West I, Wodike O (2019) Environmental degradation effects of urbanization in Port Harcourt. Jam Sci 15(9):1–7. ISSN 1545-1003 (print); ISSN: 2375-7264 (online). http://www.jofamericanscience.org

Wokekoro E (2020) The impact of erosion on real estate. Int J Res Appl Nat Soc Sci 8(5):35–40. https://www.impactjournals.us/download/archieves/27-05-2020

World Bank (2012) Nigeria – erosion and watershed management project. http://documents.worldbank.org/curated/en/728741468334143813/Nigeria-Erosion-and-Watershed-Management-Project

Yousuf A, Anil B, Sukhdeep S, Vishnu P (2022) Application of WEPP model for runoff and sediment yield simulation from ungauged watershed in Shivalik foot-hills. In: Computers in earth and environmental sciences. https://doi.org/10.1016/B978-0-323-89861-4.00028-2

Yusuf MB, Abba UJ, Isa MS (2019) Assessment of soil degradation under agricultural land use sites: emerging evidence from the savanna region of north eastern Nigeria. Ghana J Geograp 11(2):243–263

10 Strategies for Compound Urban and Climate Hazards: Linking Climate Adaptation and Sustainability to Address Risk in Environmental Justice Communities

Dalia Munenzon and Maria Noguera

Abstract

In 2022, Houston, TX, recorded the most days with unsafe Ozon levels. In combination with extreme heat, air pollution increases and exacerbates the health impacts on vulnerable communities. Bullard's environmental justice research demonstrated how a history of urban development and systemic disinvestment resulted in inequitable exposure to environmental hazards and extreme climate risk in communities of color (Bullard, Invisible Houston: the black experience in boom and bust, 1st edn. Texas A&M University Press, Collage Station, 1987). Climate change will exacerbate existing environmental and health hazards. The local government developed long-term climate action plans and adaptation strategies. However, these investments focus mainly on flood mitigation and air quality monitoring, disregarding the effects of heat and compounding environmental hazards.

We propose to integrate climate adaptation actions with sustainability goals to establish a resilient framework that addresses climate and health risks from compound hazards. Our focus is on the Eastern neighborhoods, where we first use a new model for multi-hazard vulnerability analysis to identify climate scenarios that worsen health disparities. This chapter presents a case study of how transformative climate adaptation projects can address compound climate impacts and improve public health, environmental justice issues, and community well-being while serving the best sustainability goals.

Keywords

Compound hazards · Environmental justice · Climate adaptation · Extreme heat · Health inequities

10.1 Introduction

As cities worldwide have developed climate action plans in response to natural disasters, they often fail to address the compounded impact of climate hazards, particularly in communities of color that have lacked investment historically. Houston, TX, recorded the most days with unsafe ozone levels in 2022, exacerbating health impacts on vulnerable communities facing extreme heat and compounded environmental hazards (Douglas 2022). The Texas Tribune report high-

D. Munenzon (✉) · M. Noguera
Hines College of Architecture and Design, University of Houston, Houston, TX, USA
e-mail: dmunenzo@central.uh.edu

U. Chatterjee et al. (eds.), *Climate Crisis: Adaptive Approaches and Sustainability*, Sustainable Development Goals Series, https://doi.org/10.1007/978-3-031-44397-8_10

lights the need for more comprehensive and equitable climate action plans that address flooding, air quality management, and the effects of heat and other environmental hazards (Douglas 2022). Despite the region's thriving economy, numerous chemical plants threaten public health. A Houston Chronicle investigation in 2016 found that on average, chemical fires or explosions occur every 6 weeks in the area (Collette et al. 2016). These hazardous facilities and infrastructure are more likely to be located near minority communities and low-income communities (Bullard 1987). Meanwhile, affluent communities with better access to parks and open spaces exacerbate environmental injustice.

This chapter proposes an equity-oriented approach for indicators to address the compound impacts of heat and other environmental hazards on vulnerable populations. The local government's long-term climate action plans and adaptation strategies have primarily focused on flood mitigation and air-quality monitoring and have neglected the effects of heat and compounding environmental hazards. Therefore, this chapter proposes linking climate adaptation actions to sustainability goals to establish a framework for resilience that addresses climate and health risks through the lens of compound hazards. The proposed method integrates elements of sustenance and welfare as essential components of adaptation and employs aggregated policy and strategy analysis to address adaptive capacity, community sensitivity, and environmental hazards. Localized spatial equity indications provide specific challenges that can drive policy or design actions. This chapter aims to promote climate justice and reduce health inequalities by examining how compound climate impacts can be addressed through transformative climate adaptation projects.

10.1.1 Background

This study aims to measure vulnerability as a function of future adaptation actions to develop proposals for climate risk, adaptation, and mitigation. The proposed research highlights the importance of considering the complex and interrelated factors that contribute to experiences of vulnerability to build resilient and equitable communities. Heat and environmental hazards are prominent risks in active urban areas that significantly impact human well-being. Urban Heat Islands (UHI) cause thermal discomfort, reducing life quality and amplifying heat waves (Leal Filho et al. 2021). The consequences of heat risk can be clustered under different aspects, including health and well-being, energy consumption, air quality, water management, and urban ecology (Leal Filho et al. 2021).

The literary review supports the argument that resilience and vulnerability are socially constructed and directly affected by pre-event social, economic, and cultural factors that can result in a lack of access to resources and exclusion. The evolution of risk indicators used to map vulnerability indices started with a variety of approaches used in the past, such as expert judgment, multicriteria decision analysis, and statistical techniques (Cutter et al. 2003).

Levison et al. (2018) identify gaps in understanding the local impacts of global climate change and highlight the need to expand local knowledge of health vulnerabilities. Yu et al. (2021) argue that current vulnerability and resilience literature need data-driven methodologies. They conclude that future research should focus on selecting appropriate indicators and weights based on references and considering local contextual information and relevant health outcomes. The focus on heat resilience and compound environmental hazards introduces long-term and slowly evolving impacts closely connected to decision-based spatial conditions. Therefore, to assess vulnerability we need to aggregate and evaluate it at a granular level to understand the causes and potential solutions (Heaton et al. 2014). A better understanding of intersectional population vulnerabilities during emergencies should be leveraged to drive public and municipal investment in spatial investments and better access to services (Heaton et al. 2014). The application of intersectional methods to define indicators and the profound localized outline of the individual (demographics) and root (institutionalized) causes is emphasized as a valuable tool to build adaptive capacity effectively (Lotfata and Munenzon 2022). Thus, vulnerability prioritizes urgency to address risks across sites and communities. Intersectional methods are necessary to define indicators and the profound localized outline of the individual and root causes.

In analyzing the intersectional nature of vulnerability and resilience in the face of climate change and natural hazards, this case study aims to contextualize the analysis and promote a paradigm shift towards compound hazards of urban heat as a public health issue. It is grounded in the understanding that urban design, spatial characteristics, and components are deeply connected to heat risk (Kim and Ryu 2015). Hinkel (2011) identified the limitations of the standard of vulnerability assessment in providing comprehensive adaptation strategies. This study aims to expand how we analyze adaptive capacity and vulnerability to inform adaptation actions and program activities to reduce adverse health impacts and increase resiliency and public well-being by combining social and physical factors in local-level planning. The neighborhood-scale compound hazard vulnerability analysis conducted in this study has the potential to assist community leaders and municipal agencies in the creation of local policies that implement spatial and long-term actions to protect public health and improve social well-being. It is worth noting that the outcomes of such policies are interconnected with achieving the shared goals, political priorities, and ambitions of the UN Sustainable Development Agenda 2030.

10.1.2 The Gap Between SDG and Actionable Adaptation Strategies

This chapter highlights the need to evolve vulnerability analysis to address systemic complexities (Hinkel 2011; Tonmoy et al. 2014) and incorporate intersectional and spatial challenges (Heaton et al. 2014; Kim and Ryu 2015; Lotfata and Munenzon 2022). This refined approach combines individual and social risk factors into a comprehensive vulnerability index.

Investments in adaptation need to consider spatiotemporal characteristics of populations and thermal adaptation measures (Cheng et al. 2021). Neighborhood-specific geospatial analysis can offer granular data to address inequities, inform urban planning, and ensure policies do not disproportionately affect vulnerable groups (Chakraborty et al. 2019). Such actions, sensitive to poverty and gender, are essential for achieving multiple Sustainable Development Goals (SDGs) (Paaske 2021; Salvia et al. 2019; Stuart and Woodroffe 2016). Social policy measures and spatial actions, such as tree canopy analysis and green space creation, can help reduce inequalities and localized warming (Lotfata and Munenzon 2022).

While the SDGs recognize the need for inclusive growth and regulatory capabilities in disaster management (SDGs 11 and 16), they do not directly address local environmental and social injustices (Menton et al. 2020). Challenges including inadequate funding, corruption, and the exclusion of marginalized groups hinder progress (Hope Sr. 2020). To truly build just societies, the SDGs must directly address justice, power dynamics, and structural conditions impeding environmental and social justice (Chakraborty et al. 2019). Thus, a more comprehensive approach that integrates these issues with other development goals is necessary.

SDGs 11 and 10, aimed at making cities sustainable and reducing inequalities, respectively, can mitigate heat risk and environmental justice (EJ) issues in Houston. Under SDG 11, such as developing sustainable transport, enhancing green spaces, and improving air quality, can alleviate heat risk. Strengthening local governance and increasing community engagement can ensure equitable environmental benefits (Vardoulakis et al. 2020). Under SDG 10, addressing systemic discrimination, redlining, and urban planning disparities that led to environmental hazards is crucial. Shifting decision-making power to affected communities and investing in essential services can promote a sustainable, equitable future for Houston.

However, the SDGs' indirect approach to environmental and social injustices calls for a more comprehensive strategy. Granular, site-specific data is needed to avoid disproportionate impacts on vulnerable populations, and poverty- and gender-sensitive interventions can reduce inequalities and contribute to multiple SDGs.

In urban adaptation, considering compound hazards and root causes allows for long-term strategies promoting public well-being. Transformative adaptation, if thoughtfully designed, can alter unsustainable development pathways and mitigate future vulnerabilities to climate change, yielding long-term benefits for human well-being and ecosystems (Fedele et al. 2020). Urban policymakers and designers can create sustainable, appealing cities by adopting deep transformational adaptation strategies co-developed with diverse stakeholders (Swart et al. 2023). This systemic approach uncovers opportunities for transformative urban change, leading to more effective, sustainable urban development.

10.2 Environmental Justice, Heat Risk, and Air Quality in Houston

Recent studies on environmental justice, heat risk, and air quality in Houston emphasize addressing these concerns to protect vulnerable populations and enhance the city's resilience against climate change impacts. The 2016 "Double Jeopardy in Houston" report demonstrates that marginalized communities, particularly low-income and ethnically diverse communities, face higher levels of toxic chemicals, extreme heat, and poor air quality, exacerbating existing health disparities (Union of Concerned Scientists and Texas Environmental Justice Advocacy Services 2016).

Literature reveals Houston's intricate environmental justice, air quality issues, and impacts on vulnerable communities (Bullard and Wright 2009; Johnston et al. 2016). Li et al. (2019) highlight environmental racism, while Linder et al. (2008) urge stronger pollution regulations. Hendricks and Van Zandt (2021) analyze the link between infrastructure, risk, and disinvestment in Houston, especially during Hurricane Harvey. The city's oil refineries, near predominantly Hispanic or Black residences, pose significant risks, exacerbated by the hurricane's flooding.

The Houston Ship Channel is an essential trade and petrochemical processing hub in neighborhoods with low-income, racially diverse communities and limited healthcare (Understanding Houston 2021). These communities face environmental hazards and sociodemographic stressors, heightening their pollution vulnerability. Sansom et al. (2018) detected higher polycyclic aromatic hydrocarbon levels in a Houston EJ neighborhood, necessitating further research. Post-Hurricane Harvey, Horney et al. (2018) found increased contaminant levels in the same area, emphasizing effective disaster response for vulnerable communities. Hendricks and Van Zandt (2021) highlight the need for EJ research focusing on social vulnerability due to unreported contaminant exposures during the hurricane.

Houston's substantial metal recycling industry, with over 100 facilities, poses potential health risks through torch cutting, which generates inhalable particles containing toxic heavy metals (Symanski et al. 2020). From 2006 to 2011, nearly 200 air-quality complaints about metal recycling facilities were lodged with the city (Symanski et al. 2020).

Despite high awareness, Semenza et al. (2008) found little behavior change related to hot weather and air pollution, indicating a need for targeted communication and policy interventions. Hayden et al. (2017) identified factors influencing vulnerability to heat-related illnesses, emphasizing the need for targeted interventions. Marsha et al. (2018) suggested that rising temperatures and population growth could significantly increase heat-related deaths in Houston, stressing the importance of urban planning and public health measures. Conlon et al. (2016) projected exacerbated heat exposure in urban areas, emphasizing the importance of urban planning and green infrastructure.

In 2008, the State Implementation Plan (SIP) was designed to curb air pollution in the Houston-Galveston-Brazoria metropolitan region, focusing on regulating pollution sources and promoting mass transportation (Texas Commission on Environmental Quality (TCEQ) 2008). However, proposed revisions must meet the Clean Air Act and EPA rules, causing concerns among local activists and EJ groups (Texas Commission on Environmental Quality (TCEQ) 2008).

Despite Houston's efforts to mitigate air pollution, such as investing in technology and proposing new fine-particulate matter monitors (Li et al. 2019), air quality remains concerning, leading to calls for stricter regulations (Linder et al. 2008). Recent studies highlight the pressing need for comprehensive strategies to enhance air quality and address EJ issues in Houston (Johnston et al. 2016; Bullard and Wright 2009; Li et al. 2019).

These comprehensive EJ challenges are partly attributed to infrastructural integrity issues, such as low-income and minority communities inheriting older and poorer quality housing due to historical neighborhood planning forces and persistent disinvestment. In light of Houston's complex and extreme climate events, a systematic approach to infrastructure management is crucial for addressing risks, hazard exposures, and disaster outcomes across various social groups.

10.2.1 Compound Impacts

Heatwaves and extreme heat events can worsen air pollution-related health risks, as high temperatures accelerate the formation of ground-level ozone, a primary component of smog (AghaKouchak et al. 2020; Tibbetts 2015). This can lead to respiratory illnesses and aggravate cardiovascular conditions (AghaKouchak et al. 2020; Baldwin et al. 2019). Extreme heat events also cause air stagnation, trapping pollutants near the ground and exacerbating the health impacts of particulate matter and other harmful pollutants (Ebi et al. 2021).

Urban heat island (UHI) effects can magnify city pollution, creating urban pollution islands (UPIs) (Ulpiani 2021). This heightens temperatures and pollution levels in cities, disproportionately impacting vulnerable populations (Ulpiani 2021). A study on ground-level O_3 in Houston found a decrease in high PM days due to increased southeast winds caused by atmospheric pressure differences, leading to higher O_3 concentrations north of the city (Liu et al. 2015; Johnston et al. 2016).

Historical redlining practices have contributed to present-day inequalities in air pollutant emissions, disproportionately affecting marginalized communities (Cushing et al. 2022). Climate change projections indicate increased extreme heat events, which could worsen air pollution-related health risks (U.S. Climate Resilience Toolkit n.d.). Elevated temperatures and pollutant levels can impact ecosystems, agriculture, and water resources (Tibbetts 2015).

Houston confronts climate change-related difficulties, including increasing temperatures and a higher frequency of extreme weather occurrences (Understanding Houston 2021). The 2020 Houston Climate Impact Assessment projects increased average temperatures, frequent and intense heatwaves, and heavier precipitation events, exacerbating existing vulnerabilities and amplifying the urban heat island effect (City of Houston 2020d).

Hoffman, Shandas, and Pendleton's study (2020) emphasizes the effects of past housing policies on resident exposure to urban heat, necessitating targeted adaptation strategies for the most vulnerable populations and areas. Moreover, Oliveira Santos et al. (2023) highlight the intricate relationship between air quality, extreme heat, and the built environment, offering insights into mitigating the health risks of climate change and air pollution in Houston. Yu et al. (2021) stress the relevance of geospatial indicators for assessing neighborhood vulnerability to climate-linked health hazards, reinforcing the need for an all-inclusive adaptation approach.

Synthesizing these sources, it is evident that confronting Houston's climate change challenges requires a comprehensive approach. This approach should consider the spatial distribution of vulnerabilities, the compounded effects of heat and air pollution, and the necessity for targeted adaptation strategies (Table 10.1). Such a strategy will better prepare the city for ongoing and future climate change impacts, promoting EJ and enhancing community resilience and well-being.

The comprehensive table above highlights the intricate connections between extreme heat, air

Table 10.1 Summary of compound hazards literary review, links between heat and air quality as health risks

Climate driver	Climate change impact	Health outcomes	Counter adaptation
Extreme heat	Urban Heat Island (UHI) effect	Heat-related illnesses, exacerbation of chronic diseases, dehydration, and heatstroke	Green spaces, cool roofs, and urban planning for heat mitigation
Ground-level ozone (O_3) formation	Increased O_3 concentrations	Respiratory illnesses, cardiovascular conditions, asthma, and COPD	Reduction of VOC and NOx emissions, air quality monitoring, and stricter regulations
Biogenic emissions (VOCs)	Formation of secondary organic aerosols	Respiratory issues and cardiovascular problems	Vegetation management and emission controls
Stagnation of air	Trapping of pollutants	Respiratory issues, cardiovascular problems, increased exposure to PM, and other harmful pollutants	Improved ventilation, air filtration systems, and public transportation
Changes in wind patterns	Redistribution of O_3 and other pollutants	Increased respiratory and cardiovascular issues in areas with higher O_3 concentrations	Monitoring wind patterns and adjusting pollution control measures

quality, and EJ hazards, emphasizing the need for a two-level evaluation of vulnerabilities and adaptive strategies for urban climate adaptation. Assessing the spatial patterns of heat, the urban heat island effect, and environmental hazards like pollution is crucial for identifying areas with high vulnerability to climate impacts and developing targeted interventions to reduce these risks (Ulpiani 2021).

Houston faces significant challenges as climate change intensifies, with rising temperatures and more frequent extreme weather events becoming increasingly prominent. As global temperature patterns change and Houston experiences increased heat and humidity, the long-term impacts of these climate drivers on air quality and public health become even more significant (AghaKouchak et al. 2020; Conlon et al. 2016). Integrating the evaluation of climate impacts and EJ hazards in urban planning and policymaking can facilitate a more comprehensive understanding of the risks faced by communities, particularly those historically marginalized and disproportionately exposed to pollution (Cushing et al. 2022).

10.3 Methodology

Since the IPCC's Third Assessment Report introduced the initial vulnerability framework, our understanding of vulnerability has deepened, encompassing its social, political, and economic dimensions (Houghton and Intergovernmental Panel on Climate Change 2001), this evolution, fueled by two decades of extreme weather and additional studies, has led to a more comprehensive framework that includes social and environmental factors. For our localized multi-hazard vulnerability analysis model, we rely on the latest IPCC AR6 risk framework (IPCC 2021).

The IPCC AR6's risk framework is flexible and comprehensive, accounting for climate change's complex and cascading impacts. It recognizes that climate variations and shifts in extreme events, whether sudden or gradual, are influenced by hazards. The severity of these impacts hinges on the vulnerability, sensitivity, and adaptive capacity of the systems affected. Risk disparities arise across communities due to intersecting inequalities and context-specific factors such as culture, gender, religion, ability, and ethnicity. The risk propeller concept (IPCC, p. 145) identifies intersecting elements in hazard, vulnerability, and exposure relationships, facilitating the recognition of critical determinants and possible responses.

Hendricks and Van Zandt (2021) posit that infrastructure-related hazard risks, intensified by environmental extremes like flooding, disproportionately impact low-income and ethnically diverse communities. They underline the need to consider EJ and social vulnerability in infrastructure planning. The built environment and neighborhood quality significantly shape environmental hazards' economic toll, injuries, and fatalities. Social vulnerability, a multifaceted concept, affects community support during hazard exposure. Their work advocates for an anti-environmental racism approach, using policy and practice to ensure fair resource distribution while addressing infrastructure and climate change challenges, thus promoting resilience and justice for minority communities.

The compounding effects of heat on populations directly correlate with increased mortality rates. Heaton et al. (2014) pinpointed sociodemographic factors, such as elderly and disadvantaged populations, as critical contributors to heat vulnerability in Houston. We propose to build upon the socioeconomic processes identified by O'Lenick et al. (2019) as impacting vulnerability and exposure, introducing intersectional complexity and systemic factors at the core of Spatial equity (Reckien et al. 2017). Health risks from extreme heat and environmental hazards arising from individual and social factors (sensitivity), spatial equity (adaptive capacity), and embedded socioeconomic inequities lead to greater exposure, as depicted in Fig. 10.1.

Outlining the determinants of the root causes and risk factors provides a greater understanding of the spatial issues that needed to be addressed and adapted toward resilience. For example, historic discriminatory land use practices that allowed certain contaminating industrial uses adjacent to neighborhoods, lack of investment in urban infrastructure and public transit, building regulations favorable to developers with limited energy and performance requirements, or limited indoor employment options (Reckien et al. 2017). This intersectional approach to determinant mapping allows us to respond to the transformative potential of adaptation actions and respond to the three types of EJ distributive justice, sense of justice, and procedural justice (Lotfata and Munenzon 2022; Svarstad and Benjaminsen 2020; Hendricks and Van Zandt 2021).

10.3.1 Compound Spatialized Analysis Toward Equitable Planning Strategies

This chapter addresses the gap in comprehending compound urban risks and vulnerability assessment, as underscored in various studies. It introduces the "compound urban impact" concept, focusing on a more just and equitable approach to urban climate adaptation. This concept captures the interplay of environmental justice impacts and urban planning disparities. It encourages a holistic strategy to address these interconnected challenges, including air pollution, lack of green space, and social vulnerability. It advocates for a unified strategy allowing policymakers and urban planners to tackle multiple crises simultaneously (Yazar et al. 2022).

The link between multiple and amplifying impacts can be analyzed using the complex adaptive systems (CAS) concept, which focuses on interconnected components interacting non-linear, dynamic manner (Naylor et al. 2020). The challenge lies in the political and economic powers that shape cities and reinforce systemic impacts. Brenner (2009) presents urban space as politically mediated and socially contested. Westman et al. (2022) propose combining CAS and critical urban studies (CUS), identifying overlapping concepts to bridge interdisciplinary discussions and highlight power relations. Boundary concepts such as unsettlement, unevenness, and *unbounding* can illuminate these relations. *Unsettlement* and *unevenness* are related to power dynamics and evolving policies, while *unbounding* pertains to uncertainty and the potential for crises to evolve and expand. Examining adaptive actions in response to compound crises allows the identification of root causes associated with spatial equity and embedded inequities. This chapter uses Westman's framework to connect compound urban crises and adaptive capacity root causes, leading to a conceptual diagram (Fig. 10.2 and Table 10.2).

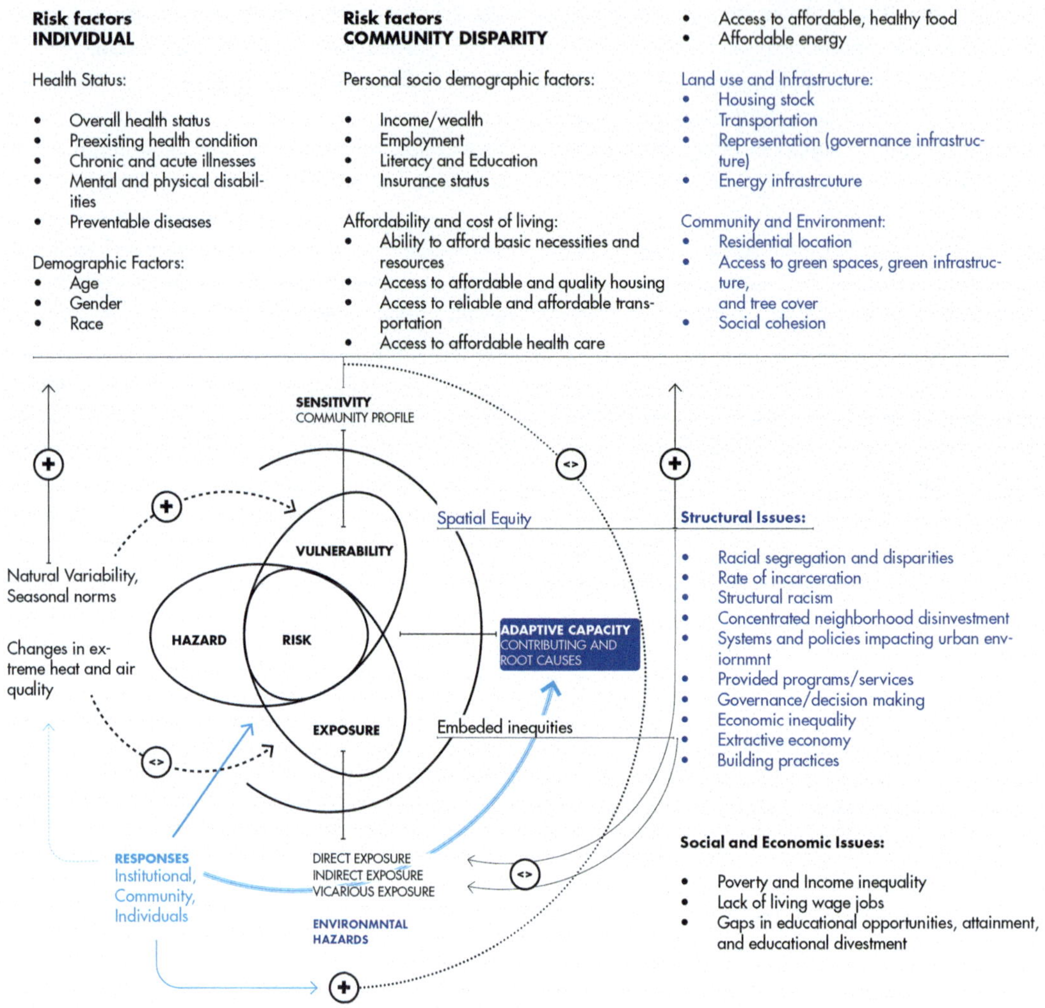

Fig. 10.1 Risk evaluation framework adapted from the IPCC AR6 report, integrating contributing causes that can translate from single action-based solutions to transformative actions

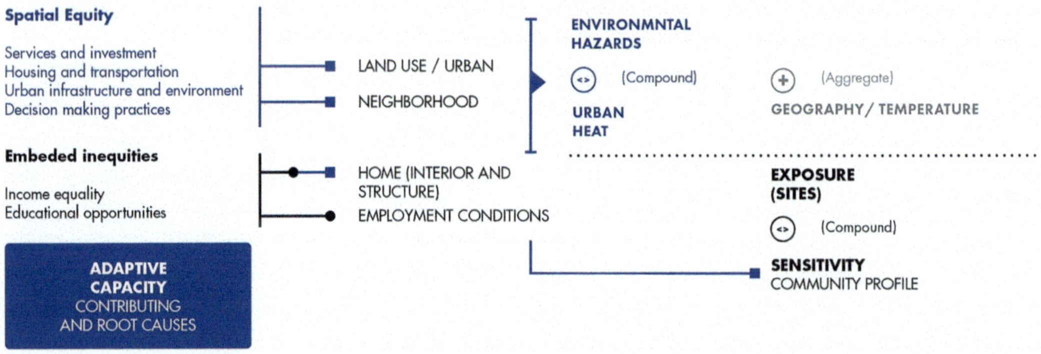

Fig. 10.2 Conceptual framework for the examination of actions and proposed solutions

Table 10.2 Link between compound urban crisis and adaptive capacity root causes as identified in literary review towards a conceptual diagram

Compound urban crises (Naylor et al. 2020; Westman et al. 2022; Yazar et al. 2022)	Spatial equity (adaptive capacity) (Lotfata and Munenzon 2022)	Embedded inequities (adaptive capacity) (Hendricks and Van Zandt 2021; Svarstad and Benjaminsen 2020)
Unsettlement is the unpredictable and unintended consequences of entangled interactions between human and natural systems	Market-driven actions that impact land use and locations of an environmental hazard	Available access to support systems/and public services
Unevenness refers to entrenched ways of thinking and doing that limit the ability to deal with crises, such as path dependencies in institutions and spatial forms that contribute to the cementation of inequalities over time	Institutional decision-making practices and resource distribution for infrastructure	Power relations and ability to participate in decision-making; historic discriminations; and intersectional experiences of various individuals and groups
Unbounding refers to emergence as a critical property of complex adaptive systems. Climate externalities	Spatial compound impacts of air quality, pollution, and heat. How the lack of vegetation impacts the long-term spatial capacity to reduce heat	Cross-cutting, evolving, uncertainty connected to the health impacts of heat and cascading cost of living

The interaction between different forms of disruption creates entirely new and unpredictable phenomena, making it challenging to observe or reduce the system's behavior to its parts. By prioritizing the needs and voices of those most affected by urban crises in policy-making processes and understanding the intertwined social-ecological problems, the compound urban impact can help address urban heat and equity.

10.3.2 Equity-Oriented Approach for Indicators

This theoretical framework highlights the significance of urban design, regulations, and management in fostering equitable resilience. To maintain the approach of successful transformative measures and solutions that will generate lasting advantages for society, we aim to adopt the framework established in the literature review and integrate elements of sustenance and welfare as essential components of adaptation. Therefore, adaptive capacity is the element we approach with aggregated policy and strategy analysis. At the same time, community sensitivity and environmental hazards will provide neighborhood risk indicators.

Adaptive capacity's influence on spatial inequities and exposure to heat and environmental hazards is depicted in Fig. 10.3. Indicators like age, health, and gender, coupled with community disparity and environmental hazards, help identify heat risk vulnerability. In contrast, localized spatial equity indicators present specific policy or design challenges (Cheng et al. 2021). Older adults and outdoor workers, mainly from marginalized communities, are more vulnerable to heat-related issues due to age and work conditions (Jessel et al. 2019; Licker et al. 2022).

Addressing heat risk and climate justice involves power equity, home ownership, and housing quality, mitigating energy access and weatherization disparities in rental properties (Hayden et al. 2017). Energy-efficient programs and targeted interventions help address energy insecurity, promoting public health and racial equity (Hernández 2013; Reames 2016; Sovacool et al. 2016).

Historic redlining policies contribute to intra-urban heat patterns and climate change inequalities. Higher temperatures in underserved urban areas are linked to a lack of tree canopy and urban features like roadways and buildings (Hoffman et al. 2020). Past discriminatory policies have led to greater heat risk vulnerability in communities

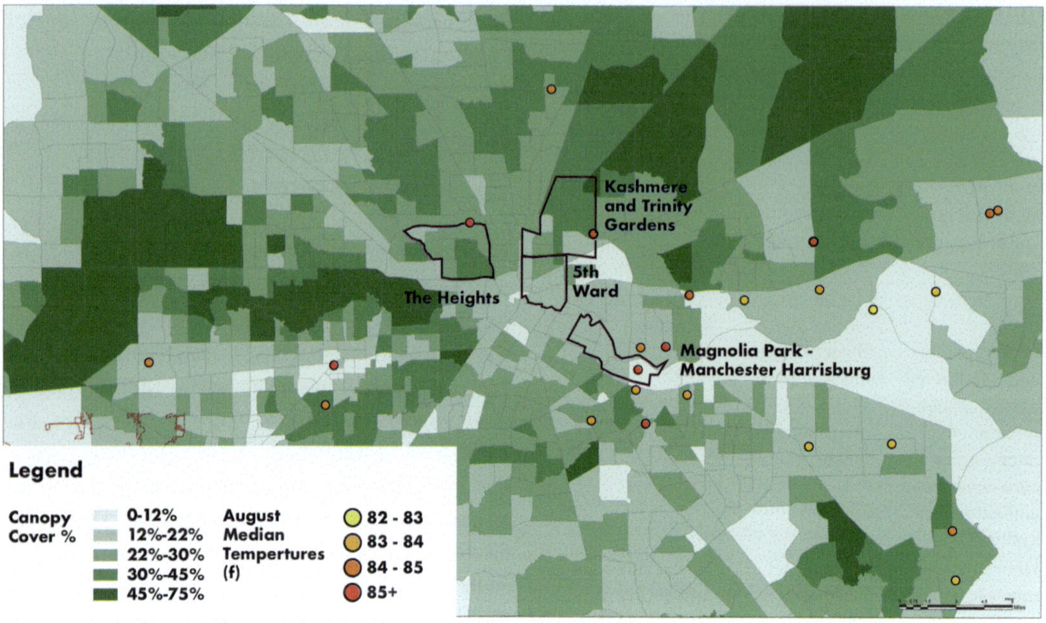

Fig. 10.3 Selected neighborhood in city context with tree canopy and location of monitored August temperatures

of color and temperature variation across US regions (Manware et al. 2022; Reckien et al. 2017).

Localized risk factors of heat and local vulnerability to compound impacts of heat and air quality require addressing social and environmental inequities. However, several worldwide heat vulnerability analysis and adaptation projects by municipalities needed more localized specificity and understanding of a place's spatial and social conditions. To address this issue, Rinner et al. (2021) conducted a study in Toronto to support policymakers and recommended using maps as tools for local data. Their study emphasized combining individual and social risk factors into a composite vulnerability index. Similarly, Inostroza et al. (2016), Yu et al. (2021), and Cheng et al. (2021) propose that employing geospatial data to analyze the combined aspects of exposure, sensitivity, adaptive capacity, and heat vulnerability indexes can guide urban planning and adaptation tactics while also serving as a foundation for public discourse. Even so, a more localized understanding of vulnerabilities cannot be found in census data or large-scale surveys. Granular information is needed through local public health data and qualitative community studies (Turek-Hankins et al. 2020).

Past research has utilized qualitative and quantitative data to examine social vulnerability in data-scarce contexts, identifying spatial and systemic inequities as key areas of adaptive capacity. These insights can guide decision-makers toward localized, hazard-specific mitigation strategies. For instance, in Lesotho, rural highland community vulnerability is attributed to factors such as poverty, resource access, family structure, and rain-dependent agriculture (Letsie and Grab 2015).

This study supports using adaptive capacity as the primary indicator for vulnerability analysis, facilitating the identification of actionable strategies to mitigate compound heat risk. Socioeconomic and built environment vulnerability sub-categories are grouped under adaptive capacity to transform vulnerability components into actionable strategies (Yu et al. 2021), assisting planners in identifying key determinants for community adaptive capacity and reducing vulnerability variability.

10.4 Neighborhood Analysis

Our study initiated by outlining key themes for data comparison across neighborhoods, namely Environmental Hazards, Community Disparity, and Community Capacity (Inostroza et al. 2016; Reckien et al. 2017; Cheng et al. 2021). The scope of these themes was extended to include potential neighborhood-specific remediation and community empowerment strategies.

We utilized environmental and demographic data from 2020, a year that, alongside 2016, recorded the highest summer temperatures by NASA and the NOAA. The Texas Commission on Environmental Quality (TCEQ) provided our air quality data, which included hourly and daily records from various Houston locations. The heat island data were sourced from the Houston and Harris County urban heat island mapping campaign of August 2020 and TCEQ. Community demographic data were retrieved from the American Community Survey 2020 (5-year estimate). We relied on the Houston State of Health portal for community health data, which amalgamates health data from several sources, including the National Cancer Institute, the Centers for Disease Control, the American Community Survey, and numerous state-specific sources. Land cover and land use data were obtained from the Houston Galveston Area Council raster imagery and GIS portal.

We examined the data at the census tract level, except health data, land use mapping, and environmental data, which were reviewed at the neighborhood level using Houston's Super Neighborhood (SNG) boundaries. SNGs serve as planning areas where residents can identify, discuss, and prioritize community needs and concerns. Boundaries are drawn based on geographic and physical features to group residents with similar identities, histories, or characteristics.

We categorized each indicator into one of the three central themes upon data source identification. Environmental Hazards encompassed built environment indicators, which have emerged due to years of deficient urban development and would necessitate large-scale remediation efforts for rectification. Community disparities included both community and individual disparities arising from long-term exposure to environmental hazards. Their rectification would call for infrastructural investment in the community to enhance amenities and programs. Finally, community capacity involves indicators reflecting the current level of city investment in the community and the strength of the social infrastructure, which contributes to quicker recovery following climate events. Enhancing these indicators would demand more specific aid from city plans and the fostering of grassroots community-building initiatives.

- Environmental hazard: Land use; land cover/greenery; morning heat index; evening heat index; air quality index; brownfield proximity; superfund site proximity; % of trees in the neighborhood; predominant industries in neighborhood; predominant occupations in neighborhood.
- Community disparity: Community percentage of cancer cases; median year of houses built; community percentage of high blood pressure cases; community percentage of children with deformity cases; community racial demographics; community highest educational attainment; number of cars available per household; community means of transportation; community median household income; number of after school programs; community percentage of health insurance; community percentage of stroke cases; community percentage of obesity cases; community percentage of coronary heart disease cases; community percentage of asthma cases; community percentage of diabetes cases; community percentage of kidney disease; community life expectancy; percent of persons older than 65 living alone; percent of persons older than 65 with independent living difficulty.
- Community capacity: Median year of houses built; number of community anchors; locations of cooling centers; number of community plans; transit connectivity; federal community clinics; number of community organizations; number of after-school programs.

We set a quantitative benchmark for each indicator using Houston's city-wide values, providing a standardized reference to compare across communities and identify significant trends. This replicable approach aids city planners in statistical analysis. Environmental data, such as heat index and air quality, were compared with city-wide records from the same day.

Our study focused on Houston's six super neighborhood groups (SNG), including Kashmere and Trinity Gardens, Magnolia Park and Manchester Harrisburg, 5th Ward, and the Heights. Most of these SNGs are in industrial Eastern Houston, while the heights, in North-West Houston, is Texas' earliest planned community. The study aimed to delve into the long-term effects of pollution on these communities, an area only recently explored by researchers (UnderstandingHouston.org 2021; Tessum et al. 2019; Sexton et al. 2007; Horney et al. 2018; Sansom et al. 2018; Oliveira Santos et al. 2023).

Moreover, recent research indicates these communities face graver environmental hazards than previously understood (Sansom et al. 2018). With increased health risks from compounded environmental and urban factors and climate change-induced disasters, our study sought to probe these neighborhoods' environmental hazards, community disparities, and adaptive capacity (Adepoju et al. 2021; Grineski et al. 2022).

10.4.1 Magnolia Park: Manchester Harrisburg

Geospatial analysis shows that Magnolia Park—Manchester Harrisburg is predominantly Latino, with 94% of the population identifying as such. Education levels could be higher, with 45% not finishing high school and only 8% obtaining a bachelor's degree or higher. The primary industries are construction (20%), manufacturing, and retail (12%). Notably, 25% of residents lack health insurance. The neighborhood's buildings, predominantly built around 1950, may need more support with rising temperatures due to outdated construction and energy inefficiency. Around 11% of residents do not own cars and a similar number carpool. Health complications are prevalent, with 70% of residents on high blood pressure medication, 44.6% classified as obese, and a life expectancy 4 years below the Houston average (74 vs. 78) (Fig. 10.4).

Tree canopy coverage varies from 19% to 21%, dropping to 7% in the southern part of Manchester Harrisburg. Heat indexes are higher than in other Houston areas, with morning temperatures around 81 °F and some industrial regions reaching 88 °F. The lower relative humidity (58%) results in lower evening heat indexes. The neighborhood contends with considerable industrial land use, posing environmental hazards near residential areas and community spaces. Yet, the area maintains a strong social network, with three community centers but limited after-school programs and healthy eating options, as exemplified by the federal food desert designation in southern Manchester Harrisburg.

10.4.2 5th Ward

The 5th Ward, home to a diverse population of 47% Black and 48% Hispanic or Latino residents, presents lower education levels than Houston at large, with 31% lacking a high school education and only 11% owning a bachelor's degree. The primary occupations are construction, extraction, maintenance (17%), transportation, and material moving (15%). The neighborhood's housing, primarily constructed around 1956, and the high rental rate (65%) might exacerbate vulnerabilities to rising temperatures due to limited capacity for energy efficiency improvements. Poverty affects one-third of families with children, with 57% being single-mother households. Health concerns are evident with high obesity (48.9%), high blood pressure medication rates (74.6%), and an 8-year lower life expectancy than Houston's average. Furthermore, 22% of elderly residents report difficulties with independent living, increasing their heat-related injury risk.

The area faces environmental hazards, including industrial land use and a Union Pacific railway yard linked to increased cancer, asthma,

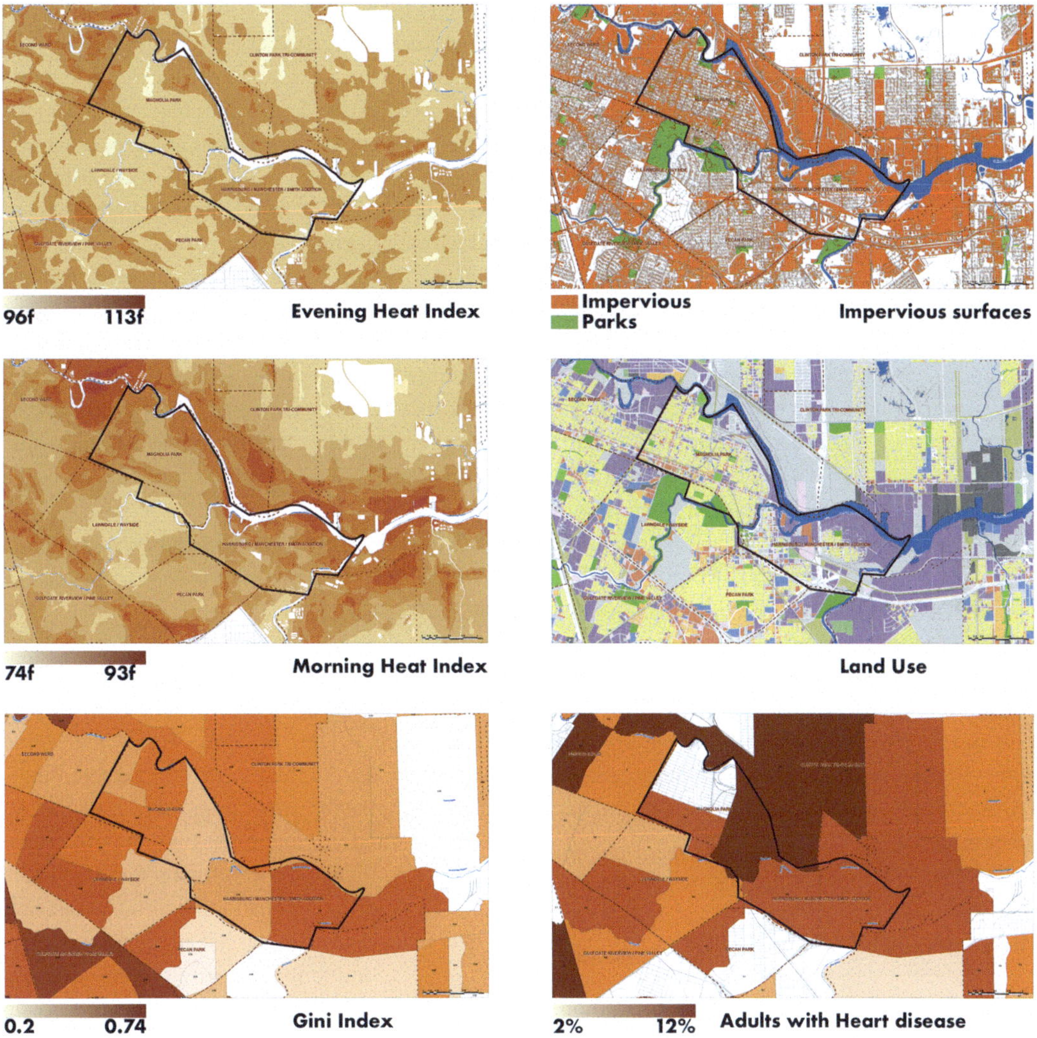

Fig. 10.4 Selected maps of Magnolia Park—Manchester Harrisburg showing the limited ability of evening temperatures to drop, intense distribution of industrial use and impermeable surfaces, and social and individual vulnerability of the neighborhoods

and childhood deformities. Tree coverage ranges from 15% to 21%, and the highest morning heat indexes are found in industrialized regions near the railway and bayou, reaching 84–88 °F. Evening temperatures often exceed 99 °F, with some areas hitting 103 °F. Access to healthy food is limited, with no major grocery stores in the area. Despite three community centers, no schools offer state-funded after-school programs. The 5th Ward confronts significant educational, health, and environmental challenges.

10.4.3 Kashmere and Trinity Gardens and the Heights

Kashmere Gardens and Trinity Gardens, located on either side of the I-610 Loop, are historically black neighborhoods with predominantly communities of color, where 58% of residents are Black, and 37% are Hispanic or Latino. The median year of housing in these neighborhoods is 1941, with a quarter of families living below the poverty line and limited vehicle access for 13%

of families. Health issues are prevalent among residents, with 27% lacking health insurance, a life expectancy of 69 years, and high rates of obesity (49%), high blood pressure (74.6%), asthma (11.2%), and disability among children (6%). Trinity Gardens has high tree canopy coverage (34–44%), while Kashmere Gardens has 19–34% coverage, influenced by industrial and residential land use. Kashmere Gardens has four superfund sites, while Union Pacific rail yard has leaked hazardous toxins, leading to increased rates of asthma and childhood disabilities.

In contrast, The Heights is a master-planned community with a rich investment history, where 65% of residents are white, and 39% hold bachelor's degrees. The area has high tree canopy coverage (25–35%), and morning heat indexes range from 78 to 81 °F, with evening heat indexes dropping to 98–99 °F. The neighborhood is not a food desert, has minimal industrial or transportation land use, and has healthy food options. The Heights has lower percentages than Houston in most figures except for the 65+ living alone category. Higher heat indexes are found near industrial areas and areas with high impervious land cover.

10.4.4 Summary of Findings

Focusing on Environmental Hazards, Community Disparity, and Community Capacity, we compared three Houston neighborhoods: Magnolia Park - Manchester Harrisburg, 5th Ward, Kashmere, and Trinity Gardens, using environmental, demographic, and health data from 2020.

All neighborhoods exhibited significant Environmental Hazards. Magnolia Park—Manchester Harrisburg and 5th Ward were most vulnerable due to their industrial occupations, low tree canopy cover, and high heat indexes. Kashmere and Trinity Gardens had four superfund sites and toxins from a rail yard, posing severe environmental threats. Community Disparity was apparent across all neighborhoods. Magnolia Park—Manchester Harrisburg and 5th Ward had low educational attainment and high rates of health issues. Kashmere and Trinity Gardens faced high rates of poverty and health issues, including asthma and childhood disabilities. In terms of Community Capacity, Magnolia Park—Manchester Harrisburg had a robust social network but needed healthy eating options and after-school programs. 5th Ward and Kashmere, and Trinity Gardens faced similar challenges but had varying degrees of tree coverage.

These findings underscore the need for targeted interventions to address these communities' significant disparities and vulnerabilities, focusing on remediation strategies, community infrastructure investment, and grassroots community-building efforts (Table 10.3).

10.4.5 Policy Analysis Framework

The geospatial analysis presents clear links between spatial decision-making, embedded inequities in land use policy, public work investments, the heat index in certain areas, and air pollution levels. The findings from the three neighborhoods, combined with the selected indicators, emphasize the importance of urban design, regulations, and management in fostering equitable resilience for developing targeted interventions. Addressing spatial inequities, exposure to heat and environmental hazards, and the intersection of community sensitivity and environmental hazards is crucial. Adaptive capacity in policy and strategy analysis can identify challenges driving policy or design actions, such as power equity, homeownership, housing quality, and localized risk factors. Historical redlining policies and current practices have led to disparities in temperature variation and heat risk vulnerability in underserved neighborhoods, necessitating strategies addressing social and environmental inequities. Utilizing geospatial data to create composite vulnerability indices and focusing on localized understandings of vulnerabilities can support urban planning and adaptation strategies, aiding decision-makers in producing more hazard-specific mitigation strategies. This study's use of adaptive capacity as the lead indicator source for vulnerability analysis allows for identifying actionable strategies to

Table 10.3 Compiled findings from the geospatial analysis

Neighborhood	Key demographics	Housing	Health	Environmental factors
Magnolia Park—Manchester Harrisburg	Predominantly Hispanic (94%), low educational attainment (45% less than high school, 8% bachelor's degree), and main occupation—construction (20%)	Median construction year: 1950, 11% without a car	High blood pressure (70%), obesity (44.6%), and life expectancy lower than Houston average	Tree canopy: 19–21%, heat indexes: 81–88 °F, one brownfield, and considerable industrial land use
5th Ward	Diverse (47% Black, 48% Hispanic/Latino), low education (31% without high school, 11% bachelor's degree), and main occupations—construction and transportation	Median construction year: 1956 and 65% rent homes	High rates of obesity (48.9%), high blood pressure (74.6%), and life expectancy 8 years lower than Houston	Tree canopy: 15–21%, heat indexes: 80–88 °F morning, 99–103 °F evening, industrial, and transportation land use
Kashmere and Trinity Gardens	Predominantly communities of color (58% Black, 37% Hispanic/Latino) and 25% families below poverty line	Median construction year: 1941 and 13% families with limited vehicle access	Life expectancy: 69 years, high rates of obesity (49%), high blood pressure (74.6%), asthma (11.2%), and childhood disability (6%)	Tree canopy: 19–44%, four superfund sites and leaked hazardous toxins from union Pacific rail yard

address compound heat risk and guiding planners in defining crucial determinants for community adaptive capacity (Table 10.4).

The challenges posed by increased temperatures, urban heat, air pollution, and health issues necessitate transformative adaptation. Swart et al. (2023) define transformative adaptation as addressing the underlying causes of climate risk and avoiding lock-ins for unsustainable development. This approach involves fundamental changes to city systems and leveraging the natural system's ability to mitigate climatic extremes. Transformative adaptation differs from reactive adaptation, which deals with immediate consequences of weather and climate extremes, and incremental adaptation, which involves preventive measures within city systems. We suggest using Hendricks and Van Zandt's (2021) research as the basis for an evaluation framework for solutions and actions toward transformative adaptation. Their emphasis on linking infrastructure with environmental justice challenges and social vulnerability can inform our approach to addressing the root causes of vulnerability. By reframing how infrastructure is explored within contexts and considering embedded inequities such as racism, classism, sexism, and ableism, we can develop policies and practices that prevent race and racism from determining the distribution of assets and resources, including infrastructure. These interventions are a fundamental right to public safety and protection and can support a more resilient and just future for marginalized communities.

10.5 Analysis of Adaptation Strategies

The geospatial analysis presented in the previous section revealed stark contrasts in public realm investment and high levels of risk in specific neighborhoods adjacent to industrial uses. These neighborhoods also face challenges such as outdated housing stock, heightening their vulnerability to rising temperatures, and environmental hazards. The limited availability of municipal investment and public infrastructure projects suggests a historical pattern of discrimination. This section evaluates Houston's efforts to address

Table 10.4 The indicators mapped and studied as adaptive capacity opportunities

Adaptive capacity	Spatial equity	Embedded inequities
Environmental hazard	Land cover/ greenery (impervious land cover and tree canopy); morning heat index; predominant industries in neighborhood;	Land use as policy (industrial and residential land use); superfund site proximity; predominant occupations in neighborhood;
	SDG 11.a economic development SDG 11.7 public and green space	SDG 11.3 capacity building; SDG 11.6 air quality
Community capacity	Median year of houses built; community anchors; community facilities;	Community means of transportation; community clinics and services; access to health and community services;
	SDG 11.1 housing affordability	1SDG 1.b energy affordability; SDG 11.2 transportation

these issues through spatial and policy interventions. The method presented in Fig. 10.5 goes beyond identifying hazards and indicators and aims to evaluate the root causes and systemic apparatus leading to socio-spatial conditions.

Our approach fosters comprehensive strategies over a direct hazard-solution method, promoting divergent thinking that leads to specific actions. Assessing solutions' time horizons and aspects like multi-benefits, policy, and governance helps discern if actions spur equitable investment and long-term transformation, as advocated by Swart et al. (2023) and Lotfata and Munenzon (2022).

Effective heat and hazard mitigation needs a holistic understanding of city systems and diverse stakeholder collaboration. This research underscores the importance of coordination among urban systems (Wang et al. 2021). Emphasizing public participation, education, and decision-making is crucial. An opportunity-focused vision promotes broader engagement, potentially increasing financing for heat mitigation strategies (Swart et al. 2023).

Houston, a Gulf Coast urban center, grapples with significant environmental challenges, including climate change, pollution, urban heat, and environmental justice. To address these issues and build resilience, Houston has undertaken numerous steps. This section introduces a policy review to identify opportunities to address community disparities and environmental hazards in Houston, using a matrix to analyze plans in three categories: environmental hazards, community disparity, and adaptive capacity. It recommends a table to evaluate projects, goals, and policies on their ability to address issues like heat, health, pollution, spatial challenges, and long-term action. It reviews five community and climate-related plans developed in Houston to address aspects of the built environment and well-being improvement (Table 10.5).

Houston's Climate Action Plan provides a roadmap for reducing greenhouse gas emissions and adapting to climate change impacts with equity measures. Air monitoring programs and initiatives like Complete Communities have been expanded to identify environmental justice issues. H3at.org and research collaborations are raising awareness and modeling extreme heat risks. Air pollution remains challenging in the Houston-Galveston-Brazoria metropolitan region, requiring stricter regulations and spatial actions. The Resilient Houston Plan strongly focuses on environmental justice. It aims to create prepared and thriving Houstonians, safe and equitable neighborhoods, healthy and connected bayous, accessible and adaptive cities, and innovative and integrated regions. The city has plans to improve neighborhoods, transportation, sustainability, and water conservation. Some of these plans are still only on paper and are awaiting further steps and funding for implementation. However, more investment is needed to protect vulnerable populations' health, including emissions reduction, pollution controls, and enforcement measures. These strategies can serve as a

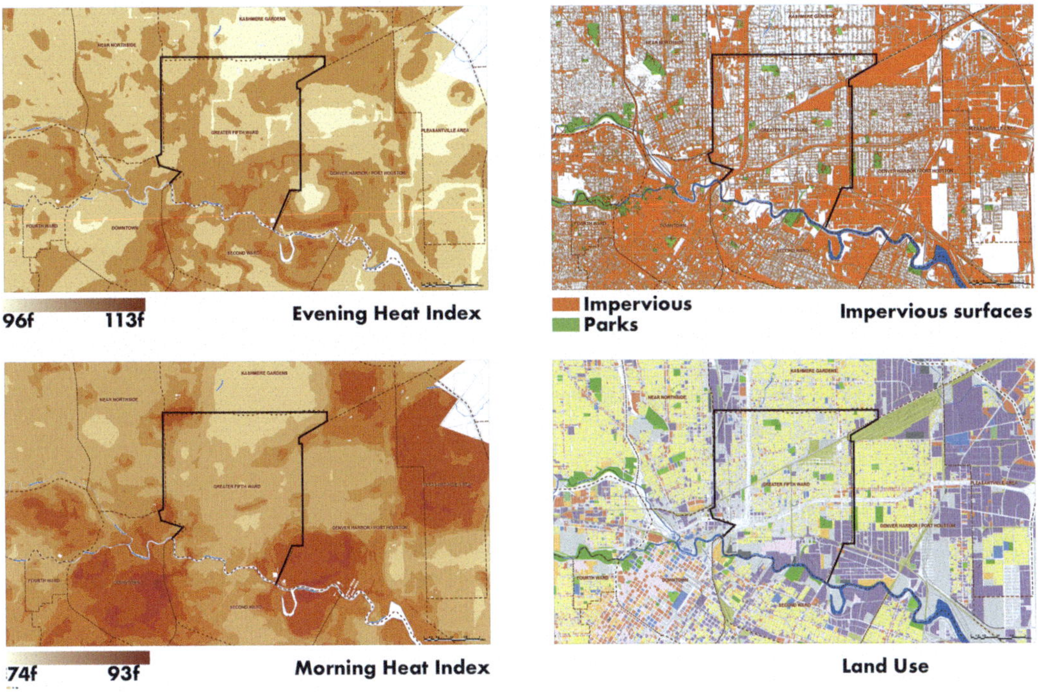

Fig. 10.5 Selected maps of Greater 5th Ward showing the limited ability of evening temperatures to drop, intense distribution of industrial use, and impermeable surfaces

Table 10.5 Review of SDGs targets as indicators across the selected five Houston city plans

	SDG 11 targets and indicators						
Plan	11.1 Housing affordability	11.b Energy affordability	11.a Economic development	11.3 Capacity building	11.2 Transportation	11.6 Air quality	11.7 Public and green space
(a)	X	X	X	X	X	X	X
(b)	X	X	X		X	X	
(c)	X		X	X	X		X
(d)	X		X		X		
(e)				X	X		X

(a) Resilient Houston (City of Houston 2020a); (b) Houston Climate Action Plan (City of Houston 2020b); (c) Complete Communities (City of Houston 2018); (d) Plan Houston (City of Houston 2015); (e) Vision Zero Action Plan (City of Houston 2020c)

model for other cities facing similar challenges (Table 10.6).

Investment is key to shielding vulnerable populations from air pollution and heat-related illnesses. The IPCC AR6 report advocates for system transition strategies including green infrastructure, sustainable land use and urban planning, and resilient water management. Coordinated government action is necessary for resilient power systems, health systems adaptation, and livelihood diversification. These strategies address the compound effects of heat risk and environmental hazards, potentially serving as a blueprint for other cities facing similar challenges.

Table 10.6 Review of the selected five Houston city plans across the compound crisis categories and the scale of proposed actions. The plans evolve from visions toward implementable actions

Plan	Urban crisis			Scale		Spatial impact	Embedded inequities
	Unsettlement/ policy	Unevenness/ governance and power	Unbounding/ compound and climate	Responsive	transformative		
(a)	X		X		X	X	
(b)			X	X			X
(c)		X		X		X	X
(d)			X	X	X	X	
(e)				X		X	

(a) Resilient Houston (City of Houston 2020a); (b) Houston Climate Action Plan (City of Houston 2020b); (c) Complete Communities (City of Houston 2018); (d) Plan Houston (City of Houston 2015); (e) Vision Zero Action Plan (City of Houston 2020c)

With escalating climate and environmental impacts on public health, the evaluation of adaptation interventions is becoming increasingly vital. As data on climate vulnerability and resilience grow more accessible, and communities demand effective evidence, municipal planners face a wider range of options in question, design, and method selection (Joseph et al. 2023). Assumptions within the evaluation process must be clearly understood, with adaptation success defined on a case-by-case basis, informed by local context and stakeholders. Future policy reviews can benefit from a clearly articulated vision and implementation roadmap, and consideration of just, equitable, or transformative adaptation levels.

10.6 Discussion and Conclusion

The discourse surrounding SDG emphasizes the importance of addressing the inequitable distribution and environmental burden on disadvantaged communities through a climate justice lens, particularly in cities such as Houston (Menton et al. 2020; Chu et al. 2017). To achieve inclusive and equitable benefits for all urban residents, comprehensive climate adaptation transformation across all sectors of urban influence is necessary, as market pressures, local policies, and investment patterns disproportionately impact low-income communities of color. The emergence of a new urban climate economy and the idea of climate urbanism establish cities as capable institutions to address climate change, reprioritize policy priorities, and promote new investment frameworks, which are essential in achieving climate adaptation (Long and Rice 2019).

Responding to climate change requires a shift towards new economic development and energy transition types, prioritizing distributive justice and access to decision-making spaces for marginalized communities within neighborhood economic development. Green infrastructure and public spaces effectively address urban heat and air pollution (Hewitt et al. 2020). Still, implementation requires resolving property ownership and jurisdictional issues if it is outside the public right of way. Neighborhood-scale actions that promote stewardship, provide employment opportunities, and build social cohesion can foster transformative politics grounded in critical self-reflection, collective action, and intersectional awareness (Leal Filho et al. 2021; Wang et al. 2021; Sultana 2022).

According to Symanski et al. (2020), a successful partnership involving the community, academic, government, and industry sectors was established to address environmental health concerns related to metal air pollution in underserved communities in Houston. The partnership employed a community-based participatory research (CBPR) approach and conducted community air monitoring, health risk assessments, and a community survey to develop an action plan based on the findings. The project resulted in voluntary actions from industry partners to reduce emissions, developing a colloquial version of key research findings, and broader outcomes, such as the Houston Health Department developing a program to train volunteers to communicate environmental health concerns to the department and promote environmental health programs.

To address the compound risks of urban and climate hazards in EJ communities, localized vulnerability analysis is critical in identifying climate scenarios that amplify existing health disparities and environmental hazards. The proposed framework for resilience provides a way forward for transformative climate adaptation projects that address the compound risks of urban and climate hazards. By integrating climate adaptation and sustainability goals, all communities can access essential services, green infrastructure, affordable housing, and reliable power, with decision-making power shifted to the communities most affected by these issues.

In conclusion, our study highlights the need for a comprehensive approach to address the compound risks of urban and climate hazards in EJ communities, particularly in Houston, TX. Our proposed framework for resilience can provide transformative solutions that address environmental injustice and improve public health and community well-being. By addressing these challenges, we can work towards achieving sustainability goals and promoting environmental justice in Houston and beyond. Local governments and communities must actively participate in climate adaptation and sustainability planning, emphasizing transformative politics, social equity, and community engagement. We hope this chapter contributes to ongoing efforts to address EJ and climate risks in Houston and other urban areas facing similar challenges.

References

Adepoju OE, Han D, Chae M, Smith KL, Gilbert L, Choudhury S, Woodard L (2021) Health disparities and climate change: the intersection of three disaster events on vulnerable communities in Houston, Texas. Int J Environ Res Public Health 19(1):35. https://doi.org/10.3390/ijerph19010035

AghaKouchak A, Chiang F, Huning LS, Love CA, Mallakpour I, Mazdiyasni O, Moftakhari H, Papalexiou SM, Ragno E, Sadegh M (2020) Climate extremes and compound hazards in a warming world. Annu Rev Earth Planet Sci 48(1):519–548. https://doi.org/10.1146/annurev-earth-071719-055228

Baldwin JW, Dessy JB, Vecchi GA, Oppenheimer M (2019) Temporally compound heat wave events and global warming: an emerging Hazard. Earth's Future 7(4):411–427. https://doi.org/10.1029/2018EF000989

Brenner N (2009) What is critical urban theory? City 13(2–3):198–207. https://doi.org/10.1080/13604810902996466

Bullard RD (1987) Invisible Houston: the black experience in boom and bust, 1st edn. Texas A&M University Press, Collage Station

Bullard RD, Wright B (eds) (2009) Race, place, and environmental justice after hurricane Katrina: struggles to reclaim, rebuild, and revitalize New Orleans and the Gulf coast. Westview Press, Boulder, CO

Chakraborty T, Hsu A, Manya D, Sheriff G (2019) Disproportionately higher exposure to urban heat in lower-income neighborhoods: a multi-city perspective. Environ Res Lett 14(10):105003. https://doi.org/10.1088/1748-9326/ab3b99

Cheng W, Li D, Liu Z, Brown RD (2021) Approaches for identifying heat-vulnerable populations and locations: a systematic review. Sci Total Environ 799(December):149417. https://doi.org/10.1016/j.scitotenv.2021.149417

Chu E, Anguelovski I, Roberts D (2017) Climate adaptation as strategic urbanism: assessing opportunities and uncertainties for equity and inclusive development in cities. Cities 60:378–387. https://doi.org/10.1016/j.cities.2016.10.016

City of Houston (2015) Plan Houston. https://www.houstontx.gov/planhouston/sites/default/files/plans/Final_Plan_Houston.pdf; https://www.houstontx.gov/mayor/Climate-Impact-Assessment-2020-August.pdf

City of Houston (2018) Complete communities. https://www.houstoncc.org/our_communities/acres_home/index.php

City of Houston (2020a) Resilient Houston. https://www.houstontx.gov/mayor/Resilient-Houston-20200518-single-page.pdf

City of Houston (2020b) Houston Climate Action Plan. http://greenhoustontx.gov/climateactionplan/

City of Houston (2020c) Houston Vision Zero Action Plan. https://houstontx.gov/visionzero/pdf/VZAP-Final-Report.pdf

City of Houston (2020d). Climate impact assessment for the City of Houston. https://www.houstontx.gov/mayor/Climate-Impact-Assessment-2020-August.pdf

Climate.gov (n.d.) U.S. Climate Resilience Toolkit. https://toolkit.climate.gov/

Collette M, Dempsey M, Carroll S (2016) Chemical breakdown. Houston Chronical, May. https://www.houstonchronicle.com/local/texas/chemical-breakdown/

Conlon K, Monaghan A, Hayden M, Wilhelmi O (2016) Potential impacts of future warming and land use changes on intra-urban heat exposure in Houston, Texas. PLoS One 11(2):e0148890. https://doi.org/10.1371/journal.pone.0148890

Cushing LJ, Li S, Steiger BB, Casey JA (2022) Historical red-lining is associated with fossil fuel power plant siting and present-day inequalities in air pollutant emissions. Nat Energy 8(1):52–61. https://doi.org/10.1038/s41560-022-01162-y

Cutter SL, Boruff BJ, Shirley WL (2003) Social vulnerability to environmental hazards. Soc Sci Q 84(2):242–261. https://doi.org/10.1111/1540-6237.8402002

Douglas E (2022) Smog levels in Texas surge during heat wave, bringing worst summer air quality in a decade. Texas Tribune, July 14. https://www.texastribune.org/2022/07/14/texas-summer-smog-ozone-pollution/

Ebi KL, Vanos J, Baldwin JW, Bell JE, Hondula DM, Errett NA, Hayes K, Reid CE, Saha S, Spector J, Berry P (2021) Extreme weather and climate change: population health and health system implications. Annu Rev Public Health 42(1):293–315. https://doi.org/10.1146/annurev-publhealth-012420-105026

Fedele G, Donatti CI, Harvey CA, Hannah L, Hole DG (2020) Limited use of transformative adaptation in response to social-ecological shifts driven by climate change. Ecol Soc 25(1):art25. https://doi.org/10.5751/ES-11381-250125

Grineski SE, Collins TW, Chakraborty J (2022) Cascading disasters and mental health inequities: Winter Storm Uri, COVID-19 and post-traumatic stress in Texas. Soc Sci Med 315:115523. https://doi.org/10.1016/j.socscimed.2022.115523

Hayden MH, Wilhelmi OV, Banerjee D, Greasby T, Cavanaugh JL, Nepal V, Boehnert J, Sain S, Burghardt C, Gower S (2017) Adaptive capacity to extreme heat: results from a household survey in Houston, Texas. Weather Clim Soc 9(4):787–799. https://doi.org/10.1175/WCAS-D-16-0125.1

Heaton MJ, Sain SR, Greasby TA, Uejio CK, Hayden MH, Monaghan AJ, Boehnert J, Sampson K, Banerjee D, Nepal V, Wilhelmi OV (2014) Characterizing urban vulnerability to heat stress using a spatially varying coefficient model. Spatial Spatio-Temporal Epidemiol 8:23–33. https://doi.org/10.1016/j.sste.2014.01.002

Hernández D (2013) Energy insecurity: a framework for understanding energy, the built environment, and health among vulnerable populations in the context of climate change. Am J Public Health 103(4):e32–e34. https://doi.org/10.2105/AJPH.2012.301179

Hendricks MD, Van Zandt S (2021) Unequal protection revisited: planning for environmental justice, hazard vulnerability, and critical infrastructure in communities of color. Environ Justice 14(2):87–97. https://doi.org/10.1089/env.2020.0054

Hewitt CN, Ashworth K, MacKenzie AR (2020) Using green infrastructure to improve urban air quality (GI4AQ). Ambio 49(1):62–73. https://doi.org/10.1007/s13280-019-01164-3

Hinkel J (2011) "Indicators of vulnerability and adaptive capacity": towards a clarification of the science–policy interface. Glob Environ Chang 21(1):198–208. https://doi.org/10.1016/j.gloenvcha.2010.08.002

Hoffman JS, Shandas V, Pendleton N (2020) The effects of historical housing policies on resident exposure to intra-urban heat: a study of 108 US urban areas. Clim 8(1):12. https://doi.org/10.3390/cli8010012

Horney JA, Casillas GA, Baker E, Stone KW, Kirsch KR, Camargo K, Wade TL, McDonald TJ (2018) Comparing residential contamination in a Houston environmental justice neighborhood before and after Hurricane Harvey. PLoS One 13(2):e0192660. https://doi.org/10.1371/journal.pone.0192660

Houghton JT, Intergovernmental Panel on Climate Change (eds) (2001) Climate change 2001: the scientific basis. Contribution of working group I to the third assessment report of the intergovernmental panel on climate change. Cambridge University Press, Cambridge/New York

Hope Sr. KR (2020) Peace, justice and inclusive institutions: overcoming challenges to the implementation of sustainable development goal 16. Global Change, Peace & Security 32(1):57–77. https://doi.org/10.1080/14781158.2019.1667320

Inostroza L, Palme M, de la Barrera F (2016) A heat vulnerability index: spatial patterns of exposure, sensitivity and adaptive capacity for Santiago de Chile. PLoS One 11(9):e0162464. https://doi.org/10.1371/journal.pone.0162464

IPCC (2021) Summary for policymakers. In: Masson-Delmotte V, Zhai P, Pirani A, Connors SL, Péan C, Berger S, Caud N et al (eds) Climate change 2021: the physical science basis. Contribution of working group I to the sixth assessment report of the intergovernmental panel on climate change. Cambridge University Press, Cambridge/United Kingdom/New York/USA. https://doi.org/10.1017/9781009157896.001

Jessel S, Sawyer S, Hernández D (2019) Energy, poverty, and health in climate change: a comprehensive review of an emerging literature. Front Public Health 7(December):357. https://doi.org/10.3389/fpubh.2019.00357

Johnston JE, Werder E, Sebastian D (2016) Wastewater disposal Wells, fracking, and environmental injustice in southern Texas. Am J Public Health 106(3):550–556. https://doi.org/10.2105/AJPH.2015.303000

Joseph HA, Mallen E, McLaughlin M, Grossman E, Holmes TJ, Locklear A, Powell E, Thie L, Uejio CK, Vacca K, Williams C, Bishop T, Jeffers C, Siegel H, Austin C (2023) Evaluating public health strategies for climate adaptation: challenges and opportunities from the climate ready states and cities initiative. PLOS Clim 2(3):e0000102. https://doi.org/10.1371/journal.pclm.0000102

Kim S, Ryu Y (2015) Describing the spatial patterns of heat vulnerability from urban design perspectives. Int J Sustain Dev World Ecol 22(3):189–200. https://doi.org/10.1080/13504509.2014.1003202

Leal Filho W, Wolf F, Castro-Díaz R, Li C, Ojeh VN, Gutiérrez N, Nagy GJ, Savić S, Natenzon CE, Quasem Al-Amin A, Maruna M, Bönecke J (2021) Addressing the urban heat islands effect: a cross-country assessment of the role of green infrastructure. Sustainability 13(2):753. https://doi.org/10.3390/su13020753

Letsie MM, Grab SW (2015) Assessment of social vulnerability to natural hazards in the mountain Kingdom of Lesotho. Mt Res Dev 35(2):115–125. https://doi.org/10.1659/MRD-JOURNAL-D-14-00087.1

Levison M, Butler A, Rebellato S, Armstrong B, Whelan M, Gardner C (2018) Development of a climate change vulnerability assessment using a public health lens to determine local health vulnerabilities: an Ontario health unit experience. Int J Environ Res Public Health 15(10):2237. https://doi.org/10.3390/ijerph15102237

Licker R, Dahl K, Abatzoglou JT (2022) Quantifying the impact of future extreme heat on the outdoor work sector in the United States. Elem Sci Anth 10(1):00048. https://doi.org/10.1525/elementa.2021.00048

Li Z, Konisky DM, Zirogiannis N (2019) Racial, ethnic, and income disparities in air pollution: a study of excess emissions in Texas. Edited by Matthew Kahn. PLoS One 14(8):e0220696. https://doi.org/10.1371/journal.pone.0220696

Linder SH, Marko D, Sexton K (2008) Cumulative cancer risk from air pollution in Houston: disparities in risk burden and social disadvantage. Environ Sci Technol 42(12):4312–4322. https://doi.org/10.1021/es072042u

Liu L, Talbot R, Lan X (2015) Influence of climate change and meteorological factors on Houston's air pollution: ozone a case study. Atmos 6(5):623–640. https://doi.org/10.3390/atmos6050623

Long J, Rice JL (2019) From sustainable urbanism to climate urbanism. Urban Stud 56(5):992–1008. https://doi.org/10.1177/0042098018770846

Lotfata A, Munenzon D (2022) The interplay of intersectionality and vulnerability towards equitable resilience: learning from climate adaptation practices. In: The Palgrave encyclopedia of urban and regional futures. Springer, pp 1–16. https://doi.org/10.1007/978-3-030-51812-7_324-1

Manware M, Dubrow R, Carrión D, Ma Y, Chen K (2022) Residential and race/ethnicity disparities in heat vulnerability in the United States. GeoHealth 6(12). https://doi.org/10.1029/2022GH000695

Marsha A, Sain SR, Heaton MJ, Monaghan AJ, Wilhelmi OV (2018) Influences of climatic and population changes on heat-related mortality in Houston, Texas, USA. Clim Chang 146(3–4):471–485. https://doi.org/10.1007/s10584-016-1775-1

Menton M, Larrea C, Latorre S, Martinez-Alier J, Peck M, Temper L, Walter M (2020) Environmental justice and the SDGs: from synergies to gaps and contradictions. Sustain Sci 15(6):1621–1636. https://doi.org/10.1007/s11625-020-00789-8

Naylor A, Ford J, Pearce T, Van Alstine J (2020) Conceptualizing climate vulnerability in complex adaptive systems. One Earth 2(5):444–454. https://doi.org/10.1016/j.oneear.2020.04.011

Oliveira Santos V, Costa Rocha PA, Scott J, Van Griensven Thé J, Gharabaghi B (2023) Spatiotemporal air pollution forecasting in Houston-TX: a case study for ozone using deep graph neural networks. Atmos 14(2):308. https://doi.org/10.3390/atmos14020308

O'Lenick CR, Wilhelmi OV, Michael R, Hayden MH, Baniassadi A, Wiedinmyer C, Monaghan AJ, Crank PJ, Sailor DJ (2019) Urban heat and air pollution: a framework for integrating population vulnerability and indoor exposure in health risk analyses. Sci Total Environ 660(April):715–723. https://doi.org/10.1016/j.scitotenv.2019.01.002

Paaske DM (2021) Impact of climate change, intergenerational justice, and human rights on SDGs. In: Filho WL, Azul AM, Brandli L, Salvia AL, Özuyar PG, Wall T (eds) Reduced inequalities. Encyclopedia of the UN Sustainable Development Goals. Cham: Springer International Publishing. https://doi.org/10.1007/978-3-319-95882-8_109

Reames TG (2016) Targeting energy justice: exploring spatial, racial/ethnic and socioeconomic disparities in urban residential heating energy efficiency. Energy Policy 97(October):549–558. https://doi.org/10.1016/j.enpol.2016.07.048

Reckien D, Creutzig F, Fernandez B, Lwasa S, Tovar-Restrepo M, Mcevoy D, Satterthwaite D (2017) Climate change, equity and the sustainable development goals: an urban perspective. Environ Urban 29(1):159–182. https://doi.org/10.1177/0956247816677778

Rinner C, Paychuk D, Bassil K, Nasr S, Gower S, Campbell M (2021) The role of maps in neighborhood-level heat vulnerability assessment for the City of Toronto [Other]. doi:https://doi.org/10.32920/14636997.v1

Sansom GT, Kirsch KR, Stone KW, McDonald TJ, Horney JA (2018) Domestic exposure to polycyclic aromatic hydrocarbons in a Houston, Texas, environmental justice neighborhood. Environ Justice 11(5):183–191. https://doi.org/10.1089/env.2018.0004

Salvia AL, Filho WL, Brandli LL, Griebeler JS (2019) Assessing research trends related to sustainable development goals: local and global issues. J Clean Prod 208(January):841–849. https://doi.org/10.1016/j.jclepro.2018.09.242

Semenza JC, Wilson DJ, Parra J, Bontempo BD, Hart M, Sailor DJ, George LA (2008) Public perception and behavior change in relationship to hot weather and air pollution. Environ Res 107(3):401–411. https://doi.org/10.1016/j.envres.2008.03.005

Sexton K, Linder SH, Marko D, Bethel H, Lupo PJ (2007) Comparative assessment of air pollution–related health risks in Houston. Environ Health Perspect 115(10):1388–1393. https://doi.org/10.1289/ehp.10043

Sultana F (2022) Resplendent care-full climate revolutions. Polit Geogr 99:102785. https://doi.org/10.1016/j.polgeo.2022.102785

Sovacool BK, Heffron RJ, McCauley D, Goldthau A (2016) Energy decisions reframed as justice and ethical concerns. Nat Energy 1(5):16024. https://doi.org/10.1038/nenergy.2016.24

Stuart E, Woodroffe J (2016) Leaving no-one behind: can the sustainable development goals succeed where the millennium development goals lacked? Gend Dev 24(1):69–81. https://doi.org/10.1080/13552074.2016.1142206

Svarstad H, Benjaminsen TA (2020) Reading radical environmental justice through a political ecology lens. Geoforum 108:1–11. https://doi.org/10.1016/j.geoforum.2019.11.007

Swart R, Timmermans W, Boon E, Van Ginkel M, Goosen H, Van Veldhoven F, Cilliers J, Ndaguba E (2023) Can managing climate risks be a catalyst for broader transformative change? Soc Sci 12(3):158. https://doi.org/10.3390/socsci12030158

Symanski E, An Han H, Hopkins L, Smith MA, McCurdy S, Han I, Jimenez M, Markham C, Richner D, James D, Flores J (2020) Metal air pollution partnership solutions: building an academic-government-community-industry collaboration to improve air quality and health in environmental justice communities in Houston. Environ Health 19(1):39. https://doi.org/10.1186/s12940-020-00590-1

Tessum CW, Apte JS, Goodkind AL, Muller NZ, Mullins KA, Paolella DA, Polasky S et al (2019) Inequity in consumption of goods and services adds to racial–ethnic disparities in air pollution exposure. Proc Natl Acad Sci 116(13):6001–6006. https://doi.org/10.1073/pnas.1818859116

Texas Commission on Environmental Quality (TCEQ) (2008) Texas State implementation plan. https://www.tceq.texas.gov/airquality/sip/hgb/hgb-status

Tonmoy FN, El-Zein A, Hinkel J (2014) Assessment of vulnerability to climate change using indicators: a meta-analysis of the literature. WIREs Clim Change 5(6):775–792. https://doi.org/10.1002/wcc.314

Tibbetts JH (2015) Air quality and climate change: a delicate balance. Environ Health Perspect 123(6):A148. https://doi.org/10.1289/ehp.123-A148

Turek-Hankins LL, Hino M, Mach KJ (2020) Risk screening methods for extreme heat: implications for equity-oriented adaptation. PLoS One 15(11):e0240841. https://doi.org/10.1371/journal.pone.0240841

Vardoulakis S, Salmond J, Krafft T, Morawska L (2020) Urban environmental health interventions towards the sustainable development goals. Sci Total Environ 748(December):141530. https://doi.org/10.1016/j.scitotenv.2020.141530

Ulpiani G (2021) On the linkage between urban heat island and urban pollution island: three-decade literature review towards a conceptual framework. Sci Total Environ 751:141727. https://doi.org/10.1016/j.scitotenv.2020.141727

Understanding Houston (2021) Understanding Houston, air and water quality, 2021. https://www.understandinghouston.org/topic/environment/air-water-quality

Union of Concerned Scientists, & Texas Environmental Justice Advocacy Services (2016) Double Jeopardy in Houston. https://www.ucsusa.org/sites/default/files/attach/2016/10/ucs-double-jeopardy-in-houston-full-report-2016.pdf

Wang C, Wang Z-H, Kaloush KE, Shacat J (2021) Perceptions of urban heat island mitigation and implementation strategies: survey and gap analysis. Sustain Cities Soc 66:102687. https://doi.org/10.1016/j.scs.2020.102687

Westman L, Patterson J, Macrorie R, Orr CJ, Ashcraft CM, Castán Broto V, Dolan D, Gupta M, van der Heijden J, Hickmann T, Hobbins R, Papin M, Robin E, Rosan C, Torrens J, Webb R (2022) Compound urban crises. Ambio 51(6):1402–1415. https://doi.org/10.1007/s13280-021-01697-6

Yazar M, Haarstad H, Drengenes LL, York A (2022) Governance learning from collective actions for just climate adaptation in cities. Front Sustain Cities 4:932070. https://doi.org/10.3389/frsc.2022.932070

Yu J, Castellani K, Forysinski K, Gustafson P, Lu J, Peterson E, Tran M, Yao A, Zhao J, Brauer M (2021) Geospatial indicators of exposure, sensitivity, and adaptive capacity to assess neighbourhood variation in vulnerability to climate change-related health hazards. Environ Health 20(1):31. https://doi.org/10.1186/s12940-021-00708-z

Part III

Climate Crisis and Smart Agriculture and Food Security

11 The Role of Indigenous Climate Forecasting Systems in Building Farmers' Resilience in Nkayi District, Zimbabwe

Joram Ndlovu, Mduduzi Ndlovu, and Douglas Nyathi

Abstract

Providing context-specific and user-friendly weather forecast information is essential in determining the resilience and variability of farmers and pastoralists to climate change. Climate change and variability have negatively affected farmers' livelihoods who depend on rain-fed agriculture in the Nkayi District. This chapter seeks to assess the capacity of farmers who rely on access to climate forecast information to adapt to climate change. The chapter examines the role of indigenous climate forecasting systems in developing adaptation strategies to mitigate the effects of climate change in the agriculture sector. A qualitative study was conducted with 150 farmers and two focus group discussions in Ward 23 of Nkayi District. The findings show limited reliable modern weather information services specific to the local area. Therefore, farmers use indigenous climate forecasting information to develop adaptation strategies through managing crop and livestock production systems. They have adopted indigenous weather forecasting methods to make significant strategic and tactical agricultural practices and decisions, such as using different species of plants, domestic and wild animals, and atmospheric indicators to predict crop production, climate conditions, and social safety nets. Other indigenous sources include knowledge from agricultural services, weather interpretation by elders, weather reports from radios, and other weather service providers. However, climate forecasting information is dependent on how farmers perceive the climate forecasting system. The chapter concludes that climate variability has limited the scope and reliability of indigenous weather knowledge for forecasting and decision-making. The chapter recommends the incorporation of indigenous climate forecasting systems in the production of scientific climate forecasting information.

Keywords

Climate change · Climate forecasting · Climate resilience · Climate risk · Adaptation strategies

11.1 Introduction

Climate change and variability significantly impact social, economic, and environmental systems. According to IPCC (2019), climate change is affecting precipitation patterns. The impact of

J. Ndlovu (✉) · D. Nyathi
School of Social Sciences, Howard College,
University of KwaZulu-Natal, Durban, South Africa
e-mail: Ndlovuj1@ukzn.ac.za

M. Ndlovu
Faculty of Humanities and Social Sciences,
Lupane State University, Bulawayo, Zimbabwe

U. Chatterjee et al. (eds.), *Climate Crisis: Adaptive Approaches and Sustainability*, Sustainable Development Goals Series, https://doi.org/10.1007/978-3-031-44397-8_11

the changes in precipitation has been observed in Africa. Much of the African continent is in the tropics, which makes it more vulnerable to the effects of climate change (Brazier 2017). Sub-Saharan countries are now prone to droughts, resulting in decreased food production and severe hunger (Sarr 2012). According to Boko et al. (2007), the Sub-Saharan region is severely and unequally affected by climate change and is vulnerable to climate variability because of limited adaptive capacity. About 70% of smallholder farmers in Zimbabwe rely on rain-fed agriculture as a source of food and livelihood (FAO 2009). However, climate change and variability affect their agricultural activities and production. Smallholder farmers largely depend on climate-related information such as total seasonal rainfall, the onset of rains, the length of the rainy season, and the frequency of dry spells (Winsemius and Werner 2013). This information enables farmers to make informed decisions regarding agricultural activities and reduce their vulnerability to climate-related stresses. Makaudze (2016) purports that economic value is contributed by having access to seasonal climate forecasting. Farmers with access to seasonal climate information record 28% higher yield gains during drought years compared to farmers who do not have access to seasonal climate information (Makaudze 2016).

The Meteorology Services Department of Zimbabwe (MSD) is the national designated authority on meteorology, climate, and seismology in Zimbabwe. The authority provides scientific climate forecasts and science-based warnings to protect life and property. MSD disseminates climate forecasting information through various platforms such as television, radio, the Department of Agricultural, Technical and Extension Services (AGRITEX), and social media. Chisadza et al. (2013) state that farmers in rural communities can not readily access scientific rainfall forecasts disseminated through different channels. As a result, farmers' ability to make farm decisions due to climate change is affected. The demarcation of the country into three spatially extensive homogenous rainfall zones means the information is not site-specific, while rainfall distribution is uneven. This hurts the accuracy and usability of climate forecasting information from MSD.

Smallholder farmers use local indicators to predict climate to make decisions and mitigate the effects of climate change and variability. The effects of climate change and variability can be mitigated by developing informed adaptation strategies and building smallholder farmers' resilience. Farmers use indigenous climate forecasting methods to develop adaptation strategies at the local level (Jiri et al. 2015). Indigenous knowledge can assist in building farmers' resilience which provides an enabling capacity for them to adapt to environmental changes and improve their livelihoods. If accessed by the farmers' scientific climate forecast information can be merged with indigenous knowledge to make appropriate responses to climate change (Andersson et al. 2020). There is potential for the two climate forecast systems to be integrated to produce accessible and usable climate information forecasts. The aim is to identify indigenous climate forecasting indicators and determine how they contribute to building farmers' resilience to climate-related stresses. The objectives were to identify indigenous knowledge systems used to forecast seasonal climate and weather in the Nkayi district; the study sought to determine farmers' perceptions of scientific climate forecasting systems and how they integrate them with indigenous climate forecasting; and lastly, to identify adaptation strategies developed by the farmers in response to indigenous climate forecasting systems to mitigate the effects of climate change.

11.2 Literature Review

11.2.1 Climate Change and Impacts on Africa

The IPCC's fourth assessment (2007) reports that it is unequivocal that anthropogenic climate change is happening. Due to climate change, extreme temperatures and rainfall variability have been noted over the years globally. Hope (2009) states that Africa is warmer than a century ago. Pachauri and Reisinger (2007) report that extreme weather events, notably floods, drought, and tropical storms, are also expected to increase in frequency and intensity across the globe. This is already putting a strain on the African agricultural sector, which depends on climatic conditions. Due to this process, Africa experiences increased water stress, decreased yields from rain-fed agriculture, increased food insecurity and malnutrition, sea level rise, and increased arid and semi-arid land (IPCC 2012). Sub-Saharan countries are now prone to droughts resulting in a decline in food production and severe hunger (Sarr 2012). Rainfall variability has been high, and drought years have become more frequent in Africa due to climate change. In Tanzania, rainfall distribution coupled with drought periods and dry spells has increased the problem of moisture stress. It is estimated that 20–30% of people in semi-arid areas are at risk (Mongi et al. 2010).

11.2.2 Climate Change Adaptation

Eriksen and Brown (2011) purport that human-induced climate change will continue to occur for an unknown period as countries develop and grow in population. Global climate mitigation efforts will take time to reduce greenhouse emissions. Therefore, climate change adaptation is required to minimise the impact of climate change on people's livelihoods (Dube 2015). According to Thornes (2001), adaptation is 'an adjustment in natural or human systems in response to actual or expected stimuli or their effects. The adjustments enable natural or human systems to survive in a changing environment'. Climate change adaptation involves acting to reduce the impact of climate-related stresses after accessing information. Adaptation is seen as a crucial component of actions to reduce the effects of climate change as it enables people to adjust and cope (Dube 2015).

11.2.3 Climate Change Adaptation Strategies in the Agriculture Sector

Farming communities in response to expected or actual changes in climate to reduce adverse impacts of climate change and take advantage of the opportunities presented by climate change (Parry et al. 2005). Climate change can present new opportunities as farmers adjust to new conditions. Due to extreme climatic events farmers have experienced over the years, farmers have adopted adaptation strategies to cope with climatic events that have the potential to affect their agricultural productivity and production (Mutekwa 2009). According to Mutekwa (2009), adaptation strategies undertaken by farmers in Zimbabwe involve re-examining land use, management practices, and farm infrastructure. Planting short-season varieties, staggering planting dates, and crop diversification are the most popular strategies farmers adopt in semi-arid areas (Mutekwa 2009). Farmers plant short-season varieties to reduce the risk of crop failure in case the rainy season becomes short. They vary planting dates to reduce the risk of germination failure due to dry spells.

Dube et al. (2018) point out adaptation strategies such as farming drought and heat-tolerant crops, water harvesting techniques, irrigation, conservation farming, stream bank cultivation, and transhumance of livestock. Crop diversification is a crop management practice that improves household food by planting different crops affected by climatic conditions. Farmers diversify by growing drought-tolerant crop varieties such as pearl millet, sorghum, and rapoko (Mutekwa 2009). Maize is a staple crop in Zimbabwe, but yields are low in semi-arid regions. However, the farmers adapt to maize by cropping short-season hybrid varieties instead of traditional varieties that take longer to mature. Mutekwa (2009) found that almost all farmers employed at least one of these adaptation strategies in the Zvishavane district.

Farmers use technological innovations and adapt to climate change through crop irrigation (Karki and Gurung 2012). Irrigation presents opportunities for farmers who have access to the technology. Irrigation farmers are found to have higher incomes during drought years (Pittman et al. 2011). However, there are technology and economic limits to irrigation technology as an adaptation strategy for smallholder farmers. Dube et al. (2018) purports that the available technology needs to be improved to sustain irrigation as an adaption strategy, and a few farmers have the resources to access the technology. Adaptation actions should adopt sustainable pathways because what might be successful adaptation strategies undermine other objectives associated with sustainable development. Sustainable adaptation should contribute to social, economic, and environmental development (Eriksen and Brown 2011). People and the environment might be negatively affected by adaptation strategies.

11.2.4 Climate Forecasting Information Systems

These early warning data can be based on science or indigenous knowledge (Soropa et al. 2015). They are used to predict season quality for better decision-making. The purpose of producing climate forecasting information is to equip and prepare stakeholders for imminent climate-related hazards that can be prepared for in anticipation (Dube 2015). Lives and property can be saved by issuing the information. Farmers can use the information to prepare based on the expected outlook of the season regarding anticipated rainfall and temperature patterns. Unganai et al. (2013) state that farmers who receive quality and timely seasonal forecasting information use the information to lessen the agricultural risks induced by climate change and variability.

11.2.4.1 Scientific Climate Forecasts (SCF)

SCF refers to climate-related information produced and disseminated by national meteorology departments. It is information produced by using scientific instruments to measure atmospheric processes and conditions to calculate the probability of the future state of the climate of a given area (Dube 2015). Scientific climate information is generated using Geographical Information Systems (GIS) and remote sensing in meteorology and climatology (Oni et al. 2015). In Zimbabwe, scientific climate forecasts are produced by the Meteorological Services Department of Zimbabwe (MSD 2021). Scientific climate information is usually produced for a given large area. In Zimbabwe, the country is divided into only three homogenous rainfall zones according to climate drivers, and the climate forecasts are produced for these spatially extensive zones (MSD 2021; Mushore 2013). In Africa, two climate patterns over the Pacific Ocean predict the quality of the rainy season probabilistically: El Nino and La Nina (ENSO). El Nino events over the Pacific Ocean are associated with drought conditions in Southern Africa, while La Nina events are associated with mild and wet seasons (Pomposi et al. 2018). Analogue methods and linear regression models are used in scientific forecasting systems to get the probability of weather events' occurrence (Murphy et al. 2001). The homogenous regions change in January, as shown in Fig. 11.1.

Annually, climate scientists from the SADC National Meteorological and Hydrological Services and the SADC Climate Services Centre convene a Southern Africa Regional Climate Outlook Forum (SARCOF) and produce a seasonal outlook for the SADC region based on atmospheric processes and conditions scientifically measured. The outlook is presented in overlapping three monthly periods covering October to March: October–November–December; November–December–January; December–January–February; and January–February–March (SARCOF-24, 2020). Following the SARCOF, Zimbabwe, through the facilitation of MSD, convenes a National Climate Outlook Forum (NACOF) meeting with various stakeholders to present the outlook for the country, categorised into three homogenous rainfall regions (MSD 2021).

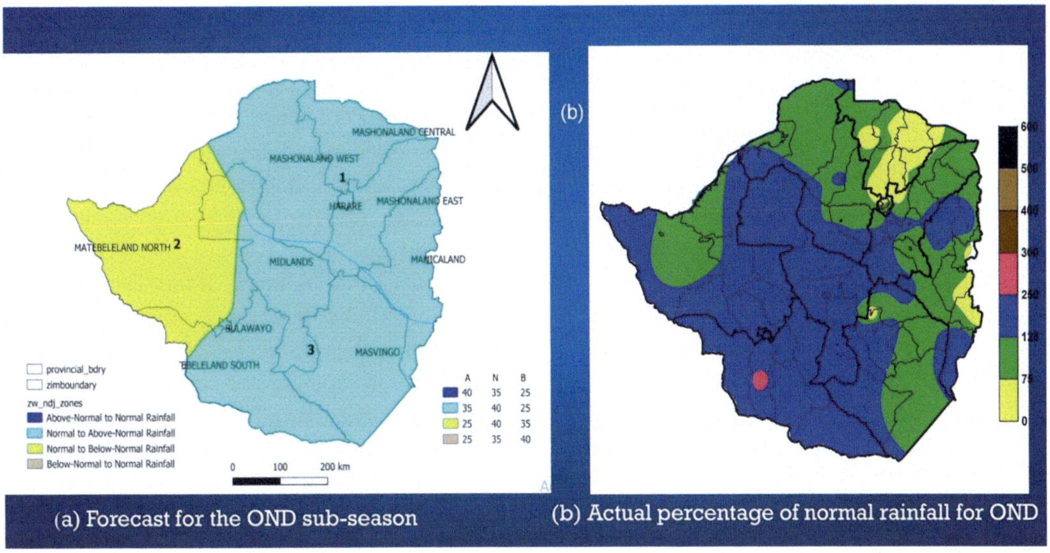

Fig. 11.1 Homogenous rainfall zone in Zimbabwe: (**a**) Oct–Nov–Dec and (**b**) Jan–Feb–Dec (MSD 2021). (Source: Meteorological Services Department, Seasonal Forecast: 2021/2022)

Accuracy and Farmers' Perception of Scientific Climate Forecasting

Mushore et al. (2017) concludes that ENSO-based climate forecasts are accurate and reliable in Zimbabwe, although accuracy was low (at 55%) in homogenous rainfall region 3. Access to information is not enough if it is not used in decision-making. Farmers should respond by changing their way of doing if the information received is of value to them. Jiri et al. (2016) note that 'inaccurate scientific climate forecasts remain a major challenge to effective use of seasonal forecasts by farmers and other users in Southern Africa'. Farmers tend to lose credibility in using the information due to inaccuracy.

11.2.5 Indigenous Knowledge Systems-Based Climate Forecasting

11.2.5.1 Indigenous Knowledge Systems

Indigenous knowledge systems (IKS) are a body or bodies of knowledge of a given geographical area in which indigenous peoples have stayed for a very long time. Ifejika Speranza et al. (2010) defines the term IKS as

> .... local ecological knowledge refers to knowledge that is location specific, acquired through long-term observation of (and interaction with) the environment, and transferred through oral traditions from generation to generation.

IKS is developed through experiences by people who live in that particular location. Since indigenous knowledge systems are based on practical experiences, they can be documented and preserved by communities living in that particular location and future generations (Mapira and Mazambara 2013).

11.2.5.2 Indigenous Climate Forecasting Knowledge

Indigenous knowledge is now recognised as a critical tool in developing adaptation strategies for various threats and problems faced in the world (Kelman and Glantz 2014). Dube (2015) notes that the concept of IKS is essential because it addresses some challenges that scientific climate forecasting methods cannot address. Ifejika Speranza et al. (2010) argue that IKS is essential to adaptation as it is at the local level where the people who have to adapt to the impacts of climate change live. Jiri et al. (2016) note that farmers use tree phenology, animal behaviour, wind cover, and other social rains and season indicators to predict rains and season quality. Farmers are always eager to know the onset of the rainy season and the quality of the season so that they make some preparations and make decisions on crop choices. Ethno-meteorology knowledge plays a vital role in smallholder farmers' ability to develop adaptation measures to reduce the effects of climate change. Kolawole et al. (2014) posit that farmers prefer using indigenous forecasts f compared to scientific ones.

Challenges of Using Indigenous Climate Forecasting Systems

One of the challenges in using IK in climate forecasting is that it risks being lost. There needs to be systematic documentation of indigenous climate forecasting knowledge. Some knowledge is lost as knowledge is passed from one generation to the next. Climate change and urbanisation are some factors that threaten the accuracy and usage of IK in predicting climate and weather. Climatic conditions disrupt traditional indicators, affecting their reliability in forecasting climate change. IK is also at risk of being eroded by modernisation (Radeny et al. 2019). The other challenge with using IKS in predicting weather is that it takes work to quantify indigenous indicators or norms. Some indicators are considered too many, normal or little without quantifiable values. IK is for a given location, limiting its application to large areas (Makwara 2019).

11.2.6 Integration of Indigenous Knowledge and Scientific Climate Forecasting

There is potential for integrating indigenous climate forecasting systems with scientific climate forecasting systems to increase the farmers' ability to prepare and make appropriate decisions to reduce the impact of climate change (Kolawole et al. 2014). Jiri et al. (2016) argue that indigenous knowledge is helpful for scientific knowledge, especially in tracking change. Jiri et al. (2016) further note that both forms of knowledge are produced through observation, experimentation, and validation. Therefore, there is a need for suitable platforms where farmers and scientists can work together and develop adaptation strategies against climate change and variability. These assist climate scientists in knowing farmers' priorities and designing forecast information that will be useful to farmers. Rural farmers only accept scientific climate forecasting information for reasons such as farmers' lack of sense of ownership. Therefore, there is a need to have participatory outreach programs and incorporate indigenous knowledge into scientific climate forecasting. This helps change the view that climate scientists are the sources of knowledge and the farmers are the recipients of knowledge (Orlove et al. 2010). Both scientists should be involved in the production of climate change information.

11.2.7 Description of the Study Area

The study was done in the Nkayi district in Matabeleland North province in Zimbabwe. The study site was selected based on climatic conditions, weather, food security, and accessibility. The district has a high number of food-insecure households. According to ZimVAC (2021), 51% of the households were food insecure. Nkayi district is in natural agroecological region IV, and climate change negatively affects agricultural production. Therefore, farmers must adapt to adverse climate change effects, especially dry spells and drought. Farmers in the area have always had crop and livestock farming as their primary land use. The agro-ecological conditions in Nkayi are characterised by unreliable rainfall between 450 and 700 mm per year and periodic dry spells experienced during the rainy season (MSD unpublished records).

Nkayi district is divided into two zones mainly characterised by different soil and vegetation types. Shangani River demarcates the district into these two zones. The southern zone and areas along the Shangani River have red soils. The vegetation in the southern part of the Nkayi district is dominated by mopane trees and acacia tree species, especially in areas where soils are sandy-loamy. Along big rivers such as Shangani and Gweru Rivers, the dominating tree species include Diamond-leaved nuclear (Umtshekesane in Ndebele), acacia species (ameva), Chinese lantern (ugagu), and Bitter albizia (Umbola). The northern part of Nkayi is a sand veld zone characterized by diverse woodlands. Jubernadia globiflora (umkusu) and Branchstegia speciformis (igonde) dominate tree species. Trees long liveive trees (isigangatsha), wild velvet medlar (umviyo), and false wild medlar (umthofi) are also found across the district (Senda et al. 2020).

11.2.8 Data Collection Methods

Data were collected from ward 23 communal farmers in the Nkayi district through questionnaires and focus group discussions. Key informant interviews were done with community leaders and government officials. The research study used the primary data collected from the smallholder farmers and key stakeholders in the study sample. Rainfall records were collected from the Meteorological Services Department in the Nkayi district. Survey questionnaires were distributed to 150 households in the selected study area and administered face-to-face to the respondents. The questionnaires provided a quantitative description of trends and perspectives of the sampled population. Key informants were chosen based on their length of stay in the district, expertise, and role in the community. A total of seven key informant interviews were done; three of the interviews were done with community village heads, and the other interviews were done with AGRITEX, the Forestry Commission officer, the EMA officer, and the District Development Coordinator's assistant. The village heads interviewed had Nkayi district as their origin, and government officials interviewed had stayed there for over 15 years. Seasonal forecasting information for the 2021/22 rainy season was collected from the Meteorological Services Department in the Nkayi district. Daily rainfall records for the 2021/22 season were collected for October–November-December.

11.3 Research Findings

11.3.1 Tree Phenology Indicators

Table 11.1 shows tree phenology indicators used by farmers in Nkayi and their climate prediction. The study found that farmers in the Nkayi district have a vast knowledge of indigenous climate forecasting systems. For some trees, the bearing of fruits or flowers in abundance indicates a good season, whilst, for some bearing, more fruits predict a drought. Trees such as wild velvet medlar, false wild medlar, long live, mobola plum, and snot apple indicate an excellent rainy season if they have too many fruits. The monkey orange trees, donkey berries, baobab, and batoka plum are some of the trees mentioned to indicate drought conditions if they have more flowers and fruits. Like other studies done by Kolawole et al. (2014), Mavhura et al. (2013), and Jiri et al. (2015), this study found that farmers use tree phenology indicators to predict the climate. However, there are different tree species and animals in the Nkayi district due to the difference in the environment and vegetation. These previous studies should have mentioned tree species such as the Long live trees, Mobola plum, and false wild medlar as plant indicators for predicting the climate.

11.3.2 Animal Behaviour

Animals and insects behave in specific ways before the onset of rain and during the rainy season. Farmers pointed out several certain behaviours that animals display to predict the climate or weather, as shown in Table 11.2. Nkayi district has extensive forests where wild animals are also found. The common duiker (impunzi) is used to predict the onset of the rains. When the duikers start giving birth, it shows that the onset of the rains is close. If terrestrial frogs croak during the day, it is known that rains are expected in the next 2 or 3 days. The behaviour of birds is also being used to predict climate and weather. The Southern ground hornbill is one of the common indicators mentioned by farmers. If the bird sings early in

Table 11.1 Plant indicators and climate prediction

Plant indicators	Description	Prediction
Mopane (iphane)	Foliage	The start of leaf development of mopane trees means that the rainy season is about to start
Long live tree (*Isigangatsha*)	Fruits	The abundance of fruits indicates that there will be more rain during the season. Fewer fruits indicate that drought is expected
Velvet wild medlar (umviyo)	Flowers and fruits	More flowers and fruits indicate that the rainy season will be good
False wild medlar (umthofu)	Fruits	The abundance of fruits shows that there will be more rain during the season. If the trees bear fewer fruits, then a drought should be expected
Branchstegia speciformis (igonde)	Foliage	The rainy season is about to start when the tree grows new leaves
Baobab (Umkhomo)	Fruits	More fruits indicate that less rainfall is expected
Sweet monkey orange (Umwawa/ umgwadi)	Fruits	The abundance of fruits indicates drought conditions
Batoka plum (Umthunduluka)	Fruits	The abundance of fruits indicates a good rainy season
Snot apple (Uxakuxaku)	Fruits	The abundance of flowers predicts a good season ahead
Mobola plum (umkhuna)	Fruits	Less rains are expected if there are fewer fruits
Donkey berry (Ubhunzu olukhulu)	Fruits	Less rains are expected if there are more fruits

the morning and late afternoon, rain is expected in the next 2 or 3 days. In the Nkayi district, farmers keep domesticated guinea fowls, and when they start laying eggs, the rainy season is about to start. They only lay before and during the rainy season. Another important and most common indicator is a Jacobin cuckoo (inkaku). It is known that if it starts singing, then the rains are expected to fall within 2–4 days. How the bird sings was noted as an essential indicator. The longer the singing, the more rain will fall.

Table 11.2 Animal behaviour and climate or weather prediction

Animal	Indicator	Prediction
The common duiker (Impunzi)	When they start giving birth	The onset of the rainy season is close
Guinea fowls (amathendele)	When they start laying eggs	The rainy season is about to start
Southern ground hornbill (Insingizi)	When the birds sing in the morning and late afternoon	Rains are expected in the next 2–3 days It also means that temperatures are high
Barn swallow (Inkonjane)	When they are out playing around	Incessant rains are expected in the next few days
Bank swallow (Inkonjane zomfula)	When they are seen at homesteads	The good season is expected if they are seen during spring
Termites (amatheza)	When they are busy moving around and carrying grass	Heavy rains are expected within a week
Terrestrial frogs (amoxoxo ahlala egangeni)	Croaking	Rains are expected in a few days
Jacobin cuckoo (inkanku)	Bird singing	Rains are expected in the next 2–4 days
Butterflies (amavevane)	If there are too many white butterflies flying around	Rainfall is expected on the day
Dragon fly (umavikinduku)	Flying around	Rainfall is expected on the day
Cicadas (inyeza)	Appearance and chirping	High temperatures and the onset of the rainy season

This study confirms several animal behaviours used to predict climate, as mentioned in Jiri et al.'s (2016) review paper. These include the appearance of sparrows, the presence of termites, the croaking of frogs, and the singing of the Jacobin cuckoo and Southern ground hornbill. However, this research found that bank swallows, guinea fowls, duikers, and dragonflies are used to forecast climate and weather.

11.3.3 Atmospheric Indicators

Temperature and wind direction are some of the parameters used by the farmers to predict the time and intensity of the rains. Although there are no scientific instruments to measure the temperature, farmers can tell the intensity of the parameter by the amount of perspiration, heat on the soil, and sunbeam irritation on the skin. Trees are used to determine wind direction. The frequency of whirlwinds between August and October indicates the quality of the coming season. The more frequently the whirlwinds occur, the more rainfall is expected. The farmers believe that the whirlwinds 'steer' the rains. High temperatures and strong winds in November indicate the fall of heavy rains and violent thunderstorms. The farmers stated that they could tell the intensity of rains from the formation and types of clouds they saw. They usually associate dark clouds (nimbostratus) with heavy rains, and stratocumulus clouds (*amaxhegu in Ndebele*) are known to have no or very little rain. Table 11.3 shows the atmospheric farmers use to predict the climate and weather.

11.3.4 Scientific Climate Forecast and Observed IKS Indicators for the 2021/22 Season

Data collected from MSD show that the 2021/22 seasonal outlook for homogenous rainfall region 2 predicted high chances of average to above normal rainfall from October to December 2021 (Fig. 11.2). The region covers parts of Matabeleland North, parts of Bulawayo, parts of Midlands, and parts of Mashonaland West. Therefore, Nkayi district is in a homogenous rainfall region 2. The figure graphically displays that the regions are spatially and extensive, considering that rainfall is unevenly distributed.

Figure 11.3 shows the rainfall records from AGRITEX for Ward 23 and MSD weather station at Nkayi centre. The figure also shows the 30-year long-term average for the Nkayi district for October, November, and December. The rainfall records from October to December 2021 for Ward 23 show that the rainfall received in November and December 2021 is below the long-

Table 11.3 Atmospheric indicators and weather prediction

Atmospheric indicator	Description	Weather prediction
Wind	Suppose the wind blows in a particular direction and then changes in the opposite direction. For instance, the wind blows from South to North and then changes from North to South	Indicates that rain will be received on the day
Moon or Sun Halo (umkhumbi)	Halo moon or sun. A circle around the moon or sun when there are cirrus clouds	Rains are expected within the next 2 weeks
Formation and types of clouds	Nimbostratus clouds start forming in the morning Formation of stratocumulus clouds (amaxhegu)	Good rains likely to be received on the day No rains are expected on the day
Whirlwinds	Frequency of whirlwinds during spring	A good rainy season expected
Dew	Dew in the morning	It indicates that no rain is to be received on the day
Mist	Mist around September or October	A good rainy season expected
Temperatures	High temperatures	Indicates the likelihood of violent thunderstorms

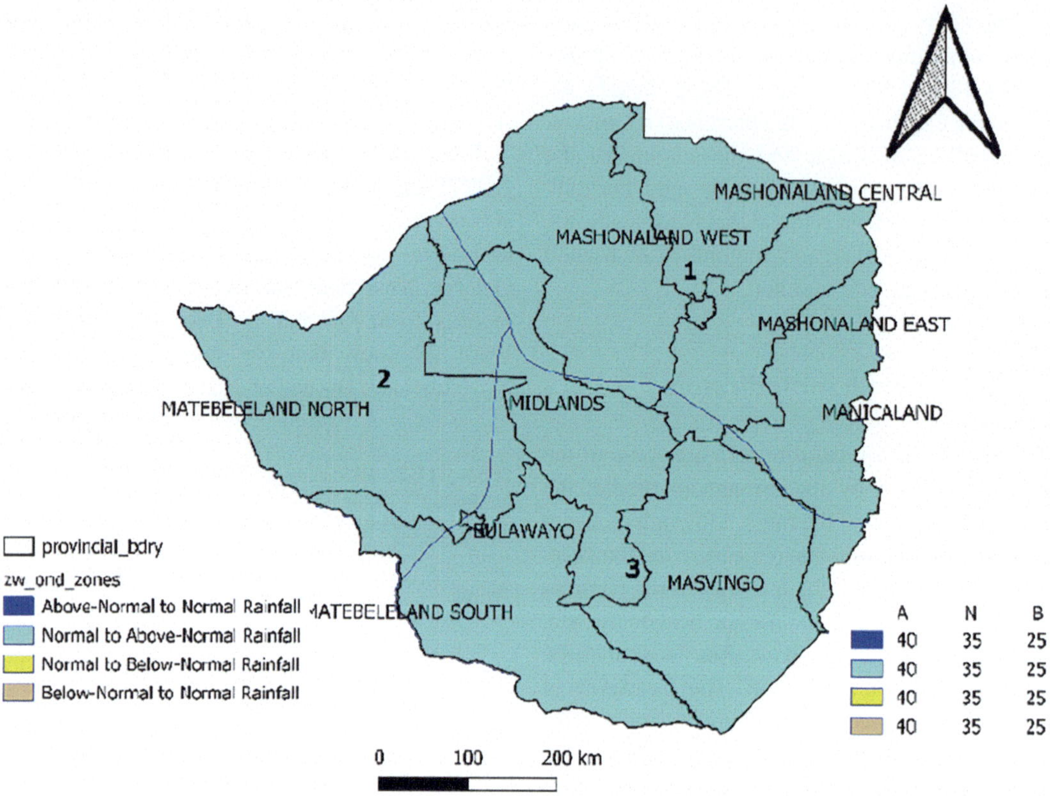

Fig. 11.2 The 2021/22 seasonal outlook for October to December 2021. (Source: Meteorological Services Department, Seasonal Forecast: 2021/2022)

Fig. 11.3 The 2021/22 rainfall records from October to December 2021 at Nkayi centre and in Ward 23

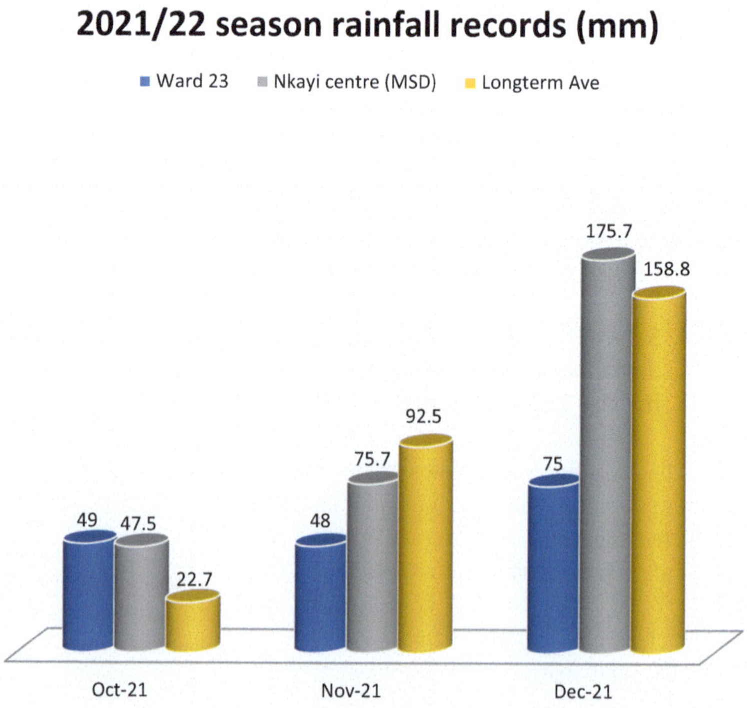

term averages. The data also show that rainfall is unevenly distributed as rainfall received in Nkayi centre far exceeds that in ward 23. From October to December, the 2021/22 season outlook predicted a normal to the above-regular season for homogenous rainfall zone 2. The prediction was incorrect for Ward 23.

Farmers observed mainly plant indicators to predict the quality of the season for 2021/22. Farmers in Ward 23 observed that long-lived trees (isigangatsha) and false wild medlar (umthofu) had many flowers, but most dropped, leading to few fruits developing. This was a prediction that the rainy season would not be good in the area. The rainfall received in Ward 23, as shown in Fig. 11.3, was relatively low, and the rainfall data shows that there were dry spells longer than 8 days from October to December 2021. Data from AGRITEX show more days without rain in ward 23 from October to December 2021 (Fig. 11.4).

Contrary to Mushore et al. (2017), this study found that scientific climate forecasting is inaccurate. The research corresponds with Kolawole et al. (2014) in that farmers prefer using indigenous knowledge systems to scientific climate forecasting information. The study also concurs with Jiri et al. (2016) in that climate forecast information produced by MSD for spatially extensive homogenous rainfall zones affects its accuracy and usability by farmers.

11.3.5 Farmers' Perception of SCF and Integration with IK Climate Forecasting Methods

The primary sources of scientific climate forecast information are AGRITEX, radios, and information sharing among farmers. Access to information could be better due to poor radio signal coverage, and 42% of the farmers need radios. Mobile smartphone penetration is low because the mobile network is inferior in the ward. As a result, farmers rely mainly on information shared by AGRITEX and other farmers. Farmers stated that this information is unusable as they get part of the information produced by MSD. Farmers indicated that they are more interested in knowing scientific climate information such as the amount of rainfall, expected days of rainfall, and length of the season. Farmers find scientific seasonal climate information somewhat reliable. Seasonal outlook forecasts sometimes come true, but sometimes the prediction does not. Farmers stated that they find the information more reliable for short-period forecasts.

Farmers highlighted that they use their location-specific indigenous knowledge systems to predict the season quality and then rely on scientific climate forecasts for the length of the season and the amount of rain expected. This helps them make appropriate decisions to mitigate the effects of climate change. Forty-two percent of

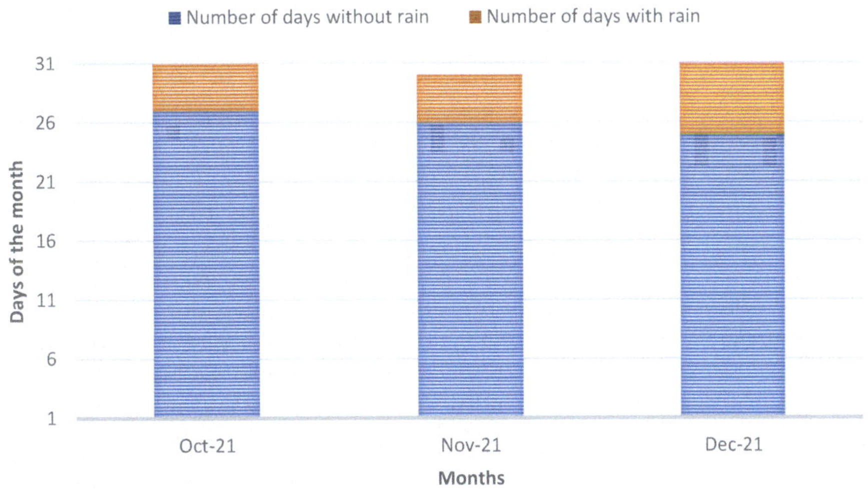

Fig. 11.4 The number of days with or without rain from October to December 2021

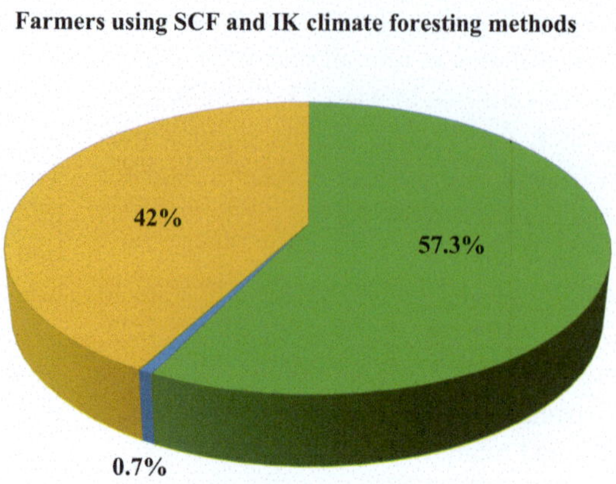

Fig. 11.5 Methods of forecasting climate and weather

the farmers rely on indigenous knowledge systems only to predict the climate, while 57.3% use indigenous knowledge systems and scientific forecasting systems. Only 0.7% of the respondents rely solely on scientific climate forecasting systems to make farming-related decisions. Those who use both information systems stated that they find the information more reliable in improving their ability to adapt to climate change effects.

The pie chart in Fig. 11.5 depicts that almost all the farmers use indigenous climate forecasting knowledge. The majority use indigenous knowledge and scientific climate information to support decision-making. Farmers find scientific climate forecast information more accurate and reliable when used solely. This contradicts a study by Andersson et al. (2020), which states that scientific climate is more reliable than IKS. However, the study corresponds with both Jiri et al. (2016) and Andersson et al. (2020) in that climate information is more reliable and usable if both IKS and scientific climate forecast information are integrated.

11.3.6 IKS and Climate Change Adaptation

The study found that indigenous climate forecasting systems play a significant role in supporting farmers' adaptation strategies. Indigenous climate indicators are location-specific, and farmers find them more reliable than scientific knowledge. Farmers take several measures based on indigenous climate indicators. However, farmers who rely on indigenous knowledge systems stated that the indicators are becoming less reliable due to climate change and variability. Farmers noted changes in temperature and rainfall patterns over the years. The temperatures they are experiencing now are higher than they were years ago, leading to a high rate of crop wilting. Farmers ranked the late onset of adequate rains as one of the significant changes in rainfall patterns. Late onset of rains is associated with short seasons in the area leading to crop failure as the season ends before the lengthy season crop varieties reach physiological maturity. Dry spells are also affecting farmers' crop production in Ward 23. The sandy-loamy soils that are dominant in the ward have high drainage capacity, and, hence, they lose moisture content relatively faster than clayey soils. Dry spells affect germination rates and crop performance in general.

Farmers in the Nkayi district adapt to climate change by managing crop and livestock production. Crop management includes changing crop varieties, diversification, and staggering planting dates. Farmers mentioned that they had changed the crop varieties they used to grow. The farmers used to grow more traditional maize varieties (*ibhabhadla* in Ndebele), which take a more extended period to mature. Farmers now grow more early-maturing varieties and small grain crops. A crop mix of early-maturing and late-maturing varieties was recorded as one of the adaptation strategies employed by the farmers. This is because there will be little or no information about the rainy season's length. The maize crop is grown by all the farmers in Nkayi. However, they stated that they diversify crop production by growing maize and drought-tolerant crops such as small grains and legumes, especially when climate forecasts indicate drought conditions. Climate forecast information is crucial in deciding which varieties should be grown and the crop mix. Farmers use indigenous knowledge systems that are easily accessible to them before the onset of the rainy season.

The farmers keep at least goats, sheep, or cattle as their livestock. Farmers plant fodder crops such as velvet beans and lab for feeding livestock during the dry season from July to November if drought conditions are expected. Plant pods such as money bread trees and acacia species are collected by the farmers when they are abundant in the forests for later use as stock feed when pastures have deteriorated. Farmers have easy access to these raw materials since they are locally available. The farmers highlighted that these low-cost adaptation strategies help reduce the number of livestock deaths during the dry season in the area. According to the farmers, dry mid-season periods have become longer due to climate change, and there is a need for the farmers to adapt. Table 11.4 shows the adaptation strategies developed by the farmers in response to climate change based on indigenous climate information accessed.

The study recognises the role of IKS in climate change adaptation. Farmers were found to use indigenous climate forecasting systems to develop adaptation strategies to mitigate the effects of climate change. This concurs with Kolawole et al. (2014), where it was found that 90% of farmers relied on IKS to generate climate and weather forecasts. Farmers adapt to these extreme climate-related stresses by managing livestock and crop production systems. Indigenous climate forecasts play an essential role in making appropriate information. These decisions include a selection of drought-tolerant crops, crop diversification, and formulation of low-cost feed for livestock, mitigating the effects of climate-related stress and reducing losses. This concurs with Jiri et al. (2015). However, this study includes adaptation strategies farmers employ, such as collecting and using raw materials found in the forests, such as the camel foot (*amahabahaba*) and acacia pods (*massage)* for stock feed during the lean season.

Table 11.4 Forecast-based adaptation strategies developed by farmers

	Period		
Level	Before the rainy season	During the rainy season	After the rainy season
Crop production	Land preparation and field fencing Crop variety selection Choosing a crop mix (diversification) Conservation agriculture – includes holing, manure application, and stocking of mulch	Planting dates staggered Re-planting with early maturing crops after germination failure due to dry spells Top dressing fertilizer application	Stocking crop residues for use as mulch in conservation the next coming season Engaging in non-agriculture activities for livelihood
Livestock production	Acquiring seed for fodder crops Use of low-cost and locally available feed to improve livestock body condition of draft animals	Give access to good pastures Production of fodder crops	Fodder processing Collection of locally available raw materials for low-cost feed production such as monkey bread, acacia pods Stocking of crop residue

11.4 Conclusion and Recommendations

There are noticeable changes in temperature and rainfall patterns due to climate change. High temperatures, dry spells, and the late onset of rains were significant climate-related stresses farmers face in Nkayi. Climate change adaptation has the potential to build farmers' resilience facing the brunt of climate change. The farmers' capacity to adapt is affected by several factors, and access to information is one of the factors. Climate-related information is crucial in supporting farmers to make appropriate decisions to mitigate the effects of climate change and subsequently reduce losses and maximise agricultural output. Farmers need climate forecast information such as the onset of rains, length of season, season quality, and amount of rainfall. MSD produces and disseminates climate forecast information. However, farmers in rural areas in the Nkayi district have limited access to scientific climate forecast information, and it needs to be delivered on time. In Zimbabwe, the homogenous rainfall zones are spatially extensive, affecting the accuracy and usability of climate forecasting information. Rainfall data for the 2021/22 season from October to December 2021 show a big difference between rainfall recorded at Nkayi centre and in Ward 23 of Nkayi. This shows that rainfall is unevenly distributed. Hence, there is a need for location-specific climate forecast information. Indigenous knowledge systems are significant in providing location-specific and timely climate forecasting information.

Farmers have indigenous knowledge methods they use to predict the climate. They use indicators of certain tree species, animal behaviour, and atmospheric phenomena to forecast future climate and weather conditions. They use indigenous knowledge climate indicators to predict rainy season quality, the onset of rains, and when to expect rains during the rainy season. However, indigenous knowledge systems need to predict the quantity of rain to be received, the degree of hotness, and the length of the season. Farmers use indigenous knowledge climate forecasting systems that are location specific and accessible to develop adaptation strategies. The information is helpful in crop variety selection and choosing a crop mix in that particular season. Indigenous climate forecasts help farmers decide which fodder crops to grow for livestock. Crop residues and locally available raw materials found in the bush are stocked based on indigenous climate indicators they observe. The low-cost feed helps reduce the number of livestock deaths during the dry season. Indigenous climate forecasting information systems enhance farmers' adaptive capacity. The ability of farmers to adapt is essential in building farmers' resilience to climate-related shocks and stresses. However, farmers indicated that they find climate information more usable and reliable if they use both scientific and indigenous knowledge systems.

The rainfall distribution was uneven, and the homogenous rainfall zones were spatially extensive. Therefore, MSD should improve location specificity when producing and disseminating climate forecasts. Licensing of community radios in Nkayi should be explored to increase access to information by the local people. Traditional leaders can be invited by the community radio stations and share indigenous climate forecasting information. There should be platforms where scientists and local people produce and share climate forecasts. Government and mobile network providers should improve network and Internet connectivity in the district. Therefore, MSD should develop an Android application for mobile phones that will incorporate indigenous climate forecasting systems in producing climate forecasts for easy and timely access to information. This will improve timely access to climate information by rural farmers. Indigenous knowledge systems used in Nkayi to predict climate should be documented for easy dissemination and to preserve local knowledge. There is a need for a bottom-up approach in formulating climate change policies. Local people have capacities, and policymakers should build upon them. Indigenous knowledge systems should be incorporated in the formulation of policies.

References

Andersson L, Wilk J, Graham LP, Wikner J, Mokwatlo S, Petja B (2020) Local early warning systems for drought – could they add value to nationally disseminated seasonal climate forecasts? Weather Clim Extremes 28:100241. https://doi.org/10.1016/j.wace.2019.100241

Boko M, Niang I, Nyong A, Vogel C, Githeko A, Medany M, Osman-Elasha B, Tabo R, Yanda P (2007) Africa. Climate change 2007: impacts, adaptation, and vulnerability. In: Parry ML, Canziani OF, Palutikof JP, van der Linden PJ, Hanson CE (eds) Contribution of Working Group II to the fourth assessment report of the Intergovernmental Panel on Climate Change. Cambridge University Press, Cambridge, pp 433–467

Brazier A (2017) Climate change in Zimbabwe: A guide for planners and decision-makers. Konrad- Adenaeur-Stiftung, Harare

Chisadza B, Tumbare MJ, Nhapi I, Nyabeze WR (2013) Useful traditional knowledge indicators for drought forecasting in the Mzingwane Catchment area of Zimbabwe. Disaster Prev Manag 22(4):312–325

Dube T (2015) Climate change information usage for adaptation purposes: the case of smallholder farmers in Matobo District, Zimbabwe (Unpublished Doctoral dissertation), University of Fort Hare

Dube, T., Mlilo, C., Moyo, P., Ncube, C. and Phiri, K. (2018). Will adaptation carry the future? Questioning the long-term capacity of smallholder farmers' adaptation strategies against climate change in Gwanda District, Zimbabwe

Eriksen S, Brown K (2011) Sustainable adaptation to climate change. Clim Dev 3(1):3–6

FAO (2009) Declaration of the World Summit on food security. Rome 7, Available online from: ftp://ftp.fao.org/docrep/faoMeeting/018/k605e.pdf

Hope KR Sr (2009) Climate change and poverty in Africa. Int J Sustain Dev World Ecol 16(6):451–461

Ifejika Speranza C, Kiteme B, Ambenje P, Wiesmann U, Makali S (2010) Indigenous knowledge related to climate variability and change: insights from droughts in semi-arid areas of former Makueni District, Kenya. Clim Change 100(2):295–315

IPCC (2007) Climate change 2007: impacts, adaptation, and vulnerability. Contribution of Working Group II to the fourth assessment report of the Intergovernmental Panel on Climate Change, Parry ML, Canziani OF, Palutikof JP, van der Linden PJ, Hanson CE (eds), Cambridge University Press, Cambridge, 1–976

IPCC (2012) Summary for policymakers. In: Field CB, Barros V, Stocker TF, Qin D, Dokken DJ, Ebi KL, Mastrandrea MD, Mach KJ, Plattner G-K, Allen SK, Tignor M, Midgley PM (eds) Managing the risks of extreme events and disasters to advance climate change adaptation, A special report of Working Groups I and II of the Intergovernmental Panel on Climate Change. Cambridge University Press, Cambridge/New York, pp 3–21

IPCC (2019) The IPCC's special report on climate change and land: What's in it for Africa? Available online from: https://www.africaportal.org/publications/ipccs-special-report-climate-change-and-land-whats-it-africa/

Jiri O, Mafongoya PL, Chivenge P (2015) Indigenous knowledge systems, seasonal 'quality and climate change adaptation in Zimbabwe. Clim Res 66(2):103–111

Jiri O, Mafongoya P, Mubaya C, Mafongoya O (2016) Seasonal climate prediction and adaptation using indigenous knowledge systems in agriculture systems in Southern Africa: a review. Can Center Sci Educ 8(5):156–172

Karki R, Gurung A (2012) An overview of climate change and its impact on agriculture: a review from the least developing country, Nepal. Int J Ecosyst 2(2):19–24

Kelman I, Glantz MH (2014) Early warning systems defined. In: Reducing disaster: early warning systems for climate change. Springer, Dordrecht, pp 89–108

Kolawole OD, Wolski P, Ngwenya B, Mmopelwa G (2014) Ethno-meteorology and scientific weather forecasting: small farmers and scientists' perspectives on climate variability in the Okavango Delta, Botswana. Clim Risk Manag 4:43–58. https://doi.org/10.1016/j.crm.2014.08.002

Makaudze EM (2016) Assessing the economic value of El Niño-based seasonal climate forecasts for smallholder farmers in Zimbabwe. In: Climate change and multi-dimensional sustainability in African agriculture. Springer, Cham, pp 591–612

Makwara EC (2019) Unpacking the functionality of indigenous knowledge systems (IKSs) in weather forecasting in Turwi Basin, Zimbabwe. J Agric Sustain 13(2020):1–26

Mapira J, Mazambara P (2013) Indigenous knowledge systems and their implications for sustainable development in Zimbabwe. J Sustain Dev Africa 15(5):90–106

Mavhura E, Manyena SB, Collins AE, Manatsa D (2013) Indigenous knowledge, coping strategies and resilience to floods in Muzarabani, Zimbabwe. Int J Disaster Risk Reduct 5:38–48. https://doi.org/10.1016/j.ijdrr.2013.07.001

Meteorological Services Department (2021) 2021/22 Seasonal rainfall forecast for Zimbabwe. The Zimbabwe National Climate outlook forum, 02 September 2021. Ministry of Environment, Climate, Tourism, and Hospitality Industry, Harare

Mongi H, Majule AE, Lyimo JG (2010) Vulnerability and adaptation of rain-fed agriculture to climate change and variability in semi-arid Tanzania. Afr J Environ Sci Technol 4(6):371–381

Murphy SJ, Washington R, Downing TE, Martin RV, Ziervogel G, Preston A, Todd M, Butterfield R, Briden J (2001) Seasonal forecasting for climate hazards: prospects and responses. Afr J Environ Sci Technol 23(2):171–196

Mushore TD (2013) Uptake of seasonal rainfall forecasts in Zimbabwe. IOSR J Environ Sci Toxicol Food Technol 5(1):31–37

Mushore TD, Mutanga O, Odindi J, Dube T (2017) Linking major shifts in land surface temperatures to long term land use and land cover changes: a case of Harare. Zimbabwe Urban Climate 20:120–134. https://doi.org/10.1016/j.uclim.2017.04.005

Mutekwa VT (2009) Climate change impacts and adaptation in the agricultural sector: the case of smallholder farmers in Zimbabwe. J Sustain Dev Africa 11(2):237–256

Oni F, Lawal BA, Adejimi O (2015) Evaluation of different meteorological data sources for agricultural uses in Ogbomoso. VEF J Agric Rural Commun Dev 2(1):27–35

Orlove B, Roncoli C, Kabugo M, Majugu A (2010) Indigenous climate knowledge in southern Uganda: the multiple components of a dynamic regional system. Clim Chang 100(2):243–265

Pachauri RK, Reisinger A (2007) IPCC fourth assessment report. IPCC, Geneva

Parry M, Rosenzweig C, Livermore M (2005) Climate change, global food supply and risk of hunger. Philos Trans R Soc B 360(1463):2125–2138

Pittman J, Wittrock V, Kulshreshtha S, Wheaton E (2011) Vulnerability to climate change in rural Saskatchewan: a case study of the Rural Municipality of Rudy No. 284. J Rural Stud 27(1):83–94

Pomposi C, Funk C, Shukla S, Harrison L, Magadzire T (2018) Distinguishing southern Africa precipitation response by strength of El Niño events and implications for decision-making. Environ Res Lett 13(7):074015. https://doi.org/10.1088/1748-9326/aacc4c

Radeny M, Desalegn A, Mubiru D, Kyazze F, Mahoo H, Recha J, Kimeli P, Solomon D (2019) Indigenous knowledge of seasonal weather and climate forecasting across East Africa. Clim Chang 156(4):509–526

Sarr B (2012) Present and future climate change in the semi-arid region of West Africa: a crucial input for practical adaptation in agriculture. Atmos Sci Lett 13(2):108–112

Senda TS, Kiker GA, Masikati P, Chirima A, van Niekerk J (2020) Modeling climate change impacts on rangeland productivity and livestock population dynamics in Nkayi District, Zimbabwe. Appl Sci 10(7):2330. https://doi.org/10.3390/app10072330

Soropa G, Gwatibaya S, Musiyiwa K, Rusere F, Mavima GA, Kasasa P (2015) Indigenous knowledge system weather forecasts as a climate change adaptation strategy in smallholder farming systems of Zimbabwe: a case study of Murehwa, Tsholotsho and Chiredzi districts. Afr J Agric Res 10(10):1067–1075

Thornes JE (2001) Climate change 2001: impacts, adaptation, and vulnerability, contribution of Working Group II to the third assessment report of the Intergovernmental Panel on Climate Change, McCarthy JJ, Canziani OF, Leary NA, Dokken DJ, White KS (eds). Cambridge University Press, Cambridge/New York

Unganai LS, Troni J, Manatsa D, Mukarakate D (2013) Tailoring seasonal climate forecasts for climate risk management in rainfed farming systems of southeast Zimbabwe. Clim Dev 5(2):139–152

Winsemius H, Werner M (2013, April) The value of forecasting key-decision variables for rain-fed farming. In: EGU general assembly conference abstracts (pp EGU2013-7466)

ZimVAC (2021) Zimbabwe vulnerability assessment committee (ZimVAC) 2021 Mashonaland Central rural livelihoods assessment report. Available online from: https://fnc.org.zw/wpcontent/uploads/2021/08/Mashonaland-Central-Province-2021-ZimVAC-Rural-Livelihoods-Assessment-Report.pdf

Agroforestry Practices: A Sustainable Way to Combat the Climate Crisis and Increase Productivity

12

Sushil Kumar, Badre Alam, Sukumar Taria, Priyanka Singh, Ashok Yadav, R. P. Dwivedi, and A. Arunachalam

Abstract

Along with local climate and terrestrial ecosystems, the effects of global warming, climate change, and their impacts, jointly referred to as the "climate crisis," are severely harming human, livestock, and environmental health. If mitigation measures, which could partly mitigate the effects of climate change, are not implemented quickly, sensibly, and responsibly, it is assumed and predicted that the impacts of climate change and global warming will soon become a global emergency. Several strategies are believed to be successful in halting climate change and global warming while also promoting sustainable agricultural production. Agroforestry is one of them and is thought to have promising potential to address climate change and global warming by reducing greenhouse gas emissions (GHGs) from the soil and storing carbon in woody biomass. Agroforestry is a distinctive and extensive agriculture system in which woody perennials are appropriately integrated with crop and livestock components to increase output while sustainably and effectively using the available natural resources. Agroforestry systems have the capacity to mitigate climate extremes like high temperatures, intra-annual climatic changes, rainfall events, floods, and droughts, among others. Agroforestry practices also avert the climate crisis by mediating microclimate, conserving resources, sequestering carbon, increasing and preserving soil fertility, and biodiversity. As a result, agroforestry has a great potential to not only avert the climate crisis but also increase productivity to ensure food, fodder, and nutritional security. The potential of agroforestry to combat the climate crisis and sustainably increase productivity is discussed in this book chapter.

Keywords

Agroforestry · Climate change · Carbon sequestration · Global warming · Microclimate · Productivity

12.1 Introduction

The term "climate crisis" refers to global warming and climate change and their impacts. Global warming and climate change are badly affecting human, livestock, and environmental health besides local climate and terrestrial ecosystems. Now, the effects of these changes in climate are visible everywhere and to everyone. It has become a widespread issue that is not limited to

S. Kumar (✉) · B. Alam · S. Taria · P. Singh
A. Yadav · R. P. Dwivedi · A. Arunachalam
ICAR-Central Agroforestry Research Institute, Jhansi, Uttar Pradesh, India

U. Chatterjee et al. (eds.), *Climate Crisis: Adaptive Approaches and Sustainability*, Sustainable Development Goals Series, https://doi.org/10.1007/978-3-031-44397-8_12

state boundaries and affects all the nations across the globe in some way. It is assumed and predicted that in the coming days, if mitigation measures, which may partially offset the effects of climate change, are not adopted timely, sensibly, and responsibly, the impacts of climate change and global warming are going to become a global emergency. The impacts of climate change and global warming will further impose severe challenges on farmers to maintain not only agricultural production but also conserve natural resources, which are very much needed for the survival of present and future generations. The climate problem will have a greater impact on developing nations than on the rest of the world (Tefera et al. 2019). Potential adverse impacts of climate change include sea-level rise, an increase in the frequency and intensity of wildfires, floods, droughts, and tropical storms, modifications to the amount, timing, and distribution of rain, snow, and runoff, and disruption of coastal marine and other ecosystems (Mbow et al. 2014).

Many approaches, including climate-resilient and smart agriculture practices, are thought to be effective in combating climate change and global warming while also assisting in the production of agricultural products in a sustainable manner. Climate-smart agriculture practices seek to solve the intertwined problems of food security and climate change by focusing on areas: sustainable agriculture production, adaptation to climate change, and migration of climate change and global warming through the reduction of greenhouse gases (GHGs). Among climate-resilient and smart agriculture practices, agroforestry is viewed as a promising sector. Agroforestry is a unique and extensive agriculture system in which woody perennials are suitably integrated with crop and livestock components to increase productivity while sustainably and effectively using available natural resources. Fast-growing, nitrogen-fixing trees and shrubs can also improve soil fertility and decrease erosion by being planted in agricultural fields. Additionally, trees have crucial roles in lowering vulnerability, boosting the resilience of farming systems, and protecting households from threats related to the climate (Meragiaw 2017). Practicing agroforestry may decrease the demand for deforestation by providing wood products from farms, reduce the need for fertilisers by improving soil quality, fertility, and nutritional balance and also strengthen agricultural resilience by increasing crop yields and offering a better environment for farm animals that all help in combating climate change (FAO 2013; Shi et al. 2013; Tefera et al. 2019). Depending on the crop, agroforestry changes the microclimate in ways that increase crop yields by 6–56%, improve soil fertility, and reduce soil erosion from wind and water (Tefera et al. 2019). In addition to the above, agroforestry also has the potential to restore degraded lands, provide a wider range of ecosystem goods and services, namely high biodiversity and carbon sequestration, and increase soil fertility and ecosystem stability through improved microclimate, erosion prevention, and additional carbon input from trees (FAO 2013; Murthy et al. 2013; Tefera et al. 2019). Agroforestry is regarded as a potential approach to address climate change because it not only reduces greenhouse gas emissions from the soil but also stores carbon in woody biomass. Agroforestry systems also have the potential to moderate climate extremes such as high temperatures and intra-annual climatic fluctuations (Mbow et al. 2014), besides mediating micro-climate and making resilient ecosystems for better crop production (DeSouza et al. 2012). Agroforestry helps to ensure not only food security but also reduces income variability in the semi-arid tropic part of central India (Singh et al. 2023). As a result, agroforestry has recently attracted a lot of interest for its potential to tackle the climate problem and boost productivity.

12.2 Climate Change Risks

The effects of climate change are becoming more obvious and real. Although its impact is visible everywhere, but the risks associated with climate change are not similar across the globe. Some countries are more and some are less susceptible to the risks of climate change due to their geographical locations. Rao et al. (2007) projected some of the changes in climate and their impact on agriculture are listed in Table 12.1. Extreme weather, a lack of fresh water, and the failure to adapt to and prevent climate change are just a few of the threats that the globe faces today. Hence, now all the stakeholders (public, policymakers, and experts) are becoming more aware of the need to work together and communicate to address risks associated with climate change. The major risks associated with the climate are elaborated in Table 12.1.

Table 12.1 Projected changes in climate and their impact on agriculture

S. no.	Phenomena and direction of change	Likelihood of occurrence	Major projected impacts on agriculture
1	Warmer and fewer cold days and nights; warmer/more frequent hot days and nights over most land areas	Virtually certain (>99% chance)	Increased yields in colder environments; decreased yields in warmer environments; increased insect outbreaks
2	Warm spells/ heat waves: frequency increases over most land areas	Very likely (90-99% chance)	Reduced yields in warmer regions due to heat stress; wildfire danger increase
3	Heavy precipitation events: frequency increases over most areas	Very likely	Damage to crops; soil erosion, inability to cultivate land due to water logging of soil
4	Area affected by drought: increases	Likely (66–90% chance)	Land degradation, lower yields/crop damage and failure; increased livestock deaths; increased risk of wildfire
5	Intense tropical cyclone activity increases	Likely	Damage to crops; windthrow (uprooting) of trees; damage to coral reefs
6	Increased incidence of extreme high sea level (excludes tsunamis)	Likely	Salinization of irrigation water, estuaries, and freshwater systems

Adopted from Rao et al. (2007)

12.2.1 Aberrations in Rainfall Events

The changes in climate are witnessed everywhere and by everyone. The aberrations in rainfall depicted in Fig. 12.1 due to climate change are more conspicuous and have become the new normal across the globe. India has experienced some of its most damaging effects from alterations in precipitation events that have generally harmed the lives of millions of people across the length and width of the country. The extreme precipitation events have reportedly become more frequent and intense in India during the past few decades. Furthermore, between the mid-twentieth century and the late twentieth century, it is predicted that the frequency of precipitation extremes will rise in southern and central India (Mukherjee et al. 2018). As Indian agriculture, wherein 55 percent of the net sown area is rainfed, completely depends on the monsoon, changes in its pattern, onset and withdrawal timings, number of rainy days, and frequency of rainfall have devastating effects by affecting food security and livelihoods. The unpredictability of rainfall has been found to alter the traditional crop calendar of many agricultural crops (Datta et al. 2022) that are considered the backbone of the food and nutritional security of the country. The aberrations in rainfall events are found to negatively affect the agriculture crops from sowing to harvesting, which leads to poor crop performance and productivity. Additionally, aberrations in rainfall are said to increase the amount of environmental stress that interferes with the system's ability to maintain productivity (Tisdell 1996). The frequency and severity of droughts along with increased rainfall unpredictability will have a severe impact on crop production (IPCC 2014).

12.2.2 Alterations in Temperature

Temperature fluctuations (Fig. 12.2) due to climate change and global warming have a significant impact on agricultural productivity. To secure sustainable crop production and global food security, agronomists, decision-makers, and

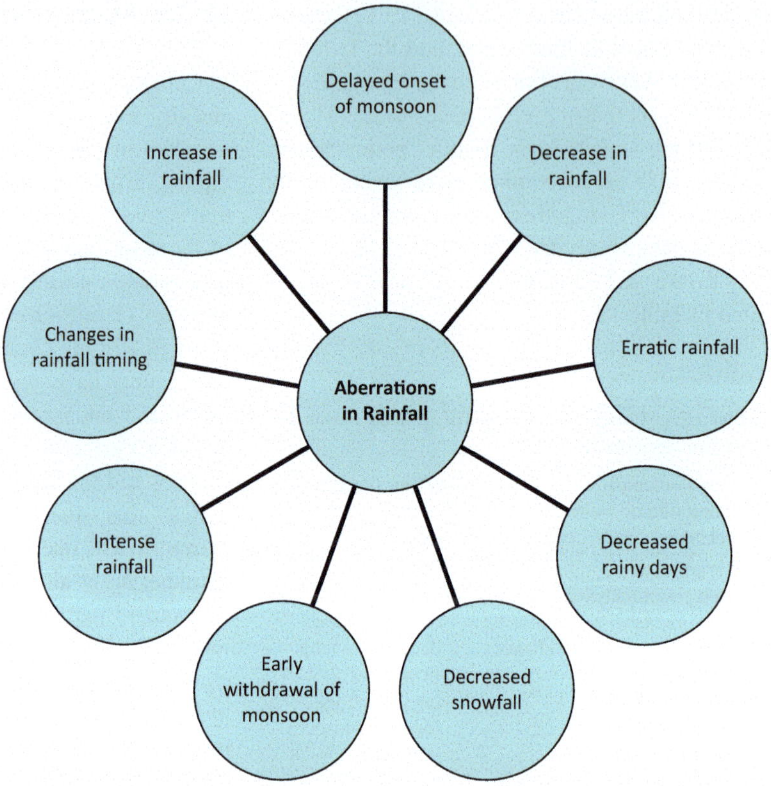

Fig. 12.1 Aberrations in rainfall due to climate change

farmers must be aware of temperature fluctuations brought on by climate change. The fluctuations in temperature at any growth stage of a crop negatively affect its performance by altering physiological processes.

These physiological processes (photosynthesis, transpiration, and respiration) ultimately influence the performance and productivity of crops. All agricultural crops require different optimum temperatures for various growth stages for better performance. The minimal fluctuations in the optimum temperature not only lead to poor crop yield but also alter the duration of developmental and reproductive stages. Among the fluctuations in temperature depicted in Fig. 12.2, the rise in temperature over optimum is known to have more detrimental effects on the growth and performance of crops as compared to others. Parthasarathi et al. (2022) reported that high temperatures affect the performance of crops, as presented in Fig. 12.3. The reproductive stage of the crop is regarded as the most sensitive phase in a crop life cycle for high temperatures and exposure of the crop to high temperatures during this phase causes a significant reduction in seed set and crop yield (Jagadish 2020). The prevalence of high temperatures at the reproductive part of the crops reduces pollen viability which results in a reduction in crop yields. Furthermore, it has been reported that exposure to temperature extremes at the start of the reproductive period has a significant impact on the production of fruit or grain in all species (Hatfield and Prueger 2015).

12.2.3 Increased Frequency and Intensity of Droughts

A prolonged break in rainfall during the monsoon season that leads to a shortage of water in ponds and soil moisture is termed drought. The hydrological cycle, which is responsible for the continuous movement of water from the earth to the atmosphere and vice versa, is disrupted by temperature changes caused by climate change. The changes in the hydrological cycle ultimately contribute to drought. The changes in climatic parameters such as temperature, wind velocity,

Fig. 12.2 Aberrations in temperature due to climate change

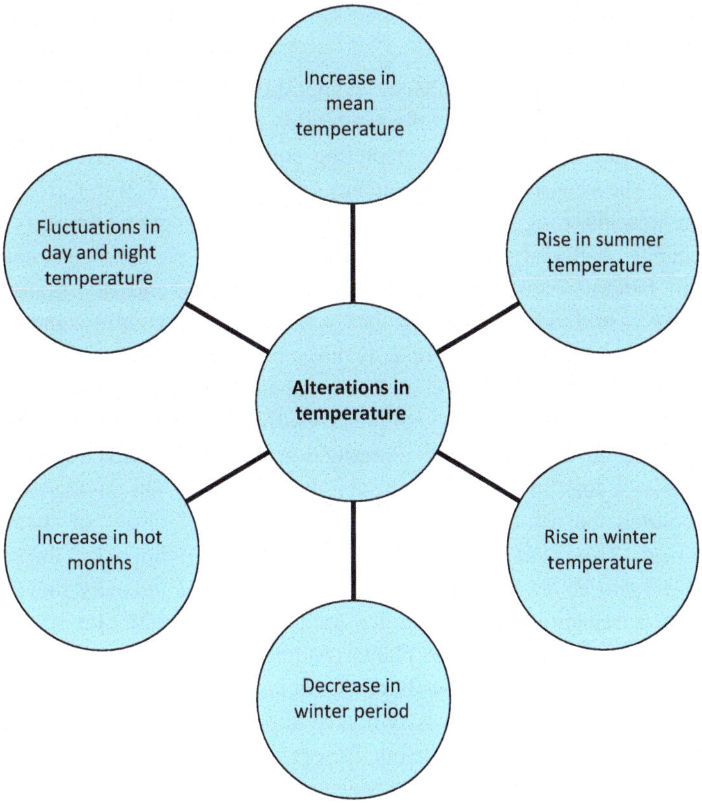

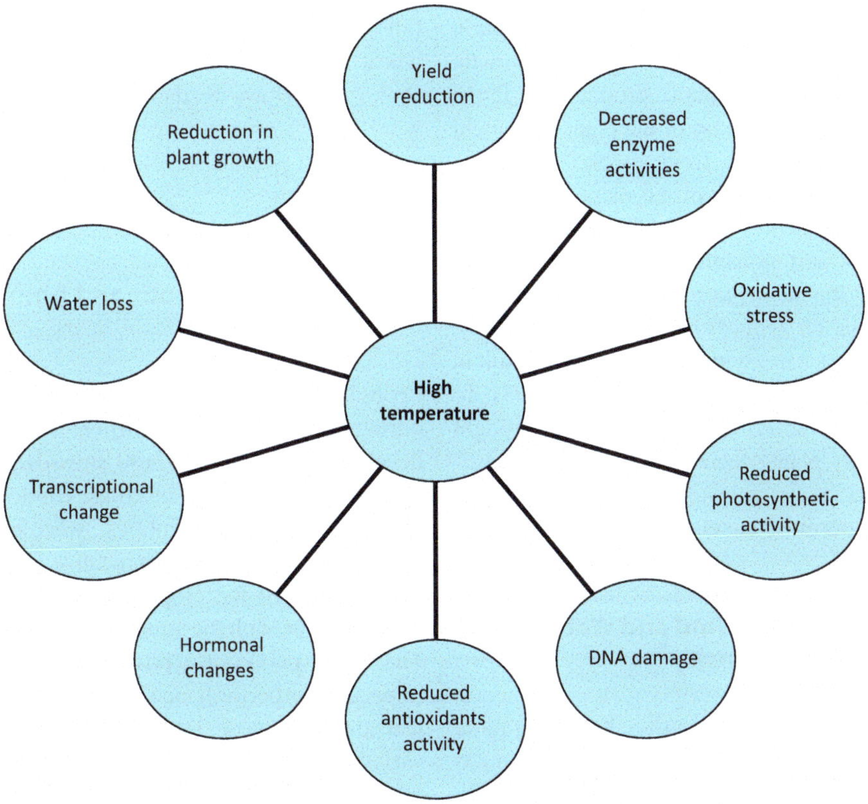

Fig. 12.3 Effect of high temperature on crops. (Adopted from Parthasarathi et al. 2022)

and relative humidity enhance evaporation and transpiration, which affect the availability of surface water besides quickly depleting soil moisture. The occurrence and frequency of droughts are now alarming across the world due to climate change. Reduced precipitation and altered rainfall patterns cause frequent droughts and heat stress worldwide, which together weaken agricultural production and ultimately have an impact on food security (Lobell et al. 2011; Fahad et al. 2017). Drought at any stage of the crop affects its growth, development, and reproduction, which causes a loss in crop productivity. It has been reported that drought results in a reduction of wheat (21%) and maize (40%) yields globally (Daryanto et al. 2016). The limited supply and higher transpiration of water are used to create drought conditions for the plants (Anjum et al. 2011). However, the intensity of the damage the drought causes to plants is typically unpredictable since it depends on many factors, namely rainfall patterns and frequency, stage of the crop, soil water holding capacity, and water losses through evapotranspiration. Drought is known to have an impact on photosynthesis, assimilation partitioning, nutrient and water relations, growth, and ultimately agricultural production (Praba et al. 2009). The response of the plant to drought stress typically varies from species to species depending on the stage of the plant growth and other environmental factors (Demirevska et al. 2009). Low soil moisture supply due to drought decreases the absorption of photosynthetically active radiation, reduces radiation use efficiency, and decreases harvest index, all of which result in low crop productivity (Earl and Davis 2003). To adapt to the severe consequences of drought stress, many plant species termed "climate resilient" alter their physiological processes and growth patterns (Duan et al. 2007).

12.2.4 Increased Wind and Water Storm Intensity

Wind and water storm events are becoming more common across the world due to changes in climatic parameters such as temperature, rainfall, and relative humidity. The unusual and sudden onset of these extreme events affects human and animal lives besides having a greater impact on the agriculture system. The hot winds cause loss of water from the plants by evapotranspiration which impacts their physiological and biochemical process namely, photosynthesis, respiration, and dry matter partitioning along with pollen viability and seed setting. These physiological and biochemical processes are very vital for the overall performance of the crop. The change in the duration of *Rabi* season crops is being witnessed due to the sudden upsurge of temperature. The sudden upsurge in the temperature causes onset of heat storms. Climate change has impacted the pattern of rainfall, its distribution, intensity, and duration. The number of rainy days and rainfall periods have decreased, but high-intensity rainfall events have been increasing. The high-intensity rainfall events not only cause the loss of productive soil and nutrients through soil erosion but also damage the standing crops and their production. Wind and water storms have caused widespread and severe crop damage around the world (Elahi et al. 2022). Heavy rainfall and ensuing waterlogging have affected wheat yield drastically (Malik et al. 2002; Marti et al. 2015). However, damage to the crops due to wind and water storm intensity depends mostly on the crop growth stage, its management, and soil conditions.

12.2.5 Increased Biotic and Abiotic Stresses

Crops face a range of stresses, biotic, and abiotic, during their life cycle at different growth and developmental stages. These stresses affect the performance and yield of crops. Biotic stress is caused by biological entities, whereas abiotic stress has an environmental component (Fig. 12.4). All the crops have genetic mechanisms to cope with the stress for a limited period. The longer period of exposure of crops to stress impacts their performance by altering physiological and biological processes. Many crops and varieties are referred to as tolerant crops because of their unique ability to withstand extreme stress. Moreover, agronomic management prac-

Fig. 12.4 Classification of stresses

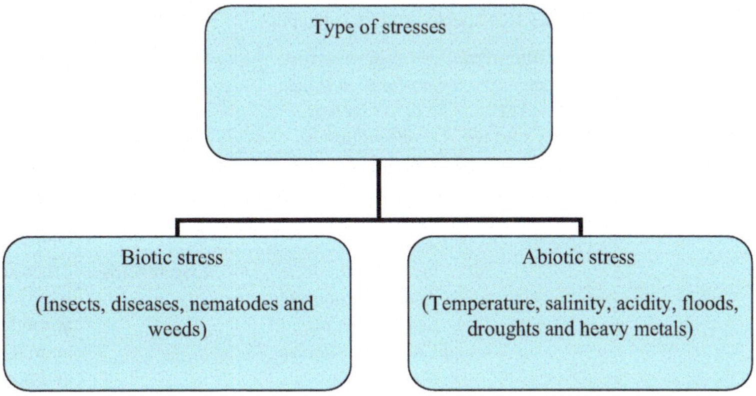

tices are also adopted to cope with stress and avoid crop losses. In the recent past, increased biotic and abiotic stresses in the agricultural crops have been reported due to climate change. High temperature due to global warming emerges as one of the gravest threats to rice production, reducing number of spikes and number of florets per plant (Fahad et al. 2016). In another study, Yadav et al. (2004) reported complete sterility in pearl millet due to drought stress at the flowering stage. Water-limiting conditions significantly reduce the growth attributes (plant height, leaf size, and stem girth) of the maize (Khan et al. 2015).

12.3 Role of Agroforestry in Combating Climate Crisis

In changing climate scenarios, the security of food, nutrition, and the environment are equally vital. Agroforestry is considered the best strategy and the only agricultural system that can solve food, nutrition, and environmental security in a sustainable manner. In India, it is practiced in a variety of ways, including planting trees on field boundaries, block plantations, alley cropping, scattered trees in fields, home gardens, homestead gardens, and so on. Agroforestry has developed over time and has a complex relationship with human life that is passed on from generation to generation. Traditional methods to secure livelihoods include maintaining trees near habitats or on field borders. Approximately 25 million hectares (Mha) of land are currently used for agroforestry in India, supporting about 50% of the country's demand for fuel wood, two-thirds of the demand for small timber, 70-80% of the demand for wood for the plywood industry, 60% of the demand for raw materials for paper pulp, and 9–11% of the demand for green feed for livestock. Although India's average biomass productivity is currently less than 2 t/ha/year, it can be further increased to 10 t/ha/year by carefully choosing compatible tree crop combinations. This will not only close the gap between supply and demand but also result in a surplus of plywood, paper pulp, and small timber for the nation, besides increasing the green cover of the country. Additionally, the production of timber on farms currently generates 450 employment days per hectare per year in India, supporting the notion that agroforestry can help reduce rural unemployment. As narrated by the farmers, tree species under agroforestry provide a myriad of benefits which are listed in Table 12.2 (Singh et al. 2023). The climate change mitigation and adaptation benefits of agroforestry are well known. Nowadays, extreme climatic events are very frequent and common across the globe due to climate change and global warming, which are proving destructive from time to time for agriculture production. Therefore, agroforestry is opined to be the way not only to optimize farm productivity to feed the burgeoning population while fulfilling their fodder, fuel, and nutritional requirements but also to sustain the natural resource base and environmental security. According to (IPCC 2019), agroforestry is one of the greatest options for simultaneously addressing land degradation, desertification, climate change adaptation, and food security.

Table 12.2 Agroforestry tree species and their benefits

Fruit	Fruits from trees are either consumed at home or sold for additional income	*Mangifera indica, Psidium guajava, Citrus sinensis, Phyllanthus emblica, Punica granatum, Syzygium cumini, Ziziphus mauritiana, Annona reticulate, Cordia species, Carissa carandas, Buchanania lanzan, Madhuca indica, Aegle marmelos, Tamarindus indica,* and *Pithecellobium dulce*
Construction material	Parts such as poles are obtained from the entire stem of young trees or from cut-offs from the stems of bigger trees. Small branches and twigs are also used for house construction (walling)	*Eucalyptus globules, Dalbergia sissoo, Tectona grandis, Bambusa* spp., *Artocarpus heterophyllus, Madhuca indica, Albizia lebbek, Acacia nilotica,* and *Azadirachta indica*
Firewood	Obtained either by pruning and drying harvested branches or by collecting dry fallen-off branches	*Eucalyptus globules, Dalbergia sissoo, Mangifera indica, Carissa carandas, Psidium guajava, Prosopis juliflora, Acacia nilotica, Pongamia pinnata, Butea monosperma,* and *Anogeissus* spp.
Timber	Timber is usually the processed wood often sold to a customer at the farm or at a market. Usually, the owner of the tree is responsible for harvesting, cutting, and transportation	*Eucalyptus globules, Dalbergia sissoo, Tectona grandis, Mangifera indica, Madhuca indica,* and *Albizia* spp.
Shade	Trees that are planted for shade in the compound and other farm components	*Azadirachta indica, Ficus religiosa, Ficus benghalensis, Ficus infectoria,* and *Cassia fistula*
Charcoal	Wood fuel is used for charcoal burning	*Bambusa* spp., and *Prosopis juliflora*
Live fencing	Trees act as a live fence around crops, animal shelters, and homesteads for protection	*Acacia catechu, Acacia Senegal, Acacia nilotica, Pithecellobium dulce, Leucaena leucocephala, Bambusa* spp., *Prosopis juliflora,* and *Anogeissus* spp.
Medicine	Trees with medicinal values	*Moringa oleifera, Azadirachta indica, Butea monosperma, Cassia fistula, Capparis deciduas, Tamrind indica, Madhuca indica, Aegle marmelos,* and *Phyllanthus emblica*
Fodder	Leaves and branches can be used as fodder	*Moringa oleifera, Leucaena leucocephala, Ziziphus mauritiana, Alianthus excels, Acacia Nilotica, Azadirachta indica, Albizia* spp., and *Dalbergia sissoo*
Vegetable	Trees planted for vegetable	*Moringa oleifera*

Adapted from Singh et al. (2023)

Agroforestry practices avert the climate crisis in many ways, which are listed in Table 12.2.

12.3.1 Microclimatic Modification

Global warming and climate change have made extreme occurrences common and frequent, which has an impact on crop performance in general and agriculture and related industries in particular. Trees grown with crops in agroforestry systems act as a buffer against these climatic extremes. These extreme events have a greater impact on the growth and development of the crop. The improvement in microclimate through agroforestry has a significant impact on agricultural performance. Microclimate climate factors include temperature, humidity, wind and turbulence, dew, frost, heat balance, and evaporation. The presence of trees in agroforestry systems helps in the moderation of microclimate to enhance the performance of understory crops in changed climate scenarios (Alam et al. 2014). In another study, it is reported that an alley-cropping agroforestry system can increase the resilience of cropping systems against extreme climate events by modifying the microclimate and water balance of croplands (Jacobs et al. 2022). Another agroforestry system option that is frequently utilised to improve microclimates is the use of shelterbelts, which increase surface roughness to slow down wind speed and reduce wind erosion and evapotranspiration (Rao et al. 2007). Trees on farms influence radiation flux, air temperature, wind speed, and the saturation deficit of understorey crops; together, these significantly affect photosynthesis' rate and duration, plant growth, transpiration, and soil water use (Monteith et al. 1991; Uthappa et al. 2017). As a result, trees on farms bring about favourable changes in the microclimate, which not only offsets the effects of climate extremes on the performance of crops but also helps to retain and maintain their productivity. Alam et al. (2018) suggested that *Dalbergia sissoo* alters microclimate factors such as incident photosynthetic photon flux density, air temperature, leaf temperature, canopy temperature depression, and soil surface in an agroforestry system, which favours the growth and yield of the understory cowpea crop. Among tree species, specifically shade trees are used to moderate the temperature for the cultivation of temperature-sensitive crops, namely coffee, cocoa, ginger, and cardamom to get higher quality produce. These shade trees protect crops from higher temperatures by altering the maximum and minimum daily temperatures, which leads to a decrease in the mean daily temperature of up to 4 °C (Beer et al. 1998). Gomes et al. (2020) opined that agroforestry systems have the potential, which depends strongly on altitude, to partly mitigate the impact of climate change on coffee suitability. The increase in temperature due to a prolonged dry season can not only affect photosynthesis and cause the abortion of flowers in coffee but also reduce its yields and quality (DaMatta and Ramalho 2006).

12.3.2 Conservation of Resources

Agroforestry is a natural resource management strategy with an ecological foundation that supports productivity and provides advantages to everyone. It is well established and widely accepted that agroforestry not only conserves natural resources while enhancing their efficiencies but also provides wood, food, fodder, fuel, fibre, and medicines. Agroforestry systems reduce run-off and soil loss and increase soil fertility levels compared to agriculture or cultivated fallow land use systems under various agro-climatic situations (Palsaniya and Ghosh 2016). Agroforestry is also well known for managing soil health since it greatly increases nutrient cycling efficiency compared to agriculture. It was reported that agroforestry in the form of alley cropping significantly reduces soil loss, mostly because of its dense canopy cover, which lowers the kinetic energy of rain falling (Tomar et al. 2021). In another study conducted in Nigeria, it is reported that alley cropping with maize and Leucaena hedge reduces the soil loss as compared to sole cropping (Lal 1990). A study conducted in the Budelkhand region of India reported reduced soil and nutrient loss in the silvopasture system as compared to bare land (Tomar et al. 2021). All forms of agroforestry practices, namely home gardens, alley cropping, silvipastoral systems, and bioengineered structures are known to overcome losses of soil and water by effectively arresting runoff and erosion (Kaushal et al. 2021). In addition to providing multiple co-benefits to farmers, agroforestry techniques are also known to improve ecosystems by storing carbon, preventing deforestation, conserving biodiversity, cleaning up the water, and controlling erosion while allowing agricultural lands to endure natural disasters like floods, droughts, and climate change (Ntawuruhunga et al. 2023). The adoption of agroforestry systems on sloping lands in Vietnam significantly reduces loss of soil, soil organic carbon, and nutrients (N, P, and K) as compared with a sole system (Do et al. 2023). By providing permanent cover and increasing organic carbon content, agroforestry systems significantly contribute to stopping and reversing land degradation, enhancing soil fertility and biological activity, increasing infiltration, and improving soil structure.

12.3.3 Carbon Sequestration

Besides providing economic and social benefits, agroforestry also plays an important role in mitigating climate change by sequestering more atmospheric carbon in plants and soil than conventional agriculture. The agroforestry system serves as an atmospheric carbon sink, and, in the process of carbon sequestration, carbon is removed from the atmosphere and deposited in carbon sinks like soils, vegetation, and seas through certain biological and physical processes. More atmospheric carbon is captured by agroforestry systems than by crop or pastureland (Sharrow and Ismail 2004; Tomar et al. 2021). However, the carbon sequestration ability of agroforestry systems depends on a myriad of factors such as tree species, age, density, climate, location, and management techniques. As compared to mono-cropping systems, agroforestry systems play a pivotal role in sustainable agroecosystems by improving atmospheric carbon assimilation (by soils and plants as well), nutrient cycling, soil biodiversity, and reducing soil disturbance, which subsequently checks soil erosion (Sauer et al. 2021). In addition to the above, agroforestry systems also provide many other benefits namely, agricultural diversification, genetic preservation, carbon capture, catchment protection, and rehabilitation, as well as improved wildlife habitat, strengthened agricultural infrastructure, increased self-sufficiency in timber and bioenergy, decreased reliance on food imports, poverty reduction, and improvements in people's nutritional status and associated health benefits (Ntawuruhunga et al. 2023). An agroforestry system not only increases agricultural productivity overall but also enriches the soil through litter fall, sequesters both above- and below-ground carbon, and maintains environmental services (Ghale et al. 2022). Agroforestry systems can sequester significant quantities of carbon in soil and tree biomass, which helps to lower the concentration of carbon dioxide (CO2) in the atmosphere. Cardinael et al. (2021) suggested that woody perennials boost above- and below-ground carbon stocks on agricultural land while reducing the crop-level temperature, under certain circumstances, and water supply risks. Agroforestry is now very important due to its multifunctionality, which includes its potential for carbon sequestration in various plant species as well as in soil, the socioeconomic aspect, and its numerous benefits for farmland.

12.3.4 Soil Fertility Management

The agroforestry systems' contribution to improving and preserving soil productivity, fertility, and sustainability is well known and established. Trees that do not fix nitrogen improve the physical characteristics of the soil, which aids in crop growth. For the sake of regional and global food security, soil fertility levels need to be maintained and improved. Increases in organic matter, whether in the form of surface litter fall, soil carbon, or root exudates in the rhizosphere, provide a substrate for a variety of organisms involved in soil biological activity and interactions, with significant effects on soil nutrients and fertility, and are closely associated with improved soil under trees and agroforestry systems. Saha et al. (2010) conducted research in the north-eastern hill region of India and concluded that agroforestry systems and multipurpose tree species improve soil physical health by lowering soil erosion and runoff, maintaining soil organic matter, enhancing soil chemical and biological properties, adding nitrogen input from trees and shrubs, and assisting in the extraction of minerals from lower horizons by roots and their recycling through litter fall on the ground. A multitiered agroforestry system involving the drumstick fruit tree, gliricidia hedgerow, and ginger: pigeon pea performed better over sole finger millet with respect to resource conservation, soil fertility and crop productivity, reduced runoff and soil loss, and enhanced soil organic carbon, phosphorus, and potash in sloppy land (Jakhar et al. 2017). The alley cropping system improves the physical, chemical, and biological characteristics of the soil by increasing nutrient recycling through the addition of clipped leafy biomass, reducing nutrient loss through erosion control, and minimising leaching losses (Hombegowda et al. 2022). Furthermore, it may also encourage the regeneration of deteriorated topsoil by improving soil physical qualities such as aeration, aggregate stability, and infiltration rate, which may help to produce more stable aggregates and provide a favourable soil medium for crop growth. In China, an agroforestry system based on *Bombax ceiba* improves potash and organic matter levels in addition to microbial diversity, increasing the fertility of rice fields (Wang et al. 2022). In comparison to monocultures, agroforestry practices improve soil health by reducing soil erosion, increasing microbiological diversity, and increasing soil organic carbon, nitrogen, and other essential nutrients. Consequently, it has enormous potential to maintain and improve environmental health and soil fertility.

12.3.5 Biodiversity Conservation

Biodiversity is very important for sustaining life and livelihood on the planet. Hence, there is an urgent need for agroecological farming methods that not only improve the ecosystem services provided by biodiversity, such as biological pest control, pollination, and nutrient cycling but also increase understanding that agrobiodiversity is essential to agricultural output and food security (Leroux et al. 2022). The faulty agricultural practices of the recent past seem to have had detrimental and alarming effects on diversity. There are a good number of good agricultural systems and practices, including agroforestry, which is touted as being best for the conservation of biodiversity. Agroforestry practices through their influence on the soil as a habitat for soil biota have the unique ability to not only encourage favourable changes in the number and diversity of soil organisms but also improve their functionality. Thus, agroforestry systems are known as a tangible example of diverse agricultural systems (Leroux et al. 2022). In tropical land use mosaics, where agriculture is mixed with natural habitat remnants, agroforestry practices aid in biodiversity conservation, besides increasing levels of wild biodiversity on farmland (McNeely and Schroth 2006). As biodiversity is a cornerstone for sustainable agriculture, increasing biodiversity in agricultural landscapes can increase farm productivity by providing ecosystem benefits and can also immediately lead to increased income and food security (Bommarco et al. 2013; Leroux et al. 2022).

12.4 Enhanced Productivity

Through mulching, trees in an agroforestry system can reduce soil moisture loss and boost crop productivity. Under trees, soil moisture availability is higher than in open regions, and the agroforestry system improves the soil's infiltration properties, trapping more water and raising the soil water content. Agroforestry systems in various agroecological zones of developing countries ensure a number of ecosystem services, such as increased system productivity, improved soil health, and climate change mitigation through carbon sequestration and related biomass production. A study conducted in the West African Sahel reported that *ziziphus*-based agroforestry with low-input cropping systems improves agricultural productivity and farmers' income, besides being affordable and compatible with low-input farming systems and ensuring sustainable management of soils, land resources, and ecosystem services (Bado et al. 2021). They also believed that incorporating fruit trees into farming systems through agroforestry could increase revenue generation, boost farming systems' resilience and productivity, and enhance the security of food and nutrition. Despite variations in soil types, environmental zones, crop types, production systems, and management practices, agroforestry systems outperformed monoculture in terms of crop and tree yields (Lehmann et al. 2020). In another study, it is opined that the adoption of integrated agroforestry systems (karonda, moringa, and phalsa with mung bean, cowpea in the rainy season, potato, and mustard during the winter season) improved farm productivity, profitability, and soil health while also doubling carbon sequestration and minimizing environmental footprints in semi-arid areas of India as compared to conventional agriculture (Rathore et al. 2022). Agroforestry systems are also known to improve agricultural output per unit area, soil health by enhancing soil fertility and halting land degradation, and environmental quality by reducing pollution through a variety of mechanisms. Furthermore, the selection and design of appropriate tree-crop combinations in agroforestry systems enhance system and resource productivity while increasing co-benefits. However, the selection and designing of appropriate tree-crop combinations always depend on many factors, such as the availability of resources, the social and economic status of the farmer, and environmental conditions. The proper combination of suitable tree types and annual agricultural crops increases system resilience by reducing negative effects (Yadav et al. 2021).

12.5 Agroforestry Practices in India

For better and holistic development of the agriculture system, India has been divided into 15 agro-climatic zones based on the soil type, climate, temperature, rainfall, and its variation, as well as the availability of water resources. Table 12.3 lists the various specialised agroforestry methods and systems that have been created over time and are currently being used in India's various agroclimatic zones. These methods include home gardens, block plantations, energy plantations, shelterbelts, and improvements to or substitutes for shifting cultivation.

Table 12.3 Major agroforestry practices followed in India

S. no.	Agroforestry practices	Agroecological region/states
1.	Boundary planting and live hedges	In all regions
2.	Farm woodlots	Throughout the country
3.	Home gardens	Mainly tropical west coast region, especially Kerala, southern Karnataka, and Andaman & Nicobar Islands
4.	Industrial plantations with crops	Intensively cropped areas in northern and northwestern regions: Haryana, Himachal Pradesh, Punjab, and Uttar Pradesh; also in southern states (Andhra Pradesh, Karnataka, Kerala, and Tamil Nadu)
5.	Scattered trees on farms, parklands	All regions, especially semiarid and arid regions
6.	Shaded perennial systems with plantation crops	Mainly humid tropical region in the southern region; also Assam and West Bengal
7.	Shelterbelts and windbreaks	In wind-prone areas, especially coastal, arid, and alpine regions
8.	Fodder trees	Throughout the country
9.	Intercropping/ grasses with fruit trees	Subtropical and tropical; orchards in hilly regions
10.	Seasonal forest grazing	Semiarid and mountainous ecosystem
11.	Taungya	Eastern region
12.	Tree planting for reclamation of saline soils and wastelands	Semiarid and canal-irrigated regions, mostly in the northern and northwestern regions
13.	Woodlots for soil conservation	In hilly areas, along the sea coast and ravine lands

Adapted from Puri and Nair (2004)

In addition to the above, the other major agroforestry systems practiced in different agroecological regions of India are listed in Table 12.4 and detailed here as agri-silviculture, boundary plantation, block plantation, energy plantation (trees + crops during initial years), alley cropping (hedges + crops), agri-horticulture (fruit trees +

Table 12.4 Major agroforestry systems practiced in different climatic regions of India

Agro-climatic zone	Agroforestry systems	Tree component	Crop/grass
Western Himalayas	Silvi-pasture (RF)	*Grewia optiva*	*Setarias* spp.
		Morus alba	*Setaria* spp.
	Agri-horticulture	*Malus pumila*	Millets and wheat
	Agri-horticulture	*Prunus persica*	Maize and soybean
Eastern Himalayas	Agri-silviculture	*Anthocephalus cadamba*	Rice
	Agri-horticulture	*Alnus nepalensis*	Large cardamom/coffee
	Silvi-pasture	Bamboos, *Parkia roxburghii, Morus alba*	–
	Silvi-pasture	*Bauhinia variegata, Ficus* spp., *Morus alba*	Napier
Lower Gangetic Plains	Agri-silviculture (Irri)	*Eucalyptus* spp., *Albizia lebbeck*	Rice
	Agri-horticulture (Irri)	Mango/Banana/Litchi	Wheat, paddy, and maize
	Silvi-pasture	*Morus alba, Albizia lebbeck*	*Dicanthium* and *Pennisetum*
Middle Gangetic Plains	Agri-silviculture (Irri)	*Populus deltoids*	Sugarcane and wheat
	Agri-silviculture (Irri)	*Eucalyptus* spp.	Rice and wheat
	Agri-silviculture	*Dalbergia sissoo*	Sesame
	Agri-horticulture (Irri)	Mango/Citrus	Rice and wheat
	Silvi-pasture	*Albizia lebbeck*	*Chrysopogan* and *Dicanthium*
Trans Gangetic Plains	Agri-horticulture (Irri)	*Emblica officinalis*	Black gram/green gram
	Agri-silviculture	*Azadirachta indica*	Black gram and wheat/ Mustard
	Silvi-pasture	*Bauhinia variegata, Albizia lebbeck*	*Cenchrus* and *Pennisetum*
Upper Gangetic Plains	Agri-silviculture (Irri)	*Populus deltoids*	Wheat and bajra fodder
	Agri-silviculture (Irri)	*Eucalyptus* spp.	Rice and wheat
	Silvi-pasture	*Bauhinia variegata, Albizia lebbeck*	*Chrysopogon, Poa*
Eastern Plateau & Hills	Agri-silviculture	*Gmelina arborea*	Rice and linseed
	Agri-silviculture	*Acacia nilotica*	Rice
	Silvi-pasture	*Acacia mangium, A. nilotica, bamboos*	–
	Silvi-pasture	*Leucaena leucocephala*	*Chrysopogon, Pennisetum,* and *Dicanthium*
Central Plateau & Hills	Agri-horticulture (Irri)	*Psidium guajava*	Bengal gram/groundnut
	Agri-horticulture (RF)	*Emblica officinalis*	Black gram/green gram
	Agri-silviculture	*Acacia nilotica/Leucaena leucocephala/ Azadirachta indica/Albizia lebbeck*	Soybean, black gram-mustard/wheat
	Silvi-pasture (RF- and degraded lands)	*Albizia amara, Leucaena leucocephala, Dichrostycuscinerea*	*Chrysopogon, Stylosanthes hamata,* and *S. Scabra*
	TBOs (RF)	*Jatropha curcas*	–

(continued)

Table 12.4 (continued)

Agro-climatic zone	Agroforestry systems	Tree component	Crop/grass
Western Plateau & Hills	Agri-horti-silviculture (Irri)	*Tectona grandis, Achrus zapot*	Rice and maize
	Agri-horticulture	*Areca catechu*	Black pepper and cardamom
	Silvi-culture	*Prosopis juliflora, Ailanthus excels*	–
	Silvi-pasture	*Acacia mangium, Albizia amara*	*Cenchrus*
Southern Plateau & Hills	Agri-silviculture (RF)	*Eucalyptus, Casuarina equisetifolia, Ailanthus excels*	Cotton and groundnut
	Agri-silviculture (Irri)	*Eucalyptus tereticornis, Melia dubia*	Chilli
	Silvi-culture (RF)	*Leucaena leucocephala, Acacia leucopholea*	–
		Eucalyptus	–
	Agri-horticulture	*Tamarindus indica*	Chilli
	TBOs	*Pongamia pinnata*	–
East Coast Plains & Hills	Agri-silviculture (RF)	*Ailanthus excelsa, Acacia leucophloea*	Cowpea
	Silvi-culture	*Casuarina equisetifolia, Leucaena leucocephala*	–
	TBOs	*Pongamia pinnata*	–
	Silvi-pasture	*Artocarpus* spp.	*Chrysopogon, Napier,* and *Cenchrus*
West Coast Plains & Hills	Agri-silviculture (RF)	*Acacia auriculiformis*	Black pepper
	Agri-horticulture (RF)	*Artocarpus heterophyllus*	Black pepper
	Agri-silviculture (RF)	*Acacia auriculiformis*	Rice
	Agri-horticulture	*Cocos nucifera/Areca catechu*	Rice
	Agri-silviculture	*Casurina equisetiofolia*	Rice
	Silvi-pasture	*Hardwickia binnata, Albizia lebbeck*	*Cenchrus*
Gujarat Coast Plains & Hills	Agri-silviculture	*Azadirachta indica, Ailanthus excels*	Cow pea and green gram
	Silvi-culture	*Prosopis juliflora, Acacia nilotica*	–
	Silvi-pasture	*Leucaena leucocephala*	*Cenchrus* and *Setaria*
Western Dry Region	Agri-silviculture	*Prosopis cineraria, Tecomella indica, Acacia nilotica, Azadirachta indica*	Pearl millet
	TBOs	*Jatropha curcas*	–
	Silvi-pasture	*Albizia lebbeck, Hardwickia binnata*	*Cenchrus*
All Islands	Agri-horticulture	*Cocos nucifera*	Rice
	Silvi-pasture	*Bauhinia* spp.*, Erythrinai indica, Leucaena leucocephala*	*Cenchrus* and *Pennisetum*

Irri irrigated, *RF* rainfed, *TBOs* tree-borne oilseeds

crops), agri-silvi-horticulture (trees + fruit trees + crops), agri-silvi-pasture (trees + crops + pasture or animals), silvi-olericulture (tree + vegetables), horti-pasture (fruit trees + pasture or animals), horti-olericulture (fruit tree + vegetables), silvi-pasture (trees + pasture/ animals), forage forestry (forage trees + pasture), shelter-belts (trees + crops), wind-breaks (trees + crops), live fence (shrubs and under- trees on boundary), silvi or horti-sericulture (trees or fruit trees + sericulture), horti-apiculture (fruit trees + honeybee), aqua-forestry (trees + fishes), and homestead (multiple combinations of trees, fruit trees, and vegetable) (Dhyani et al. 2009). Furthermore, the

All India Co-ordinated Research Programme on Agroforestry (AICRP-AF) and ICAR-Central Agroforestry Research Institute (ICAR-CAFRI), Jhansi, have also developed 35 specific agroforestry models for different agro-climatic zones of the country (Chaturvedi et al. 2016). Agroforestry models suggested for various agro-climatic regions of the country are namely: *Morus* and *Grewia* based for the western Himalayas; *Alder*-based for the North Eastern Hill (NEH) region; Poplar-based for the Indo-Gangetic Plain region (IGP); Aonla and Khejri-based for the semi-arid and arid regions; Teak-based for the tropical region; and *Gmelina* and *Acacia*-based for humid and sub-humid regions. Due to the prevalence of small land holdings, home gardens are popular in Kerala and other coastal states. Home gardens support diversity and food security while also providing for fundamental needs like food, fuel wood, fodder, plant-based medicines, and income. In the subcontinent, it is also common practise to intentionally plant trees on field bunds and in agricultural fields as scattered trees, as well as to use the open spaces between freshly planted orchards and forests for field crop cultivation.

12.6 Conclusion

Global warming and climate change and their impacts are collectively known as the climate crisis. The effects of global warming and climate change are visible everywhere to everyone. Sea-level rise, an increase in the frequency and severity of wildfires, floods, droughts, and tropical storms, alterations to the amount, timing, and distribution of rain, snow, and runoff, and disruption of coastal marine and other ecosystems are some of its potential negative effects. Together, these affect human, livestock, and environmental health besides local climate, agriculture systems, and terrestrial ecosystems. Agroforestry is viewed as an effective approach to combating climate change and global warming while also promoting sustainable agricultural production. Agroforestry practices reduce climate extremes that arise due to climate change and global warming by regulating microclimates, conserving resources, storing carbon, boosting and maintaining soil fertility, and safeguarding biodiversity, which results in enhancing productivity.

References

Alam B, Chaturvedi M, Singh AK, Ram N, Dhyani SK (2014) Microclimatic modification through agroforestry in the context of climate change: an ecophysiological appraisal. Indian J Agrofor 16(1):1–8

Alam B, Singh R, Uthappa AR, Chaturvedi M, Singh AK, Newaj R, Handa AK, Chaturvedi OP (2018) Different genotypes of *Dalbergia sissoo* trees modified microclimate dynamics differently on understory crop cowpea (*Vigna unguiculata*) as assessed through ecophysiological and spectral traits in agroforestry system. Agric For Meteorol 249:138–148

Anjum SA, Wang LC, Farooq M, Hussain M, Xue LL, Zou CM (2011) Brassinolide application improves the drought tolerance in maize through modulation of enzymatic antioxidants and leaf gas exchange. J Agron Crop Sci 197:177–185

Bado BV, Whitbread A, Manzo MLS (2021) Improving agricultural productivity using agroforestry systems: performance of millet, cowpea, and ziziphus-based cropping systems in West Africa Sahel. Agric Ecosyst Environ 305:107175. https://doi.org/10.1016/j.agee.2020.107175

Beer J, Muschler R, Kass D, Somarriba E (1998) Shade management in coffee and cacao plantations. In: Directions in Tropical Agroforestry Research: adapted from selected papers presented to a symposium on Tropical Agroforestry organized in connection with the annual meetings of the American Society of Agronomy, 5 November 1996, Indianapolis, Indiana, USA. Springer Netherlands, pp 139–164

Bommarco R, Kleijn D, Potts SG (2013) Ecological intensification: harnessing ecosystem services for food security. Trends Ecol Evol 28(4):230–238. https://doi.org/10.1016/j.tree.2012.10.012

Cardinael R, Cadisch G, Gosme M, Oelbermann M, Noordwijk MV (2021) Climate change mitigation and adaptation in agriculture: why agroforestry should be part of the solution. Agric Ecosyst Environ 319:107555. https://doi.org/10.1016/j.agee.2021.107555

Chaturvedi OP, Sikka AK, Handa AK, Bajpai CK (2016) Agroforestry technologies for different agro-climatic zones of the country. All India Coordinated Research Project on Agroforestry, ICAR-CAFRI, Jhansi

DaMatta FM, Ramalho JDC (2006) Impacts of drought and temperature stress on coffee physiology and production: a review. Braz J Plant Physiol 18:55–81

Daryanto S, Wang L, Jacinthe PA (2016) Global synthesis of drought effects on maize and wheat production. PLoS One 11:e0156362. https://doi.org/10.1371/journal.pone.0156362

Datta P, Behera B, Rahut DB (2022) Climate change and Indian agriculture: a systematic review of farmers' perception, adaptation, and transformation. Environ Chall 8:100543

Demirevska K, Zasheva D, Dimitrov R, Simova-Stoilova L, Stamenova M, Feller U (2009) Drought stress effects on RuBisCO in wheat: changes in the RuBisCO large subunit. Acta Physiol Plant 31:1129–1138. https://doi.org/10.1007/s11738-009-0331-2

DeSouza HN, DeGoede RGM, Brussaard L, Cardoso IM, Duarte EMG, Fernandes RBA, Gomes LC, Pulleman MM (2012) Protective shade, tree diversity and soil properties in coffee agroforestry systems in the Atlantic Rainforest biome. Agric Ecosyst Environ 146:179–196

Dhyani SK, Newaj R, Sharma AR (2009) Agroforestry: its relation with agronomy, challenges and opportunities. Indian J Agron 54(3):249–266

Do VH, La N, Bergkvist G, Dahlin AS, Mulia R, Nguyen VT, Öborn I (2023) Agroforestry with contour planting of grass contributes to terrace formation and conservation of soil and nutrients on sloping land. Agric Ecosyst Environ 345:108323. https://doi.org/10.1016/j.agee.2022.108323

Duan B, Yang Y, Lu Y, Korpelainen H, Berninger F, Li C (2007) Interactions between drought stress, ABA and genotypes in Picea asperata. J Exp Bot 58:3025–3036. https://doi.org/10.1093/jxb/erm160

Earl H, Davis RF (2003) Effect of drought stress on leaf and whole canopy radiation use efficiency and yield of maize. Agron J 95:688–696. https://doi.org/10.2134/agronj2003.0688

Elahi E, Khalid Z, Tauni MZ, Zhang H, Lirong X (2022) Extreme weather events risk to crop-production and the adaptation of innovative management strategies to mitigate the risk: a retrospective survey of rural Punjab, Pakistan. Technovation 117:102255

Fahad S, Hussain S, Saud S, Khan F, Hassan S, Nasim W, Huang J (2016) Exogenously applied plant growth regulators affect heat-stressed rice pollens. J Agron Crop Sci 202(2):139–150

Fahad S, Bajwa AA, Nazir U, Anjum SA, Farooq A, Zohaib A, Sadia S, Nasim W, Adkins S, Saud S, Ihsan MZ, Alharby H, Wu C, Wang D, Huang J (2017) Crop production under drought and heat stress: plant responses and management options. Front Plant Sci 8:1147. https://doi.org/10.3389/fpls.2017.01147

FAO (2013) Advancing agroforestry on the policy agenda: a guide for decision makers. In: Buttoud G, in collaboration with Ajayi O et al. (eds) Agroforestry working paper no 1. Food and Agriculture Organization of the United Nations, Rome, pp 48

Ghale B, Mitra E, Sodhi HS, Verma AK, Kumar S (2022) Carbon sequestration potential of agroforestry systems and its potential in climate change mitigation. Water Air Soil Pollut 233(228). https://doi.org/10.1007/s11270-022-05689-4

Gomes LC, Bianchi FJJA, Cardoso IM, Fernandes RBA, Fernandes Filho EI, Schulte RPO (2020) Agroforestry systems can mitigate the impacts of climate change on coffee production: a spatially explicit assessment in Brazil. Agric Ecosyst Environ 294:106858

Hatfield JL, Prueger JH (2015) Temperature extremes: effect on plant growth and development, Weather Clim Extremes 10:4–10. https://doi.org/10.1016/j.wace.2015.08.001

Hombegowda HC, Adhikary PP, Jakhar P, Madhu M (2022) Alley cropping agroforestry system for improvement of soil health. In: Shit PK, Adhikary PP, Bhunia GS, Sengupta D (eds) Soil health and environmental sustainability, Environmental science and engineering. Springer, Cham. https://doi.org/10.1007/978-3-031-09270-1_23

IPCC (2014) Climate change 2014: synthesis report. Contribution of Working Groups I, II and III to the fifth assessment report of the Intergovernmental Panel on Climate Change. IPCC, Geneva

IPCC (2019) Summary for policymakers. In: Shukla PR, Skea J, Buendia EC, Masson-Delmotte V, Portner H-O, Roberts DC, Zhai P, Slade R, Connors S, van Diemen R, Ferrat M, Haughey E, Luz S, Neogi S, Pathak M, Petzold J, Pereira JP, Vyas P, Huntley E, Kissick K, Belkacemi M, Malley J (eds) Climate change and land: an IPCC special report on climate change, desertification, land degradation, sustainable land management, food security, and greenhouse gas fluxes in terrestrial ecosystems. IPCC, Geneva

Jacobs SR, Webber H, Niether W, Grahmann K, Lüttschwager D, Schwartz C, Breuer L, Bellingrath-Kimura SD (2022) Modification of the microclimate and water balance through the integration of trees into temperate cropping systems. Agric For Meteorol 323:109065. https://doi.org/10.1016/j.agrformet.2022.109065

Jagadish SVK (2020) Heat stress during flowering in cereals – effects and adaptation strategies. New Phytol 226(6):1567–1572. https://doi.org/10.1111/NPH.16429

Jakhar P, Dass A, Adhikary PP, Sudhishri S, Naik BS, Hombegowda HC, Madhu M, Lenka NK, Chaudhary PR, Panda RK (2017) Multitier agroforestry system for integrated resource conservation on uplands of Eastern Ghats region in India. Agrofor Syst 91:697–712

Kaushal R, Mandal D, Panwar P, Rajkumar, Kumar P, Tomar JMS, Mehta H (2021) Soil and water conservation benefits of agroforestry. In: Shit PK, Pourghasemi HR, Adhikary PP, Bhunia GS, Sati VP (eds) Forest resources resilience and conflicts. Elsevier, pp 259–275

Khan MB, Hussain M, Raza A, Farooq S, Jabran K (2015) Seed priming with CaCl2 and ridge planting for improved drought resistance in maize. Turk J Agric For 39(2):193–203

Lal R (1990) Agroforestry systems to control erosion on arable tropical steep lands. IAHS Publ 192:338–346

Lehmann LM, Smith J, Westaway S, Pisanelli A, Russo G, Borek R, Sandor M, Gliga A, Smith L, Ghaley BB (2020) Productivity and economic evaluation of agroforestry systems for sustainable production of food and non-food products. Sustainability 12(13):5429. https://doi.org/10.3390/su12135429

Leroux L, Faye NF, Jahel C, Falconnier GN, Diouf AA, Ndao B, Tiaw I, Senghor Y, Kanfany G, Balde A, Dieye M, Sirdey N, Loison AS, Corbeels M, Baudron F, Bouquet E (2022) Exploring the agricultural landscape diversity-food security nexus: an analysis in two contrasted parklands of Central Senegal. Agric Syst 196:103312. https://doi.org/10.1016/j.agsy.2021.103312

Lobell DB, Schlenker W, Costa-Roberts J (2011) Climate trends and global crop production since 1980. Science 333:616–620. https://doi.org/10.1126/science.1204531

Malik AI, Colmer TD, Lambers H, Setter TL, Schortemeyer M (2002) Short-term waterlogging has long-term effects on the growth and physiology of wheat. New Phytol 153(2):225–236

Marti J, Savin R, Slafer GA (2015) Wheat yield as affected by length of exposure to waterlogging during stem elongation. J Agron Crop Sci 201(6):473–486

Mbow C, Noordwijk MV, Luedeling E, Neufeldt H, Minang PA, Kowero G (2014) Agroforestry solutions to address food security and climate change challenges in Africa. Curr Opin Environ Sustain 6:61–67

McNeely JA, Schroth G (2006) Agroforestry and biodiversity conservation – traditional practices, present dynamics, and lessons for the future. Biodivers Conserv 15:549–554. https://doi.org/10.1007/s10531-005-2087-3

Meragiaw M (2017) Role of agroforestry and plantation on climate change mitigation and carbon sequestration in Ethiopia. J Tree Sci 36(1):1–15

Monteith JL, Ong CK, Corlett JE (1991) Microclimatic interactions in agroforestry systems. For Ecol Manag 45:31–44

Mukherjee S, Aadhar S, Stone D, Mishra V (2018) Increase in extreme precipitation events under anthropogenic warming in India. Weather Clim Extremes 20:45–53

Murthy IK, Gupta M, Tomar S, Munsi M, Tiwari R, Hegde GT, Ravindranath NH (2013) Carbon sequestration potential of agroforestry systems in India. J Earth Sci Clim Change 4(1):1–7

Ntawuruhunga D, Ngowi EE, Mangi HO, Salanga RJ, Shikuku KM (2023) Climate-smart agroforestry systems and practices: a systematic review of what works, what doesn't work, and why. Forest Policy Econ 150:102937. https://doi.org/10.1016/j.forpol.2023.102937

Palsaniya DR, Ghosh PK (2016) Agroforestry for natural resource conservation, livelihood security and climate change mitigation in Himalayan agroecosystems. In: Bisht J, Meena V, Mishra P, Pattanayak A (eds) Conservation agriculture. Springer, Singapore. https://doi.org/10.1007/978-981-10-2558-7_8

Parthasarathi T, Firdous S, Mariya David E, Lesharadevi K, Djanaguiraman M (2022) Effects of high temperature on crops. Adv Plant Def Mech. https://doi.org/10.5772/intechopen.105945

Praba ML, Cairns JE, Babu RC, Lafitte HR (2009) Identification of physiological traits underlying cultivar differences in drought tolerance in rice and wheat. J Agron Crop Sci 195:30–46. https://doi.org/10.1111/j.1439-037X.2008.00341.x

Puri S, Nair PKR (2004) Agroforestry research for development in India: 25 years of experiences of a national program. Agrofor Syst 61:437–452

Rao KPC, Verchot LV, Laarman J (2007) Adaptation to climate change through sustainable management and development of agroforestry systems. SAT eJ 4(1):1–30

Rathore SS, Babu S, El-Sappah AH, Shekhawat K, Singh VK, Singh RK, Upadhyay PK, Singh R (2022) Integrated agroforestry systems improve soil carbon storage, water productivity, and economic returns in the marginal land of the semi-arid region. Saudi J Biol Sci 29(10):103427. https://doi.org/10.1016/j.sjbs.2022.103427

Saha R, Ghosh PK, Mishra VK, Majumdar B, Tomar JMS (2010) Can agroforestry be a resource conservation tool to maintain soil health in the fragile ecosystem of North-East India? Outlook Agric 39(3):191–196

Sauer TJ, Dold C, Ashworth AJ, Nieman CC, Hernandez-Ramirez G, Phillip D, Gennadiev AN, Chendev YG (2021) Agroforestry practices for soil conservation and resilient agriculture. In: Udawatta RP, Jose S (eds) Agroforestry and ecosystem services. Springer, Cham. https://doi.org/10.1007/978-3-030-80060-4_2

Sharrow SH, Ismail S (2004) Carbon and nitrogen storage in agroforests, tree plantations, and pastures in western Oregon, USA. Agrofor Syst 60:123–130

Shi S, Zhang W, Zhang P, Yu Y, Ding F (2013) A synthesis of change in deep soil organic carbon stores with afforestation of agricultural soils. For Ecol Manag 296:53–63. https://doi.org/10.1016/j.foreco.2013.01.026

Singh P, Choudhary BB, Dwivedi RP, Arunachalam A, Kumar S, Dev I (2023) Agroforestry improves food security and reduces income variability in semi-arid tropics of central India. Agrofor Syst. https://doi.org/10.1007/s10457-023-00806-6

Tefera Y, Hailu Y, Siraj Z (2019) Potential of agroforestry for climate change mitigation through carbon sequestration: review paper. Agric Res Technol 22(3):556196. https://doi.org/10.19080/ARTOAJ.2019.22.556196

Tisdell C (1996) Economic indicators to assess the sustainability of conservation farming projects: an evaluation. Agric Ecosyst Environ 57:117–131

Tomar JMS, Ahmed A, Bhat JA, Kaushal R, Shukla G, Kumar R (2021) Potential and opportunities of agroforestry practices in combating land degradation. In: Agroforestry – small landholder's tool for climate change resiliency and mitigation. Intech. https://doi.org/10.5772/intechopen.97843

Uthappa AR, Chavan SB, Handa AK, Newaj R, Kumar D, Sridhar KB, Chaturvedi OP (2017) Agroforestry - a sustainable solution to address climate change challenges. In: Gupta et al (eds) Agroforestry for increased production and livelihood security. New India Publishing Agency, New Delhi, pp 1–22

Wang WJ, Wen J, Xiang W-Q, Malabrigo PL Jr, Ren M-X (2022) Soil bacterial and fungal communities respond differently to *Bombax ceiba* (Malvaceae) during reproductive stages of rice in a traditional agroforestry system. Plant Soil 479:543–558. https://doi.org/10.1007/s11104-022-05542-x

Yadav RS, Hash CT, Bidinger FR, Devos KM, Howarth CJ (2004) Genomic regions associated with grain yield and aspects of post-flowering drought tolerance in pearl millet across stress environments and tester background. Euphytica 136:265–277

Yadav GS, Das A, Kandpal BK, Babu S, Lal R, Datta M, Chakraborty M (2021) The food-energy-water-carbon nexus in a maize-maize-mustard cropping sequence of the Indian Himalayas: an impact of tillage-cum-live mulching. Renew Sust Energ Rev 151:111602

13 Climate Crisis and Adoption of Climate-Smart Agriculture Technologies

Trisha Roy, Justin George Kalambukattu, Abhijit Sarkar, I. Rashmi, Rama Pal, Vibha Singhal, Deepak Singh, and Suresh Kumar

Abstract

The growing population projected to be 9 billion by 2050 indicates an increased food demand and pressure on natural resources including soils, water, and biodiversity that are already stretched dangerously thin. To feed this population, there is a need to increase agricultural production by 70% and agriculture sustainability plays a key role in this. However, the risk of climate change is a major threat towards the sustainability of agricultural systems across the world. The climate change problem is manifested as extreme weather events, erratic rainfall patterns, land degradation, and subsequent loss in land productivity. To manage this crisis, the evolution of Climate-Smart Agriculture (CSA) occurred during 2010 and has been a key driver to tackle the problems of agriculture associated with climate change. CSA is based on three main components that are (i) enhancement in crop productivity through sustainable agricultural practices; (ii) building climate resilient agricultural systems; and (iii) cutting down emissions to zero or reducing overall emission from production systems. The CSA practices are not entirely new practices; rather, it is an amalgamation of the already existing systems, like conservation agriculture, cover crops, integrated nutrient management, agroforestry system, diversification of crops, and introduction of local varieties and breeds that are more climate resilient. This chapter brings forth a compilation of all such CSA tools and various policy measures to improve the adoption of CSA in our country for climate proofing of agriculture.

Keywords

Climate change · Climate-smart agriculture · Sustainable agricultural practices · Resilience · Land degradation

T. Roy · R. Pal · V. Singhal · D. Singh
ICAR-Indian Institute of Soil and Water Conservation, Dehradun, Uttarakhand, India

J. G. Kalambukattu (✉)
Agriculture and Soils Department, Indian Institute of Remote Sensing, Indian Space Research Organization (ISRO), Dehradun, Uttarakhand, India
e-mail: justin@iirs.gov.in

A. Sarkar
ICAR-Indian Institute of Soil Science, Bhopal, Madhya Pradesh, India

I. Rashmi
ICAR-Indian Institute of Soil and Water Conservation, Research Centre, Kota, Rajasthan, India

S. Kumar
Agriculture, Forestry and Ecology Group, Indian Institute of Remote Sensing, Indian Space Research Organization (ISRO), Dehradun, Uttarakhand, India

U. Chatterjee et al. (eds.), *Climate Crisis: Adaptive Approaches and Sustainability*, Sustainable Development Goals Series, https://doi.org/10.1007/978-3-031-44397-8_13

13.1 Introduction: Brief About the Climate Crisis and Its Link to Agriculture

Agriculture and food security are intricately linked to climate change (Anderson et al. 2020). Climate change refers to the considerable long-term alterations in the average weather parameters over a long period of time (generally 30 years and above) (Mahato 2014). Human species over the time have evolved to adjust and acclimatize to the day-to-day variations in climatic conditions that occur naturally. However, the variations and changes in the long-term climatic parameters pose an enormous threat to all life forms on this planet and control the availability and quality of different natural resources vital for the survival and sustenance of Earth's ecosystems (Roy et al. 2023).

The release of carbon dioxide into the atmosphere is considered as the major driver of climate change. The involvement of the human developmental activities takes a center stage in the release of greenhouse gases into the atmosphere. The Intergovernmental Panel on Climate Change (IPCC) has pointed out that "majority of the warming evidenced during the past 50 years is likely due to human activities" (IPCC 2001). The IPCC, assessment Report 6 (AR6), provides a very explicit explanation of different climate projections and scenarios. The Global Surface Air Temperature (GSAT) is likely to rise by 0.2–1.0 °C during 2081–2100 compared to 1995–2014 under a low-emission scenario of Shared Socioeconomic Pathway (SSP)1 1.9. Considering all projection scenarios, an increase of 1.5 °C is expected by 2030 in comparison to 1850–1900 with a probability of 0.4–0.6 (Lee et al. 2021). As the GSAT changes, the precipitation pattern across the globe will exhibit regional variations and the annual global land precipitation will increase during the twenty-first century. The likely changes in the temperature and precipitation will affect the natural resources, productivity, and sustainability of the agricultural ecosystems. Besides being reflected in the form of changing global temperature and precipitation patterns, the climate change phenomenon also expresses itself as increased frequency of extreme climate events like floods, droughts, flash floods, melting of glaciers, aberrations in the hydrological cycles, rise in sea levels, and increased soil erosion and land degradation in different forms (Nema et al. 2012). The biodiversity of various ecosystems is also threatened by the changing climate and various flora and fauna species are on the verge of extinction while dealing with the vagaries of climate change.

The agricultural sector is one of the most susceptible sectors impacted by the ripples of climate change. Since at any place, the selection of crop and allied agricultural activities are dependent on the climate and agro-ecology, the change in climatic parameters has a direct impact on the agricultural activities. The impact of climate on agricultural activities can be summarized as given by Kim et al. (2009). The changing climatic parameters can impact the arable/livestock sector and the related hydrological cycle. Shift in climatic variables impacts the livestock production, the dynamics of insect and pests, biodiversity changes, alteration in the plant life cycle like seeding, blooming season, changes in quality traits, changes in areas suitable for cultivation, etc. which impacts the biological productivity. In the hydrology sector, the changes in water resources, river flow, water quality, etc. influence the agricultural production. In general, though the impact of climate change on agriculture is negative, there are instances where the increased CO_2 concentration in the atmosphere will enhance the photosynthetic rate and increase crop productivity. The generalized impact of climate change (positive or negative) is summarized in Fig. 13.1.

The climate change impact on agriculture is region specific, crop specific, and ambivalent in nature. This is indeed a hopeful scenario because prudent judgment and evidence-based decision-making supported by real-time data to formulate strategies for agricultural sustainability will help to shield from the negative impacts of climate change. The development of strategies to sustain agricultural production in the milieu of climate change is a daunting task and needs efforts from different sectors including researchers, farmers,

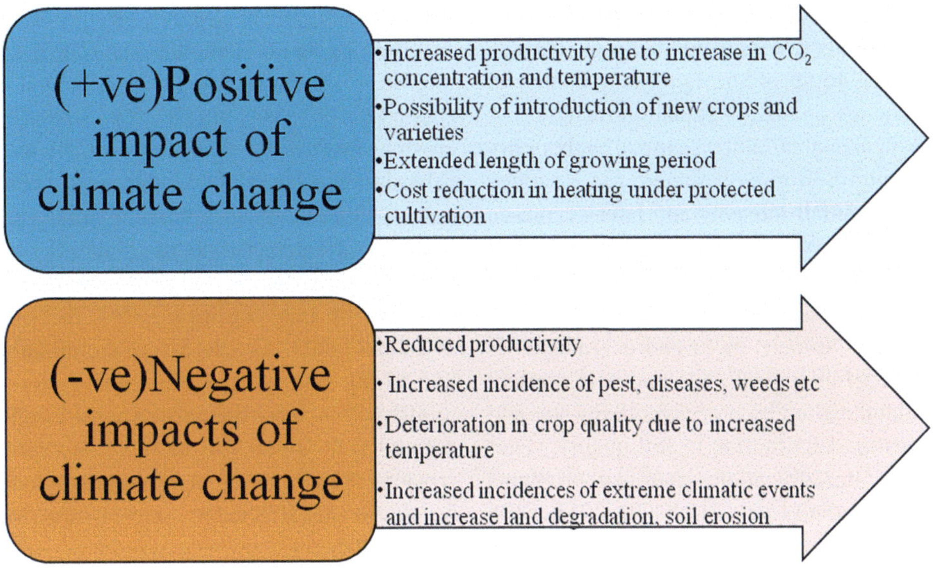

Fig. 13.1 Impacts of climate change on agriculture. (Adapted from Kim et al. 2009)

policymakers, climate change experts, and all other stakeholders. This chapter will discuss in detail about the following issues:

- The impact of climate change on agriculture.
- The different climate resilient/climate-smart agriculture (CSA) technologies.
- Popularizing climate-smart agriculture (CSA) and how it can be used for achieving the Sustainable Development Goals (SDG).

13.2 Impact of Climate Change on Agricultural Land Degradation in the Country

The occurrence of extreme climate events and erratic rainfall patterns has posed serious threats to all farmers working across the different agro-climatic zones in the country. Agricultural land degradation and climate change are intensifying challenges causing a reduction in agricultural productivity and threatening the food security of populations. Agricultural land degradation is a complex process threatening food security and rural livelihood, severely affecting the country's economy and well-being of citizens. Land degradation by soil erosion, soil salinity, nutrient loss, acidity, and other constraints are considered as a major obstacle for agricultural production and sustainability. Soil, the most important component of land essential for all terrestrial life, is degrading at an alarming rate. Population pressure is considered as a major global challenge, exerting tremendous strain on natural resources. The complexity of various pressures on land resources due to the globalization process along with the potential impact of climate change and climate extreme events further add to the higher level of uncertainties. In the developing countries, already 40% of the land area is affected by some form of degradation and this is expected to reach 78% under drylands globally with 50% population growth by the twenty-first century (Huang et al. 2015). Climate change and land degradation are interlinked and governed by various biophysical factors, chemical factors, and human interferences with impact and responses. Current climate change patterns possess several challenges in understanding ecosystem processes/services, soil resilience, and mitigation strategies to restore and improve soil health for reducing agrarian crises. The terrestrial biosphere and soil system are the interface for all the greenhouse gases (GHG) such as CO_2, N_2O, and CH_4

through tillage, deforestation, livestock rearing, biomass burning, rice cultivation, etc. (Lal 2012). Interestingly, some of the important factors of climate change on land degradation include (i) a decrease in annual rainfall causing drought or dry spells during crop growth stages; (ii) a sudden increase in rainfall intensity and later decreased rainfall; (iii) high evaporation; (iv) erratic rainfall patterns, and increased duration between two consecutive rainfall incidents; (v) enhanced runoff, low water storage, etc. Land degradation is the decline in biological and economic productivity of land caused by processes like water and wind erosion, deterioration in soil quality, vegetation loss, desertification, chemical problems like acidity, salinity, etc. (UNCCD 2015). The major impact of such degradation is observed on soil organic carbon (SOC) which is exacerbated by the erosion process resulting in the deterioration of soil properties, nutrient loss, and soil moisture depletion. Soil is the second largest carbon sink which could store four times more than aboveground vegetation and three times more than the atmosphere (Jobbagy and Jackson 2000; Houghton 2007). The importance of soil as global C bank particularly with the present climate change and land use change patterns has been emphasized in many studies (Amundson et al. 2015).

Soil erosion is accelerated by an increase in rainfall erosivity, soil erodibility, slope, crop management practices, etc. (Sharda and Ojasvi 2016). Since climate influences all these factors, the changes in climatic parameters have direct consequences on the magnitude of soil erosion. The rainfall erosivity increases by 2% for 1% increase in rainfall intensity (Sharda and Ojasvi 2016). Soils with low SOC are vulnerable to soil erosion with poor aggregation and reduced aggregate strength. It is estimated that almost 20–30% of the eroded SOC is mineralized while being transported with the eroded deposits (Lal 1995, 2005; Jacinthe and Lal 2001; Polyakov and Lal 2008). Mandal et al. (2020) estimated the erosion-induced C loss to be 115.36 Tg C year^{-1} in Indian conditions based on the existing databases. Assuming that 30% of the displaced C is mineralized, the C emitted per year is 34.61 Tg C year^{-1} which is a significant amount. Bal et al. (2016) reported a general change in the annual rainfall intensity of India by 15–24% by the end of the century (2100). Assuming a 1.5% increase in soil erosion due to 1% change in rainfall intensity (Sharda and Ojasvi 2006), the mean change in soil erosion due to an increase in annual rainfall intensity is calculated to be 29.25%. The projected soil loss by the end of the century (2100) amounts to 6.29 Pg year^{-1} and the erosion-induced soil C loss concomitantly is 215.53 Tg ha^{-1} year^{-1}. All these factors conjointly will increase the erosion hazard from agricultural land which will be more susceptible to erosion. In India by 2050, 66 Mha cropland will be additionally affected by water erosion (Mandal et al. 2020).

In addition, soil salinization will increase at an alarming rate due to increased potential evapotranspiration. Moreover, natural events such as flash floods, El Nino and La Nina, alternate wet and dry years, long dry spells, or drought have a profound effect on soil salinity and salt accumulation/leaching in the root zone (Hassani et al. 2020). Water is the most important resource for agriculture which is scarce in dryland/arid regions. Dryland and rainfed agriculture are most severely affected by climate change, resulting in low water availability for crops. According to Reibsame et al. (1995), there is a close link between water and climate change. Different drought forms such as (i) meteorological drought is caused by long-term withdrawal of rainfall; (ii) hydrological drought commonly occurs with reduced surface runoff, low groundwater level, and aquifer depletion; (iii) pedologic/edaphic drought is caused by a decrease in the soil available water; (iv) agriculture drought is commonly observed due to poor availability of soil moisture at the critical crop growth stages (Street and Findlay 1981; Maybank et al. 1995; Williams and Balling 1996). Agriculture drought will be severely affected by climate change by increasing the tendency of desertification. With high CO_2 concentrations, soil wetness during summer months would reduce thus depleting soil moisture levels (Mahato 2014). The problems of drought and soil/land degradation are more likely to be aggravated by problems of climate change (Hermans and McLeman 2021). The increase in temperature and evapotranspiration under a changing climate scenario would adversely affect and degrade the vegetation cover of land. Further, long-term drought situations modify soil moisture, albedo, temperature, surface roughness, and other soil properties (Mahato 2014). Climate change patterns of rising temperature, erratic rainfall patterns, rising sea levels, and increased extreme events severely impact agricultural production and livelihood security of rural households in developing countries. Thus, adapting to agricultural practices according to climate change is necessary for food security and identifying climate-smart practices has the potential to curb emissions.

13.2.1 Climate Change Projections and Its Impact on Agriculture

Climate change is reflected through abnormal changes in the environment and its consequences affecting other areas of the Earth. Climate change is a natural phenomenon and a sort of Earth's adjustive response to the alterations in atmospheric quality which could require a significant time span in years. However, increased anthropogenic activities such as industrialization, urbanization, deforestation, agriculture, and changes in land use patterns emitted more GHGs into the atmosphere and consequently accelerated the rate of climate change. The potential effects of climate change include higher Globally averaged Surface Air Temperature (GSAT) and Global Mean Sea Level (GMSL), altered precipitation patterns and amount, decreased cryosphere area, increased acidity of the oceans, increased frequency, intensity, and/or duration of extreme events, shift in ecosystem characteristics, increased risks to agricultural productivity and human health (https://climatechange.chicago.gov/climate-impacts/climate-impacts-human-health).

The earth's temperature has increased by 1.1 degrees Celsius since the late 1800s, which indicates that the rate of emission has been highest. The IPCC released the first part of its Sixth Assessment Report on Climate Change on August 9, 2021, and alarmed the world that global warming is occurring very rapidly and irreversibly. The temperature of our planet is projected to rise by 1.5 °C by 2030. The projected temperature rises under a set of five scenarios that are based on the framework of the shared socioeconomic pathways (SSPs) (Table 13.1).

13.2.1.1 Indian Scenario of Climate Change

According to the Sixth Assessment Report of IPCC, failure to take immediate action to reduce emissions or adjust to climate change could have disastrous effects, particularly in India. The northern regions of India may experience a greater degree of warming. The rise in maximum and minimum temperature extremes is anticipated resulting in disturbances in rainfall patterns and inequitable distribution of moisture. A 20% increase in summer monsoon rainfall is anticipated across all states in India, with the exception of Punjab, Rajasthan, and Tamil Nadu, which show a slight decline on average. The number of rainy days may decrease (e.g. in MP), but the

Table 13.1 Estimation of IPCC for increase in globally averaged surface air temperature under scenarios of shared economic pathways

SSP	Scenario	Estimated warming in °C (2041–2060)	Estimated warming in °C (2081–2100)	Projected change in temp in °C
SSP1-1.9	Net zero emission by 2050	1.6	1.4	1.0–1.8
SSP1-2.6	Net zero emission around 2075	1.7	1.8	1.3–2.4
SSP2-4.5	Intermediate GHG emissions	2.0	2.7	2.1–3.5
SSP3-7.0	Net CO_2 emissions doubled by 2100	2.1	3.6	2.6–4.6
SSP5-8.5	Net CO_2 emissions triple by 2075	2.4	4.4	3.3–5.7

Source: IPCC Sixth Assessment Report 2021

intensity is anticipated to increase across most of India (e.g. North East). From 2050, annual coral bleaching will almost certainly occur due to a rise in summer temperature. Many districts in Tamil Nadu (Nellore and Nagapattinam), Odisha (Jagatsinghpur and Kendrapara), and Gujarat (Junagadh and Porbandar) are expected to be most vulnerable to the effects of higher intensity and frequency of cyclones in India (NATCOM 2004). Sudha Rani (2017) reported the rise in sea level in India during 1973–2010 as 1.353 mm/year on the east coast and 0.372 mm/year on the west coast. By the middle of this century, the sea surface temperature adjacent to India is expected to rise by 1.5–2.0 °C. India will lose around 7.1 million persons, 5764 km^2 of land, and 4200 km of roads (NATCOM 2004).

13.2.1.2 Impact on World's Agriculture

The impacts of climate change on agriculture are unevenly distributed worldwide. Crop yield in low-latitude nations could be negatively impacted, whereas northern latitude impacts could be either positive or negative. Some vulnerable nations will face loss of arable lands, for instance, South America (1–21%), Africa (1–18%), Europe (11–17%), and India (20–40%) which could raise the risk of food insecurity. The yields will be severely affected in regions where temperatures are already close to the physiological maxima for crops (IPCC 2007). Global agricultural production is anticipated to decline between 3% and 16% by 2080. The majority of losses are centered in the lower latitudes and gains tend to be toward higher latitudes. Even if carbon fertilization benefits are considered few nations may benefit but South Asia, Latin America, and Africa will still be adversely affected (Cline 2007). Without carbon fertilization, the United States could face a 6% decline to an 8% rise in agricultural output with C fertilization. Similarly, Canada exhibits minimal losses in the absence of carbon fertilization and modest benefits in its presence. Russia could face an 8% decline in the productivity without C fertilization while rising by 6% with it. However, India would have losses varying between 30% and 40% irrespective of C fertilization. China would expect between a 7% loss and a 7% gain, similar to those of the United States, the average effects are essentially neutral (https://www.imf.org/external/pubs/ft/fandd/2008/03/cline.htm).

13.2.1.3 Impact on Indian Agriculture

The IPCC 2007 and in-house studies on the Climate Vulnerability Index (CVI) both identify India as one of the world's geographical and socio-economically vulnerable hotspots. The rate of climate change and adaptive capability will determine the impact of climate change on Indian agriculture. Since 60% of the net sown area is rainfed in India, therefore, a monsoon trend shift has a significant impact on agriculture. The wheat crop in the Indo-Gangetic Plain will be mainly impacted by the pre-monsoon changes (>0.5 °C rise between 2010 and 2039; IPCC 2007). An average 40% loss of total rice production during severe droughts (about 1 year in five) in the states of Jharkhand, Odisha, and Chhattisgarh is projected, with an estimated cost of $800 million (Pandey et al. 2007). The yields of wheat, rice, oilseeds, and legumes rise by 10–20% under the scenario of 550 ppm CO_2 levels. The yields of mustard, soybean, groundnut, wheat, and potato may decline by 3–7% for every 1 °C rise in temperature. The majority of crop productivity will drop by 10–40% by 2100 as a result of rising temperatures, erratic rainfall, and dwindling water resources. A 0.5 °C increase in winter temperature is predicted to reduce 0.45 tonnes per hectare production for rainfed wheat (Lal et al. 1998). There may have been some increase in the yields of coconut on the West Coast, as well as rabi maize, sorghum, and millets. Potato, mustard, and vegetable losses in northwestern India are lower as a result of reduced frost damage (Datta et al. 2022). Under the National Innovations in Climate Resilient Agriculture program, the effects of climate change on Indian agriculture were also investigated (NICRA) in 2011 by the Indian Agricultural Research Institute (IARI). The yield of irrigated rice will increase by 7% in 2050 and 10% in 2080 scenarios, while that of rainfed rice will reduce lightly (2.5% in each scenario). Wheat and maize crop yields are predicted to drop by 6–25% and 18–23% in 2100, respectively, whereas chickpea productivity is predicted to rise by 23–54% (Rao et al. 2019). A projected 2 °C rise in Rajasthan will result in a 10–15% decline in pearl millet production. An increase in maximum and minimum temperatures by 3 °C and 3.5 °C, respectively, will decrease soybean yields in Madhya Pradesh by 5% compared to 1998 (Singh et al. 2017). Agriculture in Gujarat's and Maharashtra's coastal regions will be most negatively impacted as arable areas are susceptible to flooding and salinization (Rao et al. 2019).

13.2.2 Agriculture/Land Use Changes as a Driver for Climate Crisis

Land use changes and shift in land use patterns have been a major driver of climate change. The link between climate and land use is complex. Land plays a dual role with respect to climate change: it acts as a sink of GHGs and also it acts as a source of the GHGs (Arneth et al. 2019). The C cycle, albedo of the Earth's surface, and consequently the radiative forcing all get impacted by the changes in land use/land cover. Thus, land use change is ranked second among the anthropogenic sources causing climate change, while burning of fossil fuels is ranked number one (Li et al. 2021). Increased demands for food, fiber, fuel, fodder, etc., which are directly supported by the land masses, have resulted in conversion of more area into agriculture and commercial forests. These changes have directly contributed to the enhanced emission of GHGs and the loss of natural ecosystems and biodiversity. Because land use changes strongly affect soil carbon stocks and trigger soil organic C depletion as CO_2 to the environment, hence, indirectly causing biodiversity loss due to substrate and habitat loss and ultimately underpins global food production (Prestele et al. 2017; Bayer et al. 2017). The impact of land use change on soil biodiversity loss is more than 75%; whereas 25% of soil biodiversity get affected by climate change (de Chazal and Rounsevell 2009). Within the last millennium, about 75% of the Earth's terrestrial area has been altered by human to perform different activities for better industrialization and civilization at the expense of environmental sustainability (https://www.unep.org/facts-about-nature). Winkler et al. (2021) reported that ~0.8 million km^2 of forest area has already been abolished on a global scale, but land use change shifted to agriculture and range land. About 1.0 million km^2 of agricultural land and 0.9 million km^2 of pasture/range land have come under recent land use change database from 1960 to 2019. Land use change and climate change together form a deadly concoction for the survival and sustenance of biodiversity in different ecosystems. According to de Chazal and Rounsevell (2009), land use change has more than 90% impact on soil biodiversity than that of climate change in tropical forests, whereas in warm mixed forests and temperate deciduous forests, the effect was more than 80%. In savannas, shrubs, and grasslands, the effect of soil biodiversity was affected to the extent of 60% due to changes in land use. Conversely, in temperate climates, typically in boreal forest, cool conifer, and tundra, land use change has only a 30% impact on soil biodiversity (de Chazal and Rounsevell 2009).

According to the National Oceanic and Atmospheric Administration (NOAA, US Dept. of Commerce), it is estimated that the global temperature rose 1.8 °F (1 °C) during the last 12 decades, which resulted rise in sea level at the rate of 1.7 mm to 3.2 mm per year (Grassi et al. 2017). However, the impact of climate change on land surface temperature is more prominent as reported by IPCC (2019), where the mean land surface air temperature has increased considerably (by 1.53 ° C) compared to the global mean surface temperature (land and ocean) (by 0.87 °C). In the Paris Agreement, land use was placed as a key component of discussion during the policy debates on climate change (Grassi et al. 2017). Since the industrial revolution, the level of atmospheric CO_2 has increased by 40% and, since 1958, the result is about 25%; that resulted in shrinkage in ice-covered areas of the Arctic region. An estimate shows that due to changes in the combined area under forestry, agriculture, and other land uses (collectively termed as AFOLU), its contribution to the total net anthropogenic emission during 2007–2016 is around 23%. Figure 13.2 illustrates the source

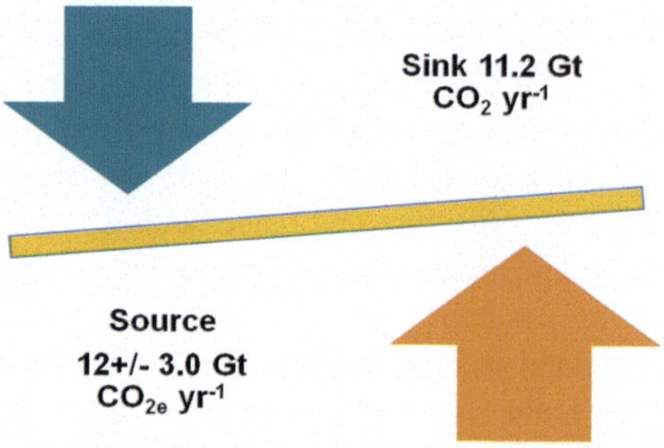

Fig. 13.2 Land (AFOLU) acting as both a source and sink of CO_2. (Adapted from IPCC (2019), Special Report)

and sink possibility of land mass with respect to the period 2007–2016 as impacted by AFOLU.

The role of land either as a sink or as a source of GHG remains very uncertain and is guided by multiple factors. For example, increased CO_2 concentration, N deposition, and an increase in the length of the growing period were able to remove atmospheric CO_2 to the tune of 11.2 +/– 2.6 Gt CO_2 year^{-1} for the period 2006–2017. However, this offset is highly uncertain because of the possibility of increased CO_2 emission from the vegetation and soil as influenced by climate change. The thawing of permafrost will increase the emission of CO_2 and C losses. Land degradation induced by climate change is also a major driver for the removal of C and CO_2 emissions from soil. Almost 30% of the C removed through soil erosion is lost to the atmosphere due to mineralization (Mandal et al. 2020) making the land mass a source of GHGs. The impact of climate change is far-flung and felt across the ecosystems. The climate crisis affects almost 80% of the land area in the Indian subcontinent and a human population to the tune of 50 million residing in these areas (Brahmananda Rao et al. 2021).

13.3 Climate-Smart Agriculture and Its Components

The climate-smart agriculture (CSA) approach increases productivity, improves livelihood, and mitigates the climate change effect. The major components of CSA are (i) increasing agricultural productivity sustainably to increase farm income and maintain food security and development; (ii) adapting and building resilience of food system to climate change; (iii) reducing GHG emission from agriculture and livestock production (FAO 2013). Climate-smart technologies are based on these outcomes and interventions should meet the above-mentioned goals. All agricultural interventions involving climate change services/solutions at field management have the potential to achieve such goals (Nyasimi et al. 2017). CSA is a holistic approach to improve crop production and food security under climate change patterns, thus mitigating climate change and associated development goals. Various CSA practices have been developed all over the world (FAO 2013). Farm-based agricultural interventions such as conservation agriculture, agroforestry, and crop residue management are adopted by stakeholders at community levels. Presently, upscaling of such CSA practices to different agro-ecological regions is required so that they have a positive impact on food security, climate change resilience, and mitigation. Climate-smart village (CSV) approach is a community approach in this direction involving researchers, farmers, line department, and other stakeholders' collaboration in selecting suitable conservation measures which are developed on a local perspective with reference to a global knowledge scale, livelihood security, animal husbandry, and achieve climate resilience and its mitigation (The Asia Foundation 2022). The CSV approach helps the farming community to adopt all smart agriculture management practices for optimal and efficient use of resources starting from seed to soil. Most CSA approaches deliver two or three climate-smart components. Other CSA practices include crop diversification, soil and water conservation practices, efficient fertilizer and manure management, tank silt application, reduced tillage practices, organic farming/natural farming practices, bio-fertilizer or microbial-mediated fertilizers, promoting bio-pesticides, integrated farm system models, efficient irrigation methods, etc. The CSA practices are aimed at reducing climate change impact on crop productivity and improving sustainable production on one hand and reduced GHG emissions on the other hand (World Bank 2016). Simple practices like integrated nutrient management, crop rotation, diversification of crops, and intercropping practices are among the most coveted CSA practices and their rate of adoption is guided by the size of farm holding, accessibility to market, extension services, weather advisories, etc. (Kifle et al. 2022). It is a well-known fact that environment-friendly farm practices such as CSA have shown a reduction in GHG up to 20% in just 1 year (USAID 2017). Thus, by reducing GHG emissions, agricultural practices such as reduced or zero tillage systems could adapt to climate change. Recently, developing countries have shown more inclination towards CSA practices for their capacity to reduce GHG emissions and ensure food security.

The CSA practices are worldwide promoted as cleaner technologies due to minimum resource exploitation through increased use efficiency, waste reduction, and recycling, and reduced gas emission and soil loss with energy savings (Athira et al. 2019; Mwalupaso et al. 2019). Many such technologies such as crop residue mulching guarantee soil cover, reduce soil moisture loss through evaporation, minimize weed problems, and soil erosion, and improve re-cycling of residues by diminishing emission due to burning; crop rotation of cereals with legumes is a common practice providing multiple benefits of nitrogen fixation saving N fertilizers, reducing pest and disease incidence; organic farming encourages efficient and effective use of available local manures encouraging waste recycling and reusing; conservation tillage reduces land disturbance, increase SOC and moisture, reducing yield loss due to rainfall variability and poor soil health. Thus, innumerable CSA approaches or farming practices are being tried in farmers fields which has both environmental and economic benefits to humans and ecosystems (Aggarwal et al. 2018; Raghuvanshi et al. 2018; Branca and Perelli 2020; The Asia Foundation 2022).

13.3.1 Climate-Smart Agricultural Technologies to Enhance Agricultural Productivity

According to the Hague Conference on Agriculture, Food Security and Climate Change (FAO of the UN 2010), CSA is an integrated approach to achieve sustainable agricultural development for food security under climate change to develop technical, policy, and investment conditions. The CSA approach is an integration of social, economic, and environmental dimensions to address food security and climate crisis. The CSA has three pillars: (i) sustainable increase of crop productivity and income, (ii) building and adapting climate resilient technologies, and (iii) reducing greenhouse gas emissions (Lipper et al. 2014). Hence, climate-smart agriculture techniques could be anything that can reduce/remove the greenhouse gas emission as well as improve crop and environmental productivity in a sustainable way. Basically, CSA encourages synchronization among researchers, producers (i.e. farmers), society, government (i.e. policymakers), and private sectors (Lipper et al. 2014).

Some of the CSA techniques could be listed as follows: (a) conservation agriculture, (b) mulching, (c) intercropping, (d) crop rotation, (e) agroforestry, (f) improved grazing, (g) integrated crop-livestock management, (h) improved water management, (i) better weather forecasting, (j) early warning system, and (k) insurance against crop failure (Totin et al. 2018; Kurgat et al. 2020). Though we are moving towards CSA but the implementation of CSA technologies is way far behind its planning. The implementation of CSA technologies is needed for strengthening local and national institutions, supporting and enabling frameworks, field-level implementation of practices, expanding evidence base, and enhancing financing options (Lipper et al. 2014). A comparative depiction of conventional and climate-smart agriculture is presented in Fig. 13.3.

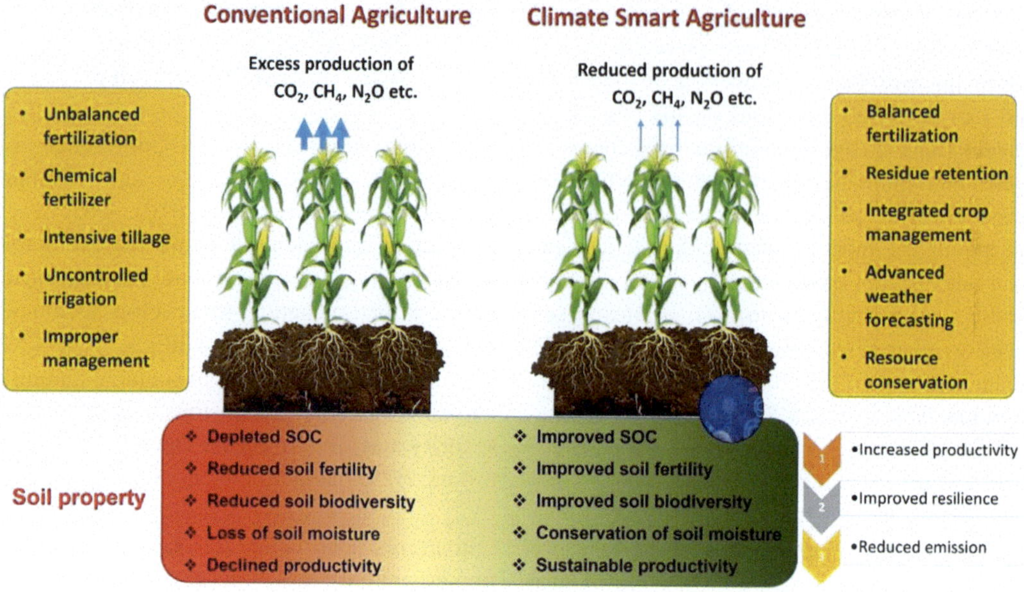

Fig. 13.3 Comparative depiction of conventional and climate-smart agriculture

13.3.1.1 CSA Technologies for Efficient Nutrient and Water Management

Conservation Agriculture and Mulching

Conservation agriculture (CA) is the combination of permanent soil cover, minimum soil disturbance, and crop rotation which is being generously promoted to improve soil quality and crop productivity at the same time (Saha et al. 2021a). In general, CA is expected to improve soil health as it provides permanent soil cover with crop residue, which retains soil nutrients to a large extent and restricts soil moisture evaporation losses. Mulching which is an important component of CA can be done via crop residue, plastic sheets, geotextiles, etc. Indeed, the use of crop residue for mulching is most beneficial and eco-friendly, which improves the microbiome (Sarkar et al. 2021a). Similar to CA, mulching with crop residue has the following benefits. Compared to no-till bare soil, it is reported that crop residue at 8 t ha^{-1} may decrease soil moisture evaporation by 30%. Crop residue mulch provides a physical barrier that reduces solar energy penetration to soil and reduces evaporation from the soil surface (Scopel et al. 2004; Lal 2008; Balwinder-Singh et al. 2011). However, the effects of mulching are more prominent when the soil is wet and plant leaves are too small to cover the entire soil surface (Lal 2008; Balwinder-Singh et al. 2011). At least, 2 t ha^{-1} of crop residue is required to gain a positive impact on infiltration of soil water and concurrent reduction in sediment loss through runoff (Ranaivoson et al. 2017). Particularly nutrient-supplying capacity of retained crop residue is directly proportional to the quantity; with an increase in the amount of crop residue retained the soil nutrient supply is also increased. A similar trend is also observed for soil organic C improvement. It was estimated that the application of crop residue at the rate of 4–5 t ha^{-1} is capable of increasing ~0.4 t ha^{-1} $year^{-1}$ soil organic C in a reduced till soil. In addition, weed emergence also decreased with the retention of crop residue. Mulching with crop residue at least at 1 t ha^{-1} can curtail the weed emergence by 50% compared to no-till bare soil (Ranaivoson et al. 2017). Effects on microflora/fauna are obvious with the retention in crop residue, even in a limited quantity. However, meta-analysis indicated that macro- and mesofauna abundance significantly improved with the incorporation of 10 t ha^{-1} of residues (Ranaivoson et al. 2017). Minimum soil disturbance along with crop residue cover improves soil biodiversity, maintains soil organic C and fertility status, and also decreases weed infestation (Corbeels et al. 2006; Teasdale and Mohler 2000; Bilalis et al. 2003; Liu et al. 2016). Though there are several benefits in terms of resource conservation using climate-smart agricultural technologies, but some authors reported that a high quantity of applied crop residues leads to slow germination, reduced crop growth, and increased pest attack (Schneider and Gupta 1985; Swanson and Wilhelm 1996).

Crop Rotation and Intercropping

Crop species, crop rotation, cropping sequence, and intercropping play a major role in sustainable soil health management and soil C sequestration (Ghosh et al. 2020). Even multi-cropping with minimal soil disturbance has reported better soil C sequestration than the mono-cropping. Particularly, root architecture differs with the species, and different species cultivated at a time resulting in soil–water and nutrient harvesting in multiple layers of soil (Biswas et al. 2022). In addition, a few crop species are capable of symbiosis with arbuscular mycorrhizae (AM) and some are compatible with N-fixing microbes in root nodules (Rhizobium spp.) (Sarkar et al. 2017, Roy et al. 2021a). Continuous retention of cereal crop residue may cause limited residue decomposition, where lignin content plays the limiting factor in SOM mineralization. On the other hand, a low C: N ratio of legumes promotes faster residue decomposition as well as improves soil nutrient status and biodiversity (Sarkar et al. 2017; Ghosh et al. 2019, 2020). Due to different crop growth cycles in intercropping, it is beneficial to grow an economical harvest without leaving the soil as fallow during any cropping season. Additionally, intercropping provides better soil cover, which ultimately reduces soil moisture and nutrient depletion (Ranaivoson et al. 2017). Apart from crop rotation, sowing time is also an important factor that helps crop establishment. Saha et al. (2020) reported that sowing rainfed lentils a fortnight before their normal sowing resulted in good soil moisture content, better crop growth, reduced phenol contents, and improved soil enzymatic activities.

Tillage and Balanced Fertilization

Rapid decomposition of initial SOC and depletion of soil moisture and nutrients are associated with improper agronomic management like tillage and unbalanced fertilization. Typically, tillage plays the biggest role in SOC restoration. Method of tillage, depth of tillage, and frequency of tillage determine the rate of SOC depletion (Triberti et al. 2016). Intensive and more frequent tillage accelerates SOC depletion. Several authors reported that with conventional tillage, exponential organic C declines were reported. According to Janzen et al. (1998) and Follett (2001), intense soil disturbance promotes exponential SOC loss, but interruption stabilizes it to an equilibrium phase where net input and output of SOC balance each other and subsequently a gradual improvement in SOC. The equilibrium phase may come approximately after 40–60 years after interruption (West and Post 2002; Jarecki and Lal 2003). In temperate Europe, the steady state of SOC generally comes after 100 years when a forest cover turns to agriculture (Jenkinson 1988; Smith et al. 1996).

In a given cropping sequence along with tillage, fertilization (i.e. rate and quantity of fertilizer application) also triggers SOC depletion and GHG emission (Ghosh et al. 2020). The availability of N, P, and S along with the substrate C is essential to build up SOC and sequester C in any soil (Roy et al. 2021b). Though higher biomass production inorganic fertilizer improves SOC content and also promotes soil C sequestration. In experiments without initial fertilization, the SOC content remains low; however, balanced fertilization with conventional tillage even slightly increased SOC (Ghosh et al. 2020). The effects of balanced fertilization on SOC restoration improved crop production and can further be achieved by using organic manures. The application of organic manures like FYM, compost, enriched compost, and municipal sludge hastens soil microbial growth and improves soil fertility and biodiversity (Triberti et al. 2016; Ghosh et al. 2019; Roy et al. 2019). A recommended dose of

K-fertilizer along with zero-tillage in rainfed lentils is capable of escaping from terminal droughts and improving soil fertility status (Saha et al. 2021b). The application of biochar (a C-rich material developed from the pyrolysis) is also beneficial to revamp weathered tropical soils, maintain soil fertility, and improve soil C sequestration (Basak et al. 2022).

Controlled-Release Fertilizer

CRFs are gaining popularity in the present era of climate change. The CRFs are the fertilizers that have controlled release of nutrients from the fertilizer in a manner to synchronize with crop nutrient demands (Trenkel 2010). Hence, it reduces nutrient losses. Sarkar et al. (2018) developed acidulated rock phosphate-based CRFs which resulted in increased wheat growth and improved residual soil phosphorus content. Similar findings were reported by Roy et al. (2015, 2018) when organic acid-loaded nano clay polymer composite improved the solubilization of Indian rock phosphate and ultimately increased wheat production in sub-tropical climatic conditions. Other than that, several biodegradable polymer-based coating agents and oil-based coating agents are also capable of reducing nutrient release from the fertilizers (Sarkar et al. 2021b, c).

13.3.1.2 Precision Agriculture as a Tool for CSA

The concept of precision agriculture (PA) is not a new one and is based on the age-old agricultural principles of applying the right input at the right place and at the right time in measured quantity (Roy and George 2020). The optimization of inputs be it fertilizers, other agrochemicals, water, seed, energy, etc. considering that the heterogeneity of the land surface is the background of PA. The variable rate technology (VRT) that aids in varying the quantity of inputs helps to reduce the overuse of fertilizers, pesticides, and other agrochemicals, thereby optimizing the energy expenditure and reducing GHG emissions. Precision agriculture has been identified as one of the key CSA technologies by the International Food Policy Research Institute to achieve yield gain in major stable crops like wheat, rice, and maize in South Asia, Latin America, and Africa, respectively (IFC 2017). The agriculture sector will face major water scarcity and, by 2025, two-thirds of the world's population will be witnessing a water crisis (Boretti and Rosa 2019). Far-sightedness should be employed to use the limited water resources more efficiently with the use of variable rate irrigation techniques, and the adoption of micro-irrigation systems like sprinklers, drips, etc. will allow farmers to save water and energy, and use nutrients more effectively. "Water Hand", an innovation under the project "Green Innovation Ventres for Agriculture and Food Sector" is one such device. It collects data on crop life cycle, soil moisture, and local weather advisory and waters the crop as per its need. This can lead to a yield increase of 30% while saving up to 60% water (https://farmingfirst.org/2020/02/solving-the-water-crisis-by-saving-money-in-agriculture/).

Emissions from both the livestock and the crop sector are taken care of through PA. However, methane which is the major GHG of concern emitted from the livestock sector is 10 times less potent in causing global warming than the nitrous oxide emitted from crop fields. Thus, precision nitrogen management and precision nutrient management can go a long way in reducing the impact of agriculture on climate change. Bates et al. (2009) observed that the application of variable rate nitrogen had the potential to reduce N_2O emission by 5% compared to the baseline emission obtained through adding optimal chemical N fertilizer. About 1.19% of the total N added to soil is released as N_2O and decreasing the quantity of N input reduces N_2O release to the tune of 1.25% of the quantity of N saved (Babcock et al. 2004). The application of the VRNA technology combining the fertilizer and irrigation systems (through drip irrigation) can result in a net benefit of up to 310 Euro per ha per year in addition to the rewards of reduced GHG emission (Balafoutis et al. 2017). The VRT can reduce the overall GHG gas emission by 5–10% and plays a pivotal role in dealing with the climate crisis in the agriculture sector.

13.3.1.3 Soil Conservation Measures for Adopting Climate Change

Soil conservation is an essential practice that helps maintain the fertility of the soil and supports the growth of crops. Climate change has made it necessary to adopt new techniques that take into account the changing weather patterns and changing soil conditions. Climate-smart soil conservation techniques are designed to help farmers, landowners, and other stakeholders conserve soil in an environmentally sustainable and economically viable manner.

To mitigate the negative impacts of climate, it is essential to adopt soil conservation techniques that not only protect the soil but also enhance its productivity, resilience, and carbon sequestration potential. In this context, the concept of "climate-smart soil conservation practices" (CSSC) has emerged as a promising approach to ensure sustainable soil management. CSSC involves the integration of climate change considerations into soil conservation practices, thereby promoting sustainable land use and supporting the resilience of ecosystems and communities (https://www.worldbank.org/en/topic/climate-smart-agriculture).

CSSC is defined as a soil conservation approach that aims to enhance the resilience of soil and ecosystems, as well as support the adaptation of communities and livelihoods to the impacts of climate change. The objective of CSSC is to promote sustainable land use and management practices to combat/reduce the impact of climate change.

CSSC integrates the principles of soil conservation with those of climate change adaptation and mitigation, which is achieved through the integration of knowledge and best practices from both fields. The approach recognizes the interdependence among soil conservation, sustainable land use, and climate change, and seeks to promote mutually reinforcing actions that address these issues in a holistic manner.

Principles of Climate-Smart Soil Conservation Practices (World Bank 2016).

The principles of CSSC include the following:

(i) Integration of climate change considerations into soil conservation practices: CSSC recognizes that soil degradation and erosion are exacerbated by climate change and that soil conservation measures must be adapted to address the changing climate.
(ii) Promotion of sustainable land use and management practices: CSSC supports the adoption of sustainable land use practices that contribute to the reduction of greenhouse gas emissions and enhance the resilience of ecosystems and communities.
(iii) Support for community-based and participatory approaches: CSSC recognizes the importance of local communities and stakeholders in soil conservation efforts and supports the integration of their knowledge and perspectives into soil conservation planning and implementation.
(iv) Enhancement of soil productivity and resilience: CSSC aims to promote soil conservation practices that enhance the productivity and resilience of soil and ecosystems, as well as support the adaptation of communities and livelihoods to the impacts of climate change.

Climate-Smart Soil Conservation Techniques

There are several CSSC techniques that can be adopted to conserve and enhance the soil, as well as support the adaptation of communities and livelihoods to the impacts of climate change (https://www.nrcs.usda.gov/conservation-basics/natural-resourceconcerns/climate/climate-smart-mitigation-activities; FAO, Climate-Smart Agriculture Source Book 2013; Cerri et al. 2021). These techniques include the following:

(i) Agroforestry: Agroforestry is a land-use system that integrates trees and crops into a single management unit, which provides multiple benefits such as soil conservation, water management, and carbon sequestration.

(ii) Terrace farming: Terrace farming involves the creation of terraces on sloping lands, which helps to reduce soil erosion, improve water retention, and enhance soil fertility and productivity.
(iii) Conservation tillage: Conservation tillage involves reducing tillage operations and leaving crop residues on the soil surface, which helps to reduce soil erosion and improve soil health by increasing organic matter and promoting the growth of beneficial soil microbes.
(iv) Cover cropping: Cover cropping involves planting cover crops between main crops, which helps to reduce soil erosion, conserve soil moisture, and improve soil health by adding organic matter and nutrients to the soil.
(v) Rainwater harvesting: Rainwater harvesting involves collecting and storing rainwater for later use, which helps to conserve soil moisture and improve soil fertility by reducing water stress and promoting the growth of crops.
(vi) Integrated soil fertility management: Integrated soil fertility management involves the use of a combination of organic and inorganic fertilizers, as well as other soil management practices, to enhance soil fertility and productivity.
(vii) Soil erosion control: Soil erosion control involves the use of physical and vegetative methods to control soil erosion, such as contour plowing, intercropping, and the planting of cover crops.

Benefits of climate-smart soil conservation: The adoption of CSSC has several benefits, including the following:

(i) Improved soil health and productivity: CSSC helps to conserve and enhance the soil, thereby improving soil health and productivity and supporting the growth of crops.
(ii) Increased resilience to climate change: CSSC helps to enhance the resilience of ecosystems and communities to the impacts of climate change, including droughts, floods, and extreme weather events.
(iii) Improved food security: CSSC supports the sustainable production of food, thereby contributing to food security and reducing hunger.
(iv) Enhanced carbon sequestration: CSSC promotes the growth of crops and vegetation, which helps to sequester carbon and reduce greenhouse gas emissions.
(v) Improved livelihoods and income: CSSC supports the development of sustainable land use practices, which can contribute to improved livelihoods and income for local communities.

Challenges in the implementation of climate-smart soil conservation: The implementation of CSSC is not without its challenges, including:

(i) Lack of awareness and understanding: There is a lack of awareness and understanding of CSSC among communities and stakeholders, which can hinder the adoption of CSSC practices.
(ii) Limited resources: CSSC requires resources such as funding, technical expertise, and equipment, which can be scarce in many communities and countries.
(iii) Resistance to change: There may be resistance to the adoption of CSSC practices, particularly among traditional farmers who are used to traditional methods of soil conservation.
(iv) Institutional barriers: Institutional barriers, such as conflicting policies and regulations, can hinder the implementation of CSSC.

13.3.1.4 Crop Simulation Models for CSA

Crop simulation models or crop growth simulation models (CSM) are computer-based programs making use of different mathematical equations to simulate crop/plant growth, their development processes, and yield, in relation to or as a function of soil conditions, weather as well as crop management (Hoogenboom et al. 2004). Such models are also widely being employed to assess the variations in growth parameters as well as yield of different cultivars of a crop under the same set of soil, weather, and management parameters making use of the different equations describing plant physiology. CSMs use different sets of differential equations to calculate rate as well as state variables related to plant growth over time, usually ranging from planting until the harvesting stage or final harvest (Murthy 2004).

Due to their inherent ability to effectively assess the impact of various weather parameters on crop growth/development as well as yield, CSMs form an important component in our efforts towards practicing climate-smart agriculture. For studying climate change-related yield response variations of major crops, CSMs are coupled with various global or regional climate change models and run under a diverse range of widely accepted emission scenarios. Projected crop yield changes of various major crops (namely rice, wheat, maize, and soy) due to the impact of climate change over the twenty-first century are mainly generated using simulations from various CSMs (IPCC 2014, 2022). Such projections involve coupling of different CSMs like CERES (Crop Environment Resource Synthesis), APSIM (Agricultural Production Systems Simulator), etc. to suit a large number of widely accepted advanced GCMs and impact the evaluation under the standard range of various IPCC emission scenarios involving mitigation and adaptation strategies (Defra 2005). Such attempts not only address the direct impact of climate change on crop growth/yield but also simulate the interaction effects of changes in various climatic parameters (namely, rainfall, temperature, and radiation) with pest infestation, irrigation practices, soil fertility, agronomic management strategies, etc. Similarly, yield response variations of a particular crop under a specific climate change scenario under different geographical/climatic regions could also be assessed in detail by the use of CSMs (Hodson and White 2010). Such simulation results/findings provide vital information helping in future decision-making processes in a much better way.

Many CSMs could assess the impact on agricultural systems too in addition to individual crops. Subash and Ram Mohan (2012) have evaluated the impact of climatic trends and variability on the productivity of rice–wheat cropping system over the Indo-Gangetic Plains of India employing Decision Support System for Agrotechnology Transfer (DSSAT) crop model. Gaydon et al. (2017) evaluated the ability of the APSIM model to simulate the performance of various cropping systems in the Asian region, considering different dimensions such as crop phenology, water use, soil dynamics, and climate change-associated impacts. Apart from assessing crop responses to abiotic factors, CSMs like CERES-RICE could be used to assess the impact of biotic factors on crop growth under future climate scenarios (Luo et al. 1998). These attempts will aid us in planning and implementation of different management strategies well in advance, thus reducing the adverse impacts of pests and diseases.

CSMs capable of genetic simulation play an important role in developing improved crop varieties that are capable of performing better under climate change scenarios (Jha et al. 2023). Such CSMs provide vital genetic data of different crops to the plant breeding community, enabling them in the identification of numerous genes responsible for various desirable traits with respect to climate change impacts and identifying best breeding methodologies for taping their potentialities. Model-derived data will provide insights into genetic responses in the form of various biochemical mechanisms as well as gene interactions under differing micro/macroenvironmental conditions (Wang 2012). Such information are an integral component of our efforts towards the development of newer climate-resilient varieties with various desirable characteristics such as drought tolerance, increased pest/disease resistance, high yielding, and improved tolerance to abiotic stress conditions like salinity. Models capable of genetic simulation could also be used for studying the genetic response of different wild varieties/relatives of major crops towards long-term changes in climatic parameters, for prospective screening of beneficial genetic characteristics.

CSMs also provide a useful tool for identifying the environmental niches as well as priority areas of wild relative diversity with respect to major cultivated crops, and their variations in response to changing climate scenarios. Such attempts could also provide vital information regarding the likely extinction or habitat reduction of various related species in the future by the incorporation of future climate data into CSMs. Jones and Beebe (2001) studied the impact of climate change on wild bean habitat in Central America using the FloraMap™ model with HadCM3 model-derived climate data for 2055. The model simulations indicated the virtual disappearance of suitable habitats by 2055, in five out of the seven countries included in the study. By 2055, similar trends like high chances of extinction, increased fragmentation of suitable habitats, reduction in range sizes, etc. were reported for related wild species of different crops such as groundnut, potato, and cowpea under future climate scenarios (Jarvis et al. 2001; Jarvis et al. 2008).

13.4 CSA for Achieving Sustainable Development Goals

Adoption of the CSA practices can successfully help in achieving all 17 Sustainable Development Goals (SDGs). Since agriculture is an important component of the SDGs and most of the developing and least developing nations still have a major workforce engaged in agriculture and almost 30% of the Gross Domestic Product (GDP) is attributed to it (Stromquist 2019), thus, strengthening the agricultural sector as an important priority. The different SDGs and their associated targets make it very clear that the societal developmental aspect cannot be separated from the sustainable development of the ecosystem as a whole. Uplifting the agricultural sector will have a key role in achieving the targets of SDG.

The Nationally Determined Contributions (NDCs) of various countries were analyzed by the Food and Agricultural Organization (FAO), which indicated that 131 countries highlighted the adaptation and/or mitigation in the agricultural sector for achieving the targets. Most of the SDGs prioritize climate change, agricultural production, natural resources and ecosystems, and income/food security, and CSA dwells at the nexus of all these targets. The CSA is primarily based on three pillars:

- to increase agricultural production and income sustainably,
- to develop resilience against climate change/climate proofing, and
- to bring down the emission of GHG or curb GHG emission in totality.

These three pillars are again categorized on the basis of some actionable points or from the implementation point of view which is presented in Fig. 13.4. This figure gives the framework for mapping and assessing the CSA-SDG interlinkages.

All three strategies of CSA allow a triple-win situation; however, not at all times, the three objectives will be simultaneously met. So, it is the vision of the implementer and expert to balance between the advantages and trade-offs so that the trade-offs are minimized. India has set the target of achieving Land Degradation Neutrality (LDN) by 2030 through rehabilitation of 26 Mha lands which is one of the prime objectives of SDG. The adoption of CSA will also help in achieving this goal for the nation.

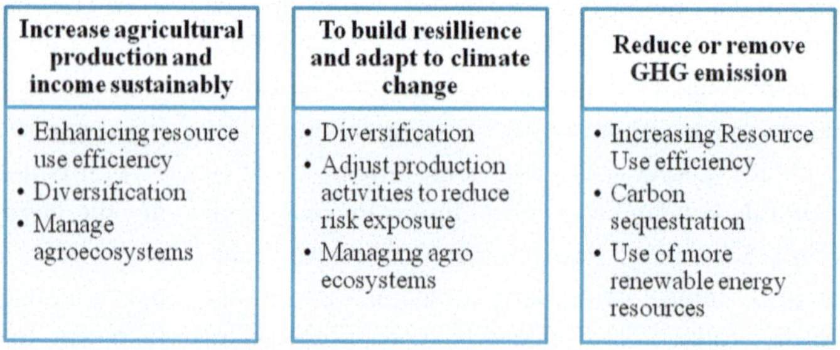

Fig. 13.4 The pillars of CSA and the action points to achieve them and set linkage with the SDGs

13.5 Strategies to Popularize CSA for Better Adoption Among Farmers

Climate change has a direct consequence on the productivity and sustainability of all agricultural production systems. In India, agriculture contributes to 16% of the country's GDP; in which, 4.5–9% of a negative effect of crop production indicates that the cost of climate change can be up to 1.5% every year (Raghuvanshi et al. 2018). Therefore, to alleviate such complex challenges, agriculture should become "climate-smart" so as to sustain crop productivity and economic security, adapt and build climate-resilient system, and reduce GHG emissions. The CSA practices are flexible options that provide context-specific interventions, by preserving the ecosystem and improving production. However, successful implementation and achievement of CSA approaches require collaboration among farmers, researchers, consumers, policymakers to identify monitorable, and achievable indicators for overcoming competing resource demands (World Bank 2016).

Scaling-up CSA technologies bring more benefits to more people over large areas which could be more lasting and equitable. Moreover, there is a need to involve a large number of farmers from developing countries to adopt CSA practices which would also involve new and other innovative approaches for large-scale adoption. According to Wambugu et al. (2014) and Gikunda et al. (2021), creating awareness among the farming community and capacity building among farmers are the important steps to ensure the success of CSA adoption and its dissemination. In India, climate changes have shown a damaging effect on agriculture and crop production. The Indian farming community is slowly adopting new approaches for transforming their farming practices to survive the climate change effect. Various advanced systems such as artificial intelligence (AI), drones and sensor-based technology enhance climate services by providing timely and accurate information to farmers, aiding them to take up climate-resilient agriculture management practices and decisions. Such information can make both agriculture and livestock production remain viable under changing climate patterns (The Asia Foundation 2022). Strong extension programs and outreach activities directly involving the farming community could provide necessary information thereby enhancing their capacity to improve crop production, so as to adapt to climate change effects (Raghuvanshi et al., 2018). Although it has been demonstrated through various studies that the adoption of CSA approaches improves natural resources and ecosystem services and ensures sustainable food security, but in many developing countries, the adoption is low. Some of the major constraints associated with lower adoption are scarcity and ownership of resources such as land and water, policy and implementation strategy, poor infrastructure, demographic, and social and other cultural factors (USAID 2017; Kifle et al. 2022). The economic constraints are highlighted as a major setback towards proactive adoption of CSA technologies (Mbow et al. 2014; Aggarwal et al. 2018; Raghuvanshi et al. 2018). A few strategies for better adoption and implementation of CSA practices at the farm level (Aggarwal et al. 2018; Raghuvanshi et al. 2018) are as follows:

(i) *Creating awareness*: Awareness campaigns about the latest technological interventions in agriculture and livestock is the most important step in popularizing CSA practices among small and resource-poor farmers. Sensitizing the farming community about climate change and its consequences on agricultural production. For example, creating awareness about CSA practices such as the SRI method of rice cultivation, preparation of seasonal crop calendar, efficient water management, encouraging rural women for goat and poultry farming, and resource mapping.

(ii) *Capacity building programs*: Regular training programs on the latest technologies of CSA could be organized at the farm level to sensitize farmers. This would provide adaption and mitigation practices to farmers, and training field extension workers is also done to make them aware of CSA practices and their implementation.

Skill training programs may be conducted for farmers, including farm women through national institutions, KVKs, NGOs, and SAUs across the country.

(iii) *Contingency planning:* Village or community level land use and contingency planning is essential for current and future climate change patterns based on agro-ecological and socio-economic conditions. Contingency planning for agriculture and allied sectors is an important step for the successful adoption of CSA practices for crop production, horticulture, agroforestry, and livestock. Such planning includes information on erratic weather patterns such as heat waves, cold waves, droughts, floods, hailstorms, untimely and extreme rainfall events, long dry spells, pest and disease incidences, etc.

(iv) *Planning and participation*: Strategic guidelines before (based on seasonal weather forecast) and during the adoption of CSA practices, services, processes, and other related information on market, labor, capital, etc. should be provided in advance and during the implementation of programs. Such activities are encouraged with local community participation, self-help groups, water user associations, extension systems, local private sectors, etc.

(v) *Weather information*: Timely and accurate real-time weather information and value-added information and communication technology-based advisories such as warning of extreme events, dry spells, use of good quality inputs, use of organic amendments, improving nutrient and water use efficiency, minimizing pesticide usage, and risk transfer through insurance schemes for crop and livestock losses.

(vi) *Field visits and exposure*: Farmers should be exposed to climate-smart technologies such as drought-tolerant varieties, mulching, less water-intensive cropping, intercropping systems, integrated farming system models, poultry rearing, polyhouse cultivation, etc. should be promoted. Field visits at progressive farmer's sites to explain the benefit of in situ soil moisture conservation, successful farm models, etc. will encourage farmers through a "learning by doing" approach.

(vii) *Field demonstration*: Front line and on-farm demonstration of SRI method, conservation agriculture and zero tillage with Happy seeder, organic fertilizer, vermicompost preparation, silage making, etc. will demonstrate the adoption and mitigation strategies of climate change.

(viii) *Gender equity and social inclusion*: This technique involves women in addressing issues on drought conditions such as groundwater depletion, low crop productivity, irregular cash flow, and increased climate risk. This is a community-based approach that focuses on building resilience among women farmers through awareness, knowledge sharing, and facilitating various government schemes for improving farming and allied livelihood.

(ix) *Custom hiring*: Village-level custom hiring approaches empower farmers to manage the shortage of labor and improve the efficiency of various agricultural operations.

(x) *Policy issues*: Policy guidance on the options and obstacles of CSA practices at local and national levels, including financial assistance for scaling-up practices at the field level. Various government initiatives in this direction are now in place in our country. Programs like the National Mission for Sustainable Agriculture (NMSA), Pradhan Mantri Krishi Sinchayi Yojna (PMKSY), per drop more crop, Soil Health Card, etc. These programs implemented in the right place with clear-cut objectives will help make the agriculture climate resilient.

13.6 Conclusions

Shielding agriculture from the uncertainties of climate change and, at the same time, making it more resilient and climate positive is a challenge. Tackling this challenge effectively depends upon the integration and amalgamation of various climate-smart practices which can be a real troubleshooter. The three pillars of CSA are often interlinked and provide a system approach for tackling the climate challenge. The projections through different SSPs indicate a wide range of scenarios and to deal with those the CSA measures need to be effectively implemented. The various CSA techniques discussed in this chapter are not some out-of-the-box techniques but it only involves the prudent judgment and use of the existing technologies region-wise for climate-proofing of agriculture at a regional scale. The farmers need to be at the center stage of implementing the CSA techniques and, for this evidence-based data, awareness programs, incentives, community approaches, and linking of CSA with existing government programs like PMKSY, NMSA, etc. should be done. Holding the hand of CSA, the country can achieve the targets of SDGs and land degradation neutrality as well which has been laid down by the UN for 2030.

References

Aggarwal PK, Jarvis A, Campbell BM, Zougmoré RB, Khatri-Chhetri A, Vermeulen SJ, Loboguerrero A, Sebastian LS, Kinyangi J, Bonilla-Findji O, Radeny M, Recha J, Martinez-Baron D, Ramirez-Villegas J, Huyer S, Thornton P, Wollenberg E, Hansen J, Alvarez-Toro P, Aguilar-Ariza A, Arango-Londoño D, Patiño-Bravo V, Rivera O, Ouedraogo M, Tan Yen B (2018) The climate-smart village approach: framework of an integrative strategy for scaling up adaptation options in agriculture. Ecol Soc 23(1):14. https://doi.org/10.5751/ES-09844-230114

Amundson R et al (2015) Soil and human security in the 21stcentruy. Science 348:1261071. https://doi.org/10.1126/science1261071

Anderson R, Bayer PE, Edwards D (2020) Climate change and the need for agricultural adaptation. Curr Opin Plant Biol 56:197–202

Arneth A, Barbosa H, Benton T, Calvin K, Calvo E, Connors S et al (2019) IPCC special report on climate change, desertification, land degradation, sustainable land management, food security, and greenhouse gas fluxes in terrestrial ecosystems, Summary for policy makers. Intergovernmental Panel on Climate Change (IPCC), Geneva, Switzerland

Athira G, Bahurudeen A, Appari S (2019) Sustainable alternatives to carbon intensive paddy field burning in India: a framework for cleaner production in agriculture, energy, and construction industries. J Clean Prod 236:117598. https://doi.org/10.1016/j.jclepro.2019.07.073

Babcock BA, Paustian K, Hatfield J, Kling CL, Lal R, McCarl B, McLaughlin S, Mosier A, Post W, Robertson GP (2004) Climate change and greenhouse gas mitigation: challenges and opportunities for agriculture. Council on Agricultural Science and Technology (CAST), Ames, IA, USA

Bal PK, Ramachandran A, Palanivelu K, Thirumurugan P, Geetha R, Bhaskaran B (2016) Climate change projections over India by a downscaling approach using PRECIS. Asia-Pac J Atmos Sci 52:353–369

Balafoutis A, Bert B, Fountas S, Vangeyte J, Wal T, Soto I, Gómez-Barbero M, Barnes A, Eory V (2017) Precision agriculture technologies positively contributing to GHG emissions mitigation, farm productivity and economics. Sustainability 9(8):1339

Balwinder-Singh, Eberbach PL, Humphreys E, Kukal SS (2011) The effect of rice straw mulch on evapotranspiration, transpiration and soil evaporation of irrigated wheat in Punjab, India. Agric Water Manag 98:1847–1855. https://doi.org/10.1016/j.agwat.2011.07.002

Basak BB, Sarkar B, Saha A, Sarkar A, Mandal S, Biswas JK, Wang H, Bolan NS (2022) Revamping highly weathered soils in the tropics with biochar application: what we know and what is needed. Sci Total Environ 822:153461. https://doi.org/10.1016/j.scitotenv.2022.153461

Bates J, Brophy N, Harfoot M, Webb J (2009) Sectoral emission reduction potentials and economic costs for climate change (SERPEC-CC). In: Agriculture: methane and nitrous oxide. Ecofys Netherlands, Utrecht, The Netherlands

Bayer AD, Lindeskog M, Pugh TAM, Anthoni PM, Fuchs R, Arneth A (2017) Uncertainties in the land-use flux resulting from land-use change reconstructions and gross land transitions. Earth Syst Dynam 8(1):91–111. https://doi.org/10.5194/esd-8-91-2017

Bilalis D, Sidiras N, Economou G, Vakali C (2003) Effect of different levels of wheat straw soil surface coverage on weed flora in Viciafaba crops. J Agron Crop Sci 189:233–241

Biswas SS, Biswas DR, Ghosh A, Sarkar A, Das A, Roy T (2022) Phosphate solubilizing bacteria inoculated rock phosphate can supplement P fertilizer to grow wheat in sub-tropical inceptisol. Rhizosphere 23:100565. https://doi.org/10.1016/j.rhisph.2022.100556

Boretti A, Rosa L (2019) Reassessing the projections of the world water development report. NPJ Clean Water 2(1):15

Brahmananda Rao V, Ashok K, Govardhan D (2021) Unprecedented climate change in India and a three-pronged method for reliable weather and climate prediction. Front Clim 3:716507. https://doi.org/10.3389/fclim.2021.716507

Branca G, Perelli C (2020) Clearing the air: common drivers of climate-smart smallholder food production in Eastern and Southern Africa. J Clean Prod 270:121900

Cerri CEP, Cherubin MR, Damian JM, Mello FFDC, Lal R (2021) Fact sheets: best practices for a climate smart agriculture, Inter-American Institute for Cooperation on Agriculture (IICA)

Cline WR (2007) Global warming and agriculture: Impact estimates by country, Peterson Institute Press: All Books, Peterson Institute for International Economics, number 4037, July

Corbeels M, Scopel E, Cardoso A et al (2006) Soil carbon storage potential of direct seeding mulch-based cropping systems in the Cerrados of Brazil. Glob Chang Biol 12:1773–1787. https://doi.org/10.1111/j.1365-2486.2006.01233.x

Datta P, Behera B, Rahut D, B. (2022) Climate change and Indian agriculture: a systematic review of farmers' perception, adaptation, and transformation. Environmental Challenges 8:100543

De Chazal J, Rounsevell MDA (2009) Land-use and climate change within assessments of biodiversity change: a review. Glob Environ Chang 19(2):306–315. https://doi.org/10.1016/j.gloenvcha.2008.09.007

Department for Environment, Food and Rural Affairs (Defra) (2005) Climate change impacts on agriculture in India. Available at: http://www.defra.gov.uk/environment/climatechange/internat/devcountry/pdf/india-climate-6-agriculture.Pdf

FAO (2013) Statistical yearbook: world food and agriculture. Food and Agriculture Organization of the United Nations (FAO), Rome

Follett RF (2001) Soil management concepts and carbon sequestration in cropland soils. Soil Tillage Res 61:77–92

Food and Agriculture Organization of the United Nations (2010) "Climate-Smart" agriculture, policies, practices and financing for food security, adaptation and mitigation

Gaydon DS, Balwinder-Singh, Wang E, Poulton PL, Ahmad B, Ahmed F, Akhter S, Ali I, Amarasingha R, Chaki AK, Chen C, Choudhury BU, Darai R, Das A, Hochman Z, Horan H, Hosang EY, Kumar PV, Khan ASMMR, Laing AM, Liu L, Malaviachichi MAPWK, Mohapatra KP, Muttaleb MA, Power B, Radanielson AM, Rai GS, Rashid MH, Rathanayake WMUK, Sarker MMR, Sena DR, Shamim M, Subash N, Suriyagoda LDB, Wang G, Wang J, Yadav RK, Roth CH (2017) Evaluation of the APSIM model in cropping systems of Asia. Field Crop Res 204:52–75. https://doi.org/10.1016/j.fcr.2016.12.015

Ghosh A, Kumar S, Manna MC, Singh AK, Sharma P, Sarkar A, Saha M, Bhattacharyya R, Misra S, Biswas SS, Biswas DR, Gautam K, Kumar RV (2019) Long-term in situ moisture conservation in horti-pasture system improves biological health of degraded land. J Environ Manag 248:109339. https://doi.org/10.1016/j.jenvman.2019.109339

Ghosh A, Misra S, Bhattacharyya R, Sarkar A, Singh AK, Tyagi VC, Kumar RV, Meena VS (2020) Agriculture, dairy and fishery farming practices and greenhouse gas emission footprint: a strategic appraisal for mitigation. Environ Sci Pollut Res 27:10160–10184

Gikunda RM, Lawver DE, Baker M, Boren-Alpizar AE, Guo W (2021) Extension education needs for improved adoption of sustainable organic agriculture in Central Kenya. Am J Geogr Inf Syst 10:61–71

Grassi G, House J, Dentener F, Federici S, den Elzen M, Penman J (2017) The key role of forests in meeting climate targets requires science for credible mitigation. Nat Clim Chang 7(3):220–226. https://doi.org/10.1038/nclimate3227

Hermans K, McLeman R (2021) Climate change, drought, land degradation and migration: exploring the linkages. Curr Opin Environ Sustain 50:236–244

Hassania A, Azapagica A, Shokri N (2020) Predicting long-term dynamics of soil salinity and sodicity on a global scale. PNAS 117(52):33017–33027

Hodson D, White J (2010) GIS and crop simulation modelling applications in climate change research. In: Climate change and crop production. CABI Publishing, pp 245–262. https://doi.org/10.1079/9781845936334.0245

Hoogenboom G, White JW, Messina CD (2004) From genome to crop: integration through simulation modelling. Field Crop Res 90(1):145–163

Houghton RA (2007) Balancing the global carbon budget. Annu Rev Earth Planet Sci 35:313–347

Huang J, Yu H, Guan X et al (2015) Accelerated dryland expansion under climate change. Nat Clim Change 6:166–171

International Finance Corporation, IFC, (2017) https://www.ifc.org/wps/wcm/connect/

IPCC (2001) In: Houghton JT, Ding Y, Griggs DJ, Noguer M, van der Linden PJ, Dai X, Maskell K, Johnson CA (eds) Climate change 2001: the scientific basis. Contribution of working group I to the third assessment report of the intergovernmental panel on climate change. Cambridge University Press, Cambridge, United Kingdom and New York, NY, USA, 881pp

IPCC (2007) Climate change 2007: the physical science basis. Contribution of working group I to the fourth assessment report of the intergovernmental panel on climate change. Cambridge University Press, Cambridge

IPCC (2014) Climate change 2014: synthesis report. Contribution of working groups I, II and III to the fifth assessment report of the intergovernmental panel on climate change [Core Writing Team, R.K. Pachauri and L.A. Meyer (eds.)]. IPCC, Geneva, Switzerland, 151 pp

IPCC (2019) In: Shukla PR, Skea J, Buendia EC, Masson-Delmotte V, Pörtner H-O, Roberts DC, Zhai P, Slade R, Connors S, van Diemen R, Ferrat M, Haughey E, Luz S, Neogi S, Pathak M, Petzold J, Pereira JP, Vyas P, Huntley E, Kissick K, Belkacemi M, Malley J (eds) Climate change and land: an IPCC special report on climate change, desertification, land degradation, sustainable land management, food security, and greenhouse gas fluxes in terrestrial ecosystems

IPCC (2021) Climate change 2021: the physical science basis. Contribution of working group I to the sixth assessment report of the intergovernmental panel on climate change. Cambridge University Press, Cambridge

IPCC (2022) Annex VI: acronyms. In climate change 2022: mitigation of climate change. Contribution of working group III to the sixth assessment report of the intergovernmental panel on climate change [P.R. Shukla, J. Skea, R. Slade, A. Al Khourdajie, R. van Diemen, D. McCollum, M. Pathak, S. Some, P. Vyas, R. Fradera, M. Belkacemi, A. Hasija, G. Lisboa, S. Luz, J. Malley, (eds.)]. Cambridge University Press, Cambridge and New York

Jacinthe PA, Lal R (2001) A mass balance approach to assess carbon dioxide evolution during erosional events. Land Degrad Dev 12(4):329–339

Janzen HH, Campbell CA, Izaurralde RC, Ellert BH, Juma N, McGill WB, Zentner RP (1998) Management effects on soil C storage on the Canadian prairies. Soil Tillage Res 47:181–195

Jarecki MK, Lal R (2003) Crop management for soil carbon sequestration. Crit Rev Plant Sci 22(6):471–502

Jarvis A, Jones P, Mottram G, Williams D, Guarino L, Ferguson M (2001) Predicting the impact of climate change on the distribution of plant genetic resources in wild peanuts. Poster presented at the conference on challenges of a changing earth – global change Open Science conference, 10–13 July, Amsterdam

Jarvis A, Lane A, Hijmans RH (2008) Impacts of climate change on crop wild relatives. Agric Ecosyst Environ 126(1–3):13–23

Jenkinson DS (1988) Soil organic matter and its dynamics. In: Wild A (ed) Russell's soil conditions and plant growth, 11th edn. Longman, London, pp 564–607

Jha PK, Prasad PVV, Araya A, Ciampitti IA (2023) Estimation of crop genetic coefficients to simulate growth and yield under changing climate. In: Global agricultural production: resilience to climate change. Springer, Cham, pp 283–309

Jobbagy EG, Jackson RB (2000) The vertical distribution of soil organic carbon and its relation to climate and vegetation. Ecol Appl 10:423–436

Jones PG, Beebe S (2001) Predicting the impact of climate change on the distribution of plant genetic resources in wild common bean (Phaseolus vulgaris L.) in Central America. Paper presented at the III international conference on geospatial information in agriculture and forestry, 5–7 July, Denver, Colorado

Kifle T, Ayal DY, Mulugeta M (2022) Factors influencing farmers adoption of climate smart agriculture to respond climate variability in Siyadebrina Wayu District, Central Highland of Ethiopia. Clim Serv 26:100290

Kim C-G, et al (2009) Impacts and countermeasures of climate change in Korean agriculture. (in Korean). Research Report No. 593. Korea Rural Economic Institute

Kurgat BK, Lamanna C, Kimaro A, Namoi N, Manda L, Rosenstock TS (2020) Adoption of climate-smart agriculture Technologies in Tanzania. Front Sustain Food Syst 4:55. https://doi.org/10.3389/fsufs.2020.00055

Lal R (1995) Erosion-crop productivity relationships for soils of Africa. Soil Sci Soc Am J 59(3):661–667

Lal R (2005) Forest soils and carbon sequestration. For Ecol Manag 220(1–3):242–258

Lal R (2008) Managing soil water to improve rainfed agriculture in India. J Sustain Agric 32:51–75. https://doi.org/10.1080/10440040802121395

Lal R (2012) Climate change and soil degradation mitigation by sustainable management of soils and other natural resources. Agricultural Research 1:199–212

Lal M, Whettori PH, Pittodi AB, Chakraborty B (1998) The greenhouse gas induced climate change over the Indian sub-continent as projected by GCM model experiments. Terr Atmos Ocean Sci 9(4):663–669

Lee J-Y, Marotzke J, Bala G, Cao L, Corti S, Dunne JP, Engelbrecht F, Fischer E, Fyfe JC, Jones C, Maycock A, Mutemi J, Ndiaye O, Panickal S, Zhou T (2021) Future global climate: Scenario-based projections and near-term information. In Climate change 2021: The physical science basis. Contribution of working group I to the sixth assessment report of the intergovernmental panel on climate change [Masson-Delmotte, V., P. Zhai, A. Pirani, S.L. Connors, C. Péan, S. Berger, N. Caud, Y. Chen, L. Goldfarb, M.I. Gomis, M. Huang, K. Leitzell, E. Lonnoy, J.B.R. Matthews, T.K. Maycock, T. Waterfield, O. Yelekçi, R. Yu, and B. Zhou (eds.)]. Cambridge University Press, Cambridge and New York pp 553–672, https://doi.org/10.1017/9781009157896.006

Li J, Zhang C, Zhu S (2021) Relative contributions of climate and land-use change to ecosystem services in arid inland basins. J Clean Prod 298:126844

Lipper L, Thornton PK, Campbell B, Baedeker T, Braimoh A, Bwalya M, Caron P, Cattaneo A, Garrity D, Henry K, Hottle R, Jackson L, Jarvis A, Kossam F, Mann W, McCarthy N, Meybeck A, Neufeldt H, Remington T, Sen PT, Sessa R, Shula R, Tibu A, Torquebiau EF (2014) Climate smart agriculture for food security. Nat Clim Chang 4:1068–1072

Liu T, Chen X, Hu F et al (2016) Carbon-rich organic fertilizers to increase soil biodiversity: evidence from a meta-analysis of nematode communities. Agric Ecosyst Environ 232:199–207. https://doi.org/10.1016/j.agee.2016.07.015

Luo Y, Teng PS, Fabellar NG, TeBeest DO (1998) The effects of global temperature change on rice leaf blast epidemics: a simulation study in three agroecological zones. Agric Ecosyst Environ 68(3):187–196

Mahato A (2014) Climate change and its impact on agriculture. Int J Sci Res Publ 4(4):1–6. ISSN 2250-3153

Mandal D, Giri N, Srivastava P (2020) The magnitude of erosion-induced carbon (C) flux and C-sequestration potential of eroded lands in India. Eur J Soil Sci 71(2):151–168

Maybank J, Bonsal B, Jones K, Lawford RG, O'Brien EG, Ripley EA, Wheaton E (1995) Drought as a natural disaster. Atmos-Ocean 33:195–222

Mbow C, Smith P, Skole D, Duguma L, Bustamante M (2014) Achieving mitigation and adaptation to climate change through sustainable agroforestry practices in Africa. Curr Opin Environ Sustain 6:8–14

Murthy VRK (2004) Crop growth modeling and its applications in agricultural meteorology. In: Satellite remote sensing and GIS applications in agricultural meteorology, pp 235–261

Mwalupaso GE, Korotoumou M, Eshetie AM, Essiagnon Alavo JP, Tian X (2019) Recuperating dynamism in agriculture through adoption of sustainable agricultural technology – implications for cleaner production. J Clean Prod 232:639e647. https://doi.org/10.1016/j.jclepro.2019.05.366

NATCOM (2004) India's initial national communication to the UNFCCC, Ministry of Environment and Forests, Government of India

Nema P, Nema S, Roy P (2012) An overview of global climate changing in current scenario and mitigation action. Renew Sust Energ Rev 16(4):2329–2336

Nyasimi M, Kimeli P, Sayula G, Radeny M, Kinyangi J, Mungai C (2017) Adoption and dissemination pathways for climate-smart agriculture technologies and practices for climate-resilient livelihoods in Lushoto, Northeast Tanzania. Climate 5:2–22. https://doi.org/10.3390/cli5030063

Pandey S, Bhandari H, Hardy B (2007) Economic costs of drought and rice farmers' coping mechanisms: a cross-country comparative analysis. International Rice Research Institute, Los Baños (Philippines), p 203

Polyakov VO, Lal R (2008) Soil organic matter and CO2 emission as affected by water erosion on field runoff plots. Geoderma 143(1–2):216–222

Prestele R, Arneth A, Bondeau A, de Noblet-Ducoudré N, Pugh TAM, Sitch S et al (2017) Current challenges of implementing anthropogenic land-use and land-cover change in models contributing to climate change assessments. Earth Syst Dynam 8(2):369–386. https://doi.org/10.5194/esd-8-369-2017

Raghuvanshi R, Raj S, Bhattacharjee S (2018) Climate smart agriculture and advisory services: approaches and implications for future, research report brief 1, MANAGE-Centre for Agricultural Extension Innovations, Reforms and Agripreneurship, National Institute for Agricultural Extension Management (MANAGE)

Ranaivoson L, Naudin K, Ripoche A, Affholder F, Rabeharisoa L, Corbeels M (2017) Agro-ecological functions of crop residues under conservation agriculture. A review Agronomy for sustainable development 37:1–17

Rama Rao CA, Raju BMK, Islam A, Subba Rao AVM, Rao KV, Ravindra Chary G, Nagarjuna Kumar R, Prabhakar M, Sammi Reddy K, Bhaskar S, Chaudhari SK (2019) Risk and vulnerability assessment of Indian agriculture to climate change. Hyderabad, ICAR-Central Research Institute for Dryland Agriculture, p 124

Reibsame WE, Strzepek KM, Wescoat JL Jr, Perritt R, Gaile GL, Jacobs J, Leichenko R, Magadza C, Phiem H, Urbiztondo BJ, Restrepo P, Rose WR, Saleh M, Ti LH, Tucci C, Yates D (1995) Complex river basins. In: Strzepek KM, Smith JB (eds) As climate changes. Cambridge University Press, Cambridge

Roy T, George KJ (2020) Precision farming: a step towards sustainable, climate-smart agriculture. In: Global climate change: Resilient and smart agriculture, pp 199–220

Roy T, Biswas DR, Datta SC, Dwivedi BS, Bandyopadhyay KK, Sarkar A, Agarwal BK, Shahi DK (2015) Solubilization of Purulia rock phosphate through organic acid loaded nanoclay-polymer composite and phosphate solubilizing bacteria and its effectiveness as P-fertilizer to wheat. J Indian Soc Soil Sci 63:327–338. https://doi.org/10.5958/0974-0228.2015.00043.2

Roy T, Biswas DR, Datta SC, Sarkar A, Biswas SS (2018) Citric acid loaded nano clay polymer composite for solubilization of Indian rock phosphates: a step towards sustainable and phosphorus secure future. Arch Agron Soil Sci 64:1564–1581. https://doi.org/10.1080/03650340.2018.1444275

Roy T, Biswas DR, Ghosh A, Patra AK, Singh RD, Sarkar A, Biswas SS (2019) Dynamics of culturable microbial fraction in an Inceptisol under short-term amendment with municipal sludge from different sources. Appl Soil Ecol 136:116–121. https://doi.org/10.1016/j.apsoil.2018.12.024

Roy T, Mandal U, Mandal D, Yadav D (2021a) Role of arbuscular mycorrhizal fungi in soil and water conservation: a potentially unexplored domain. Curr Sci 120(10):1573

Roy T, Biswas AK, Sarkar A, Jha P, Sharma NK, Mishra PK, Patra AK (2021b) Impact of varied levels of N, P, and S stoichiometry on C mineralization from three contrasting soils with or without wheat straw amendment: a laboratory study. J Soil Sci Plant Nutr 22:501–514

Roy T, Kalambukattu JG, Biswas SS, Kumar S (2023) Agro-climatic variability in climate change scenario: adaptive approach and sustainability. In: Ecological footprints of climate change: adaptive approaches and sustainability. Springer, Cham, pp 313–348

Saha M, Bandyopadhyay PK, Sarkar A, Nandi R, Singh KC, Sanyal D (2020) Understanding the impacts of sowing time and tillage in optimizing the micro-environment for rainfed lentil (Lens culinaris Medik) production in the lower indo-Gangetic plain. J Soil Sci Plant Nutr 20:2536–2551. https://doi.org/10.1007/s42729-020-00319-6

Saha M, Das M, Sarkar A (2021a) Distinct nature of soil organic carbon pools and indices under nineteen years of rice based crop diversification switched over from uncultivated land in eastern plateau region of India. Soil Tillage Res 207:104856. https://doi.org/10.1016/j.still.2020.104856

Saha M, Sarkar A, Bandyopadhyay PK, Nandi R, Singh KC (2021b) Tillage and potassium management for improving yield, physiological, and biochemical responses of rainfed lentil under moisture stressed rice-fallow. J Soil Sci Plant Nutr 20:2536–2551

Sarkar A, Saha M, Meena VS (2017) Plant beneficial rhizospheric microbes (PBRMs): prospects for increasing productivity and sustaining the resilience of soil fertility. In: Meena V, Mishra P, Bisht J, Pattanayak A (eds) Agriculturally important microbes for sustainable agriculture. Springer, Singapore, pp 3–29. https://doi.org/10.1007/978-981-10-5589-8_1

Sarkar A, Biswas DR, Datta SC, Roy T, Moharana PC, Biswas SS, Ghosh A (2018) Polymer coated novel controlled release rock phosphate formulations for improving phosphorus use efficiency by wheat in an Inceptisol. Soil Tillage Res 180:48–62. https://doi.org/10.1016/j.still.2018.02.009

Sarkar A, Saha M, Biswas SS, Roy T (2021a) Preface of phytobiome in nutrient recycling, biogeochemistry, and spatial dynamics. In: Solanki MK, Kashyap PL, Ansari RA, Kumari B (eds) Microbiomes and plant health: Panolpy and their applications. Academic/Elsevier, pp 243–266. https://doi.org/10.1016/B978-0-12-819715-8.00008-2

Sarkar A, Biswas DR, Datta SC, Dwivedi BS, Bhattacharyya R, Kumar R, Bandyopadhyay KK, Saha M, Chawla G, Saha JK, Patra AK (2021b) Preparation of novel biodegradable starch/poly(vinyl alcohol)/bentonite grafted polymeric films for fertilizer encapsulation. Carbohydr Polym 259:117679. https://doi.org/10.1016/j.carbpol.2021.117679

Sarkar A, Biswas DR, Datta SC, Dwivedi BS, Kumar R, Bandyopadhyay KK, Bhattacharyya R, Saha M, Biswas SS, Patra AK (2021c) Double-boiled linseed and mustard oil-based formulations to prepare oil-coated controlled release fertilizers. Indian J Agric Sci 91:310–314

Schneider EC, Gupta SC (1985) Corn emergence as influenced by soil temperature, matric potential, and aggregate size distribution. Soil Sci Soc Am J 49:415–422. https://doi.org/10.2136/sssaj1985.03615995004900020029x

Scopel E, Da Silva FAMM, Corbeels M et al (2004) Modelling crop residue mulching effects on water use and production of maize under semi-arid and humid tropical conditions. Agronomie 24:383–395. https://doi.org/10.1051/agro:2004029

Sharda VN, Ojasvi PR (2006) Development of daily rainfall erosivity model for sub-humid climate of outer Himalayas. Indian Journal of Soil Conservation 34(3):199–203

Sharda VN, Ojasvi PR (2016) A revised soil erosion budget for India: role of reservoir sedimentation and land-use protection measures. Earth Surf Process Landf 41(14):2007–2023

Singh P, Boote KJ, Kadiyala MDM, Nedumaran S, Gupta SK, Srinivas K, Bantilan MCS (2017) An assessment of yield gains under climate change due to genetic modification of pearl millet. Sci Total Environ 601:1226–1237

Smith P, Powlson DS, Glendining MJ (1996) Establishing a European soil organic matter network (SOMNET). In: Powlson DS, Smith P, Smith JU (eds) Evaluation of soil organic matter models using existing, long-term datasets, vol 38. Springer-Verlag, Berlin, pp 81–98. NATO ASI Series I

Street RB, Findlay BF (1981) An objective climatological study of prolonged dry spells (meteorological drought) in the Canadian prairies (pp 29) report #81-10, Canadian Climate Center, AES, Downsview

Stromquist, N. P. (2019). World Development Report 2019: The changing nature of work: By the World Bank. Washington, DC, World Bank, 2019, 151 pp. ISBN 978–1–4648-1342-9 (hbk). https://doi.org/10.1596/978-1-4648-1328-3

Subash N, Ram Mohan HS (2012) Evaluation of the impact of climatic trends and variability in rice–wheat system productivity using cropping system model DSSAT over the indo-Gangetic Plains of India. Agric For Meteorol 164:71–81

Sudha Rani NNV, Satyanarayana ANV, Bhaskaran PK (2017) Assessment of climatological trends of sea level over the Indian coast using artificial neural network and wavelet techniques. Pure Appl Geophys 174(4):1527–1546

Swanson S, Wilhelm W (1996) Planting date and residue rate effects on growth, partitioning, and yield of corn. Agron J 88:205–210

Teasdale JR, Mohler CL (2000) The quantitative relationship between weed emergence and the physical properties of mulches. Weed Sci 48:385–392. https://doi.org/10.1614/0043-1745(2000)048[0385:TQRBWE]2.0.CO;2

The Asia Foundation (2022) India's experience with climate smart agriculture: opportunities for triangular cooperation in the Indo-Pacific. The Asia Foundation, New Delhi

Totin E, Segnon A, Schut M, Affognon H, Zougmoré R, Rosenstock T, Thornton P (2018) Institutional perspectives of climate-smart agriculture: a systematic literature review. Sustainability 10(6):1990. https://doi.org/10.3390/su10061990

Trenkel ME (2010) Slow- and controlled-release and stabilized fertilizers: an option for enhancing nutrient use efficiency in agriculture. International Fertilizer Industry Association, Paris

Triberti L, Nastri A, Baldoni G (2016) Long-term effects of crop rotation, manure and mineral fertilisation on carbon sequestration and soil fertility. Eur J Agron 74:47–55. https://doi.org/10.1016/j.eja.2015.11.024

UNCCD (United Nations Convention to Combat Desertification) (2015) Land degradation neutrality: resilience at local, national and regional levels. http://bit.ly/2vJ0ZJm. Viewed 25 Jan 2023

USAID (2017) Greenhouse gas emissions in Ghana. In: USAID factsheet; USAID, Washington, DC, pp 1–2

Wambugu SW, Chomba SW, Atela J (2014) Institutional arrangements for climate-smart landscapes. In: Climate-smart landscapes: multifunctionality in practice; world agroforestry Centre, Nairobi, pp 257–274
Wang J (2012) Modelling and simulation of plant breeding strategies. In: Abdurakhmonov I (ed) Plant breeding. InTech Press, London, p 19
West TO, Post WM (2002) Soil organic carbon sequestration rates by tillage and crop rotation: a global analysis. Soil Sci Soc Am J 66:1930–1946
Williams MAJ, Balling RC (1996) Interactions of desertification and climate. WMO, Geneva, p 230
Winkler K, Fuchs R, Rounsevell M, Herold M (2021) Global land use changes are four times greater than previously estimated. Nat Commun 12(1). https://doi.org/10.1038/s41467-021-22702-2
World Bank (2016) Climate smart agriculture indicators, World Bank Group Report No. 105162-LB. World Bank, Washington, DC

14

Farming Technologies and Carbon Sequestration Alternatives to Combat Climate Change Through Mitigation of Greenhouse Gas Emissions

Knight Nthebere, M. R. Apoorva, Mandapelli Sharath Chandra, M. Bhargava Narasimha Yadav, and T. Ram Prakash

Abstract

Land use changes from forestry to agriculture result in an increase in the mineralization of soil organic matter (SOM) and a decrease in woody biomass, which serves as a source of carbon dioxide (CO_2). It is imperative to adopt mechanisms that act as carbon sinks in order to reduce atmospheric CO_2 emissions emanating from the sources through recycling into terrestrial pool. To minimize the negative effects of climate change on the quality and quantity of soil resources and land degradation, an increase in food production must always be supported by sustainable management of agricultural land. As farming, forestry, and land use change can contribute for up to 25% of human-induced GHG emissions, these practices must be adopted in order to decline the climate change effects on agriculture. The potential for soil C storage in India is significant, with estimates ranging from 39 to 49 (44 ± 5) Tg C $year^{-1}$ on average for restoring damaged ecosystems and soils, preventing erosion, and implementing recommended management practices (RMPs) on agricultural soils and secondary carbonates. In this chapter, we discuss the functions and effects of these various GHGs, agricultural farming technologies that can be used to reduce GHG emissions under different cropping systems and their role on carbon sequestration under different land uses, decarbonization strategies towards meeting the growing demand for carbon-free agriculture, as well as the potential to sequester C under diverse ecological conditions for sustainability of soil resources.

Keywords

Farming technologies · C sequestration · Climate change · GHG emissions

K. Nthebere (✉) · M. R. Apoorva
Department of Soil Science and Agricultural Chemistry, Jayashankar Telangana Agricultural University, Hyderabad, India

M. S. Chandra
AICRP on Integrated Farming System, Jayashankar Telangana Agricultural University, Hyderabad, Telangana, India

M. B. N. Yadav
Department of Soil Science and Agricultural Chemistry, University of Agricultural Sciences, Dharwad, Karnataka, India

T. R. Prakash
AICRP on Weed Management, Jayashankar Telangana Agricultural University, Hyderabad, Telangana, India

U. Chatterjee et al. (eds.), *Climate Crisis: Adaptive Approaches and Sustainability*, Sustainable Development Goals Series, https://doi.org/10.1007/978-3-031-44397-8_14

14.1 Introduction

The incoming and outgoing flow of energy from the surface of the earth through the atmosphere and oceans serves as a function of the earth's climate. It has been documented that accumulation of these GHGs into the atmosphere sets up fluctuations, which tend to trap greater amount of solar radiation in the form of heat energy within the airspace and oceans (Sadatshojaei et al. 2022). A spacious range of novel technologies and their utilizations that were initiated in modern years are of utmost importance in enhancing the living calibre as well as the effective ingress in making use of available energy resources such as enhanced oil recovery, technologies for fossil energy, corrosion, biofuels, and CO_2 (Sadatshojaei et al. 2018, 2019, 2020a, 2020b; Choubineh et al. 2017; Dehshibi et al. 2019; Nezhad et al. 2020; Seddiqi et al. 2019; Jafari et al. 2020), solvents (Kozinski et al. 2018; Wood 2020), manipulating and usage, and sewage treatments (Heidari et al. 2017, 2019; Shokri et al. 2015). In contempt of these cutting-edge technologies, CO_2 levels in the atmosphere and oceans grow alarmingly. The rise in the average yearly temperature, global warming, cyclones, storms, and other extreme weather phenomena are all caused by entrapped energy in the atmosphere and seas. Changes in the CO_2 concentration impact the ecosystems on prodigious degree and universal climate also (Sadatshojaei et al. 2022). According to Sadatshojaei et al. (2022), the increase in atmospheric CO_2 concentration from 1750 to 2015 was caused by human activity and provoked into a long-term greenhouse gas rise of 37%, CO_2 record of 79% growth roughly since 1990–2015 (Ussiri and Lal 2017). Consumption of fossil fuels and fluctuations in land use type, specifically growing urbanization and large portion of land conversion into a variety of agriculture, are the two main fundamental sources of CO_2 emissions. Smart farming technologies and carbon sequestration are potential approaches for mitigating GHG emissions. Various strategies that could be utilized to implement carbon capturing and storage (CCS) technology that would aid in combating the climate change are explored in this chapter.

14.2 Greenhouse Gas Effect

14.2.1 Natural Greenhouse Effect

The energy balance between the sun's rays striking the earth's surface and the heat radiation backward into the air space maintains the earth's surface temperature at constant. Since gases like oxygen, argon, and nitrogen do not emit or absorb thermal irradiance, if they were the only gases present in earth's atmosphere, we would have no clouds and no greenhouse effects. CO_2, methane (CH_4), nitrous oxide (N_2O), and water vapour are the primary trace gases in the atmosphere that contribute to greenhouse effects. These gases (CO_2, N_2O, CH_4, O_3, *etc.*) behave like greenhouse glass panels or closed vehicle window glass when their atmospheric quantities rise. The sun's rays can pass easily through them on their way to earth, but the longer-wavelength (infrared) radiations emitted by earth are blocked before they can reach space. Due to the increased concentration and blanketing effect of the aforementioned gases, the temperature of the earth rises, resulting in the melting of the surface of the earth (Fig. 14.1). This is known as a global warming brought about by the effect of greenhouse gases. The higher the concentrations of these gases in the earth's surface and more is the global warming. Atmospheric clouds play a significant role by cooling the higher atmosphere by reflecting some of the sun's incident radiation.

The utilization of fossil fuels, emissions from vehicles, forest fires, tree cutting, and other human activities all contribute significantly to the release of carbon dioxide into the atmosphere. When atmospheric CO_2 levels rise by 50%, surface temperatures increase by 3 °C on average globally. According to IPCC (2001), an occurrence of 1 °C increase brings about the prediction of the following bad effects:

Fig. 14.1 A symbolic illustration of global warming giving rise to melting of the earth

1. The ice caps of the polar region melting thereby increasing the level of the sea water by 90 cm.
2. The rate of evaporation of water from the seas, rivers, and ponds increases due to global warming, which leads to extreme weather events like untimely rains, hurricanes, and cyclones.
3. Agricultural part being impacted owing to rapid evaporation of water on the surface. Consequently, lack of water supply for agricultural purposes occur.

The relative efficacy of greenhouse gases entrapping heat over a specific time is shown by their Global Warming Potentials (GWPs). The Global Warming Potential (GWP) of carbon dioxide is taken as one (IPCC 2013). For instance, SF_6 has a GWP of 23,500 over a 100-year time period (IPCC 2013) *i.e.*, its impact radiative on mass basis is 23,500 times that of CO_2 over the same time. Commonly, greenhouse gas emissions are expressed in terms of CO_2 mass equivalents, calculated by multiplying the mass of emissions by the gas-specific global warming potential (EPA 2022).

14.2.2 Anthropogenic Greenhouse Effect

Climate change is being driven by man-made greenhouse gas emissions that change the energy balance of the earth between coming solar radiation from the sun and radiation that is discharged into space (IPCC 2013). It was reported recently that global warming caused by human beings came to nearly 1 °C in 2017, greater than pre-industrial levels and by 2006–2015, human-induced activities had increased the temperature of the world by 0.87 °C (±0.12 °C) in comparison to pre-industrial era (1850–1900) which means that the whole world would reach human-induced global warming of 1.5 °C around 2040 (Fig. 14.2), if the current warming rates prevails (IPCC 2018). According to Allen et al. (2019) estimates, planetary-scale temperatures rise is increasing by about 0.2 °C per decade.

Herzog (1998) and Mohajan (2011) reported that GHG emissions have been shown to be strongly associated with this process of global warming since the early 1990s (In addition to the water vapour, CO_2, CH_4, and N_2O, the upper atmosphere also accumulates man-made halogenated compounds like chlorofluorocarbons (CFC) (Sadatshojaei et al. 2022). A graphical distribution of these greenhouse gases is depicted in Fig. 14.3. These gases contribute negatively to the worsening effects of global warming. They are continuously raising the temperature of the planet. These GHGs in combination, including the anthropogenic, are known to cause climate change and damage the ecosystem (Bloomfield and Pearson 2000; Bolin and Doos 1989; Botkin et al. 2007; Moore 1998; Shao et al. 2014). As a result, stabilizing the concentrations of GHGs in the air at the level that could fend off hazardous influence of anthropogenic with the climate the United Nations Framework Convention on

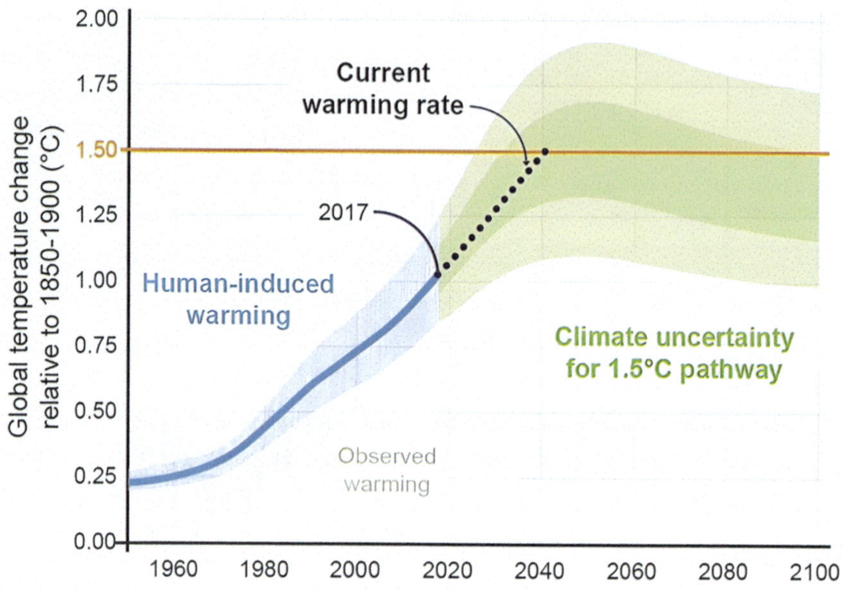

Fig. 14.2 A graphical distribution of greenhouse gases indicating the current status on the levels of universal temperature expected to approach 1.5°C near the year 2040 and the distinct 1.5°C line demonstrated here includes the emission decrements starting instantly, and atmospheric CO_2 release approaching to zero by 2055. (Source: IPPC 2018)

Fig. 14.3 Distribution of greenhouse gases. (Source: Marc, 2015)

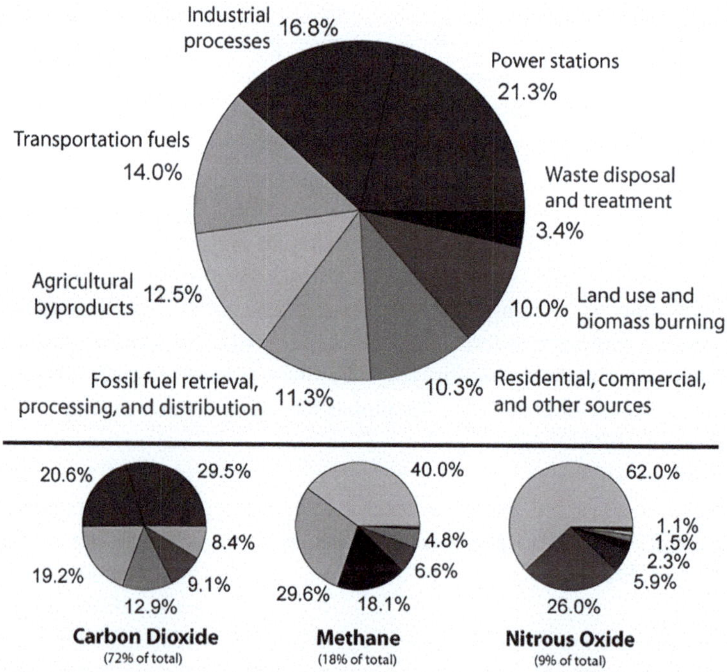

Climate Change (UNFCCC) was established as ultimate goal. According to this concept, policy-makers updated the Kyoto Protocols from 1997 to include a declared objective to keep the increase in the world average temperature to 2 °C. (Meinshausen et al. (2009) and Van Der Ploeg and Rezai (2017) established calculations, which can be accomplished by limiting total anthropogenic CO_2 emissions from 2011 to 2050 to 1100 gigatonnes. The natural carbon cycle (NCC) serves as one among other processes for preventing GHG buildup in the atmosphere (Sadatshojaei et al. 2022).

14.2.3 Atmospheric Greenhouse Gas Emissions

The concentrations of greenhouse gases such as carbon dioxide (CO_2), methane (CH_4), and nitrous oxide (N_2O) in the atmosphere have risen by 149%, 262%, and 123%, respectively, since 1750, to the level unseen in the past 800,000 years (IPCC 2013; WMO 2021). CO_2 concentrations were relatively stable at roughly 280 ppm (IPCC 2014a, b) before the industrial revolution. On monthly basis, mean concentration across the globe rose to 418.28 ppm in March 2022, an increase of around 2.8 ppm from 2021 (Loa 2015).

14.2.4 Sources of Greenhouse Gas Emissions

The combustion of fossil fuels is the primary source of anthropogenic CO_2 emissions. Other major contributors to CO_2 emissions include the manufacturing of iron and steel, cement, and petrochemicals (EPA 2022). Both natural and anthropogenic processes contribute to CH_4 and N_2O emissions. The most common human-made sources of CH_4 are livestock, landfills, and natural gas infrastructure. Seventy-four percent of all manmade N_2O comes from fertilizer use in agriculture. Livestock and other forms of mobile and stationary combustion also contribute significantly (EPA 2022). The category of greenhouse gases (GHGs) that rise very rapidly are hydrofluorocarbons (HFCs) and are used for cooling, refrigeration, and as solvents in replacement of ozone disruption chlorofluorocarbons (CFCs) (Centre for Climate and Energy Solutions 2021).

14.2.5 Universal Emissions and Trends

The sum of worldwide human-induced emissions of greenhouse gases (GHGs) in 2019 were reported to be 51.7 Gt CO_2. Since 1990, annually, these GHGs emissions rose by 57%. The increase in 2019, was by 0.57 Gt CO_2. From 1970 to 2000 an average surge of 0.4 Gt CO_2 per year was exhibited (IPCC 2014a, b; PBL Netherlands Environmental Assessment Agency 2021). Emissions that emanate from fossil fuel combustion contribute for a large portion (73%) of universal human-induced emissions (PBL Netherlands Environmental Assessment Agency 2021). EIA (2022) stated out that the total global emissions of CO_2 from energy use was 35.5 Gt CO_2 in 2019. From 2000 to 2019, the percentage increase of these global CO_2 emissions from energy use was 47.

14.2.6 Impact of GHG Emissions on Agricultural Resources and Food Production

Although India produced more than 219 million tonnes (MT) of cereals in 2000, 50 MT in 1947 (Swaminathan 2000; FAO 2011), there are still main connected issues. One, the 300 MT of cereals in predicted food demand by 2050 must come from a base of land resources that is running out. Two, there are serious issues with soil and water resource degradation that result in decreased input efficiency (such as fertilizer, irrigation, and tillage), surface and groundwaters contamination, and GHG emissions from ecosystem(s) (terrestrial and aquatic) into airspace. In order to improve soil and water quality, boost production per unit of area, time, and input, and store carbon

(C) in the ecosystems *i.e.*, terrestrial and aquatic, there should be sustainable development aims in order to achieve these goals (soil, water, and atmosphere).

Carbon is a key component present in all living things that exist on planet. There are many different types of carbon, but mostly found in biomass from the plant, soil organic matter (SOM), atmospheric CO_2, and dissolved in seawater. The most significant components of the soil, which are both soil organic carbon (SOC) and soil inorganic carbon (SIC), influences fertility status of the soil, soil water, environment, functional diversity, and other soil properties to define ecosystem and agroecosystem functions. These play a pivotal role to the global carbon cycle and the decrease of greenhouse gas (GHG) levels in the atmosphere, particularly CO_2 (Pal et al. 2015; Bhattacharyya et al. 2017). Earth's average temperature and precipitation can be affected by fluctuations in the concentrations of atmosphere as a result of carbon dioxide (CO_2) (a shift from 280 ppm in the pre-industrial era to 390 ppm during the year 2010 and 412 ppm at present) and other greenhouse gases *i.e.*, N_2O and CH_4. Therefore, there is a growing interest in devising methods to offset human-caused CO_2 emissions in the atmosphere and to slow down the rate of enrichment of CO_2 in the atmosphere.

It is well established and documented that the United Nations Framework Convention on Climate Change (UNFCCC) was the first international treaty to deal with the issue on climate change, operating to prevent human-caused emissions that account for climate disturbances. UNFCCC was adopted in Rio de Janeiro in the year 1992, with the major objective to stabilize atmospheric GHGs at levels that will hinder harmful anthropogenic interference with the climate. It has become nearly impossible for rich and developing countries to reach consensus on how to adapt the significant legislation necessary to combat climate change, as seen by current discussions on climate change and global warming. India is among the list of countries over the world, which suffer from climate change impacts. Carbon sequestration in terrestrial sinks can be employed to compensate for GHG emissions in accordance with the Kyoto Protocol (KP) (Jandl et al. 2007). The concern, thereof, is the solution to lower elevated CO_2 concentrations to the levels that cannot pose any threat to the climate.

Reduction of global energy consumption, the implementation of low- or no-carbon fuel, and utilization of engineering and natural protocols to capture CO_2 from point sources are the three alternatives available for reducing CO_2 emissions to combat climate change. Depending on the processes and technological innovations, C sequestration can be broken down into three categories: (a) methods that rely on natural processes such as photosynthesis and CO_2 to biomass conversion, SOM, and other components of the terrestrial biosphere; (b) methods that use approaches of engineering; and (c) methods that involve chemical transformations.

14.3 Carbon Sequestration

Carbon sequestration refers to the process of using biotic and abiotic (engineered) techniques to capture and store CO_2 in the atmosphere in stabilized pools. Combining these efforts can lower CO_2 levels in the atmosphere and reduce the impacts of global warming.

14.3.1 Abiotic Strategies

These entail the deep injection of CO_2 into geological basalts and the ocean as well as the separation, capture, compression, transport, and injection of industrial gases and effluents (Lal 2004a, b, c). Carbon sequestration using abiotic strategies is still in its infancy, and it takes a lot of money, chemicals, and equipment to put into practice because these methods are wholly based on cutting-edge engineering.

14.3.1.1 Consequences of Abiotic Carbon Sequestration

The oceans are becoming more acidic as a result of chemical changes brought on by the absorption of CO_2 from the atmosphere. About 50% of the carbon dioxide released by burning of fossil fuels as well as cement manufacturing are taken up in oceans over 200 years ago. The ocean's surface pH has dropped from around 8.16–8.05, a drop of about 0.1 units since the beginning of the industrial revolution, approximately 200 years back. Feeley et al. (2009) estimated that by 2100, the ocean's average pH might drop by 0.5 units (three times corresponding rise in hydrogen ions concentration) only if worldwide production of emission (CO_2) produced by man-made activities scale-up on the modern path. That condition is known as ocean acidification, and it is predicted to have negative repercussions for fishing, tourism, and the marine economy over time. The implications of injecting carbon dioxide into the ocean are uncertain, however, because so little is known about the deep sea's ecology, chemistry, and geology. Further, risks to people, ecosystems, and groundwater may arise if the CO_2 escapes from its storage.

14.3.2 Biotic Strategies

Through the process of photosynthesis, fixed atmospheric CO_2 is converted into plant biomass and pools of soil organic matter (SOM), which serves as the basis for biotic strategies to sequester terrestrial carbon in contrast with marine and geological carbon storage techniques (Kishwan et al. 2009; Kleber and Johnson 2010). Soil organic matter (SOM) is any remains of animals, plants and microbes undergoing various decomposition stages altogether with different rates of turnover. Soil organic matter acts as a major sink for carbon all over the globe. During ecosystem development, biota (including both autotrophs and heterotrophs) interact with environmental controls (including temperature and moisture) to accumulate SOM, of which about 58% is carbon (Post et al. 2001). The inputs of litter and its decomposition determine the rates at which various natural ecosystems accumulate SOM. The rate and build-up of soil organic matter is, of course, directly influenced by the productivity of the present and former vegetation, the soil conditions (physical and biological), and land management history and practices. As a result, natural ecosystems and artificial ecosystems have different rates of SOM accumulation. Organic carbon (OC) forms found in nature come from the breakdown of dead organisms. Anthropogenic components, such as pesticides and municipal trash, also contribute to OC in soils (Post et al. 2001). Soil has a wide range of organic carbon (OC) forms, from recently fallen litter like leaves and twigs to highly degraded forms (humus). Litter from the plants and microbial biomasses are the major parent material for soil organic matter production. The plants absorb carbon dioxide (CO_2) during photosynthesis and use some of that carbon to build their tissues and organs. Therefore, C sequestration through perennials has been demonstrated as financially viable strategy to alleviate the effects of climatic change on a wider scale.

14.4 Soils of India and Their Soil Organic Carbon Pool

The soils are referred to as "tropical soil" in this region due to the fact that a significant portion of India's geographical area is located between the Tropics of Cancer and Capricorn. For a long time, it was believed that tropical soils could only be found in the hot humid tropics where they were deeply coloured and severely eroded. Additionally, they are frequently mistakenly believed to be either poor at agriculture or nearly useless (Eswaran et al. 1992). According to Bhattacharjee and Dewar (1982), there are five different bioclimatic systems in India with variation in mean annual rainfall (MAR). *Alfisols*, *Ultisols*, *Aridisols*, *Inceptisols*, and *Entisols*, are the principal soil types of India, which account for 8.1%, 0.5%, 12.8%, 2.6%, 4.1%, 39.4%, and 23.9% of the thermogravimetric analysis (TGA), respectively (Bhattacharyya et al. 2009).

Global SOC estimates was an issue of interest for most researchers in the early 1970s to the early 1990s in which two various approaches were followed for estimation of SOC stock globally. (Post et al. 1982). These are based on ecosystem estimates and soil taxonomy, respectively. Based on the area/extent and mean of carbon content in each of the world's main soil taxonomic groups, SOC stocks were tallied, using the first approach (the taxonomic approach) (Eswaran et al. 1993). The world soil map produced by the food and agricultural organizations (FAO) and the United State Department of Agriculture's (USDA) soil taxonomy were the two main soil categorization or mapping techniques utilized in the global carbon tabulation. The SOC stocks of the various soil orders around the world were computed as 1576 Pg C by Eswaran et al. (1993). SOC stock was concentrated in *Histosols* and depleted in *Vertisols* among different soil orders. SOC stock was assessed using a technique based on an ecosystem-based approach, based on overall ecological life zones that are clustered according to some average annual precipitation and average annual temperature combinations (Brown and Lugo 1990). The method put total amount of carbon stored in soil as 2542.30 Pg C. According to a modern research conducted by Jobbagy and Jackson (2000), the global SOC sequestration in the upper surfaces of the soil (3 m soil layers) is 2344.0 Pg C on average, with 1502, 491 and 351 Pg C in the first, second and third meters, respectively.

Batjes (1996) compared the SOC stock for tropical regions and the world to that of Indian soils accumulated in the top 0.3 m and 1.5 m soil layers (Table 14.1) and reported no significant contribution by Indian soils. The lack of organic matter (OM)-rich soils like *Histosols*, *Spodosols*, *Andosols*, and *Gelisols* in India is the primary reason for this. Also, *Mollisols* don't cover a huge area, and Andosols can't be mapped at a 1:250,000 scale. Further, Indian soils cover only 11% of the total area of the world with sufficient

Table 14.1 The main greenhouse gases

Compound	Pre-industrial concentration (ppmv*)	Concentration in 2020 (ppmv)	Lifetime of atmosphere (years)	Major human-induced activity source	GWP**
Carbon-dioxide (CO_2)	278	413	Variable	Fossil fuels, cement production, land use change	1.0
Methane (CH_4)	0.722	1.889	12	Fossil fuels, rice paddies, waste dumps, livestock	28
Nitrous oxide (N_2O)	0.27	0.333	121	Fertilizers, combustion industrial processes	265
HFC 23 (HFC_3)	0	0.000024***	222	Electronics, refrigerants	12,400
HFC 134_a (CF_3CH_2F)	0	0.000062***	13	Refrigerants	1300
HFC 152_a (CH_3CHF_2)	0	0.0000064***	1.5	Industrial processes	138
Perfluoromethane (CF_4)	0.00004	0.000079***	50,000	Aluminium production	6630
Perfluoroethane (C_2F_6)	0	0.0000041***	10,000	Aluminium production	11,100
Sulphur Hexafluoride (SF_6)	0	0.0000073***	3200	Electrical insulation	23,500

Source: IPCC (2013), WMO (2021)

ppmv* parts per million by volume, *GWP* 100-years global warming potential, ***concentration in 2011, exclusive to aqueous vapour potential to store organic C even under environmental conditions that are not conducive for OC-rich soils (Pal et al. 2000). Approximately 50% of the total area of the TGA of India is covered by arid, semi-arid, and dry sub-humid climatic zones, where organic C accumulates at a lower rate (1%) (Pal et al. 2000; Velayutham et al. 2019). Soils in humid tropical (HT) climates, on the other hand, have OC greater than 1% (Velayutham et al. 2019). These soils are not restricted to a certain farming method, but play a vital role in India's increasing food security (Pal et al. 2015; Bhattacharyya et al. 2014).

14.5 SOC Sequestration in Agricultural Soils

Restoration of the ecological community and reduction of the universal climate change are higher priorities for C sequestration in degraded and arable lands. Around 1.6 billion hectares of land are used for farming globally, according to FAO. This account for 12% of the global total area (13.2 billion hectares). Since 1961, this amount of land used for farming has increased by 159 million hectares (FAO 2011). In the first, second, and third metres, agricultural soils worldwide stored 157, 210, and 248 gigatonnes of carbon, respectively. Carbon sequestration potential can be measured in relation to the amount of carbon lost from agricultural soil (Paustian et al. 1997). Cropland, degraded or desertified land, irrigated soil, and grazing/rang lands all have different potentials for carbon sequestration, with estimates ranging from 0.4 to 0.8, 0.2 to 0.4, and 0.01 to 0.03 Gt year^{-1}, respectively (Lal 2004a, b, c). About half of India's food grains and the diets of nearly 40% of the country's population come from the Indo-Gangetic plains (IGP), which occupy only 13% of the country's total land area. Using the Global Environment Facility Soil Organic Carbon (GEFSOC) model, Bhattacharya et al. (2007a, b) accurately estimated that IGP regions will store 1.27 Pg of carbon in 1990, 1.32 Pg in 2000, and 1.27 Pg in 2030. They also reported that by 2030, the soil in IGP could hold as much as 1.1 Pg C.

14.5.1 Factors Affecting the Storage of SOC in Soils

Carbon storage varies greatly amongst ecosystems due to the following factors:

(a) *Climate*

It has been assumed for a very long time that climate has a role on how much carbon build-up in the soil. Poor oxygen supply and colder climatic conditions in wet soils may slow the rate of decomposition process, which tends to increase soil carbon stocks are higher in these regions of cold and damp biomes and low in heat and dry biomes (Schimel 1995). Soil carbon storage is greatly impacted by climatic conditions including temperature and moisture. Precipitation, cooler temperatures, and a lower evapotranspiration-to-precipitation ratio are all associated with higher SOC levels (Jobbagy and Jackson 2000). The rapid decrease in organic carbon (OC) within the clay soils, *Haplusterts* from a damp to arid ecological community (Fig. 14.4) is also noteworthy (Pal et al. 2003; Pal et al. 2015). Table 14.2 indicates that a rainfall regime of 500–1000 mm/year have continuously lowered SOC concentration.

(b) *Altitude*

Carbon is distributed between plants and soil in an approach that varies with latitude. According to a previous study, high-latitude forests are home to a sizable portion of the world's vegetative (25%) and soil (59%) carbon stores. It has been estimated that 59% of worldwide forest vegetation and 27% of SOC on global scale are found in low-latitude tropical forests (Dixon et al. 1994). In general, low-latitude forest soils have a higher carbon density than their mid-latitude counterparts. Hobbie et al. (2000) found that permafrost dynamics and drainage contributed to the high SOC stock in high latitudes. At higher elevations, carbon stock in India's Himalayan mountain areas' various forest types tend to decline (Sheikh et al. 2009). Furthermore, elevation significantly influenced species richness, which dropped by even a hundred metres with an increase in

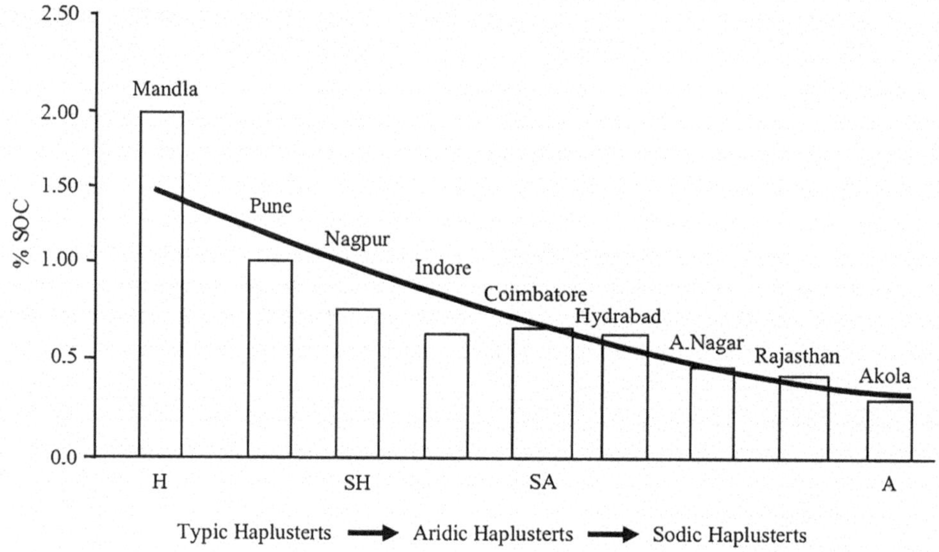

Fig. 14.4 Negative impacts of climate on soil organic carbon (SOC) accumulated in top soil (0–30 cm) of Vertisols. *H* humid, *SH* sub-humid, *SA* semi-arid, *A* arid. (Adapted from Pal et al. 2015)

Table 14.2 Total SOC stock in India, tropical regions and the world (Pg)

SOC	0–0.3 m	0–1.5 m
Soil organic carbon		
India[a]	9.55	29.92
Tropical region[a]	201–203	616–640
World[b]	684–724	2376–2456
% SOC in the tropical region of India	4.72	4.77
Stock		
% SOC stock in the world	1.4	12

Source: [a] Bhattacharyya et al. (2000), [b] Batjes (1996)

latitude. Because of the normal decrease in plantation that happens at higher altitudes, there is less litter to gather and less organic carbon introduced into soils.

(c) *Plant Cover*

The classification of plantation that shields the surface of the earth is known as soil cover. Universally, the volume and the distribution of SOC are mostly related with plant cover than climate, according to numerous researches on SOC dynamics. Because different plant life forms have different rooting depths, detrital input patterns, and dominant plant species, these differences showcase the impact on SOC content (Gill and Burke 1999). According to a few studies, vegetation life forms often differ in topography and spread of their root penetration, which impacts the amount and SOC distribution (Jackson et al. 1996). Through changes in carbon input, vegetation species have the potential of affecting SOC pools and its dynamics by impacting the loss of carbon, including the decomposition of SOM.

(d) *Soil Texture*

Soil comprises two various proportion of particles: (1) coarse particle size (2 mm) (gravels, cobbles, boulders and other soil fragments; (2) fine particle size fractions *i.e.*, sand (0.05–2 mm), silt (0.002–0.05 mm) and clay (< 0.002 mm). A large number of studies have put forward that C sequestration rely on soil texture and highly related to the proportion of fine soil materials (Bosatta and Agren 1997; Hassink 1997). Recently added organic carbon was temporarily stored in silt-sized particles, and proportions of sand materials that can serve as markers detective to deviations in the status of soil carbon as a result of land use management. SOC content of some soils of India is depicted in Table 14.3.

Table 14.3 Relationships between the content of soil organic carbon (SOC) and temperature and rainfall patterns in India

Rainfall (mm)	Mean annual temperature (°C)	SOC content (g kg^{-1})	
		Surface soil	Sub-surface soil
<500	25.90–26.70	1.20–8.00	1.20–4.00
500–100	23.60–27.90	1.80–12.50	0.70–11.70
>1000	24.40–27.20	2.00–9.00	2.30–8.40

Adapted from Sekhon et al. (1994)

14.5.2 Inorganic Carbon Sequestration

Soil inorganic carbon accounts for 38% of the world's total carbon stores (Luo 2007), making it to be among the most important C storages in proximity of the earth's surface. Soil carbonate, derived from the weathering of silicate rock, is the largest soil carbon reservoir in dry and semi-arid regions because of its rapid accumulation rate and sensitivity to changes in climate, hydrology, and other factors. Soil inorganic carbon comprises CO_2 in the gaseous state, HCO_3^- and in a liquid solution state, and CO_3^{2-} in the solid state. As per the source, the soil CO_3^{2-} entails the primary carbonate and the pedogenic carbonate. A type of soil or parent rock that has not weathered and never been in contact with the soil habitat is known as the source of principal. The parent CO_3^{2-} disintegrating into soil and dissolves in the air and water, gives rise to pedogenic carbonate (IGBP Terrestrial Carbon Working Group 1998).

The soil of arid places of the world frequently contains carbonate minerals. The excessive production of calcium carbonate ($CaCO_3$) caused by too much salt and sodicity in the soil has a negative impact on soil characteristics (Bhattacharyya et al. 2000). In the arid-semiarid region, the amount of soil inorganic carbon (SIC) is twice to thrice more than SOC up to a depth of 1 m (Sahrawat 2003). India has calcareous soil covering almost 229 million hectares, and SIC contributing to C storage essential for maintaining soil nutrient status (Sahrawat et al. 2005). Bioclimatic regions (semi-arid, sub-humid, and humid) in the Indo-Gangetic Plain (IGP) of India had soil inorganic carbon (SIC) stocks of about 124.48, 12.28, and 24.11 Tg per lakh ha, respectively. Interestingly, previous studies have discovered a connection that lies in between the SIC and the development of soil organic carbon (SOC). Small substrate additions as well as slow rate of decomposition are responsible for nadir levels of soil organic carbon (SOC) and microbial biomass in problem soil known as salt-affected soils (Wong et al. 2009). Future studies on the role of soil SIC sequestration would be stimulated by the basal computation of SIC in arid zones of India.

In India, calcareous soils can be found in majority of the country's regions (54%); however, they are most common various states of India (Rajasthan, Gujarat, Punjab, Haryana, Uttar Pradesh, Maharashtra, Karnataka, Tamil Nadu, Andhra Pradesh, and even some regions of Madhya Pradesh and Bihar). Indian soils are estimated to contain about 196 Pg at 1 m soil profile, the total of SIC pool (Pal et al. 2000). Globally, it is estimated that there is a SIC pool of 722 Pg at a depth of 1 m in soils (Batjes 1996). Therefore, approximately 27% of the SIC pool in soils worldwide can be found in India. Particularly important for C sequestration in irrigated settings is the development of pedogenic or secondary carbonates. According to Pal et al. (2000), secondary carbonates formation rates have been estimated between 30 and 130 kg ha^{-1} $year^{-1}$.

The contributing factors for the differences in SIC stocks of the soils of India in relation to different land uses and productivity were recently remarked in a review on Indian soils and their potential to store SIC in seven soil orders, as well as the factors that are conducive for sequestering carbon amidst subtleties of pedogenesis and polygenesis as a result of tectonic, atmospheric, and geomorphic episodes during the Holocene (Pal et al. 2015). Soil organic carbon stocks are twice more predominant as opposed to soil inorganic carbon (SIC) stocks at the 0–30 cm soil layer in five bioclimatic regions of India (Table 14.4). Pal et al. (2000) reported that the principal pedogenic mechanism that captures atmospheric CO_2 for the creation of pedogenic $CaCO_3$ in soils, fast calcification, is the key contributor to inorganic SIC storage in dry and SAT soils (Table 14.5).

Table 14.4 Soil organic carbon (SOC) content of some soils of India

Site	Texture of the soil	Soil type	Content of SOC (g kg^{-1})
Bangalore	Sandy loam	*Haplustalf*	5.5
Barrackpore	Sandy loam	*Eutrochrept*	7.1
Bhubaneshwar	Sandy	*Haplaquept*	2.7
Coimbatore	Clay loam	*Vertic Ustochrept*	3.0
Delhi	Sandy loam	*Ustochrept*	4.4
Hyderabad	Sandy clay	*Tropaquept*	5.1
Jabalpur	Clayey	*Chromustert*	5.7
Ludhiana	Loamy sand	*Ustochrept*	2.1
Palampur	Silty clay loam	*Hapludalf*	7.9
Pantnagar	Silty clay loam	*Hapludoll*	14.8

Adapted from Nambiar (1994)

Table 14.5 Different bioclimatic zones of Indian soils' organic and inorganic carbon (SOC and SIC) stocks in Pg, at 0–30 cm soil depth

		SOC		SIC	
Bio-climatic regions	Area (mha)	SOC stocks	SOC in %	SIC stocks	SIC in %
Cold-arid	15.20	0.60	6	0.7	17
Hot-arid	36.80	0.40	4	1.0	25
Semi-arid	116.40	2.80	30	2.0	47
Sub-humid	105.00	2.40	26	0.33	8.0
Humid to per humid	34.90	2.00	21	0.04	1.0
Coastal	20.40	1.30	13	0.07	2.0

Adapted from Bhattacharyya et al. (2000, 2008)

14.5.3 Principles of Soil Organic Carbon (SOC) Sequestration

Soil organic carbon content or stock is decadally renewable resource that serves a crucial part in assessing the quality of soil or function as well as ecosystem services. For this reason, two fundamental approaches for improving the slowly renewable SOC resources are (i) to achieve a productive ecological community and SOC budget, biomass-C input without moving beyond the losses due to its decay, removal by erosion, and leaching (ii) to set an ultimate goal for preventing SOC from re-emission into the atmosphere by extending its mean residence time (MRT) (Lal 2018).

In the light of this context, the six "Rs" of SOC stewardship herein must always be followed:

R1: Soil reduction losses from and SOC stock preservation in existence.

R2: Biomass-C recycling processes.

R3: Reinstating exhausted and degraded soils.

R4: Restore the inputs and outputs of biomass-C.

R5: Re-evaluate soil functionality in the light of the important SOC limitations that are unique to each soil rhizosphere.

R6: Re- define and modernize the application of fertilizer inputs with special reference to C: N: P: K rather than N: P: K.

Lal (2010, 2015) coined these six "Rs" of SOC restoration in accordance with the notion of eco-intensification (Fig. 14.5).

14.5.4 Major Techniques for SOC Storage Based on Restoration of Land Use and Recommended Management Practices to Be Adopted

These practices are grouped into two main categories:

(i) C-sequestration at soil surface (0–30 cm).

(ii) C-sequestration at sub-surface (>30 cm).

Conservation Agriculture (CA) with its components as depicted in Fig. 14.6, should be practised as to aid in SOC sequestration in the top soil. Deep injection of biomass-C into the sub-soil via mechanical methods or natural systems, as well as deep-injection of plant debris and other bio-wastes are important for C storage

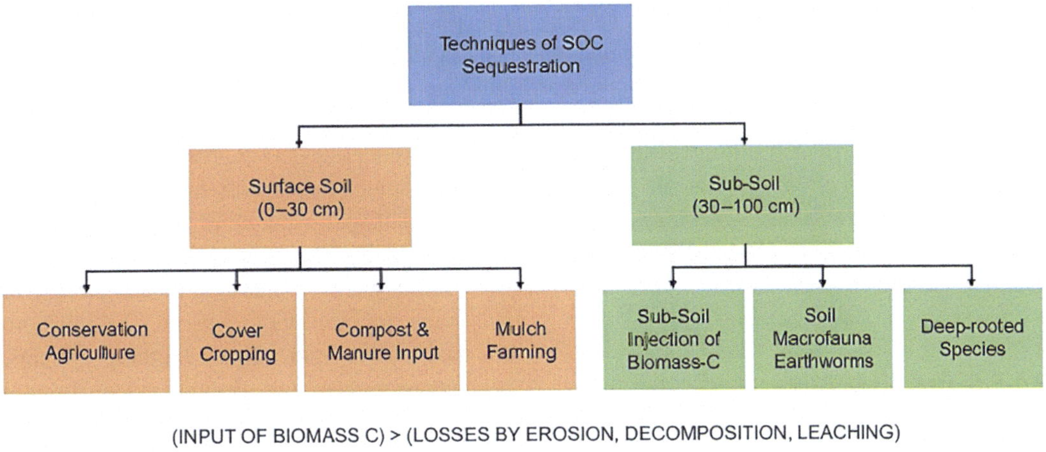

Fig. 14.5 Major C-sequestration methods in restorating land use and advised management approaches to be adopted

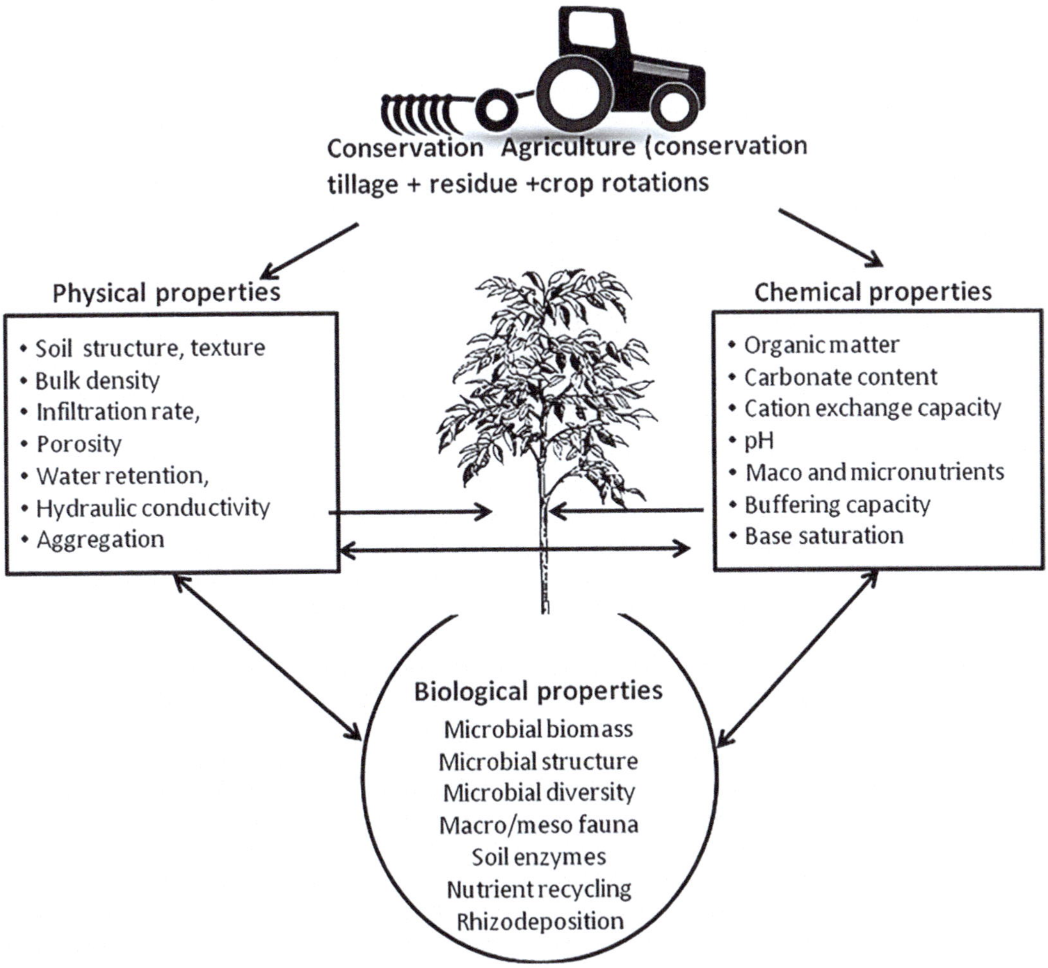

Fig. 14.6 Fundamental aspects of CA on soil parameters and their impact on soil health

in lower profile depth (>30 cm) (Hammond et al. 1993; Kalisz et al. 1994). The use of mulching methodology such as vertical has been viewed as utmost importance for boosting crop yields in both rain-fed and irrigated settings (Meyer et al. 1992), as well as to lower run-off of water (Denardin et al. 2008) and watersheds particularly micro-sheds management (Fairbourn and Gardner 1972).

14.5.5 Practices Advocated to Enhance Soil Carbon (C) Sequestration and to Remove Net Carbon Dioxide (CO_2)

The new National Academies paper categorizes soil C storage management approaches under two major groups as to enhance soil C stocks (NASEM 2019). For sites with current crops and management practices, the first group contains well-known tested conservation management approaches that can raise soil C levels. These are techniques that, while not (yet) extensively used, are being used by more farmers that care about conservation and could eventually be adopted on a much larger scale. For improving soil carbon storage, these management strategies are referred to as BMPs ("Best Management Practices"). Second, "frontier technologies" are methods that have not yet overcome significant technological and/or economic hurdles, and hence are still considered experimental. As a result, they stand for technologies and methods that are still largely in the developmental stages, as they are rarely used in production agricultural systems.

14.5.6 Conventional Conservation Practices to Store Soil Carbon

(a) *Improved Crop Rotations and Cover Cropping*

According to CAS. (2004), cultivation of high-residual crops, seasonal green leaf manure cover crops, continuous cropping with short fallow frequency, and cultivation of involving rotation of perennial grasses in the cropping system are viable options that farmers can take advantage of to improve inputs of C into soils. A wide modern evaluation of cover crops, for instance, found that on average, they sequester 0.32 tC per hectare per year, with some studies even finding rates higher than 1 tC per hectare per year (Poeplau and Don 2015). The farmers follow the practice of fallowing croplands on yearly basis in many dry climates to preserve soil moisture and stable grain output. In these systems, intensification and diversification through crop rotations are believed to raise mean annual carbon inputs and result in larger soil carbon stocks compared to an increase of fallow periods (West and Post 2002; Sherrod et al. 2005). Row crop rotations with a period between 2 and 3 years with forage crops addition can enhance C inputs having fine rooting system and raise soil organic carbon reserves in more humid conditions.

(b) *Manure and Compost Addition*

Compost and manure additions can improve soil C contents by adding C for the amendment as well as by improving the physical properties as well as soil nutrient availability, which ultimately increases crop productivity and residual C inputs (Paustian et al. 1997).

(c) *Tillage*

Tillage is the primary cause of soil destruction especially in annual cropping utilized by the farmers for management of plant debris and creation of conditions necessary for seed bed. In recent decades, improvements in till technologies and agronomic techniques have created flexibility for the farmers to lower tillage intensity, and occasionally even to stop it completely with a method known as "no-till" (NT). To lessen soil erosion, many farmers are reducing their tillage practices. According to some studies conducted earlier, soil erosion has been reported to be largely decreased under NT, with reductions of up to 90% (Ghidey and Alberts 1998; Williams et al. 2011).

The degradation of stability of soil aggregation that "shield" organic materials from decomposition is facilitated by tillage operations

(Six et al. 2002). Higher C storage is expected to occur under NT because of the dramatic enhancement of aggregation and aggregate stability that occurs under NT (Six and Paustian 2014). Increases of about 0.25 tC ha^{-1} $year^{-1}$ and 0.29 t C ha^{-1} $year^{-1}$ were expected by Ogle et al. (2005) under no-till on a sandy and without sandy soils, respectively. However, there are certain cases where zero tillage (ZT) does not enhance OC in relation to conventional tillage (CT), such as soils having high surface C contents already and frequent cool and wet conditions where productivity in terms of the crop and carbon inputs may slow down under ZT, for instance, due to germination delayed (Ogle et al. 2012). Ten sites of Germany were investigated by Alcantara et al. (2016); overall having undergone a one deep tillage manipulation between the years 1965 and 1978 with the goal of reducing subsurface layer compaction. It had been discovered that the SOC stocks at deep-tilled locations were on average, 42 t ha^{-1} higher (up to 1.5 m depth) than those at the control sites. Crop yields were about comparable in both heavy-tilled and untreated fields. Over a 45-year time period, the rate of soil C 0.96 tC ha^{-1} $year^{-1}$ (3.5 t CO_2 eq ha^{-1} $year^{-1}$) increased after implying deep tillage (Fig. 14.7).

(d) *Nutrient management*

Gaseous greenhouse gases (GHGs) can be released through synthetic fertilizers containing N_2O associated with fertilizer production and distribution. The mineralization of SOC enhances crop yields as well as profitability when fertilizers are utilized properly and farmlands had added of CO_2 approximately (50 Pg) to the environment (Jandl et al. 2007). Nitrogen fertilization has been shown to decrease soil microbial activity; nonetheless, the use of fertilizers has greatly increased crop productivity. Applying the right amounts of fertilizers is essential for maintaining soil fertility and crop yield (Bhattacharyya et al. 2000). Up to 21.3–32.5% of carbon is sequestered by crop residues and nutrients, particularly N (Bhattacharyya et al. 2007a, b).

Liu et al. (2013) conducted ongoing long-term study in Northwest China, which was taken up since 1979 to observe the influence brought by fertilization on soil organic carbon and soil organic carbon fractions at various soil depths from (0–100 cm). The trial comprises six different tests: unfertilized, nitrogen fertilizer, nitrogen and phosphorus (NP), straw with N and P fertilizers (NP + S), farmyard manure (FYM), and NP fertilizer + FYM. As compared to the control treatment, the results showed that SOC storage in the 0–60 cm increased by 41.5, 32.9, 28.1, and 17.9% in the NP + FYM, NP + S, FYM, and NP treatments, respectively. The labile pool of carbon in the upper surface of the soil (0–60 cm) was also improved by the addition of organic manure and inorganic fertilizer. These observations demonstrate that among the evaluated kinds of fertilization, organic manure treatments imposed over an extended period of time had the most beneficial effects on building carbon pools.

(e) *Crop rotations*

Crop rotation refers to the practice of growing different crops in the same field over the course of two or more consecutive years. Changes in crop management, soil composition, temperature, and crop rotation can all have an effect on carbon sequestration. Despite the fact that continuous cropping systems lead to the loss in soil organic matter, amendments of soil nutrients containing 10.7–18% C, such as balanced NPK fertilization, organic supplements, and likewise, the supply of plant debris, can improve C-sequestration rates up to 5–10 Mg ha^{-1} $year^{-1}$ (Luo et al. 2010). Other sources of nitrogen include peas, lentils, alfalfa, chickpeas, *sesbania*, and other legume crops. Soil organic carbon can be stabilized through crop rotations, particularly employment of cover crops particularly legumes, which include C molecules that can withstand microbial degradation (Lal 2004a, b, c). Tillage in combination with the cropping sys-

Fig. 14.7 Relationship between soil tillage and soil carbon dynamics. (Kimble 1997)

tems has recently gained more attention as a way to reduce agricultural CO_2 emissions (Ellert and Gregorich 1995; Bhadwal and Singh 2002). Different kinds of cropping systems (cover, ratoon, and companion cropping) have the ability to aid store carbon in the soil. Inter-row cropping, strip intercropping, mixed cropping, and relay intercropping are all types of intercropping that can boost soil fertility and increase yields (Shekhawat et al. 2012). Wheat and mustard are two examples of intercropping, as well as cotton intercropping with peanut, peanut with sunflower, wheat with chickpea, and so on.

(f) *Rewetting Organic Soils*

The methods have been discussed thus far, for working with "mineral soils," which are those where inorganic matters constitute for a relatively low percentage of the overall mass and minerals, like sand, silt, and clay, makes up the majority. Organic soils, on the other hand, are formed from peat and muck and include a high concentration of organic materials (often termed as "histosols" in the system of soil classification). Waterlogged circumstances (therefore, very low CO_2 concentrations) severely slow decomposition processes, resulting to the formation of these soils, which are characterized by the accumulation of thick layers of only partially digested plant matter. Organic soils, contrary to mineral soils, do not have susceptibility to saturation, meaning that organic matter can continue to build, resulting in a greater soil "depth," so long as the conditions that prevent decomposition persist.

(g) *Soil biota management*

Several reports indicated that soil microbes, especially bacteria, had improved soil physical, chemical, and biological properties, which may aid in biological C-sequestration. The living component of soils is called the soil biota, which is made up of a variety of micro-organisms in enormous numbers. They interact with plants and one another, directly supplying nutrients and other advantages. A significant portion of the soil structure is aided by their physical makeup and products. Moreover, they are in charge of transforming the minerals and nitrogen that are bonded to organic materials into forms that plants can use. Through bio-control agents, these micro-organisms control their own inhabitants and that of micro-organism's interfering with the ecosystem function, which are significantly impacted by various agronomic management practices.

(h) *Agroforestry*

The practice of growing perennial trees and shrubs alongside agricultural crops is known as agroforestry, which combines agriculture with forestry. Improved soil carbon sequestration can result from the planting of various tree species, such as orchards, fruit trees, and forests, in agriculture. Dijkstraa et al. (2009) reported that agroforestry systems have a carbon sequestrations potential of 12–228 Mg ha^{-1}, which means that during the next 50 years, an overall of 1.1–2.2 Pg C can be stored in farm lands all over the world's total area, appropriate for crop production. Once land was converted to agroforestry, a meta-analysis of 53 published research on the subject of changes in soil organic carbon (SOC) stocks at depths of 0–15, 0–30, 0–60, 60–100, and 0–100 cm indicated substantial declines in SOC stores of 26 and 24% at 0–15 and 0–30 cm, respectively. The SOC stock increased dramatically by 26, 40, and 34% at different soil depths of 0–15, 0–30, and 0–100 cm, respectively, when land use for agriculture was converted to agroforestry. A range of 0.29–15.21 Mg ha^{-1} year^{-1} in above-ground plant biomass and 30–300 Mg ha^{-1} year^{-1} in below-ground plant parts up to a depth of 1.0 m has been estimated for C storage capacity of AFS in various ecosystem and environmental management (Nair et al. 2010). Assuming that half of the biomass is composed of C (Nair et al. 2010), above-ground biomass is a direct measure of C sequestration. Under AFS, both above- and below-ground C sequestration are significantly higher than on treeless croplands under comparable ecological and management conditions. Silviculture, ally cropping, forest farming, windbreaks, home gardens, riparian buffers, woodlots, *etc.* are all examples of agroforestry techniques. It has been projected that the world forest system contributes for about 90% of the annual SOC pool in between the soil and atmospheric carbon (Wani and Qaisar 2014). Ibrahim et al. (2013) had discovered that the forest system is intended to capture 3 Pg C equivalence per year. According to a recent study by Minasny et al. (2017), agroforestry has the highest ability to store carbon of any land use form.

(i) *Irrigation Management*

Soil carbon sequestration could be greatly increased through the use of irrigation water. Therefore, in arid and semiarid ecosystems, SOC stock can be increased through careful irrigation water management by increasing biomass production and the quality of above- and below-ground plant parts returned to the soil. SOC sequestration also necessitates efficient water recycling and appropriate management of the water table, like the use of irrigation methods (drip or sprinkler). According to experimental findings, soil may sequester between 0.05 and 0.15 t SOC per hectare per year. (Conant et al. 2001) and 0.05 to 0.10 t SIC per hectare per year (Nordt and Dees 2000). C additions to the soil ecosystem is controlled by the availability of soil moisture, which in turn controls vegetative growth and NPP (Yuste et al. 2007). There are both significant and adverse impacts of irrigation on SOC accumulation over a long period of time in cropland. Enhancement of

C input into the soil in the form of root exudates, rhizodeposition, roots decay, and other plant components are facilitated by the improved water availability (Kochsiek et al. 2009). Contrarily, irrigation supports the accumulation of soil moisture and the related microbial activities. Increase in SOM decomposition process as well as atmospheric CO_2 emissions are the outcomes for that (Trost et al. 2013; Gogoi et al. 2018).

14.6 Frontier Mechanisms for Soil Carbon Sequestration

Numerous management techniques that are not conventional showcase more potential for reducing emissions, but a large number of studies are necessitated to create the required recent techniques and/or better control projections of costs and life-cycle emissions under widespread adoption. In this chapter, we cover a variety of approaches, including biochar application to agricultural soils, the cultivation of perennial, and annual crops with increased root depth and root area attributes to boost C inputs.

(i) *Biochar Additions*

A thermochemical conversion method called pyrolysis is used to create biochar, a carbon-rich material, from biomass. The development of solid biochar is favoured by lower temperatures and longer residence times during pyrolysis, while more fraction of gases and liquid bio-oil and less char is produced at higher temperatures and shorter residence times (Tripathi et al. 2016). Moreover, biochar comprises up to 35% of the total organic carbon (TOC) in numerous burning habitats, such as grasslands, savannas, and woods (generally known as pyrogenic carbon) (Skjemstad et al. 2002; Glaser and Amelung 2003; Bird et al. 2015). Hence, biochar is a natural component in many soils, and its addition in high amounts (*i.e.,* 100 t ha^{-1} or more) generally has no negative effects on soil function. As a result, majority of soils have a potential storage capacity size for biochar. In three separate ways, biochar supplements have an effect on soil C sequestration and atmospheric net CO_2 removals. Santos et al. (2012) and Wang et al. (2016) first demonstrated that the majority of pyrogenic carbon produced as by-product of bio-fuel pyrolysis operations (80–95%) is exceptionally resistant to microbial breakdown when added to soils and has a mean residence life of 100 years or more. As a result, the carbon storage that the biochar represents has the potential to last a very long time once it has been incorporated into the soil. Second, the addition of biochar can potentially interfere with natural SOM, which is already present in the soils to accelerate or decrease the pace of decomposition. These interactions could be caused by a variety of elements, such as changes in soil moisture and water holding capacity, pH variations, nutrient availability, and direct effects of adding biochar on the composition and activity of microbial communities. Following biochar additions, both having favourable and unfavourable impacts on the native SOM decay process have been explored (Song et al. 2016; Wang et al. 2016), although the effects soil C balance are typically small in a long run (Wang et al. 2016). Last but not the least, the use of biochar might affect plant production and, consequently, C present in the soil as vegetation wastes. Incorporation of biochar can have a wide range of effects on plant production, depending on the biochar's properties as well as those of the soil or the plant.

(ii) *Deployment of Perennial Grain Crops*

For the past three decades, scientists have been working on breeding programmes to develop perennial varieties of both cereal and annual crops. Cultivars of perennial grasses such as intermediate wheatgrass, used as breeding stocks, have root penetration systems and larger percentage of dry mass dedicated under the ground than conventional annual crops. Soil C inputs are substantially higher than those from annual crops, supporting higher SOC stock levels. Moreover, the necessity for soil manipulation and its detrimental impacts on soil organic carbon stocks and soil removal would be significantly reduced by perennial crops. Bigger and deeper roots may also lower N_2O emissions to the atmosphere and nitrate leaching losses to waterways (Abalos et al. 2016; Crews and Rumsey 2017; Glover et al. 2010; Pimentel et al. 2012).

14.7 Benefits for Soil Carbon Storage Entail the Following

- Lowering of the emissions of GHGs, particularly of CO_2.
- Usefulness in bringing down atmospheric temperatures.
- Reduces nutrient losses and aids in the maintenance of an appropriate biotic ecosystem.
- Raises the fertility and production of the soil.
- That could also lead towards conserving more water.
- Encourage and support root growth development.
- Prevent soil removal through erosion.

14.8 Conclusions

Different methods having feasibility were extensively assessed to conduct carbon sequestration (CS), and in a wider-scale as an attempt to reduce elevated atmospheric carbon dioxide (CO_2) build-up and to mitigate its adverse environmental effects. The complex carbon cycling that affects the atmosphere, soils *etc.*, results in equilibrium CO_2 concentrations that change all over geological periodic scales as situations change among its various components. Large and growing man-made emissions of CO_2 and other greenhouse gases (GHGs) have damaged the natural carbon cycle and threatened permanent climate change and ecological community disturbance in few decades if remained unstopped. If rapidly adopted on a large scale, the approaches offered by CS and farming technology could overcome these problems. Effective CS alternatives entails expansion of agroforests; shrinking agricultural land use and adopting agricultural cultivation practices that are more efficient, eco-friendly, and conservation tillage practices; cropping system diversification; enhancing soil carbon pools through augmenting production of biomass; accumulation of soil carbonates and stopping soil removal; geological CO_2 sequestration in lower profile of the reservoirs; and deeper CO_2 sequestration in the oceans. These technological methods altogether can create significant input into man-made carbon sinks. Some of the techniques, which are of significance, can be developed as CS systems that can bring both sustainability and economically valuable results. These include carbon capture and sequestration, change in the land use combined with burning of fossils from industrial processes, like Torrefaction, which can slow down wastes from agriculture and forests. Awareness campaigns in the form of placards, posters and logos can be advisable to showcase how climate change occurs and on how to protect and save the planet. They are a really effective approach to show why global warming is bad for the environment. There is necessity also for international support policies and agencies, which are fully boosted by the governments and industries to exploit the various paths for atmospheric CO_2 reduction.

References

Abalos D, Brown SE, Vanderzaag AC, Gordon RJ, Dunfield KE, Wagner-Riddle C (2016) Micrometeorological measurements over 3 years reveal differences in N_2O emissions between annual and perennial crops. Glob Chang Biol 22(3):1244–1255

Alcántara V, Don A, Well R, Nieder R (2016) Deep ploughing increases agricultural soil organic matter stocks. Glob Chang Biol 22(8):2939–2956

Allen M, Antwi-Agyei P, Aragon-Durand F, Babiker M, Bertoldi P, Bind M Zickfeld K (2019) Technical summary: global warming of 1.5 C. An IPCC Special Report on the impacts of global warming of 1.5 C above pre-industrial levels and related global greenhouse gas emission pathways, in the context of strengthening the global response to the threat of climate change, sustainable development, and efforts to eradicate poverty

Batjes NH (1996) Total carbon and nitrogen in the soils of the world. Eur J Soil Sci 47(2):151–163

Bhadwal S, Singh R (2002) Carbon sequestration estimates for forestry options under different land-use scenarios in India. Curr Sci 83(11):1380–1386

Bhattacharjee A, Dewar RL (1982) Energy principle with global invariants. The Physics of Fluids 25(5):887–897

Bhattacharyya T, Pal DK, Mandal C, Velayutham M (2000) Organic carbon stock in Indian soils and their geographical distribution. Curr Sci 79(5):655–660

Bhattacharyya T, Pal DK, Easter M, Batjes NH, Milne E, Gajbhiye KS, Powlson DS (2007a) Modelled soil organic carbon stocks and changes in the Indo-Gangetic Plains, India from 1980 to 2030. Agric Ecosyst Environ 122(1):84–94

Bhattacharyya T, Chandran P, Ray SK, Pal DK, Venugopalan MV, Mandal C, Wani SP (2007b) Changes in levels of carbon in soils over years of two important food production zones of India. Curr Sci 93(12):1854–1863

Bhattacharyya T, Pal DK, Chandran P, Ray SK, Mandal C, Telpande B (2008) Soil carbon storage capacity as a tool to prioritize areas for carbon sequestration. Curr Sci 95(4):482–494

Bhattacharyya T, Sarkar D, Sehgal JL, Velaytham M, Gajbhiye KS, Nagar AP, Nimkhedkar SS (2009) Soil taxonomic database of India and the states (1: 250, 000 scale). NBSS Publication

Bhattacharyya T, Chandran P, Ray SK, Mandal C, Tiwary P, Pal DK, Sahrawat KL (2014) Processes determining the sequestration and maintenance of carbon in soils: a synthesis of research from tropical India. Soil Horizons 55(4):1–16

Bhattacharyya T, Wani SP, Pal DK, Sahrawat KL (2017) Soil as source and sink for atmospheric CO_2. In: Goel M, Sudhakar M (eds) Carbon utilization. Green energy and technology. Springer, Singapore. https://doi.org/10.1007/978-981-10-3352-0_4

Bird MI, Wynn JG, Saiz G, Wurster CM, McBeath A (2015) The pyrogenic carbon cycles. Annu Rev Earth Planet Sci 43:273–298

Bloomfield J, Pearson HL (2000) Land use, land-use change, forestry, and agricultural activities in the clean development mechanism: estimates of greenhouse gas offset potential. Mitig Adapt Strateg Glob Chang 5:9–24

Bolin, B, Doos, BR (1989) Greenhouse effect

Bosatta E, Ågren GI (1997) Theoretical analyses of soil texture effects on organic matter dynamics. Soil Biol Biochem 29(11–12):1633–1638

Botkin DB, Saxe H, Araújo MB, Betts R, Bradshaw RH, Cedhagen T, Stockwell DR (2007) Forecasting the effects of global warming on biodiversity. Bioscience 57(3):227–236

Brown S, Lugo AE (1990) Tropical secondary forests. J Trop Ecol 6(1):1–32

Centre for Climate and Energy Solutions (2021) Short-lived climate pollutants

Choubineh A, Ghorbani H, Wood DA, Moosavi SR, Khalafi E, Sadatshojaei E (2017) Improved predictions of wellhead choke liquid critical-flow rates: modelling based on hybrid neural network training learning based optimization. Fuel 207:547–560

Conant RT, Paustian K, Elliott ET (2001) Grassland management and conversion into grassland: effects on soil carbon. Ecol Appl 11(2):343–355

Council for Agricultural Science (CAS) (2004) Climate change and greenhouse gas mitigation: challenges and opportunities for agriculture

Crews TE, Rumsey BE (2017) What agriculture can learn from native ecosystems in building soil organic matter: a review. Sustainability 9(4):578

Dehshibi RR, Sadatshojaie A, Mohebbi A, Riazi M (2019) A new insight into pore body filling mechanism during waterflooding in a glass micro-model. Chem Eng Res Des 151:100–107

Denardin JE, Kochhann RA, Faganello A, Sattler A, Manhago DD (2008) "Vertical mulching" como prática conservacionista para manejo de enxurrada em sistema plantio direto. Rev Bras Ciênc Solo 32:2847–2852

Dijkstra JPM, Reinds GJ, Kros H, Berg B, de Vries W (2009) Modelling soil carbon sequestration of intensively monitored forest plots in Europe by three different approaches. For Ecol Manag 258(8):1780–1793

Dixon RK, Solomon AM, Brown S, Houghton RA, Trexier MC, Wisniewski J (1994) Carbon pools and flux of global forest ecosystems. Science 263(5144):185–190

Ellert BH, Gregorich EG (1995) Management-induced changes in the actively cycling fractions of soil organic matter. In: Carbon forms and functions in forest soils. Wiley, London, pp 119–138

EPA (2022) "Emissions and Generation Resource Integrated Database (eGRID) 2020"

Eswaran H, Kimble J, Cook T, Beinroth FH (1992) Soil diversity in the tropics: implications for agricultural development. In: Myths and science of soils of the tropics, vol 29. Soil Science Society of America: American Society of Agronomy, Madison, pp 1–16

Eswaran H, Van Den Berg E, Reich P (1993) Organic carbon in soils of the world. Soil Sci Soc Am J 57(1):192–194

Fairbourn ML, Gardner HR (1972) Vertical mulch effects on soil water storage. Soil Sci Soc Am J 36(5):823–827

FAO (2011) The state of the worlds land and water resources for food and agriculture (SOLAW) managing systems at risk. United Nations, London

Feeley RA, Doney SC, Cooley SR (2009) Ocean acidification: present conditions and future changes in a high CO_2 world. Oceanography 22:36–47

Ghidey FESSEHAIE, Alberts EE (1998) Runoff and soil losses as affected by corn and soybean tillage systems. J Soil Water Conserv 53(1):64–70

Gill RA, Burke IC (1999) Ecosystem consequences of plant life form changes at three sites in the semiarid United States. Oecologia 121:551–563

Glaser B, Amelung W (2003) Pyrogenic carbon in native grassland soils along a climosequence in North America. Glob Biogeochem Cycles 17(2)

Glover JD, Reganold JP, Bell LW, Borevitz J, Brummer EC, Buckler ES, Xu Y (2010) Increased food and ecosystem security via perennial grains. Science 328(5986):1638–1639

Gogoi N, Baruah KK, Meena RS (2018) Grain legumes: impact on soil health and agroecosystem. In: Legumes for soil health and sustainable management. Springer, Singapore, pp 511–539

Hammond B, Mills L, Johnson W (1993) Vertical mulch for healthier trees and shrubs. University of Nevada, Cooperative Extension, Reno

Hassink J (1997) The capacity of soils to preserve organic C and N by their association with clay and silt particles. Plant Soil 191:77–87

Heidari S, Jafarzadeh Y, Seyfollahi Samarin M, Yegani R (2017) Study on the fouling behaviour of HDPE/PE-g-MA/EVA blend membrane fabricated via thermally induced phase separation method. Polyolefins J 4(2):235–251

Heidari S, Amirinejad M, Jahangirian H (2019) Investigation of fouling mechanisms using surface morphology and physicochemical membrane features. Chem Eng Technol 42(6):1310–1320

Herzog HJ (1998) Ocean sequestration of CO_2: An overview. In: Proceedings of the AWMA's second international specialty conference, 13–15 October 1998, Washington DC

Hobbie SE, Schimel JP, Trumbore SE, Randerson JR (2000) Controls over carbon storage and turnover in high-latitude soils. Glob Chang Biol 6(S1):196–210

Ibrahim MA, Guerra L, Casasola Coto F, Neely C (2013) Importance of silvo-pastoral systems for mitigation of climate change and harnessing of environmental benefits. Programa Agroambiental Mesoamericano (MAP). Fase I

IGBP Terrestrial Carbon Working Group, Steffen W, Noble I, Canadell J, Apps M, Schulze ED, Jarvis PG (1998) The terrestrial carbon cycle: implications for the Kyoto Protocol. Science 280(5368):1393–1394

IPCC (2001) IPCC third assessment report, WGII

IPCC (2013) Climate change 2013: the physical science basis. TF Stocker, *et al* Cambridge University Press, Cambridge, United Kingdom and New York, NY, USA

IPCC (2014a) Climate change 2014 synthesis report. IPCC, Geneva

IPCC (2014b) Mitigation of climate change. Contribution of working group III to the fifth assessment report of the intergovernmental panel on climate change. Cambridge University Press, Cambridge

IPCC (2018) Special report: global warming of 1.5 C

Jackson RB, Canadell J, Ehleringer JR, Mooney HA, Sala OE, Schulze ED (1996) A global analysis of root distributions for terrestrial biomes. Oecologia 108:389–411

Jafari A, Esmaeilzadeh F, Mowla D et al (2020) New insights to direct conversion of wet microalgae impregnated with ethanol to biodiesel exploiting extraction with supercritical carbon dioxide. Fuel

Jandl R, Lindner M, Vesterdal L, Bauwens B, Baritz R, Hagedorn F, Byrne KA (2007) How strongly can forest management influence soil carbon sequestration? Geoderma 137(34):253–268

Jobbágy EG, Jackson RB (2000) The vertical distribution of soil organic carbon and its relation to climate and vegetation. Ecol Appl 10(2):423–436

Kalisz PJ, Stringer JW, Wells RJ (1994) Vertical mulching of trees: effects on roots and water status. J Arboric 20(3):141–145

Kimble LR (1997) JM Conservation tillage for carbon sequestration/JM Lal R. Kimble. Nutr Cycl Agroecosyst 49:243–253

Kishwan J, Pandey R, Dadhwal VK (2009) India's forest and tree cover

Kleber M, Johnson MG (2010) Advances in understanding the molecular structure of soil organic matter: implications for interactions in the environment. Adv Agron 106:77–142

Kochsiek AE, Knops JM, Walters DT, Arkebauer TJ (2009) Impacts of management on decomposition and the litter-carbon balance in irrigated and rainfed no-till agricultural systems. Agric For Meteorol 149(11):1983–1993

Kozinski J, Łukasik RM, Matsuda T et al (2018) Supercritical and other high-pressure solvent systems: for extraction, reaction and material processing. Royal Society of Chemistry, Cambridge

Lal R (2004a) Carbon sequestration, terrestrial. In: Encyclopedia energy, vol 1, pp 289–298

Lal R (2004b) Soil carbon sequestration impacts on global climate change and food security. Science 304(5677):1623–1627

Lal R (2004c) Soil carbon sequestration to mitigate climate change. Geoderma 123:1–22

Lal R (2010) Enhancing eco-efficiency in agro-ecosystems through soil carbon sequestration. Crop Sci 50:S-120

Lal R (2015) Managing soil carbon through sustainable intensification of agro- ecosystems. Agric Dev 24:13–18

Lal R (2018) Digging deeper: a holistic perspective of factors affecting soil organic carbon sequestration in agroecosystems. Glob Chang Biol 24(8):3285–3301

Liu E, Yan C, Mei X, Zhang Y, Fan T (2013) Long-term effect of manure and fertilizer on soil organic carbon pools in dryland farming in northwest China. PLoS One 8(2):e56536

Loa M (2015) Trends in atmospheric carbon dioxide (2022-06-13). https://gml.noaa.gov/ccgg/trends/global.html

Luo Y (2007) Terrestrial carbon–cycle feedback to climate warming. Annu Rev Ecol Evol System 38:683–712

Luo Z, Wang E, Sun OJ (2010) Soil carbon change and its responses to agricultural practices in Australian agro-ecosystems: a review and synthesis. Geoderma 155(3–4):211–223

Marc L (2015) Greenhouse gas emissions: causes and sources. http://www.livescience.com/37821-greenhousegases.html

Meinshausen M, Meinshausen N, Hare W, Raper SC, Frieler K, Knutti R, Allen MR (2009) Greenhouse gas emission targets for limiting global warming to 2 C. Nature 458(7242):1158–1162

Meyer JH, Van Antwerpen R, Henry PC, Leibbrandt N (1992) Improved cane yields from vertical mulching under rainfed and irrigated cane conditions. The South African Sugar Technologists' Association, pp 89–94

Minasny B, Malone BP, McBratney AB, Angers DA, Arrouays D, Chambers A, Winowiecki L (2017) Soil carbon 4 per mille. Geoderma 292:59–86

Mohajan H (2011) Greenhouse gas emissions increase global warming

Moore TG (1998) Health and amenity effects of global warming. Econ Inq 36(3):471–488

Nair PKR, Nair VD, Kumar BM, Showalter JM (2010) C sequestration in agroforestry systems. Adv Agron 108:237–307

Nambiar KKM (1994) Soil fertility and crop productivity under long-term fertilizer use in India

NASEM (2019) Negative emissions technologies and reliable sequestration: a research agenda. National Academies of Sciences, Engineering, and Medicine/ The National Academies Press, Washington, DC

Nezhad SM, Wood DA, Sadatshojaei and E. et al (2020) New insight to experimental study of ionic solutions with a nonionic surfactant on wettability, interfacial tension and micro-model flooding. Fuel 285:119126

Nordt LC, Wilding LP, Drees LR (2000) Pedogenic carbonate transformations in leaching soil systems: Implications for the global C cycle. In: Global climate change and pedogenic carbonates. Lewis, Boca Raton, pp 43–64

Ogle SM, Breidt FJ, Paustian K (2005) Agricultural management impacts on soil organic carbon storage under moist and dry climatic conditions of temperate and tropical regions. Biogeochemistry 72:87–121

Ogle SM, Swan A, Paustian K (2012) No-till management impacts on crop productivity, carbon input and soil carbon sequestration. Agric Ecosyst Environ 149:37–49

Pal DK, Dasog GS, Vadivelu S, Ahuja RL, Bhattacharyya T (2000) Secondary calcium carbonate in soils of arid and semi-arid regions of India. In: Global climate change and pedogenic carbonates. Lewis, Boca Raton, pp 149–185

Pal DK, Bhattacharyya T, Ray SK, Bhuse SR (2003) Developing a model on the formation and resilience of naturally degraded black soils of the peninsular India as a decision support system for better land use planning. NRDMS, Department of Science and Technology (Govt. of India) Project Report, NBSSLUP (ICAR), Nagpur, p 144

Pal DK, Wani SP, Sahrawat KL (2015) Carbon sequestration in Indian soils: present status and the potential. Proc Natl Acad Sci India B Biol Sci 85:337–358

Paustian KAOJH, Andren O, Janzen HH, Lal R, Smith P, Tian G, Woomer PL (1997) Agricultural soils as a sink to mitigate CO_2 emissions. Soil Use Manag 13:230–244

PBL Netherlands Environmental Assessment Agency (2021) Trends in global CO_2 and total greenhouse gas emissions

Pimentel D, Cerasale D, Stanley RC, Perlman R, Newman EM, Brent LC, Chang DTI (2012) Annual vs. perennial grain production. Agric Ecosyst Environ 161:1–9

Poeplau C, Don A (2015) Carbon sequestration in agricultural soils via cultivation of cover crops–a meta-analysis. Agric Ecosyst Environ 200:33–41

Post WM, Emanuel WR, Zinke PJ, Stangenberger AG (1982) Soil carbon pools and world life zones. Nature 298(5870):156–159

Post WM, Izaurralde RC, Mann LK, Bliss N (2001) Monitoring and verifying changes of organic carbon in soil. In: Storing carbon in agricultural soils: A multipurpose environmental strategy. Springer, Dordrecht, pp 73–99

Sadatshojaei E, Esmaeilzadeh F, Fathikaljahi J, Barzi SEH, Wood DA (2018) Regeneration of the midrex reformer catalysts using supercritical carbon dioxide. Chem Eng J 343:748–758

Sadatshojaei E, Jamialahmadi M, Esmaeilzadeh F, Wood DA, Ghazanfari MH (2019) The impacts of silica nanoparticles coupled with low-salinity water on wettability and interfacial tension: experiments on a carbonate core. J Dispers Sci Technol 41:1159

Sadatshojaei E, Mowla D, Wood DA (2020a) Review of progress in microalgal biotechnology applied to wastewater treatment. In: Sustainable green chemical processes and their allied applications. Springer, Cham, pp 539–557

Sadatshojaei E, Wood DA, Mowla D (2020b) Third generation of biofuels exploiting microalgae. In: Sustainable green chemical processes and their allied applications. Springer, Cham, pp 575–588

Sadatshojaei E, Heidari S, Edraki Z, Wood DA, Ismail AF (2022) Carbon sequestration alternatives for mitigating the accumulation of greenhouse gases in the atmosphere. In: Bioresource technology: concept, tools and experiences. Wiley, Hoboken, pp 443–465

Sahrawat KL (2003) Importance of inorganic carbon in sequestering carbon in soils of the dry regions. Curr Sci 84(7):864–865

Sahrawat KL, Bhattacharyya T, Wani SP, Chandran P, Ray SK, Pal DK, Padmaja KV (2005) Long-term lowland rice and arable cropping effects on carbon and nitrogen status of some semi-arid tropical soils. Curr Sci 89(12):2159–2163

Santos F, Torn MS, Bird JA (2012) Biological degradation of pyrogenic organic matter in temperate forest soils. Soil Biol Biochem 51:115–124

Seddiqi H, Sadatshojaie A, Vaferi B et al (2019) Mathematical model for iron corrosion that eliminates chemical potential parameters. Chin J Chem Eng 28(2):603–612

Schimel DS (1995) Terrestrial ecosystems and the carbon cycle. Glob Chang Biol 1(1):77–91

Sekhon GS, Meelu OP, Katyal JC, Eswaran H, Abrol IP (1994) Organic matter management in relation to crop production in stressed rainfed systems. In: Stressed ecosystems and sustainable agriculture

Shao W, Keim BD, Garand JC, Hamilton LC (2014) Weather, climate, and the economy: explaining risk perceptions of global warming, 2001–10. Weather Clim Soc 6(1):119–134

Sheikh MA, Kumar M, Bussmann RW (2009) Altitudinal variation in soil organic carbon stock in coniferous subtropical and broadleaf temperate forests in Garhwal Himalaya. Carbon Balance Manag 4:1–6

Shekhawat NS, Phulwaria M, Rai MK, Kataria V, Shekhawat S, Gupta AK, Modi R (2012) Bioresearches of fragile ecosystem or desert. Proc Natl Acad Sci India B Biol Sci 82:319–334

Sherrod LA, Peterson GA, Westfall DG, Ahuja LR (2005) Soil organic carbon pools after 12 years in no-till dryland agroecosystems. Soil Sci Soc Am J 69(5):1600–1608

Shokri E, Yegani R, Heidari S, Shoeyb Z (2015) Effect of PE-g-MA compatibilizer on the structure and performance of HDPE/EVA blend membranes fabricated via TIPS method. Chem Eng Res Des 100:237–247

Six J, Paustian K (2014) Aggregate-associated soil organic matter as an ecosystem property and a measurement tool. Soil Biol Biochem 68:A4–A9

Six J, Conant RT, Paul EA, Paustian K (2002) Stabilization mechanisms of soil organic matter: implications for C-saturation of soils. Plant Soil 241:155–176

Skjemstad JO, Reicosky DC, Wilts AR, McGowan JA (2002) Charcoal carbon in US agricultural soils. Soil Sci Soc Am J 66(4):1249–1255

Song X, Pan G, Zhang C, Zhang L, Wang H (2016) Effects of biochar application on fluxes of three biogenic greenhouse gases: a meta-analysis. Ecosyst Health Sustain 2(2):e01202

Swaminathan MS (2000) Science in response to basic human needs. Science 287(5452):425–425

Tripathi M, Sahu JN, Ganesan P (2016) Effect of process parameters on production of biochar from biomass waste through pyrolysis: a review. Renew Sust Energ Rev 55:467–481

Trost B, Prochnow A, Drastig K, Meyer-Aurich A, Ellmer F, Baumecker M (2013) Irrigation, soil organic carbon and N_2O emissions. A review. Agron Sustain Dev 33:733–749

U.S. Energy Information Administration (EIA) (2022) International emissions by fuel

Ussiri DA, Lal R (2017) Carbon sequestration for climate change mitigation and adaptation. Springer, Cham, pp 287–325

Van der Ploeg F, Rezai A (2017) Cumulative emissions, unburnable fossil fuel, and the optimal carbon tax. Technol Forecast Soc Chang 116:216–222

Velayutham M, Pal DK, Bhattacharyya T (2019) Organic carbon stock in soils of India. In: Global climate change and tropical ecosystems. CRC Press, Boca Raton, pp 71–95

Wang J, Xiong Z, Kuzyakov Y (2016) Biochar stability in soil: meta-analysis of decomposition and priming effects. GCB Bioenergy 8(3):512–523

Wani NR, Qaisar KN (2014) Carbon percent in different components of tree species and soil organic carbon pool under these tree species in Kashmir valley. Curr World Environ 9(1):174

West TO, Post WM (2002) Soil organic carbon sequestration rates by tillage and crop rotation: a global data analysis. Soil Sci Soc Am J 66(6):1930–1946

Williams JD, Wuest S, Williams JD, Wuest SB (2011) Tillage and no-tillage conservation effectiveness in the intermediate precipitation zone of the inland Pacific Northwest, United States. J Soil Water Conserv 66(4):242–249

Wong VN, Dalal RC, Greene RS (2009) Carbon dynamics of sodic and saline soils following gypsum and organic material additions: a laboratory incubation. Appl Soil Ecol 41(1):29–40

Wood G (2020) Fossil fuels in a carbon-constrained world. The Palgrave Handbook of Managing Fossil Fuels and Energy Transitions:3–23

World Meteorological Organization (WMO) (2021) Greenhouse gas bulletin

Yuste JC, Baldocchi DD, Gershenson A, Goldstein A, Misson L (2007) Microbial soil respiration and its dependency on C inputs, soil temperature and moisture. Glob Chang Biol 13:2018–2035

15 Nature-based Solutions (NbS) for Dryland Agriculture in Semi-Arid Regions of Maharashtra, India: A Short Review with Possible Approaches for Building Climate Resilience

Wasim Ayub Bagwan

Abstract

In this chapter, we will discuss the environmental issues related to semi-arid zones in Maharashtra and explore possible solutions that are easily accessible at the local level. The semi-arid ecosystem poses challenges for long-term residents due to its harsh conditions, making it unsuitable for habitation. To create a proper and habitable environment and prevent migration resulting from climate change in these regions, it is crucial to implement scientific and effective remedies to restore the natural ecosystem. There is an urgent need to implement measures to protect natural resources, reverse resource depletion, and ensure sustainability. Increased resource consumption and declining quality have significantly impacted the environment worldwide, jeopardizing livelihoods and economic well-being. To address these challenges arising from human activities and climate change, sustainable solutions are imperative. Nature-based Solutions (NbS) offer a promising approach to mitigate the impacts of extreme weather events and human influence. Through the adoption of nature-based climate-resilient technologies, we can achieve multiple benefits from socioecological and environmental perspectives. NbS enables the development of local ecosystem-based resilience, effectively addressing climate change while tackling social issues such as poverty and resource depletion. Ecosystem-based Adaptation (EbA) further enhances this approach by incorporating cost-effective natural solutions into green infrastructure. Given that the semi-arid region of Maharashtra covers 67% of the total geographical area and is highly vulnerable to climate change, NbS emerges as the most effective and affordable approach to manage present and future climatic variability.

Keywords

Climate action · Climate change · Ecosystem-based adaptation (EbA) · Ecosystem services (ES) · Nature-based solutions · Semi-arid Maharashtra

15.1 Introduction

Nature-based Solutions (NbS) are defined by IUCN as "actions to protect, sustainably manage, and restore natural or modified ecosystems, that

W. A. Bagwan (✉)
School of Rural Development, Tata Institute of Social Sciences, Tuljapur, Maharashtra, India

U. Chatterjee et al. (eds.), *Climate Crisis: Adaptive Approaches and Sustainability*, Sustainable Development Goals Series, https://doi.org/10.1007/978-3-031-44397-8_15

address societal challenges effectively and adaptively, simultaneously providing human well-being and biodiversity benefits" (IUCN 2022). According to the NbS concept, infrastructure should be built in harmony with nature, and performance evaluations should include efforts to reduce the negative environmental effects of these initiatives (Stefanakis et al. 2021). Drylands cover 41.3% of the global terrestrial area and are located at tropical and temperate latitudes. They are divided into four categories: dry sub-humid, semi-arid, arid, and hyper-arid environments, which cover a wide range of habitats (IUCN 2022c). Drylands are ecosystems that cover more than 40% of the terrestrial surface and are characterized by high temporal and geographical rainfall variability. They include rangelands, grasslands, and woods. Grasslands dominate the drylands, which account for more than a fifth of the planet's geographical area (IUCN 2019a). NbS are an important part of the larger global effort to meet the goals of the Paris Climate Agreement. They are an important part of decarbonization, minimizing climate change risks, and building climate-resilient societies. They promote human–nature harmony as well as ecological progress, and they offer a comprehensive people-centered solution to climate change. They are efficient, effective, long-term, and globally scalable (UNCA 2019).

Semi-arid regions (SARs) face extreme climates, environmental degradation, and limited resources, while also being affected by poverty, inequality, and rapid socioeconomic changes. Climate change further complicates these vulnerabilities and amplifies existing pressures. It is vital to understand how to empower individuals, organizations, and governments to adapt to these challenges, minimize vulnerability, and promote long-term resilience (Pillai and Bendapudi 2019). The ability to perform numerous operations simultaneously to supply a set of connected Ecosystem Services (ES) is a major feature of NbS. The NBS strategy is based on the idea that improving and maintaining some ecosystems can help to mitigate the negative effects of climate change while also delivering many environmental, economic, and social advantages (Gómez Martín et al. 2020). To effectively scale up climate change adaptation strategies while achieving environmental and socioeconomic development goals, countries can benefit from ecosystem-based adaptation (EbA), a nature-based and human-centered strategy to tackle the impacts of climate change (D'souza et al. 2020). A healthy ecosystem can supply a variety of critical public goods, such as clean water, nutrient cycling, climate regulation, and food security services, all of which contribute to human well-being directly or indirectly (Sonneveld et al. 2018). The current study is based on the identification of methodologies and natural solutions for adapting to climate change across the entire semi-arid Maharashtra.

15.2 Study Area

As displayed in Fig. 15.1, Maharashtra covering an area of 307,713 km^2 lies in the western part of India. And Fig. 15.2 shows the semi-arid area of Maharashtra susceptible to climate change and the adaptation measures suggested for ecological restoration under this short review. The state shares a boundary with the Arabian Sea with a coastline of 720 km. The entire state is located on the Deccan peninsula with basalt as the major geological setting. The state has diverse landforms and land covers. The state is divided into 36 districts with 6 administrative divisions. Chhattisgarh, Goa, Gujarat, Karnataka, Telangana, Madhya Pradesh, and Dadra and Nagar Haveli (Union Territory) all share the state's borders. Bhima, Godavari, Krishna, Purna, Tipi, Wainganga, and Narmada are major rivers. The rainy season is restricted to the southwest monsoon, which accounts for 80% of precipitation from June to October. The average temperature is 25–28 °C, with the maximum temperature in May and the lowest in December. Entisols, Inceptisols, and Vertisols are the most common soil types found in the state, followed by Alfisols and Mollisols (Sharma and Verma 2010). According to the

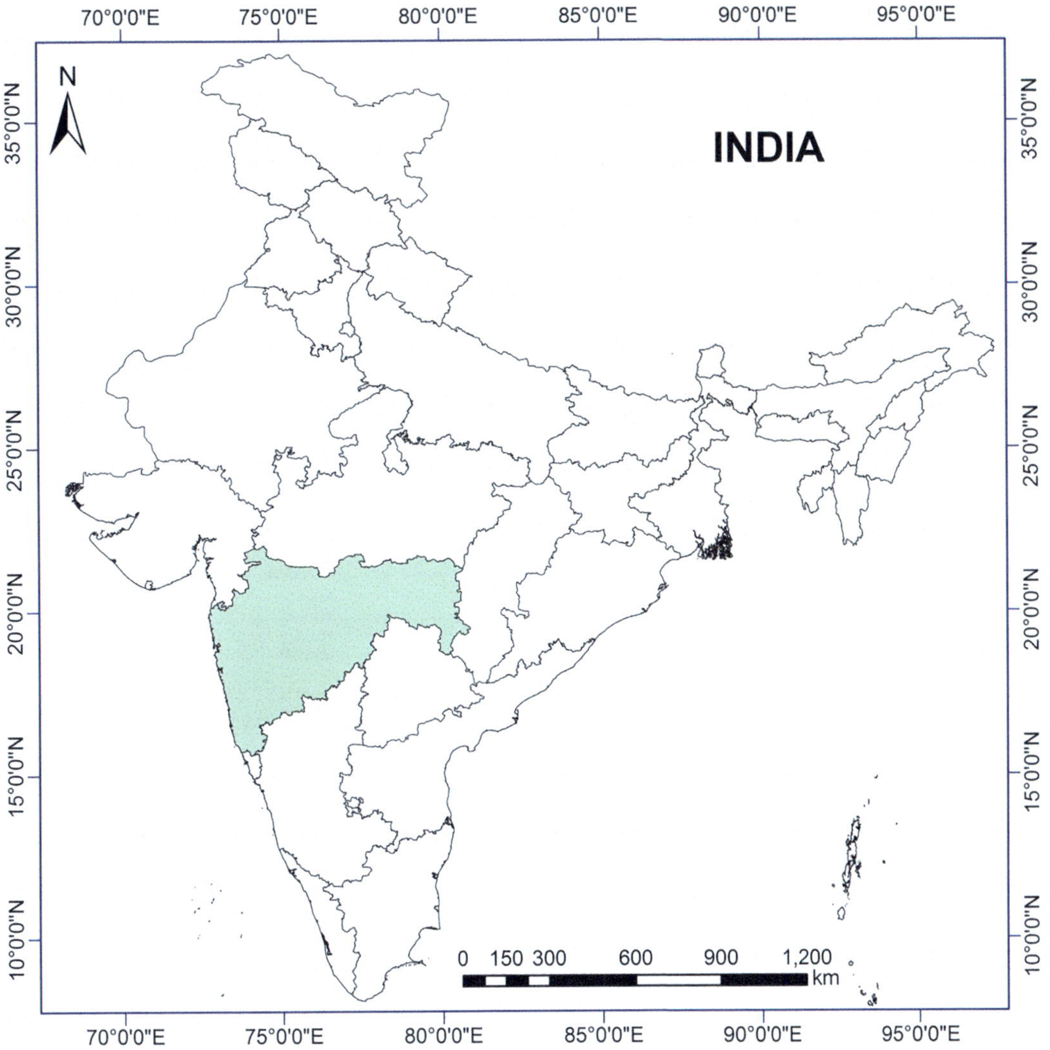

Fig. 15.1 Location of Maharashtra in India

2011 Population Census, the State's population was 11.24 crores, accounting for 9.3% of the total population of India. The state had India's second-largest population. The state's population density was 365 people/km^2 (DES 2022). According to the Köppen-Geiger climate classification, the Koyna reservoir catchment occupies Am: Tropical, monsoon, and Aw: Tropical. Savannah zone; whereas Ujjani reservoir catchment comes under BSh: Arid, steppe, hot zone, and CSa: Temperate, dry summer, hot summer can be observed across Maharashtra (Beck et al. 2018).

15.3 Data and Preprocessing

To showcase the zones of Maharashtra based on the Aridity Index AI), the high-resolution evapotranspiration (30 arc-seconds) data were downloaded from https://cgiarcsi.community/data/global-aridity-and-pet-database/ [Accessed on 20 Feb 2022]. The data are available in GeoTIFF format with the capability to achieve global sustainability, biodiversity, environmental conservation, poverty alleviation, and climate change adaptation, particularly in underdeveloped nations (Trabucco and Zomer 2019). As per the

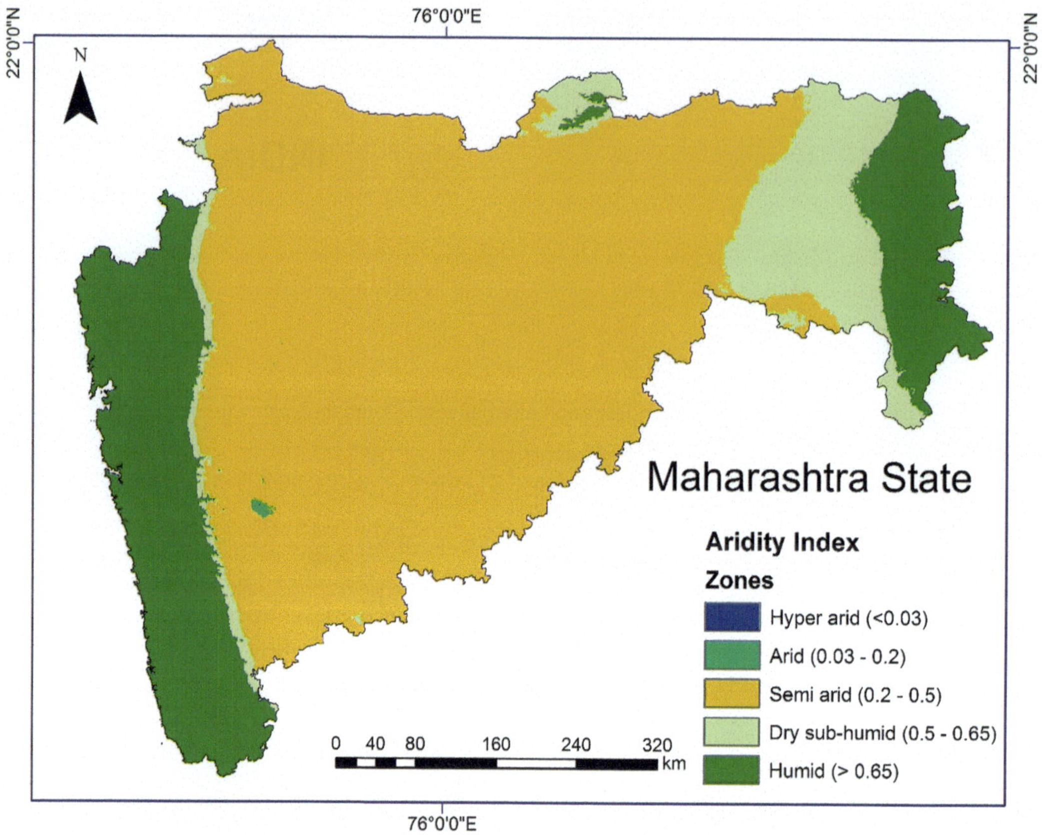

Fig. 15.2 Classification of Maharashtra based on aridity index

Table 15.1 Area under zones of Aridity Index defined by UNEP (1997)

Zones	Area (km^2)	Proportion (%)
Hyper arid	0.56	0.00
Arid	232.21	0.08
Semi-arid	208035.4	67.61
Dry sub-humid	31879.30	10.36
Humid	67565.50	21.96
	307713.00	

shape file of the study region, the raster data were extracted. The dataset was mapped into the correct unit by multiplying the raster grid by 0.0001 by using the map algebra tool in ArcGIS 10 platform. Then, reclassified it by using UNEP (1997) zones based on AI values and converted to polygon form. Table 15.1 shows the area in (km^2) under each zone.

15.4 Search for Literature

The present study is based on scientific data collection of the Nature-based Solutions (NbS) strategy adopted worldwide. References were retrieved using structured keywords from online platforms, including Scopus and Web of Science, with searching the articles published on "Nature-based Solutions", "Ecosystem-based Adaptation (EbA)", "Climate Change Adaptation", "Climate Resilient Techniques", "Semi-arid Tropics adaptation" and so on. NbS-related books, proceedings, comments, reports, review articles, and research papers from peer-reviewed journals from the publishers like Nature, Springer, Elsevier, Taylor, and Francis, etc. have been considered to carry out the literature inspection. Considering the objectives of the present study, the reports, planning reports, laws, and legislations are also covered by the respective domain of the government body.

15.5 India's Picture for NbS

The National Environment Policy (NEP) 2006 establishes a wide policy framework for the environment and climate change, promoting sustainable development while respecting ecological restrictions and social justice imperatives. The present development paradigm emphasizes long-term growth and tries to capitalize on the synergies between tackling climate change and fostering economic growth. The National Action Plan on Climate Change (NAPCC) puts a greater emphasis on the interventions that are required. Currently, the NAPCC is implemented through eight National Missions, which outline mitigation and adaptation goals in the face of climate change (UNFCC 2020).

15.6 Climate Change Has Driven Problems in Dryland Agriculture

Drylands possess complex characteristics and intertwined biophysical and socioeconomic linkages, including high temporal and geographical rainfall variability. Drylands are constantly threatened on several fronts. The distribution of resources, such as water and nutrients, and, as a result, the distribution and abundance of vegetation, is determined by these yearly and inter-annual rainfall changes. To deal with the extreme changes in rainfall and temperature, species living in these environments have developed very distinctive physiological and behavioral adaptations, allowing them to take advantage of the uneven distribution of water and nutrients throughout dryland landscapes (IUCN 2019b). Changing climate jeopardizes human development by skipping SDGs. Due to climate change slow-down or undercut, the progress achieved in fighting against hunger targets has been reversed. The attacking face of environmental change and climatic variability in more frequent and severe weather conditions has an impact on agricultural productivity and puts additional strain on already vulnerable food and ecological systems. This occurrence of changes primarily affects small landholders and the rural poor (FAO 2017a). Figure 15.2 shows the zonation of Maharashtra based on the aridity index. Based on the zonation, Table 15.1 shows the area of dryness and wetness with respected proportion.

15.7 Ecosystem Management Through NbS

Nature-based solutions that improve agricultural land management often increase productivity while also providing climatic benefits, contributing to reduced land conversion pressure (UNEP 2021). This can be attained through ecosystem protection and restoration can benefit in developing resilience, this collection helps to mitigate and buffer the consequences of droughts, floods, and extreme weather events (Mijatović et al. 2013). The concept of these solutions has only lately developed in the context of our failure to stabilize the climate or halt the loss of biodiversity. High-level pledges for 'nature', on the other hand, translate into afforestation targets, often monocultures with species, which can result in climate change maladaptation, carbon storage compromises, and negative impacts on biodiversity and sustainable development (Seddon et al. 2019).

15.8 Knowledge of Indigenous People

Indigenous knowledge contains comprehensions of how to deal with and adapt to environmental change and operates at smaller spatial and temporal scales than science (UNESCO 2022). NbS provides an opportunity for developing countries to not only demand a fair and cost-effective share of financial resources from the developed world but also to take the lead in demonstrating how traditional technologies may be used to combat climate change consequences (Gokhale 2021). Many Indigenous peoples have strong ties to the "land," which necessitates special considerations when it comes to comprehending and responding to environmental change (Ford et al. 2020). It refers to a variety of activities that have been used for decades, are based on indigenous knowledge, or were simply referred to as "conservation agriculture" (Hallstein and Iseman 2021).

15.9 Protection Against Evaporation

Sustainable water resource management and water conservation technologies are required to meet socioeconomic demands for water resources while sustaining healthy dryland ecosystems, as future water deficits will be driven mostly by increased water demand (Lian et al. 2021). In dryland cropping systems, evaporation accounts for the majority of water loss, and management strategies such as no-till have been used to reduce evaporative losses (Hansen et al. 2012). Additionally, the locally available material has a great contribution to nutrient recycling and resource value enhancement. In dry and semi-arid regions, mulching has emerged as a crucial water conservation technique in current agricultural production. By retaining soil water and adjusting soil temperature, mulch material minimizes evaporation by shielding the soil surface from sunlight (Kader et al. 2019).

15.10 Circular Economy in Agriculture

In a circular economy-based agriculture system, all steps of the food system, including producing, harvesting, packaging, processing, transporting, marketing, consuming, and disposing of food, are designed to support sustainable development. An important component of a circular agriculture model that attempts to minimize CO_2 emissions, use natural resources more efficiently, and reduce input use significantly is the combination of mixed crop-livestock and organic farming, agroforestry, water recycling, and wastewater reuse (Kristinn et al. 2021). A circular economy reduces energy and material leakage from the system by recirculating them throughout production, whereas a linear economy relies on external inputs to generate outputs and waste. Improved cattle integration in the circular economy will result in fewer losses and lower GHG emissions (FAO 2021).

15.11 Role of Forests and Grasslands in Semi-Arid Dryland

To optimize the promising potential of forests and agroforestry programs in arid areas, they must be completely integrated into the REDD+ framework, with a particular focus on co-benefits (UNCCD 2013). Dryland biodiversity is critical in the fight against poverty, climate change, and desertification (Davies et al. 2012). Nature-based solutions (NbS) include ecosystem-based adaptation (EbA) and ecosystem-based disaster risk reduction (Eco-DRR). These phrases refer to the application of biodiversity and ecosystems to social adaptation to global change (UNFCCC 2021). To show the Land Use and Land Cover (LULC) of Maharashtra in 2020, Sentinel-2 data of 10 m spatial resolution is used (Karra et al. 2021). The source of the website is https://livingatlas.arcgis.com/landcoverexplorer. The raster data comes in GeoTIFF files for each year of the Sentinel-2 Land Use/Land Cover map produced by ESRI, Microsoft, and Impact Observatory. The area of interest is extracted from provided data using the 'Extract by mask' tool in ArcGIS software. Next, a raster to polygon operation was performed to compute the area under each land cover type and its proportion calculated (Table 15.2).

The dominance of agriculture as the most extensive land use type (covering approximately

Table 15.2 Land use/land cover in 2020 extracted from the map provided by ESRI

Land cover	Area (km^2)	Proportion (%)
Shrubs	4423.67	1.44
Herbaceous vegetation	9733.6	3.16
Agriculture	229,942	74.73
Urban	4135.31	1.34
Bare	217.79	0.07
Permanent water bodies	1768.3	0.57
Herbaceous wetland	609.45	0.20
Closed forest, needle leaf needle leaf	527.73	0.17
Closed forest, evergreen, broad leaf	2003.4	0.65
Closed forest, deciduous broad leaf	30144.3	9.80
Closed forest	2265.24	0.74
Open forest, evergreen	0.632	0.00
Open forest, evergreen	2.626	0.00
Open forest, deciduous	5163.88	1.68
Open forest	16507.7	5.36
Open sea	267.582	0.09
	307713.3	100.00

74.73% of the total area) indicates a heavy reliance on intensive farming practices that could lead to environmental degradation. Such practices may involve the use of chemical fertilizers and pesticides that can contribute to soil and water pollution, as well as the clearing of forests and other natural habitats for agricultural land conversion, leading to the loss of biodiversity. Moreover, the expansion of urban areas (accounting for 1.34% of the total area) may result in the conversion of agricultural lands and natural habitats, leading to habitat fragmentation and loss of wildlife. Urbanization also poses challenges such as increased air and noise pollution, increased demand for water and energy, and the generation of waste and sewage. On the other hand, the presence of forests, wetlands, and permanent water bodies in the state (albeit in relatively smaller proportions) can provide various ecosystem services, such as carbon sequestration, water regulation, and biodiversity conservation, which are critical for environmental sustainability. Therefore, preserving and managing these natural ecosystems can help to maintain the balance of the local environment, reduce greenhouse gas emissions, and ensure the continued provision of ecosystem services. Overall, the LULC status in Maharashtra in 2020 highlights the need for a more sustainable approach to land use planning and management, which considers the impacts of human activities on the environment and seeks to maintain a balance between economic development and environmental protection. This may involve promoting more sustainable farming practices, protecting natural habitats, and promoting urban planning that prioritizes green spaces and environmental sustainability (Fig. 15.3).

15.12 Agriculture Practices for Conservation of Nature

Many NbS are applied largely by farmers or producers in the context of agricultural productivity and pasture management. These actions may provide a direct economic advantage to the producer in the form of higher yields or lower costs, as well as societal benefits. Technical help and transition financing may be sufficient to make long-term reforms if the advantages to the landowner are significant. Many of these approaches are in line with the burgeoning area of 'regenerative agriculture' (Iseman and Miralles-Wilhelm 2021). Conservation tillage, cover crops, biochar application to agricultural fields, and strategic use of synthetic and organic fertilizers have all been proposed as ways to minimize GHG emissions from agriculture. Agricultural management practices can be improved to reduce soil disturbance by reducing the frequency and scope of cultivation to reduce soil C loss and/or increase soil C storage (Zaman et al. 2021). Drought resilience is increased by maintaining vegetation cover in dryland areas and agricultural methods such as the use of shadow crops, nutrient-rich plants, and vegetation litter. In arid terrain, prescribed burning and the development of physical firebreaks minimize fuel loads and the risk of large-scale fires (Sudmeier-Rieux et al. 2019).

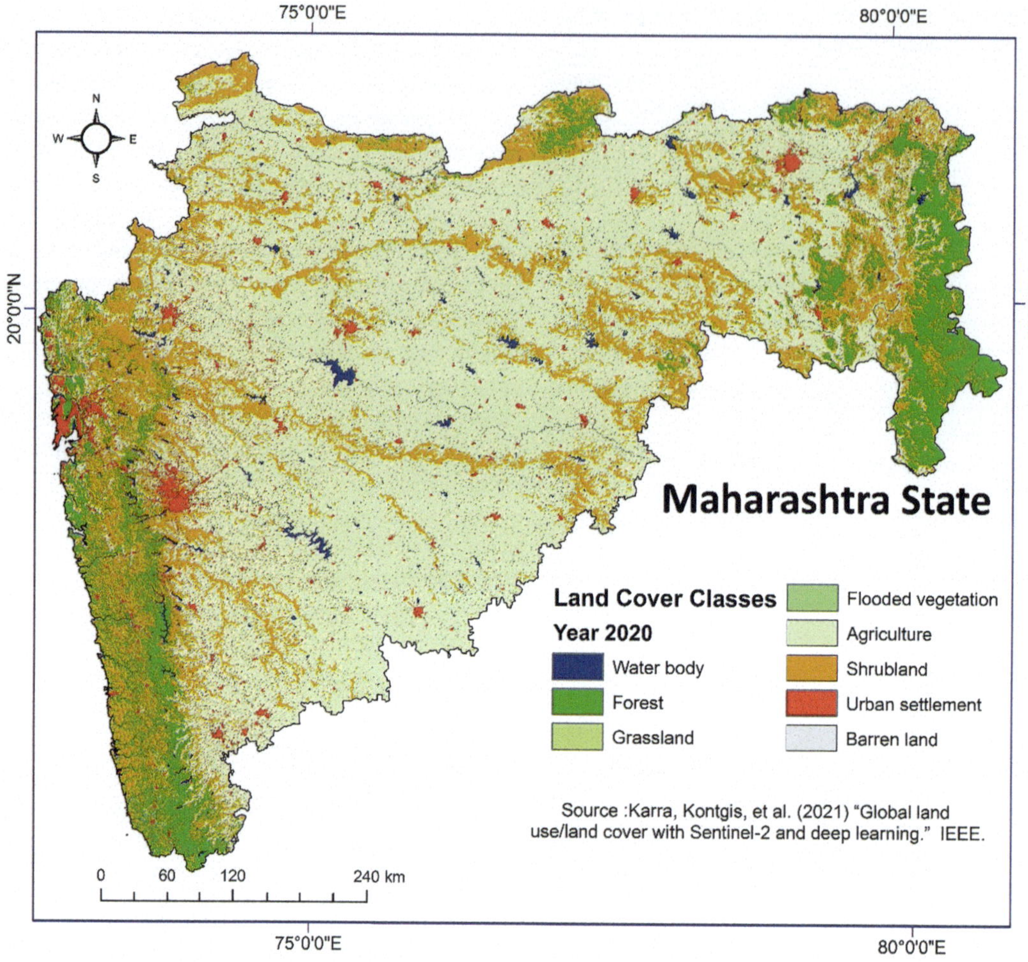

Fig. 15.3 Land use/land cover of 2020 by using Sentinel-2 satellite. (Source: Karra et al. (2021) "Global land use/land cover with Sentinel-2 and deep learning." IEEE)

15.13 Agroforestry for Ecological Restoration

Soil fertility can be improved by agroforestry. The increase in soil organic matter and biological nitrogen-fixing by leguminous trees are the key reasons for this (Mbow et al. 2014). The pigeon pea is a leguminous shrub from *the Fabaceae* family. Its life cycle can span anywhere from 1 to 5 years and it can reach a height of 4 m. A woody stem and a taproot that can reach depths of 2 m aid in loosening compacted soil. After 4–5 months, it begins to flower and produce pods with edible seeds that can be white, yellow, brown, black, or purple in color, depending on the type, and may even be spotted with brown or purple. Pigeon peas naturally self-pollinate, but they also have 20% cross-pollination and are heavily visited by bees (Miccolis et al. 2016).

15.14 Ecosystem Restoration Through NbS

The UN declared this decade intended for ecosystem restoration. Ecological restoration at a small scale for recovering native ecosystems for plants and soil has a great help in the reduction of land degradation. There are many ways to restore the natural ecosystem and it includes many aspects of safeguard. Ecosystem restoration comprises assisting in the recovery of ecosystems that have been degraded or destroyed, as well as the preservation of intact ecosystems (decadeonrestoration.org 2022). For the delivery of ecosystem services, restoring functions, flows of energy, nutrients, and other subsidies through the landscape or seascape may be as vital as, if not more important than, restoring composition and structure. Understanding the landscape or seascape environment is critical for prioritizing restoration priorities such as where, what, how, and with whom to work, as well as the sorts of management actions required (Valderrábano et al. 2021).

15.15 Soil Health Management

Fewer carbon emissions and more carbon capture are the hallmarks of eco-friendly and sustainable land management. Soil organic carbon adds to the soil's fertility and water-holding capacity, and hence controls the soil's ability to generate food and support other species to a considerable extent. When soil productivity, and hence carbon stock, increases, so does the resilience of societies and ecosystems (IUCN 2015).

15.16 Water Resource Management

Water is the fundamental component in defining dryland social-ecological systems; aridity in drylands is strongly intertwined with vegetation, climate, and humans, all of which fluctuate over various spatiotemporal scales (Lian et al. 2021). Climate change has had an impact on agricultural production, prompting individuals to travel in search of jobs. As a result, women's agricultural workload has expanded dramatically. Water sources have begun to dry up, and unexpected monsoons and floods have begun to wreak havoc on crop output (Sonneveld et al. 2018).

15.17 Forest Landscape Management

Forest ecosystem deterioration in drylands causes a loss of biomass and biodiversity, as well as water supplies and carbon storage capacity. This, however, can be avoided. Forest resource protection, restoration, and sustainable management can aid minimize and adapt to climate change and drought impacts, as well as prevent land degradation/desertification (UNCCD 2013). Maharashtra is not new to the NbS activities. One such activity is 'Seed ball'. The seeds of various fruits are dried in the sun with a thick layer of black soil to create a form of "seed ball" that farmers then toss at the foot of hills and along the roadside during celebrations before the onset of the seasonal rains. These seed balls eventually sprout trees after being wet with rain, and over time, vast forests are formed. These all fall under the category of "Nature-based solutions" to numerous issues with life and a means of subsistence (Haider 2022). The India State of Forest Report (ISFR) 2021 revealed that Maharashtra has a total forest cover of 50,798 km^2, an increase of 20 km^2 from the 2019 report. The Forest Department, Forest Development Corporation of Maharashtra (FDCM), and Revenue Department manage 55,827.96 km^2, 3461.91 km^2, and 1434.31 km^2 of the forest area, respectively. The Forest Department has also taken over 1182.90 km^2 of private forest. Despite this, the forest cover in the state is only 16.5% of the geographical area, falling short of the National Forest Policy's target of 33%. The forest cover consists of 17.2% very dense forest, 40.5% moderately dense forest, and 42.3% open forest (DES 2023).

15.18 Biochar for Soil Health Enrichment

Biochar production utilizing low-tech and thus accessible technologies can significantly reduce climate change in a distinctive developing-world setting. The carbon footprint of manufacturing biochar was lower than that of composting, and both compost and biochar had much lower carbon footprints than landfilling (Aquije et al. 2021). A modeling study to manage the agricultural soil as a carbon sink through the adoption of negative emission strategies conducted in the five districts namely, Jalna, Dhule, Ahmednagar, Amravati, and Yavatmal the state to deal with this global burning issue. The study found that how the correct proportion of irrigation, biochar, and fertilizer might potentially boost soil carbon by up to 300% and aid in reducing climate change (vigyanprasar.gov.in 2022).

15.19 River Rejuvenation

NbS can be used in several areas of the riverine system, including the river channel, the wetlands, and the riparian zone. Allowing the river to meander, incise, and flood its floodplains in the channel creates a biodiverse ecosystem that slows high water flows and so mitigates flood hazards, which is a typical strategy in river restoration. In terms of the benefits it can provide, riparian vegetation serves a dual purpose. It not only benefits the riverine habitat, but it also slows the flow of water and retains silt from the hill slopes that must travel through the riparian zone before entering the channel (Keesstra et al. 2018).

15.20 Restoration of Green Infrastructure

The term "Green Infrastructure" (GI) refers to the interconnected network of natural areas and pathways within a specific location. This type of infrastructure encompasses a range of open spaces, such as parks, gardens, fields, and woodlands, as well as water features like lakes, ponds, and rivers. Additionally, infrastructure that incorporates water resources may be referred to as "blue infrastructure," but is still considered part of the larger green infrastructure network (devon.gov.uk 2023). One of the examples of the restoration of GI is located near Solapur city named Ekrukh Lake; a constructed wetland through rejuvenation in 2019 (Fernandes et al. 2022). This restoration helps to combat water scarcity and acts as a habitat for birds and aquatic fauna. Also, it helps to control the temperature in close vicinity.

Artificial and natural reservoirs are crucial for securing water for people's lives and livelihoods. About 580,000 tanks of various sizes are scattered out across India, 150,000 of which are found in the semi-arid Deccan plateau. The majority (42%) of irrigation dams are located in Maharashtra alone. Tank systems are very helpful for recharging groundwater, giving cattle access to water, and irrigating crops. Tanks are complex ecological systems that are influenced by a variety of factors including urbanization, agricultural patterns, land use, and managerial institutions nearby. They are also a useful source of silt for fertilization and construction material. Community-based tank rejuvenation is vital in drought-prone regions to conserve water for irrigation. These systems require continuous maintenance, repair, and monitoring. In the pre-British era, local communities played a role in upkeep, but ownership was limited to the wealthy. However, post-independent India saw a decline due to limited community involvement and a lack of an integrated approach as tanks came under state ownership. State governments are now taking steps to revitalize small dams and tanks. In Maharashtra, initiatives like tank construction and silt removal are regularly carried out during drought years through schemes like the Employment Guarantee Act. The Jalayukt Shivar Yojana, launched in 2016, focuses on desilting and rejuvenating water bodies. These efforts aim to improve the utility and effectiveness of these irrigation systems (Zade et al. 2020).

15.21 Participatory Rural Appraisal (PRA) as a Tool for Community Engagement

The needs of rural communities and target groups can help local people to analyze their condition and communicate with outsiders. This enables communities to propose demands dealing with development agencies and institutions. With the help of PRA, the immediate support necessary for community strengthening is done by decision-making (FAO n.d.). PRA acts as a stimulant for the active participation of local people. This is one of the best methods to gather people and to summarize the situation in real time prevailing at ground level at the village level. The community engagements by conducting PRA help all age groups to involve in it with enthusiasm. Public participation and active involvement is a key to success in any planning and development of projects and schemes initiated by the government.

As documented by Pretty and Vodouhê (1997), extensionists and researchers can collaborate on the same team when PRA and other participatory methodologies are applied. They converse with farmers, share their knowledge and experiences, and come to an understanding of what is most important. All sides are brought together as a result. Farmers have more faith that experts can assist them without forcing solutions on them.

PRA has great potential to locate and safeguard community resources. By the PRA technique, we can enhance the water resources, and identify the area for rehabilitation and restoration. A manual prepared by (FAO 2006) states that in the identification of physical and biophysical characteristics from an agroecological focus, the local community involvement could assess in a better way. The local community knows better resources and natural issues continuing in the village so, the management with the help of technology could be easy. The different sessions conducted during PRA execution such as brainstorming, transect mapping, resource mapping, crop calendar, opportunity matrix, stakeholder analysis, Venn diagram, time budgeting, etc. play a vital role in comprehensive investigation by multiple expertise engagement and prepare the clear image of the current situation with threats and possibilities. This social tool not only helps researchers and academicians but also all age groups and gender. From the aspect of the sustainable livelihood approach (SLA), PRA helps to analyze all the pillars of SLA i.e. social, financial, human, natural, and physical capital of the village. PRA helps to study the anthropogenic effect on the socioeconomic dynamics of the village, natural ecosystem study, the impact of resources on the village development, etc. Figure 15.4 displays PRA conduction in Jawlga Mesai village, Osmanabad district.

From the achievement of sustainable development goal (SDG) point of view, PRA touches on every goal in a way or other, either direct or indirect based on their locality. It is an effective way to combat climate change, build resilience, attitudes and behavioral change, community engagement, and Strength, Weakness, Opportunity, and Threat (SWOT) analysis done at a village scale. By conducting PRA, capacity enhancement, teamwork, and creating a development framework and blueprint of plans can be done by sustainable resource consumption and protection. In a way, it is the tool necessary for every village for their holistic growth.

15.22 Control of Soil Erosion

Degradation caused by erosion processes is not limited to soil loss in detachment areas; it also has an impact on transportation and deposition zones (Olsson et al. 2019). A vegetative barrier is a continuous strip of stiff, dense vegetation planted along the slope's general contour. Vegetative barriers have great potential in reducing sheet and rill erosion, controlling water flow, stabilizing steep slopes, and trapping sediment. All eroding regions, including cropland, grazing land, forest land, farmsteads, mined ground, and construction sites, are subject to this activity. Below the barrier, a proper outlet is required. On slopes of less than 10%, barriers are the most effective way to control soil loss (USDA 2015). Gliricidia plants grown on the bunds could effectively reduce soil loss and also strengthen

Fig. 15.4 Participatory Rural Appraisal conducted at Jawlga Mesai (Geographic coordinates: 17.9805 N, 76.1922 E), by Tata Institute of Social Sciences, Tulapur in October 2022

the bunds. The list of few species given below for vegetative barriers: Sewan: *Lasiurus sindicus*, Sania: *Crotalaria burhia*, Kair: *Capparis aphylla* are the exemplary species that are effective in controlling the erosion of soil and maintaining soil health condition in good condition.

15.23 Government Schemes: Groundwater Improvement

Atal Bhujal Yojana (ATAL JAL) is a Central Sector Scheme aimed at promoting sustainable groundwater management. This scheme is being implemented by the government in 13 districts of Maharashtra with the aim of community participation and demand-side interventions to promote sustainable groundwater management in water-stressed areas across seven states in India. The scheme aims to enhance the sustainability of water sources for the Jal Jeevan Mission and contribute to the government's objective of doubling farmers' income. Additionally, the scheme aims to foster behavioral changes within communities to encourage optimal water utilization (jalshakti-dowr.gov.in 2022).The scheme strengthens the community by availing the groundwater, to combat water scarcity due to changing climate scenarios. The various components covered under this action scheme such as water budgeting, and a water security plan through a participatory approach assist stakeholders in the judicious and sustainable utilization of groundwater.

15.24 Watershed Management

A watershed is distinguished by the interconnection of the local population as well as the water that runs through and links to the same river or wetland. So, during the past 10 years, watershed management has gained popularity as a comprehensive strategy for protecting water, land, and biodiversity resources while preserving human livelihoods. (FAO 2017b). In the Tarai districts of the Himalayas, a traditional community practice of excavating a succession of conservation ponds for groundwater table recharging was also identified as a nature-based solution to watershed management (IUCN BRIDGE GBM 2018). Infiltration expansion, water-holding volume extension, soil erosion control, and recharge techniques are all common watershed operations. Strip cropping, pasture cropping, and the utilization of grasslands and woods for agriculture are all examples of control strategies. Engineering solutions that support the effects of vegetation cover, soil conservation practices, and surface runoff on erosion, surface runoff, and nutrient losses include contour bunding, terracing, earth embankment, check dams, farm ponds, and rock dams (Pande 2020). In integrated watershed management, one of the leading and well-known non-governmental organizations (NGO) named *Paani Foundation* founded in 2016 actively works for drought control and mitigation throughout Maharashtra. In 2016, the foundation initiated the inaugural Satyamev Jayate Water Cup competition. It commenced with a pilot phase in three talukas, encompassing 116 villages during its first year. Over the next 2 years, the competition expanded its reach to include numerous talukas and thousands of villages across the state. Initially, the foundation selected only three tehsils for this endeavor, but it subsequently scaled up its efforts to cover a staggering 4032 villages situated in 75 drought-affected tehsils across four regions of Maharashtra: Marathwada, Vidarbha, North Maharashtra, and Western Maharashtra, as of 2018. During the years 2016 and 2017, the participating villages collectively established a storage capacity of 100 billion liters of water. Furthermore, the foundation actively encourages water-sufficient villages to embrace permaculture practices (Wikipedia 2022). The foundation has implemented a comprehensive approach to motivate, train, and educate communities. They utilized various methods such as mass and digital media to educate people through entertainment. Additionally, they provided technical and leadership training to five individuals per village, reaching a total of 150,000 individuals across 30,000 drought-vulnerable villages in Maharashtra. To facilitate these efforts, the foundation developed a digital platform that offered technical information and also facilitated access to crowd-sourced funds, government support, and volunteers. As a result, the perception of rural residents, who previously viewed walking long distances for water as an inevitable part of their lives, has shifted. They now understand the importance of measuring rainfall in millimeters and have embraced a new perspective on water management (mowr.gov.in 2018). The NGO has a great role in uniting the villages for their benefits from socioeconomic, environmental, and sustainability perspectives. In the collective challenges of local needs and to prioritize and strengthen the community, NGOs have played a crucial role.

15.25 Conclusion

The present review highlighted the bundle of solutions available in nature to execute at the individual or community level to build resiliency for climate change. NbS is an effective way to boost the people's mindset towards the conservation of resources and their protection too. It will also insights into the long-term outcome of your current actions along with its co-benefits. From the ecosystem service point of view, each mentioned activity had an important role in achieving any of the Sustainable Development Goals by safeguarding the environmental and social management framework. The ground-based action plans are one of the best methods to restore degrading natural conditions in faster ways. Along the same line, the initiatives taken by the government schemes trigger the community to combat the new emerging challenges arising due to climate change in more cost-effective ways. The possible ways and methods covered in this chapter are the solutions to environmental and socioeconomic issues that are the closest to the community, and simple motivation to stimulate the people can create a huge change in the resource availability to them in terms of quality and quantity in the long term. Finally, the abovementioned methodologies help in the policy-making process to provide long-lasting solutions with the harmony of the environment, promotion of climate resilient technologies, and awareness of people.

Conflict of Interest I wish to confirm that there are no known conflicts of interest associated with this publication and there has been no significant financial support for this work that could have influenced its outcome.

Code Availability Not Applicable.

References

Aquije C, Schmidt H-P, Draper K et al (2021) Low-tech biochar production could be a highly effective nature-based solution for climate change mitigation in the developing world. Plant Soil. https://doi.org/10.1007/s11104-021-05159-6

Beck HE, Zimmermann NE, McVicar TR et al (2018) Present and future Köppen-Geiger climate classification maps at 1-km resolution. Sci Data 5:180214. https://doi.org/10.1038/sdata.2018.214

D'souza M, Srinidhi A, Banerjee S, Indurkar A et al (2020) Scaling ecosystem-based adaptation to climate change in Maharashtra, India an analysis of policies and programmes

Davies J, Poulsen L, Schulte-Herbrüggen B, Mackinnon K, Crawhall N, Henwood, WD, Dudley N, Smith J, Gudka M (2012) Conserving dryland biodiversity. xii +84p

decadeonrestoration.org (2022). https://www.decadeonrestoration.org/what-ecosystem-restoration

DES (2022) Economic survey of Maharashtra 2021–22. Directorate of Economics and Statistics, State Government of Maharashtra. https://mahades.maharashtra.gov.in/ESM1920/chapter/English/esm2122_e.pdf

DES (2023) Economic survey of Maharashtra 2022–23. Directorate of Economics and Statistics, State Government of Maharashtra. http://mls.org.in/PDF2023/BUDGET/ESM_2022_23_Eng_Book.pdf

devon.gov.uk (2023) What is green infrastructure and why is it important? https://www.devon.gov.uk/greeninfrastructure/what-is-it-and-why-is-it-important

FAO (2006) Guidelines for soil description, 4th edn. https://www.fao.org/3/a0541e/a0541e.pdf

FAO (2017a) FAO strategy on climate change. https://www.fao.org/3/i7175e/i7175e.pdf

FAO (2017b) Watershed management in action – lessons learned from FAO field projects

FAO (2021) Climate-smart livestock production. A practical guide for Asia and the Pacific region. FAO, Bangkok

FAO (n.d.) Participatory rural appraisal. https://www.fao.org/3/w2352e/W2352E06.htm

Fernandes R, Panwar V, Sen M (2022) Nature-based solutions for urban climate resilience in South Asia: cases from Bangladesh, India and Nepal

Ford JD, King N, Galappaththi EK et al (2020) The resilience of indigenous peoples to environmental change. One Earth 2:532–543. https://doi.org/10.1016/j.oneear.2020.05.014

Gokhale Y (2021) Nature-based solutions: convenient ignorance of the inconvenient truth. https://www.downtoearth.org.in/blog/environment/nature-based-solutions-convenient-ignorance-of-the-inconvenient-truth-77783. Accessed 9 Mar 2022

Gómez Martín E, Giordano R, Pagano A et al (2020) Using a system thinking approach to assess the contribution of nature based solutions to sustainable development goals. Sci Tot Environ 738:139693. https://doi.org/10.1016/j.scitotenv.2020.139693

Haider M (2022) How mangrove afforestation can fight the adverse effects of climate change. The Business Standard

Hallstein E, Iseman T (2021) Nature-based solutions in agriculture: project design for securing investment. FAO and TNC

Hansen NC, Allen BL, Baumhardt RL, Lyon DJ (2012) Research achievements and adoption of no-till, dryland cropping in the semi-arid U.S. Great Plains. Field Crop Res 132:196–203. https://doi.org/10.1016/j.fcr.2012.02.021

Iseman T, Miralles-Wilhelm F (2021) Nature-based solutions in agriculture: the case and pathway for adoption. FAO and TNC

IUCN (2015) Land degradation and climate change. The multiple benefits of sustainable land management in the drylands. https://www.iucn.org/downloads/land_degradation_issues_brief_cop21_031215.pdf

IUCN (2019a) Drylands and climate change. https://www.iucn.org/sites/dev/files/iucn_issues_brief_drylands_and_climate_change_sept_2019.pdf

IUCN (2019b) Drylands and climate change

IUCN (2022) Nature-based solutions

IUCN (2022c) Global drylands initiative. https://www.iucn.org/theme/ecosystem-management/our-work/global-drylands-initiative

IUCN BRIDGE GBM (2018) Nature-based solutions and their application in river basin management: BRIDGE GBM CSO dialogue report. IUCN, Bangkok, 20pp

jalshakti-dowr.gov.in (2022) Atal Bhujal Yojana. https://jalshakti-dowr.gov.in/atal-bhujal-yojana/

Kader MA, Singha A, Begum MA et al (2019) Mulching as water-saving technique in dryland agriculture: review article. Bull Natl Res Cent 43:147. https://doi.org/10.1186/s42269-019-0186-7

Karra K, Kontgis C, Statman-Weil Z, et al (2021) Global land use/land cover with Sentinel 2 and deep learning. In: 2021 IEEE International geoscience and remote sensing symposium IGARSS. IEEE, Brussels, pp 4704–4707

Keesstra S, Nunes J, Novara A et al (2018) The superior effect of nature based solutions in land management for enhancing ecosystem services. Sci Tot Environ 610–611:997–1009. https://doi.org/10.1016/j.scitotenv.2017.08.077

Kristinn SVH, Kenneth I, Alex J (2021) Development Research Branch, Economic Analysis and Policy Division, UN DESA. This policy brief is based on world social report 2021: reconsidering rural development, available at http://bit.ly/wsr2021

Lian X, Piao S, Chen A et al (2021) Multifaceted characteristics of dryland aridity changes in a warming world. Nat Rev Earth Environ 2:232–250. https://doi.org/10.1038/s43017-021-00144-0

Mbow C, Smith P, Skole D et al (2014) Achieving mitigation and adaptation to climate change through sustainable agroforestry practices in Africa. Curr Opin

Environ Sustain 6:8–14. https://doi.org/10.1016/j.cosust.2013.09.002
Miccolis A, Pereira AVB, Peneireiro FM, Marques HR, Vieira DLM, Arco-Verde MF, Hoffmann MR, Rehder T, Pereira AVB (2016) Agroforestry systems for ecological restoration
Mijatović D, Van Oudenhoven F, Eyzaguirre P, Hodgkin T (2013) The role of agricultural biodiversity in strengthening resilience to climate change: towards an analytical framework. Int J Agric Sustain 11:95–107. https://doi.org/10.1080/14735903.2012.691221
mowr.gov.in (2018) Success story the water cup by Paani Foundation. https://www.mowr.gov.in/sites/default/files/PaaniFdn_Success_Story_2018.pdf
Olsson L, Barbosa H, Bhadwal S, Cowie A, Delusca K et al (2019) Land degradation. In: Climate change and land: an IPCC special report on climate change, desertification, land degradation, sustainable land management, food security, and greenhouse gas fluxes in terrestrial ecosystems
Pande CB (2020) Watershed management and development. In: Sustainable watershed development. Springer, Cham, pp 13–26
Pillai S, Bendapudi R (2019) Inclusion of local aspirations in the village development plans: A case study of the semi-arids in Maharashtra. CARIAA-ASSAR working paper. International Development Research Centre/UK Aid, Ottawa/London. Available online at: www.assar.uct.ac.za
Pretty J, Vodouhê S (1997) Chapter 6 – using rapid or participatory rural appraisal. https://www.fao.org/3/w5830E/w5830e00.htm#Contents. In: Improving agricultural extension. A reference manual. FAO
Seddon N, Turner B, Berry P et al (2019) Grounding nature-based climate solutions in sound biodiversity science. Nat Clim Chang 9:84–87. https://doi.org/10.1038/s41558-019-0405-0
Sharma RP, Verma SA (2010) Degraded and wastelands of India, status and spatial distribution. ICAR and NAAS, New Delhi
Shivanandappa T, Rajashekar Y (2014) Mode of action of plant-derived natural insecticides. In: Singh D (ed) Advances in plant biopesticides. Springer India, New Delhi, pp 323–345
Sonneveld BGJS, Merbis MD, Alfarra A, Ünver O, Arnal MA (2018) Nature-based solutions for agricultural water management and food security. http://catalogue.unccd.int/1037_ca2525en.pdf
Stefanakis AI, Calheiros CSC, Nikolaou I (2021) Nature-based solutions as a tool in the new circular economic model for climate change adaptation. Circ Econ Sust 1:303–318. https://doi.org/10.1007/s43615-021-00022-3
Sudmeier-Rieux K, Nehren U, Sandholz S, Doswald N (2019) Disasters and ecosystems, resilience in a changing climate – source book. UNEP and Cologne: TH Köln – University of Applied Sciences, Geneva
Trabucco A, Zomer R (2019) Global aridity index and potential evapotranspiration (ET0) climate database V2. Figshare. Dataset. https://doi.org/10.6084/m9.figshare.7504448.v3
UNCA (2019) Compendium of contributions nature-based solutions. UN Climate Action Summit 2019
UNCCD (2013) UNCCD thematic factsheet series No. 7. REDD+ and desertification. https://catalogue.unccd.int/28_Factsheet%207_redd.ENGweb.pdf
UNEP (1997) World atlas of desertification. United Nations Environment Programme
UNESCO (2022) Indigenous knowledge and climate change
UNFCC (2020) India's intended nationally determined contribution: working towards climate justice. https://www4.unfccc.int/sites/ndcstaging/PublishedDocuments/India%20First/INDIA%20INDC%20TO%20UNFCCC.pdf
UNFCCC (2021) Scoping paper on knowledge gaps in integrating forest and grassland biodiversity and ecosystems into adaptation strategies, Bonn, 100 pp.
United Nations Environment Programme, and International Union for Conservation of Nature, UNEP (2021) Nature-based solutions for climate change mitigation. Nairobi and Gland
USDA (2015) Conservation practice standard overview. Vegetative barrier (601)
Valderrábano M, Nelson C, Nicholson E, et al (2021) Using ecosystem risk assessment science in ecosystem restoration: a guide to applying the Red List of Ecosystems to ecosystem restoration. IUCN, International Union for Conservation of Nature
vigyanprasar.gov.in (2022) Soil carbon sequestration can help fight climate change
Wikipedia (2022) Paani Foundation. https://en.wikipedia.org/wiki/Paani_Foundation
Zade D, Kale E, Sood A, et al (2020) Tank desiltation scheme in Maharashtra: policy concerns and way forward. https://www.indiawaterportal.org/articles/tank-desiltation-scheme-maharashtra-policy-concerns-and-way-forward
Zaman M, Kleineidam K, Bakken L et al (2021) Climate-smart agriculture practices for mitigating greenhouse gas emissions. In: Zaman M, Heng L, Müller C (eds) Measuring emission of agricultural greenhouse gases and developing mitigation options using nuclear and related techniques. Springer, Cham, pp 303–328

Smart Farming and Carbon Sequestration to Combat the Climate Crisis

16

K. R. Sooryamol, Suresh Kumar, Anu David Raj, and M. Sankar

Abstract

Agricultural lands are a major source of greenhouse gas emission that causes global warming. Carbon sequestration in agricultural sector can be achieved by assimilating atmospheric carbon dioxide in biomass and as soil organic carbon in soil through sustainable management practices. The concept of smart farming emerged to support sustainable agricultural production and food security to overcome the challenges of climate change. It helps in reducing greenhouse gas emission, increase production, resilience, and the possibilities of carbon sequestration at the same time. The smart farming approaches involves various elements that are site specific such as proper management of soil, water, crops and livestock, use of high yielding and drought-resistant varieties, conservation agricultural practices, mulching and crop residues, practicing of agroforestry and systems, etc. Smart farming can help to capture the CO_2 and reduce the climate change impact. Agriculture sectors can effectively reduce emissions and boost carbon storage in soils and plant biomass in balancing the world's climate.

Keywords

Smart farming · Crop diversification · Agricultural management · Climate change · Carbon sequestration

16.1 Introduction

Present food production and supply systems are under pressure to withstand the growing global population and their diverse dietary needs. Hence agriculture production needs to be effectively doubled from the current level to feed the population in the coming decades. Climate change has emerged as a crisis of our time and the devastating consequences are affecting the agricultural sector, including forestry, livestock and fisheries. Climate change due to natural causes are something that always happens on earth. But the studies show that human influences somehow accelerated these effects and causes long-term changes.

K. R. Sooryamol (✉) · M. Sankar
Indian Institute of Soil & Water Conservation (ICAR-IISWC), Dehradun, Uttarakhand, India
e-mail: sooryamolkr96@gmail.com

S. Kumar
Agriculture, Forestry and Ecology Group, Indian Institute of Remote Sensing, Indian Space Research Organization (ISRO), Dehradun, Uttarakhand, India

A. David Raj
Agriculture and Soils Department, Indian Institute of Remote Sensing, Indian Space Research Organization (ISRO), Dehradun, Uttarakhand, India

Forest Research Institute, Dehradun, Uttarakhand, India

U. Chatterjee et al. (eds.), *Climate Crisis: Adaptive Approaches and Sustainability*, Sustainable Development Goals Series, https://doi.org/10.1007/978-3-031-44397-8_16

Variability in global temperature and rainfall pattern, extreme weather events, increased emission of greenhouse gases (GHGs), sea level rise, glacier reduction, ocean acidification, and degradation of various ecosystems are the expected common impacts due to change in climatic conditions in recent years. Various organizations around the world are trying to combat the eventual warming due to global greenhouse gas emission to limit further impacts. Faster melting of glaciers and sea level rise makes the population lives in the coastal region at danger. Those areas may go under water within our life time if it continues. Climate change directly affects soil degradation. It limits soil's capacity to contain carbon and affects about 500 million people who lives in erosion-affected areas (UN, 2020).

Predicting impacts of climate change on agriculture have various levels of uncertainties. They may vary over time, region and will affect people depending on their vulnerability and capacity to adapt. People in developing countries who are poor and economically unstable may face several challenges to meet their food security. They will be unable to buy food, feed the hunger and reduce malnutrition. Thus, the necessity to increase current agricultural output is the key goal of food security. Climate change may seriously affect land degradation, water availability and other natural resources. Nearly 33% of world's land is under the threat of moderate to high degradation because of soil erosion, chemical pollution, compaction and salination (FAO 2011). The scope of expansion of agriculture area is very less as the additional land available is not fit for cultivation. Making those areas suitable for production will consume more time and high expenditure. Thus, the food to satisfy the growing demands need to be produced from the lands that are currently under cultivation using the same resources. Therefore, the soil quality of the available land must be managed. Besides the availability of land, agriculture sector consumes a reasonable amount of water for irrigation purposes. Inefficient use of water resource for crop production may cause serious consequences to crop production. Impacts of climate change may exaggerate the further climate crisis. Variability in temperature and precipitation may adversely affects crop production, soil quality, and epidemics of pests and diseases. Some areas may become more drought-prone, while others will witness more altered rainfall patterns. Extreme weather events like floods and hurricanes may affect the people's access to food and market volatility. Indian agriculture system is at high risk due to inadequate arable land, monsoon-depended farming and limited technological and financial development (Birthal et al., 2014). It makes India as the seventh most vulnerable country to climate change (Eckstein et al. 2019).

16.2 Climate Change Impacts on Agriculture and Vice Versa

Climate change induced risks are significantly interlinked with food security as it directly and negatively affects the productivity. All crops have specific conditions such as ideal temperature and adequate water in order to thrive. Warmer temperatures may aid the growth of certain crops in some parts of the world to a limit. But exceeding or lagging to the crop's critical thresholds will reduce the yields. Similarly, elevated levels of atmospheric carbon may increase the vegetative growth of crops. However, crops may eventually lose the nutritional value, proteins and nutrient contents significantly (Myers et al. 2014). Growing crops in regions where temperature is projected to increase and rainfall to decrease is most challenging. Increased temperature, atmospheric carbon and wet climate may help the growth of weeds, insects and pests. Even the physical properties of soil may change in response to increased temperature. In regions where heat waves are likely to form more often will threaten livestock. Heat stress will reduce the feed intake by the livestock, meat and dairy production, fertility and will make them vulnerable to various parasites, pathogens and diseases (FAO 2017). Increased soil temperature can accelerate the decomposition rate of soil organic matter (SOM) (Bai et al. 2010). Dramatic changes in rainfall events can affect the soil's capacity to make nutrients and water available to plants. Erratic rainfall causes the risk of soil erosion and damage of above-ground plant biomass. Water logging condition by floods may affect the gas exchange between soil and plant roots. Climate change can be the major cause for the loss of genetic resource of plants, animals, fishes, forests and micro-organisms. Concisely, projected changes in climate are responsible for 20% of reduction of agricultural outputs from developing countries by 2080 (Khor 2009). Reduction of 3.8% of maize and 5.5% of wheat yields was estimated globally because of climate change. Decrease in crop production and lowering of people's income challenges in coping ability of the farming community (Lipper et al. 2014). About 750 million extremely poor people lives in the rural area depend only on the benefits from agriculture and are smallholder farmers. Thus, the eradication of poverty by coming decade will be challenging (Olinto et al. 2013).

Emission from agriculture including forest and other land use shares 24% of the GHG emission in the world (IPCC 2014). Approximately 75% of world total anthropogenic nitrous oxide (N_2O) and 50% of methane (CH4) emissions are contributed from agriculture system (FAO 2021). Nearly 6% of global carbon is stored by the agriculture farm lands (IPCC 2001). However, nearly 80 tonnes per hectare soil organic carbon is lost due to the conversion of natural ecosystem to agriculture (Lal 2004). Concisely, one third of global anthropogenic GHG emissions released from agri-food systems (Crippa et al. 2021).

Soil holds 81% of the global terrestrial carbon stock. The soil carbon pool (pedologic pool) estimated as 2500 Gt up to 2 m depth, which comprises soil organic, inorganic and elemental carbon pools (Batjes 1996). Soil carbon pools link between atmosphere, vegetation and ocean. Thus, they contain more carbon than atmospheric and biotic pools. The amount of carbon stored vary with different ecosystems such as forests, wetlands, crop lands, grasslands, and deserts. Among the various biomes around the world, boreal forests have the highest carbon stocks (471 Gt). One-meter soil of crop land biome contains 128 Gt of carbon (IPCC 2001). Even though the soils are active sink of atmospheric carbon, their efficiency to contain carbon declined over time under disturbed conditions such as erosion, tillage, leaching and decomposition of organic matter. Approximately 1.6 Gt carbon per year emitted due to land use change (deforestation) for the last decade (Friedlingstein et al. 2020). After 60 years of cultivation, temperate humid grasslands lost nearly 30% of their soil organic carbon (SOC) stocks (Tiessen and Stewart 1983; Guo and Gifford 2002). Pasture lands converted from Amazon rain forest emits around 8–12 tonnes of carbon per hectare (Fearnside and Barbosa 1998; Cerri et al. 2007). More than 60% of SOC lose can caused by cultivation in tropical forest soils (Brown and Lugo 1990). However, very little exposure is given to the carbon sequestration regarding the mitigation of climate change, increase the agriculture production and food security. Several agricultural practices adapted to diminish the risks of climate change and have a limited potential for carbon sequestration for a long run and some of them are not suitable for diverse environmental conditions (Stavi and Lal 2013).

Practice of high-yielding varieties, irrigation methods and higher amount of chemical inputs may boost the production. Agriculture intensification without growing farm area may save forest land from conversion to agricultural lands, ecosystem services and carbon stored in the soil. But still the situation is not a guarantee for the future food security until they are sustainable. Sustainable agriculture can be attained through the conservation of water, land, plant and animal and their genetic resources. It is environmentally non-degrading, technically appropriate, economically viable and socially acceptable. (FAO 1988). Achieving sustainable food production from the depleted sources needs changes in agricultural systems. Transformation of agriculture systems in a sustainable manner to meet the sustainable development goals is the need of this century for a better future.

16.3 Climate Smart Agriculture (CSA)

Climate smart agriculture is not a new agricultural system with a set of standard practices that can be applied universally. It is a novel strategy to guide the agricultural sector to sustainably increase agriculture production and income, building capacity for adaptation and resilience towards climate change and reduction of GHG from agriculture system (Fig. 16.1). The approach needs site-specific assessments to identify suitable land management practices, policies, investments and technologies to increase the agriculture production (FAO 2013). That is the reason why CSA approach is more sustainable.

Increased productivity will support the food security by making food available for households and local markets. Increased income helps the accessibility of food. This objective can be achieved through increased resource use efficiency, diverse production system and managing agro-ecosystems. Judicious application of fertilizers and pesticides, high-quality livestock feed, use of high-yielding seeds and animal varieties will all increase production without adding external inputs. Choosing healthy plants and animals that can utilize the fertilizers and feed more efficiently can increase the productivity. Regular disease surveillance, integrated pest and disease management and maintenance can also support to meet the objective. Recycling the by-products and waste such as manure as fertilizers and crop residue and its by-products as feed to livestock etc. within the system will further reduce the use of external input. Reducing food loss by adapting good post-harvesting practices, appropriate storage and processing methods to increase the product durability is also an important aspect of effective use of resource. It can also help with the reduction of GHG emission as the loss of food through the agriculture value chain is a wastage of resources and sustainable income (FAO 2019a, b).

Enabling the adaptation capacity and build resilience of agriculture system to long-term changes in climate can be achieved by diversifying production system. Diversification of production systems could be a way of economic growth. Integrated farming systems that incorporate livestock systems and multiple crops, agroforestry, silvopastoral systems and mixed crop-aquaculture-livestock systems etc. are better ways for diversification. It can create favourable environment for crop growth and reduce GHG emission compared to monoculture systems. Resilience can be developed through precautionary measures to climate extremes such as careful design of production sites, use of early warning systems, selection of adapted varieties and species etc. (FAO 2019a, b). Crossbreeding of local varieties and improved varieties can help to develop new varieties that have tolerance to climate risks, greater resilience, higher productivity and can use limited resource efficiently (FAO 2017). Use of climate and weather information services can support the farmers to manage the timing and operations of crop and livestock. Management practices that improve soil structure

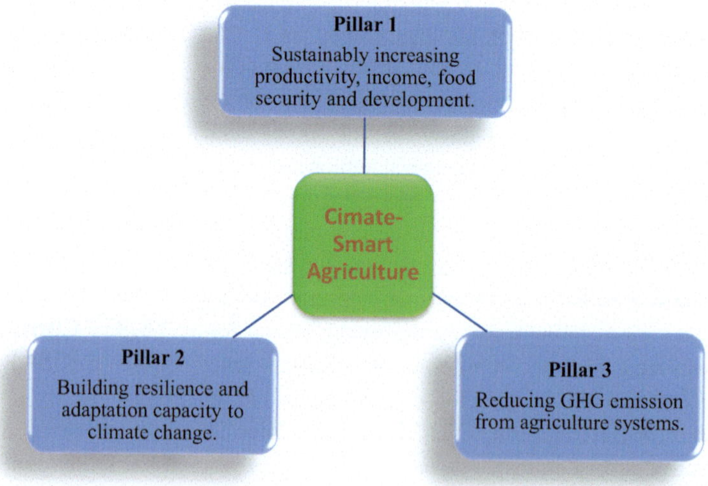

Fig. 16.1 Pillars of climate-smart agriculture. (Source: FAO 2019a, b)

increases water infiltration and water retention. Higher water infiltration and retention enhances the recharge of groundwater. Thus, they build resilience agro-ecosystems to face climate change.

Reducing GHG emissions in terms of absolute emissions and emission intensities is the third objective of CSA. GHG emissions is way of loss of nutrients such as nitrogen and carbon from agriculture soils. Assimilation of carbon in agriculture soils and its long-term storage can be achieved through conservation tillage practices, mulching cover crops, afforestation integration of crop, livestock and trees and restoration of ecosystems etc. Use of efficient technology can also reduce GHG emission through the wastage of resources and energy. Producing bioenergy from agriculture by-products can reduce the use of fossil fuel utilization, which in turn can decline the GHG emission.

To obtain the resilience to climate change, a concept of climate resilient agriculture (CRA) was developed to enhance the capacity of agriculture system to cope with the various climate crises (e.g., drought, flood and extreme weather conditions) by reducing the damage and fast recovery. Indian Council of Agricultural Research (ICAR) formed a project network named National Initiative in Climate Resilient Agriculture (NICRA) to develop novel strategies and demonstrate suitable management practices to combat the impacts of climate change on Indian agriculture. NICRA primarily aimed towards the transfer of resilient technologies to the farmers. They have implemented various practices to address various climate-related issues in various states of India. Table 16.1 shows the climate resilient practices implemented in NICRA villages across the country (Prasad et al. 2014).

Table 16.1 Climate-resilient practices implemented in NICRA villages

Climate vulnerability	Smart practices	Impacts
Drought, heat stress	Zero till drill Direct seeding (Punjab, Chhattisgarh)	Minimum soil disturbance Reduces the cost and time of tillage operations and labours Reduces moisture stress in the crucial stages of crop growth and increased productivity.
Flood, cyclone	Use of flood tolerant varieties which can stay in water-logged condition for 2 weeks and produce better crop yield. Use of lodging resistant varieties. (Andhra Pradesh, West Bengal, Bihar and Maharashtra)	Minimum crop damage Minimum grain loss Increased dry fodder yield
Intensive rainfall, prolonged dry spell and water logging	Construction of broad bed furrow Construction of furrow irrigated raised bed (West Bengal, Rajasthan, Tamil Nadu)	Proper drainage and aeration Efficient management of irrigation water, increased crop yield.

16.3.1 Climate-Smart Crop Production

Climate smart crop production is part of sustainable crop production to address various impacts of climate change through adaptation and mitigation strategies. Management practices and technologies may differ with different cropping systems. Maintaining healthy soil is the basic necessity of climate smart crop production to achieve sustainable crop production. Climate smart crop production involves use of high-quality seeds and practices that minimize soil disturbance, cover the soil to reduce carbon emissions, and promote crop diversity at the landscape and farm levels over time. Additionally, managing soil, land, and livestock sustainably, selecting suitable and sustainable mechanization, managing nutrients and water, and implementing integrated pest and disease management can help sustainable crop production. Undisturbed soil with higher organic matter concentration can support higher growth of soil organisms. Those soil fauna will produce more soil organic matter, stable aggregates and improves soil structure, water storage, etc., which in turn support the growth of plants. Introduction of crops, trees or shrubs through crop rotation, cover crop, green manure, etc. to these soils will encourage bio-tillage and thereby sustainable production. (FAO 2009, 2012). Integrating various production in a single farming system to support and depend mutually by using products, by-products or services all together. By optimizing efficiency in the use of resources and mitigating climate change by reducing GHS emission support the objectives of climate-smart agriculture. Agroforestry systems, silvopastoral systems and integrated crop livestock systems are some of the promising practices that can be adapted to address climate change. It improves intensity and diversity of the production and enhance resilience. Emission intensities of integrated systems are lower than those of specialized systems. Higher yields add greater amount of plant residue and biomass on the ground, which may increase the biological activity and the carbon storage (FAO 2013). Livestock is also a part of crop production system since they contribute to nutrient cycling in the soil. Integrating livestock in the crop system would increase the efficiency of the soil to contain carbon. The impact of manure on soil carbon sequestration can be found in those integrated systems. Integrated crop livestock system reduces the cost or organic manure for crop production. It is a way to diversify the production system and manage soil organic carbon through the management of organic matter.

16.3.2 Smart Practices for Carbon Sequestration

Rainfall affects the moisture availability to plants. Higher temperature may accelerate the decomposition of surface soil organic matter that contains about 58% of organic carbon. It will particularly reduce the soil's capacity to sequester carbon and will not be adequate to balance the GHG releases. Soil carbon has its direct relation with soil quality. It can help the soil to become more sustainable and resilient against the degradation processes. Sequestrating carbon in soil to balance between the gains and losses will enhance the soil fertility and limit the GHG emission. Thus, carbon sequestration has the highest mitigation potential against climate change from agriculture lands. Conserving large volume of carbon stocks stored in natural ecosystems and incorporating carbon in agriculture soils through land management practices can help to acquire carbon benefits. However, intensive conventional agriculture practices helped the sequestrated carbon to escape. Continuous cropping without rotations, repetitive tillage, nutrient mining, over grazing, burning of crop residues, etc. are unsustainable management practices that reduce the productivity of lands. Approximately 1–3% of SOC declined in many regions because of the tillage-based agriculture system and associated degradation processes over the last century (FAO 2013). Inter-governmental panel on climate change (IPCC) estimates that 89% of global technical mitigation potential for agriculture assumed to be from carbon sequestration in soils by the year of 2030 (Venkateswarlu and Shanker 2009). Concisely, mitigation potential of agriculture is allied not with lessening of GHG emission but with the sequestration of carbon in soils. Suitable land and farm management practices can improve the capacity of soil to capture the carbon and, therefore, multiple benefits such as soil health, fertility, higher production, fostering climate change adaptation and reduction of GHGs. Climate of a region influences the patterns of soil carbon sequestration. Soil type that has higher clay content retains carbon the most. Potential soil carbon sequestration occurs within the first 20–30 years of implementing better land management techniques in most of the cases.

16.3.3 Tillage Practices

Tillage activities that use devices and machinery for seed germination, planting, etc. has important influence on carbon sequestration and deposition.

Tillage exposes the carbon molecules trapped in the aggregates to mineralization by disrupting the untilled soil (World Bank 2012). Tillage systems are categorized into conventional, reduced, and conservation tillage (Fig. 16.2). Traditional tillage techniques used man power or animals for tillage operations. The conventional tillage method is also known as intensive tillage due to the use of motor-powered numerous farm operations such as mold board, disk, plough, and harrow for soil preparation. This method leaves the minimum residue on the soil surface after operations. Ploughing breaks the topsoil and facilitates aeration and seedling growth. Minimum tillage uses less machinery for the land preparation and leaves 15–30% residue after the operation.

No tillage/zero tillage, strip tillage, ridge tillage, mulch tillage, rotational tillage, etc. are the tillage practices that come under conservation tillage. Zero tillage retains 100% soil cover and does not disturb the soil for operations. Thus, the soil aggregates remain unbroken and protect the carbon from emission and improve soil organic matter (World Bank 2012). Continuously practicing no tillage can maintain carbon gains in soil. It will sequester more soil organic carbon near the surface soil compared to conservation tillage. Strip tillage uses minimum tillage. Only the seed rows are tilled for better seedbed. Ridge tillage performed on the ridges, hills or bund formed from crop residue. Rows are maintained and ridges are levelled at planting in each season. During mulch tillage soil is disrupted only before planting and incorporate residues using chisel, sweeps and field cultivars. Drills and seeders are used for planting (World Bank 2012). In rotational tillage system, different tillage operation is conducted at specific intervals to begin different crops during crop rotations. Conservation tillage systems retain the most crop residues (more than 30%) on the soil. Conservation agriculture includes minimum soil disturbance along with crop residues, cover crops, and their diversification and enhancing SOM and mitigate carbon emissions through carbon sequestration up to 0.16–0.49 Mg C ha^{-1} $year^{-1}$ (Farooqi et al. 2018).

16.3.4 Crop Residue Management

Incorporating residues such as biomass from trees, sugarcane, rice, residues from previous crop and other grain crops into the soil is an important renewable resource for agroecosystems (World Bank 2012). Proper management of residues can result in soil organic carbon accumulation and reduced carbon emission. Residues converted into humus compounds have higher residence time. The amount of carbon sequestered depends on the quality and quantity of residues. The amount of residue formed depends on the area of cropland and agronomic practices. Cereals can sequester carbon two to three times better than legumes (World Bank 2012).

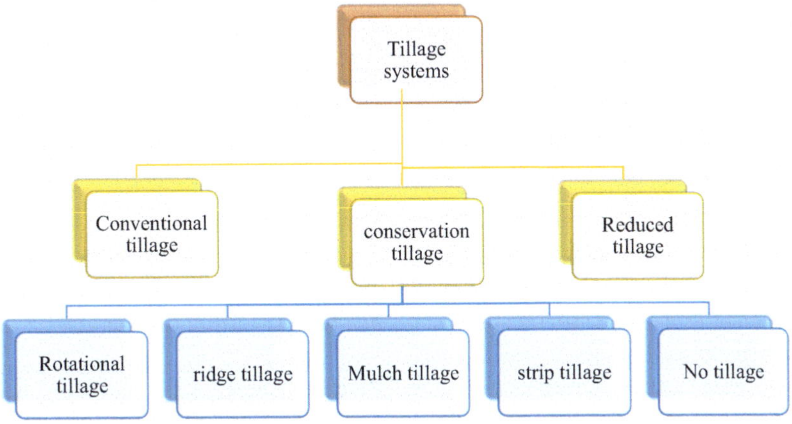

Fig. 16.2 Crop residue management-based tillage systems. (Source: World Bank 2012)

Burning and inadequate management of crop residues are responsible for loss of organic carbon and emission of CO_2 to the atmosphere. Nearly 9.2 Mt. of carbon is lost due to the burning of rice residues in the northwest region of India each year (Sing 2018). Application of residues shortly before the beginning of rainy season will bring out best results since the large amount of carbon loss happens during wet conditions.

16.3.5 Cover Crops and Crop Rotation

Cover crops are commonly known as green manure. They enhance soil quality by increasing soil organic carbon by incorporating biomass. Covering soil surface will protect the soil from exposure to climatic factors such as rainfall and temperature and thereby reduces the loss due to decomposition of soil organic carbon (Shahzad et al. 2022). Green manures are commonly crops with high nitrogen content such as leguminous crops, for example ground nuts and cow pea. Soil carbon sequestration rates in cover cropping are more than the soils with low or no cover cropping (Vicente-Vicente et al. 2016). Cover crops and crop rotation can successfully implement no-tillage practice.

Crop rotation is the order of the incorporation of different crops in a cycle on the same field and the rotation may planned for 2 or more years. The previous crop may be followed by a variety of the same crop or a different species. The following crop may be of a different species or a variety of the previous crop. They bring larger diversity in the cropping system (World Bank 2012). Intensifying and diversifying crop rotations are part of crop rotation system. Even though the variation is high, tendency towards higher carbon sequestration rates is found under the triple cropping systems. Soils, climate, and cropping systems can affect carbon sequestration under crop rotation. Diversified cropping systems are much effective in increasing C contents of soil as compared to monoculture systems. Diverse crop rotations that include cover crops in their rotation have higher soil carbon and soil microbial biomass than less diverse systems (McDaniel et al. 2014). Incorporating legume crops in the rotation increased the carbon contents in the soil as compared to cropping system where no legume crop was involved (Becker et al. 2017).

16.3.6 Mulching

Mulching can enhance the soil aggregation and thereby increase the carbon content in soil (Senjam Jinus et al. 2021). Different mulch practices such as straw mulch, plastic mulch, and no mulch are present. Highest organic carbon found in the straw mulch plots (Scopel et al. 2005). Under no-tillage mulch-based (NTM) cropping systems, the SOC stocks in the 30 cm soil layer in the fields ranges from 4.2 and 6.7 kg C m^{-2} and increased on average with 0.19 kg C m^{-2} $year^{-1}$ (Neto et al. 2010).

16.3.7 Agroforestry

Agroforestry is a conservation farming practice that integrates land-use system with trees, shrubs, crops and livestock. This system also maintains diversity and thereby augment long-term carbon sequestration. They can sequester more carbon than field crops and pastures (Shahzad et al. 2022). Tree-crop farming store average carbon about 1.4 t C ha/year. Promoting this system will increase above-ground and below-ground biomass and are a rich source of soil organic carbon. The capacity of agroforestry system to store carbon widely varies with the climate, soil, management practices physical and biotic conditions, etc. (World Bank 2012) carbon sequestration through agroforestry and conservation agriculture can offset 25–75% of the annual carbon release globally (Lal 2005; Silver et al. 2000).

16.3.8 Biochar

Biochar production is an effective way for carbon sequestration with minimum leakage for a long term. It can contain twofold carbon than the original material. Residues of trees, crops, grass and poultry wastes can be converted into biochar through pyrolysis (Shahzad et al. 2022). Biochar form of carbon will be stable for a long time since they are resistive to microbial attack and decomposition rate and higher retention time of 108 and 556 years (Wang et al. 2016). They remain in the soil nearly 10–1000 times longer than most of the soil organic matter (World Bank 2012). Thus, they can reduce the terrestrial carbon emission.

16.3.9 Land-Use Changes

Studies showed that conversion of land use highly affects the carbon storage capacity of soils around the world. Replacing annual crops with perennials, conversion of cultivated land to forests and plantations etc. increased significant amount of carbon sequestration. However, conversion of native grasslands including savannahs resulted in a net loss of soil carbon. Studies have reported that grassland soils may not store carbon when converted to forests and some soils may even lose carbon (World Bank 2012). While converting crop lands into perennial grasslands, they may increase soil carbon below the ground and reduce the soil disturbance. Afforestation, reforestation and revegetation will increase carbon stocks. Converting low productive forest to high productive, converting conventional logging to reduced logging and managing forest land for wood products are also some ways to increase carbon sequestration. Conversion of temperate forest/prairie to cropland can reduce 30–50% of native SOC in 50-year period. Clearing tropical forest can degrade 75% of SOC over 25 years (Brady and Weil 2008; Olson et al. 2016). Approximately 516 billion tons of biosphere carbon was emitted to the atmosphere as CO_2 since the establishment of agriculture (Lal 2005).

16.3.10 Improvement of Pastures and Grasslands

Nearly 40% of land is occupied by grasslands/grazing lands around the world. Thus, improving grassland systems is a major measure of smart farming through smart management of grasslands and pastures for carbon sequestration. They can retain higher amount of carbon through return of biomass. Controlled grazing can minimize the loss of soil carbon. Overgrazing may expose the soil and increase mineralization of carbon in the surface soil. Higher total organic carbon content was observed in time-controlled grazing than in controlled grazing and no grazing lands (Sanjari et al. 2008). Grazing management supports continuous supply of forages. Converting crop lands to pastures and use of pasture face as a part of cropping system can increase carbon content in soil. Managing pasture land similar to agriculture land can improve the soil fauna and biomass production. Cultivation of highly productive species of grasses can boost the carbon sequestration. Improved pasture productivity can avoid the land conversion and associated carbon loss. Using higher quality forage, adequate application of nutrients reducing the fire extent, etc. are the pasture and grassland managements that can incorporates higher amounts of carbon to the soil.

16.3.11 Restoring Degraded Soils

Soil holds 30% more carbon than the biotic and atmospheric pool. Half of the net loss of carbon is caused by the soil erosion, which degrades the soil (Harden et al. 2018). It makes soil vulnerable to serious loss of soil organic matter. Conservation and sequestration of carbon can be effectively done through appropriate land and management practices. Partial restoration of carbon storage in degraded lands can be achieved by this (Lal 2004). Continuous cropping with reduced rotations, intensive tillage and soil nutrient mining, overgrazing and burning of range land are some unsustainable land management practices that degrade world's croplands, rangelands and forests. Maximizing vegetative cover, cover cropping, mulch, green manure, use of biochar, and minimized tillage are some restoration practices, which also can enhance the carbon content in soils.

16.3.12 Water Management

Availability of adequate amount of water and its efficient utilization is necessary to soil organic carbon management through enhanced above-ground and below-ground biomass production. Improved irrigation can sequester significant amount of carbon. Stable forms of carbon can contain seven times more water than their weight. Proper irrigation in grassland can enhance soil organic carbon concentration (Conant et al. 2001).

16.3.13 Nutrient Management

Judicious application of nutrients is vital to carbon sequestration. Use of compost and organic manures improves the carbon content efficiently than application of the same amount of inorganic fertilizer (Gregorich et al. 2001). The amount of biomass produced/returned to the soil from crop residues and nitrogen fertilization management. Supply of N and other essential nutrients can enhance biomass production. Smart management practices for the incorporation of nitrogen in soil rises the production of soil organic matter, and yield biomass. Continuous application of organic manure increases the potential of conservation tillage to retain carbon (Hao et al. 2002).

16.3.14 Conservation Agriculture

Primary objective of conservation agriculture is to minimize soil disturbance, permanent soil cover, crop rotation and diversification through mechanical tillage, residue management, and using legumes and green manure or cover crops, respectively (Senjam Jinus et al. 2021). Conservation agriculture thus promises an agriculture carbon sink. Undisturbed soils decrease the mineralization of carbon. Cover crops and mulches continuously supply biomass for incorporating carbon into soil (World Bank 2012). Use of machinery is very less under conservation tillage. It can help to cut the emission from the use of fuel and energy. Increasing the conservation agriculture practices will positively balance the carbon input and output. The carbon content in the soils is much higher if the crop yields and residue recycling are greater in the soils, that the supply of organic matter can balance the outgoings (FAO 2013) (Fig. 16.3).

16.4 A Case Study

Bhattacharyya et al. (2009) conducted experiment for 5 years from 1999 to 2003 at the experimental farm of the Vivekananda Institute of Hill Agriculture. They examined the effects of conventional, minimum, and zero tillage systems (CT, MT, and ZT, respectively) and crop rotations systems, soybean–lentil (S–L), soybean–wheat (S–W), and soybean–pea (S–P)] on SOC and TSN storage and their distribution within aggregate-size fractions in a field where the major soil type was sandy clay loam. Surface (0–15 cm)and subsurface samples from each plot was collected using core sampler during 1999 (surface samples collected) and 2003 (surface and subsurface samples collected) and created composite sample for each plot. Total soil organic carbon (SOC), total soil nitrogen (TSN), aggregate size distribution, aggregate stability, bulk density were analyzed and calculated for collected samples.

The main objective of this study is to determine how aggregate distribution and aggregate associated carbon (C) and nitrogen (N) affected by short-term tillage and crop rotation practices.

Salient Findings A significant variation in bulk density was observed from surface layer of plots under tillage practices as higher to lower in the order of ZT> MT> CT. Cropping system had no marked impact on bulk density in surface and subsurface soils. Higher microaggregate was found in the surface layer. Macroaggregate was found greater under ZT and MT than CT. Zero-tilled plots stored higher SOC than CT in the surface soil layer and found no statistical difference in the subsurface layer. Correspondingly, considerable influences on SOC content in the surface soil layer was found due to crop rotation.

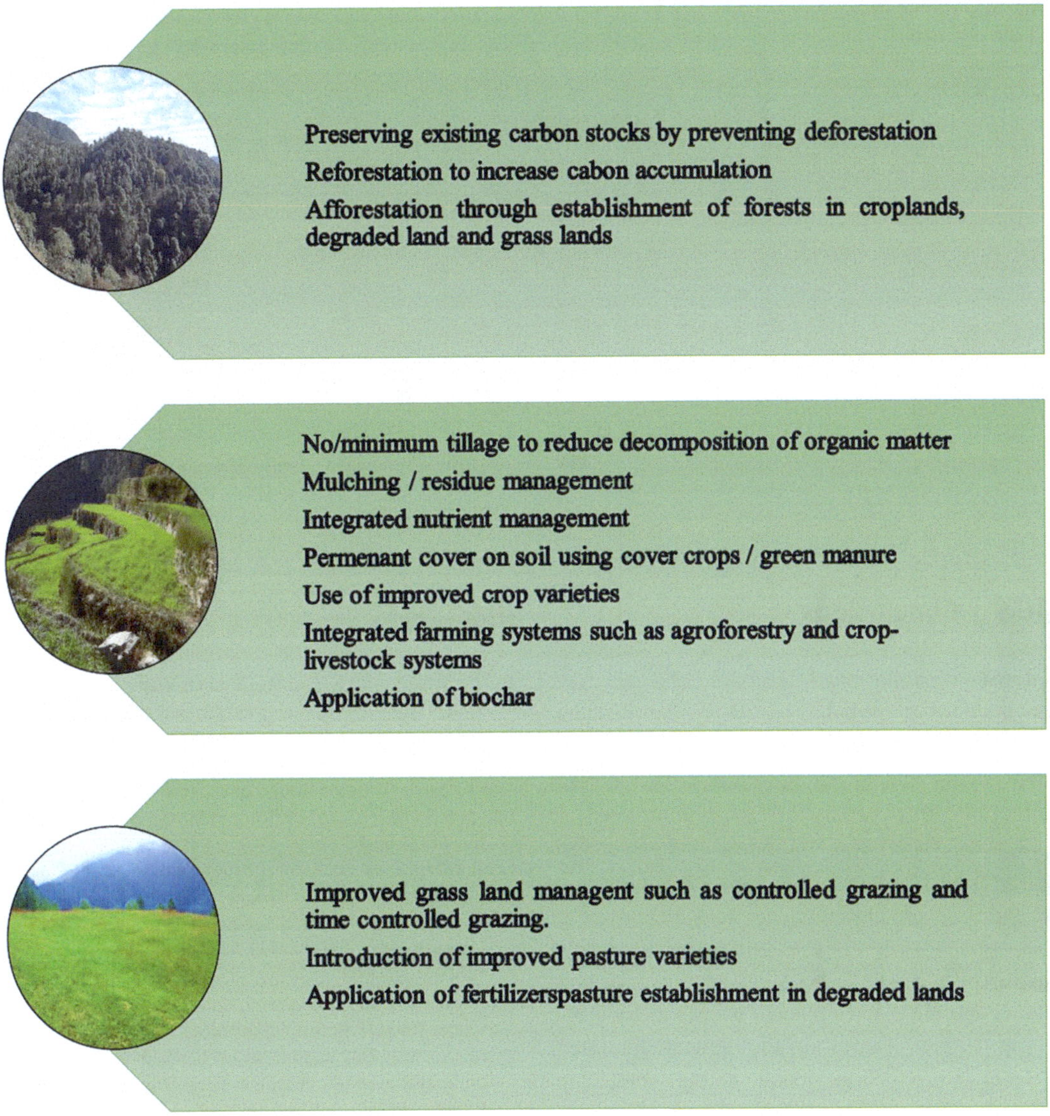

Fig. 16.3 Practices that sequester carbon in forest, cropland and grassland. (Source: World Bank 2012)

Soybean-based cropping system increased the SOC content from the initial soil. Adoption of ZT and continous cropping significantly increased SOC after the experiment.

Higher TSN was found under ZT and MT in the surface soil layer. S-L and S-P cropping systems found to have more TSN content than S-W. To understand the influence of tillage system mean grain yield (MGY) of soyabean-based cropping system was calculated (Fig. 16.4). Higher SOC and TSN found in the ZT system was may be due to reduced tillage. Increased macroaggregates and associated C and N concentration are due to less mechanical desruption on the surface soil. ZT increased bulk density and great proportion of macroaggregates. Higher level of SOC and TSN supported the growth of crops and higher MGY was observed from the plot under ZT practice. Thus, the study clearly indicates that the higher organic carbon in soil can increase the production and improve soil's ability to incorporate more carbon in a sustainable way for better crop production.

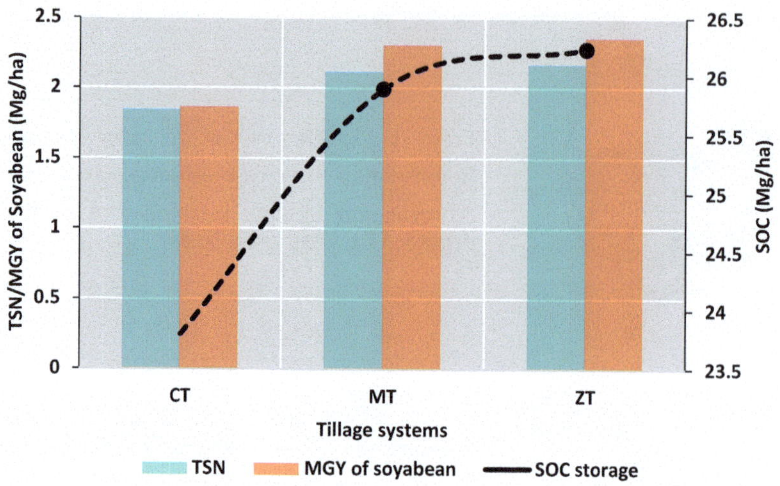

Fig. 16.4 Influence of tillage system on SOC,TSN and mean grain yield (MGY)

16.5 Conclusions

Global warming and climate change have emerged as a global crisis in recent decades. Agriculture is the prime sector suffering from climate crisis and needs to mitigate the adverse effect of climate crisis through climate-resilient practices. The effects of climate change on agricultural sector may begin with crop failures in response to temperature and rainfall variations but can extend up to the stability of food security by affecting all its dimensions such as availability, economical and physical access to food and its utilization. Therefore, the demand for intensive crop production, adequate food, and adaptive measure to cope with climate change is increasing simultaneously. Thus, the agricultural system needs to transform to hurdle the challenges interlinked with feeding a growing population. Climate smart agriculture is an effective way to sustainable production and mitigation of climate change effects through reduced carbon and other GHG emissions. Understanding the potential of each land and adopting suitable farming operation to increase the capacity of soil to incorporate more carbon while conserving the carbon stocks is the way to sequester carbon through climate smart agriculture.

References

Bai W, Wan S, Niu S, Liu W, Chen Q, Wang Q, Zhang W, Han X, Li L (2010) Increased temperature and precipitation interact to affect root production, mortality, and turnover in a temperate steppe: implications for ecosystem C cycling. Glob Change Biol 16(4):1306–1316

Batjes NH (1996) Total carbon and nitrogen in the soils of the world. Eur J Soil Sci 47:151–163

Becker JN, Gütlein A, Sierra Cornejo N, Kiese R, Hertel D, Kuzyakov Y (2017) Legume and non-legume trees increase soil carbon sequestration in Savanna. Ecosystems 20:989–999. https://doi.org/10.1007/s10021-016-0087-7

Bhattacharyya R, Prakash V, Kundu S, Srivastva AK, Gupta HS (2009) Soil aggregation and organic matter in a sandy clay loam soil of the Indian Himalayas under different tillage and crop regimes. Agric Ecosyst Environ 132(1–2):126–134

Birthal PS, Khan TM, Negi DS, Agarwal S (2014) Impact of climate change on yields of major food crops in India: implications for food security. Agric Econ Res Rev 27(2):145–155

Brady NC, Weil (2008) The nature and properties of soils, 14th edn. Pearson Education, Upper Saddle, 992 pp

Brown S, Lugo AE (1990) Effects of forest clearing and succession on the carbon and nitrogen content of soils in Puerto Rico and US Virgin Islands. Plant Soil 124:53–64

Cerri CEP, Easter M, Paustian K, Killian K, Coleman K, Bernoux M, Falloon P, Powlson DS, Batjes NH, Milne E, Cerri CC (2007) Predicted soil organic carbon stocks and changes in the Brazilian Amazon between 2000 and 2030. Agric Ecosyst Environ 122(1):58–72

Conant RT, Paustian K, Elliott ET (2001) Grassland management and conversion into grassland: effects on soil carbon. Appl Ecol 11:343–355

Crippa M, Solazzo E, Guizzardi D, Monforti-Ferrario F, Tubiello FN, Leip AJNF (2021) Food systems are responsible for a third of global anthropogenic GHG emissions. Nat Food 2(3):198–209

Eckstein D, Künzel V, Schäfer L, Winges M (2019) Global climate risk index 2020: who suffers most from extreme weather events? Weather-related loss events in 2018 and 1999–2018

FAO (1988) Sustainable agricultural production: implications for international agricultural research. TAC Secretariat, Rome

FAO (2009) The state of food and agriculture: livestock in the balance. Rome

FAO (2011) Energy-smart food for people and climate. Issue paper. http://www.fao.org/docrep/014/i2454e/i2454e00.pdf

FAO (2012) Stability of food security in a green economy environment. Greening the economy with agriculture. Rio+20 working paper no. 3, Rome

FAO (2013) Climate-smart agriculture sourcebook

FAO (2017) Climate-smart agriculture sourcebook, 2nd edn. Rome. http://www.fao.org/climate-smart-agriculture-sourcebook/en

FAO (2019a) Agriculture and climate change – challenges and opportunities at the global and local level – collaboration on climate-smart agriculture. Rome. 52 pp. Licence: CC BY-NC-SA 3.0 IGO

FAO (2019b) Climate-smart agriculture and the sustainable development goals: mapping interlinkages, synergies and trade-offs and guidelines for integrated implementation. Rome

FAO (2021) Land statistics. Global, regional and country trends, 1990–2018. FAOSTAT analytical brief series no. 15. Rome. http://www.fao.org/3/cb2860en/cb2860en.pdf

Farooqi ZUR, Sabir M, Zeeshan N, Naveed K, Hussain MM (2018) Enhancing carbon sequestration using organic amendments and agricultural practices. In: Carbon capture, utilization and sequestration. InTech

Fearnside PM, Barbosa RI (1998) Soil carbon changes from conversion of forest to pasture in Brazilian Amazonia. For Ecol Manag 108(1–2):147–166

Friedlingstein P, O'sullivan M, Jones MW, Andrew RM, Hauck J, Olsen A, Peters GP, Peters W, Pongratz J, Sitch S, Le Quéré C (2020) Global carbon budget 2020. Earth Syst Sci Data 12(4):3269–3340

Gregorich EG, Drury CF, Baldock JA (2001) Changes in soil carbon under long-term maize in monoculture and legume-based rotation. Can J Soil Sci 81:21–31

Guo LB, Gifford RM (2002) Soil carbon stocks and land use change: a meta-analysis. Glob Change Biol 8(4):345–360

Hao Y, Lal R, Owens LB, Izaurralde RC, Post M, Hothem D (2002) Effect of cropland management and slope position on soil organic carbon pools in the North Appalachian experimental watersheds. Soil Till Res 68:133–142

Harden JW, Hugelius G, Ahlström A, Blankinship JC, Bond-Lamberty B, Lawrence CR, Loisel J, Malhotra A, Jackson RB, Ogle S, Phillips C (2018) Networking our science to characterize the state, vulnerabilities, and management opportunities of soil organic matter. Glob Change Biol 24(2):e705–e718

IPCC (2001) Climate change 2001. In: Houghton JT, Ding Y, Griggs DJ, Noguer M, van der Linden PJ, Xiaosu D (eds) The scientific basis, contribution of working Group I to the third assessment report of the intergovernmental panel on climate change (A). Cambridge University Press, Cambridge, p 944

IPCC (2014) Summary for policymakers. In: Climate change 2014: Mitigation of climate change. Contribution of working Group III to the fifth assessment report of the intergovernmental panel on climate

Khor M (2009) The food crisis, climate change and the importance of sustainable agriculture. Penang Malaysia Third World Network

Lal R (2004) Soil carbon sequestration impacts on global climate change and food security. Science 304(5677):1623–1627

Lal R (2005) World crop residues production and implications of its use as a biofuel. Environ Int 31:575–584

Lipper L, Thornton P, Campbell BM, Baedeker T, Braimoh A, Bwalya M, Caron P, Cattaneo A, Garrity D, Henry K, Hottle R (2014) Climate-smart agriculture for food security. Nat Clim Chang 4(12):1068–1072

McDaniel MD, Tiemann LK, Grandy AS (2014) Does agricultural crop diversity enhance soil microbial biomass and organic matter dynamics? A meta-analysis. Ecol Appl 24(3):560–570

Myers SS, Zanobetti A, Kloog I, Huybers P, Leakey AD, Bloom AJ, Carlisle E, Dietterich LH, Fitzgerald G, Hasegawa T, Holbrook NM (2014) Increasing CO2 threatens human nutrition. Nature 510(7503):139–142

Neto MS, Scopel E, Corbeels M, Cardosoc AN, Douzetd JM, Fellere C, Piccoloa MC, Cerria CC, Bernouxe M (2010) Soil carbon stocks under no-tillage mulch-based cropping systems in the Brazilian Cerrado: an on-farm synchronic assessment. Soil Till Res 2815:1–9

Olinto P, Beegle K, Sobrado C, Uematsu H (2013) The state of the poor: where are the poor, where is extreme poverty harder to end, and what is the current profile of the world's poor. Economic premise series no. 125

Olson KR, Al-Kaisi M, Lal R, Cihacek L (2016) Impact of soil erosion on soil organic carbon stocks. J Soil Water Conserv 71(3):61A–67A. https://doi.org/10.2489/jswc.71.3.61A

Prasad YG, Maheswari M, Dixit S, Srinivasarao Ch, Sikka AK, Venkateswarlu B, Sudhakar N, Prabhu Kumar S, Singh AK, Gogoi AK, Singh AK, Singh YV, Mishra A (2014) Smart practices and technologies for climate resilient agriculture. Central Research Institute for Dryland Agriculture (ICAR), Hyderabad, 76 p

Sanjari G, Ghadiri H, Ciesiolka CAA, Yu B (2008) Comparing the effects of continuous and time-controlled grazing systems on soil characteristics in

Southeast Queensland. Aust J Soil Res 46:348–358. https://doi.org/10.1071/SR07220
Scopel E, Findeling A, Guerra EC, Corbeels M (2005) Impact of direct sowing mulch-based cropping systems on soil carbon, soil erosion and maize yield. Agron Sustain Dev 25:425–432
Senjam Jinus S, Meinam T, Melanglen K, Potsangbam M, Devi AR, Nongthombam L, Bhaigyabati T, Shephrou HD, Tilotama K, Thokchom DS (2021) The climate smart agriculture for carbon capture and carbon sequestration: the challenges and opportunities. In: Advances in carbon capture and utilization. Springer, Singapore, pp 195–233
Shahzad K, Sintim H, Ahmad F, Abid M, Nasim W (2022) Importance of carbon sequestration in the context of climate change. In: Building climate resilience in agriculture: theory, practice and future perspective. Springer, Cham, pp 385–401
Silver WL, Ostertag R, Lugo AE (2000) The potential for carbon sequestration through reforestation of abandoned tropical agricultural and pasture lands. Restor Ecol 8:394–407
Sing JS (2018) Crop residues management in agro-environmental sustainability. Climate Change 4:653–660
Stavi I, Lal R (2013) Agroforestry and biochar to offset climate change: a review. Agron Sustain Dev 33:81–96
Tiessen HJWB, Stewart JWB (1983) Particle-size fractions and their use in studies of soil organic matter: II. Cultivation effects on organic matter composition in size fractions. Soil Sci Soc Am J 47(3):509–514
United Nations (2020) The climate crisis – a race we can win. https://www.un.org/en/un75/climate-crisis-race-we-canwin
Venkateswarlu B, Shanker AK (2009) Climate change and agriculture: adaptation and mitigation strategies. Indian J Agron 54(2):226–230
Vicente-Vicente JL, García-Ruiz R, Francaviglia R, Aguilera E, Smith P (2016) Soil carbon sequestration rates under Mediterranean woody crops using recommended management practices: a meta-analysis. Agric Ecosyst Environ 235:204–214
Wang J, Xiong Z, Kuzyakov Y (2016) Biochar stability in soil: meta-analysis of decomposition and priming effects. GCB Bioenergy 8(3):512–523
World Bank (2012) Carbon sequestration in agricultural soils. World Bank, Washington, DC. https://openknowledge.worldbank.org/handle/10986/11868

17 Alleviation of Climate Catastrophe in Agriculture Through Adoption of Climate-Smart Technologies

D. T. Santosh, Subhankar Debnath, Sagar Maitra, Masina Sairam, La Lichetti Sagar, Akbar Hossain, and Debojyoti Moulick

Abstract

Present-day agriculture is facing enough burden to feed the future human population; since climate change is considered a vital negative factor for agricultural sustainability. It has been already well established that farming is the most vulnerable sector that has been affected by climatic aberrations. Anthropogenic interventions in agriculture are one of the major causes of the emission of greenhouse gases (GHGs) release to atmosphere and global warming impacts crop growth leading to a negative impact on farm output. Therefore, mitigation as well as adaptation of climatic aberration impacts should be considered, as the world population is increasing day by day. The adaptation of climate-smart technologies has enough potential to reduce the ill effects of changing climate on agriculture. In this regard, the adoption of efficient water management practices, climate-resilient crops and cropping systems, agroforestry systems, carbon sequestration, lowering the GHGs emission, integrated farming systems and appropriate inputs delivery by precision and smart agriculture technologies can show the arena of climate-smart agriculture. The present chapter is focused on climate-resilient agricultural technologies that are precisely relevant to the consequences of climate change.

Keywords

Agriculture · Climate change · GHGs emission · Smart technologies · Food security

17.1 Introduction

Climatic changees are major environmental challenges that the earth is currently facing, with various effects on human societies and ecosystems (Choudhury and Moulick, 2022). The Earth's climate is changing rapidly because

D. T. Santosh · S. Debnath
School of Agriculture and Bioengineering, Centurion University of Technology and Management, Paralakhemundi, Odisha, India
e-mail: santosh.dt@cutm.ac.in; subhankar.debnath@cutm.ac.in

S. Maitra · M. Sairam · L. L. Sagar
M.S. Swaminathan School of Agriculture, Centurion University of Technology and Management, Paralakhemundi, Odisha, India
e-mail: sagar.maitra@cutm.ac.in; sairam.masina@cutm.ac.in; lalichetti.sagar@cutm.ac.in

A. Hossain
Division of Soil Science, Bangladesh Wheat and Maize Research Institute, Dinajpur, Bangladesh

D. Moulick (✉)
Department of Environmental Science, University of Kalyani, Nadia, West Bengal, India

U. Chatterjee et al. (eds.), *Climate Crisis: Adaptive Approaches and Sustainability*, Sustainable Development Goals Series, https://doi.org/10.1007/978-3-031-44397-8_17

of the enhancement of greenhouse gases (GHGs) in the atmosphere, which primarily results from anthropogenic causes, namely, burning fossil fuels and deforestation. According to the Intergovernmental Panel on Climate Change (IPCC), global temperatures have already raised by around 1 °C above preindustrial levels, and the impacts of climatic aberrations are getting more severe (Masson-Delmotte et al. 2021). Climate change (CC) has various and complex impacts, such as aberrant temperatures, and more frequent and intense weather events such as heatwaves, droughts, floods, and storms. These changes have significant implications for natural systems, including the timing and pattern of rainfall, vegetation distribution, and species abundance and distribution. Climatic fluctuations have enormous influence on growth and productivity of all major crops, and thus hindrances in flourish of crops are commonly observed under climatic extremes (Raza et al. 2019). Besides environmental effects, climate change also has considerable social and economic impacts, with the possibility of worsening poverty, food insecurity, and social inequality. In this chapter, our aim will be to present (i) the consequences of climate changes on agriculture as well as make (ii) an assessment on the efficacy of climate smart technologies to ameliorate the adverse impact of climatic fluctuations.

17.1.1 Climate Change

Climate change (CC) is a crucial ecological concern of the twenty-first century. It is characterized by persistent alterations in the statistical distribution of weather factors for extended period, ranged between decades and millions of years. Climate change can result from both internal processes and external forcing, including natural influences such as solar radiation and volcanism, as well as anthropogenic factors, including changes in atmospheric composition due to human activities during the industrial revolution (Bhadra et al. 2022). As a result, climate change refers to long-term changes in climate, including mean temperature and precipitation.

Climatic change refers to a long-term shift in the average, variability, and extremes of meteorological variables that persist for at least 30 years. One of the important reasons of CC is fossil fuels burning, which emits greenhouse gases, especially carbon dioxide (CO_2), into the atmosphere (Mohammad et al. 2020). These gases trap heat and can cause various ecological impacts, such as sea level rise, extreme weather events, and droughts that increase the risk of wildfires in certain terrains. Human interventions such as farming as well as, also responsible for enhancement of GHGs into the atmosphere causing CC. The Fifth Assessment Report (AR5) of the IPCC provides extensive evidence on the contribution of human activities to CC, including sea level rise and its causes over the last few decades. It warns that continued GHGs emissions would result in an increased warming and long-term variations in all aspects of the climate system, with drastic and irretrievable influences on both human being and and ecology.

17.1.2 Climate Change Scenarios

The IPCC created four different families of emission pathways in 2000, called A1, B1, A2, and B2, in their report on emission scenarios (SRES), on the basis of different assumptions about socioeconomic development (Nakicenovic et al. 2000). The IPCC's fourth assessment report (AR4) published in 2007 used these SRES-based emission scenarios and climate predictions to describe projected climate change and its effects on society and ecosystems (Solomon et al. 2007). In 2014, the IPCC adopted a new set of emission scenarios called Representative Concentration Pathways (RCPs) that reflect radiative forcing pathways rather than entire socioeconomic scenarios. These pathways are defined by their cumulative measure of human emissions of greenhouse gases expressed in W/m^2 and their level by the year 2100. There are four RCP scenarios, including RCP 2.6, RCP 4.5, RCP 6.0, and RCP 8.5. Van Vuuren et al. (2011) provided a summary of the RCPs' evolution and key traits, which have radiative forcing values ranging from 2.6 to 8.5 W/m^2 for the year 2100. The RCP 2.6

Table 17.1 List of popular climate models and their horizontal resolution

Sl. no.	Climate model name	Organization	Horizontal Resolution (Latitude × longitude)	Type of climate model
1	CanESM2	Canadian Centre for Climate Modeling and Analysis, Canada	2.8° × 2.8°	GCM
2	MPI-ESM-LR	Max Planck Institute for Meteorology, Germany	1.8° × 1.8°	
3	IPSL-CM5A-LR	Institute Pierre Simon Laplace,France	1.9° × 3.75°	
4	CNRM-CM5	Centre national de Recherches Meteorologiques, France	1.4° × 1.4°	
5	HadGEM3-RA	National Institute of meteorological sciences, Korea	0.44° × 0.44°	RCM
6	RegCM4	Kongju National University, South Korea		
7	SNU-RCM	Ulsan National Institute of Science and Technology, South Korea		

scenario reaches a radiative forcing level of about 3.1 W/m^2 by the middle of the century, dropping back to 2.6 W/m^2 by 2100. The RCP 4.5 scenario aims to stabilize total radiative forcing before 2100, using a range of GHG emission reduction technologies and tactics. The RCP 6.0 scenario also seeks to stabilize total radiative forcing after 2100, without overshoot, through various strategies to lower greenhouse gas emissions. Meanwhile, the RCP 8.5 scenario is characterized by increasing greenhouse gas emissions over time, leading to high concentration levels.

Climate models are used to investigate the impacts of climate change based on data availability in a given region. Two main categories of climate models are global circulation models (GCMs) and regional climate models (RCMs). RCMs provide a more accurate simulation of future meteorological data compared to GCMs (Mishra et al. 2013; Debnath et al. 2021a, b). Table 17.1 lists the various climate models that are frequently used in impact assessment studies.

17.1.3 Effect of Climate Change on Crop Production

Crop production is facing several challenges due to climate change and extreme weather events, leading to reduced yields and lower quality. Land use patterns and soil degradation also contribute to adverse effects on crop productivity. Research indicates that maize and wheat crops have experienced losses of up to 10% over the last three decades due to climate change (Sagar et al. 2022; Sairam et al. 2023). Extreme weather conditions, namely, flood, heatwaves, and drought, cause severe harm to crop plants, impacting planting and harvesting schedules and reducing productivity (Elahi et al. 2022). Modification in land use pattern, overgrazing and deforestation contribute to soil degradation, nutrient depletion, and reduced soil health, thus limiting productivity of crops. Therefore, it is crucial to create climate-resilient cropping systems that can be appropriate to the changing climate conditions and withstand under harsh weather conditions. Sustainable land use practices, including conservation agriculture (CA), integrated pest management (IPM), adoption of appropriate cropping system, and agroforestry, can mitigate soil degradation, maintain soil health, and enhance crop productivity and resilience (Mbow et al. 2014).

To achieve climate-resilient crop production, it is vital to focus on scientific soil management as the effects of climate change become more severe. This can be achieved by implementing sustainable practices that can enhance soil health and fertility. The study carried out by Wezel et al. (2014) showed the implementation of CA like reduced tillage, cover cropping, and mixed stand of crops can recover soil quality as well as enhanced crop productivity. These practices can the moisture holading and infiltration capacity of soil, making crops more resilient to drought conditions. Incorporating organic matter into the soil can also increase soil fertility and replace chemical fertilizers. In drought-prone areas, Jat et al. (2019) found that incorporating organic matter into the soil significantly increased crop yields and improved soil quality. Similarly, soil management practices such as crop rotation and the use of soil amendments can improve soil health and fertility, leading to enhanced crop productivity and resilience to climatic variability.

To address the challenges of climate change in agriculture, the United Nations has developed a framework called climate-smart agriculture (CSA). This approach aims to upsurge crop yield and income, create resilience to climatic burdens, and reduce GHGs emissions. Additionally, CSA promotes the sustainable organization of natural resources and ecosystem services. The influence of CC on agriculture is likely to be significant for vulnerable population, as around 80% of the world population lives in rural areas with poverty and farming-based livelihoods. In some regions, losses of up to 25% of crop yields are estimated by 2050 (da Cunha Dias et al. 2021), which is compounded by the increasing world population and growing demand for food, feed, and fiber. The UN's framework for CSA integrates three key pillars to address these challenges:

1. Sustainable increases in crop yield and farm income.
2. Adaptation as well as building resilience to climate change.
3. Reduction and/or removal of GHHs release to the atmosphere, where possible.

This chapter delves into the topic of climate change, examining its origins and how it impacts on crop productivity, as well as approaches to alleviate its direct and indirect consequences on agriculture. The principles and approaches utilized to manage the negative effects of CC on agriculture are also discussed.

17.2 Climate-Resilient Cropping Sequences

Climate-resilient cropping sequences can help farmers adapt to changing climate conditions, such as increased temperatures and changing rainfall patterns. Climate-resilient crops are developed through genetic modification or breeding programs to be more tolerant to extreme weather conditions. Furthermore, appropriate cropping sequences can help maintain soil health, reduce pests and diseases, and improve overall crop productivity in a changing climate.

17.2.1 Cropping Systems

Cropping systems are a combination of agricultural practices and techniques that farmers employ to manage their lands and optimize crop yield and sustainability. These systems often include crop rotation, intercropping, cover cropping, and conservation tillage, among others. In modern agriculture, these systems are increasingly crucial for farmers seeking to reduce environmental impact while maintaining profitability. Recent studies have shown that conservation agriculture, which utilizes cropping systems such as cover crop cultivation and reduced tillage, can facilitate improvement of soil quality, cut soil erosion, and productivity of crops and cropping systems (Zhang et al. 2021). Similarly, intercropping is commonly known to increase soil fertility, reduce pest damage, and enhance overall productivity of crops in smallholder farming systems (Maitra et al. 2019, 2021). Cropping systems not only offer environmental benefits but also economic advantages for farmers. For instance, crop diversification, a critical component of many cropping systems, has been shown to increase farm incomes and reduce vulnerability to market fluctuations (Xu and Du 2022). Similarly, conservation agriculture practices, which are often incorporated into cropping systems, have been found to lead to higher profits for smallholder farmers in developing countries.

17.2.2 Recent Trends in Climate Change

CC is an ongoing global problem, and it is evident that the Earth's climate is continuing to warm at an extraordinary rate. Recent investigations provided evidence of the ongoing trend in the CC. A study conducted by Liao et al. (2021a, b) revealed that the 2010s were the hottest decade, with the past five years being the hottest globally. The study also found that the global temperature increase in 2020 was about 1.2 °C more than pre-industrial levels. In another study, Swapna et al. (2020) examined the temperature records of more than 3000 meteorological stations worldwide and discovered that the increased temperature became threefold in the past few decades. The research also suggested extrimity of high temperature, such as heatwaves, has increased tenfold since the 1950s. Furthermore, the impact of CC is being experienced in other aspects of the Earth's system. A study by Sherwood et al. (2020) recorded that atmospheric humidity has increased by roughly 7% per degree of warming, resulting in more intense and frequent rainfall events. The consequences of climate change are being felt in different regions worldwide. Moreover, Hu et al. (2021) found that in the Arctic region, temperature is increasing at the rate of around 0.6 °C per decade, which is double than the global average. This caused the loss of sea ice, permafrost thawing, and changes in ecosystems.

17.2.3 Impact of Climate Change on Crops and Cropping System

CC is a significant concern for global food and nutritional security in developing countries that rely heavily on farming. Crop yields are declining, agricultural productivity is decreasing, and food prices are increasing. Recent studies have highlighted this issue and projected intensification of the impact of CC. For example, Lobell et al. (2014) found that global maize and wheat yields have decreased by 4% and 5%, respectively, due to CC since the 1980s, and low-income countries may suffer due to its continuation. Similarly, Challinor et al. (2014) projected that global maize, wheat, and rice yields could decline by 10–25% by 2050 due to climate change, with tropical regions being more vulnerable to extreme weather events. Climate change is also altering the timing of crop growth and development, as reported by Anwar et al. (2015), who found that crop phenolgy, namely, flowering and maturity has shifted by an average of 2.5–5 days per decade over the past 30 years. This has implications for crop management practices, such as planting and harvesting dates. Moreover, CC is impacting the distribution of pests, as noted by Dudney et al. (2021), who found that the global distribution of plant pathogens is shifting due to climate change, resulting in some pathogens moving towards higher latitudes and altitudes.

17.2.3.1 Phenology and Physiology of Crops

The timing of various plant growth stages is referred to as phenology, while the internal functioning of the plant is referred to as physiology. Climate change is causing alterations in the timing of crop growth stages, as per a study by Easterling et al. (2000), which discovered that increasing temperatures in the United States led to earlier maturity of maize and soybean crops. This shift in timing has resulted in an increase in yields for these crops. Furthermore, climate change is affecting the physiology of crops by disrupting their metabolic processes. For instance, Sarkar et al. (2020) found that increased temperatures cause variations in the photosynthetic rate and soil moisture utilization of crops like wheat and corn, resulting in lower crop yields. Moreover, climate change is affecting the interplay between phenology and physiology in crops. According to Wang et al. (2020a), rising temperatures can expedite the growth and development of wheat while decreasing its photosynthetic capacity, indicating that a better understanding of these intricate interactions is important to anticipate the CC impacts on crop yields. On the other hand, continuous efforts are provided to develop crop cultivars suitable to the abiotic stresses (Billah et al. 2021). Kothari and Lachowiec (2021) found that rice varieties acclimatized to higher temperatures have a higher photosynthetic rate and produce greater yields under heat-stress conditions.

17.2.3.2 Crop Yield

Crop yields may suffer because of climate change, which can impact crop growth and productivity through changes in heat, rainfall and other extrimity of climatic factors. Recent studies have explored the link between climate change and crop yields. For example, Yu et al. (2021) discovered that corn and soybean yield in the United States declined by 1–2% per degree celsius of warming above a specific temperature threshold. Song et al. (2022) observed that high temperatures during the reproductive stage of rice caused significant yield loss in China. Shi et al. (2021) found that aberrant weather events such as soil moisture scarcity and flood had a significant negative impact on wheat yields in China, and Wu et al. (2021a, b) reported that warming and precipitation changes led to significant reductions in maize yields in China. Zhai and Zhuang (2012) found that warming and precipitation changes also caused significant reductions in soybean yields in the United States, but elevated carbon dioxide levels partially offset the negative impacts. These findings highlight the importance of comprehending the relationship between climate change and crop yields, as well as the necessity for adaptation strategies to minimize adverse effects on agriculture.

17.2.4 Climate-Resilient Crops and Varieties

CC is expected to influence the global food production, and efforts are underway to develop climate-resilient crops and varieties that can withstand changing climate conditions. Recent research has focused on identifying genetic traits, biotechnological tools and breeding strategies for developing the climate resilient crop cultivars. For instance, Kumar et al. (2021) identified genetic markers associated with heat tolerance in wheat, which could be used to breed more resilient wheat varieties. Similarly, Waheed et al. (2021) identified drought-tolerant traits in rice, such as root architecture and water use efficiency, which could be incorporated into breeding programs. Other studies have focused on the use of genetic engineering to develop crops with enhanced resilience to climate stressors. For example, Hassan et al. (2021) developed genetically modified cotton plants with enhanced drought tolerance. Excluding genetic approaches, strategies like developing climate-resilient cropping systems that integrate crops, livestock, and other elements of the agroecosystem can be adopted. Apart from these, cutting edge technologies like application of metabolomics (Choudhury et al. 2021a), potentials of crop distant/wild relatives with desired agronomic traits (Hossain et al. 2022), efficacy of osmoprotectants and secondary metabolites (Hossain et al. 2021a, b, Choudhury et al. 2021b) are being examined costantly by the reseachers for developing climate resilience in a wide range of crops.

17.2.4.1 Rainfed Ecosystems

Rainfed ecosystems are exposed to CC, as changes in heat stress and unpredicted and erratic precipitation can have significant influences on crop yields and livelihoods. However, there is a growing interest in developing climate-resilient cropping systems that can enhance crop yields and resilience of rainfed cropping systems.

Recent research has focused on identifying climate-resilient crops and cropping systems for rainfed ecosystems. For example, Anantha et al. (2021) evaluated the performance of various drought-tolerant crops, including pearl millet, sorghum, and chickpea, in rainfed areas of India. They found that pearl millet and sorghum performed well under drought conditions and that intercropping these crops with legumes such as pigeonpea and mungbean could further enhance productivity and resilience. Other studies have focused on the use of CA, such as minimum tillage along with cover crops to improve soil health and water use efficiency in rainfed ecosystems. For example, Mishra et al. (2021) found that conservation agriculture practices improved the productivity and resilience of rainfed maize systems in Vietnam, while also reducing greenhouse gas emissions.

17.2.4.2 Heat Stress

The negative impacts of heat stress are expected to increase with climate change. The research was conducted to understand the mechanisms of heat stress in plants and develop strategies to mitigate its negative impacts. Huang et al. (2021) carried out a research on the molecular mechanisms of heat burdens in rice, and identified genes involved in heat tolerance that could be used in breeding programs. Similarly, dos Santos et al. (2022) reviewed the strategies used by plants to handle heat stress, including the activation of heat shock proteins and the regulation of metabolic pathways, and highlighted the potential of genetic engineering to enhance heat tolerance in crops. Appart from the genetic approaches, there is increasing interest in developing cropping systems that can buffer the negative impacts of heat stress. Li et al. (2022) studied the effects of intercropping maize and soybean on heat stress tolerance and found that intercropping improved soil moisture and reduced temperature stress in the maize canopy. Similarly, Hobbs et al. (2008) showed that CA practices can improve soil moisture storage, and thus reduced high temperature stress in wheat fields.

17.2.4.3 Cold Stress

Cold stress can significantly impact crop growth and productivity, particularly in temperate and cold regions. With CC, extremity of cold is becoming more common and severe in some areas, making it important to develop climate-resilient crops and cropping systems that can withstand cold stress. Recent research has focused on identifying genetic traits and molecular mechanisms that can enhance the cold tolerance of crops. For instance, Yang et al. (2021) identified a gene in rice that adjusts the countenance of cold-responsive genes, which could be used to develop rice varieties with enhanced cold tolerance. Similarly, Fritsche-Neto et al. (2021) identified key genes involved in cold tolerance in maize, which could be targeted in breeding programs. In addition to genetic approaches, there is also interest in developing cropping systems that can mitigate the impacts of cold stress. Earlier, Huang et al. (2021) found that intercropping of corn and soybean improved the cold tolerance of maize by altering soil properties and microbial communities.

17.2.4.4 Waterlogging Areas

Waterlogging is a key challenge in many agricultural areas, particularly in low-lying regions where excess water can accumulate and reduce crop productivity. Climate change is expected to exacerbate this problem, as changing precipitation patterns and risky weather conditions can lead to more frequent and severe waterlogging.

Recent studies have focused on identifying genetic traits and breeding strategies that can improve the resilience of crops to waterlogging. For example, Najeeb et al. (2021) found genetic markers linked to waterlogging tolerance in rice, which can be used to breed more resilient rice varieties. Similarly, Nguyen et al. (2021) identified genetic traits in maize, such as root length and porosity, which can be targeted in breeding programs to enhance waterlogging tolerance. Besides genetic approaches, raised beds, improved drainage systems, and alternate wetting and drying irrigation strategies are adopted to minimize the risk of waterlogging that results in improved crop productivity. Crop rotation and intercropping strategies can also be used to diversify the agroecosystem and reduce crop failure due to waterlogging. Ahmed et al. evaluated the effectiveness of different cropping systems, including intercropping and crop rotation, in mitigating the impact of waterlogging on wheat and lentil crops in Pakistan. The study found that intercropping was more effective than crop rotation in improving crop yield and reducing the negative impact of waterlogging.

17.2.5 Climate-Resilient Cropping Systems

Climate-resilient cropping systems are developed to cope with the negative effects of climatic variations. They typically consist of various techniques for crop management, such as utilizing stress tolerant crop cultivars that show resilience to climatic aberrations, employing proper crop rotations, and practicing soil conservation methods.

17.2.5.1 Rainfed Agro-ecosystems

Rainfed agro-ecosystems are highly susceptible to the effects of climate change, particularly due to their reliance on rainfall for crop production. Unpredictable rainfall patterns and droughts can lead to reduced crop yields and even crop failure. A key strategy for building climate-resilient cropping systems is the use of tolerant crop cultivars to soil moisture stress. For instance, the International Crops Research Institute for the Semi-Arid Tropics (ICRISAT) has successfully developed drought-tolerant varieties of crops, such as various millets, and pigeon pea, which have demonstrated promising results in improving yield under drought conditions (Kumar et al. 2022; Maitra et al. 2022).

Further, the use of CA practices such as minimum tillage, intercropping, and mulching, which can enhance soil health, increase water retention, and reduce soil erosion, thus improving the resilience of rainfed agro-ecosystems to climate variability (Alam et al. 2014). Crop diversification and rotation can also enhance the resilience of rainfed agro-ecosystems to CC. Crop diversification enables farmers to minimize the impact of any single crop failure by cultivating different crop species in a particular area, thereby spreading their risk. On the other hand, crop rotation can help break pest and disease cycles, improve soil health, and increase crop yields (Thakur and Kumar 2021). Moreover, climate information services can play a vital role in aiding farmers to anticipate and respond to climate variability. By providing weather forecasts and advice on crop management practices, farmers can adjust their planting and harvesting schedules to curtail the impact of CC on their crops (Kumar et al. 2020).

17.2.5.2 Irrigated Agro-Systems

Irrigated agro-ecosystems are highly productive, but also highly susceptible to CC because of their reliance on water resources. Crop diversification is a primary approach to enhance resilience in such systems. This technique involves growing various crops on the same land, which can improve soil health, manage pest population dynamics, and increase the agroecosystem's resilience to climate variability. In India, for instance, crop diversification reduced water use by 30% and increased yields by up to 28% (Sharma et al. 2021). Precision irrigation is another strategy that can enhance resilience. This system provides water to crops at the right time and amount, based on real-time weather information, soil moisture sensors, and crop water requirements. It enhances judicious use of water, improves water use efficiency, and enhances crop resilience to water scarcity (Liao et al. 2021a, b). Conservation agriculture practices such as minimum tillage, intercropping, and mulching are also effective strategies to make climate-resilient agro-ecosystems to combat against CC. These practices improve soil health, and water retention, and reduce soil erosion, thereby enhancing the overall productivity and resilience of the agro-ecosystem (Sarkar et al. 2021). Providing farmers with weather forecasts and advice on crop management practices can help them adjust their irrigation schedules and reduce the ill effects of CC on crops (Kumar et al. 2021).

17.2.6 Cropping Systems Management

Effective cropping systems management is essential for adjusting to CC and developing climate-resilient cropping systems. This includes the use of climate-smart agriculture practices such as crop diversification, precision irrigation, conservation agriculture, and climate information services.

17.2.6.1 Crop Planning as per Climate-Soil-Site Suitability

Climate change is expected to affect the suitability of different crops and cropping systems in various regions, making it imperative to plan crops according to their suitability to the local climate, soil, and site conditions. With the use of climate modeling, determining the projected changes in temperature, precipitation, and other climate variables, farmers can plan crops that are suitable adapted to CC. Different crops have different nutrient requirements and grow best in soils with specific physicochemical properties. Soil testing and analysis can help farmers determine which crops are most suitable for their particular soil conditions. Site suitability, including factors such as topography and water availability, also plays a significant role in crop planning for climate-resilient cropping systems. For instance, crops that require high water availability may be unsuitable for regions with limited water resources, while crops that require well-drained soils may be unsuitable for areas prone to flooding.

17.2.6.2 Seed of Resilient Crop Varieties

To achieve climate resilience in cropping systems, the seed of resilient crop varieties must possess traits such as tolerance to several abiotic burdens and capability to use nutrients efficiently. Some of the climate-resilient varieties of seed developed are mentioned in Table 17.2; however, there are many more varieties developed and released by different organizations across the world.

17.2.6.3 Weed Management

CC has been observed to influence on weed growth and distribution, with modifications in temperature, rains, and atmospheric CO_2 concentrations affecting the timing and intensity of weed emergence and growth. Effective weed management strategies, including cultural, mechanical, and chemical control methods (commonly known as Integrated Weed Management or IWM), must be tailored to local climate conditions and crop management practices to ensure maximum crop productivity and yield stability under changing climate conditions (Ghosh et al. 2020a, b, 2021, 2022a, b).

Research studies have shown that variations in climate variables such as temperature, rainfall patterns, and atmospheric carbon dioxide concentrations can affect weed growth, distribution, and composition. For instance, higher temperatures and increased atmospheric carbon dioxide concentrations have been found to favor the growth of certain weed species such as pigweed and ragweed, which can compete with crops for resources and reduce crop yields (Ziska et al. 2011). On the other hand, changes in rainfall patterns and soil moisture regimes can affect weed emergence and distribution, with wetter conditions favouring the growth of certain weed species such as waterhemp and nutsedge (Norsworthy et al. 2012).

To effectively manage weeds in climate-resilient cropping systems, farmers and researchers must employ a combination of cultural, mechanical, and chemical management methods tailored to local climate conditions and crop management practices. Some examples of climate-smart weed management practices include the use of cover crops and intercropping to reduce weed growth and competition with crops, mechanical control methods such as hand weeding and tillage, and the use of herbicides in combination with other management practices to reduce weed populations (Shekhawat et al. 2020). These strategies must be implemented as part of an IWM approach to ensure sustainable and productive cropping systems in the face of CC.

17.2.6.4 Water Management

As climate patterns continue to change, it becomes increasingly important to conserve and make the most of available water resources. By employing climate-smart irrigation practices, such as deficit irrigation and microirrigation, farmer can achieve better water-use efficiency and boost crop yields. In addition, optimization of water use can help reduce the carbon footprint of crop production by minimizing energy consumption for irrigation. Recent research has demonstrated that efficient irrigation water management can also mitigate the adverse effects of climate change on crop yields (Daryanto and Christata 2021; Kaur and Singh 2019).

17.2.6.5 Nutrient Management

Nutrient management is an essential component of sustainable crop production, especially under changing climate conditions. Adequate and balanced nutrient supply can enhance crop growth and resilience to climate stress. CC is anticipated to impact on soil nutrient cycling and availability,

Table 17.2 Some of the crop varieties developed to achieve resilience for different climatic extremes

Crops	Drought tolerance	Heat tolerance	Pest and disease resistance
Rice	1. NERICA (new Rice for Africa) 2. Sahbhagi Dhan 3. DroughtGard 4. Swarna Sub1 5. Samba Mahsuri 6. IR64 Sub1 7. IRRI Dhan 28 8. MAS26 9. Vandana 10. CSR10	1. Swarna-Sub1 2. Sahbhagi dhan 3. N22 4.TDK1 5. Supanburi 1 6. LAC23 7. MR219–4 8. PSBRc82 9. IR64-Sub1 10. TNI	1. Swarna-Sub1 2. IR64-Sub1 3. NERICA-L-19 4. NERICA-L-44 5. BG300 6. Curinga 7. AG8 8. Sahbhagi Dhan 9. TDK1 10. Luna Sufaid
Wheat	1. Creso 2. Seri M82 3. BRS 331 4. KAUZ 5. NARC 2011 6. N22 7. Pavon 76 8.. Raj 3765 9. RAC875 10. Sahel 108	1. Krichauff 2. Drysdale 3. Wyalkatchem 4. Excalibur 5. H45 6. Pavon 76 7. Jupateco 8. Katana 9. Huapeño 10. CENCOSUD 1	1. Kukri 2. Roblin 3. Sava 4. Caledonia 5. Lee 6. Yecora Rojo 7. Sumai 3 8. Gladius 9. Catbird 10. Norin 10
Maize	1. Drought Tego (DT) 2. DroughtGard 3. IITA Drought Tolerant 4. SC Drought Tolerant 5. Ua Kayongo 6. DKC 80-17 7. Oaxaca 169 8. Water Efficient Maize for Africa (WEMA) 9. B73 10. ZM 521	1. DKC 63-42 VT3PRO 2. P 3394 VT3PRO 3. P 3253H 4. P 2958H 5. P 2822H 6. P 2539YH 7. SC 647 8. SC 533 9. SC 719 10. ACH1415	1. Tegemeo 2. Longe 5 3. DK 8033 4. SC 513 5. WE2108 6. SAWAHYBRID 8 7. IITA 937 8. Suwan 1 9. Katumani Composite A 10. Viptera
Millets	1. CR Dhan 310 (Pearl Millet) 2. GPU-28 (Finger Millet) 3. ICTP 8203 (sorghum) 4. HB 3 (pearl millet) 5. Raj 171 (sorghum) 6. Raj 379-2 (sorghum) 7. PSH 104 (pearl millet) 8. Dharwad Sankar (sorghum) 9. S35 (Sorghum) 10. Pusa Baisakhi (Proso millet)	1. Pearl millet 863B (pearl millet) 2. Phule G-0815 (green gram) 3. ICTP 8203 (sorghum) 4. Prasad (Foxtail Millet) 5. Souna 3 (sorghum) 6. Okashana 1 (cowpea) 7. B35 (bean) 8. Tift 23D2B1 (sorghum) 9. Landrace (common bean) 10. PSB-SM 706 (soybean)	1. B35 (maize) 2. ICTP 8203 (pearl millet) 3. PAVAN (chickpea) 4. LDTM-1 (cowpea) 5. GPU-28 (Finger Millet) 6. AHP 9714 (sorghum) 7. IPCA 5043 (pigeon pea) 8. KARI-Mtama 1 (sorghum) 9. ICMV 221 (mung bean) 10. RACER (cowpea)
Legumes	1. PUSA-44 (green gram) 2. PUSA-105 (black gram) 3. PUSA-307 (green gram) 4. RIL-204 (cowpea) 5. RIL-122 (cowpea) 6. Pusa-372 (pigeon pea) 7. RIL-72 (cowpea) 8. PUSA-256 (green gram) 9. EC-404911 (chickpea) 10. N-1380 (cowpea)	1. Pusa Vishal (black gram) 2. Pant Urd (black gram) 3. Rajendra Soybean-1 (soybean) 4. PKM-1 (cluster bean) 5. RIL-41 (cowpea) 6. JG-74 (pigeon pea) 7. K-851 (green gram) 8. Phule G-0815 (green gram) 9. Maruti (red gram) 10. RG-283 (chickpea)	1. PDM-11 (pigeon pea) 2. ICTA-L812 (dolichos bean) 3. IT-97 K-556-6 (cowpea) 4. PBCV-0301 (common bean) 5. MIB-289 (moth bean) 6. V-146 (mung bean) 7. BG-406-4 (black gram) 8. Seredo (chickpea) 9. Njavara (rice bean) 10. KAT B1 (lablab bean)

thereby impacting crop nutrient uptake and productivity. Integrated nutrient management (INM) inclusive of organic and inorganic fertilizers with soil amendments and conservation agriculture techniques can optimize soil fertility and crop yield while mitigating greenhouse gas emissions. Recent studies have suggested that nutrient management strategies should be tailored to specific agroecological zones and cropping systems to enhance available nutrients efficiently and reduce environmental risks (Chen et al. 2019).

17.2.7 Various Adaptation Options for Rainfed Agro-ecosystems

Various adaptation options are available to mitigate the ill effects of CC on rainfed agriculture. Some important methods available for rainfed agro-ecosystems to mitigate the adverse effects of climate change are listed below:

- Soil and water conservation measures.
- Use of drought-tolerant crop varieties and species diversification.
- Conservation agriculture practices like zero-tillage and residue retention.
- Agroforestry practices.
- Efficient water use like rainwater harvesting and microirrigation.
- Use of organic fertilizers and soil amendments to improve soil fertility and water-holding capacity.
- Intercropping and crop rotations to improve soil health and reduce pest pressure.
- Adoption of weather forecasting and climate information services to optimize crop management practices.

These methods can be combined and tailored to suit the specific needs and conditions of different rainfed agro-ecosystems. Research has shown that the adoption of these adaptation options can lead to increased productivity, improved soil health, and enhanced resilience to climate change in rainfed agro-ecosystems (Jat et al. 2019; Malyan et al. 2021). Additionally, policies and programs that support the adoption of these adaptation options, along with capacity-building initiatives for farmers, can help to further enhance the resilience of rainfed agriculture to climate change (Kumar et al. 2021).

17.3 Temperature and Rainfall Variability Due to Climate Change

17.3.1 Change in Temperature and Rainfall

To study the effects of climate change and extreme weather in China, Chen et al. (2020) utilized the outputs of climate variables from 17 General Circumstance Models (GCMs) in the Coupled Model Inter-comparison Project phase five (CMIP5) dataset to analyze the effects of extreme weather and CC in China. The study found that during the growing season, the average temperature for rice cultivation increased from a baseline level of 1.14 to 1.33 °C under the RCP4.5 scenario and from 1.22 to 1.48 °C in the RCP8.5 scenario (1961–2005). For the research area, the study projected that the growing season cumulative rainfall for rice will increase from 1.48 to 104.08 mm under the RCP4.5 scenario and from 4.57 to 95.82 mm under the RCP8.5 scenario.

CC is predictable to increase annual rainfall in Indonesia by 2 to 3%, with the Moluccas region undergoing a significant transformation that will increase rainfall across the country. The higher rainfall is likely to persist, shortening the rainy season and increasing the risk of flooding. Rahman and Lateh (2017) found that Bangladesh's climate is warming more rapidly than the global average, with an average temperature increase of 0.20 °C per decade versus 0.13 °C globally. During the period 1971–2010, the northern, northwestern, northeastern, central, and central southern regions of Bangladesh experienced the greatest warming in T_{min}, while the south, southeast, and northeast regions witnessed the greatest warming in T_{max} (1.20–2.48 °C range). Pre-monsoon and post-monsoon rainfall trends declined, whereas yearly rainfall increased by 7.13 mm per year and annual rainfall decreased by 0.75 mm per year. Schmidt-Thome et al. (2015) reported that the annual mean temperature in Vietnam has increased by 0.5 °C nationwide, with annual precipitation decreasing in the north and increasing in the south. Tmax in Vietnam ranged between −3 and 3 °C, while Tmin varied between −5 and 5 °C. Both Tmax and Tmin have increased in recent years following the global warming trend, with minimum temperatures rising more rapidly than maximum temperatures. In the past 50 years, rainfall during the dry season (November–April) has significantly increased in the southern regions, while it has remained relatively stable in the northern parts. During the monsoon period (May–October), precipitation has decreased by 5 to over 10% in most of northern Vietnam, while it has increased by 5–20% in the southern regions. Similarly, annual rainfall varies across regions, with the south-central area experiencing the greatest increase in annual precipitation, up to 20% in some locations, compared to other areas of the country.

17.3.2 Change in Temperature and Rainfall in India

Due to climate change, extreme weather events including droughts, periods of extremely high temperatures, and torrential rain are predicted to happen more frequently in the future. The global impact of extreme weather events and climate change has been the subject of an increasing number of research. According to Allen et al. (2014), the global surface temperature climbed by 0.5–1.3 °C between 1951 and 2010 and is anticipated to rise by an additional 3.7 °C by the end of the century. As a result, it is projected that rainfall patterns, frequency, and intensity may alter in the future and intensify globally. At 125 sites dispersed throughout India, Arora et al. (2005) detected the changes in average maximum temperatures (T_{max}) and minimum temperatures (T_{min}) at yearly and seasonal scales. According to the study, during the years 1901–2000, the annual mean temperature increased at a pace of 0.42 °C, 0.92 °C, and 0.09 °C, respectively. India has witnessed considerable warming of 0.5 °C over the past 100 years between 1901 and 2007, with accelerated warming over the last 40 years (1971–2010), and intense warming in the most recent 10 years (1998–2007), according to the Ministry of Earth Sciences (2010). Up to the year 2100, climate change projections for India show an overall increase in temperature of 2–4 °C along with an increase in rainfall, especially during the monsoon season (Kumar 2009; Chaturvedi et al. 2012). By the end of the century, however, it is predicted that the frequency of heavy rainfall events will increase while that of low and medium rainfall events will decrease (Kundu et al. 2014). Also, there are significant geographical variations across the nation (Ministry of Environment and Forests 2010).

17.4 Carbon Management

Carbon (C) is an essential element for soil fertility and crop growth. The amount of C present in the soil indicates soil quality, as it affects physical, chemical and biological properties of soil. Several studies have shown that an increase in soil carbon content can improve crop yields, increase soil organic matter, and enhance soil microbial activity (Lal 2004; Kamal et al. 2020). C sequestration through the use of efficient land management approaches, namely, CA agroforestry, intercropping and alley cropping and crop rotation can help to minimize GHGs emission, enhance soil health quality, and increase the resilience of crops to CC. Additionally, the use of C credits and other financial incentives can help to promote the adoption of carbon management practices by farmers. Recent studies have demonstrated the potential of carbon management practices to reduce the carbon footprint of agricultural production while increasing crop yield (Ashfaq et al. 2020; Qin et al. 2020).

17.4.1 Basic Principles of Soil Carbon Management

Climate change and extreme weather events have direct and indirect impacts on soil carbon content and crop production. Changes in temperature and precipitation patterns can alter the decomposition of organic matter, which affects soil carbon sequestration. Extreme weather events such as floods and droughts can also affect soil carbon storage by altering soil moisture and temperature conditions. The indirect impact of changes in soil carbon content on crop production can result from changes in nutrient cycling, water availability, and soil structure. The basic principles of soil carbon management include maximizing plant inputs, minimizing carbon losses, and enhancing soil organic matter stability. Practices such as cover cropping, crop rotation, reduced tillage, and organic amendments can increase plant inputs and soil organic matter. Moreover, management practices such as avoiding compaction, maintaining soil moisture, and reducing erosion can help minimize carbon losses.

Recent studies have demonstrated the potential of carbon management practices to reduce the carbon footprint of agricultural production while increasing agricultural productivity (Ashfaq et al. 2020; Qin et al. 2020). However, promoting microbial activity, selecting appropriate crop and soil management practices, and reducing the use of synthetic fertilizers can enhance soil organic matter stability. De Moraes Sa et al. (2020) estimates that soil carbon sequestration could mitigate up to 20% of global greenhouse gas emissions. Similarly, Smith et al. (2020) argue that sustainable land management practices, including soil carbon management, can contribute significantly to meeting the goals of the Paris Agreement.

The potential indirect effects of climate change on crop productivity via CO_2 fertilization may lead to an increase in plant growth and associated yield due to photosynthesis being the largest transfer of CO_2. The increase in daytime temperature may also enhance photosynthesis and plant growth, as long as water and nutrient requirements are met, and a rise in night temperature may reduce respiration. Elevated CO_2 levels often increase the root:shoot ratio, favouring the accumulation of soil organic matter. Both C3 and C4 crops benefit from the CO_2 fertilization effect, resulting in increased assimilation of plant biomass. Studies by Meena et al. have shown a strong response to elevated CO_2 levels in terms of increased total and root biomass of pulse crops, and elevated CO_2 levels have been found to significantly increase total oil yield. Additionally, the risks of crop damage due to air pollutants, such as nitrogen oxide (NOx), sulphur dioxide (SO_2), and ozone (O_3), are reduced under elevated CO_2 levels due to the partial closure of stomata and other physiological changes. (Reddy et al. 2010).

17.4.2 Carbon Management in Various Production Systems

Increasing soil organic carbon (SOC) is important for improving soil health, water retention, and nutrient availability, as well as reducing greenhouse gas emissions (Lal 2020). Conservation tillage practices such as no-till and reduced tillage are effective ways to increase SOC levels in various cropping systems, such as corn-soybean rotations, cotton, and wheat (Smith et al. 2020). Cover crops are another strategy to improve soil quality, reduce erosion, and increase SOC levels in cash crop rotations (Poeplau and Don 2015). In addition, adding animal manure and organic amendments can enhance SOC levels and improve soil fertility (Morugán-Coronado et al. 2020). Research conducted in China shows that the application of composted animal manure to rice paddies can increase SOC levels and rice yields while decreasing greenhouse gas emissions (Nayak et al. 2020). Agroforestry systems, where crops are grown with trees, can also increase SOC levels and improve the resilience of cropping systems to climate change (Ivezić et al. 2022).

17.4.3 Carbon Management for Adaptation

The concept of 'adaptation to climate change' refers to the adjustments made in response to the current or projected effects of climate change to reduce adverse impacts or take advantage of beneficial opportunities (Robinson, 2020). Conservation of soil and water resources, along with effective crop management, is considered important pillar of adaptation strategies. Given that human activities play a significant role in the concentration, distribution, and life cycle of greenhouse gases (GHGs) (Stocker 2014), evaluating the impact of anthropogenic activities is crucial. For instance, land use change from forests or grasslands to agriculture can lead to a reduction in soil organic carbon (SOC) stocks of up to 50% within 10–15 years, depending on factors such as soil type, temperature, moisture, natural vegetation, and cropping systems (Zingore et al. 2007). The recommended management practices (RMPs) for carbon sequestration in soils involve improving aggregate stabilization, deep translocation into the soil, and the formation of recalcitrant material. These practices include conservation tillage, diverse crop rotations, cover cropping, agroforestry, perennial pasture, use of biochar and other amendments, integrated nutrient management (INM), and supplemental irrigation. Some RMPs are discussed briefly below.

17.4.3.1 Recycling Crop Residue

Crop residues, such as straw, leaves, and stems, are valuable sources of organic matter that can be recycled back into the soil. The practice of leaving crop residues on the soil surface after harvest, also known as conservation agriculture, can help reduce soil erosion and increase water infiltration. When crop residues are incorporated into the soil, they can contribute to the build-up of SOC, which can increase the soil's ability to store carbon and improve soil fertility.

Research has shown that recycling crop residues can increase SOC levels in various cropping systems. In a study conducted in Pakistan, the incorporation of wheat straw into the soil increased SOC levels by 4.4% compared to the control treatment (Zahid et al. 2020). Another study in India found that the addition of rice straw to the soil increased SOC levels by 0.12% per year over a period of 4 years (Benbi et al. 2015). In addition to crop residues, other organic materials such as animal manure, compost, and biochar can also be recycled to enhance SOC levels. The application of composted animal manure has been shown to increase SOC levels and crop yields while reducing greenhouse gas emissions (Li et al. 2020). However, it is important to consider the potential trade-offs between the use of crop residues for carbon management and other purposes, such as livestock feed and bioenergy production. Sustainable management practices that balance the competing demands of different stakeholders are necessary for the effective implementation of recycling residues as a carbon management technique for adaptation.

17.4.3.2 Cover Crops

Crop cover management is an effective approach to enhance soil organic carbon (SOC) levels, mitigate climate change impacts, and improve crop productivity. This technique involves growing crops between the main crops or during the fallow period, which includes cover crops, intercropping, and relay cropping. Cover crops, particularly legumes, contribute to increased SOC levels by increasing carbon input, reducing soil erosion, and improving nutrient cycling. Similarly, intercropping of multiple crops has been found to increase crop productivity, reduce soil erosion, and improve SOC levels. Relay cropping of a second crop after the main crop harvest can increase SOC levels by lengthening the growing season, adding more organic matter to the soil, and reducing soil erosion.

Several studies have shown the positive impact of crop covers on SOC levels. A study conducted in the United States reported that cover crops increased SOC levels by up to 19% in a corn-soybean rotation system (Cavigelli et al. 2013). Similarly, a study conducted in Brazil found that intercropping maize and soybean increased SOC levels by up to 34% (Crusciol et al. 2013). Another study conducted in China reported that relay cropping of maize and soybean increased SOC levels by up to 11% (Wang et al. 2020b).

17.4.3.3 Conservation Agriculture

Conservation agriculture (CA) is a sustainable approach to land management that has the goal of improving soil health, increasing crop productivity, and reducing greenhouse gas emissions. The approach is based on three main principles, including minimum soil disturbance, permanent soil cover, and crop rotation or intercropping. By reducing soil disturbance, carbon losses are minimized and soil organic carbon (SOC) levels are increased under CA practices. Additionally, the use of permanent soil covers, such as mulch or cover crops, helps to reduce soil erosion and increase SOC levels by adding organic matter to the soil. Crop rotation and intercropping can also contribute to SOC enhancement by promoting crop diversity and reducing crop residues (Pretty et al. 2018).

According to research, implementing CA practices has a beneficial impact on SOC levels. For instance, studies carried out in South Africa, Brazil, and Zimbabwe found that SOC levels increased by up to 43%, 30%, and 61%, respectively, compared to conventional tillage systems (Bationo et al. 2018; Almeida et al. 2019; Nyamangara et al. 2014). Furthermore, the adoption of CA practices has been shown to decrease greenhouse gas emissions. In Mexico, a study revealed that carbon dioxide emissions were reduced by up to 33% compared to conventional tillage practices (Díaz-José et al. 2016). Similarly, in Brazil, CA was found to lower nitrous oxide emissions by up to 69% compared to conventional tillage practices (Monteiro et al. 2021). In addition, CA enhances soil water retention, reduces nutrient leaching, and increases crop productivity, thus increasing resilience to climate change (Derpsch et al. 2010).

17.4.3.4 Nutrient Management

The appropriate application of nutrients, such as nitrogen and phosphorus, can increase crop yields and, therefore, enhance carbon sequestration potential in soils. For example, a study conducted in Zimbabwe found that combining conservation agriculture practices with the appropriate use of nitrogen fertilizer increased soil carbon stocks by up to 5.5 tons per hectare per year (Chikowo et al. 2014). Similarly, a study conducted in China reported that balanced nutrient management practices increased soil carbon sequestration by up to 45% compared to unbalanced nutrient management (Shehu et al. 2019). However, the overapplication of nutrients can have negative impacts on soil health and lead to greenhouse gas emissions. Therefore, it is important to adopt precision nutrient management practices that aim to optimize nutrient use efficiency while minimizing environmental impacts. A study conducted in Brazil reported that precision nutrient management practices reduced nitrous oxide emissions by up to 78% compared to conventional nutrient management practices (Borges et al. 2018).

One approach to nutrient management is the use of integrated nutrient management (INM) practices, which combine organic and inorganic fertilizers to improve nutrient availability, enhance soil fertility, and increase crop yields (Shrestha et al. 2021). Studies have shown that the use of INM practices can increase SOC levels and reduce greenhouse gas emissions. For example, a study in Zimbabwe found that the use of INM practices increased SOC levels by up to 27% compared to conventional practices (Thierfelder et al. 2018). Similarly, a study in India reported that the use of INM practices reduced nitrous oxide emissions by up to 30% (Fagodiya et al. 2019). Moreover, nutrient management practices can also contribute to the production of bioenergy crops, which can further enhance carbon sequestration potential. A study conducted in Argentina found that the production of bioenergy crops, such as sweet sorghum, can increase soil carbon stocks by up to 4 tons per hectare per year, depending on fertilization and tillage management practices (Button et al. 2023).

17.4.3.5 Soil Erosion Control

Soil erosion control measures aim to reduce soil erosion and improve soil health, which can in turn enhance carbon sequestration and climate change adaptation. One approach to soil erosion control is the use of conservation tillage practices, such as reduced tillage, minimum tillage, and no-till farming. These practices reduce soil disturbance and leave crop residues on the soil surface, which can help to reduce erosion and improve soil health. Studies have shown that conservation tillage practices can increase soil organic carbon (SOC) levels and reduce greenhouse gas emissions. For example, a study in Australia found that the adoption of no-till farming practices increased SOC levels by up to 22% compared to conventional tillage practices. Similarly, a study in the United States reported that the use of minimum tillage practices reduced greenhouse gas emissions by up to 39% compared to conventional tillage practices (Dabney et al. 2010).

Another approach to soil erosion control is the use of cover crops, which can help to protect soil from erosion and improve soil health. Cover crops are planted between main crops or during fallow periods and can include grasses, legumes, and other plant species. Studies have shown that the use of cover crops can increase SOC levels and reduce greenhouse gas emissions. For example, a study in Brazil found that the use of cover crops increased SOC levels by up to 25% compared to conventional practices (Silva et al. 2020). Similarly, a study in the United States reported that the use of cover crops reduced greenhouse gas emissions by up to 34% compared to conventional practices (Qin et al. 2020). In addition to conservation tillage and cover crops, other soil erosion control measures include the use of terracing, contour farming, and vegetative barriers. These practices can help to reduce the flow of water and wind over the soil surface, which can in turn reduce erosion and improve soil health (Keesstra et al. 2018).

17.4.3.6 Crop Season Management

Crop season management involves the planning and implementation of farming practices throughout the growing season to optimize crop growth and yield while minimizing the negative impact on the environment. These practices can also contribute to carbon management and climate change adaptation. One approach to crop season management is the use of crop residue management techniques such as crop residue retention, incorporation, or removal. The management of crop residues can have a significant impact on soil carbon sequestration and greenhouse gas emissions.

A study in China reported that crop residue retention increased SOC levels by up to 14% compared to conventional practices (Zhao et al. 2019). Another study conducted in Australia found that incorporating crop residues into the soil increased SOC levels by up to 30% compared to removal practices (Dhaliwal et al. 2020). Another important aspect of crop season management is the timing and amount of fertilizer application. Overuse of fertilizers can lead to nutrient loss, greenhouse gas emissions, and reduced soil carbon sequestration (Gaikwad et al. 2022). Therefore, optimizing fertilizer application through precision agriculture techniques can enhance crop productivity while minimizing negative environmental impacts.

17.4.3.7 Agroforestry

The integration of trees with crops and/or livestock to create a more sustainable agricultural landscape is known as agroforestry. This land use management system is believed to enhance carbon sequestration, reduce greenhouse gas emissions, and promote climate change adaptation. Trees are capable of sequestering carbon through photosynthesis, and transferring it into the soil through root systems and leaf litter, thereby increasing the amount of carbon stored in vegetation and soils. Agroforestry practices such as alley cropping and silvopasture have been found to improve SOC levels and reduce greenhouse gas emissions. For example, a study conducted in Nigeria showed that alley cropping increased SOC levels by up to 30% compared to conventional agriculture practices (Kuyah et al. 2022). Likewise, a study in Brazil found that silvopasture reduced greenhouse gas emissions by up to 68% compared to traditional pasture systems (Cerri et al. 2010).

The incorporation of trees in agroforestry practices can also improve the ability of agricultural systems to adapt to climate change by making them more resilient to environmental stressors, including pests, diseases, and drought. Trees provide shading, shelter, and microclimatic regulation, which can enhance the diversity and productivity of crops and livestock. Several agroforestry practices, such as multiple strata agroforestry and agroforestry parklands, have been shown to improve soil fertility and increase crop yields. For instance, a study conducted in Cameroon showed that agroforestry parklands increased maize yields by up to 36% when compared to monoculture systems (Pemunta and Mbu-Arrey 2013).

17.4.3.8 Biochar

Biochar is a form of charcoal generated by heating organic matter like agricultural waste in a low-oxygen atmosphere. It has been recognized as a tool for carbon management as it has the potential to sequester carbon in the soil for an extended period, ranging from hundreds to thousands of years (Lehmann and Joseph 2015). The inclusion of biochar in soil has been observed to boost soil organic carbon (SOC) levels and enhance soil fertility. An analysis in China found that the application of biochar to soil increased SOC levels by up to 33% in contrast to soils without biochar. Likewise, a study conducted in Sweden found that biochar improved crop yields by up to 20% (Aviles et al. 2020). Additionally, biochar has been found to minimize greenhouse gas emissions, particularly nitrous oxide emissions. A report in New Zealand showed that biochar application to soil reduced nitrous oxide emissions by up to 80% (Clough et al. 2013).

Biochar can be produced from various raw materials, including agricultural waste, forestry residues, and animal manure. Utilizing biochar as a soil amendment can also help to reduce the amount of organic waste burned or disposed of in landfills, which can reduce greenhouse gas emissions further (Sohi et al. 2010). Even though biochar has shown substantial potential benefits for carbon management and soil health, additional research is necessary to comprehend its full potential benefits and limitations. The effectiveness of biochar as a soil amendment can be influenced by several factors, such as feedstock selection, production methods, and application rates.

17.5 Reducing the Greenhouse Gases Footprint in Agriculture

Agriculture significantly contributes to global climate change through its greenhouse gas (GHG) emissions. The IPCC has reported that AFOLU activities account for around 23% of global GHG emissions. Agricultural processes such as enteric fermentation, manure management, and synthetic fertilizers result in emissions of carbon dioxide (CO_2), methane (CH_4), and nitrous oxide (N_2O) (Shukla et al. 2019).

To address the challenges posed by climate change while improving food security and increasing agricultural productivity, the concept of climate-smart agriculture (CSA) has been introduced. The implementation of CSA practices focuses on three main objectives: reducing GHG emissions, increasing carbon sequestration, and enhancing the resilience of agricultural systems to climate change. Climate-smart crop management practices can be a potential solution to reduce GHG emissions from agriculture. Conservation agriculture practices such as reduced tillage and crop rotation can minimize soil disturbance and increase soil organic carbon (SOC) levels, thus reducing GHG emissions. Climate-smart crop varieties can also help to reduce GHG emissions. For example, some crop varieties are more efficient at using nitrogen fertilizer, which can help to reduce N_2O emissions. In China, the use of nitrogen-efficient rice varieties reduced N_2O emissions by up to 40% compared to conventional varieties in a study (Iqbal et al. 2023). However, some important techniques and methods can be used to reduce the greenhouse gas footprint in agriculture.

1. Conservation Agriculture: Conservation agriculture (CA) is a farming system that involves minimal soil disturbance, crop rotations, and the use of cover crops. It can help to reduce GHG emissions by increasing soil organic

carbon (SOC) levels and reducing soil erosion (Six et al. 2002).
2. Nutrient Management: Nitrogen fertilizers are a significant contributor to GHG emissions in agriculture. However, the use of precision agriculture techniques such as variable rate fertilization and site-specific nutrient management can help to reduce GHG emissions (Balafoutis et al. 2017).
3. Livestock Management: Livestock is a significant source of GHG emissions, mainly due to enteric fermentation and manure management. However, management techniques such as dietary manipulation, manure management, and biogas production can help to reduce GHG emissions (Zhang et al. 2013).
4. Renewable Energy: The utilization of renewable energy sources such as solar and wind power can help to reduce GHG emissions in agriculture.

17.6 Integrated Farming System for Developing Countries

Integrated Farming System (IFS) is a sustainable and climate-resilient farming system that integrates multiple agricultural components such as crops, livestock, poultry, fish, and agroforestry on the same farm (Panda et al. 2022). IFS is designed to increase productivity, reduce environmental degradation, and improve the resilience of farming systems to climate change. The integration of multiple components in IFS provides a diverse and stable income source, reduces risk, and increases the efficiency of resource use. IFS has been proven to be an effective climate-resilient technology in many parts of the world.

According to research conducted in India, implementing IFS led to a 20% boost in crop yield and a 30% increase in livestock productivity, as well as a decrease in greenhouse gas emissions of up to 50% (Nayak et al. 2020). Similarly, a study conducted in Vietnam showed that adopting IFS raised farm income by up to 50%, lowered the use of synthetic fertilizers and pesticides, and enhanced soil fertility (Nguyen et al. 2021). The utilization of IFS in agriculture can also decrease the carbon footprint of farming systems. A study in China demonstrated that the use of IFS lowered carbon emissions by up to 51% compared to traditional farming practices (Ullah et al. 2020). The incorporation of crops and livestock in IFS can support nutrient recycling and minimize the need for synthetic fertilizers. In addition, the use of agroforestry in IFS can help to store carbon in the soil and diminish greenhouse gas emissions.

Furthermore, IFS is a low-cost and low-input farming system that is accessible to small-scale farmers in developing countries. It is a sustainable alternative to conventional farming practices that rely heavily on synthetic inputs and are often not suitable for small-scale farmers. By integrating multiple components and reducing reliance on synthetic inputs, IFS can provide a sustainable and climate-resilient farming system that can help to alleviate poverty and improve food security.

17.6.1 Role of IFS in Climate Change Adaptation

17.6.1.1 Resilience to Reduce Vulnerability

The implementation of IFS practices can enhance the adaptive capacity of farming systems to climate change, which can lead to reduced vulnerability. By integrating various components, such as agroforestry, IFS practices can aid in carbon sequestration in the soil and mitigate greenhouse gas emissions. According to a study conducted in Vietnam, the adoption of IFS practices led to a reduction in the use of synthetic fertilizers and pesticides, which can increase the vulnerability of farming systems to input availability and price fluctuations (Behera and France 2016). Furthermore, by integrating livestock and crops, IFS practices can help to recycle nutrients and decrease the reliance on synthetic fertilizers, which can further enhance the resilience of farming systems to climate change (Yadav et al. 2020).

17.6.1.2 Flexibility to Enhance Adaptive Capacity

The flexibility of IFS can play a crucial role in enhancing the adaptive capacity of farming systems to changing market demands and environmental conditions. Farmers can adjust their production strategy based on changing consumer preferences for organic or free-range products, and IFS provides a diverse and stable income source, reduces the risk of crop failure, and increases the efficiency of resource use.

According to Sundari et al. (2021), an Indonesian study found that IFS adoption can enhance the resilience of farming systems to climate change by increasing the adaptive capacity of farmers to changing environmental conditions. The study emphasized the role of IFS in providing flexibility to farmers to choose crops and livestock species based on changing weather patterns and market demands, which can improve the overall resilience of the farming system. Similarly, Gautam et al. (2021) reported that the integration of crops, livestock, and trees in IFS in India improved the adaptive capacity of farming systems to climate change. The study revealed that IFS helped farmers cope with climate change impacts, such as droughts and floods, by providing a flexible and resilient farming system that can withstand and recover from environmental shocks. Moreover, IFS reduces vulnerability by offering a diverse and stable income source, decreasing the risk of crop failure, and enhancing the efficiency of resource use.

17.6.1.3 Diversity to Cope with Variability

The diversity of crops and livestock in IFS can help to cope with the variability of climate and market conditions. This diversity also improves the efficient use of resources, as different components of IFS utilize different resources and have different production cycles. A study conducted in Ethiopia found that the adoption of IFS increased farm productivity and reduced the vulnerability of farming systems to climate change by enhancing diversity and resilience (Egziabher et al. 2013). Similarly, another study conducted in Brazil found that IFS practices helped to cope with the variability of rainfall by increasing the water-use efficiency of farming systems. The study also found that the integration of livestock and crops in IFS improved soil fertility and reduced the need for synthetic fertilizers. The diversity of IFS components can also provide additional benefits such as biodiversity conservation, soil conservation, and water conservation (Paramesh et al. 2022).

17.7 Agroforestry Systems

Agroforestry systems (AFS) are a type of land-use system that combines the cultivation of trees with crops or livestock on the same land. AFS has been recognized as a promising approach for climate change adaptation and mitigation, as it offers several environmental, social, and economic benefits. These benefits include improved productivity, soil conservation, biodiversity conservation, carbon sequestration, and better livelihoods for farmers. AFS can help farmers adapt to climate change by providing a diverse and resilient farming system that can cope with extreme weather conditions, such as droughts and floods.

Studies have shown the potential of AFS in mitigating climate change by sequestering carbon in above- and below-ground biomass and soil. According to Zougmoré et al. (2018), AFS can help to enhance soil fertility, increase water-use efficiency, and improve the resilience of farming systems to climate change. Similarly, Lasco et al. (2014) reported that AFS can help to diversify income sources, improve food security, and reduce the vulnerability of farming systems to extreme weather events. These findings highlight the potential of AFS as a sustainable land-use option that can contribute to both adaptation and mitigation efforts to address climate change.

17.8 Smart Farming for Precise Input Delivery

Smart farming technologies, such as precision agriculture, enable farmers to deliver inputs precisely, thereby reducing waste and increasing resource efficiency in agriculture. Precision agriculture uses advanced technologies such as GPS, sensors, and drones to monitor and analyze soil and crop conditions, allowing farmers to apply inputs such as water, fertilizers, and pesticides only where and when they are needed. This can significantly reduce the use of inputs, such as water and fertilizers, and thereby reduce the environmental impact of agriculture while maintaining or even increasing yields. A study conducted in China found that precision agriculture practices resulted in a 20–30% reduction in the use of fertilizers, a 30–50% reduction in the use of pesticides, and a 20–30% increase in crop yield. Another study conducted in the United States found that precision agriculture practices reduced nitrate leaching into groundwater by up to 57% (Li et al. 2018).

The use of advanced farming technologies can enhance the resilience of farming systems to climate change by providing real-time monitoring of changing environmental conditions and the ability to adapt accordingly. Precision agriculture technologies, for example, can aid farmers in monitoring soil moisture levels and adjusting irrigation schedules during droughts to decrease the risk of crop failure and maintain productivity even in variable weather conditions. In addition, smart farming technologies can also contribute to reducing the carbon footprint of farming systems by decreasing the use of energy-intensive inputs like fertilizers and pesticides. By reducing the use of these inputs, precision agriculture can aid in mitigating climate change by lowering greenhouse gas emissions.

17.9 Conclusion

The impact of climate change is being felt in various regions of the world, leading to food insecurity, land degradation, and water scarcity. Climate-smart agriculture offers a promising solution to address these challenges while ensuring sustainable food production, livelihoods, and environmental conservation. Integrated Farming Systems (IFS), agroforestry, precision farming, and other climate-smart technologies can enhance the resilience, adaptation, and mitigation capacity of farming systems. Numerous research studies conducted globally have proven the efficacy of climate-smart agriculture technologies in augmenting agricultural productivity, curtailing greenhouse gas emissions, and elevating the adaptive capacity of farming systems to changing environmental conditions. For instance, the adoption of IFS practices has been shown to enhance farm productivity, reduce environmental degradation, and improve the resilience of farming systems to climate change. Integrating agroforestry in IFS can help to sequester carbon in the soil and reduce greenhouse gas emissions, thereby aiding in mitigating climate change. Precision farming technologies can improve the efficient utilization of resources and reduce the adverse impacts of synthetic inputs on the environment. The adoption of climate-smart agriculture technologies can lead to sustainable agriculture and alleviate the detrimental effects of climate change on farming systems. However, it necessitates the participation of all stakeholders, including farmers, policymakers, researchers, and extension agents, for successful adoption. Effective policies, research, and extension services are critical to promote the adoption and dissemination of climate-smart agriculture technologies and elevating sustainable agricultural production, resilience, and food security.

References

Alam MD, Islam M, Salahin N, Hasanuzzaman M (2014) Effect of tillage practices on soil properties and crop productivity in wheat-mungbean-rice cropping system under subtropical climatic conditions. Sci World J 437283. https://doi.org/10.1155/2014/437283

Allen SK, Plattner GK, Nauels A, Xia Y, Stocker TF (2014) Climate change 2013: the physical science basis. An overview of the working group 1 contribution to the fifth assessment report of the intergovernmental panel on climate change (IPCC). In EGU General Assembly Conference Abstracts, p 3544

Almeida RF, Queiroz IDS, Mikhael JER, Oliveira RC, Borges EN (2019) Enriched animal manure as a source of phosphorus in sustainable agriculture. Intl J Recycl Organic Waste Agricult 8:203–210

Anantha KH, Garg KK, Petrie CA, Dixit S (2021) Seeking sustainable pathways for fostering agricultural transformation in peninsular India. Environ Res Lett 16(4):044032

Anwar MR, Li Liu D, Farquharson R, Macadam I, Abadi A, Finlayson J, Ramilan T (2015) Climate change impacts on phenology and yields of five broadacre crops at four climatologically distinct locations in Australia. Agric Syst 132:133–144

Arora M, Goel N, Singh P (2005) Evaluation of temperature trends over India/Evaluation de tendances de température en Inde. Hydrol Sci J 50(1):81–93

Ashfaq M, Lal MH, Moghal AAB, Murthy VR (2020) Carbon footprint analysis of coal gangue in geotechnical engineering applications. Indian Geotech J 50(4):646–654

Aviles D, Berglund K, Wesström I, Joel A (2020) Effect of liming products on soil detachment resistance, measured with a cohesive strength meter. Acta Agricult Scandinavica Sect B Soil Plant Sci 70(1):48–55

Balafoutis A, Beck B, Fountas S, Vangeyte J, Van der Wal T, Soto I, Eory V (2017) Precision agriculture technologies positively contributing to GHG emissions mitigation, farm productivity and economics. Sustainability 9(8):1339

Bationo A, Fening JO, Kwaw A (2018) Assessment of soil fertility status and integrated soil fertility management in Ghana. In: Bationo et al. (eds) Improving the profitability, sustainability and efficiency of nutrients through site specific fertilizer recommendations in West Africa agro-ecosystems. Springer Nature, pp 93–138

Behera UK, France J (2016) Integrated farming systems and the livelihood security of small and marginal farmers in India and other developing countries. Adv Agron 138:235–282

Benbi DK, Kiranvir BRAR, Sharma S (2015) Sensitivity of labile soil organic carbon pools to long-term fertilizer, straw and manure management in rice-wheat system. Pedosphere 25(4):534–545

Bhadra P, Maitra S, Shankar T, Hossain A, Praharaj S, Tariq Aftab T (2022) Climate change impact on plants: plant responses and adaptations. In: Plant perspectives to global climate changes, Elsevier Inc/Academic, pp 1–24. https://doi.org/10.1016/B978-0-323-85665-2.00004-2

Billah M, Aktar S, Brestic M, Zivcak M, Khaldun ABM, Uddin MS, Bagum SA, Yang X, Skalicky M, Mehari TG, Maitra S, Hossain A (2021, 1910) Progressive genomic approaches to explore drought- and salt-induced oxidative stress responses in plants under changing climate. Plants 10. https://doi.org/10.3390/plants10091910

Borges O, Raimundo F, Coutinho J, Gonçalves B, Oliveira I, Martins A, Madeira M (2018) Carbon fractions as indicators of organic matter dynamics in chestnut orchards under different soil management practices. Agrofor Syst 92:301–310

Button ES, Marshall M, Sánchez-Rodríguez AR, Blaud A, Abadie M, Chadwick DR, Jones DL (2023) Greenhouse gas production, diffusion and consumption in a soil profile under maize and wheat production. Geoderma 430:116310

Cavigelli MA, Mirsky SB, Teasdale JR, Spargo JT, Doran J (2013) Organic grain cropping systems to enhance ecosystem services. Renew Agricult Food Syst 28(2):145–159

Cerri CC, Bernoux M, Maia SMF, Cerri CEP, Costa Junior C, Feigl BJ, Carvalho JLN (2010) Greenhouse gas mitigation options in Brazil for land-use change, livestock and agriculture. Sci Agric 67:102–116

Challinor AJ, Watson J, Lobell DB, Howden SM, Smith DR, Chhetri N (2014) A meta-analysis of crop yield under climate change and adaptation. Nat Clim Chang 4(4):287–291

Chaturvedi RK, Joshi J, Jayaraman M, Bala G, Ravindranath NH (2012) Multi-model climate change projections for India under representative concentration pathways. Curr Sci 103:791–802

Chen C, Chen HY, Chen X, Huang Z (2019) Meta-analysis shows positive effects of plant diversity on microbial biomass and respiration. Nat Commun 10(1):1332

Chen X, Wang L, Niu Z, Zhang M, Li J (2020) The effects of projected climate change and extreme climate on maize and rice in the Yangtze River basin, China. Agricult Forest Meteorol 282:107867–107881

Chikowo R, Zingore S, Snapp S, Johnston A (2014) Farm typologies, soil fertility variability and nutrient management in smallholder farming in sub-Saharan Africa. Nutr Cycl Agroecosyst 100:1–18

Choudhury S, Moulick D (2022) Response of field crops to abiotic stress: current status and future prospects, 1st edn. CRC Press. https://doi.org/10.1201/9781003258063

Choudhury S, Sharma P, Moulick D, Mazumder MK (2021a) Unrevealing metabolomics for abiotic stress adaptation and tolerance in plants. J Crop Sci Biotechnol 24(5):479–493. https://doi.org/10.1007/s12892-021-00102-8

Choudhury S, Moulick D, Mazumder MK (2021b) Secondary metabolites protect against metal and metalloid stress in rice: an in silico investigation using dehydroascorbate reductase. Acta Physiol Plant 43:1–10. https://doi.org/10.1007/s11738-020-03173-2

Clough TJ, Condron LM, Kammann C, Müller C (2013) A review of biochar and soil nitrogen dynamics. Agronomy 3(2):275–293

Crusciol CAC, Nascente AS, Mateus GP, Borghi E, Leles EP, Santos ND (2013) Effect of intercropping on yields of corn with different relative maturities and palisadegrass. Agron J 105(3):599–606

da Cunha Dias TA, Lora EES, Maya DMY, del Olmo OA (2021) Global potential assessment of available land for bioenergy projects in 2050 within food security limits. Land Use Policy 105:105346

Dabney SM, Delgado JA, Meisinger JJ, Schomberg HH, Liebig MA, Kaspar T, Reeves W (2010) Using cover crops and cropping systems for nitrogen management. Advances in nitrogen management for water quality, Soil and Water Conservation Society, USA, pp 231–282

Daryanto Y, Christata B (2021) Optimal order quantity considering carbon emission costs, defective items, and partial backorder. Uncertain Supply Chain Manage 9(2):307–316

De Moraes Sa JC, Tivet F, Lal R, de Oliveira Ferreira A, Briedis C, Inagaki TM, Romaniw J (2020) Carbon management practices and benefits in conservation agriculture systems: soil organic carbon fraction

losses and restoration. In: Advances in conservation agriculture. Burleigh Dodds Science Publishing, pp 229–266
Debnath S, Mishra A, Mailapalli DR, Raghuwanshi NS (2021a) Identifying most promising agronomic adaptation strategies to close rainfed rice yield gap in future: a model-based assessment. J Water Climate Change 12(6):2854–2874
Debnath S, Mishra A, Mailapalli DR, Raghuwanshi NS, Sridhar V (2021b) Assessment of rice yield gap under a changing climate in India. J Water Climate Change 12(4):1245–1267
Derpsch R, Friedrich T, Kassam A, Li H (2010) Current status of adoption of no-till farming in the world and some of its main benefits. Intl J Agricult Biolog Eng 3(1):1–25
Dhaliwal SS, Naresh RK, Gupta RK, Panwar AS, Mahajan NC, Singh R, Mandal A (2020) Effect of tillage and straw return on carbon footprints, soil organic carbon fractions and soil microbial community in different textured soils under rice–wheat rotation: a review. Rev Environ Sci Biotechnol 19:103–115
Díaz-José J, Rendón-Medel R, Govaerts B, Aguilar-Ávila J, Muñoz-Rodriguez M (2016) Innovation diffusion in conservation agriculture: a network approach. Eur J Dev Res 28:314–329
dos Santos TB, Ribas AF, de Souza SGH, Budzinski IGF, Domingues DS (2022) Physiological responses to drought, salinity, and heat stress in plants: a review. Stress 2(1):113–135
Dudney J, Willing CE, Das AJ, Latimer AM, Nesmith JC, Battles JJ (2021) Nonlinear shifts in infectious rust disease due to climate change. Nat Commun 12(1):5102
Easterling DR, Meehl GA, Parmesan C, Changnon SA, Karl TR, Mearns LO (2000) Climate extremes: observations, modeling, and impacts. Science *289*(5487):2068–2074
Egziabher KG, Mathijs E, Deckers JA, Gebrehiwot K, Bauer H, Maertens M (2013) The economic impact of a new rural extension approach in northern Ethiopia (no. 1067-2016-86816)
Elahi E, Khalid Z, Tauni MZ, Zhang H, Lirong X (2022) Extreme weather events risk to crop-production and the adaptation of innovative management strategies to mitigate the risk: a retrospective survey of rural Punjab, Pakistan. Technovation 117:102255
Fagodiya RK, Pathak H, Bhatia A, Jain N, Gupta DK, Kumar A, Tomer R (2019) Nitrous oxide emission and mitigation from maize–wheat rotation in the upper Indo-Gangetic Plains. Carbon Manage 10(5):489–499
Fritsche-Neto R, Galli G, Borges KLR, Costa-Neto G, Alves FC, Sabadin F, Crossa J (2021) Optimizing genomic-enabled prediction in small-scale maize hybrid breeding programs: a roadmap review. Front Plant Sci 12:658267
Gaikwad DJ, Ubale NB, Pal A, Singh S, Ali MA, Maitra S (2022) Abiotic stresses impact on major cereals and adaptation options – a review. Res Crops *23*(4):896–915. https://doi.org/10.31830/2348-7542.2022.ROC-913
Gautam R, Shriwastav CP, Lamichhane S, Baral BR (2021) The residual effect of pre-rice green manuring on a succeeding wheat crop (Triticum aestivum L.) in the rice-wheat cropping system in Banke, Nepal. Intl J Agron 2021:1–10
Ghosh D, Brahmachari K, Skalicky M, Hossain A, Sarkar S, Dinda NK, Das A, Pramanick B, Moulick D, Brestic M, Raza MA (2020a) Nutrients supplementation through organic manures influence the growth of weeds and maize productivity. Molecules 25(21):4924. https://doi.org/10.3390/molecules25214924
Ghosh D, Brahmachari K, Brestic M, Ondrisik P, Hossain A, Skalicky M, Sarkar S, Moulick D, Dinda NK, Das A, Pramanick B (2020b) Integrated weed and nutrient management improve yield, nutrient uptake and economics of maize in the rice-maize cropping system of Eastern India. Agronomy 10(12):1906. https://doi.org/10.3390/agronomy10121906
Ghosh D, Brahmachari K, Das A, Hassan MM, Mukherjee PK, Sarkar S, Dinda NK, Pramanick B, Moulick D, Maitra S, Hossain A (2021) Assessment of energy budgeting and its indicator for sustainable nutrient and weed management in a rice-maize-green gram cropping system. Agronomy 11(1):166. https://doi.org/10.3390/agronomy11010166
Ghosh D, Brahmachari K, Skalický M, Roy D, Das A, Sarkar S, Moulick D, Brestič M, Hejnak V, Vachova P, Hassan MM (2022a) The combination of organic and inorganic fertilizers influence the weed growth, productivity and soil fertility of monsoon rice. PloS One *17*(1):e0262586. https://doi.org/10.1371/journal.pone.0262586
Ghosh D, Brahmachari K, Sarkar S, Dinda NK, Das A, Moulick D (2022b) Impact of nutrient management in rice-maize-greengram cropping system and integrated weed management treatments on summer greengram productivity. https://doi.org/10.5958/0974-8164.2022.00004.1
Hassan S, Ahmad A, Batool F, Rashid B, Husnain T (2021) Genetic modification of Gossypium arboreum universal stress protein (GUSP1) improves drought tolerance in transgenic cotton (*Gossypium hirsutum*). Physiol Mol Biol Plants 27(8):1779–1794
Hobbs PR, Sayre K, Gupta R (2008) The role of conservation agriculture in sustainable agriculture. Philos Trans Royal Soc B: Biol Sci 363(1491):543–555
Hossain A, Pramanick B, Bhutia KL, Ahmad Z, Moulick D, Maitra S, Ahmad A, Aftab T (2021a) Emerging roles of osmoprotectant glycine betaine against salt-induced oxidative stress in plants: a major outlook of maize (*Zea mays* L.). In: Frontiers in plant-soil interaction. Academic, pp 567–587. https://doi.org/10.1016/B978-0-323-90943-3.00015-8
Hossain A, Ahmad Z, Moulik D, Maitra S, Bhadra P, Ahmad A, Garai S, Mondal M, Roy A, Sabagh AE, Aftab T (2021b) Jasmonates and salicylates: mechanisms, transport and signalling during abiotic stress in plants. Jasmonates and salicylates signaling in plants, pp 1–29. https://doi.org/10.1007/978-3-030-75805-9_1

Hossain A, Maitra S, Pramanick B, Bhutia KL, Ahmad Z, Moulik D, Syed MA, Shankar T, Adeel M, Hassan MM, \Aftab T (2022) Wild relatives of plants as sources for the development of abiotic stress tolerance in plants. In: Plant perspectives to global climate changes. Academic, pp 471–518. https://doi.org/10.1016/B978-0-323-85665-2.00011-X

Hu G, Zhao L, Wu T, Wu X, Park H, Fedorov A, Zou D (2021) Spatiotemporal variations and regional differences in air temperature in the permafrost regions in the Northern hemisphere during 1980–2018. Sci Total Environ 791:148358

Huang D, Sun M, Zhang A, Chen J, Zhang J, Lin C, Huang L (2021) Transcriptional changes in pearl millet leaves under heat stress. Genes 12(11):1716

Iqbal MF, Zhang Y, Kong P, Wang Y, Cao K, Zhao L, Fan X (2023) High-yielding nitrate transporter cultivars also mitigate methane and nitrous oxide emissions in Paddy. Front Plant Sci 14:433

Ivezić V, Lorenz K, Lal R (2022) Soil organic carbon in alley cropping systems: a meta-analysis. Sustainability 14(3):1296

Jat HS, Datta A, Choudhary M, Sharma PC, Yadav AK, Choudhary V, McDonald A (2019) Climate smart agriculture practices improve soil organic carbon pools, biological properties and crop productivity in cereal-based systems of North-West India. Catena 181:104059

Kamal NLM, Itam Z, Sivaganese Y, Razak NA (2020) Carbon dioxide sequestered concrete. Intl J Integr Eng 12(9):45–51

Kaur H, Singh SP (2019) Sustainable procurement and logistics for disaster resilient supply chain. Ann Oper Res 283:309–354

Keesstra S, Nunes J, Novara A, Finger D, Avelar D, Kalantari Z, Cerdà A (2018) The superior effect of nature based solutions in land management for enhancing ecosystem services. Sci Total Environ 610:997–1009

Kothari A, Lachowiec J (2021) Roles of brassinosteroids in mitigating heat stress damage in cereal crops. Int J Mol Sci 22(5):2706

Kumar KS (2009) Climate sensitivity of Indian agriculture: Madras School of Economics Chennai, working paper 43. https://www.mse.ac.in/wp-content/uploads/2016/09/working-paper-43.pdf

Kumar U, Werners S, Roy S, Ashraf S, Hoang LP, Kumar Datta D, Ludwig F (2020) Role of information in farmers' response to weather and water related stresses in the lower Bengal Delta, Bangladesh. Sustainability 12(16):6598

Kumar S, Singh VP, Saini DK, Sharma H, Saripalli G, Kumar S et al (2021) Meta-QTLs, ortho-MQTLs, and candidate genes for thermotolerance in wheat (Triticum aestivum L.). Mol Breed 41:1–22

Kumar S, Gopinath KA, Sheoran S, Meena RS, Srinivasarao C, Bedwal S, Praharaj CS (2022) Pulse-based cropping systems for soil health restoration, resources conservation, and nutritional and environmental security in rainfed agroecosystems. Front Microbiol, 13:1041124. https://doi.org/10.3389/fmicb.2022.1041124

Kundu A, Dwivedi S, Chandra V (2014) Precipitation trend analysis over eastern region of India using CMIP5 based climatic models. The international archives of photogrammetry. Remote Sens Spatial Inform Sci 40(8):1437

Kuyah S, Muoni T, Bayala J, Chopin P, Dahlin AS, Hughes K, Öborn I (2022) Grain legumes and dryland cereals for enhancing carbon sequestration in semi-arid and sub-humid agro-ecologies of Africa and South Asia. Working paper. ICRISAT, Hyderabad

Lal R (2004) Soil carbon sequestration to mitigate climate change. Geoderma 123(1–2):1–22

Lal R (2020) Managing organic matter content for restoring health and ecosystem services of soils of India. J Indian Soc Soil Sci 68(1):1–15

Lasco RD, Delfino RJP, Catacutan DC, Simelton ES, Wilson DM (2014) Climate risk adaptation by smallholder farmers: the roles of trees and agroforestry. Curr Opin Environ Sustain 6:83–88

Lehmann J, Joseph S (eds) (2015) Biochar for environmental management: science, technology and implementation. Routledge

Li S, Zhang Y, Yan W, Shangguan Z (2018) Effect of biochar application method on nitrogen leaching and hydraulic conductivity in a silty clay soil. Soil Tillage Res 183:100–108

Li J, Wang YK, Guo Z, Li JB, Tian C, Hua DW, Xu Y (2020) Effects of conservation tillage on soil physicochemical properties and crop yield in an arid loess plateau, China. Sci Rep 10(1):1–15

Li S, Zhao J, Li J, Shao R, Li H, Fang W, Liu T (2022) Inter-and mixed cropping of different varieties improves high-temperature tolerance during flowering of summer maize. Sustainability 14(12):6993

Liao R, Zhang S, Zhang X, Wang M, Wu H, Zhangzhong L (2021a) Development of smart irrigation systems based on real-time soil moisture data in a greenhouse: proof of concept. Agric Water Manag 245:106632

Liao Z, Chen Y, Li W, Zhai P (2021b) Growing threats from unprecedented sequential flood-hot extremes across China. Geophys Res Lett *48*(18):e2021GL094505

Lobell DB, Roberts MJ, Schlenker W, Braun N, Little BB, Rejesus RM, Hammer GL (2014) Greater sensitivity to drought accompanies maize yield increase in the US Midwest. Science 344(6183):516–519

Maitra S, Palai JB, Manasa P, Kumar DP (2019) Potential of intercropping system in sustaining crop productivity. Intl J Agricult Environ Biotechnol 12(1):39–45

Maitra S, Hossain A, Brestic M, Skalicky M, Ondrisik P, Gitari H, Brahmachari K, Shankar T, Bhadra P, Palai JB, Jena J, Bhattacharya U, Duvvada SK, Lalichetti S, Sairam M (2021) Intercropping–a low input agricultural strategy for food and environmental security. Agronomy 11:343. https://doi.org/10.3390/agronomy11020343

Maitra S, Praharaj S, Hossain A, Patro TSSK, Pramanick B, Shankar T, Sahoo U (2022) Small millets: the next-generation smart crops in the modern era of climate change. In: Pukade RN et al (eds) Omics of climate resilient small millets. Springer Nature Singapore, Singapore, pp 1–25

Malyan SK, Kumar SS, Fagodiya RK, Ghosh P, Kumar A, Singh R, Singh L (2021) Biochar for environmental sustainability in the energy-water-agroecosystem nexus. Renew Sust Energ Rev 149:111379

Masson-Delmotte V, Zhai P, Pirani A, Connors SL, Péan C, Berger S, Zhou B (2021) Climate change 2021: the physical science basis. Contribution of working group I to the sixth assessment report of the intergovernmental panel on climate change, p 2

Mbow C, Smith P, Skole D, Duguma L, Bustamante M (2014) Achieving mitigation and adaptation to climate change through sustainable agroforestry practices in Africa. Curr Opin Environ Sustain 6:8–14

Ministry of Earth Sciences (2010) Climate profile of India, met monograph no. In: Environment Meteorology-01/2010. Government of India, Indian Meteorological Department, Ministry of Earth Sciences

Ministry of Environment and Forests (2010) Climate change and India: a 4×4 assessment – a sectoral and regional analysis, Indian network for climate change assessment report 2, Ministry of Environment and forests, Government of India

Mishra A, Singh R, Raghuwanshi NS, Chatterjee C, Froebrich J (2013) Spatial variability of climate change impacts on yield of rice and wheat in the Indian Ganga Basin. Sci Total Environ 468:S132–S138

Mishra A, Ketelaar JW, Uphoff N, Whitten M (2021) Food security and climate-smart agriculture in the lower Mekong basin of Southeast Asia: evaluating impacts of system of rice intensification with special reference to rainfed agriculture. Int J Agric Sustain 19(2):152–174

Mohammad M, Isaifan RJ, Weldu YW, Rahman MA, Al-Ghamdi SG (2020) Progress on carbon dioxide capture, storage and utilisation. Intl J Global Warm 20(2):124–144

Monteiro RA, Camara MC, de Oliveira JL, Campos EVR, Carvalho LB, de Freitas Proenca PL, Fraceto LF (2021) Zein based-nanoparticles loaded botanical pesticides in pest control: an enzyme stimuli-responsive approach aiming sustainable agriculture. J Hazard Mater 417:126004

Morugán-Coronado A, Linares C, Gómez-López MD, Faz Á, Zornoza R (2020) The impact of intercropping, tillage and fertilizer type on soil and crop yield in fruit orchards under Mediterranean conditions: a meta-analysis of field studies. Agric Syst 178:102736

Najeeb S, Mahender A, Anandan A, Hussain W, Li Z, Ali J (2021) Genetics and breeding of low-temperature stress tolerance in rice. In: Rice improvement: physiological, molecular breeding and genetic perspectives, pp 221–280

Nakicenovic N, Alcamo J, Davis G, Vries BD, Fenhann J, Gaffin S, Zhou D (2000) Special report on emissions scenarios. IPCC, available online: https://www.ipcc.ch/site/assets/uploads/2018/03/emissions_scenarios-1.pdf. Accessed 25 Dec 2022

Nayak PK, Nayak AK, Kumar A, Kumar U, Panda BB, Satapathy BS, Pathak H (2020) Rice based integrated farming Systems in Eastern India: a viable technology for productivity and ecological security. NRRI research bulletin, p 24

Nguyen HD, Fox D, Dang DK, Pham LT, Viet Du QV, Nguyen THT, Petrisor AI (2021) Predicting future urban flood risk using land change and hydraulic modeling in a river watershed in the central province of Vietnam. Remote Sens 13(2):262

Norsworthy JK, Ward SM, Shaw DR, Llewellyn RS, Nichols RL, Webster TM, Barrett M (2012) Reducing the risks of herbicide resistance: best management practices and recommendations. *Weed Science* 60(SP1):31–62

Nyamangara J, Chikowo R, Rusinamhodzi L, Mazvimavi K (2014) Conservation agriculture in southern Africa. In: Conservation agriculture: global prospects and challenges. CABI, Wallingford, pp 339–351

Panda M, Nandi S, Sahoo U, Sairam M (2022) Integrated farming system for agricultural sustainability. Indian J Nat Sci 13(71):41311–41317

Paramesh V, Ravisankar N, Behera U, Arunachalam V, Kumar P, Solomon Rajkumar R, Rajkumar S (2022) Integrated farming system approaches to achieve food and nutritional security for enhancing profitability, employment, and climate resilience in India. Food Energy Secur 11(2):321

Pemunta NV, Mbu-Arrey OP (2013) The tragedy of the governmentality of nature: the case of national parks in Cameroon. In: Smith JB (ed) National parks: sustainable development, conservation strategies, and environmental effects. Nova Science Publishers Inc, USA), pp 1–56

Poeplau C, Don A (2015) Carbon sequestration in agricultural soils via cultivation of cover crops–a meta-analysis. Agric Ecosyst Environ 200:33–41

Pretty J, Benton TG, Bharucha ZP, Dicks LV, Flora CB, Godfray HCJ et al (2018) Global assessment of agricultural system redesign for sustainable intensification. Nature Sustain 1(8):441–446

Qin Y, Chen Z, Ding B, Li Z (2020) Impact of sand mining on the carbon sequestration and nitrogen removal ability of soil in the riparian area of Lijiang River, China. Environ Pollut 261:114220

Rahman MR, Lateh H (2017) Climate change in Bangladesh: a spatio-temporal analysis and simulation of recent temperature and rainfall data using GIS and time series analysis model. Theor Appl Climatol 128(1–2):27–41

Raza A, Razzaq A, Mehmood SS, Zou X, Zhang X, Lv Y, Xu J (2019) Impact of climate change on crops adaptation and strategies to tackle its outcome: a review. Plan Theory 8(2):34. https://doi.org/10.3390/plants8020034

Reddy KR, Hodges HF, Reddy VR (2010) Climate change and global crop productivity. CABI

Robinson SA (2020) Climate change adaptation in SIDS: a systematic review of the literature pre and post the IPCC fifth assessment report. Wiley Interdiscip Rev Clim Chang 11(4):e653

Sagar L, Praharaj S, Singh S, Attri M, Pramanick B, Maitra S, Hossain A, Tanmoy Shankar T, Palai JB, Sahoo U (2022) Drought and heat stress tolerance in field crops: consequences and adaptation strategies. In: Choudhury S, Moulick D (eds) CRC Press, pp 91–102. https://doi.org/10.1201/9781003258063-8

Sairam M, Maitra S, Praharaj S, Nath S, Shankar T, Sahoo U et al (2023) An insight into the consequences of emerging contaminants in soil and water and plant responses. In: Emerging contaminants and plants: interactions, adaptations and remediation technologies. Springer, Cham, pp 1–27

Sarkar S, Skalicky M, Hossain A, Brestic M, Saha S, Garai S et al (2020) Management of crop residues for improving input use efficiency and agricultural sustainability. Sustainability 12(23):9808

Sarkar D, Dubey PK, Chaurasiya R, Sankar A, Chatterjee N, Ganguly S, Rakshit A (2021) Organic interventions conferring stress tolerance and crop quality in agroecosystems during the United Nations decade on ecosystem restoration. Land Degrad Dev 32(17):4797–4816

Schmidt-Thome P, Nguyen TH, Pham TL, Jarva J, Nuottimäki K, Schmidt-Thome P, Nuottimäki K (2015) Climate change in Vietnam. Climate change adaptation measures in Vietnam: development and implementation, pp 7–15

Sharma G, Shrestha S, Kunwar S, Tseng TM (2021) Crop diversification for improved weed management: a review. Agriculture 11(5):461

Shehu BM, Lawan BA, Jibrin JM, Kamara AY, Mohammed IB, Rurinda J, Merckx R (2019) Balanced nutrient requirements for maize in the Northern Nigerian savanna: parameterization and validation of QUEFTS model. Field Crop Res 241:107585

Shekhawat K, Rathore SS, Chauhan BS (2020) Weed management in dry direct-seeded rice: a review on challenges and opportunities for sustainable rice production. Agronomy 10(9):1264

Sherwood SC, Webb MJ, Annan JD, Armour KC, Forster PM, Hargreaves JC, Zelinka MD (2020) An assessment of Earth's climate sensitivity using multiple lines of evidence. Rev Geophys *58*(4):e2019RG000678

Shi W, Wang M, Liu Y (2021) Crop yield and production responses to climate disasters in China. Sci Total Environ 750:141147

Shrestha J, Subedi S, Timsina KP, Subedi S, Pandey M, Shrestha A, Hossain MA (2021) Sustainable intensification in agriculture: an approach for making agriculture greener and productive. J Nepal Agricult Res Council 7:133–150

Shukla PR, Skea J, Calvo Buendia E, Masson-Delmotte V, Pörtner HO, Roberts DC, Malley J (2019) IPCC, 2019: climate change and land: an IPCC special report on climate change, desertification, land degradation, sustainable land management, food security, and greenhouse gas fluxes in terrestrial ecosystems

Silva JND, Bezerra Neto F, Lima JSSD, Chaves AP, Nunes RLC, Rodrigues GSDO, Santos ECD (2020) Sustainability of carrot-cowpea intercropping systems through optimization of green manuring and spatial arrangements. Ciência Rural 51

Six J, Feller C, Denef K, Ogle S, de Moraes Sa JC, Albrecht A (2002) Soil organic matter, biota and aggregation in temperate and tropical soils-effects of no-tillage. Agronomie 22(7–8):755–775

Smith P, Soussana JF, Angers D, Schipper L, Chenu C, Rasse DP, Klumpp K (2020) How to measure, report and verify soil carbon change to realize the potential of soil carbon sequestration for atmospheric greenhouse gas removal. Glob Chang Biol 26(1):219–241

Sohi SP, Krull E, Lopez-Capel E, Bol R (2010) A review of biochar and its use and function in soil. Adv Agron 105:47–82

Solomon S, Qin D, Manning M, Averyt K, Marquis M (eds) (2007) Climate change 2007-the physical science basis: working group I contribution to the fourth assessment report of the IPCC, vol 4. Cambridge University Press, Cambridge

Song Y, Wang C, Linderholm HW, Fu Y, Cai W, Xu J, Chen D (2022) The negative impact of increasing temperatures on rice yields in southern China. Sci Total Environ 820:153262

Stocker T (ed) (2014) Climate change 2013: the physical science basis: working group I contribution to the fifth assessment report of the intergovernmental panel on climate change. Cambridge University Press, Cambridge

Sundari RS, Arshad A, Sulistyowati L, Noor TI, Setiawan I (2021) Enhancing food security throughout Aquaponicsin urban farming development strategy. J Phys Conf Ser 1764(1):012209. IOP Publishing

Swapna P, Ravichandran M, Nidheesh G, Jyoti J, Sandeep N, Deepa JS, Unnikrishnan AS (2020) Sea-level rise. Assessment of climate change over the Indian region: a report of the Ministry of Earth Sciences (MoES), Government of India, pp 175–189

Thakur M, Kumar R (2021) Mulching: boosting crop productivity and improving soil environment in herbal plants. J Appl Res Med Aromat Plants 20:100287

Thierfelder C, Baudron F, Setimela P, Nyagumbo I, Mupangwa W, Mhlanga B, Gérard B (2018) Complementary practices supporting conservation agriculture in southern Africa: a review. Agron Sustain Develop 38:1–22

Ullah S, Ai C, Huang S, Song D, Abbas T, Zhang J, He P (2020) Substituting ecological intensification of agriculture for conventional agricultural practices increased yield and decreased nitrogen losses in North China. Appl Soil Ecol 147:103395

Van Vuuren DP, Edmonds JA, Kainuma M, Riahi K, Weyant J (2011) A special issue on the RCPs. Clim Chang 109:1–4

Waheed R, Ignacio JC, Arbelaez JD, Juanillas VM, Asif M, Henry A, Arif M (2021) Drought response QTLs in a Super Basmati× Azucena population by high-density GBS-based SNP linkage mapping. Plant Breed 140(5):758–774

Wang B, Cai W, Li J, Wan Y, Guo C, Wilkes A, Liu K (2020a) Leaf photosynthesis and stomatal conductance acclimate to elevated [CO_2] and temperature thus increasing dry matter productivity in a double rice cropping system. Field Crop Res 248:107735

Wang X, Feng Y, Yu L, Shu Y, Tan F, Gou Y, Wang J (2020b) Sugarcane/soybean intercropping with reduced nitrogen input improves crop productivity and reduces carbon footprint in China. Sci Total Environ 719:137517

Wezel A, Casagrande M, Celette F, Vian JF, Ferrer A, Peigné J (2014) Agroecological practices for sustainable agriculture: a review. Agron Sustain Develop 34(1):1–20

Wu B, Zhai B, Mu H, Peng X, Wang C, Patwary AK (2021a) Evaluating an economic application of renewable generated hydrogen: a way forward for green economic performance and policy measures. Environ Sci Pollut Res 29(10):15144–15158

Wu JZ, Zhang J, Ge ZM, Xing LW, Han SQ, Chen SHEN, Kong FT (2021b) Impact of climate change on maize yield in China from 1979 to 2016. J Integr Agric 20(1):289–299

Xu L, Du X (2022) Land certification, rental market participation, and household welfare in rural China. Agric Econ 53(1):52–71

Yadav S, Kumar R, Chandra MS, Singh S, Yadav RB, Kumar M (2020) Soil organic carbon sequestration and carbon pools in Rice based cropping Systems in Indo-Gangetic Plains, an overview. Intl Res J Pure Appll Chem 1:122–136

Yang L, Lei L, Li P, Wang J, Wang C, Yang F, Zou D (2021) Identification of candidate genes conferring cold tolerance to rice (*Oryza sativa* L.) at the bud-bursting stage using bulk segregant analysis sequencing and linkage mapping. Front Plant Sci 12:647239

Yu C, Miao R, Khanna M (2021) Maladaptation of US corn and soybeans to a changing climate. Sci Rep 11(1):1–12

Zahid A, Ali S, Ahmed M, Iqbal N (2020) Improvement of soil health through residue management and conservation tillage in rice-wheat cropping system of Punjab, Pakistan. Agronomy 10(12):1844

Zhai F, Zhuang J (2012) Agricultural impact of climate change: a general equilibrium analysis with special reference to Southeast Asia. Climate change in Asia and the Pacific: how can countries adapt, pp 17–35

Zhang WF, Dou ZX, He P, Ju XT, Powlson D, Chadwick D, Zhang FS (2013) New technologies reduce greenhouse gas emissions from nitrogenous fertilizer in China. Proc Natl Acad Sci 110(21):8375–8380

Zhang S, Wang H, Sun X, Fan J, Zhang F, Zheng J, Li Y (2021) Effects of farming practices on yield and crop water productivity of wheat, maize and potato in China: a meta-analysis. Agric Water Manag 243:106444

Zhao H, Ning P, Chen Y, Liu J, Ghaffar SA, Xiaohong T, Shi J (2019) Effect of straw amendment modes on soil organic carbon, nitrogen sequestration and crop yield on the North-Central Plain of China. Soil Use Manag 35(3):511–525

Zingore S, Murwira HK, Delve RJ, Giller KE (2007) Influence of nutrient management strategies on variability of soil fertility, crop yields and nutrient balances on smallholder farms in Zimbabwe. Agric Ecosyst Environ 119(1–2):112–126

Ziska LH, Blumenthal DM, Runion GB, Hunt ER Jr, Diaz-Soltero H (2011) Invasive species and climate change: an agronomic perspective. Clim Chang 105(1–2):13–42

Zougmoré RB, Partey ST, Ouédraogo M, Torquebiau E, Campbell BM (2018) Facing climate variability in sub-Saharan Africa: analysis of climate-smart agriculture opportunities to manage climate-related risks. Cahiers Agricultures (TSI) 27(3):1–9

18 Climate Crisis and Adoption of Climate-smart Agriculture Technologies and Models

Khadijeh Bazrafkan, Ali Karami, Naser Valizadeh, Samira Esfandyari Bayat, Hajar Zareie, and Dariush Hayati

Abstract

Climate-smart agriculture (CSA) is one of the latest generations of agricultural systems, which aims to reduce the vulnerability of the agricultural sector to uncertainties such as climate change. Adoption of the technologies of this agricultural system by farmers can be one of the first steps towards reducing the vulnerability of the agricultural sector to climate change. In this regard, introducing models and frameworks for adoption of climate-smart agriculture technologies was determined as the main goal of this chapter. For this purpose, different technology acceptance models were evaluated in terms of strengths and weaknesses. Comparison of different theoretical approaches to the adoption of CSA technologies shows that each framework has strengths and weaknesses. In other words, it can be mentioned that each of these models can be a good model to encourage the adoption of CSA technologies in different spatial and temporal situations. In this chapter, we demonstrated that Theory of Planned Behavior (TPB), Health Belief Theory (HBT), Protection Motivation Theory (PMT), and Reasoned Action Theory (RAT) are based on the presupposition that people act completely rationally in the application of CSA technologies. However, Value-Belief-Norm (VBN) theory and Norm Activation Theory (NAT) reject this assumption and claim that in some cases it is norms or moral considerations that lead people to use CSA technologies. Finally, some implications were presented that can be used by policy-makers, managers, farmers, and other stakeholders engaged in the process of encouraging adoption of climate-smart agriculture technologies.

Keywords

Climate change · Climate-smart agriculture · Adoption of technologies · Behavioral models

K. Bazrafkan · N. Valizadeh (✉) · D. Hayati
Department of Agricultural Extension and Education, School of Agriculture, Shiraz University, Shiraz, Iran
e-mail: n.valizadeh@shirazu.ac.ir

A. Karami
Department of Soil Science, School of Agriculture, Shiraz University, Shiraz, Iran

S. Esfandyari Bayat
Department of Agricultural Extension and Education, College of Agriculture, Tarbiat Modares University (TMU), Tehran, Iran

H. Zareie
Department of Agricultural Extension and Education, College of Agriculture, University of Tehran, Tehran, Iran

U. Chatterjee et al. (eds.), *Climate Crisis: Adaptive Approaches and Sustainability*, Sustainable Development Goals Series, https://doi.org/10.1007/978-3-031-44397-8_18

18.1 Introduction

Climate change is considered as an unfavorable phenomenon that can threaten agricultural production systems, people's livelihood and well-being, food security, and health of human beings. The excessive increase in the world's population and climatic variabilities result in issues such as the increase in floods, droughts, and storms. Climate changes also increase in the prevalence of pests and diseases and reduce the agricultural land. It should also be mentioned that climate changes can result in the decrease in soil fertility and the quantity and quality of yield in agricultural systems (Jayne et al. 2014; Nelson et al. 2014; Seneviratne et al. 2012). In particular, climate change affects agriculture in different ways. The first effect is rising temperatures and migrating weather. Depending on the extent of mitigation measures taken in the coming decades, 1–3 °C temperature increases are expected to occur worldwide, which is equivalent to a 300–500 km change in weather patterns from the equator to the poles. Therefore, the temperature will increase in areas with higher altitude (Lipper et al. 2017; Ohmura 2012). In addition to having a negative effect on plants, climate migration affects the migration of pests and their invasion to other agricultural areas. Second effect is rising sea levels. Sea level rise leads to the loss of agricultural land as well as important infrastructure in the field of food supply (Lipper et al. 2017). Also, with the rise of the sea level, a large area of valuable agricultural land, especially in tropical regions, will be threatened (Kurukulasuriya and Rosenthal 2013). About 10% of the world's population lives in coastal areas (altitude ≤10 m), which is estimated to produce about 14% of the world's gross domestic product (GDP) with a large variation in the share of the population to the country that the sea level rising due to climate change and global warming are a serious threat to these people (McGranahan et al. 2007).

Increased snow melting and irrigation timing is the third effect of climate change. Climate change and increase in temperature, in addition to changes in precipitation patterns, increase the melting of snow and reduce the possibility of using the water stored in the snow accumulated in the wet season for irrigation in the dry season, and with the increase in the melting of snow and ice caused by the increase. Temperatures may increase the likelihood of flooding (Lipper et al. 2017). These changes could have a significant impact on agriculture and thus food supply around the world. The fourth effect of climate change is related to the increased likelihood of severe events. Climate change also changes the climatic distribution, which leads to an increase in the probability of extreme events such as heat waves, heavy rainfall, rainstorms and coastal flooding (Lipper et al. 2017). Also, temperature changes lead to the possibility of methane gas release as a result of the loss of permanent ice and an increase in temperature up to 6 °C (Lenton et al. 2008). Such extreme events can threaten the world's agriculture and food security and destroy the agricultural system.

It should also be emphasized that since climate change has destructive effects on agriculture, food security, and the health of the world community, the management of this crisis requires immediate action by global societies, governments, regional and international organizations by the performance of a more climate-resilient and low-emissions climate-smart agriculture approach (Nyasimi et al. 2017; Ogada et al. 2020; Traore et al. 2021; Zerssa et al. 2021). In this regard, suitable agricultural technologies or practices are required to reduce restrictions, increase productivity, enhance resilience, reduce or eliminate greenhouse gas emissions from farms in order to sustainably maintain agricultural livelihoods (Teklu et al. 2023; Zerssa et al. 2021). Agriculture reduces climate change through various methods such as avoiding deforestation and converting wetlands and pastures or grassland, increasing carbon storage or carbon sequestration in soil and plants, and strategies to avoid greenhouse gas emissions such as methane and nitrous oxide (Djojodihardjo and Ahmad 2015). In many countries in the world, agriculture is more important than industry and transportation in terms of the potential to reduce climate change (Djojodihardjo and Ahmad 2015). In order to build capacity, experience and guide future choices, initial action in climate-smart agriculture is necessary.

Climate-smart agriculture (CSA) is an agricultural procedure that changes the direction of agricultural development in the face of climate changes caused by natural or anthropogenic, takes steps in the direction of mitigation and adaptation, and ultimately ensures food security in a changing climate condition (Komarek et al. 2019; Lipper et al. 2014; Teklu et al. 2023). In this context, CSA is an approach that seeks to integrate the need for adaptation and the possibility of mitigation in agricultural development strategies to support food security. In another definition of climate-smart agriculture, it can be said that it is a method to develop technical, political and investment conditions to achieve sustainable agricultural development in accordance with food security under climate change, and its goal is to continuously increase agricultural productivity and income. Adapting and creating resistance to climate change are reducing or eliminating greenhouse gas emissions. CSA actually pays special attention to social, economic, and environmental issues in order to maximize benefits and minimize trade-offs (Teklu et al. 2023). It also considers technological, political, and institutional interventions that depend on the spatiotemporal context (Lipper et al. 2017).

Among the advantages of CSA, it is worth mentioning that this method combines plants, livestock, agroforestry, improved pest and disease management, water and nutrients, landscape management, grassland and forest management, methods such as reduced plowing, and the use of varieties, diverse species and breeds, integration of trees in agricultural systems, rehabilitation of degraded lands, improving the efficiency of water and nitrogen consumption, fertilizer management including the use of anaerobic decomposers and tries to bring biodiversity, efficiency, self-sufficiency, self-regulation and self-reliance to the systems to induce agriculture and thereby improve the degree of pliability and reduce the risk of food insecurity (Barnett et al. 2008; Thornton et al. 2014). It evaluates the interaction between departments and the needs of different stakeholders. It seeks to create empowering environments through the regulation of policies, financial investments and organizational coordination. By improving access to services, knowledge, resources, financial services and markets, it prioritizes the livelihoods of smallholders, considers climate change control as one of the possible secondary benefits, especially in low-income, agricultural-oriented populations. It takes into account adaptation and creates pliability against sudden changes, especially those related to climate change, and seeks to identify opportunities to access climate-related investments and integrate them with financial resources. It is traditional in the agricultural sector. One of the disadvantages of CSA is that this method is not actually a set of operations that can be used everywhere, but rather an approach that includes various factors in local concepts and it changes strongly under the influence of location (Shirsath et al. 2017).

Generally, there are several advantages to adopt climate-smart agriculture, including increasing productivity and yield, improving water management, enhancing soil health, reducing greenhouse gas emissions, and improving farmer resilience. However, it should be mentioned that there are some challenges that farmers may face in adopting CSA, including high upfront costs, limited knowledge and awareness, market challenges, policy and regulatory barriers, and social and cultural barriers. CSA innovations include old and new practices and technologies that are integrated into agricultural systems at different scales (Mutenje et al. 2019; Zerssa et al. 2021). On the other hand, there are some strategies as examples of climate-smart practices that farmers can be used to manage climate change like crop diversification, livestock breeds, soil and land management technologies, improved fodder production and livestock feeding technologies, physical soil and water management practices (Babatunde and Qaim 2010; Burney and Naylor 2012; Karamba et al. 2011; Nyasimi et al. 2017; Ogada et al. 2020). Regarding soil and land management, we can mention practices such as creating terraces, ridges and bunds, micro-catchments, stopping the burning of crop residues, mulching, change land scale under cultivation of specific crops, compost and vermicompost, cover crops, inorganic and organic fertilizers, minimum tillage with conservative use of pesticides and herbicides, crop rotation intercropping or mixed cultivation, crop residue management, row planting, biochar, and agroforestry (Ajibade et al. 2022; Nyambo et al. 2023; Ogada et al. 2020; Simane et al. 2016; Teklewold et al. 2019; Teklu et al. 2023).

Although people and farmers can be expected to accept and use CSA technologies due to the adverse effects of climate change on their agriculture, farmer-to-farmer interactions, visits to demonstration farms, and participation in informal social networks; however, the adoption and diffusion of agricultural technologies is complicated and difficult due to uncertainty, costs and benefits of technology, farmer's gender, social capital, labor and credit constraints, and market access (Bandiera and Rasul 2006; Nordin et al. 2014; Ogada et al. 2020; Rao and Qaim 2011).

Due to the fact that CSA is considered a new approach to increase resilience and adaptability to climate change, the dimensions and methodologies necessary to develop its adoption have not yet been developed. In other words, not many studies have been conducted in the field of theoretical and scientific bases for the adoption of CSA technologies in different countries of the world. At the same time, any progress in the field of agricultural sustainability requires knowing the factors that determine the behaviors and intentions of different stakeholders such as farmers. The social and behavioral dimensions of the adoption of technologies such as CSA technologies have always received less attention than other dimensions. In this regard, the lack of sufficient knowledge of the theoretical foundations of the development of adoption of CSA technologies and its determining factors is presented as a research gap in this chapter. In order to achieve this goal, in present study, an effort was made to introduce and compare the most important models and development models for the adoption of CSA technologies.

18.2 Frameworks to Adopt CSA Technologies

Studies have shown that theoretical interventions can motivate people to change their attitudes and behaviors technologies like CSA technologies (Aquilina et al. 2004). The first step in this process is to examine the current situation and the perspectives of people involved in adoption of CSA technologies and activities. In this regard, Paul (1998) states that if people believe that they cannot do something to prevent the loss of their products, they will not do anything in that field. Therefore, understanding beliefs about climate change management and CSA technologies is the key to understand vulnerability patterns against climate change within a society and region. Therefore, farmers' attitude towards climate change and CSA technologies is closely related to behavioral management and their experience of past events (Zarafshani et al. 2007). The attitude and experience of the past can be effective in evaluating CSA technologies in the future, especially from the point of view of preventive action (Yazdanpanah et al. 2013). For this reason, environmental psychology and the theories of this field of science have a special place in CSA. In other words, these theories are considered as suitable tools for understanding people's behavior in adoption of CSA technologies. Various theories have been used in this field (Truelove et al. 2015; Keshavarz and Karami 2016) and general theories of environmental behavior often failed to identify the role of individual motivations in environmental protection and adoption of CSA technologies (Lafreniere et al. 2013; Price and Leviston 2014). However, some studies (see Bamberg and Moser 2007) show that the psychological characteristics of farmers are of vital importance in adoption of CSA technologies. In this part of this chapter, some of the most important theoretical frameworks that can be applied to encourage the adoption CSA are discussed.

18.2.1 Protection Motivation Theory (PMT)

PMT was first developed by Rogers in 1975, based on the expectation value model (Fig. 18.1). This theory provides a combination of psychological constructs (originated from the value expectancy theories and cognitive theories) to help clarify how to approach health protection (Scarpa and Thiene 2011). Although the original model was designed to study health protection behavior (Rogers 1975, 1983), its use has now been extended to several other areas, including the adoption of technologies such as CSA technologies (Bubeck et al. 2013). Therefore, PMT, as a deliberative and decision-making framework, can identify barriers and facilitators for the adoption of CSA technologies. The main idea in PMT is that people face risks such as climate change and adopt technologies such as CSA technologies through two cognitive processes "threat appraisal" and "coping appraisal" to participate in adaptive actions (Rogers 1983). Threat appraisal describes a person's assessment of the threat level of an action or technology in terms of "perceived vulnerability" and "perceived severity". Perceived vulnerability indicates a person's sensitivity to the threat of a phenomenon or technology such as CSA technology (Scarpa and Thiene 2011). Perceived severity is the amount of threat that people expect to be realized if they are resilient against that threat or adopt that technology (Dang et al. 2012). In addition, coping appraisal refers to an individual's assessment of his/her ability to respond to a perceived threat (for example, the consequences of climate change or the adoption of CSA technologies) and thereby avoid a risk (Woon et al. 2005). Coping appraisal includes three variables: "response efficacy", "self-efficacy" and "response costs" (Floyd et al. 2000). In fact, it can be said that response efficiency describes the effect of an adaptive response to reduce or avoid the risks of adopting CSA technologies (Floyd et al. 2000: Milne et al. 2000). Self-efficacy emphasizes the ability and judgment of a person to deal with the threats caused by the adoption of CSA technologies (Ifinedo 2012). Finally, response costs in this theory refer to the financial costs, time, effort, and intellectual-emotional costs, which represent all the perceived costs associated with adopting technologies such as CSA technologies (Scarpa and Thiene 2011; Bubeck et al. 2013). Therefore, the coping evaluation structure in PMT is based on the perceived ability to prevent or reduce the risks of adopting CSA technologies against the anticipated costs of protective measures (Keshavarz and Karami 2016). Applicability of this theory to encourage the behaviors and intentions of individuals to adoption technologies like CSA technologies has been approved by different studies (see Keshavarz and Karami 2016; Luu et al. 2019; Neisi et al. 2020).

18.2.2 Health Belief Theory (HBT)

The HBT is one of psychological theories that can investigate and predict preventive behaviors in the face of crises such as climate change and the adoption of CSA technologies (Yazdanpanah et al. 2013). The HBT (Fig. 18.2), which was designed based on Kurt Lewin's theory, was first

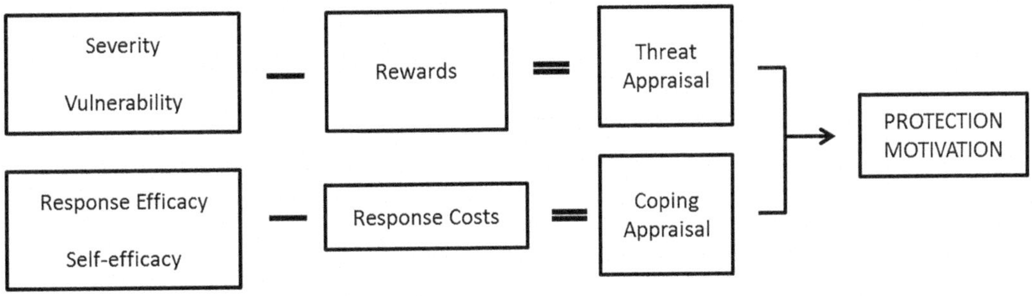

Fig. 18.1 Protection motivation theory

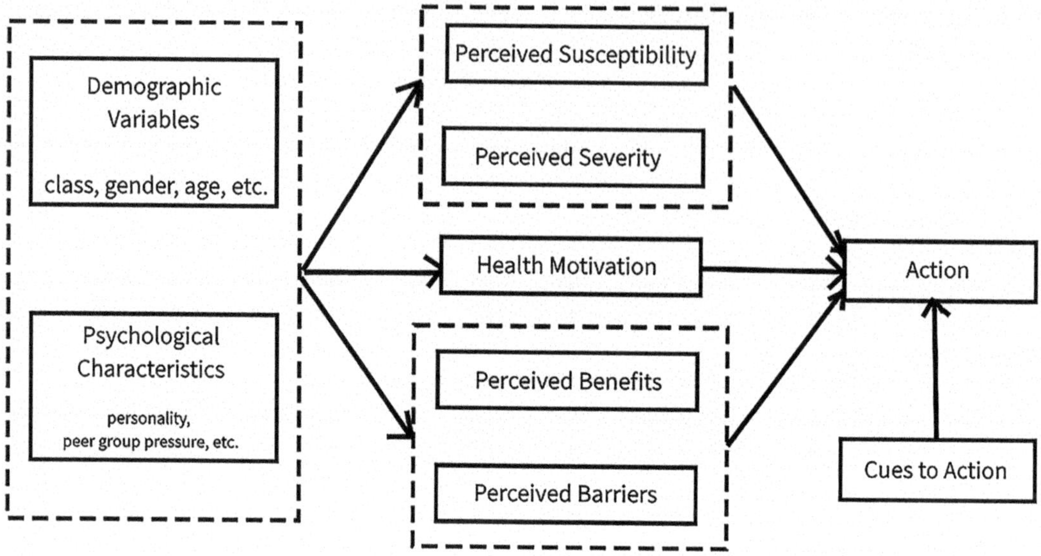

Fig. 18.2 The health belief model

proposed in 1974 by Rosentac in order to explain why people do not participate in preventive health programs (Vassallo et al. 2009). Then it was developed by Mayman and Baker (Yazdanpanah et al. 2016) to be used as a framework to explain people's future risk behaviors and adoption of different technologies in the field of climate change (Pender et al. 2006). This model focuses on the change in beliefs and assumes that a change in beliefs leads to a change in people's behavior in dealing with risks or adoption technologies (Yazdanpanah et al. 2014). The basis of this model is the motivation of people to act, and it emphasizes how a person's perception leads to the motivation of behavior in him/her; it also emphasizes the fact that human behavior is based on two major antecedents: the value given by the person to the goal and the result of the work and the person's prediction and estimate of the probability of reaching the goal if he/she performs that behavior (Delfian et al. 2016). The characteristic of the HBT is that it emphasizes "the individual" too much and only the effect of health beliefs on behavior is considered. This model is based on the assumption that a person will adopt an action or CSA technology if he/she feels that this action will save him/her from the desired harm like climate change harms. Individuals have a positive expectation from this model and that is health and prevention of its unfortunate consequences by adopting recommendations. That is, people expect that by adopting the recommendations and technologies of CSA, they will be secured against the dangers of climate change. Therefore, by adopting the recommendations, this belief and trust will be created in them that they will succeed in reaching the goal (Yazdanpanah et al. 2015).

The key constructs of the HBT in adoption CSA technologies are as follows (Mckellar and Sillence 2020):

- Degree of perceived risk of a climate change. This construct includes perceived susceptibility of individuals against climate crisis and its perceived severity.
- Perceived benefits adopting CSA adaptive strategies. Refers to the perceived effectiveness of adopting CSA adaptive strategies in reducing the threat of climate change on agricultural sector.
- Perceived barriers to adopt CSA adaptive strategies. This construct refers to the potential negative consequences that may originate from taking specific adopting CSA adaptive strategies.

- Cues to action. Events that motivate farmers and individuals to take action in changing their CSA adaptive strategies are pivotal determinants of change.
- Self-efficacy. One of the most important constructs in HBT is the belief in being able to successfully execute the CSA adaptive strategies required to produce the desired outcomes.
- Other constructs. Demographic, sociopsychological, and structural constructs influence the farmers' perceptions of changing their adaptive strategies and thus indirectly influence their ability to sustain agricultural activities.

18.2.3 Theory of Planned Behavior (TPB)

This theory (Fig. 18.3) was proposed by Ajzen in the early 1990s as a general model of conscious behavior such as the adoption of CSA technologies. This theory was expanded to solve the limitation of Reasoned Action Theory (RAT); because the RAT was limited to behavior in which people had incomplete voluntary control over their behavior (Ajzen 2006). Therefore, Ajzen proposed a third factor in the RAT, next to the two factors of attitude and subjective norms, which is called perceived behavior control. When the subjective norms and attitude are stable, the ease or difficulty of performing a behavior will affect the intention and desire towards that behavior. By adding the construct of perceived behavioral control, this theory tries to predict involuntary behaviors as well (Madden et al. 1992). Perceived behavioral control refers to people's belief that their behavior will successfully lead to the achievement of expected goals. Both TPB and RAT focus on the importance of intention to perform a specific behavior such as adopting CSA technologies (Sawitri et al. 2015). The TPB has been used in various fields such as sexual behavior (Boldero et al. 1992), mechanics (Parker 1992), health-related measures (Black and Babrow 1991). Recently, this theory has been used to explain pro-environmental behaviors (Stern et al. 1995) such as recycling behavior (Cheung et al. 1999), water conservation behavior (Trumbo and O'Keefe 2001), consumer support (Sparks and Shepherd 1992) and water management (Shaw et al. 2011). But it is easy to infer that this theory and its components can be employed to encourage the adoption of CSA technologies (Tikir and Lehmann 2011; Masud et al. 2016; Zhang et al. 2020; Shalender and Sharma 2021; Delistavrou et al. 2023). The power of this theory in explaining behaviors and intentions towards adopting CSA technologies has been proven in several studies (see Zhang et al. 2020; Delistavrou et al. 2023; Sobaih and Elshaer 2023). Therefore, this theory can be a basis for researchers and practitioners to be able to employ it to encourage the adoption of CSA technologies under climate change.

Fig. 18.3 Theory of planned behavior

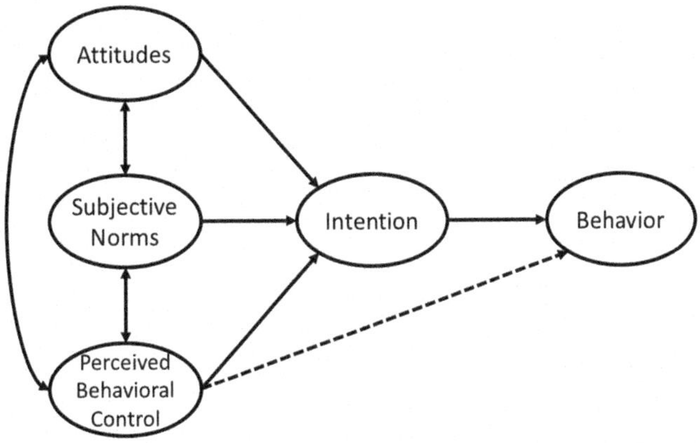

18.2.4 Norm Activation Theory (NAT)

The NAT was presented in 1973 by Schwartz. In this theory, the constructs of awareness of the consequences of behavior, responsibility, and personal norm are the three main predictors of CSA technologies adoption behavior. This theory argues that awareness of potential harmful consequences and personal responsibility activates a personal norm towards the use of CSA technologies. In this way, a person acts to prevent the harmful results of the climate change phenomenon. This theory is a good intervention framework to encourage the use of different technologies. In other words, this theory provides a framework for intervention in the adoption of CSA technologies and is used when someone believes that events such as climate change will lead to adverse consequences for the individual and society (Sawitri et al. 2015). Schwartz's theory is based on the awareness of the consequences and acceptance of the components of responsibility and the content of personal norms. This theory argues that personal norms are likely to be strengthened due to the importance or intensity of awareness of consequences and increased responsibility (Sawitri et al. 2015). If the content of personal norms prescribes an action such as adopting CSA technologies, that person will act to avoid expected harmful consequences (Sawitri et al. 2015). In other words, farmers will avoid taking actions incompatible with climate change that have harmful consequences for themselves and the agricultural community. Some studies (see Qiao and Gao 2017; Schweizer et al. 2013; Vaske et al. 2015; He and Zhan 2018; Hallaj et al. 2021) have tested the applicability of this theory to encourage adaptive behaviors with climate change. The success of this model in explaining the intention to adopt or actual adoption of CSA or climate change-adapted technologies has been confirmed in many studies (Oom Do Valle et al. 2005).

18.2.5 Value-Belief-Norm (VBN) Theory

VBN theory (Fig. 18.4) was introduced by Stern et al. (1999) and Stern (2000) based on NAT and by adding general values and environmental concern specifically to explain pro-environmental behavior. This theory provides a causal chain of general and basic values and beliefs for the formation of specific behavioral beliefs and norms, which ultimately ends in behavior (Abrahamse and Steg 2011). The theory of VBN is based on the assumption that the successive effects of values and beliefs are linked to environmental concerns, which causes individual and social norms to lead to pro-environmental behaviors (Kaida and Kaida 2016). VBN theory argues that environmental altruistic behavior occurs through the activation of norms. These norms originate from the three factors of personal values, beliefs that feel threatened towards values, and beliefs that a

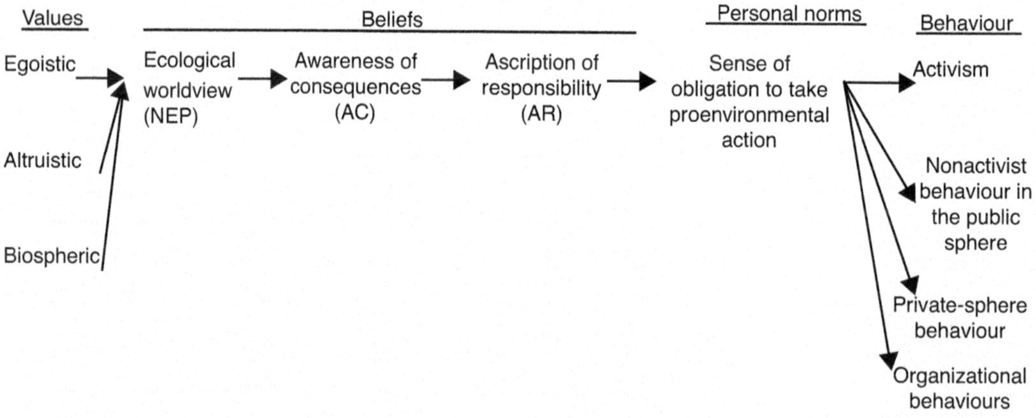

Fig. 18.4 Value-Belief-Norm (VBN) theory

person can have based on his/her ability to reduce these threats (Sawitri et al. 2015). The difference between VBN theory and NAT is that NAT only emphasizes altruistic values, while VBN theory includes other values, and also in this theory, individual beliefs are a direct evaluation of behavior (Stern et al. 1995).

18.3 Comparison of Frameworks for Adoption of CSA Frameworks

Each of the frameworks for encouraging the adoption of CSA technologies makes different assumptions about the adoption behavior of agricultural communities. PMT, TPB, and RAT are based on the assumption that farmers act rationally in adopting CSA technologies (Steg and Vlek 2009). HBT can also be placed in the category of rational or individualistic theories such as PMT, TPB, and RAT; because this theory also emphasizes more on constructs that are more effective in accelerating the achievement of individual's goals. While VBN and NAT consider the moral motivations in people's decision to use CSA technologies. The theoretical constructs used in this framework for encouraging the adoption of CSA technologies have extensive empirical support (Valizadeh et al. 2018). PMT, TPB, HBT, VBN, NAT and RAT have provided useful frameworks for the adoption and application of CSA technologies.

Very few studies have compared TPB, HBT, VBN, NAT and RAT. Kaiser et al. (2005) and Ives and Kendal (2014) state that VBN is a stronger model than TPB. They claim that the relationships between variables in VBN and NAT are fully described. Nevertheless, TPB, HBT, and RAT are considered stronger models for predicting behaviors such as adopting CSA technologies. Comparing VBN with TPB, López-Mosquera and Sánchez (2012) state that despite the fact that TPB represents the respondents' future interest in changing their attitudes and norms in response to perceived conditions, VBN refers to the values or principles that people have. In other words, unlike the VBN theory, the TPB does not fit public-sphere and altruistic behaviors (López-Mosquera and Sánchez 2012). Aguilar-Luzón et al. (2012) and Chan and Bishop (2013) mention in their studies that the validity and predictive ability of TPB was better than VBN theory. Such results in the studies show that the theoretical constructs in TPB can be good predictors of the adoption of CSA technologies. There are also studies that have supported the suitability of frameworks such as VBN and NAT compared to rational frameworks such as PMT, TPB, and RAT. Andersson et al. (2005) claim that VBN has a plausible ability to explain environmental and risk coping behaviors such as the adoption of CSA technologies.

In general, it is believed that VBN and NAT theories are comparable to TPB, HBT, PMT and RAT in terms of explaining individualistic (private-sphere) and collective (public-sphere) behaviors (Pradhananga 2014).

18.4 Conclusion and Future Directions

Comparison of different theoretical approaches to the adoption of CSA technologies shows that each framework has strengths and weaknesses. In other words, there is no model that can be claimed to be the best model for encouraging the adoption of CSA technologies in all circumstances. It can be mentioned that each of these models can be a good model to encourage the adoption of CSA technologies in different spatial and temporal situations. In addition, each framework has different defaults. TPB, HBT, PMT, and RAT are based on the presupposition that people act completely rationally in the application of CSA technologies. However, VBN and NAT reject this assumption and claim that in some cases it is norms or moral considerations that lead people to use CSA technologies.

In general, it should be mentioned that the moral and logical behavior of agricultural communities in the context of adopting CSA technologies is a completely contingent and context-specific matter. In other words, it cannot be claimed that the presuppositions of each of these theories can be true in all societies and these theories can be used in all situations. In this regard, it is suggested that before using these frameworks to encourage the adoption of CSA technologies, the context in which the frameworks are to be used should be known. It means that a preliminary study should be done to determine whether the society in question is acting more rationally or morally. Perhaps VBN and NAT are good options to encourage CSA technologies if they are moral; but if communities act more rationally, it can be said that PMT, TPB, HBT, and RAT can be appropriate frameworks for developing the use of CSA technologies. Nevertheless, it should not be forgotten that it may be appropriate to use a combination of these theories that have structures from several theories.

Each of the theories and models introduced in this chapter, with a certain number of variables, try to predict the adoption of CSA technologies. It is clear that these models represent a very simplified version of farmers' behavior in adopting CSA technologies. Therefore, it is suggested that researchers, in addition to using these theories and models and the variables in them, always try to include context-specific variables that are more compatible with the conditions of the regions and societies. This can be a great help to get a realistic idea of the adoption behavior of CSA technologies. It also provides the researchers, policy-makers, behavioral change practitioners, and other users with real and applicable implications on the behavioral change in the field of CSA adoption. Such an approach also helps the theoretical development of models.

References

Abrahamse W, Steg L (2011) Factors related to household energy use and intention to reduce it: the role of psychological and socio-demographic variables. Hum Ecol Rev 18(1):30–40

Aguilar-Luzón MDC, García-Martínez JMÁ, Calvo-Salguero A, Salinas JM (2012) Comparative study between the theory of planned behavior and the value–belief–norm model regarding the environment, on Spanish housewives' recycling behavior. J Appl Soc Psychol 42(11):2797–2833

Ajibade S, Mupambwa HA, Manyevere A, Mnkeni PNS (2022) Vermicompost amended with rock phosphate as a climate smart technology for production of organic Swiss chard (Beta vulgaris subsp. vulgaris). Front Sustain Food Syst 6:1

Ajzen I (2006) Behavioral interventions based on the theory of planned behavior. http://people.umass.edu/aizen/pdf/tpb.intervention.pdf

Andersson L, Shivarajan S, Blau G (2005) Enacting ecological sustainability in the MNC: a test of an adapted value-belief-norm framework. J Bus Ethics 59(3):295–305

Aquilina S, Amato Gauci A, Ellul M, Scerri L (2004) Sun awareness in Maltese secondary school students. J Eur Acad Dermatol Venereol 18(6):670–675

Babatunde RO, Qaim M (2010) Impact of off-farm income on food security and nutrition in Nigeria. Food Policy 35(4):303–311

Bamberg S, Moser G (2007) Twenty years after Hines, Hungerford, and Tomera: a new meta-analysis of psycho-social determinants of pro-environmental behavior. J Environ Psychol 27(1):14–25

Bandiera O, Rasul I (2006) Social networks and technology adoption in northern Mozambique. Econ J 116(514):869–902

Barnett BJ, Barrett CB, Skees JR (2008) Poverty traps and index-based risk transfer products. World Dev 36(10):1766–1785

Black DR, Babrow AS (1991) Identification of campaign recruitment strategies for a stepped smoking cessation intervention for a college campus. Health Educ Q 18:235–247

Boldero J, Moore S, Rosenthal D (1992) Intention, context, and safe sex: Australian adolescents' responses to aids. J Appl Soc Psychol 22(17):1374–1396

Bubeck P, Botzen WJW, Kreibich H, Aerts JCJH (2013) Detailed insights into the influence of flood-coping appraisals on mitigation behavior. Glob Environ Chang 23(5):1327–1338

Burney JA, Naylor RL (2012) Smallholder irrigation as a poverty alleviation tool in sub-Saharan Africa. World Dev 40(1):110–123

Chan L, Bishop B (2013) A moral basis for recycling: Extending the theory of planned behaviour. J Environ Psychol 36:96–102

Cheung SF, Chan DKS, Wong ZSY (1999) Reexamining the theory of planned behavior in understanding wastepaper recycling. Environ Behav 31(5):587–612

Dang LH, Li E, Bruwer J (2012) Understanding climate change adaptive behavior of farmers: an integrated conceptual framework. Int J Clim Change Impacts Responses 3(2):255–272

Delfian F, Yazdanpanah M, Forouzani M (2016) Investigating farmers' managerial behaviors during drought as conservation responses (the case of Dehloran County). Spat Anal Environ Risks 4(4):79–92

Delistavrou A, Tilikidou I, Papaioannou E (2023) Climate change risk perception and intentions to buy consumer

packaged goods with chemicals containing recycled CO_2. J Clean Prod 382:135215
Djojodihardjo H, Ahmad D (2015) Opportunities and challenges for climate-smart agriculture. Paper presented at the proceedings of the 3rd international conference on natural resource management (NRM'15), New Delhi, India
Floyd DL, Prentice-Dunn S, Rogers RW (2000) A meta-analysis of research on protection motivation theory. J Appl Soc Psychol 30(2):407–429
Hallaj Z, Sadighi H, Farhadian H, Bijani M (2021) Human ecological analysis of farmers' pro-environmental behaviour in the face of drought: application of norm activation theory. Water Environ J 35(4):1337–1347
He X, Zhan W (2018) How to activate moral norm to adopt electric vehicles in China? An empirical study based on extended norm activation theory. J Clean Prod 172:3546–3556
Ifinedo P (2012) Understanding information systems security policy compliance: an integration of the theory of planned behavior and the protection motivation theory. Comput Secur 31(2012):83–95
Ives CD, Kendal D (2014) The role of social values in the management of ecological systems. J Environ Manag 144:67–72
Jayne TS, Chamberlin J, Headey DD (2014) Land pressures, the evolution of farming systems, and development strategies in Africa: a synthesis. Food Policy 48:1–17
Kaida N, Kaida K (2016) Pro-environmental behavior correlates with present and future subjective well-being. Environ Dev Sustain 18:111–127
Kaiser FG, Hübner G, Bogner FX (2005) Contrasting the theory of planned behavior with the value-belief-norm model in explaining conservation behavior. J Appl Soc Psychol 35(10):2150–2170
Karamba WR, Quiñones EJ, Winters P (2011) Migration and food consumption patterns in Ghana. Food Policy 36(1):41–53
Keshavarz M, Karami E (2016) Farmers' pro-environmental behavior under drought: application of protection motivation theory. J Arid Environ 127(2016):128–136
Komarek AM, Thurlow J, Koo J, De Pinto A (2019) Economywide effects of climate-smart agriculture in Ethiopia. Agric Econ 50(6):765–778
Kurukulasuriya P, Rosenthal S (2013) Climate change and agriculture: a review of impacts and adaptations. The World Bank Environment Department, Washington, DC
Lafreniere KC, Deshpande S, Bjornlund H, Hunter MG (2013) Extending stakeholder theory to promote resource management initiatives to key stakeholders: a case study of water transfers in Alberta, Canada. Environ Manage 129(15):81–91
Lenton TM, Held H, Kriegler E, Hall JW, Lucht W, Rahmstorf S, Schellnhuber HJ (2008) Tipping elements in the Earth's climate system. Proc Natl Acad Sci 105(6):1786–1793
Lipper L, Thornton P, Campbell BM, Baedeker T, Braimoh A, Bwalya M et al (2014) Climate-smart agriculture for food security. Nat Clim Chang 4(12):1068–1072
Lipper L, McCarthy N, Zilberman D, Asfaw S, Branca G (2017) Climate smart agriculture: building resilience to climate change. Springer Nature, Cham
López-Mosquera N, Sánchez M (2012) Theory of Planned Behavior and the Value-Belief-Norm Theory explaining willingness to pay for a suburban park. J Environ Manag 113:251–262
Luu TA, Nguyen AT, Trinh QA, Pham VT, Le BB, Nguyen DT et al (2019) Farmers' intention to climate change adaptation in agriculture in the Red River Delta Biosphere Reserve (Vietnam): a combination of Structural Equation Modeling (SEM) and Protection Motivation Theory (PMT). Sustainability 11(10):2993
Madden TJ, Ellen PS, Ajzen I (1992) A comparison of the theory of planned behavior and the theory of reasoned action. Personal Soc Psychol Bull 18(3):3–9
Masud MM, Al-Amin AQ, Junsheng H, Ahmed F, Yahaya SR, Akhtar R, Banna H (2016) Climate change issue and theory of planned behaviour: relationship by empirical evidence. J Clean Prod 113:613–623
McGranahan G, Balk D, Anderson B (2007) The rising tide: assessing the risks of climate change and human settlements in low elevation coastal zones. Environ Urban 19(1):17–37
Mckellar K, Sillence E (2020) Current research on sexual health and teenagers. In: Sexual health information and the digital age. Elsevier, pp 5–23
Milne S, Sheeran P, Orbell S (2000) Prediction and intervention in health-related behavior: a meta-analytic review of protection motivation theory. J Appl Soc Psychol 30(1):106–143
Mutenje MJ, Farnworth CR, Stirling C, Thierfelder C, Mupangwa W, Nyagumbo I (2019) A cost-benefit analysis of climate-smart agriculture options in Southern Africa: balancing gender and technology. Ecol Econ 163:126–137
Neisi M, Bijani M, Abbasi E, Mahmoudi H, Azadi H (2020) Analyzing farmers' drought risk management behavior: evidence from Iran. J Hydrol 590:125243
Nelson GC, Valin H, Sands RD, Havlík P, Ahammad H, Deryng D et al (2014) Climate change effects on agriculture: economic responses to biophysical shocks. Proc Natl Acad Sci 111(9):3274–3279
Nordin SM, Noor SM, bin Md Saad, M. S. (2014) Innovation diffusion of new technologies in the Malaysian paddy fertilizer industry. Procedia Soc Behav Sci 109:768–778
Nyambo P, Zhou L, Chuma T, Sokombela A, Malobane M, Musokwa M (2023) Prospects of vermicompost and biochar in climate smart agriculture. In: Vermicomposting for sustainable food systems in Africa. Springer, pp 145–159
Nyasimi M, Kimeli P, Sayula G, Radeny M, Kinyangi J, Mungai C (2017) Adoption and dissemination pathways for climate-smart agriculture technologies and

practices for climate-resilient livelihoods in Lushoto, Northeast Tanzania. Climate 5(3):63
Ogada MJ, Radeny MA, Recha JW, Solomon D (2020) Adoption of climate-smart agricultural technologies in Lushoto Climate-Smart Villages in north-eastern Tanzania. CCAFS Working Paper
Ohmura A (2012) Enhanced temperature variability in high-altitude climate change. Theor Appl Climatol 110:499–508
Oom Do Valle P, Rebelo E, Reis E, Menezes J (2005) Combining behavioral theories to predict recycling involvement. Environ Behav 37:364–396
Parker D (1992) Intention to commit driving violations: an application of the theory of planned behavior. J Appl Psychol 77(1):94–101
Paul BK (1998) Coping mechanisms practised by drought victims (1994/5) in North Bengal, Bangladesh. Appl Geograp 18(4):355–373
Pender et al (2006) Strategies for sustainable land management in the East African highlands: conclusions and implications. In: Pender J, Place F, Ehui S (eds) Strategies for sustainable land management in the East African highlands. International Food Policy Research Institute, Washington, D.C., pp 377–416
Pradhananga AK (2014) A moral obligation model of landowner conservation norms and behavior. Ph.D. dissertation, University of Minnesota. (No Published)
Price JC, Leviston Z (2014) Predicting pro-environmental agricultural practices: the social, psychological and contextual influences on land management. J Rural Stud 34(April 2014):65–78
Qiao G, Gao J (2017) Chinese tourists' perceptions of climate change and mitigation behavior: an application of norm activation theory. Sustainability 9(8):1322
Rao EJ, Qaim M (2011) Supermarkets, farm household income, and poverty: insights from Kenya. World Dev 39(5):784–796
Rogers RW (1975) A protection motivation theory of fear appeals and attitude change. J Psychol Interdiscip Appl 91(1):93–114
Rogers RW (1983) Cognitive and physiological processes in fear appeals and attitude change: a revised theory of protection motivation. In: Cacioppo BL, Petty RE (eds) Social psychophysiology: a sourcebook. Guilford Press, London. Available at: https://www.researchgate.net/publication/229068371_
Sawitri DR, Hadiyanto H, Hadi SP (2015) Pro-environmental behavior from a social cognitive theory perspective. Procedia Environ Sci 23:27–33
Scarpa R, Thiene M (2011) Organic food choices and protection motivation theory: addressing the psychological sources of heterogeneity. Food Qual Prefer 22(6):532–541
Schweizer S, Davis S, Thompson JL (2013) Changing the conversation about climate change: a theoretical framework for place-based climate change engagement. Environ Commun J Nat Cult 7(1):42–62
Seneviratne S, Nicholls N, Easterling D, Goodess C, Kanae S, Kossin J et al (2012) Changes in climate extremes and their impacts on the natural physical environment. In: Field CB, Barros V, Stocker TF, Qin D, Dokken DJ, Ebi KL, Mastrandrea MD, Mach KJ, Plattner G-K, Allen SK, Tignor M, Midgley PM (eds) Managing the risks of extreme events and disasters to advance climate change adaptation, A special report of working groups I and II of the Intergovernmental Panel on Climate Change (IPCC). Cambridge University Press, Cambridge/New York, pp 109–230
Shalender K, Sharma N (2021) Using extended theory of planned behaviour (TPB) to predict adoption intention of electric vehicles in India. Environ Dev Sustain 23(1):665–681
Shaw BR, Radler B, Chenoweth R, Heilberger P, Dearlove P (2011) Predicting intent to install a rain garden to protect a local lake: an application of the theory of planned behavior. J Ext 49(4):204–218
Shirsath PB, Aggarwal PK, Thornton PK, Dunnett A (2017) Prioritizing climate-smart agricultural land use options at a regional scale. Agric Syst 151:174–183
Simane B, Zaitchik BF, Foltz JD (2016) Agroecosystem specific climate vulnerability analysis: application of the livelihood vulnerability index to a tropical highland region. Mitig Adapt Strateg Glob Chang 21:39–65
Sobaih AEE, Elshaer IA (2023) Risk-taking, financial knowledge, and risky investment intention: expanding theory of planned behavior using a moderating-mediating model. Mathematics 11(2):453
Sparks P, Shepherd R (1992) Self-identity and the theory of planned behavior: assessing the role of identification with green consumerism. Soc Psychol Q 55(4):388–399
Steg L, Vlek C (2009) Encouraging pro-environmental behaviour: an integrative review and research agenda. J Environ Esychol 29(3):309–317
Stern PC (2000) New environmental theories: toward a coherent theory of environmentally significant behavior. J Soc Issues 56(3):407–424
Stern PC, Dietz T, Guagnano GA (1995) The new ecological paradigm in social psychological context. Environ Behav 27(6):723–743
Stern PC, Dietz T, Abel T, Guagnano GA, Kalof L (1999) A value-belief-norm theory of support for social movements: the case of environmentalism. Hum Ecol Rev 6(2):81–98
Teklewold H, Mekonnen A, Kohlin G (2019) Climate change adaptation: a study of multiple climate-smart practices in the Nile Basin of Ethiopia. Clim Dev 11(2):180–192
Teklu A, Simane B, Bezabih M (2023) Multiple adoption of climate-smart agriculture innovation for agricultural sustainability: empirical evidence from the Upper Blue Nile Highlands of Ethiopia. Clim Risk Manage 39:100477

Thornton PK, Ericksen PJ, Herrero M, Challinor AJ (2014) Climate variability and vulnerability to climate change: a review. Glob Chang Biol 20(11):3313–3328

Tikir A, Lehmann B (2011) Climate change, theory of planned behavior and values: a structural equation model with mediation analysis: a letter. Clim Chang 104(2):389–402

Traore B, Birhanu BZ, Sangaré S, Gumma MK, Tabo R, Whitbread AM (2021) Contribution of climate-smart agriculture technologies to food self-sufficiency of smallholder households in Mali. Sustainability 13(14):7757

Truelove HB, Carrico AR, Thabrew L (2015) A socio-psychological model for analyzing climate change adaptation: a case study of Sri Lankan paddy farmers. Glob Environ Chang 31(March 2015):85–97

Trumbo CW, O'Keefe GJ (2001) Intention to conserve water: environmental values, planned behavior, and information effects. A comparison of three communities sharing a watershed. Soc Nat Resour 14(10):889–899

Valizadeh N, Bijani M, Hayati D (2018) A comparative analysis of behavioral theories towards farmers' water conservation. Int J Agric Manage Dev 9(1047-2019-3460):1–10

Vaske JJ, Jacobs MH, Espinosa TK (2015) Carbon footprint mitigation on vacation: a norm activation model. J Outdoor Recreat Tour 11:80–86

Vassallo M, Saba A, Arvola A, Dean M, Messina F, Winkelmann M, Shepherd R (2009) Willingness to use functional breads. Applying the health belief model across four European countries. Appetite 52(2):452–460

Woon IMY, Tan GW, Low RT (2005) A protection motivation theory approach to home wireless security. In: Avison D, Galletta D, DeGross JI (eds), Proceedings of the 26th international conference on information systems, Las Vegas, December 11–14, 367–380

Yazdanpanah M, Thompson M, Hayati D, Zamani GH (2013) A new enemy at the gate: tackling Iran's water super-crisis by way of a transition from government to governance. Prog Dev Stud 13(3):177–194

Yazdanpanah M, Hayati D, Hochrainer-Stigler S, Zamani GH (2014) Understanding farmers' intention and behavior regarding water conservation in the Middle-East and North Africa: a case study in Iran. J Environ Manag 135:63–72

Yazdanpanah M, Forouzani M, Hojjati M (2015) Willingness of Iranian young adults to eat organic foods: application of the health belief model. Food Qual Prefer 41(April 2015):75–83

Yazdanpanah M, Tavakoli KH, Marzban A (2016) Investigating factors influence framers' intention regarding safe use of pesticides through health belief model. Iran Agric Extens Educ J 1.11(2):21–29

Zarafshani K, Gorgievski M, Zamani GH (2007) Dealing with drought: a comparison of perceptions and coping strategies of Iranian farmers from regions with different drought intensities. J Agric Educ Extens 19(1):69–80

Zerssa G, Feyssa D, Kim D-G, Eichler-Löbermann B (2021) Challenges of smallholder farming in Ethiopia and opportunities by adopting climate-smart agriculture. Agriculture 11(3):192

Zhang L, Ruiz-Menjivar J, Luo B, Liang Z, Swisher ME (2020) Predicting climate change mitigation and adaptation behaviors in agricultural production: a comparison of the theory of planned behavior and the Value-Belief-Norm Theory. J Environ Psychol 68:101408

Part IV

Climate Crisis and Urban Health

Mainstreaming Biodiversity in Urban Habitats for Enhancing Ecosystem Services: A Conceptual Framework

19

Jayeeta Sen and Meenakshi Dhote

Abstract

Biodiversity endures life on this planet. It ensures the delivery of ecosystem services upon which human well-being reposes. The trajectory of sustainable growth of a nation is often measured by biological diversity, equitable distribution, judicious consumption with minimal effect on nature. Earlier, human settlements have primarily been ecologically sensitive and climate responsive ensuring a 'connect with nature'. However, after industrial revolution, development witnessed a 'disconnect with nature' especially in the urban realm. Such discordance has led to a diminishing quality of life and human well-being. With climate change being a reality, the cities need to be resilient and future ready. Biodiversity needs to be factored in to urban planning to "reconnect with nature". A variety of instruments enable mainstreaming biodiversity into urban planning. Here an attempt has been made to explore the available diverse instruments for biodiversity conservation across scales. Singapore has taken great strides in factoring biodiversity and Mumbai, a coastal city in India, that supports a population of above 12 million are facing gargantuan challenges. The idea is to develop a conceptual framework that enables biodiversity mainstreaming in coastal cities.

Keywords

Urban habitats · Urban biodiversity · Ecosystem services · Urban and regional planning · Sustainable development · Urban spatio-cultural-bioshed

19.1 Introduction

The expression 'Biodiversity' denotes the variety and abundance of life on earth. The veracity that Biodiversity sustains the biosphere and reinforces human civilization on earth is irrefutable. It has endowed humankind with a variety of goods and services and enabled them to lead a 'decent quality of life'. The term Biodiversity, has been coined by W.G. Rosen in 1985 and was contracted from the expression 'biological diversity'.

The Convention on Biological Diversity (CBD) is an outcome of the 1992 Conference on the Environment and Development - "Earth or the Rio Summit". The CBD was convened for the global protection, conservation and enhancement of biodiversity with triple objectives: (1) Preservation of Biological Diversity, (2) Sustainable use of its components, and (3) Fair

J. Sen (✉) · M. Dhote
Department of Environmental Planning, School of Planning and Architecture, New Delhi, India
e-mail: jayeetaplanner@gmail.com; m.dhote@spa.ac.in

U. Chatterjee et al. (eds.), *Climate Crisis: Adaptive Approaches and Sustainability*, Sustainable Development Goals Series, https://doi.org/10.1007/978-3-031-44397-8_19

and equitable sharing of its natural resources. The CBD for conservation and protection of biodiversity endorsed the precautionary environmental principle. The Article two of the Convention defines Biodiversity as "the variability among living organisms from all sources including, inter alia, terrestrial, marine and other aquatic ecosystems and the ecological complexes of which they are part; this includes diversity within species, between species and of ecosystems."

Biodiversity relies heavily on the abiotic components of the geographies they are part. Biodiversity is an integral component of a defined structural and functional unit when expressed on a spatial scale is known as the "ecosystem". The goods and services that we obtain from ecosystems are known as ecosystem services. The four major components of the ecosystem services are (a) Provisioning services, (b) Regulating services, (c) Habitat and supporting services and (d) cultural ecosystem services [Millennium Ecosystem Assessment (M.E.A. 2003), TEEB Foundations 2010]. Biodiversity is not an ecosystem service by itself but rather supports the supply of all services and builds resilience within the community. The conservation and enhancement and restoration of Biodiversity is linked with Sustainable Development Goals 2030.

19.2 Biodiversity and Sustainable Development

Humankind has always been associated with nature and hence biodiversity for sustenance. However, in the modern times the pursuit of managing global biodiversity gained grounds in the United Nations Conference on Human Environment held in Stockholm in 1972. The expression 'biological diversity' had also found a mention in the United Nation's report on "Our Common Future", which emphasized the need for conserving biodiversity and how biodiversity can benefit mankind with respect to species, ecosystem and genetic diversity. The action plan of the Earth Summit – Agenda 21, also advocated for preservation of Biodiversity.

The 20 Aichi Biodiversity Targets under the five strategic goals were adopted by the Conference of Parties to the Convention on Biological Diversity in 2010 in Japan. The Strategic Plan along with the Aichi Targets were adopted for conservation of biodiversity vide decision X/2 that provided a framework for conservation of biodiversity between 2011 and 2020, and also developed the vision for the Plan "Living in Harmony with Nature" by 2050. There exists a link between the Sustainable Development Goals 2030 and the Strategic Goals and Aichi Biodiversity Targets (2011–2020).

According to the Global biodiversity Outlook 5 (2020), some of the Strategic Goals and Aichi Biodiversity Targets were directly compatible with the targets of Sustainable Development Goals (SDG) 2030. It may be noted that SDG 14 that supports life below water directly co-related with Aichi Biodiversity Target 3 (Incentives reformed under Strategic Goal A relating to addressing of underlying causes of Biodiversity Loss and mainstreaming biodiversity across government and society), target 6 and other relevant Aichi targets. Similarly, other Aichi targets are also interlinked with other SDGs.

India, rich in biodiversity, is one of the seventeen mega diverse countries of the world. India has made an endeavour at the national level to co-relate the Aichi targets and the SDG 2030. The National Report on Implementation of India's National Biodiversity Action Plan–An Overview (2019) has illustrated the connect between the SDGs and the National Biodiversity Targets that have been derived from the Aichi Biodiversity Targets of the CBD.

19.2.1 Post-2020 Global Biodiversity Framework

The recent Conference of Parties (COP) 15 convened in December 2022 drafted the new Global Biodiversity Framework (GBF) 2022 with a vivacity to obstruct biodiversity damage (CBD, GBF 2022). The Post-2020 GBF has been devised to further augment the progress made by the Strategic Goals and the Aichi Biodiversity targets. According to Veronica Lo, Nicole Jang of the International Institute for Sustainable Development (December 2022), the Post-2020 GBF primarily emphasized on (a) Conservation of 30% World's terrestrial and aquatic ecosystems and restoration of another 30% of Degraded Ecosystems; (b) Adoption of Nature-Based Solutions and ecosystems-based approach in all climate change mitigation, adaptation and disaster risk reduction measures; (c) Ensure Human Rights-based approach by acknowledging the roles, rights and contribution of indigenous people and local communities in biodiversity conservation and sustainable utilization of the natural resources and also maintain the right to development for all; (d) emphasizes on adequate means of implementation and financial resources and necessitates the convening of a global biodiversity fund for implementation of the set goals and targets.

However, realization of these ecological targets shall necessitate, implementation across scales and integration into the wider web of economic development. The Framework of Convention on Biological Diversity supports biodiversity inclusive spatial planning. Further, it advocates enhancing the areas under blue green infrastructure in densely populated urban areas.

19.3 Biodiversity, Human Health and Human Well-Being

Biodiversity is the foundation upon which ecosystem services reposes and is intricately linked to human health and well-being. Health is one of the invaluable resources of human population as it enables achievement of all other entities that define a good quality of life and is not the mere objective for living alone.

There exists an inevitable link between biodiversity, mental health and physical health. According to a Lancet Planet Health paper (Volume 5 November 2021), the pandemic (CoVID 19 outbreak) bears an inextricable link with biodiversity damage and degraded ecosystems. The direct drivers that demonstrate such phenomena are land use/land cover change, climate change, trading in wild life and livestock management. The strategies to contain the negative impacts of pandemic must include environmental safeguards for obstructing damage to biodiversity and fragile ecosystems. The One Health Approach is a step in this direction and India too has supported the one health approach within the National Mission on Biodiversity and Human Well-being. In addition to this, biodiversity is often found to be fundamental to preservation of culture, traditions, knowledge, customs and value systems that contribute to human and cultural well-being (WHO, CBD 2015). Clark in 2014 further identified the direct and indirect linkages between health component, biodiversity and culture. The social and the cultural dimensions of biodiversity co-relate with mental health. Further arguments in favour of quality of life (QOL) and public heath also support preservation of biodiversity. Health is defined by the World Health Organization (WHO) as a state of complete physical, mental and social well-being. The environmental component of the WHO QOL included determinants like safety, security, access to resources and interaction with local environments as significant constituents. 'Quality of life' concept is a significant component of spatial planning approach. The green space management that entails a hierarchy of open spaces across urban planning scales is an instrument for integrating biodiversity. The connect with nature and biodiversity that was more direct and consumptive in traditional rural culture transmutes to a more abstract and indirect one in a contemporary urban culture. The relationship can be primarily co-related with physical and leisure activity that impact physical and mental health.

19.4 Biodiversity and Climate Change

Climate Change is a gargantuan challenge confronting human kind. The frequent exposure to climate extremes has led to the vulnerability of natural and human-made systems. This has further jeopardized the life of the underprivileged population whose livelihoods are directly dependent on biodiversity and they also are deprived of the basic infrastructure too. The report on the Ecosystems and Human Well-being, A Framework for Assessment, 2003 drafted by the MEA 2005 has already illustrated Climate Change as one of the direct drivers of biodiversity loss. According to the CBD, the climate change impact on species element of biological diversity ranges from variability in distribution of species, augmented rates of extinction of species, changes in the extent of growing season of plants and agricultural crops, changes in flowering timings and reproduction timings in flora and fauna.

Further, biodiversity plays a pivotal role in the global carbon cycle and enables maximising the mitigation and adaptation potential for addressing climate change. The recent six assessment report of Inter-governmental Panel on Climate Change (IPCC 2022b) shows that most of the anthropogenic green-house gas (GHG) emissions across the globe were from the Agriculture, Forestry and Other Land Use (AFOLU) sector and the driver responsible for such emissions is deforestation. Biodiversity also enables adaptation to climate change and develops resilience. Ecosystems like Mangroves not only act as carbon sinks but also aid in disaster risk reductions. The sixth Assessment report of IPCC (2022a) has identified almost 127 key risks.

India too has made endeavors in diminishing the impacts of Climate Change. It has drafted the National Action Plan on Climate Change (NAPCC) that has eight major missions—that focus on Solar power, Sustainable Habitat, Water, Himalayan Eco-system, Green India, etc. The NAPCC that commenced on 30th June 2008 designed the country's development trajectory by devising strategies that would enable India to adapt to climate change and enhance the national ecological security and protect livelihoods. The States in India have also developed their respective Climate Change Plans like the Madhya Pradesh State Action Plan on Climate Change (2014) that focuses on forests and biodiversity, water, energy, urban and rural development etc. The focus is primarily on adaptation measures. The District Climate Action Plans like the Climate Change and Environment Action Plan for Pune District (2022) and the City level Climate Action Plans like the Climate Resilient Cities Action Plan—Siliguri that focuses on the internal and external stressors through vulnerability and risk assessments have also been developed.

Understanding that human-made systems are equally vulnerable and are at risks to climate hazards, biodiversity conservation, enhancement and restoration with tailor-made nature-based solutions that serve as instruments for safeguarding cities need to be integrated in development planning to arrest the impacts of climate change that are currently demonstrated through climate extremes like excess rainfall, urban and coastal floods, droughts, heat, sea level rise etc. Green and ecological infrastructure in the built environment have greater capacities to provide environmental remunerations. The natural ecosystems like mangroves, coastal wetlands and marshes and sea grasses sequester carbon in their biomass and act as reservoirs of carbon and works as long-term pools for carbon. The mudflats and salt-pans act as buffers and the sand dunes and mangroves further enable reduction of climate hazards by attenuating the wind speed and wave heights.

19.5 Urban Biodiversity, Ecosystem Services and Urban Planning

The cities are "by far the largest visible expression of human activity on earth" and the most "efficient means of spatial organisation yet devised" because of their ability to accommodate a large number of people and economic activities. They are the centres of economic hub that endorse growth, innovation and development enabling social and economic sustainability. Cities are primarily modified and managed habitats with a high built-up component. Though according to Naji Akbar (2020), ancient cities were constructed in accordance with nature manifested through climate responsive architecture, compact urban forms and well-knit green infrastructure with large open spaces, farms and water bodies, the disconnect with nature was demonstrated with the advent of the Industrial Revolution in approximately 1750 A.D., urbanization played a decisive role in socio-economic, socio-cultural and socio-ecological metamorphosis of human societies and settlements. The over-reliance on technology post-industrial revolution has led to alienation of urban dwellers from nature and has caused a disconnect.

The United Nations website on Resource Efficiency and Green economy states that 80% of the global population will be urban by 2050. Even after the world population stabilizes around 2050, the urban population will continue to grow. It is certain, that our future will be increasingly urban and urbanization, therefore, is inevitable. However, cities have been highly criticized for their consumptive behaviour. According to the United Nations Environment Programme the cities occupy 2% of the earth's surface, but the inhabitants use 75% of the natural resources of the planet. This gargantuan consumption of natural resources leads to 78% of carbon emissions, 60% of residential water use, and 76% of the wood used for industrial purposes. Conventionally too, the ecological narrative has always held cities responsible for habitat destruction, degradation and fragmentation as they cause massive land use and land cover change and modify natural habitats. This narrative ought to be transformed to an argument in favour of cities. Such a case was floated by J. Wu of School of Life Sciences and Global Institute of Sustainability, Arizona State University (2013), which stated that cities enable biodiversity conservation by accommodating large number of people within a defined administrative jurisdiction that support them with livelihood and necessary infrastructure and allow in principle at least to save land for conservation of nature. Further cities harbour biodiversity within the urban fabric and every land use within the urban jurisdiction is capable to support biodiversity whether the habitats are natural, semi-natural or man-made. The cities need to be reconceptualized and replanned as biodiversity hubs that are climate resilient and are on the trajectory of sustainable development.

According to Uslu and Shakouri (2013), urban biodiversity is the variety and richness of living organisms including genetic variation and habitat diversity found in and on the edge of human settlements. This biodiversity ranges from the rural fringe to the urban core. Though traditionally, biodiversity primarily focused on cataloguing of species—flora and fauna within a habitat, in recent times, the scientific progress within the domain has enabled the lexis 'urban biodiversity' to go beyond and encompass vegetation and vegetation techniques as well as all green infrastructure that act as habitat and harbour biodiversity within the urban fabric (ICLEI Local governments for Sustainability, Canada). This enabled supporting a diverse range of habitats as mentioned in the Table 19.1 that propagate a wide variety of flora and fauna. These diverse range of habitats not only enhance biodiversity but also increase local food security and water security, enable water flow regulation and runoff mitigation, wastewater treatment, sequester carbon, purify air and reduce noise, regulate urban temperature and microclimate conditions. These enable reduction of the internal stressors reducing vulnerability to climate extremes.

Table 19.1 The natural, semi-natural and man-made habitats of cities

Biodiversity in natural areas within the cities	Biodiversity in semi-natural areas within the cities	Biodiversity in man-made areas within the cities
Rivers Floodplains Riparian Vegetation Ox Bow Lakes, Paleo channels, Marshes and Natural drainage channels Inland Waterbodies and wetlands Ridges Forests and Wilderness Mangroves Estuaries and Creeks Coastal Wetlands Hills and Mountains	Railway Yards and railway lines Urban agriculture Deserted industrial complexes Along the verges, medians, green spaces on traffic islands, bus depots Water Treatment Plants Sewage Treatment Plants Green spaces of Institutional Areas, zoos and botanical gardens. In Landfills	Residential Areas (buildings, housing plots, neighborhoods) Solid waste dumps and landfill sites Kitchen Gardens, Gardens of plotted houses, Totlots, Street Trees, Neighbourhood parks, Allotment Gardens Stormwater channels and wastewater channels Green and Biodiverse roofs Green and Living Walls Constructed Wetlands Hydroponics and Aquaponics

Source: As adapted from Dhote Meenakshi, 2005, Land use strategy for conservation of biodiversity, Ph.D. thesis, School of Planning and Architecture

According to a background paper of World Wide Fund for Nature (WWF) on Cities and Climate Change in the Indian Context, urban biodiversity enables provision of ecosystem services and aids climate change mitigation and adaptation measures. The paper cites that enhanced biodiversity increases resilience. The Intergovernmental Panel on Biodiversity and Ecosystem Services (IPBES 2019) in its Global Assessment of Biodiversity and Ecosystem Services acknowledges "Building sustainable cities" as one of the approaches for enhancing biodiversity. Urban ecosystem services delivered by the natural areas and public greens especially within planned development in cities enable people to "connect with nature" and also at times to "reconnect with nature".

Urban areas are primarily governed by the tenets of urban and spatial planning. Urban Planning guides the orderly development of space and its social allocation through adequate scientific and technical knowledge and drafting of meticulous land use plans that minimize conflicts. The discipline has been connected with nature largely through the principles and theories of environmental planning that seeks to balance social, economic, environmental, cultural, technical and financial considerations collated within the environmental governance framework. Urban Planning is demonstrated through spatial planning frameworks/physical plans/Master Plan. A master plan is a statutory document that provides a framework for a definite period to guide future urban development that includes recommendations and proposals for a city's population, economy, housing, transportation, community facilities, and open spaces. It is based on public input, surveys, planning initiatives, existing development, physical characteristics, and social and economic conditions that is expressed primarily through a land use plan (URDPFI 2015). The city Master Plan is the instrument that has the potential to integrate biodiversity within the urban space.

Spatial Planning studies too are multi-scale and environment transcend all scales ranging from the national to regional to local levels. Further at the urban level, it can descend to the zonal level, neighbourhood level and site or building level. Urban Biodiversity, however, has to be understood and studied in conjunction with the spatial planning scales that is commensurate with the specifications and norms as prescribed in the Urban and Regional Development Plan Formulation and Implementation Guidelines (URDPFI).

Further, the similarity between the two scales cannot be achieved as ecological scales rarely reconcile with the spatial planning scales. This is further emphasized as many environmental and biodiversity related issues emerge from the incompatibility of scales. The scales at which ecological processes operate are not congruent with the scales at which the socio-economic needs of human population investigated. In other words, the scale at which the decisions on ecosystems are made are not always appropriate and the results are significantly influenced by the interplay of multiple drivers at varied scales. The drivers are further dependent on the diverse socio-ecological, socio-political, socio-economic and socio-cultural milieu across space. This enunciates the fact that for environment and biodiversity-related matters, it is inadequate to focus on a single spatial unit (urban or metropolitan region) and scale as the interactions are often not fully understood. This disparity may affect potent decision-making and have detrimental impacts on the ecosystems, supply of ecosystem services and overall human well-being. This argument of scalar impact of biodiversity and ecosystems and the interaction between multiple drivers draws support from the literature inscribed in the Framework for M.E.A., Ecosystems and Human well Being (2003).

19.6 The Need for Mainstreaming Biodiversity into Urban Planning

It may be reasoned that the conservation of urban biodiversity is attainable by way of 'Mainstreaming' or 'including' biodiversity into urban planning. The need for mainstreaming biodiversity within the urban planning process found support by the decisions of the COP to the CBD in 2010 wherein the 193 parties to the Contract adopted Decision X/22 in Aichi Nagoya. The Decision X/22 of the tenth meeting of the Conference of the Parties (COP 10) provided guidelines to facilitate local governments and enable implementation of CBD at the local scale. The Scientific Advisory Technical Panel of Global Environment Facility (identified three broad approaches in this regard (a) Mainstreaming of biodiversity into economic sectors; (b) Mainstreaming into cross-sectoral policies and strategies; (c) Mainstreaming into spatial planning (CBD 2012).

Though the argument for integration of urban biodiversity within the urban development process gained momentum through the urban ecology debate, the process of mainstreaming largely remained blurred. The efforts of conservation progressed in silos and the vocabulary of the two domains often remained not understood and at times misunderstood. Though, globally, several mainstreaming projects have been conceived and operationalised, in India incorporation of biodiversity concerns in development planning and in urban land use planning is a relatively recent phenomenon. Mainstreaming can be achieved through different approaches like ecosystems approach, ecosystem services approach, spatial planning approach and also through urban biocultural divesity approach. Whilst the Ecosystems approach as advocated by the CBD for fulfilment of its objectives focuses on the conservation, and management of biodiversity through integrated management of land and water resources and all living organisms on earth, the ecosystem services approach is an utilitarian approach that treats urban nature and urban biodiversity as foundation for delivering goods and 'services' to humankind that can be valued in monetary terms and can be part of the decision making process. The urban biocultural diversity approach is an advancement over the ecosystem services approach that emphasizes the link between biodiversity and cultural diversity within cities and between city and city regions. It reconnects people with nature promoting a new urban culture that has dynamic narrative with pro-environmental behavior and creates place-based values with nature. The approach allows for conservation of biodiversity by recognizing allotment gardens and also includes modification of relationships that allows evolving of new socio-cultural patterns (Vierikko, Kati et al. 2016, Stalhammar et al. 2021).

However, in the urban realm it is urban planning that guides development and hence the mainstreaming of biodiversity into it shall necessitate taking cognizance of other approaches mentioned above. In India, Biodiversity traditionally has been governed by the environmental governance framework. This translates that the instruments that are applicable for conservation and protection of environment when implemented appropriately shall conserve biodiversity. The current state of affairs, however, does not reflect the picture envisaged thus creating the need for separate enactment and instruments for conservation of biodiversity. Whilst India has managed to enact the Biodiversity Conservation Act 2002 and the National Biodiversity Authority has been instrumental in developing the the National Biodiversity Strategy and Action Plan (NBSAP) in 2008 and 2014, the local level instruments and initiatives are still deficient. The respective states have formed their State Biodiversity Action Plans. Further, the State Biodiversity Boards and the Biodiversity Management Committees are responsible for developing the People Biodiversity Register (PBR) in both the rural and urban realm. The instrument documents biodiversity and traditional knowledge. However, such endeavors in urban settings are still deficient. The URDPFI Guidelines 2015 mentions the need for conserving biodiversity and supports the City Biodiversity Index and Local Biodiversity Strategy and Action Plan as a subset of it indicating that Biodiversity as a theme has arrived in the urban planning process. The 2019 sixth report of the country to the CBD further acknowledged the need for conservation of Urban Biodiversity. The post-2020 framework also supports biodiversity inclusive urban planning. Mainstreaming of Biodiversity into Urban Planning is advocated to create a convergence between the processes of urbanization and urban development and biodiversity conservation. The convergence will enable a dialogue between the two policy spheres, which are otherwise progressing in silos.

19.7 Case Study

The study has been primarily conducted through secondary literature following a methodical framework and has been conducted mainly in three stages. The first stage constituted of the background study, exploring the connect between biodiversity climate change and human well-being, sustainable development and the need for mainstreaming biodiversity into urban planning. The second stage assessed the biodiversity conservation framework and instruments as operated in the city of Singapore for conservation of Urban Biodiversity. Further, a case study was taken up in the Indian context for the city of Mumbai that is again located within a coastal ecosystem. This was done to explore how biodiversity can be integrated in Indian coastal cities. The third stage details out the key premises of the conceptual framework and the typology of instruments.

19.8 Singapore

In the recent past endeavors have been made by the City-State of Singapore to integrate biodiversity in its urban planning and management and enhance Urban Greens. The accomplishment has been possible through meticulous Biodiversity Conservation and Planning by the National Parks Board Singapore (NParks) and Land Use Planning by Urban Redevelopment Authority of the City-State. Singapore is located at the end of Malayan Peninsula and is a part of the Indo Malayan biogeographic Realm. Physiographically, Singapore has a very low-lying terrain and 75% of the island lies 15 m above sea level. Singapore was originally home to the tropical rain forests (Dipterocarpus forest with species of Shorea). The City-State today supports a population of approximately 6 million.

19.8.1 Biodiversity Planning in Singapore

The NParks is the organization responsible for conserving, restoring and enhancing biodiversity within all natural and human-made ecosystems in the City-State. The NParks is also the custodian of four nature reserves (Bukit Timah Nature Reserves, Central Catchment Nature Reserve, the Sunghi Buloh Nature Reserve and the Labrador Nature Reserve) and 400 parks with strong streetscapes and roadside greenery that contributes immensely to the Urban greens. The National Parks Board in 2018 managed more than two million street trees in Singapore and the city stands today as one of the greenest cities in the world (Alex Thiam Koon et al. 2019). According to NParks Singapore, the city possesses an estimated 23,000–28,000 species of terrestrial organisms and 12,000–17,000 marine organisms, making up over 40,000 kinds of nonmicrobial organisms. Further, the city boasts of having more than 390 species of birds and at least 2100 native vascular plants, of which more than 1500 species are classified as extant in Singapore. The NParks makes enduring efforts to retain the native and endemic species and natural ecosystems in a highly urbanized City-State.

The country has been one of the forerunners of biodiversity inclusive urban planning and have developed several biodiversity planning instruments and potent regulatory framework for safeguarding and strengthening biodiversity conservation. The NParks has been instrumental in developing The Nature Conservation Master Plan (NCMP) that includes the Habitat Enhancement Restoration and Recovery of species in terrestrial, freshwater and marine habitats and serves as an unswerving guidance document for conservation of Biodiversity. The document. Further, there are a variety of biodiversity planning instruments like BIOME that is an online database on local species allowed to be accessed by the public, instruments like SGBioAtlas, Community Stewardship and Outreach in Nature. The efforts are also put in by the Ministry of National Development, Urban Redevelopment Authority, National Environmental Agency Public Utilities Board, Housing Development Board, and the National Parks Board in developing Singapore into a 'city in a garden'. Singapore is an archetypal of Biophilic urbanism and the statement is supported by Dr. *Lena Chan of* N*Parks and Perter* Newman (2014).

In addition to this, NParks has developed a Park Connectors Network and a City Biodiversity Index or CBI that aids planning and monitoring of biodiversity. The City Biodiversity Index is a self-assessment tool for cities to evaluate and monitor the progress of their biodiversity conservation efforts against their own individual baselines. The Index has a total score of 112 points based on Native biodiversity in the city, ecosystem services emanating from biodiversity, and governance and management (Chan et al. 2021) . The Singapore Index on Cities Biodiversity (SICB) 2021 is a comprehensive tool for measuring the performance of cities in conserving and managing their biodiversity.

19.8.2 Urban Planning in Singapore

The Urban Planning process is guided by the formulation of a Concept Plan, a Master Plan and Development Controls. The Concept Plan is long-term strategy document developed by Urban Redevelopment Authority (URA) and is formulated under the aegis of the Ministry of the National Development of Singapore is the Department responsible for Singapore's Land Use Planning. The Concept Plan constitutes of broad strategies, identifying land for different uses and needs, and establishing Singapore's overall development pace underscored by the principles of sustainability. The broad strategies and proposals set out in the Land Use Plan are translated into detailed plans in the Master Plan, which guides development over the next 10 to 15 years that focuses on planning for inclusive, sustainable, and green neighbourhoods with community spaces and amenities. The Master Plan is to be read in conjunction with its Written Statement. The Master Plan is supported by Special and Detailed Control Plans (SDCP) that include Parks and Waterbodies, Public Spaces, Landed Housing Areas, Street Block, Envelop Control, Building Height and Urban Design, Conservation Areas, Conserved Buildings and Monuments, Connectivity, and Underground Plans, published by the competent Authority and unlike the Master Plan, SDCP are non-statutory plans.

19.8.3 Instruments Across Scales That Enable Biodiversity Mainstreaming

Singapore has plethora of instruments both for Biodiversity and Urban Planning that enables conservation of Biodiversity at different scales. According to L. L. Heng (2008) there are several environment related instruments too. The Active, Beautiful and Clean Water Programme (ABC programme) by Public Utility Board of Singapore under the aegis of National Water Agency is another effective instrument for building Blue Green Infrastructure. Other Instruments like that for 3 D Greening at the building level supported by the Landscaping for Urban Spaces and High Rises (LUSH) programme by the Urban redevelopment Authority (URA) has further made it possible to integrate biodiversity considerations at that scale. The LUSH programme supports vertical greenery towards the Landscape Replacement Area requirements (LRA), rooftop urban farming as a part of LRA. An attempt has been made to categorize some of the Biodiversity Planning Instruments and the Urban Planning Instruments based on their typology like legal instruments, planning instruments, which are applicable at different scales and congruent with the spatial planning approach has been considered (Table 19.2).

Table 19.2 Biodiversity and urban planning instruments in Singapore based on their typology

Instrument types	Name of the instrument	Scales
Legal	The National Parks Board Act (Chapter 198A)	City-State
	The Nature Reserves Ordinance was enacted in 1951, Nature Reserve Act in 1970, National Parks Act 1996	
	The Parks and Trees Act 1996 (Chapter 216)	
	Subsidiary legislation	
	Wild Animals and Birds Act 1985 (Chapter 7)	City-State
	Control of Plants Act (Chapter 57A)	
	Fisheries Act 1985	
	Other legislations	
	Planning Act (PA 1998)	City-State
	Public Utilities Act 1996	
	Carbon Pricing Act (CPA) 2018	
Policy	Climate Change Policy	City-State
	National Air Quality Policy	
	Energy Policy	
Planning and management	National Biodiversity Strategy and Action Plan	City-State
	Sixth report to the CBD (2015–2018) for implementation of the NBSAP	
	Nature Conservation Master Plan	
	Master Plan and written Statement 2019	
	Regional Plan—Park Connectors Network Plan-	Neighbourhood, District, City-State
	Biophilic town framework by the Housing Development board	Building, Neighbourhood
	Heritage Streets and Heritage Trees programme	Building, Neighbourhood, District
	Sky Rise Greening Initiative and the LUSH, Green Plot Ratio-Green Balconies, Green roofs and Green Facades, green walls	Building
Process	The Garden City campaign	Neighbourhood, District
	Tree planting campaign	City-State
	BIOME and SGBioAtlas	Building, Neighbourhood, District, City-State

Source: Developed by authors from multiple sources

19.9 Mumbai

The second case study pertains to Mumbai, located on the western coast of India. Going by the regional classifications, the city can be said to be located within the Biogeographic Zone of 5(A) Western Ghats and Malabar Plains. The temperature varies between 27 °C and 35 °C with an average mean annual rainfall of over 2000 mm and with relative humidity of 60–70%.

The urban biodiversity of Greater Mumbai Municipal Corporation may not be well understood by restricting the study to the natural and human made ecosystems exiting within the legal boundaries of Mumbai city alone. The species and the diverse ecosystems forests, costal and riverine extend beyond the legal boundaries of the Mumbai city and are part of various watersheds and larger hinterland. The rivers that drain the city and the hinterland have their origins often in the adjacent Sahyadri and the Western Ghats, which is the biodiversity hotspot of the country. The urban biodiversity too relates with the wider natural landscape.

Furthermore, it may be understood that these urban areas are largely governed by the spatial planning and urban planning approaches. The city region may serve as the functional area upon which the city of Mumbai may be dependent for flow goods and services. The city region may also absorb the excess population by providing housing and infrastructure facilities and, therefore, may influence the commuting dynamics. The insatiable demand for land may have an impact on biodiversity within terrestrial and aquatic habitats in the city region if not planned. In case of Mumbai, this city region has been delineated as the Mumbai Metropolitan Region by the Mumbai Metropolitan Regional Development Authority.

19.9.1 Mumbai Metropolitan Region

Mumbai as a city is largely contained within the Mumbai Metropolitan Region (MMR) that spreads over the districts of Palgarh, Thane, Raigad and Greater Mumbai primarily. The Mumbai Metropolitan Region (MMR) covers an area of over 4355 sq. km. approx. and according to the MMR Plan 2016–36, the metropolitan region had a total population of 22.8 million people in 2011. Whilst MMR is highly urbanized, about 27% of the population Mumbai Metropolitan region live in slums indicating that they are devoid of basic physical infrastructure. According to the Environmental Status report of Mumbai Metropolitan Region, (2015) in MMR only 59% of 4753 million litres per day of wastewater that is generated is treated through primary and secondary treatment mechanisms. This again have a negative impact on the rivers, mangroves, and the coastal ecosystems.

The total area of wetland present in MMR is around 564.7 sq.km. The MMR has three protected areas, one eco-sensitive zone and reserved and protected forests. The Sanjay Gandhi National Park, Karnala Bird Sanctuary, Karnala Panvel, Tungareshwar Wildlife Sanctuary, Tansa Wildlife Sanctuary, Phansad Wildlife Sanctuary, and the Matheran Eco Sensitive Zone.

19.9.2 Mumbai Metropolitan Regional Plan and Greater Mumbai Municipal Corporation Spatial Plans

The Mumbai Metropolitan Regional Development Authority has prepared the Mumbai Metropolitan Regional Plan 2016–2036, a statutory spatial planning instrument for Regional Planning. The Plan envisages various sectors like the Growth Centres with urbanizable zones and Industrial Zones, Blue Green Spaces and developing a network, Green Zones 1 and 2 with low intensity development to have less impact on natural asset. The Floor Space Index (FSI) for each of the zones have been designated. The intensity of development is indicated by the Floor Space Index has a bearing on Biodiversity. The instruments like Coastal Regulation Zone (CRZ) though provide protection to the immediate mangrove habitats are largely influenced by the anthropogenic activities within the matrix.

The city of Mumbai has recently notified its Development Control and Promotion Regulations (2034), a statutory spatial planning instrument to regulate and have orderly development in the city. The Plan wishes to develop Mumbai as slum free city and periodize affordable housing. The higher FSIs along creeks may inflict pressure on the mangrove and the creek ecosystem.

19.9.3 Mumbai Climate Action Plan 2022

The Mumbai Climate Action Plan 2022 (MCAP 2022) prepared by Brihan Mumbai Municipal Corporation in partnership with C40 cities and World Resources Institute shows that the total GHG emission from Mumbai city is 23.42 million tons CO_2 e with a per capita emission of about 1.8 tons CO_2 e (with base year 2019). Though Mumbai has rich biodiversity assets, the citizens are deprived of their right to greens and only a mere 3.7% open space is accessible to the people of Mumbai. The per capita availability of open space is a meagre 1.8 m^2 (mentioned in the MCAM 2022) as compared with 10–12 m^2 prescribed in the URDPFI guidelines making Mumbai the least green metro city. All these make the city highly vulnerable.

According to the MCAP 2022 the ecological assets of MCGM comprise the Protected areas of Sanjay Gandhi National Park, the Aarey Forests, coastal wetlands, Mangroves, mudflats and urban greens (trees outside forest) within the planned areas. The Plan has made an assessment of the carbon sequestration capability of the vegetation (2016–2021 period) within the legal boundaries of the city. Whilst the Protected Areas removed 3448.74 tonnes of CO_2/year (from forest remaining category), 308.80 tonnes of CO_2/year is removed because of the net gain of forest and 1067.99 tonnes of CO_2/year is emitted due to loss of forest. The Mangroves remove 87622.08 tonnes of CO_2/year (of the Mangroves remaining category) and the net gains in Mangroves remove about 4569.95 tonnes of CO_2/year and 1572.06 tonnes of CO_2/year is emitted due to loss of mangroves. In the trees outside Forest category that covers 24 wards of Mumbai the annual carbon removed is 76,991.35 tonnes of CO_2/ year while carbon emissions from the same category is to the tune of 19,640.899 tonnes of CO_2/year clearly demonstrating that the urban greens are instrumental in delivering the regulating ecosystem services.

Further, the city of Mumbai ranks fifth globally amongst world's most flood prone cities that has been demonstrated by the deluge of 2005. The city also has unauthorised colonies and slums of high densities devoid of tree cover. The Plan proposes to enhance the per capita greens to 6 m^2 per capita. In the future actions, to be taken the plan has upheld the role of multifunctional urban greens and biodiversity in imparting resilience to the city.

19.9.4 Instruments Across Scales That Are Applicable to Environment and Biodiversity Conservation in Mumbai Metropolitan Region

In India there is an abundance of instruments for protection of environment. However, there are very few instruments that are available for protection of urban biodiversity per se. Some of the instruments that are currently being used to protect the environment and Biodiversity of Mumbai Metropolitan Region and Greater Mumbai Municipal Corporation have been listed below based on their typology. The instruments that have been found to be congruent with the spatial planning approach has been considered. However, genetic biodiversity is a limitation and instruments related to the same has not been considered in the current study (Table 19.3).

The Spatial Panning instruments in Mumbai, the Regional and the Master Plans that are applicable at the Mumbai Metropolitan regional Level and Greater Mumbai Municipal Corporation needs to integrate biodiversity as layer in the plans; however, the process of integration is still wanting. Mumbai fairs poorly in terms of per capita availability of Green Space. The dense

Table 19.3 Instruments that are applicable for conservation of environment and biodiversity to Mumbai (MCGM and MMR)

Instrument type	Name of the instrument	Scale
Policy	Convention on biological diversity	Global
	Convention on migratory species	
Legal	Directive principles of state policy, part IV – Article 48, Article 51 (A) (g)3	National
	Indian Forest Act 1927	
	The Environment (Protection) Act, 1986	
	Bio-safety regulatory framework	
	Seeds Act, 1966	
	EIA notification 2006	
	Maharashtra biological diversity rules, 2008	Sub-National
	Matheran Eco Sensitive Zone Notification 2003	
	Maharashtra Groundwater Act 2009	
Policy	National Forest Policy 1988	National
	National Forestry Action Programme 1999	
	National Biodiversity Strategy 2008 and 2014	
	National River Conservation Plan 1995	
	Green India mission	

(continued)

Table 19.3 (continued)

Instrument type	Name of the instrument	Scale
Planning and Management	Vulnerability Atlas of India	National
	Composite water management index 2018	
	National Wetland Atlas Maharashtra 2010	
	Regional Plan 2016–2036 Mumbai Metropolitan region	Sub-National
	Zonal Master Plan for Matheran Eco- Sensitive Zone 2016– 2036	
	Coastal Zone Management Plan of Mumbai Sub-urban Mumbai, Raigad Thane, Palgarh 2019	Sub-National, Local
	Mumbai climate action plan 2022	Local
	City biodiversity index and local biodiversity strategy and action plan	Not Available
	Peoples biodiversity register	

Source: Developed by the authors from multiple sources.

development of Mumbai ought to take cognizance of the local biodiversity (both of the city Developed by authors from multiple sources and the region) and make endeavour to protect not only the existing natural ecosystems but also enhance biodiverse greens within the planned development through nature-based solution and 3 D Greening (biodiverse green roofs, roof top urban farming, sky gardens, in addition to flower beds). The city had confronted the Cyclone Nisarga and, therefore, needs to preserve its ecological infrastructure (the Mangroves and the coastal habitats) and enhance green infrastructure that enables building resilience (Fig. 19.1).

19.10 The Conceptual Framework

The Conceptual Framework has been developed for the city of a coastal ecosystem and it explores the interrelationship between the variables (identified from the case studies like the urban biodiversity, ecosystem services, urban culture, governance and instruments) and the preliminary outline of the research process (Tan et al. 2020). This enables developing a comprehensive understanding of the phenomenon of mainstreaming that facilitate integration of urban biodiversity within the urban planning framework. This necessitates a thorough understanding of the ecosystems, culture and governance framework within the delineated 'Urban Spatio-Cultural-Bioshed'. The key premises of the Framework are represented below.

19.11 Urban Biodiversity and Its Connect with City Region

As evident from the case studies, it is equally important to understand the 'City region' while understanding urban Biodiversity as nature does not respect anthropogenic boundaries. The Urban Biodiversity often is an outcome of 'spill over' effects of the hinterland especially in the peripheral areas adjoining the peri-urban areas. The peri-urban areas often support biodiversity and may constitute of critical terrestrial and aquatic habitats and ecological corridors within a watershed that need to be protected and conserved as "No development Zones" within a Development Plan.

Further, the city is also dependent on the hinterland for flow of goods and services, the production of which may have an indirect or direct bearing on the biodiversity. Furthermore, the city region being a functional zone is also influenced by the production processes and the urban airshed, built environment and the commuting dynamics. In the words of Jenifer Rae Pierce (2014, 2022) this constitutes an Urban Bioshed. However, this bioshed (in the context of the study) in addition to the layer on biodiversity is further governed by a set of instruments that are multiscale and are differentiated based on their typology. The instruments have the capability to modify human behavior [according to the literature inscribed in the Framework for Millennium Ecosystem Assessment, Ecosystems and Human

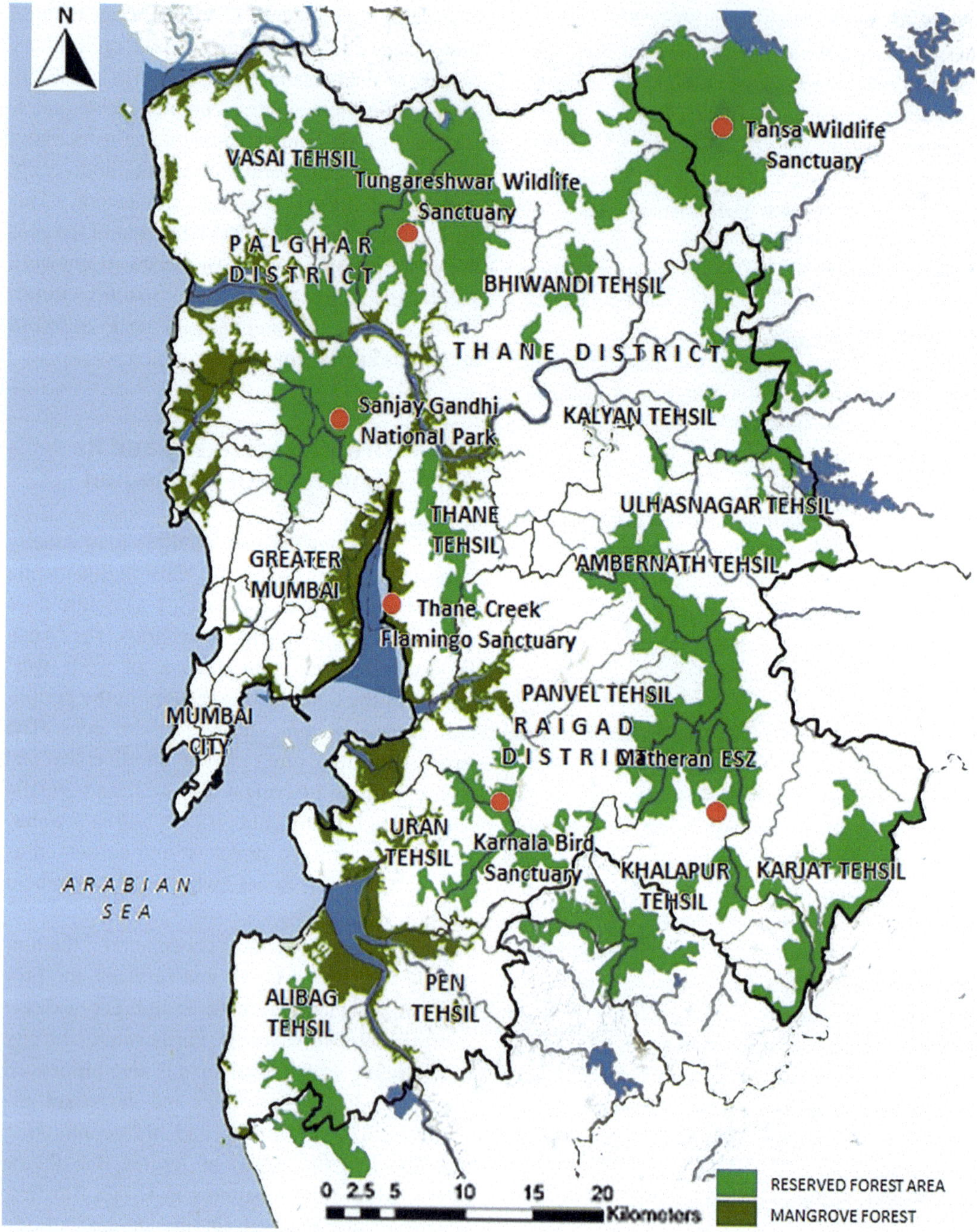

Fig. 19.1 Map showing green assets of Mumbai metropolitan region. (Source: As adapted from MMRDA Regional Plan Map 2036 https://maps-mumbai.com/mmrda-map) (Map is not accurate representative of administrative units)

well-being (2003)]. The third layer that has a bearing on the human and ecosystem interaction is traditional rural and modern urban culture. The instruments also aid in developing a 'new culture that enable biodiversity conservation. The interaction and the interconnectedness between the three spheres decide the status of Biodiversity within the urban and the rural realm. The Spatial

Urban Cultural Bioshed would necessarily be an areal unit that shall include all the critical and the core natural habitats, the ecological support areas, corridors ensuring connectivity within the matrix i.e., within the urban limits or outside in the hinterland/within the watershed and may also constitute of man-made habitats/land uses that may themselves support biodiversity (in the form of species-flora and fauna) or have spillover effects. In addition to the biophysical elements, it shall also include the socio-economic, socio-cultural and socio-political domains with the governance mechanisms and instruments within the defined areal unit. Figure 19.2 shows the interconnectedness between the different spheres within the Spatial Urban Cultural Bioshed.

19.12 Developing the Baseline for the City and Understanding the Multiscalar Dimensions of Urban Biodiversity and Ecosystem Services

Urban ecosystem services that benefit humankind are much dependent on the abiotic and biotic components. Therefore, for understanding the existing status of biodiversity and urban ecosystem services within the city the following themes of Bio-geophysical elements, Socio-economic elements, Socio-cultural elements and Planning and Governance Mechanisms should be taken care of. These will enable developing the base-

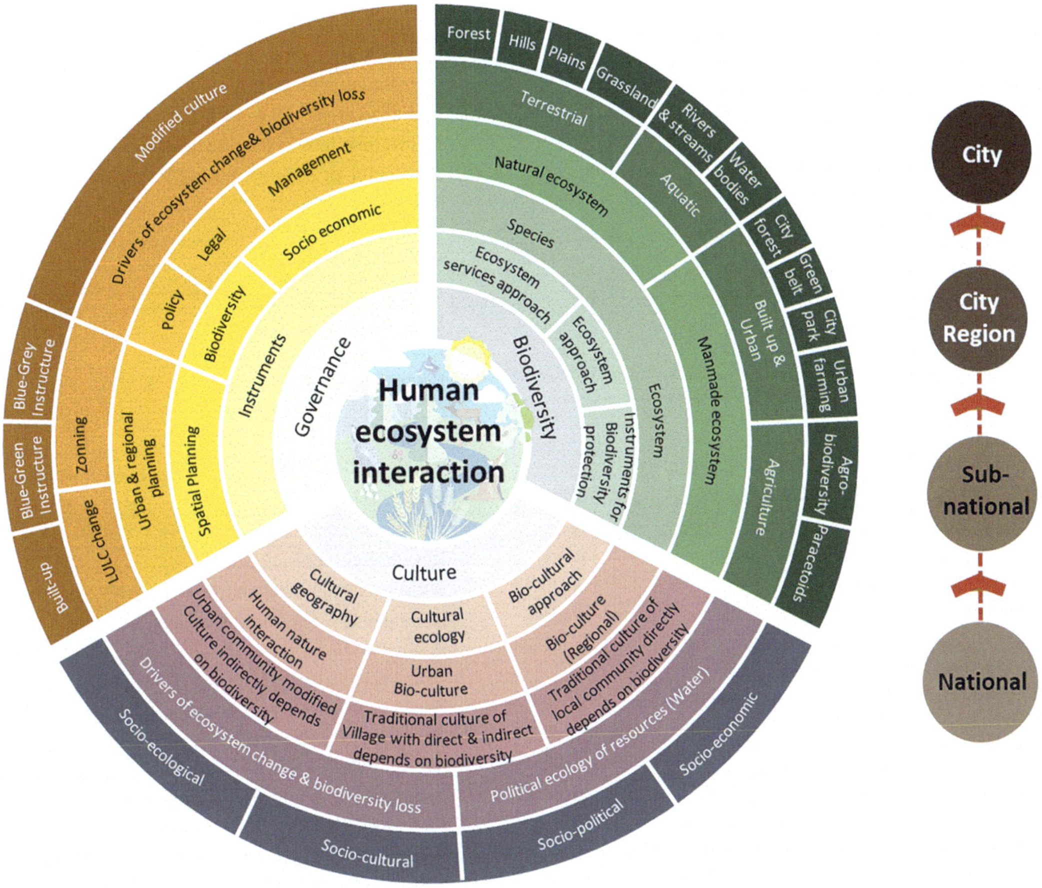

Fig. 19.2 The inter-relationship of biodiversity, culture and instruments within an urban spatio-cultural-bioshed. (Source: Developed by authors)

line for the city taking cognizance of its environmental resources. Besides the themes may also emanate from the variety of frameworks and instruments like the Singapore City Biodiversity Index and approaches that may enable the conservation of biodiversity like the Ecosystems Approach, Ecosystem Services Approach, Urban Bio-Cultural Diversity Approach and the Spatial Planning Approach (that are instruments). The selection of the frameworks also shall depend on the applicability of the frameworks to a definite ecosystem. Some of the components for the baseline area particularly for a coastal habitat is as follows (Table 19.4).

Table 19.4 Parameters for developing the baseline for cities in coastal habitats

S. No.	Component	Themes
1	Bio-physical elements	Geology and geomorphology
		Soil type and texture
		Relative relief
		Slope
		Agricultural land capability
		Surface water (rivers, inland water bodies, coastal wetlands creeks), watersheds, water quantity regulation
		Ground water (quantity and quality)
		Climatic conditions (temperature, wind-speed, wind-direction, precipitation)
		Area under natural assets
		Protected areas and ecologically sensitive zones
		Macro invertebrates in streams/ waterbodies
		Percentage of endangered and Vulnerable species
		Climate regulation
		Sea level rise and shoreline changes
		Degraded land and water bodies
2	Socio-Cultural and Socio-Economic Elements	Poverty level
		Literacy rate
		Livelihood
		Community participation
		Languages, customs and traditions
		Moral and traditional values with respect to nature
3	Urban Planning elements and Governance	Urbanised area
		Residential typology, slums
		Solid waste management
		Road density, non-motorized infrastructure, public transit
		Area under urban greens and parks
		Urban agriculture
		Air quality index
		Noise levels
		Area under green buffers for industries and industrial estates
		Waste management facilities in industrial estates
		Policies and legal frameworks for protection of biodiversity (including CRZ, Mangroves)
		Budget for biodiversity conservation

Source: Developed by Authors (based on parameters for Spatial Planning Approach, Urban Biocultural Diversity Approach and City Biodiversity Index)

19.13 Instruments: Deemed Imperative for Integration of Urban Biodiversity and Ecosystem Services and Their Applicability Across Scales

The mainstreaming of biodiversity into development planning/spatial planning/urban planning is possible through diverse set of instruments that operate at varied planning scales. Instruments can be defined as concepts, frameworks, documents, standards, codes of conduct, guidelines and certification schemes, maps, parameters, indicators, plans (spatial or non-spatial) and some of the legal devices that will enable inclusion/integration of biodiversity in the development plans and processes at different scales. They are the tools/techniques or devices that restrict/regulate/allow planning and development of human societies at large and also enable protection of environment, ecology and biodiversity that ensure human well-being. Only those instruments have been discussed that have a direct or indirect bearing on biodiversity conservation and are compatible with the spatial planning and environmental authorization approach (Dhote and Sen 2020).

These instruments can further be categorized as 'Policy Instruments' that either provide information, generate awareness on issues of biodiversity and their position within the development framework, are part of the regulatory design or have an economic overtone.

The 'Legal Instruments' are instruments governing the ownership, access, and utilization of natural resources and are particularly important for the protection of biodiversity and sustainable use and management and conservation of ecological infrastructure. Therefore, the expression legal instrument enforces the formal legal nature of the instruments in the form of existing domestic law. The rules, codes of conduct, etc. that emanate from laws or acts are the legal instruments.

The 'Planning Instruments' are primarily spatial in character that operate within a given geography and defined administrative jurisdiction at different spatial planning scales and also reflects on the governance framework and institutional mechanisms and development controls necessary to mainstream biodiversity. Spatial Planning that operates within different biogeographic region across states (perspective plans), districts (administrative regions and regions based on flow of goods and services and environmental regions, therefore, can have perspective, regional, development plans or special purpose plans), metropolitan regions (Metropolitan Regional Plans like the MMR Plans of Mumbai or NCR Plans of Delhi) and urban areas (City Master Plans). The application of these instruments to planned greenfield development enable conservation of greens within the urban fabric.

The 'Management Instruments' are the devices to direct/enforce and administer the decisions pertaining to urban and environment planning. It can also be referred as elements and methods that enable the decision-makers to develop a rationale for informed choices between alternative actions. They enable retrofitment of nature-based solutions within the existing planned and unplanned development thereby integrating urban biodiversity.

Further, the 'Process Instruments' provide ways of doing something, steps that can be taken to reach a desired goal. These can refer to top-down and highly centralized processes in executing urban environment strategies, prioritization of issues and also be highly participatory in nature. These instruments advocate for increased stakeholder participation in the decision-making process and enable building of consensus thus creating new partnerships.

The instruments have been mapped and categorized into policy, legal framework and legal instruments, planning and process instruments and management instrument within the case studies. However, depending on the nature of the instrument, there may be overlaps. An application of combination of instruments shall enable conservation of biodiversity across scales.

19.14 The Framework

The frame work is a distillation of the literature review, the case studies that reflect the ecological fragility of the coastal habitats that need preservation for biodiversity and the activities within the urban jurisdiction that has a bearing on the city's natural blue green assets.. The different attributes like the biophysical elements, socio-economic and socio-cultural elements and the planning governance elements that have a bearing on the composite variables of urban biodiversity, ecosystem services, culture and governance and their interrelationships with the set of instruments based on typology decides the status of biodiversity in the city. The potential zones are further identified with the elements of urban bio-cultural diversity approach (that encourage process instruments) and that are also congruent with the spatial planning approach. The potential zones identified are compatible with the planning zones specified in the Master Plans. The status of the biodiversity further decides the type of instruments in the form of nature-based solutions to be designed for fulfilling the specific need (Fig. 19.3).

19.15 Conclusion

The above discussion was an endeavour to explore the process of mainstreaming within urban planning. Whilst the post 2020 Global Framework supports biodiversity inclusive urban planning, the dichotomy that exists between biodiversity conservation and economic development need to be resolved and an ecologically balanced development promoted. It may further be noted that urban biodiversity may not be similar to that of the pre-urbanization levels but shall constitute of species that can thrive within modified urban habitats and continue to support ecosystem services. The natural areas within the cities that harbor greater biodiversity can be conserved as No Development Zones both within the legal boundaries of the city and in the city region with which the species have a connect. Furthermore, since a symbiotic relationship

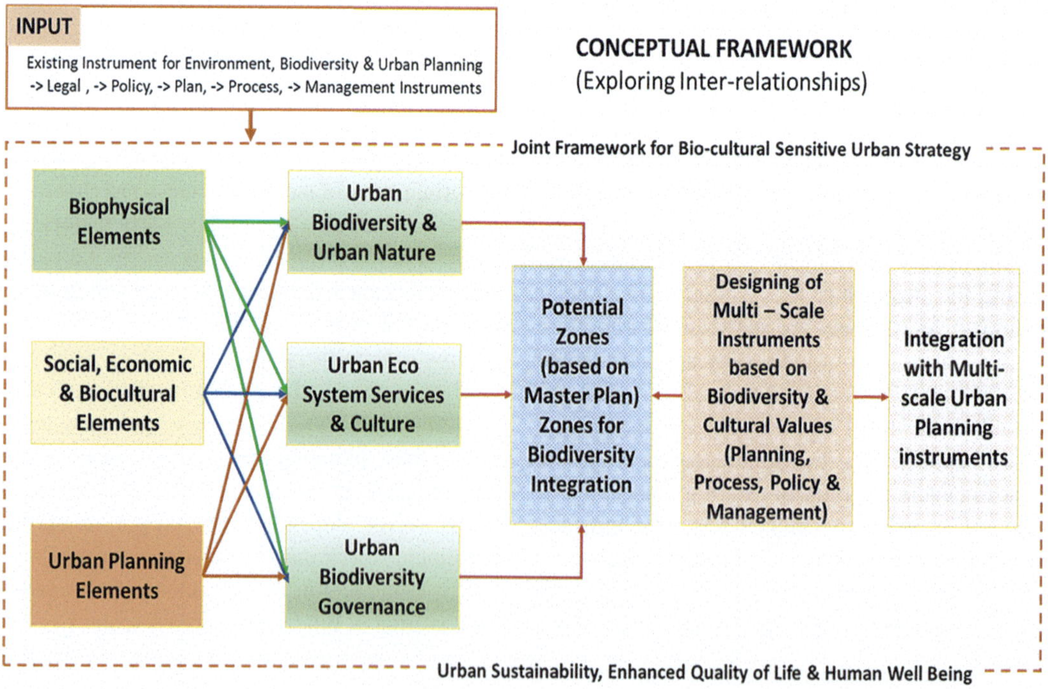

Fig. 19.3 Conceptual framework: Exploring inter-relationship. (Source: Developed by authors)

exists between the city and its region and this relationship should not be perceived as only functional but also ecological. The critical habitats and ecological support areas that sustain biodiversity in the region need to be identified by conducting scientific studies. This association is further strengthened with the aid of variety of instruments that operate across scales in order to make development in conformity with environment protection and biodiversity conservation. The integration of Biodiversity within the urban planning framework will further attenuate the impact of the internal and external stressors within the legal boundaries of the city and also in the hinterland and thereby reduce vulnerability to climate extremes.

References

Akbar N, Abubakar IR, Bouregh AS (2020) Fostering urban sustainability through the ecological wisdom of traditional settlements. Sustainability 12(23):10033. https://doi.org/10.3390/su122310033

Brihan Mumbai Municipal Corporation. C40 Cities and WRI, Mumbai Climate Action Plan 2022, Towards A Climate Resilient Mumbai.

Chan L, Hillel O, Werner P, Holman N, Coetzee I, Galt R, Elmqvist T (2021) Handbook on the Singapore index on cities' biodiversity (also known as the city biodiversity index). Secretariat of the Convention on Biological Diversity/National Parks Board, Singapore, Montreal/Singapore, 70 Pages.

Cities and Biodiversity Outlook, Action and Policy (2012) A global assessment of the links between urbanisation, biodiversity and ecosystem services

Dhote M, Dr (2005) Landuse strategy for conservation of biodiversity [Doctoral Thesis, School of Planning and Architecture, New Delhi].

Dhote M, Dr, Sen J (2020) Interface between instruments of development planning and biodiversity planning and conservation. Biodiversity and livelihoods, lessons for community research in India

Heng LL (2008) A fine city in a garden – environmental law and governance in Singapore. Singapore Journal of Legal Studies

ICLEI, Local governments for Sustainability, Canada, Biodiver Cities, A Primer on Nature in Cities https://icleicanada.org/wp-content/uploads/2019/07/biodiverCities_A-Primer-on-Nature-in-Cities.pdf

International Bank for Reconstruction and Development, The World Bank (2021), https://www.thegpsc.org/sites/gpsc/files/final_urban_nature_and_biodiversity_for_cities_0.pdf (accessed on 14 May 2022)

IPBES (2019) Global assessment report on biodiversity and ecosystem services of the intergovernmental science-policy platform on biodiversity and ecosystem services. (Eds. Brondizio ES, Settele J, Díaz S, Ngo HT). IPBES Secretariat, Bonn, 1148 pages

IPCC (2022a) Summary for policymakers (Pörtner HO, Roberts DC, Poloczanska ES, Mintenbeck K, Tignor M, Alegría, Craig M, Langsdorf S, Löschke S, Möller V, Okem A (eds)). In: Climate change 2022: impacts, adaptation and vulnerability. Contribution of working Group II to the sixth assessment report of the intergovernmental panel on climate change (Pörtner HO, Roberts DC, Tignor M, Poloczanska ES, Mintenbeck K, Alegría A, Craig M, Langsdorf S, Löschke S, Möller V, Okem A, Rama B (eds)). Cambridge University Press, Cambridge/New York, pp 3–33. 10.1017/9781009325844.001

IPCC (2022b) Summary for policymakers. In: Shukla PR, Skea J, Slade R, Al Khourdajie A, van Diemen R, McCollum D, Pathak M, Some S, Vyas P, Fradera R, Belkacemi M, Hasija, Lisboa G, Luz S, Malley J (eds) Climate change 2022: mitigation of climate change. Contribution of working Group III to the sixth assessment report of the intergovernmental panel on climate change. Cambridge University Press, Cambridge/New York. https://doi.org/10.1017/9781009157926.001

Linkage between Cities Biodiversity and Governance.: Perspectives and Challenges of the Implementation of the convention on biological diversity at the city level, UNU –IAS. ISBN: 9789280845174

Lo V, Jang N (2022) The Global Biodiversity Framework's "30x30" Target: Catchy slogan or effective conservation goal?, International Institute For Sustainable Development, https://www.iisd.org/articles/insight/global-biodiversity-framework-30x30-target. (accessed 14th May 2023)

Millennium Ecosystem Assessment (2003) Ecosystems and human well-being, a framework for assessment

Ministry of Urban Development, Government of India (2015) Urban and Regional Development Plan Guidelines (2015) Formulation and Implementation – Volume 1. January 2015

Municipal Corporation of Greater Mumbai (2018) Greater Mumbai Development Plan-2034, Development Control And Promotion Regulations-2034

Mumbai Metropolitan Region Development Authority (2021) Regional Plan 2016-2036., Mumbai Metropolitan Region

National Parks Board (NParks) Singapore, https://www.nparks.gov.sg/about-us. (accessed 14 May 2022)

National Parks Board (NParks) Singapore, Wildlife in Singapore https://www.nparks.gov.sg/biodiversity/wildlife-in-singapore. (accessed on 14 May 2023)

Newman P (2014) Biophilic urbanism: a case study on Singapore. Aust Plan 51(1):47–65. https://doi.org/10.1080/07293682.2013.790832

Pierce JR (2022) Chapter 17: cities and biodiversity. In: The Routledge handbook of sustainable cities and landscapes in the Pacific Rim. Routledge, London

Stalhammar S, Brink E (2021) 'Urban biocultural diversity' as a framework for human–nature interactions: reflections from a Brazilian favela. Urban Ecosystems. 24. https://doi.org/10.1007/s11252-020-01058-3

Tan PY, Jingyuan Z, Mahyar M, Jahson AB, Edwards, Peter, Grêt Adrienne R, Daniel R, Justine S, Xiao SP, Wei Lynn W (2020) A conceptual framework to untangle the concept of urban ecosystem services. Landsc Urban Plan 200:103837. https://doi.org/10.1016/j.landurbplan.2020.103837

The Energy and Resources Institute WRC (Western Regional Centre) (2015) Environmental Status Report of Mumbai Metropolitan Region.

UN Environment Programme, Secretariat of the Convention on Biological Diversity (2020) Global Biodiversity Outlook 5. Montreal.

UN Environment Programme, Secretariat of the Convention on Biological Diversity (CBD/WG2020/5/L.2 5 December 2022) Post 2020 Global Biodiversity Framework

UN Environment Programme, Resource efficiency & green economy, https://www.unep.org/explore-topics/resource-efficiency/what-we-do/cities/resource-efficiency-green-economy. (accessed on 14 May 2023)

Urban Redevelopment Authority, Singapore, https://www.ura.gov.sg/Corporate/Resources/Publications/Skyline/Skyline-Issue10/Biophilic-city. (accessed on 14 May 2023). Urban Redevelopment Authority, Singapore, https://www.ura.gov.sg/Corporate/Planning/Master-Plan (accessed on 14 May 2023).

Uslu A, Shakouri N (2013) Urban landscape design and biodiversity. https://doi.org/10.5772/55761

Vierikko K, Elands B, Niemelä J, Andersson E, Buijs A, Fischer L, Haase D, Kabisch N, Kowarik I, Luz A, Olafsson A, Száraz L, van der Jagt A, Konijnendijk C (2016) Considering the ways biocultural diversity helps enforce the urban green infrastructure in times of urban transformation. Curr Opin Environ Sustain 22:7–12. https://doi.org/10.1016/j.cosust.2017.02.006

World Health Organization and Secretariat of the Convention on Biological Diversity (2015) Connecting global priorities: biodiversity and human health: a state of knowledge review. ISBN: 978 92 4 150853 7

Wu J., Arizona State University, Ganlin Huang, Beijing Normal University, Chunyang He, Beijing Normal University, September 2013, Urban landscape ecology: Past, Present, and Future.

20 Climate-Resilient Agropolitan Approach Towards Sustainable Regional Development of Barddhaman District of West Bengal

Tanmoy Basu and Biraj Kanti Mondal

Abstract

In the sphere of globalization, increasing competition, and complexity, sustainability has become a global buzzword as a potential solution, especially for regional development. The agropolitan approach is a mode of strategic move that promises economic and social benefits, empowerment, and amendment for optimal services in environment-friendly development. Climate-resilient agricultural, rural, and urban development comprehensively build up an integrated development of a regional unit. The present study attempts to assemble the balance of regional heterogeneity and sustainability by applying the climate-resilient agropolitan approach in Barddhaman district (undivided) of West Bengal. To address the issues of increasing agricultural production in Purba Bardhaman and rapid urban growth in Paschim Bardhaman districts, field investigation has been done along with the consideration of secondary databases. Monthly and annual temperature, rainfall, and relative humidity have been analyzed to identify the climatic conditions of the selected blocks of the study area. All the socio-economic data have been analyzed and represented by using Principal Component Analysis (PCA), relational triangle, and predicted values selecting major crop production, rural and urban population, literacy rate, main workers, educational and medical development, roadways, and electricity as the prime indicators for the current diagnostic progression. Potential zones of agropolitan development also are delineated in some blocks of the district where the resultant standardized predicted values show their connotation. Furthermore, the geometric arrangement $A_d = R_d = U_d$ indicates sustainability equilibrium and the correlation between composite development and climatic parameters shows a positive direction, which would be functional for building agropolis in the study districts.

Keywords

Agropolitan · Climate · Resilience · PCA · Relational triangles · ZPR · Barddhaman

20.1 Introduction

Agriculture is the base of all economies in developing countries like India (Praburaj et al. 2018). To promote the overall development in a region rural, agricultural, and urban development are

T. Basu
Faculty (SACT-1), Department of Geography, Katwa College, Bardhaman, West Bengal, India

B. K. Mondal (✉)
Department of Geography, Netaji Subhas Open University, Kolkata, West Bengal, India

U. Chatterjee et al. (eds.), *Climate Crisis: Adaptive Approaches and Sustainability*, Sustainable Development Goals Series, https://doi.org/10.1007/978-3-031-44397-8_20

required in an integrative way (Li et al. 2005; Qian and Wong 2012). Sustainability for regional development implies capacity building for both natural and socio-economic resources (Neto 2003; Singh 2006). Climatic parameters temperature, rainfall, and humidity condition influence the agricultural development of a region, which also promotes rural and urban resiliency to climate change (Jat et al. 2016; Smith et al. 2017; Javadinejad et al. 2019, 2021). "The development of the agropolitan program can also be a strategic move since it can become the nation's backbone in boosting national sustainability" (Rosdiana 2014, p. 564). The Sustainable Development Goals (SDGs) 13 (United Nations 2022) which promote climate resilience are intensively linked with the focus of the present study. Thus, the agropolitan approach would be applied to balance regional heterogeneity and climate-resilient sustainability in the district of West Bengal where the economic activity is based on both rural development and the amelioration of urbanization.

20.1.1 Approaches to Climate-Resilient Agropolitan Study

The agropolitan approaches and climate resilience have been postulated by multidimensional studies. Huq et al. (2015) explained the impact of climate change on the agrarians in the rural areas of coastal Bangladesh. The land use situation and rice cultivation have been decreasing after the severe hit by the cyclones named Sidrin and Sidar in 2007 and 2008, respectively (Huq et al. 2015, p. 8447). The earlier capacities were built up to adapt to the impact of the climatic disaster on the cultivation (Huq et al. 2015, p. 8447). Viswanathan et al. (2020, p. 3) postulated the adaptation strategies of climate-resilient agriculture in India that are management of 'crop', 'land', 'water', 'livelihood', 'technological and social innovations Arnold et al. (2021) exposed the survey by Michigan State University on 'rural climate-resilience' in the U.S.A. The survey found the progression of the knowledge domain on climate change, a collaboration of the communities in capacity building, and implementation of plans and programmes of climate-resilient development of locale in the U.S.A. Arnold et al. (2021, p. 16). Kim et al. (2020) systematically reviewed the scholarly publications on rural resilience and framed that climate-change-related rural resilience is a significant part of the ecological dimensions of the appraisal of rural resilience. Leichenko (2011) researched climate change and associated resilience in urban areas. According to Leichenko (2011), "resilience is typically understood as the ability of a system to withstand a major shock and maintain or quickly return to normal function" (p. 164). In urban areas, the "promotion of urban resilience is essential for enabling both adaptation and mitigation efforts" (Leichenko 2011, p. 165). Zhong et al. (2022) studied the strategies of farmers to combat climate change effects on urbanization and its adaptation in rural Chengdu in Southwest China. Regarding this, Zhong et al. (2022) opined that 'ecosystem-based strategies' were required to contend with the extreme impact of climatic phenomena and weather catastrophes. 85%, 80%, 88%, and 84% of farmers percept about promoting the 'functioning ecosystem' strategies to climate change and urbanization in Paotong and Dantu, the two study villages, respectively (Zhong et al. 2022, p. 15). In the case of 'agrobiodiversity' strategies, the percentages are 80 (climate change), 75 (urbanization) in Paotong and 96 (climate change), 80 (urbanization) in Dantu.

In researching the agropolitan approach, Buang, et al. (2011, p. 1) postulated that the agropolitan approach provides the rural poor true monetary and psychological strength. Indah et al. (2017, p. 60) conceptualized that the idea of agropolitan development arises from the contrast between rural growth as a farming activity core and urban development as a locus of economic activity in Ponorogo, Indonesia. Prasetiya et al. (2014) asserted that the availability of natural resources, the availability of facilities, social groups, and sufficient quantity and diversity of infrastructural framework and facilities (p. 60). It provides the provision to support systems integration and farming activities are the major influencing factors of agropolitan developments in Sendang, Tulungagung in Indonesia (Prasetiya et al. 2014, p. 60). Agropolitan areas would be developed through different strategic plans. Safariah et al. (2016, p. 695) mentioned that the formulation of a development idea is necessary for rural regions to create an agropolitan region in Padang Pariaman. The other studies highlight the keys to rural, agricultural, and urban developments in West Bengal and other areas. Sakir et al. (2017, p. 357) pointed out that developing agropolitan initiatives is frequently highlighted as the key to growth in rural development issues in countries of emerging economies. Sarkar and Ghosh (2017) studied that West Bengal should aggrandize 'diversified and export-oriented' agriculture (p. 483). Farhanah and Prajanti (2015, p. 158) found the "gaps between planning and implementation in developing agropolitan areas" in Rojonoto of Wonosobo district. Thus, sustainability would be maintained to establish a homogeneous and balanced agropolitan developmental area in any region. Friedmann (1979, p. 607) thought about 'basic needs', 'agropolitan development', and 'planning corn' regarding agropolitan approach.

Poli et al. (2013, p. 35) aimed to research identifying the potential for wholesome agropolitan advancement in a region. Poli et al. (2013, p. 35) also suggested that spatial planning is concerned with the future of regional sustainability to generate agriculture production in agropolitan area growth. Roy et al. (2014, p. 1) assessed the evidence accumulating information from official sources that West Bengal's districts concerning one another on the route to sustained development. Regarding the implementation of agropolitan approach to building up an agropolitan area Shaffril et al. (2010, p. 2354) implied that Agropolitan projects and its possibilities in elevating the communities socio-economic status with the enactment of the Gahai Agropolitan Project in Kuala Lipis, Pahang in Malaysia.

The previous literature explicates the agropolitan development with sustainability building and climate resilience, but the integration of rural, agricultural, and urban developmental factors would be considered to build up a predicted agropolitan area. Besides, to show the overall sustainability relational triangle and its changing structures would also be taken into consideration. Regarding those aspects, the present study pointed out the implementation of a climate-resilient agropolitan approach to building up a comprehensive regional development in the study area. To promote the regional sustainability coalescence of rural, agricultural, and urban development has to be implemented in a consolidative way (Fig. 20.1).

(a) (b)

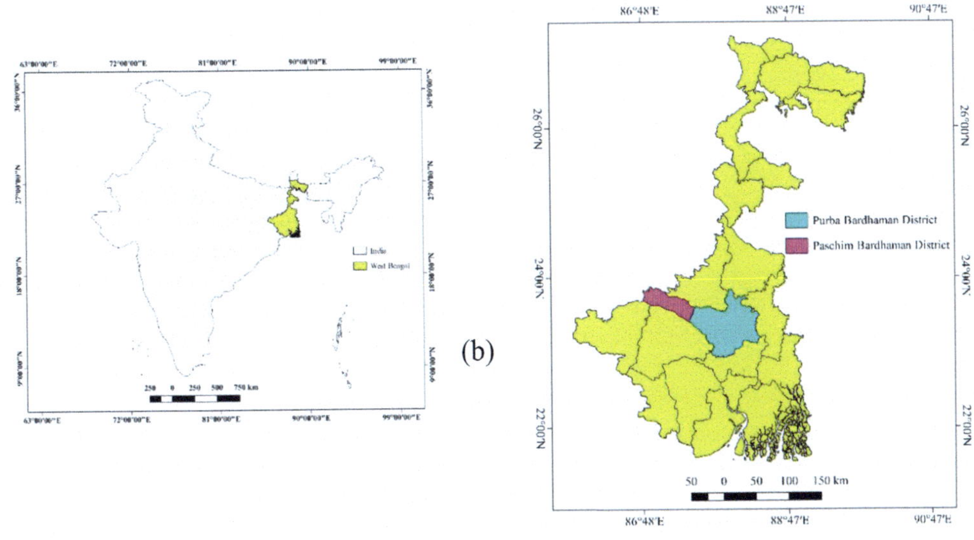

Fig. 20.1 (**a**, **b**) Location map of the study area. (**a**). West Bengal in India and (**b**). Bardhaman District in West Bengal

20.2 Objectives

The main objectives of the study are as follows:

(i) To analyze the conditions of temperature, rainfall, and humidity parameters in the study area (2011).
(ii) To identify the principal factors of rural, agricultural, and urban development in the study area (2011).
(iii) To assess the relationship between climatic parameters and composite development in the study area.

20.3 Study Area

Barddhaman district (undivided) of West Bengal, India, which was bifurcated into two separate districts in 2017 (Personnel and Administrative Reforms Department, 2017) is selected as the study area (Fig. 20.2). The locational extent of the district was 22° 56′ to 23° 56′ north latitude and 86° 48′ to 88° 25′ east longitude (Peterson 1910; HDRCC 2011). The newly formed two districts are Purba Bardhaman (in the east) and Paschim Bardhaman (in the west) (Mondal et al. 2023). Purba Bardhaman is enriched with rural and agricultural activities (Dey and Mistri 2019) where the rapid industrial and urban development has mostly occurred in Paschim Bardhaman (Dutta and Guchhait 2022). The physiographic situation of the undivided Barddhaman district varied from a part of the western plateau of West Bengal in the western portion of Barddhaman where the main river is Damodar and Barakar (Peterson 1910) and the eastern portion is the alluvial plain of river Bhagirathi-Hugli, Damodar, and Ajay, which is enriched with the fertile alluvium of the river Bhagirathi-Hugli (Peterson 1910). The district had a variation of average temperature as 30° in the hot summer and 20° in the cold winter. The temperature increased from east to west while humidity decreased due to scorching heat in the western part (near Asansol

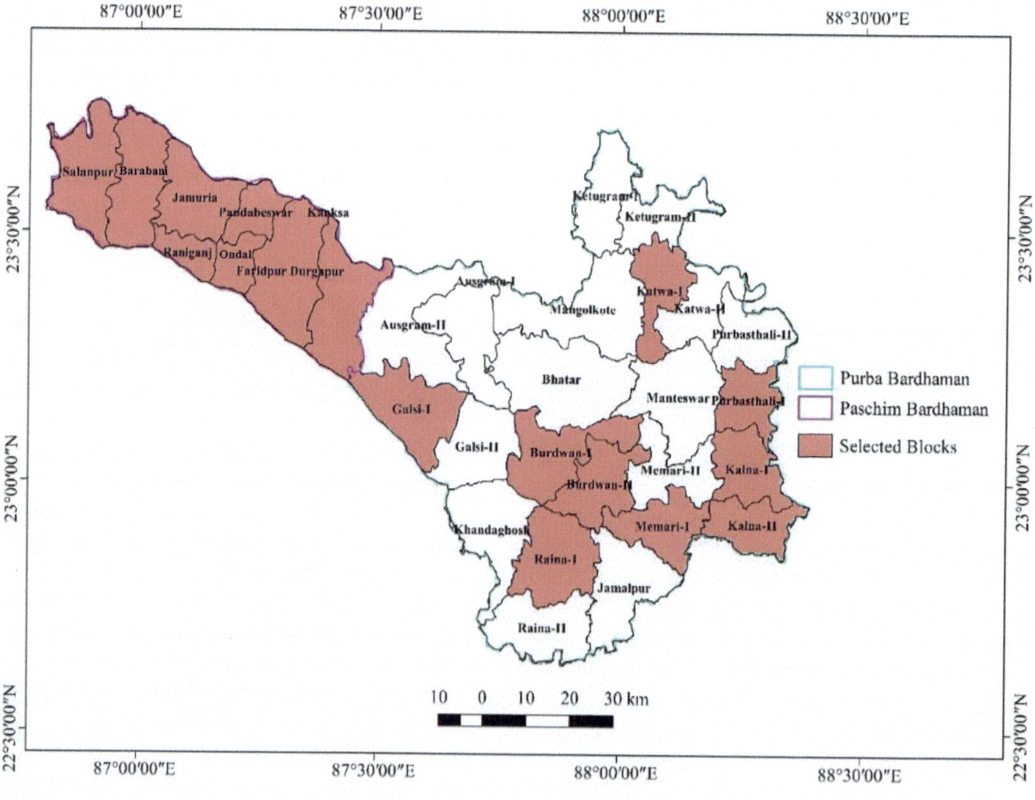

Fig. 20.2 Selected Blocks of Barddhaman (Purba Bardhaman and Paschim Bardhaman) District

and surroundings, Peterson 1910; National Informatics Centre, Burdwan District Unit 2011). The average rainfall was 150 mm in the monsoon season and it is also decreasing from the east to the western part of the district (Peterson 1910; National Informatics Centre, Burdwan District Unit 2011). About 60.11% of the total population lived in rural areas whereas 39.39% lived in urban areas, respectively (Census of India 2011) in Barddhaman district (undivided).

Paddy production is the main agricultural activity of Barddhaman district (HDRCC 2011) and the district is called the 'rice bowl of Bengal' (Dutta 2012). The average landholding size of the district was 0.97 hectares, whereas the highest individual landholding belonged to the 10.00–20.00 hectares of landholding size class in 2011 (Agriculture Census 2011). Very high cropping intensity had been found in the Community Development Blocks (C. D. blocks) named Galsi-I, Galsi-II, Jamalpur-I, Memari-I, and Kalna-I (HDRCC 2011). The undivided Barddhaman district consisted of a total of 31 C. D. blocks (Census of India 2011). In the present study, a total of 17 blocks are selected, which are Salanpur, Barabani, Jamuria, Raniganj, Ondal, Pandabeswar, Faridpur Durgapur, Kanksa (in Paschim Bardhaman, District Administration Paschim Bardhaman 2023), and Katwa-I, Purbasthali-I, Galsi-I, Burdwan- I, Burdwan- II, Memari-I, Kalna-I, Kalna-II, Raina-I (in Purba Bardhaman, District Administration Purba Bardhaman 2023). The rest of the blocks were not taken into consideration in the study as there is no urban population that consists of the blocks as per the Census of India (2011). To minimize the heterogeneity of the overall regional development, the study has analyzed and explained the agropolitan approach to be implemented in the study area.

20.4 Materials and Methods

The study was conducted based on field observation along with employing secondary databases. Climatic data for the year 2011 have been collected from the websites of Solar Radiation Data (SoDA): MERRA Project collaboration with NASA and the India Meteorological Department (Regional Meteorological Centre Kolkata). Socio-economic data have been collected from the District Census Handbook (2011). District Statistical Handbook (2010 and 2011 combined), and District Human Development Report of Barddhaman. The following methods and techniques have been applied to analyze the collected datasets.

To formulate the composite factor scores of rural, agricultural, and urban developments, the Composite indices method through Principal Component Analysis (PCA, Pearson 1901) has been used. For the Composite Factor Analysis (Principal Component Analysis) the following equation has been formulated,

$$P_1 = a_{11}.Z_1 + a_{12}.Z_2 + a_{13}.Z_3 + \ldots + a_{n1}.Z_n \quad (20.1)$$

where P_1 denotes the composite index of development of a unit study as the first factor denotes the factor loading of the '*j*'th variable and *1* indicates the factor number that is the first factor, vector of factor loadings. Z_j denotes the standardized value of the '*j*'th variable, which is expressed as

$$Z_j = \frac{x_j - x_m}{\delta_j} \quad (20.2)$$

where *Xj* denotes the original value of '*j*'th variable, *Xm* denotes the mean (Simple arithmetic mean ($\overline{x}$) of '*j*'th variable and δ*j* denotes the standard deviation (*δ*) of '*j*'th variable.

In this aspect,

$$\text{Mean}\left(\overline{x}\right) = \frac{\sum \overline{x}}{n} \quad (20.3)$$

$$\text{Standard Deviation}\left(\delta\right) = \sqrt{\left(\frac{x - \overline{x}}{n}\right)} \quad (20.4)$$

where

$\overline{x}$ is the arithmetic mean,

x Is the individual value of items,

n is the number of terms in the distribution.

Mean factor scores have been represented through the following formula.

$$\text{Mean composite factor scores} = \frac{sum\text{ of composite factor scores of development }(\text{rural, agricultural and urban})}{\text{Total Numbers of developemnt }(\text{rural, agricultural and urban})} \quad (20.5)$$

Simply principal factors of rural, agricultural, and urban development are selected to delineate agropolitan areas in the study districts by implying the agropolitan approach. The details of the indicators have been discussed later in the study.

Sustainability based on rural, agricultural, and urban development has been represented through the relational triangle. The concept of the 'relationship triangle' was postulated by Friedman (1985) to express 'any three parts of the human relational system'. The triangles are also organized to represent the pillars of agricultural sustainability by Fallah-Alipour et al. (2018) as 'AMOEBA diagrams' (p. 16).

The relational triangle (Fig. 20.3) shows the equilibrium position where Rural development (R_d) = Agricultural Development (A_d) = Urban Development (U_d). The various changes in the shapes of the triangle occur due to the changes in the relationship between the standardized predicted values of rural, agricultural, and urban development which are mentioned in the results and discussion section (Fig. 20.4).

A multiple linear regression model has been constructed to show the relationship between the factor scores as the dependent variable and indicators/factors of the rural, agricultural, urban, and composite development as independent variables based on selected factors and formulating the standardized predicted values (to delineate the predicted agropolitan zones) by the following formulas.

The formula of '*R*' (Multiple Correlation Coefficient) (Pearson 1897, 1914),

$$R = \sqrt{\frac{\left[(ry.x1)^2 + (ry.x2)^2\right] - (2ry.x1ry.x2rx1.x2)}{1-(rx1.x2)^2}} \quad (20.6)$$

where *R* is the value of the correlation coefficient

X1 is one independent variable.

X2 is another independent variable.

Y is the dependent variable.

The formula of the multiple regression model is

$$Y_c = a + b_1X_1 + b_2X_2 \ldots b_nX_n \quad (20.7)$$

where Y_c is a predicted value of *Y* (which is the dependent variable)

a is the '*Y* intercept',

b_1 is the change in *Y* for each increment change in X_1,

b_2 is the change of *Y* for each increment change in X_2,

b_n is the change of Y for each increment change in X_n,

X is the *X* score (independent variable) for which a value of *Y* is predicted.

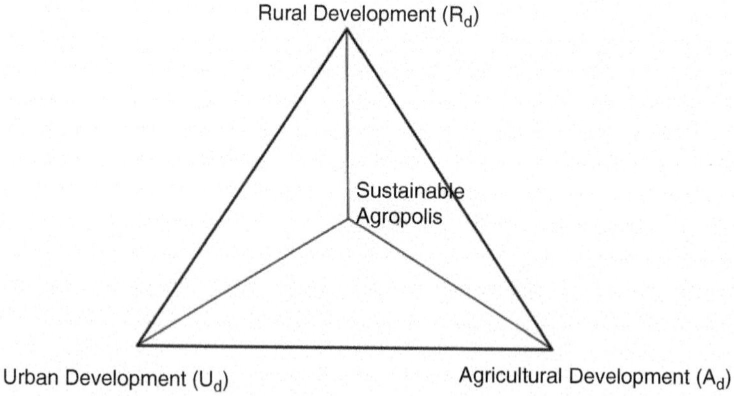

Fig. 20.3 Relational triangle of sustainability

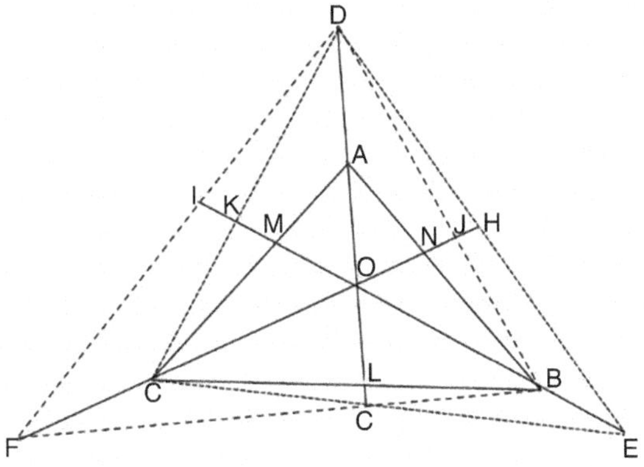

Fig. 20.4 Changing shapes of the triangle. Equilibrium of agropolitan areas

The '*F*' value in the *j*th one-way ANOVA (Fisher 1934) is calculated through the following formula:

$$F = \frac{\text{Explained variance}}{\text{Unexplained variance}} \quad (20.8)$$

or

$$F = \frac{\text{Between-group variability}}{\text{Within-group variability}} \quad (20.9)$$

The 'Explained variance', or 'Between-group variability' is.

$$\sum_{i=1}^{k} ni\left(\overline{Yi} - \overline{Y}\right)^2 / (K-1) \quad (20.10)$$

where
$\overline{Yi}$ denotes the sample mean in the *i*th group.
ni is the number of observations in the *i*th group
$\overline{Y}$ denotes the composite mean of the data.
And *K* denotes the number of groups.

The 'Unexplained variance' or 'Within-group variability is

$$\sum_{i=1}^{n}\sum_{j=1}^{ni}\left(Yij - \overline{Yi}\right)^2 / (N-K) \quad (20.11)$$

where

Yij is the observation in the *i*th out of *K* groups and *N* is the overall sample size.

This *F*-statistic follows the F-distribution with *K* – *1*, *N* – *K* degree of freedom under the null hypothesis.

The test of significance (Fisher 1925 following Student 1908) test has been adopted to observe the statistical significance of the computed correlations.

$$t = \sqrt{\frac{n-2}{1-r^2}} \quad (20.12)$$

where
t = Value of Significance,
r = Correlation Coefficient,
r^2 = Coefficient of determinants and,
n = No. of observation.

20.5 Results

20.5.1 Climatic Conditions of the Selected Blocks of Barddhaman

Based on the collected data of first day of every month in 2011 from Global Modeling and Assimilation Office (GMAO) (2015) and India Meteorological Department (2015), the temperature, rainfall, and relative humidity situation of the selected blocks have been analyzed. The average value of the mean annual temperature, mean annual rainfall, and average relative humidity of the selected C.D. blocks was 23.69 °C, 129.01 mm, and 65.72%, respectively, and the standard deviation were 0.15, 4.39, and 1.85, respectively, in 2011 (Table 20.1). Among the three variables mean annual rainfall shows more variability in nature. Figure 20.5a–q shows the month-wise temperature and rainfall conditions of the selected blocks of Barddhaman in 2011. The maximum temperature was highest in Katwa-I (31.2 °C) and lowest in Faridpur Durgapur (30.06 °C) blocks and the minimum temperature was highest in Kalna-II (16.58 °C) and lowest in Salanpur (16.06 °C) blocks. The highest and lowest temperature had been observed in June and February, respectively. The highest mean annual temperature was observed in Katwa-I (23.83 °C) and lowest in Faridpur Durgapur (23.18 °C) blocks. The annual maximum temperature condition varied from lowest to highest in the middle western portion to the western portion and eastern portion in the selected C. D. blocks and the annual minimum temperature varied inversely to the annual minimum temperature (Fig. 20.6). Figures 20.7 and 20.8 show the average annual maximum and minimum temperature in the selected blocks, respectively. Figure 20.9 shows the variation in

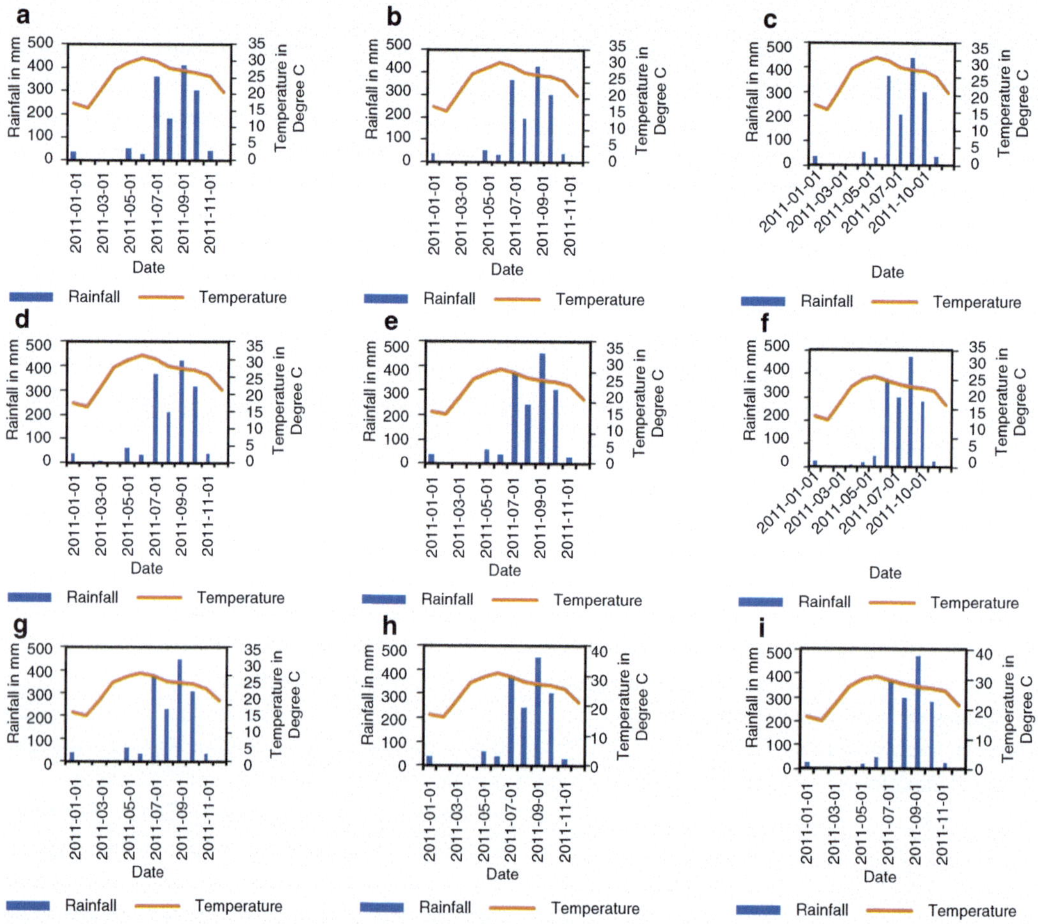

Fig. 20.5 (**a**) Rainfall-temperature graph of Salanpur block. (**b**). Rainfall-temperature graph of Barabani block. (**c**) Rainfall-temperature graph of Jamuria block. (**d**) Rainfall-temperature graph of Raniganj block. (**e**) Rainfall-temperature graph of Ondal block. (**f**). Rainfall-temperature graph of Pandabeswar block. (**g**) Rainfall-temperature graph of Faridpur Durgapur block. (**h**) Rainfall-temperature graph of Kanksa block. (**i**) Rainfall-temperature graph of Katwa-I block. (**j**) Rainfall-temperature graph of Purbasthali-I block. (**k**) Rainfall-temperature graph of Galsi-I block. (**l**) Rainfall-temperature graph of Burdwan-I block. (**m**) Rainfall-temperature graph of Burdwan-II block. (**n**). Rainfall-temperature graph of Memari-I block. (**o**) Rainfall-temperature graph of Kalna-I block. (**p**) Rainfall-temperature graph of Kalna-II block. (**q**) Rainfall-temperature graph of Raina-I block. (**a–q**) Month-wise temperature and rainfall conditions of the selected blocks of Barddhaman (2011)

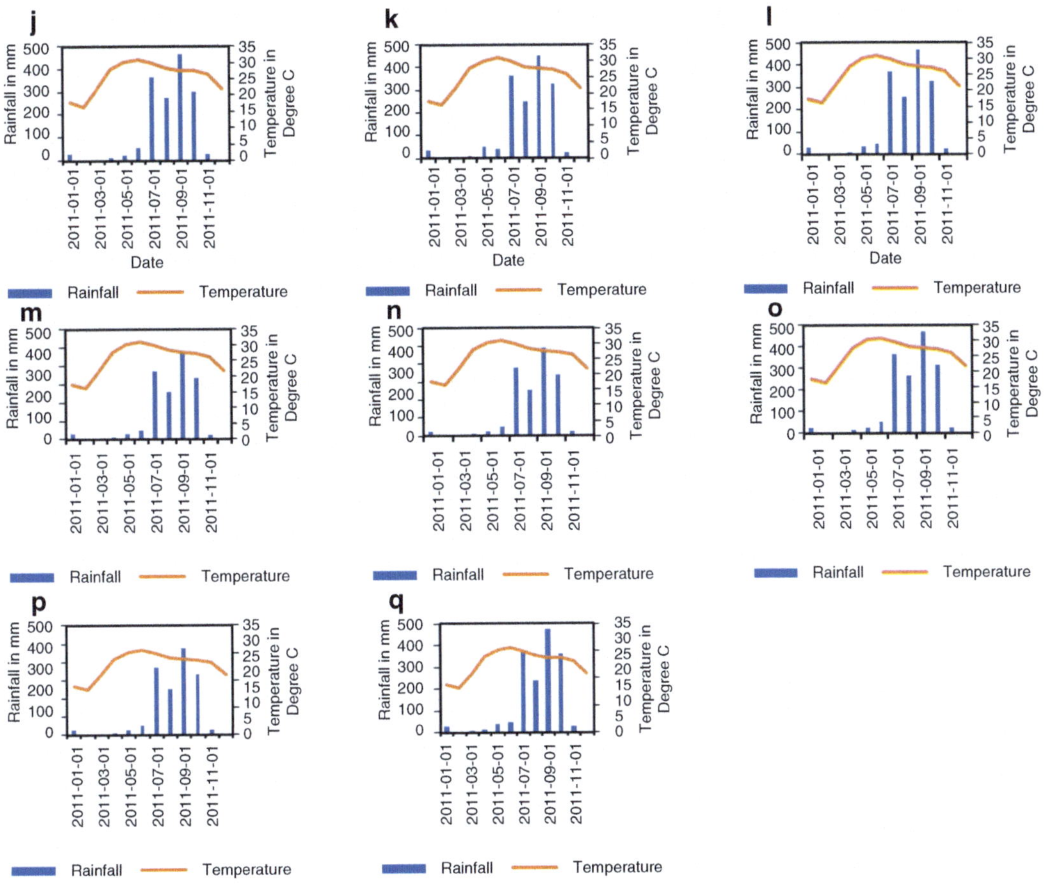

Fig. 20.5 (continued)

mean annual temperature. Based on the range of the average annual maximum and minimum temperature the mean annual temperature was comparatively higher in the blocks Katwa-I, Purbasthali-I, Galsi-I, Burdwan-I, Burdwan-II, Memari-I, Kalna-I, Kalna-II, Raina-I and the part of Raniganj, Ondal, Faridpur, Durgapur and Kanksa than the other blocks.

The impact of the tropic of cancer (Figs. 20.7, 20.8, and 20.9) has been observed on the range of temperature distribution among the eastern part of the district, whereas the western part faces highly average maximum temperature due to physiographic conditions. The average annual rainfall was highest in the C. D. block Memari-I (135.03 mm) and lowest in the C. D. block Salanpur (119.62 mm). The maximum concentration of rainfall had been identified from June to October when monsoon season hits in South Bengal. Block-wise variation of monthly rainfall (2011) has been shown in Fig. 20.5a–q. The annual average rainfall was highly concentrated in the eastern and south-eastern blocks of the district whereas the western and north-western blocks are lacking rainfall in 2011. In the case of annual average rainfall, C. D. blocks Burdwan- I, Burdwan- II, Memari-I, Kalna-I, Kalna-II, and Raina-I consisted of high rainfall, whereas Raniganj, Ondal, Pandabeswar, Faridpur Durgapur, Kanksa, Katwa-I, Purbasthali-I, and Galsi-I consisted moderate rainfall as well as Salanpur, Barabani, and part of Jamuria consisted low rainfall situation in 2011 (Fig. 20.10). Figure 20.6a–q shows the condition of relative humidity in the selected blocks of Barddhaman. The highest average relative humidity had been

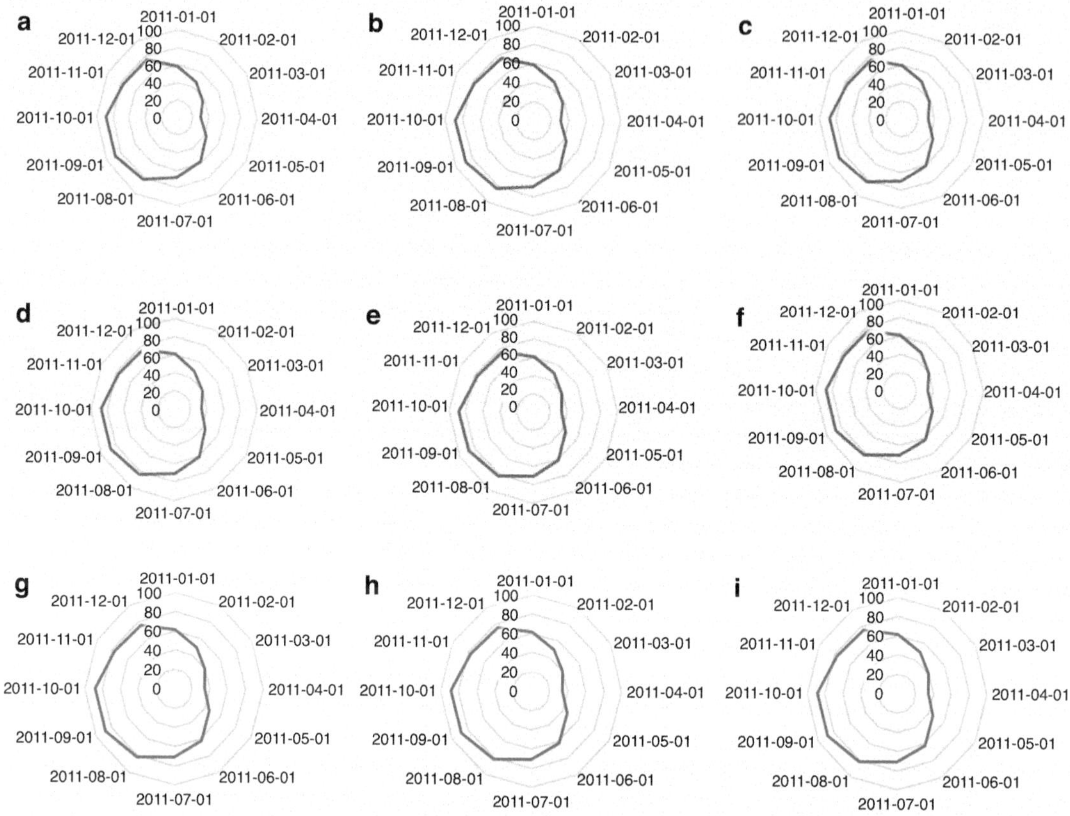

Fig. 20.6 (**a**) Humidity condition of Salanpur block. (**b**) Humidity condition of Barabani block. (**c**) Humidity condition of Jamuria block. (**d**) Humidity condition of Raniganj block. (**e**) Humidity condition of Ondal block. (**f**) Humidity condition of Pandabeswar block. (**g**) Humidity condition of Faridpur Durgapur block. (**h**) Humidity condition of Kanksa block. (**i**) Humidity condition of Katwa-I block. (**j**) Humidity condition of Purbasthali-I block. (**k**) Humidity condition of Galsi-I block. (**l**) Humidity condition of Burdwan-I. (**m**) Humidity condition of Burdwan-II Kalna-I block. (**n**) Humidity condition of Memari-I block. (**o**) Humidity condition of Kalna-I block. (**a–q**) Conditions of relative humidity (%) in the selected blocks of Barddhaman (2011)

recognized in Kalna-II (68.03%) and the lowest average relative humidity in Salanpur (62.66%) in 2011. The distribution of average annual relative humidity also varied from highest to lowest from the eastern and south-eastern to the western and north-western part of the district (Fig. 20.11) and this was also high mostly in the rainy season. The overall conditions of rainfall and relative humidity conjugate impact the crop cultivation of the district and the overall climatic situation in accordance that the temperature, rainfall, and relative humidity condition being more favourable in the eastern part that the western part of the district for the development of farming activities.

20.5.2 Developmental Factors of the Study Area

The regional development of the study area is heterogeneous. Rural and agricultural areas are mostly concentrated in the eastern part and urban-industrial areas are in the western part of the district. To show the rural, agricultural, and urban development total of 36 indicators (Tables 20.2, 20.3 and 20.4) have been selected for analysis. Tables 20.2, 20.3, and 20.4 show the component score coefficients of extracted first and second components (with suppression of small coefficients as an absolute value below 0.010) of the selected factors of rural, agricultural, and

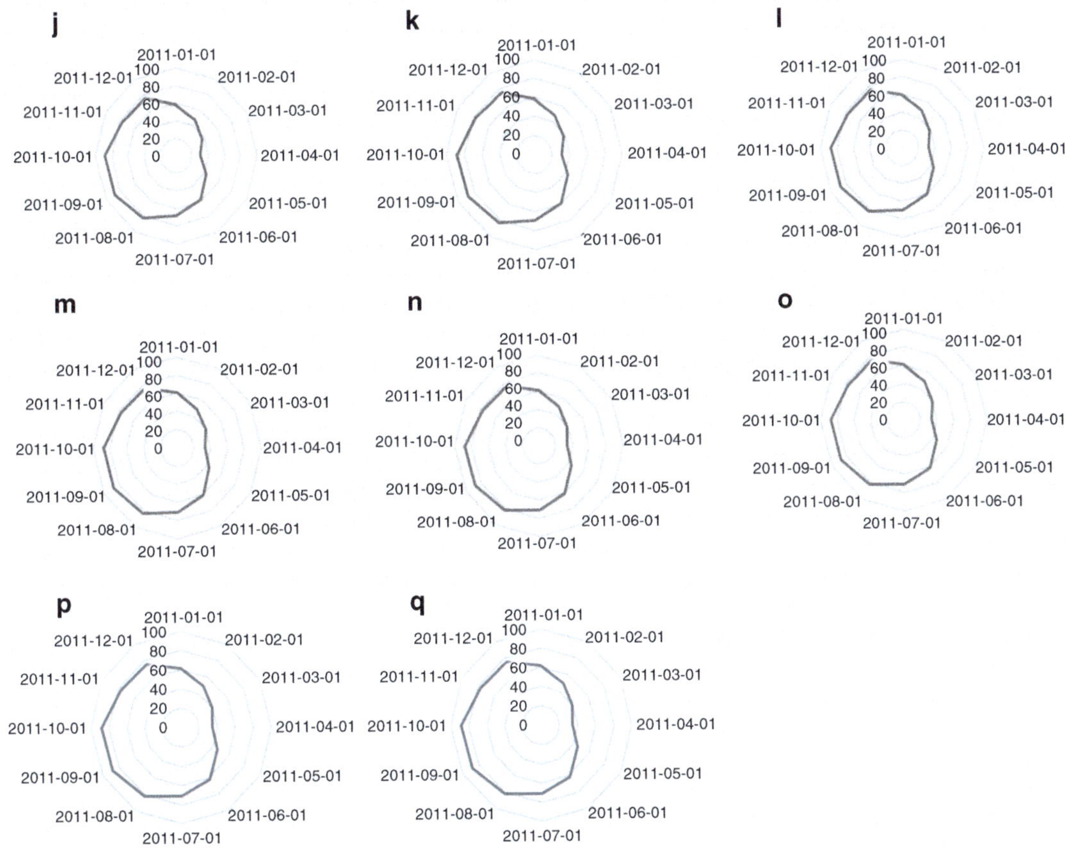

Fig. 20.6 (continued)

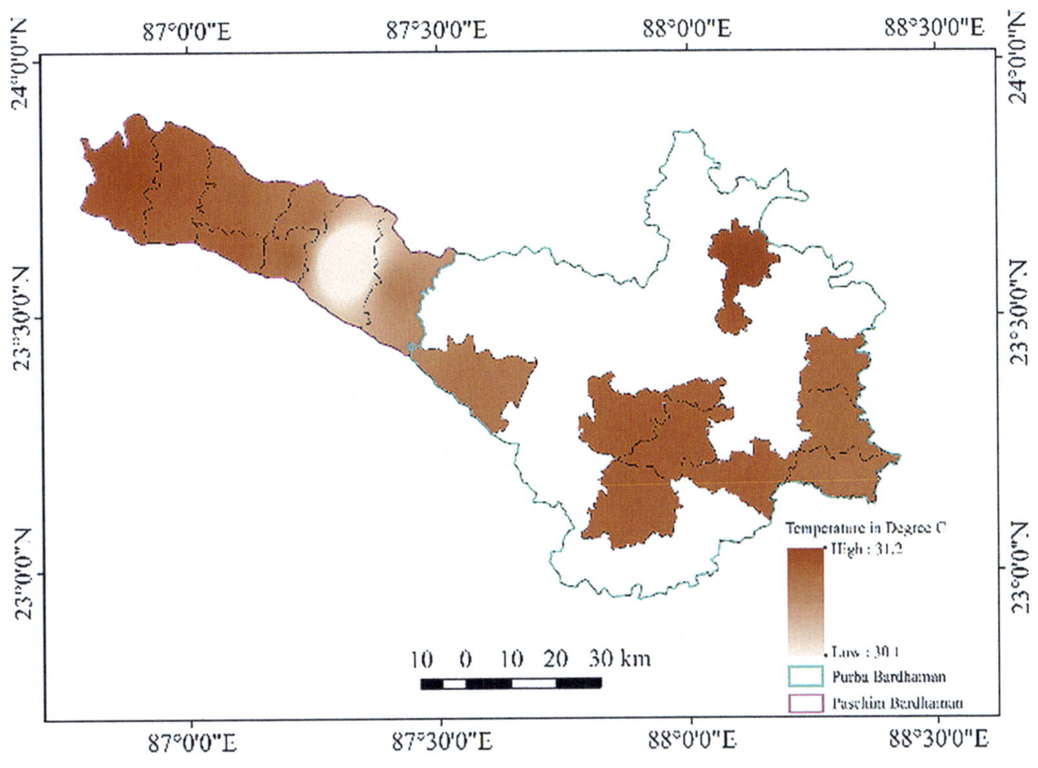

Fig. 20.7 Variation of average annual maximum temperature in the selected blocks of Barddhaman (2011)

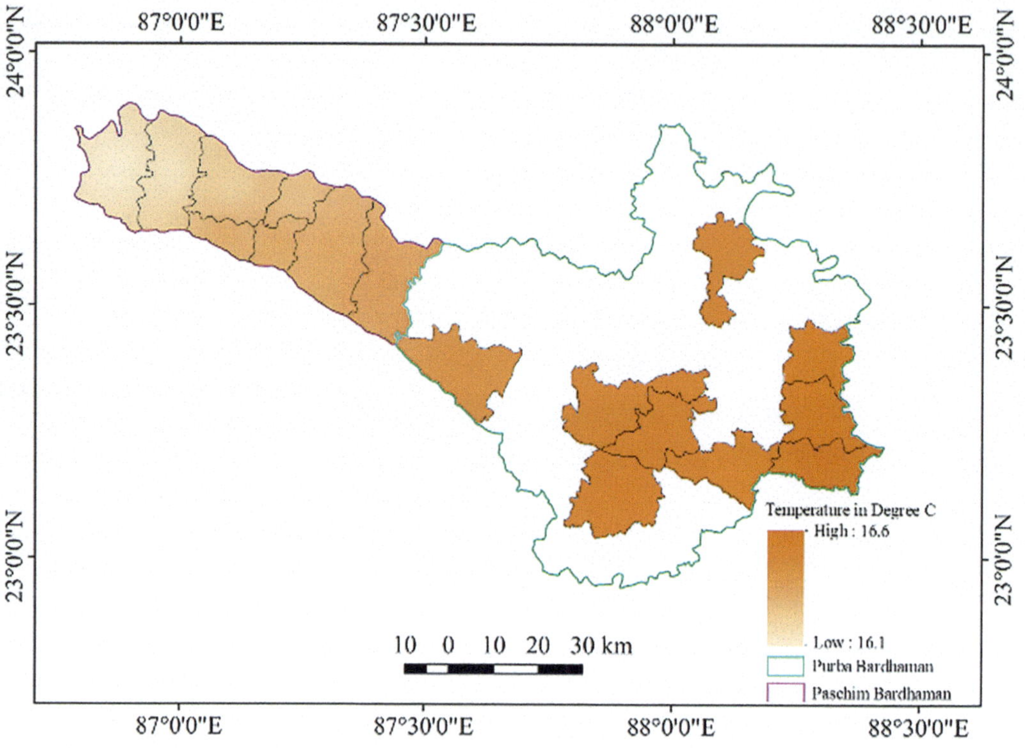

Fig. 20.8 Variation of average annual minimum temperature in the selected blocks of Barddhaman (2011)

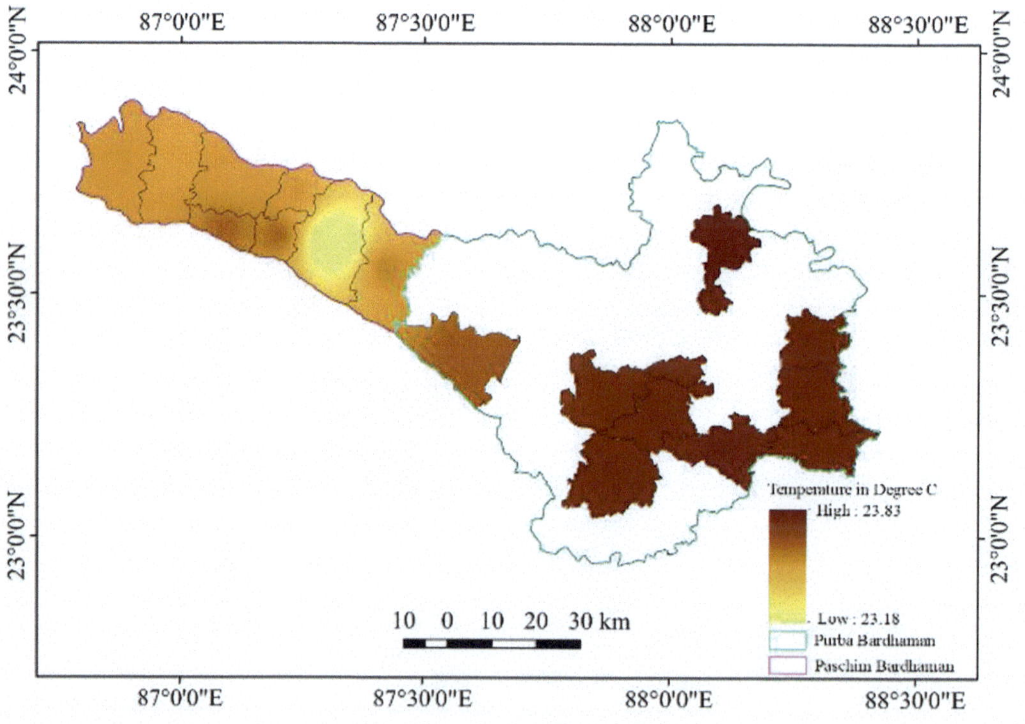

Fig. 20.9 Variation of mean annual temperature in the selected blocks of Barddhaman (2011)

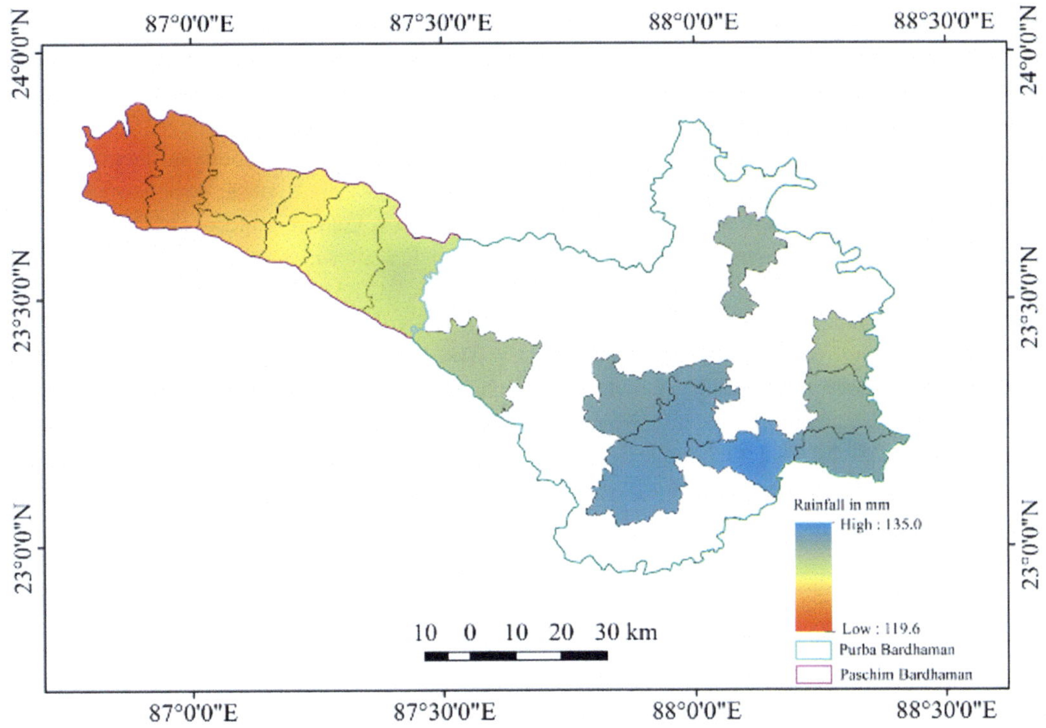

Fig. 20.10 Variation of average annual rainfall in the selected blocks of Barddhaman (2011)

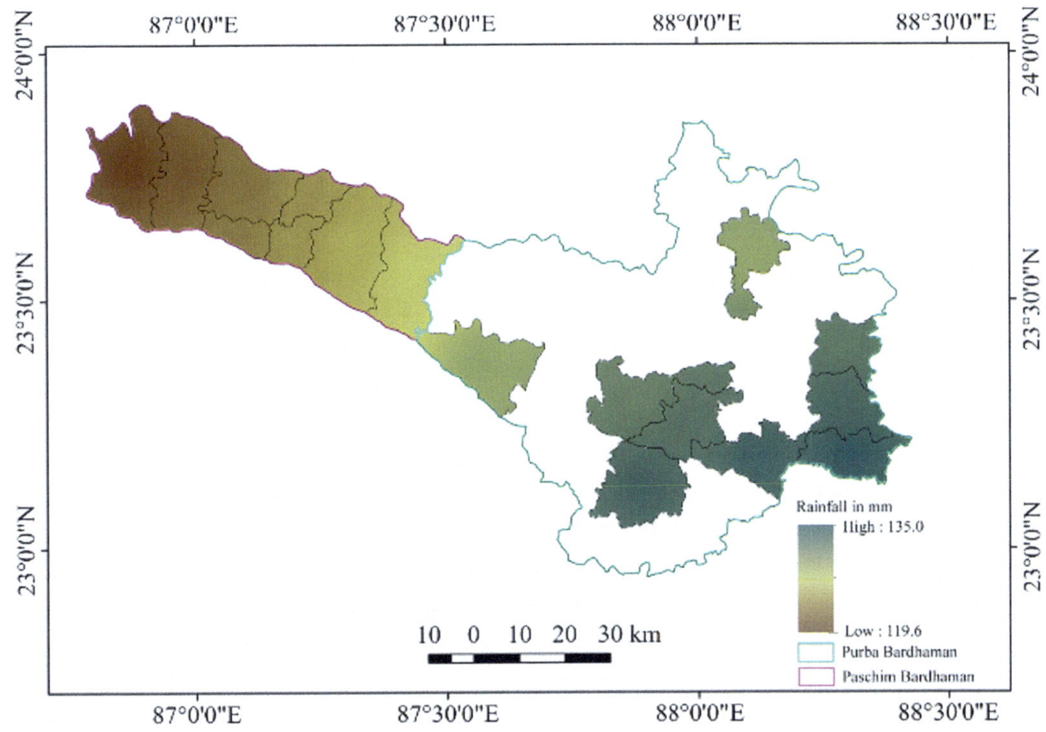

Fig. 20.11 Variation of the average annual relative humidity in the selected blocks of Barddhaman (2011)

urban development, respectively. The principal components have been recognized as total household, total Scheduled Castes (SC population), total Scheduled Tribes (ST population), total literacy, the total number of main workers and the total number of marginal workers in case of rural development; gross cropped area, net cropped area, the area under principal crops, average yield, the total number of main cultivators, the total number of marginal cultivators, the total number of main agricultural labourers, and the total number of marginal agricultural labourers in case of agricultural development and total population, total household, total ST population, total literacy, the total number of main workers, electricity (domestic connection) in case of urban development. Figure 20.12a–c shows the scree plots and Fig. 20.13a–c shows the component pilots of the principal component analysis. The data collected from the Department of Planning and Statistics (2016) show that the total number of inhabited villages in 2011 was 69 in Salanpur, 46 in Barabani, 38 in Jamuria, 12 in Raniganj, 12 in Ondal, 14 in Pandabeswar, 48 in Faridpur Durgapur, 77 in Kanksa; whereas Katwa-I, Purbasthali-I, Galsi-I, Burdwan- I, Burdwan- II, Memari-I, Kalna-I, Kalna-II, and Raina-I consisted 63, 91, 85, 75, 83, 111, 98, 112 and 110 inhabited villages, respectively, in 2011.

The data show that the concentration of rural inhabitants was greater in the eastern part of Barddhaman rather than in the western part. On the other hand, the total number of urban bodies identified in 2011 was 2 in Salanpur, 6 in Barabani, 1 in Jamuria, 1 in Raniganj, 14 (12 + 2 Part) in Ondal, 14 (11 + 3 Part) in Pandabeswar, 6 in Faridpur Durgapur, 7 in Kanksa; whereas the urban bodies were 1, 5, 3, 4, 1, 1, 3, 1 and 1,

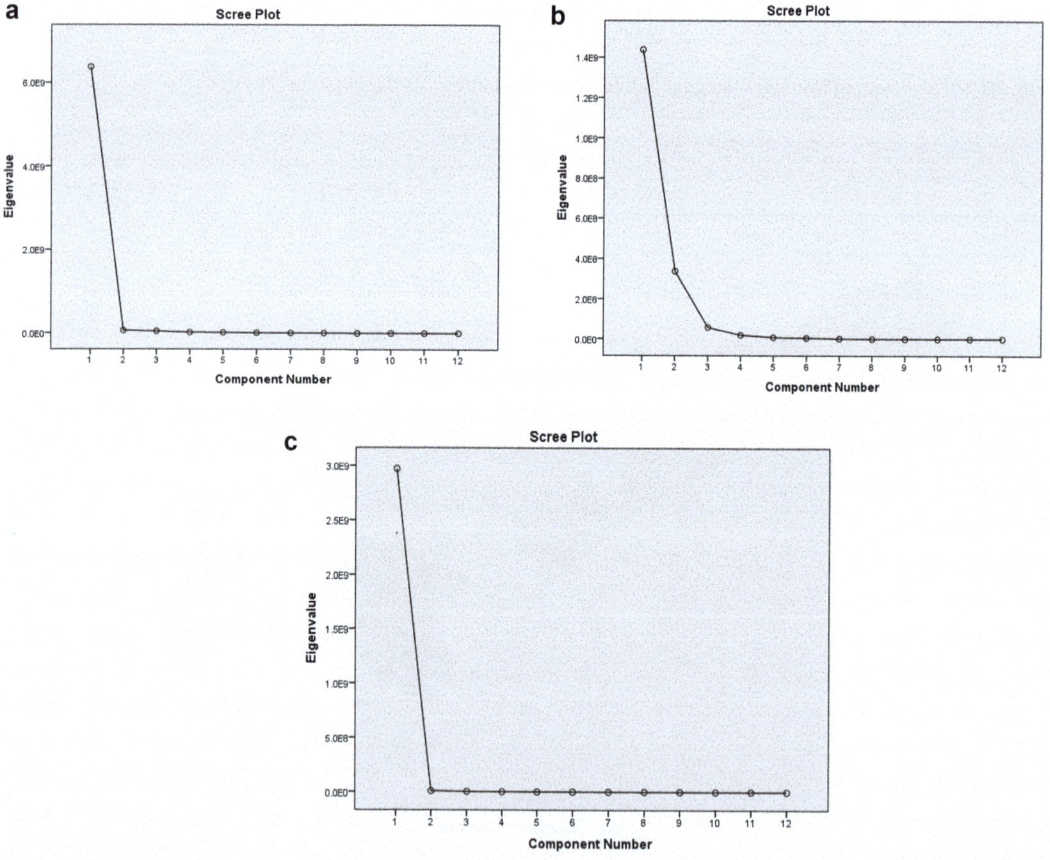

Fig. 20.12 (**a**) Scree plot of the factors of rural development. (**b**) Scree plot of the factors of Agricultural development. (**c**) Scree plot of the factors of Urban development. (**a–c**) Scree plots of the eigenvalues of component numbers of the factors of rural, agricultural, and urban development

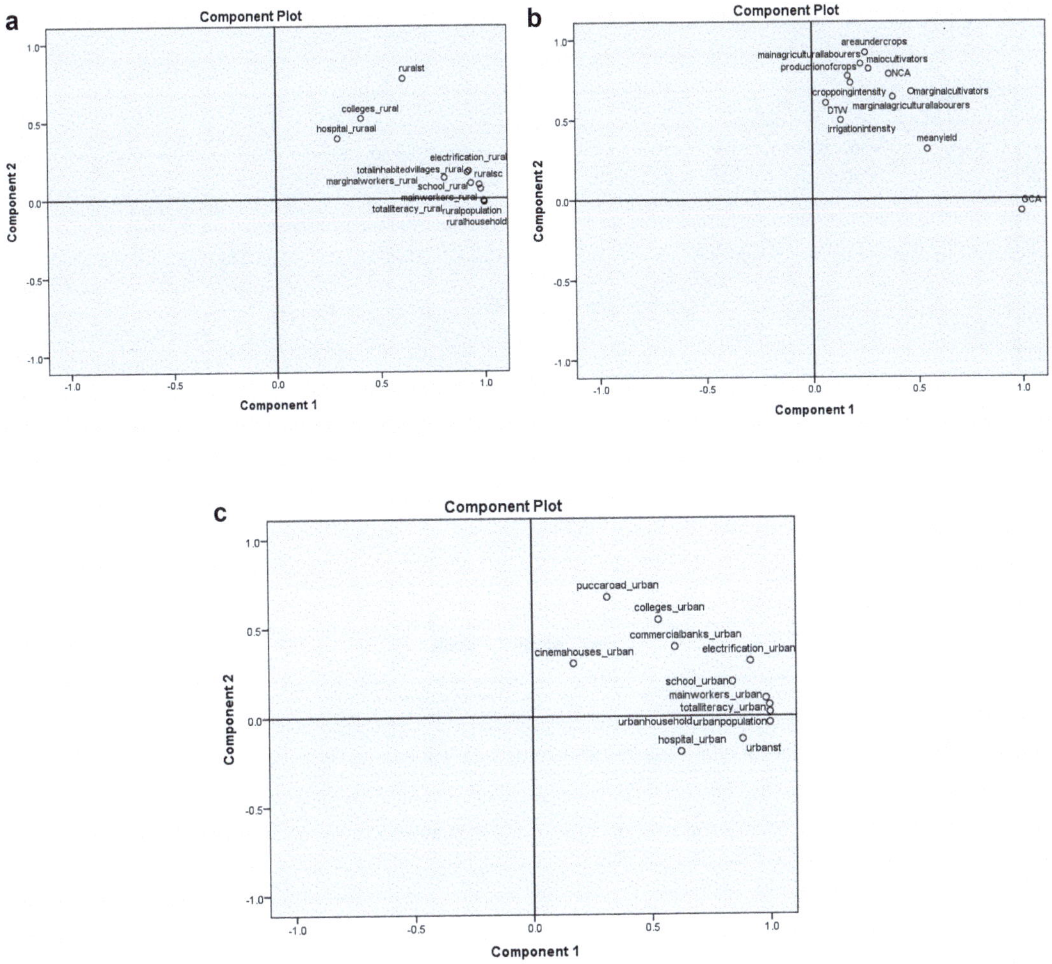

Fig. 20.13 (**a**) Component plots of the factors of rural development. (**b**) Component plots of the factors of agricultural development. (**c**) Component plots of the factors of urban development. (**a–c**) Component plots of the factors of rural, agricultural, and urban development

respectively, in Katwa-I, Purbasthali-I, Galsi-I, Burdwan- I, Burdwan- II, Memari-I, Kalna-I, Kalna-II, and Raina-I. According to Census (2011), the total number of households in the district was 1,725,511; the total population of the district was 7,717,563 where the male was 3,966,889 persons and the female was 3,750,674 persons. The total rural population of the district was 4,639,264 and the total urban population was 3,078,299. The percentage of urban population to the total pollution of the district was 39.89. The total literacy rate was 76.21%, whereas male and female literacy rates were 82.42% and 69.63%. The percentage of SC and ST population to the total population were 27.41% and 6.34%. Barddhaman district consisted of 28.08% of the main workers and 9.65% of the marginal workers to the total workers. Among the categories of main and marginal workers, the district comprised 11.75% of cultivators, 33.43% of agricultural labourers, 4.28% of workers in the household industry, and 50.54% of other workers. The calculated factor score of rural development, agricultural development, and urban development (Table 20.5) shows the spatial distribution of the developmental status (Figs. 20.14, 20.15, and 20.16) in the study area (2011).

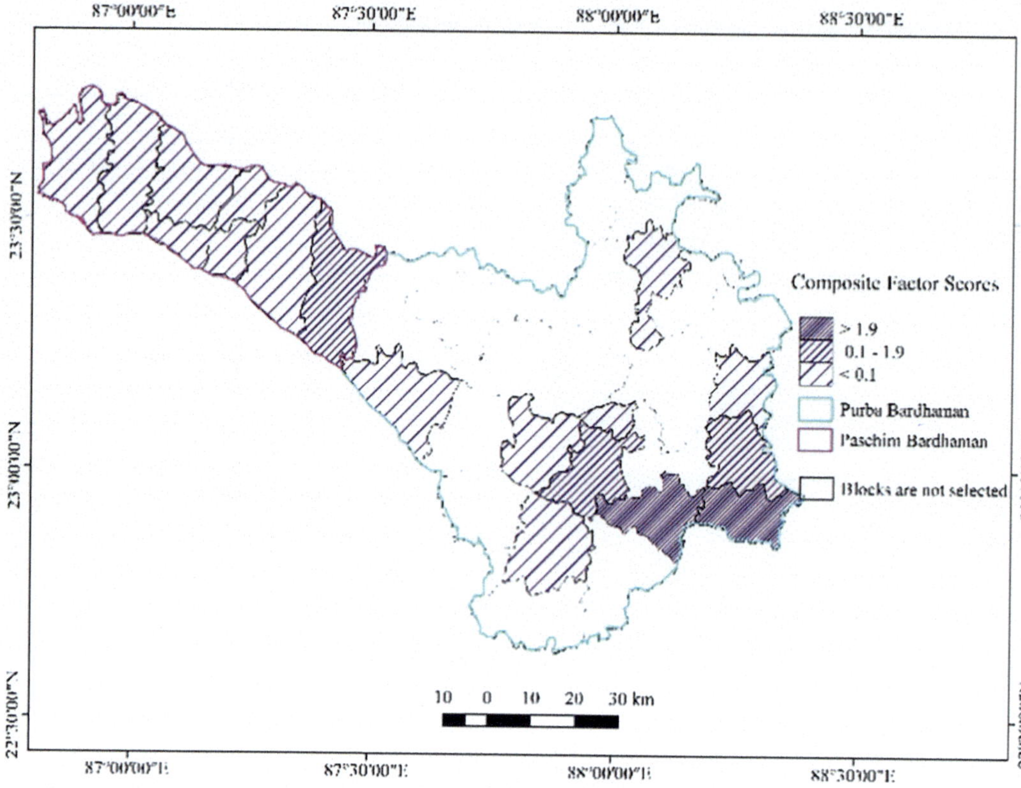

Fig. 20.14 Developmental zones of rural areas (based on composite factor scores) in the selected blocks of Barddhaman

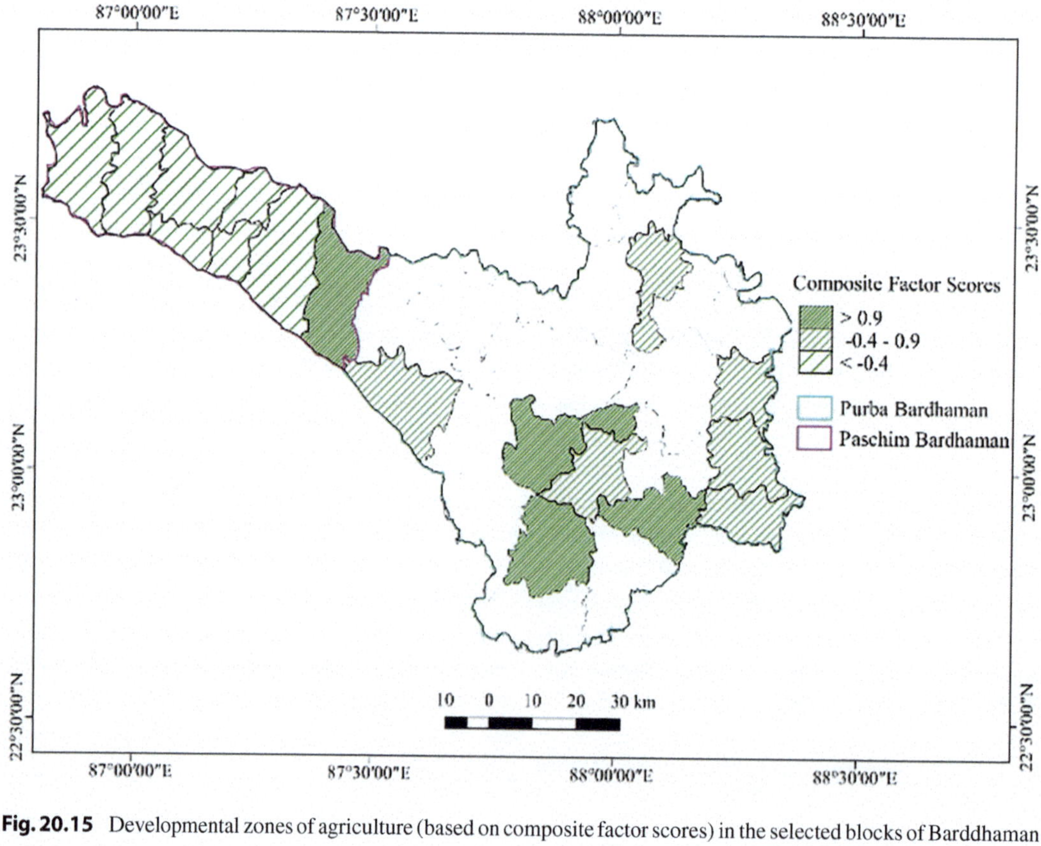

Fig. 20.15 Developmental zones of agriculture (based on composite factor scores) in the selected blocks of Barddhaman

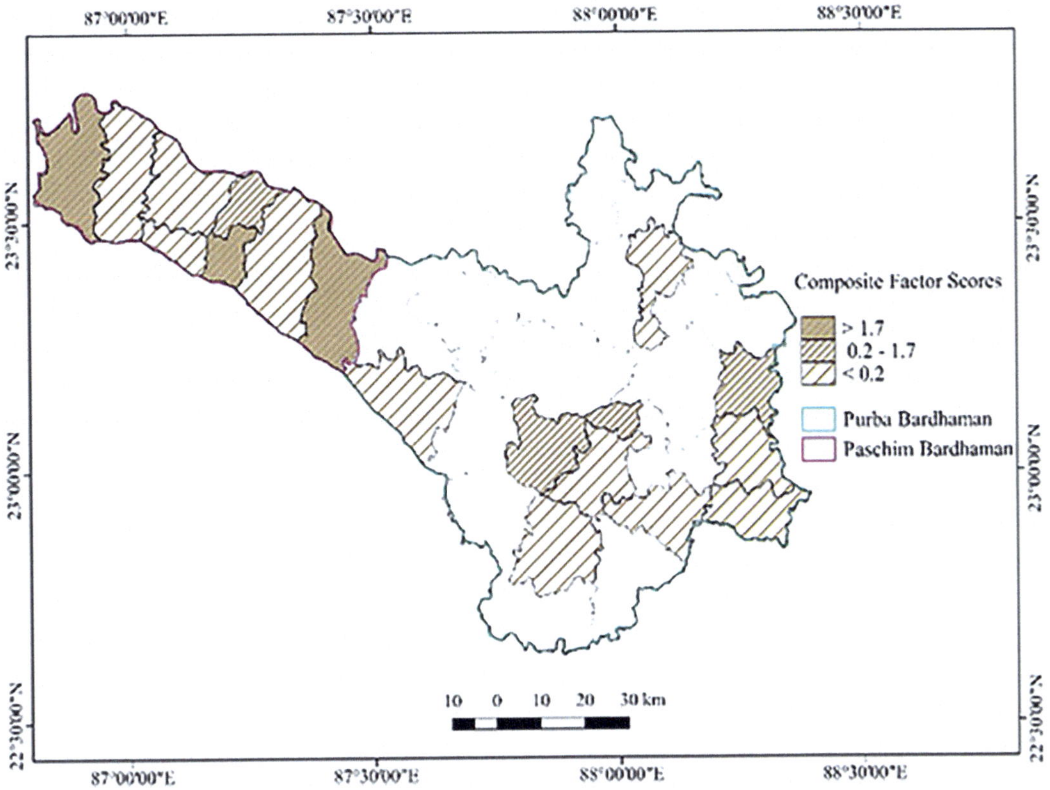

Fig. 20.16 Developmental zones of urban areas (based on composite factor scores) in the selected blocks of Barddhaman

Comparatively high developmental zones (value of factor scores are >1.9 in case of rural development, value of factor scores are >0.9 in case of agricultural development, value of factor scores are >1.7 in case of urban development) had been recognized in the C. D. blocks Memari-I and Kalna-II (rural development), Kanksa, Burdwan-I, Memari-I and Raina-I (agricultural development), Salanpur, Ondal and Kanksa (urban development); moderate developmental zones (value of factor scores are 0.1–0.9 in case of rural development, value of factor scores are −0.4 – 0.9 in case of agricultural development, value of factor scores are 0.2–1.7 in case of urban development) had been recognized in Kalna-I, Burdwan-II and Kanksa (rural development), Katwa-I, Purbasthali-I, Galsi-I, Burdwan-II, Kalna-I and Kalna-II (agricultural development), Pandabeswar, Purbasthali-I and Burdwan-I (urban development) and low developmental zones (value of factor scores are <0.1 in case of rural development, value of factor scores are <−0.4 in case of agricultural development, value of factor scores are <0.2 in case of urban development) had been recognized in Salanpur, Barabani, Jamuria, Raniganj, Ondal, Pandabeswar, Faridpur Durgapur, Katwa-I, Purbasthali-I, Galsi-I, Burdwan-I and Raina-I (rural development), Salanpur, Barabani, Jamuria, Raniganj, Ondal, Pandabeswar and Faridpur Durgapur (agricultural development), Barabani, Jamuria, Raniganj, Faridpur Durgapur, Katwa-I, Galsi-I, Burdwan- II, Memari-I, Kalna-I, Kalna-II and Raina-I (urban development) respectively in 2011. Besides, In 2011, high developmental zones (value of composite factor score is >0.6) of composite rural, agricultural, and urban development were also found in Kanksa, Burdwan-I, and Memari-I; moderate developmental zones (value of composite factor score is −0.2 – 0.6) of composite development were found in Salanpur, Purbasthali-I, Galsi-I, Burdwan-II, Kalna-I and II and Raina-I and low developmental zones (value of composite factor score is <−0.2) of composite development were found in Barabani, Jamuria, Raniganj, Ondal, Pandabeswar, Faridpur Durgapur, and Katwa-I (Table 20.5, Fig. 20.17).

The area, production, and yield of the selected blocks are also regionally varied. The data from the Department of Planning and Statistics (2016) indicated that the yield (2010–2011) of the *aman* paddy, one of the main *Kharif* crops of the district was 0 kg/hectare in Raniganj block, the yield rate of *aman* < 2000 kg/hectare was found in the blocks Ondal and Pandabeswar; yield rate between 2000–3000 kg/hectare was found in the blocks Salanpur, Barabani, Jamuria, Faridpur Durgapur, Kanksa, Burdwan-I, Burdwan-II, Memari-I, Kalna-II and Raina-I as well as yield rate > 3000 kg/hectare was found in Katwa-I, Purbasthali-I, Galsi-I, and Kalna-I. There was no cultivation of *aus* and *boro* paddy conjugately in Salanpur, Barabani, Jamuria, Raniganj, Ondal, Pandabeswar, and Faridpur Durgapur in 2011. The rest of the C. D. blocks almost seasonally cultivated the *aus* and *boro* paddy. The irrigation status (in terms of total irrigated area) in 2011 showed that there was no irrigated area in Salanpur, Jamuria, Ondal, and Pandabeswar. Contrarily, the highest irrigated area was observed in Kalna-II (21941.27 hectares), Kanksa (20422.89 hectares), Jamuria (18036.49 hectares), Raina-I (14283.12 hectares) and Burdwan-II (11720.42 hectares); whereas the irrigated area below 10,000 hectares was observed in Memari-I (8534.64 hectares) followed by Katwa-I (5760.55 hectares), Kalna-I (5233.21hectares), Burdwan-I (4958.65 hectares), Pandabeswar (2614.61 hectares) and Purbasthali-I (2147.71 hectares). In Ondal the total irrigated area is very low (354.58 hectares). Insignificant irrigation status belonging to Barabani (the total irrigated area was 22.91 hect-

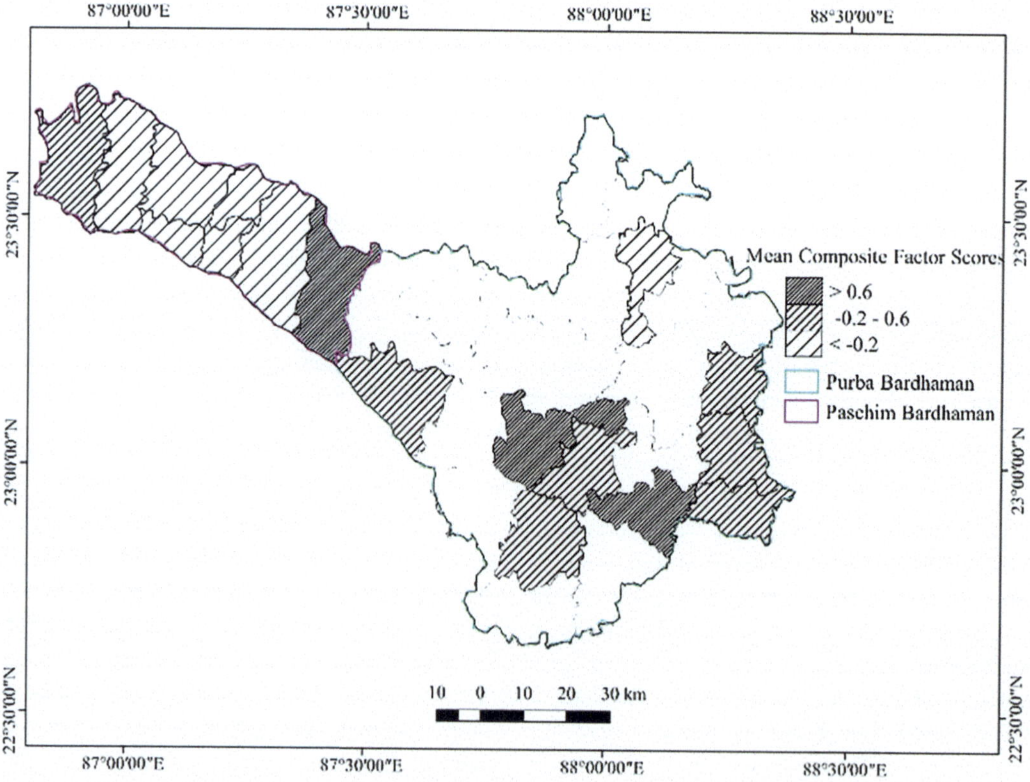

Fig. 20.17 Composite developmental zones (based on mean composite factor scores) in selected blocks of Barddhaman

ares) and Salanpur (22.01 hectares). There was also a regional disparity in the distribution of gross cropped area, net cropped area, and electricity (domestic connection). Based on the above discussion, it is summarized that the selected blocks situated in the western part of Barddhaman comparatively lag in terms of agricultural development but urban expansion was greater there than in the blocks of the eastern part. On the other hand, agricultural development and occurrences of rurality are abundant in the eastern portion of the district, which comprises a low to moderate urban expansion.

20.5.3 Relationship of Developmental Factor Scores with the Components of Development and Climatic Variables

The analysis of variance shows a 0.657 mean square value of the dependent variable as mean composite factor scores. The relationship has been predicted by the selected variables of composite development (Table 20.7), which is statistically significant ($p < 0.0005$). Figures 20.18, 20.19, 20.20 show the predicted zones of rural development, agricultural development, and urban development, respectively. The highly standardized predicted value of composite development (based on selected factors) was found in the major parts of Salanpur, Kanksa, Burdwan-I, Memari-I, Kalna-I, and II, and moderate standardized predicted value of composite development was found in the major parts of Ondal, Purbasthali-I, Galsi-I, Burdwan-II, and Raina-I and low standardized predicted value of composite development was found in the major parts of Barabani, Jamuria, Raniganj, Pandabeswar, Faridpur Durgapur and Katwa-I (Tables 20.6 and 20.7, Fig. 20.21). While predicting the climate resiliency regression coefficient has been calculated using the standardized predicted scores of the composite development as the dependent variable and mean composite Z-score as the predictor. Here, Table 20.8 shows that the correlation value between the two variables is moderate ($r = 0.538$) (Figs. 20.22, 20.23, 20.24). The correlation of determinants is 0.290 (r square = 0.290) as a 1 unit increase of mean composite Z-score increases 29% of standardized predicted scores of composite developments. The above prediction is also statistically significant ($p < 0.05$, Table 20.8). Figures 20.25 and 20.26 show the relationship between mean composite Z-scores of the climatic variables and mean composite regression standardized predicted values of composite development and the relationship between regression standardized predicted values and predicted mean composite standardized predicted values of composite development.

Here, predicted agropolitan areas are measured by formulating the standardized predicted values of the relationship between the mean composite factor scores and the selected variables of the composite development (rural, agricultural, and urban) and the areas have been delineated based on standardized predicted values of the selected factors. Figure 20.22 shows the predicted agropolitan areas in Barddhaman district. The predicted agropolitan areas are situated in Salanpur, Kanksa, Burdwan-I, Kalna-I and II, and Memari-I (where standardized predicted values composite development is high based on the selected factors). The rural, agricultural, and urban developmental status in those areas is progressing. Table 20.6 shows the comparative relationship between the predicted values of rural, agricultural, and urban development. There is no equal position between the three developmental statuses but agropolitan areas can be demarcated where agricultural development is also prioritized in the domain of urban growth and development.

20.5.4 Formulating Sustainability Equilibrium Between Rural, Agricultural, and Urban Development

The sustainability equilibrium of agropolitan areas has been represented by the relational triangles based on the predicted values of rural, agricultural, and urban development. Table 20.6 shows the relationship among the regression

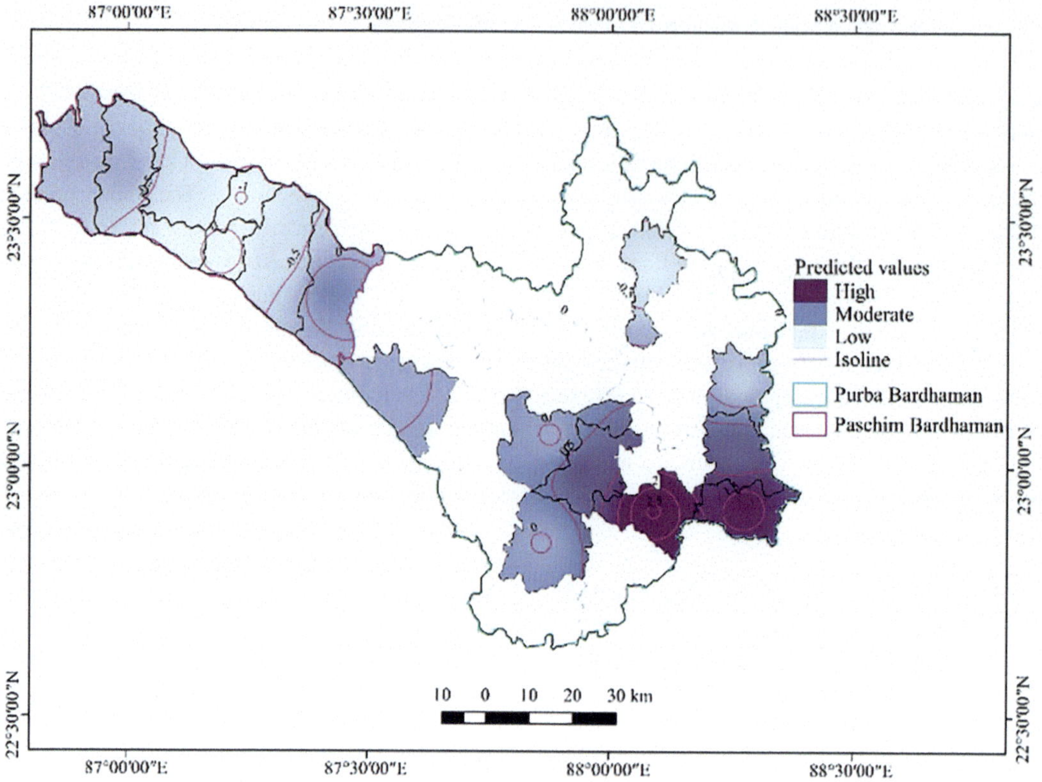

Fig. 20.18 Predicted zones of rural development (based on standardized predicted values) in the selected blocks of Barddhaman

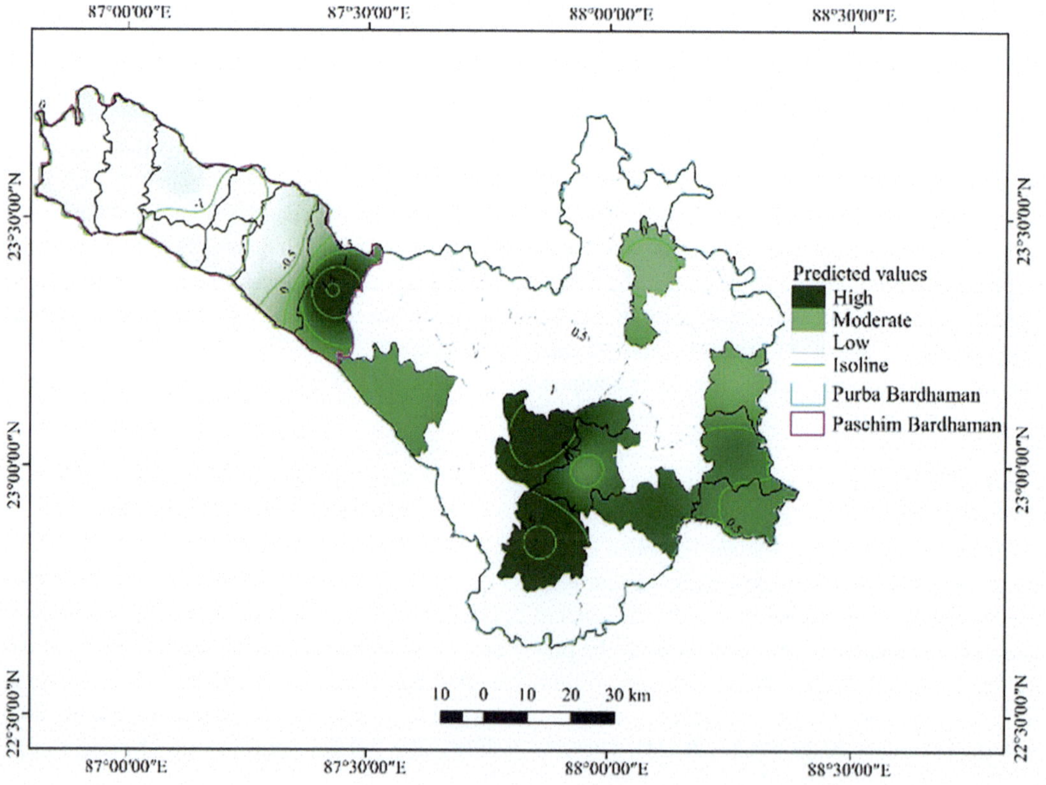

Fig. 20.19 Predicted zones of agricultural development (based on standardized values) in the selected blocks of Barddhaman

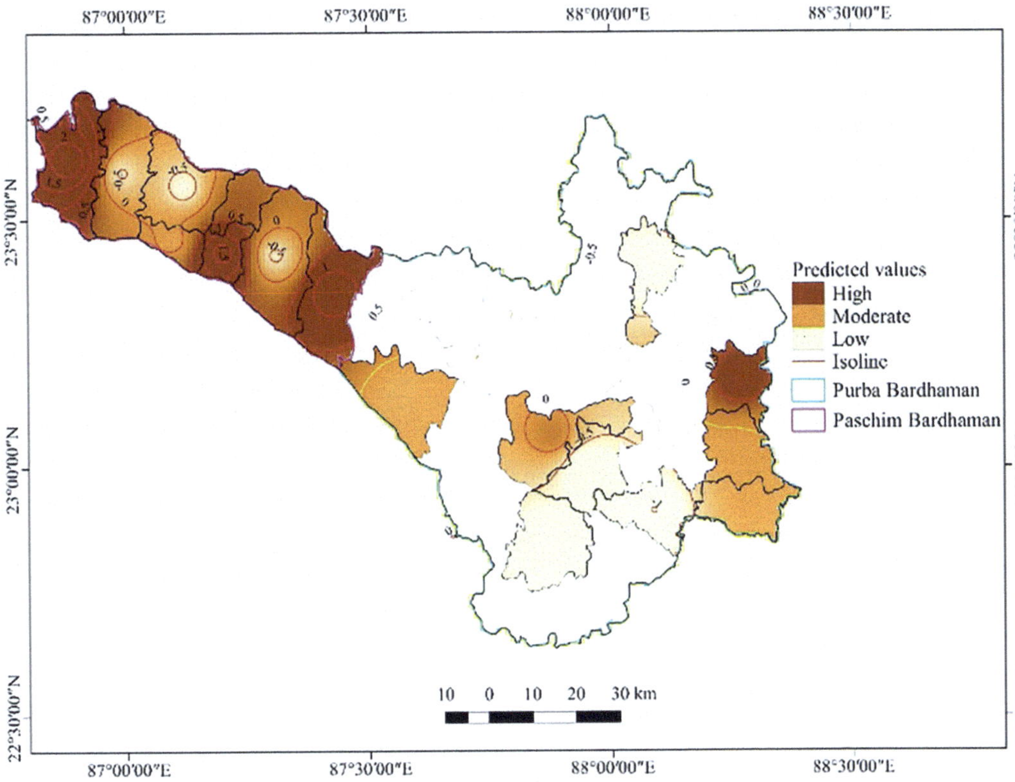

Fig. 20.20 Predicted zones of urban development (based on standardized predicted values) in the selected blocks of Barddhaman

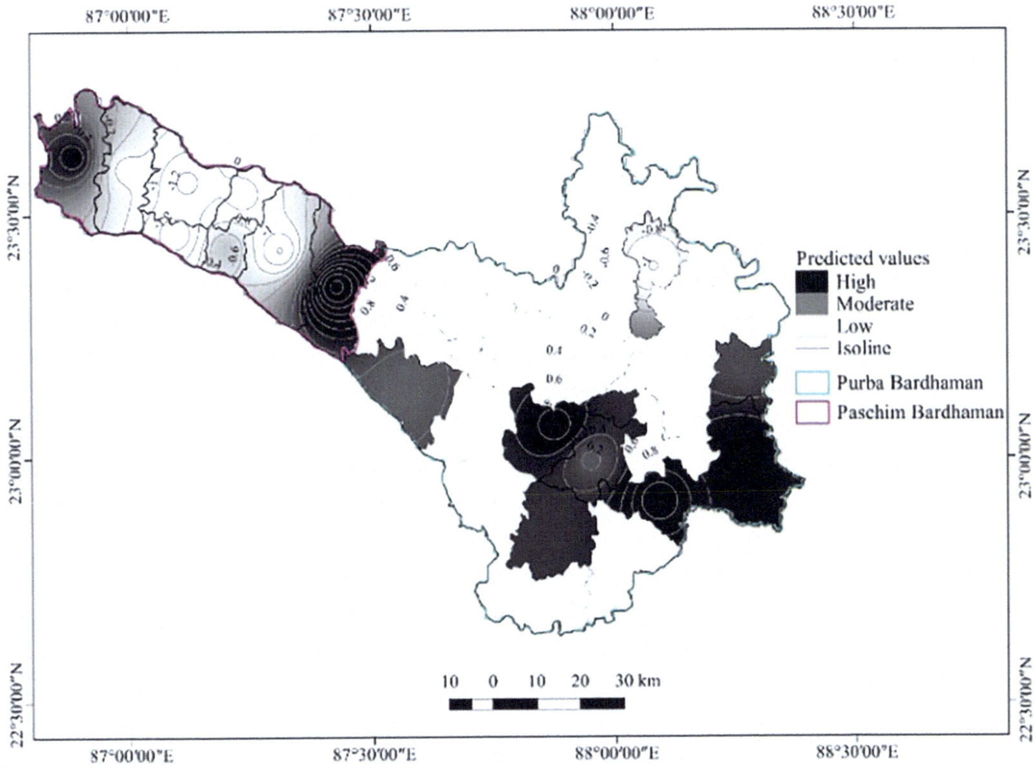

Fig. 20.21 Predicted zones of composite development by selected factors (based on standardized predicted values) in the selected blocks Barddhaman

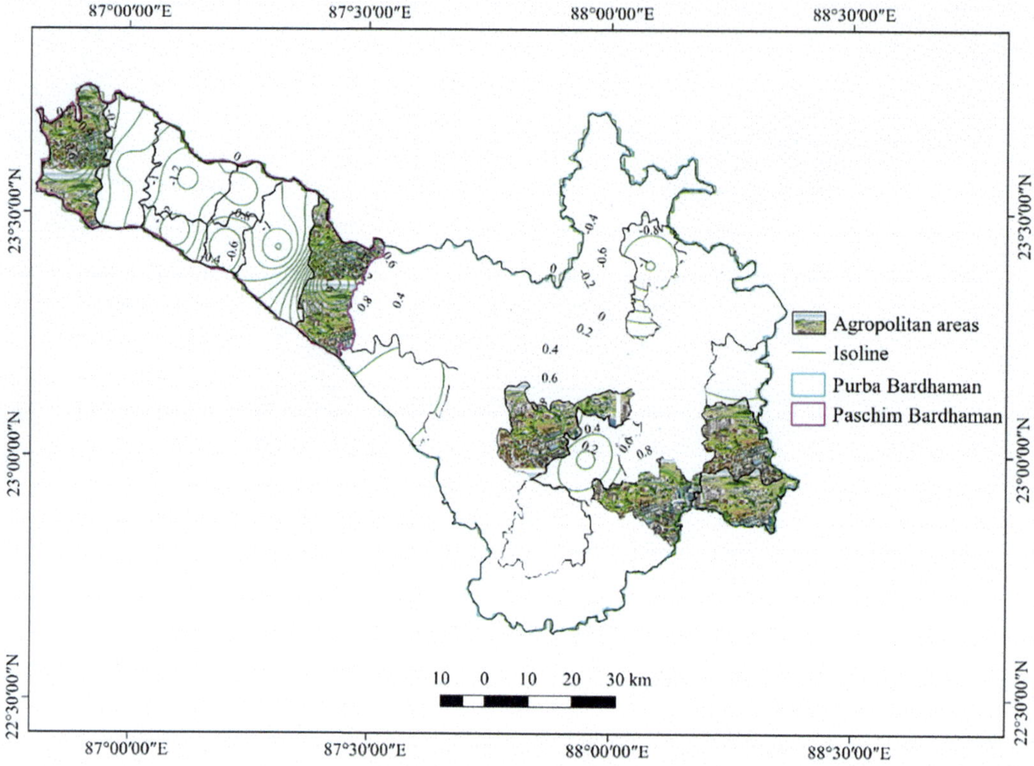

Fig. 20.22 Predicted agropolitan areas of Barddhaman (Purba Bardhaman and Paschim Barddhaman)

standardized predicted values of rural, agricultural, and urban development, and based on those relationships, the representation of different relational triangles has been figured out in Fig. 20.23a–q. The triangles have inequilateral shapes because the standardized predicted values of rural, agricultural, and urban development are different. Figure 20.24 shows the representation of relational triangles of predicted values of rural, agricultural, and urban development in the selected blocks of Barddhaman. In Salanpur, Barabani, Jamuria, Raniganj, Ondal, Pandabeswar, and Katwa-I, the predicted value of agricultural development is less than the predicted value of rural and urban development ($R_d > A_d < U_d$).

In Faridpur Durgapur, Kanksa and Galsi-I, Kalna-II, and Raina-I the predicted value of agricultural development is greater than the predicted values of rural and urban development ($R_d < A_d > U_d$). In addition, the predicted value of agricultural development is greater than the predicted value of rural development and less than the predicted value of urban development in Purbasthali-I ($R_d < A_d < U_d$) and the predicted value of the agricultural development is greater than the predicted value of urban development and less than the predicted value of rural development in Burdwan-I, Burdwan-II, Memari-I, and Kalna-I ($R_d > A_d > U_d$). To balance the regional heterogeneity of the developmental statuses, disparity among the rural, agricultural,

Fig. 20.23 (continued) (**i**) Situation of ZPR (Katwa-I block). (**j**) Situation of ZPR (Purbasthali-I block). (**k**) Situation of ZPR (Galsi-I block). (**l**) Situation of ZPR (Burdwan-I block). (**m**) Situation of ZPR (Burdwan-II block). (**n**) Situation of ZPR (Memari-I block). (**o**) Situation of ZPR (Kalna-I block). (**p**) Situation of ZPR (Kalna-II block). (**q**) Situation of ZPR (Raina-I block). (**a–q**). Relational triangles of standardized predicted values (ZPR) of rural, agricultural, and urban development of the selected blocks of Barddhaman

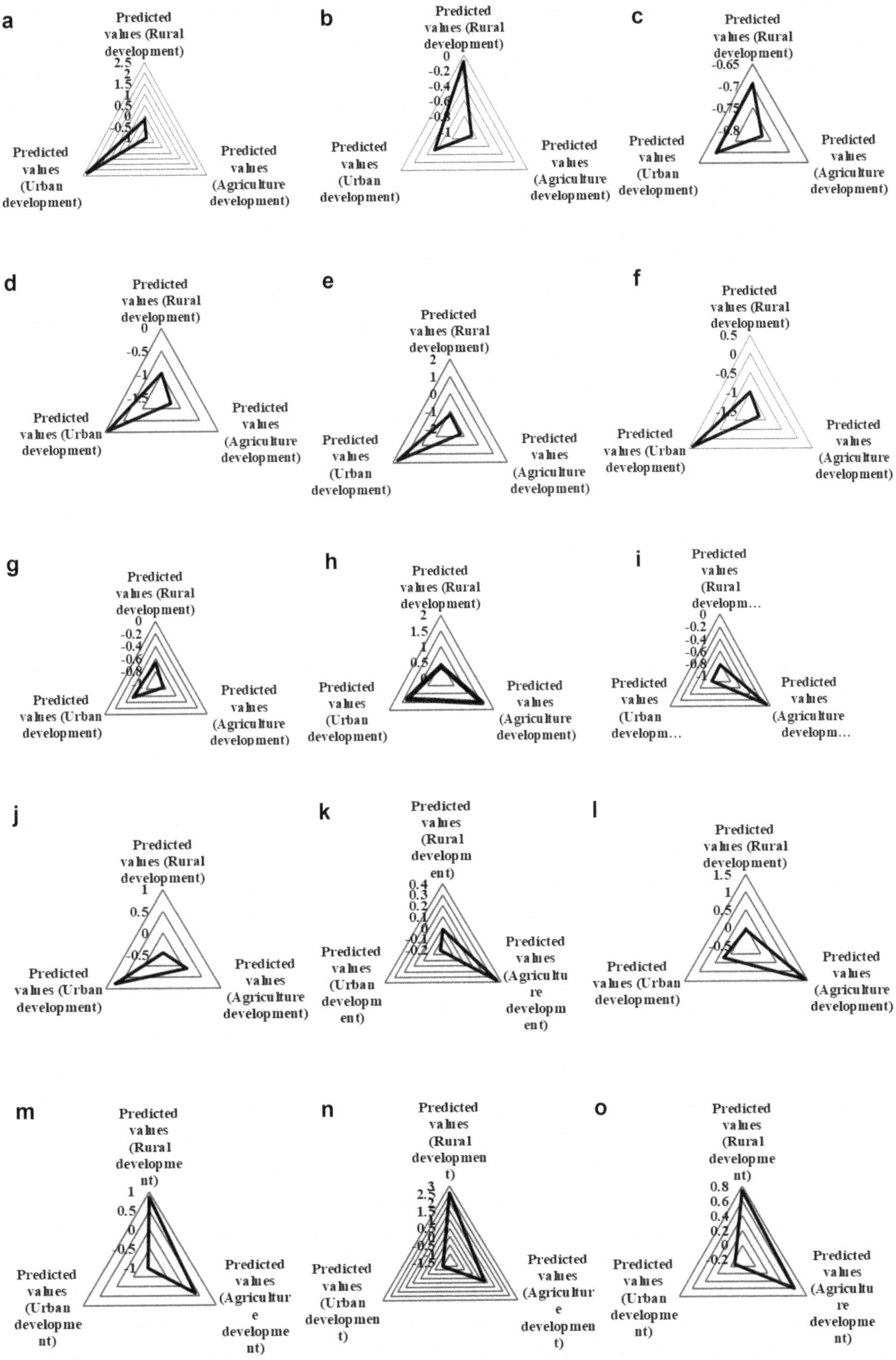

Fig. 20.23 (**a**) Situation of ZPR (Salanpur block). (**b**) Situation of ZPR (Barabani block). (**c**) Situation of ZPR (Jamuria block). (**d**) Situation of ZPR (Raniganj block). (**e**) Situation of ZPR (Ondal block). (**f**) Situation of ZPR (Pandabeswar block). (**g**) Situation of ZPR (Faridpur Durgapur block). (**h**) Situation of ZPR (Kanksa block).

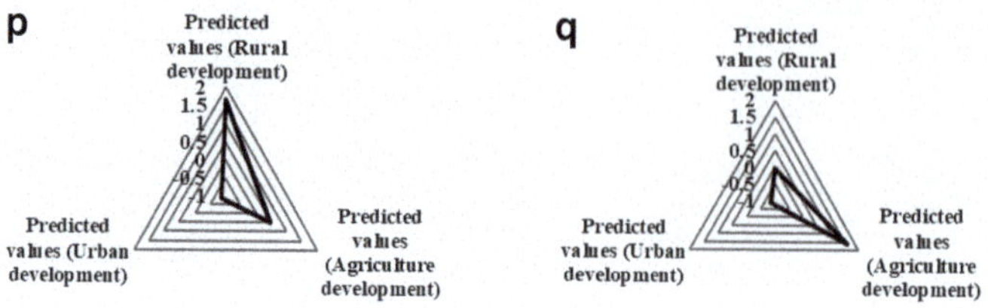

Fig. 20.23 (continued)

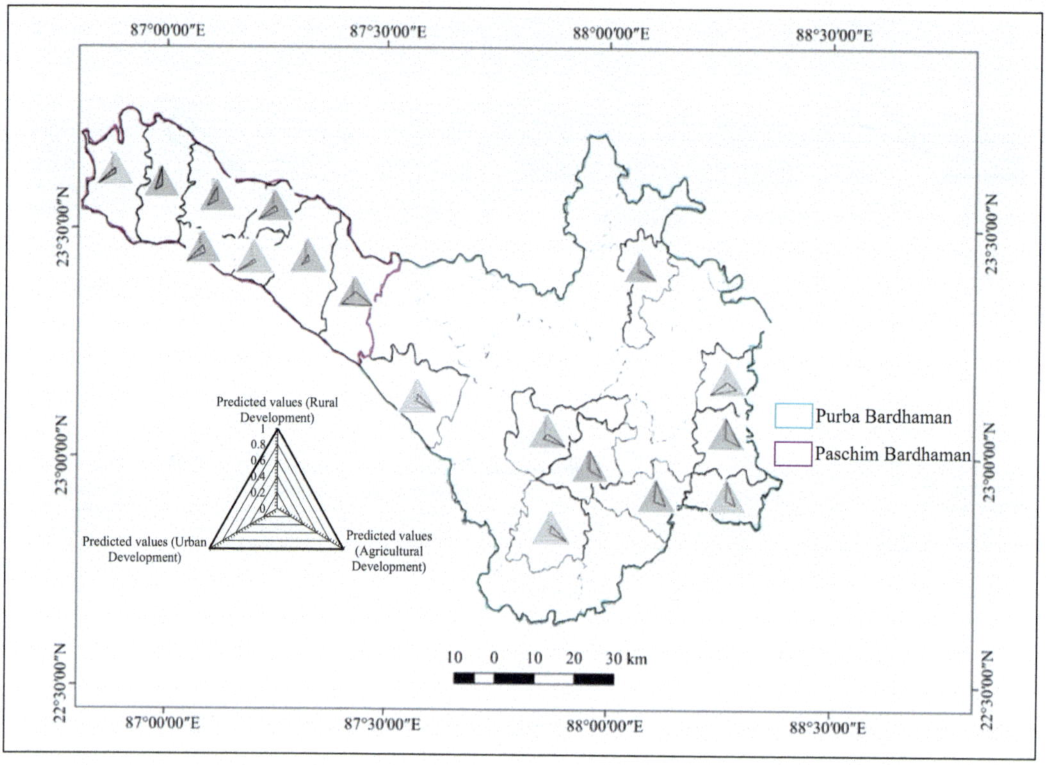

Fig. 20.24 Representation of relational triangles of predicted values of rural, agricultural, and urban development in the selected blocks of Barddhaman

and urban development would be minimized in the agropolitan areas, which have been delineated based on predicted values of composite development. The changing adjustment of every triangle nearer to the equilateral position would suggest the maximum lessening of regional inequality in terms of rural, agricultural, and urban development. As conceptually Fig. 20.27 shows that building up a sustainable agropolitan area is depended on the formation of a rural–urban continuum, delineation of the boundary of an agropolitan area, practicing cultivation process in the city fringe, implementation of an integrated rural, agricultural, and urban development, integration of the developmental statuses, the implication of the keys of climate-resilience sustainable regional development, maintaining the equilibrium of the composite development as much possible, which would tend the composite development to build up a climate-resilient sustainable agropolitan area.

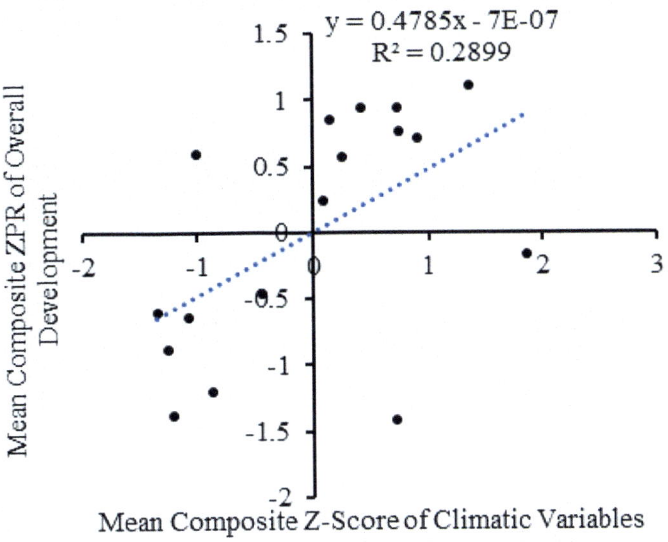

Fig. 20.25 Relationship between mean composite Z-Score of climatic variables and mean composite regression standardized predicted values of overall development (composite development)

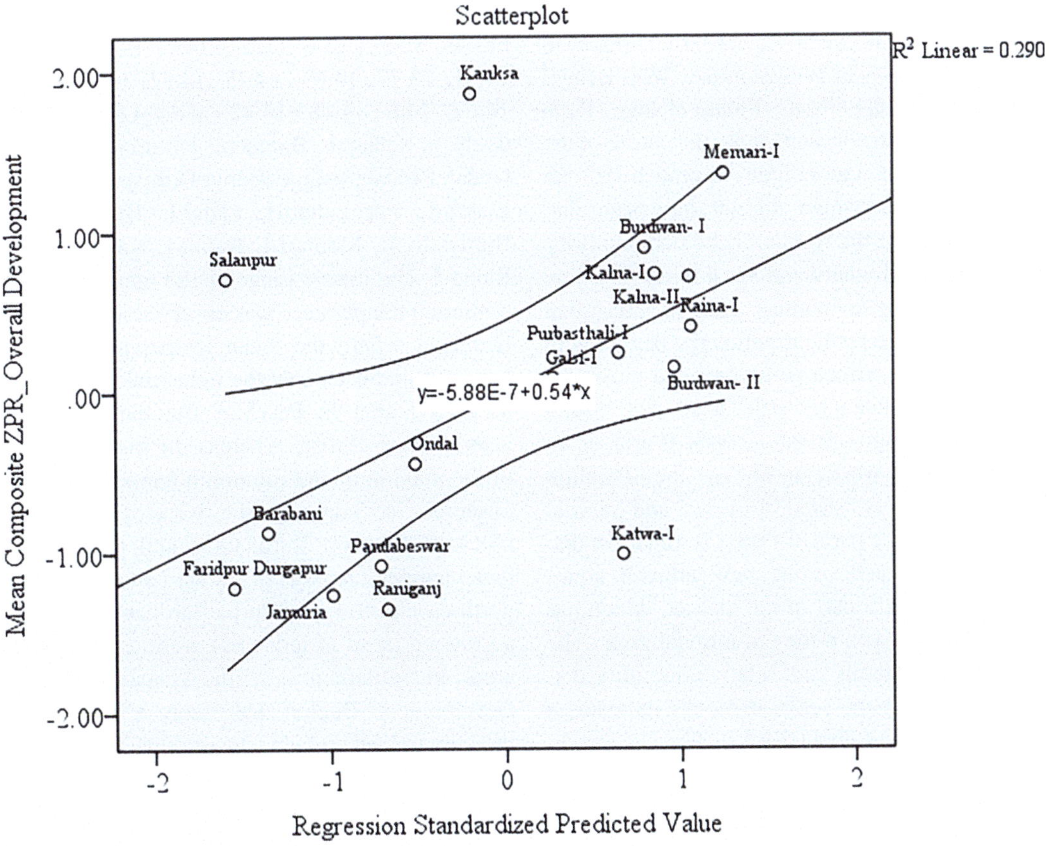

Fig. 20.26 Relationship between regression standardized predicted values and predicted mean composite regression values of overall development (composite development)

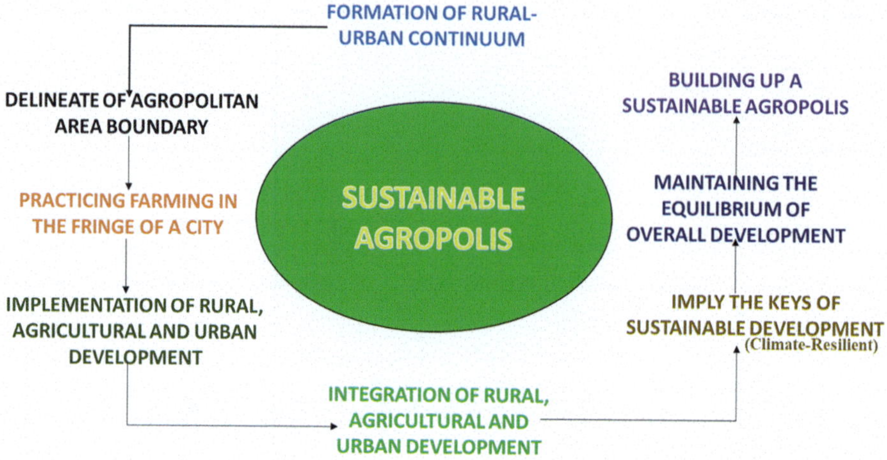

Fig. 20.27 Conceptual organization of a sustainable agropolitan area

20.6 Discussion

Being situated between the eastern Himalayan range and the Bay of Bengal makes West Bengal particularly vulnerable to climate change (Paria et al. 2022). In the present study, the nature of the climate-resilient agropolitan approach in the undivided Barddhaman district has been discussed based on the results of the data analysis. According to Mohanty and Wadhawan (2021, p. vii), organizing economies and societies that embrace climate-resilient pathways begins with identifying risk, which is the first and most crucial step. Barddhaman district is situated within three distinct agro-climatic zones (Paria et al. 2022) where Westernmost of the Paschim Bardhaman has consisted of the red and lateritic zone, most of the parts of Purba Bardhaman district has consisted of the new alluvial zones except the eastern part of the district, which has been consisted within the old alluvial zone. The temperature, rainfall, and relative humidity distribution among the selected blocks indicates a certain range throughout 2011.

According to estimates and projections made by Rama Rao et al. (2013), Barddhaman showed a very low vulnerability index in 2013, 2021–2050, and 2071–2098 for the district-level vulnerability of Indian agriculture to climate change. In 2011, the range of the maximum and minimum temperature (in °C) of the selected C.D. blocks were 15.11, 15.02, 14.9, 14.82, 14.77, 14.81, 13.77, 14.63, 14.75, 14.52, 14.6, 14.63, 14.62, 14.6, 14.48, 14.42, and 14.59, respectively, in Salanpur, Barabani, Jamuria, Raniganj, Ondal, Pandabeswar, Faridpur Durgapur, Kanksa, Katwa-I, Purbasthali-I, Galsi-I, Burdwan- I, Burdwan- II, Memari-I, Kalna-I, Kalna-II, and Raina-I. The lowest range of the maximum and minimum temperature was observed in Faridpur-Durgapur where the mean temperature is also lowest. It indicates that the industrial belt of this urbanized area in Paschim Bardhaman has a near-extreme climate whereas the highest range of the maximum and minimum temperature was observed in Salanpur, the westernmost C.D. block of Paschim Bardhaman and the highest mean temperature was in Katwa-I situated in the north-eastern fringe of Purba Bardhaman. Due to a growing trend in industrial areas, coal mining areas, and settlement areas, the Asansol-Durgapur Development Region has seen a significant increase in land surface temperature (Choudhury et al. 2019, p. 15). The rainfall situation and humidity condition had a proportional relationship in 2011 when most of the C.D. block showed a high humidity condition where the rainfall is high and vice versa. The highest mean annual rainfall (135.03 mm) was observed in Memari-I with 67.87% of average relative humidity. The lowest mean annual rainfall (119.62 mm) was observed in Salanpur with 62.66% of average relative humidity. Except for Murshidabad, Purba Bardhaman, Paschim Bardhaman, Birbhum, and Purulia, all districts in the southern parts of the state had shown a significant decrease trend in monsoon rainfall (Biswas et al. 2022, pp. 414–415). The lowest relative humidity was observed in Salanpur, the westernmost block of the district, whereas the highest relative humidity was observed in Kalna-II block situated in the south-eastern fringe of the district. Thus, it is evident that the C.D. blocks of Purba Bardhaman are less extreme than the C.D. blocks of Paschim Bardhaman.

The rural, agricultural, and urban developmental scenarios have been depicted in the present study intensively. The selected indicators of rural, agricultural, and urban development have regional heterogeneity. In totality, the highest population, household, STs, and main workers were found in the C. D. block Memari-I in 2011 whereas the lowest population, household, literacy, and main workers were found in the C. D. block Raniganj in 2011. The lowest ST population was found in Katwa-1 and the highest literacy in Burdwan-1 in the same census year. Dinda and Ghosh (2015) researched the backward blocks of Barddhaman district and found that one of the three most backward blocks was Memari-I. In the Barddhaman district, Salanpur was the least disadvantaged block, and Ondal was the next-least disadvantaged in 2011 (Dinda and Ghosh 2015). The total number of inhabited villages in Salanpur, Barabani, Jamuria, Raniganj, Ondal, Pandabeswar, Faridpur Durgapur, Kanksa, Katwa-I, Purbasthali-I, Galsi-I, Burdwan- I, Burdwan- II, Memari-I, Kalna-I, Kalna-II, and Raina-I were 69, 46, 38, 12, 12, 14, 48, 77, 63, 91, 85, 75, 83, 111, 98, 112, and 110, respectively, in 2011. There were two Municipal Corporations namely Asansol Municipal Corporation, and Durgapur Municipal Corporation situated in Asansol and Durgapur subdivisions in 2011. Asansol subdivision consisted 3 municipalities (Raniganj, Jamuria, and Kulti), Durgapur subdivision had no municipalities, Burdwan (North) subdivision consisted of 1 municipality (Burdwan), Burdwan (South) subdivision consisted of 1 municipality (Memari), Katwa subdivision consisted 2 municipalities (Katwa and Dainhat), and Kalna subdivision consisted 1 municipality (Kalna) in the district in 2011. The present study has identified that rural development is comparatively higher in the south-eastern portion of the district under Purba Bardhaman, whereas the urban development is comparatively higher in the western and north-western portions under Paschim Bardhaman of the study area. According to Ghorai et al. (2014, p. 44), agricultural cooperative societies and local Panchayats should promote rural agro-production industry-based non-farm activities, which need to be established near the source location of the primary production. Efficiency development in farm and non-farm activities of rural people and small farmers is necessary to achieve the socio-economic development benefits (p. 45). Rahaman (2018, pp. 10421–10422) opined that In West Bengal, the Barddhaman district retained the fourth-place spot (31.16%) in terms of the percentage of the urban population. In comparison to the state and national averages, this district has a higher level of urbanization (39.87%), and the urban population has grown by 20.91% between 2001 and 2011 (p. 10422).

The agricultural development has a distinctive characteristic in Barddhanman district. Paria et al. (2022) postulated that in five districts of West Bengal, namely Burdwan, Bankura, Darjeeling, Cooch Behar, and Purulia, the share of non-food grains as a percentage of the total area cultivated increased from 2004–2005 to 2013–2014. There was no production of *aus* and *boro* paddy in Salanpur, Barabani, Jamuria, Raniganj, Ondal, Pandabeswar, and Faridpur Durgapur in 2010–2011. Galsi-I, Burdwan-I, Burdwan-II, and Raina-I had not shown any production of *aus* paddy also. Raniganj and Kanksa had no production of *aman* and *boro* paddy, respectively. The highest yield of *aus*, *aman*, and *boro* was followed in Katwa-I (2975 kg/hectare), Galsi-I (3487 kg/hectare), and Raina-I (4046.79 kg/hectare), respectively, in 2010–2011. It is observed that agricultural development was higher in the eastern-side blocks in 2010–2011 than in the western-side blocks because of their locational situation, climatic condition, and soil characteristics. Dey and Mistri (2018) mentioned in their study that Purba Bardhaman (the eastern part of the undivided Barddhaman) is situated in "a flat alluvial plain area" (p. 54), which is an agriculturally developed district of West Bengal formally known as the 'granary of the West Bengal' (p. 54). Galsi-I and Kalna-II, as well as Burdwan-I, Burdwan-II, Memari-I, Raina-I, and Kalna-I, had high agricultural efficiency in the C. D. blocks in 2004–2005 (Dey and Mistri, 2018, p. 57). Similarly in 2014–2015, very high agricultural efficiency was observed in Galsi-I and Kalna-II (p. 57). Dey and Mistri (2018) also noted that from 2004–2005 to 2014–2015, Purbasthali–I, Katwa–I, and Kalna–I all displayed increased agricultural productivity (pp. 57–58). The high agricultural productivity is a result of crop rotation, a variety of crops, and multi-crop farming (p. 58). Thus, the present study also depicts that the composite rural, agricultural, and urban development situation was high in the C. D. blocks Kanksa, Burdwan-I, and Memari-I.

While establishing the climate-resilient agropolitan approach in the study area the components of development and climatic variables significantly predict the developmental factor scores ($p < 0.0005$ and $p < 0.05$, respectively). The correlation ($r = 0.538$) shows that climatic variables moderately impact the composite development of the study area. The equilateral figure of the plotted value on the triangles, i.e., rural development (R_d) = Agricultural development (A_d) = Urban development (U_d) should be the sustainability equilibrium position. Due to the heterogeneous and uneven developmental situation, the inserted triangles are inequilateral in the case of sustainability equilibrium of agropolitan area development. Only Kanksa blocks have shown a near-equilateral position among the selected C. D. blocks.

20.7 Conclusion

An accomplished development of any region depends on the integration of urban development with rural and agricultural development. A varied and progressive development in both urban and rural-agricultural sectors has been observed in the present study area. Although the Barddhaman district has been split into two separate districts—Purba Bardhaman and Paschim Bardhaman, where Purba Bardhaman is a center of rural and agricultural development and Paschim Bardhaman is a center of industrial and urban development, the regional heterogeneity, as well as potential patches, could be identified to create the predicted agropolitan areas. The district is diverse due to its physiographic and climatic conditions, as well as the fact that the distribution of rainfall, temperature, and relative humidity is not consistent. Ghosh et al. (2017, pp. 172–173) analyzed that despite experiencing a total of nine floods from 1956 to 2015, the Barddhaman district (undivided) outperformed the Agricultural Scenario Index in both 2011 and 2012. During the monsoon season, the climate was most vulnerable (Ghosh et al. 2017, p. 174) due to extreme rainfall and flood occurrences in the district. Therefore, it is quite impracticable to develop a uniform agricultural sector within a climate-resilient framework. However, rainfall dependence was reduced following the founding of Damodar Valley Corporation (DVC) in 1948 (HDRCC 2011) as a result of the implementation of irrigation projects, primarily in the western part of the district. Further analysis of climate change through institutional arrangements, sector- and livelihood-specific vulnerability, and adaptation is also suggested by Ghosh et al. (2017, p. 175). The developmental patterns of rural, agricultural, and urban areas indicate that some C.D. blocks, like Salanpur and Kanksa in the west, have the potential for both agricultural and urban development. Along with the potential for rural and agricultural development, the eastern part's blocks, including Burdwan-I, Kalna-I and II, and Memari-I, also have the potential for urban growth and development. Rahaman et al. (2020) also identified the scenarios of 'agricultural land', 'built-up land', 'forest cover', 'plantation', 'bare land', and 'waterbodies' as being in a changing situation (1987–2017) in the various areas of the Barddhaman district (p. 4). Delineating agropolitan areas with the characteristics of sustained urban growth surrounded by developing zones of villages and agricultural activities would be promising in maintaining the balance between developmental statuses. Based on the analysis of the standardized predicted values of the composite development, six blocks are appropriate for the development of the agropolitan areas or 'agropolis' (terminology adopted from Koscica 2014). According to Paul and Mehera (2016, p. 555), the contributing factors—increased per capita income and decision-making power, improved women's status and environmental protection, slowed population growth and raised living standards had a significant positive impact on sustainable development. As the relational triangles demonstrate that there are no precise equilibrium positions of climate-resilient development, the gaps between the development of rural, agricultural, and urban areas are required to be minimized in an integrated and sustainable manner to increase the regional equity of the agropolitan areas.

Appendix

Table 20.1 Descriptive statistics of the selected climatic variables

Variables	Mean	Standard deviation
Mean annual temperature (° C)	23.6888	0.15394
Mean annual rainfall (mm)	129.0100	4.38922
Average relative humidity (%)	65.7147	1.85128
Valid N: 17		

Source: Calculated by the authors

Table 20.2 Component score coefficient matrix of the factors of development of rural areas

Component score coefficient matrix[a]		
Factors of the development of rural areas	Component	
	1	2
Total population	**0.561**	−1.398
Total household	**0.033**	−0.042
Total SC population	0.069	**0.680**
Total ST population	0.008	**1.102**
Total inhabited villages	0.000	0.000
Total literacy (persons literate excluding 0–6 years)	**0.265**	−0.471
Total number of main workers	0.066	**0.453**
Total number of marginal workers	0.009	**0.156**
Total number of government schools (primary, secondary, and senior secondary)	0.000	0.000
Total number of government colleges	0.000	0.000
Total number of hospitals (allopathic)	0.000	0.000
Total number of Mouza electrified	0.000	0.000
Extraction method: principal component analysis Component scores[a]		

Source: Calculated by the authors
[a]Coefficients are standardized

Table 20.3 Component score coefficient matrix of the factors of development of agriculture

Component score coefficient matrix[a]		
Factors of the development of agriculture	Component	
	1	2
Gross cropped area (GCA in hectares)	**0.977**	−0.306
Net cropped area (NCA in hectares)	0.008	**0.073**
Cropping intensity (%)	0.000	0.000
Irrigation intensity (%)	0.000	0.000
The area under principal crops (in hectares)	0.041	**0.635**
Production of principal crops (in kg)	0.000	0.000
Average yield (kg/hectares)	0.003	**0.008**
Total number of main cultivators	0.004	**0.047**
Total number of marginal cultivators	0.000	**0.001**
Total number of main agricultural labourers	0.018	**0.281**
Total number of marginal agricultural labourers	0.013	**0.094**
Total number of deep tube wells	0.000	0.000
Extraction method: principal component analysis. Component scores[a]		

Source: Calculated by the authors
[a]Coefficients are standardized

Table 20.4 Component score coefficient matrix of the factors of development of urban areas

Component score coefficient matrix[a]		
Factors of the development of rural areas	Component	
	1	2
Total population	**0.643**	−8.218
Total household	0.028	**0.231**
Total ST population	**0.002**	−0.092
Total literacy (persons literate excluding 0–6 years)	0.291	**6.445**
Total number of main workers	0.034	**1.205**
Total number of government schools (primary, secondary, and senior secondary)	0.000	0.000
Total number of government colleges and universities	0.000	0.000
Total number of hospitals (allopathic)	0.000	0.000
Pucca road length (in km)	0.000	0.000
Electricity domestic connection (numbers)	0.004	**0.488**
Total number of commercial banks	0.000	0.000
Total number of private cinema houses	0.000	0.000
Extraction method: principal component analysis Component scores[a]		

Source: Calculated by the authors
[a]Coefficients are standardized

Table 20.5 REGR factor sores, composite factor scores and mean composite factor scores of the rural, agricultural, and urban developments

Name of the blocks	Rural development			Agricultural development			Urban development			Mean composite factor scores
	REGR factor scores1	REGR factor scores 2	Composite factor scores	REGR factor scores1	REGR factor scores 2	Composite factor scores	REGR factor scores1	REGR factor scores 2	Composite factor scores	
Salanpur	−0.39351	0.14782	−0.24569	−0.66282	−0.66194	−1.32476	0.43977	2.86788	3.30765	0.579067
Barabani	−0.75076	0.63191	−0.11885	−0.53221	−0.71807	−1.25028	−0.09245	−0.65603	−0.74848	−0.70587
Jamuria	−0.81412	−0.16676	−0.98088	−0.3641	−0.73144	−1.09554	−0.01392	−0.97173	−0.98565	−1.02069
Raniganj	−1.61511	0.22035	−1.39476	−0.7525	−1.00458	−1.75708	0.80185	−0.92427	−0.12242	−1.09142
Ondal	−1.41043	−0.27386	−1.68429	−0.76348	−1.01779	−1.78127	2.3884	0.00405	2.39245	−0.3577
Pandabeswar	−1.56573	0.13361	−1.43212	−0.74945	−0.95147	−1.70092	1.95489	−1.44212	0.51277	−0.87342
Faridpur Durgapur	−0.61976	−0.32246	−0.94222	−0.43342	−0.78681	−1.22023	−0.39162	−0.3971	−0.78872	−0.98372
Kanksa	−0.27953	0.80327	0.52374	3.59001	−1.39628	2.19373	0.67163	1.15561	1.82724	1.514903
Katwa-I	0.66587	−1.8361	−1.17023	−0.17775	0.06878	−0.10897	−0.9403	−0.23198	−1.17228	−0.81716
Purbasthali-I	0.55151	−1.1823	−0.63079	−0.17216	0.32677	0.15461	0.16625	0.91113	1.07738	0.2004
Galsi-I	0.69204	−0.71894	−0.0269	0.20785	0.30178	0.50963	−0.54179	0.28751	−0.25428	0.07615
Burdwan- I	0.87913	−0.95535	−0.07622	0.32644	1.64635	1.97279	−0.14171	0.43809	0.29638	0.730983
Burdwan- II	0.45283	0.75096	1.20379	0.02252	0.49773	0.52025	−0.94545	−0.40016	−1.34561	0.126143
Memari-I	1.59414	2.02586	3.62	0.11185	1.00814	1.11999	−0.99083	−0.43015	−1.42098	1.106337
Kalna-I	1.06039	−0.01208	1.04831	−0.01894	0.88039	0.86145	−0.59132	0.47972	−0.1116	0.599387
Kalna-II	0.652	1.70166	2.35366	0.00261	0.62966	0.63227	−0.86228	−0.37045	−1.23273	0.5844
Raina-I	0.90105	−0.94759	−0.04654	0.36554	1.90879	2.27433	−0.91113	−0.32	−1.23113	0.33222

Source: Calculated by the authors

Table 20.6 Standardized predicted values of regression of the rural, agricultural, and urban development

Selected blocks	Standardized predicted values (rural development)	Standardized predicted values (agricultural development)	Standardized predicted values (urban development)	Standardized predicted values of selected variables (composite development)	Relationship among the standardized predicted values of rural development (R_d), agricultural development (A_d), and urban development (U_d) (Dependent variables)
Salanpur	−0.17373	−0.93675	2.33886	0.71416	$R_d > A_d < U_d$
Barabani	−0.08404	−0.88408	−0.52925	−0.87055	$R_d > A_d < U_d$
Jamuria	−0.69359	−0.77466	−0.69696	−1.25882	$R_d > A_d < U_d$
Raniganj	−0.98624	−1.24244	−0.08656	−1.34605	$R_d > A_d < U_d$
Ondal	−1.19097	−1.25955	1.69172	−0.44116	$R_d > A_d < U_d$
Pandabeswar	−1.01266	−1.20273	0.36258	−1.0772	$R_d > A_d < U_d$
Faridpur Durgapur	−0.66625	−0.86283	−0.55771	−1.21323	$R_d < A_d > U_d$
Kanksa	0.37034	1.5512	1.29205	1.86834	$R_d < A_d > U_d$
Katwa-I	−0.82748	−0.07705	−0.82892	−1.00781	$R_d > A_d < U_d$
Purbasthali-I	−0.44604	0.10933	0.76182	0.24715	$R_d < A_d < U_d$
Galsi-I	−0.01902	0.36036	−0.1798	0.09392	$R_d < A_d > U_d$
Burdwan- I	−0.0539	1.39497	0.20957	0.90152	$R_d > A_d > U_d$
Burdwan- II	0.85121	0.36787	−0.95149	0.15557	$R_d > A_d > U_d$
Memari-I	2.55972	0.79195	−1.00478	1.36445	$R_d > A_d > U_d$
Kalna-I	0.74127	0.60914	−0.07891	0.73923	$R_d > A_d > U_d$
Kalna-II	1.66429	0.44708	−0.87168	0.72074	$R_d < A_d > U_d$
Raina-I	−0.03291	1.60819	−0.87054	0.40973	$R_d < A_d > U_d$

Source: Calculated by the authors

Table 20.7 Analysis of variance (ANOVA) of composite development (using selected socio-economic factors)

ANOVA[a]						
Model		Sum of squares	df	Mean square	F	Sig.
1	Regression	10.519	16	0.657	.	0.000[b]
	Residual	0.000	0	.		
	Total	10.519	16			

Source: Calculated by the authors

[a]Dependent variable: mean composite factor scores

[b]Predictors: (constant), Total number of private cinema houses_urban, neat cropped area (NCA in hectares), Total number of government colleges_rural, irrigation intensity, Total number of government colleges and Universities_urban, average yield (kg/hectares), Pucca road length (km)_urban, cropping intensity, Total number of hospitals (allopathic)_urban, deep tube well, area under principal crops (in hectares), Total number of commercial banks_urban, Total number of marginal workers_rural, Total number of main workers_urban, Total number of main cultivators, Total number of government schools (primary, secondary and senior secondary)_rural

Table 20.8 Coefficients of the linear regression of the relationship between climatic variables and composite development

Coefficients[a]						
		Unstandardized coefficients		Standardized coefficients		
Model		B	Std. error	Beta	t	Sig.
2	(Constant)	0.0000000055790	0.211		0.000	1.000
	Mean composite Z-score of climatic variables	0.606	0.245	0.538	2.474	0.026

R: 0.538
R Square: 0.290
Sig.: 0.026

Source: Calculated by the authors
[a]Dependent variable: ZPR_Composite development

References

Agriculture Census (2011) District tables (number and area of operational holdings). National Informatics Centre (NIC), Agriculture Census Division, Department of Agriculture and Farmers Welfare, Government of India. https://agcensus.dacnet.nic.in/districtsummarytype.aspx

Arnold JL, Cangelosi E, Beyea WR, Shaaban A, Kim SK (2021) Rural climate resilience through built-environment interventions: modified deliberation with analysis as a tool to address barriers to adaptive capacity. Reg Stud Reg Sci 8(1):1–24. https://doi.org/10.1080/21681376.2020.1854110

Biswas S, Verma A, Sendhil R, Dixit AK, Singh A, Ponnusamy K (2022) Rainfall pattern analysis to depict the affects of climate change in West Bengal. Environ Ecol 40(2):410–417

Buang A, Habibah A, Hamzah J, Ratnawati YS (2011) The Agropolitan way of re-empowering the rural poor. World Appl Sci J 13(Special Issue of human Dimensions of Development):1–6

Census of India (2011) District census handbook Barddhaman, village and town wise Primary Census Abstract (PCA). Directorate of Census Operations, West Bengal, Series-20, Part XII-B, 1–464. https://censusindia.gov.in/census.website/data/handbooks

Choudhury D, Das K, Das A (2019) Assessment of land use land cover changes and its impact on variations of land surface temperature in Asansol-Durgapur development region. Egypt J Remote Sens Space Sci 22(2):203–218. https://doi.org/10.1016/j.ejrs.2018.05.004

Department of Planning and Statistics (2016) District Statistical Handbook Burdwan (2010–2011, 2013 and 2014). Government of West Bengal. http://wbpspm.gov.in/publications/District%20Statistical%20Handbook

Dey CK, Mistri T (2018) Changing trends of agricultural efficiency (Ei) in Purba Bardhaman District, West Bengal, India. J Emerg Technol Innov Res 5(6):53–65

Dey CK, Mistri T (2019) Changing trends of market prices of rice in Burdwan and Memari Markets of Purba Bardhaman District, West Bengal, India. J Inf Comput Sci 9(8):492–508

Dinda S, Ghosh D (2015) A search for backward blocks in Burdwan District, West Bengal: an application of composite development index. Munich Personal RePEc Archive (MPRA). https://mpra.ub.uni-muenchen.de/67158/

District Administration Paschim Bardhaman (2023) Subdivisions and blocks. Government of West Bengal. https://paschimbardhaman.gov.in/subdivision-blocks/

District Administration Purba Bardhaman (2023) Subdivision and blocks. Government of West Bengal. https://purbabardhaman.nic.in/subdivison/

Dutta AK (2012) Rice trade in the 'rice bowl of Bengal': Burdwan 1880–1947. Indian Econ Soc Hist Rev 49(1):73–104. https://doi.org/10.1177/001946461104900103

Dutta S, Guchhait SK (2022) Assessment of land use land cover dynamics and urban growth of Kanksa block in Paschim Barddhaman District, West Bengal. GeoJournal 87(2):971–990. https://doi.org/10.1007/s10708-020-10292-3

Fallah-Alipour S, Mehrabi Boshrabadi H, Zare Mehrjerdi MR, Hayati D (2018) A framework for empirical assessment of agricultural sustainability: the case of Iran. Sustainability 10(12):4823. https://doi.org/10.3390/su10124823

Farhanah L, Prajanti SDW (2015) Strategies in developing Agropolitan areas in Indonesia. J Ekonomi Pembangunan 16(2):158–165. https://doi.org/10.23917/jep.v16i2.1460

Fisher RA (1925) Theory of statistical estimation. Proc Camb Philos Soc 22(5):700–725. https://doi.org/10.1017/S0305004100009580

Fisher RA (1934) Statistical methods for research workers, 5th edn. Oliver and Boyd, pp 1–319

Friedman EH (1985) Generation to generation: family process in church and synagogue. Guilford

Friedmann J (1979) Basic needs, Agropolitan development, and planning corn below. World Dev 7:607–613. https://doi.org/10.1016/0305-750X(79)90096-2

Ghorai D, Jana CK, Ziauddin G, Sarkar S, Singh MS, Saikia P, Ghatak S, Chatterjee J (eds) (2014) Comprehensive district agriculture plan, Burdwan, under XIIth plan. A report submitted to the District Magistrate, Burdwan, pp 1–204
Ghosh S, Mahato K, Gorain S, Das U, Mondal B (2017) Resilience of agriculture reducing vulnerability to climate change in West Bengal. Curr Adv Agric Sci 9(2):170–177. https://doi.org/10.5958/2394-4471.2017.00031.4
Global Modeling and Assimilation Office (GMAO) (2015) *MERRA-2 day and month-wise rainfall data of selected coordinate points in West Bengal, India.* Goddard Earth Sciences Data and Information Services Center (GES DISC), Accessed on 23 December, 2022. https://doi.org/10.5067/VJAFPLI1CSIV
HDRCC (2011) District human development report: Bardhaman. Development and Planning Department, Government of West Bengal. http://wbpspm.gov.in/publications/District%20Human%20Development%20Report
Huq N, Hugé J, Boon E, Gain AK (2015) Climate change impacts in agricultural communities in rural areas of coastal Bangladesh: a tale of many stories. Sustainability 7(7):8437–8460
Indah PN, Sam ZA, Damaijati E (2017) Identifying potential estate commodity for Agropolitan development in Ponorogo. Int J Agric Syst (IJAS) 5(1):60–68. https://doi.org/10.20956/ijas.v5i1.1171
India Meteorological Department (2015) Climatological table. Ministry of Earth Sciences. Government of India
Jat ML, Dagar JC, Sapkota TB, Govaerts B, Ridaura SL, Saharawat YS et al (2016) Climate change and agriculture: adaptation strategies and mitigation opportunities for food security in South Asia and Latin America. Adv Agron 137:127–235. https://doi.org/10.1016/bs.agron.2015.12.005
Javadinejad S, Eslamian S, Ostad-Ali-Askari K, Nekooei M, Azam N, Talebmorad H et al (2019) Relationship between climate change, natural disaster, and resilience in rural and urban societies. In: Handbook of climate change resilience, pp 1–25. https://doi.org/10.1007/978-3-319-71025-9_189-1
Javadinejad S, Dara R, Jafary F (2021) Analysis and prioritization the effective factors on increasing farmers resilience under climate change and drought. Agric Res 10:497–513. https://doi.org/10.1007/s40003-020-00516-w
Kim G, Kang W, Lee J (2020) Knowledge structures and components of rural resilience in the 2010s: conceptual development and implications. Sustainability 12(22):9769. https://doi.org/10.3390/su12229769
Koscica M (2014) Agropolis: the role of urban agriculture in addressing food insecurity in developing cities. J Int Aff:177–186. https://www.jstor.org/stable/24461745
Leichenko R (2011) Climate change and urban resilience. Curr Opin Environ Sustain 3(3):164–168. https://doi.org/10.1016/j.cosust.2010.12.014
Li F, Wang R, Paulussen J, Liu X (2005) Comprehensive concept planning of urban greening based on ecological principles: a case study in Beijing, China. Landsc Urban Plan 72(4):325–336. https://doi.org/10.1016/j.landurbplan.2004.04.002
Mohanty A, Wadhawan S (2021) Mapping India's climate vulnerability– a district level assessment. Council Energy Environ Water, New Delhi, pp 1–74
Mondal BK, Basu T, Das R, Mahata S, Lee MA (2023) Geospatial assessment of fluvial dynamics and associated flood vulnerability in the Bhagirathi-Hugli subbasin in West Bengal, India. In: Islam A, Shit PK, Datta DK, Islam MS, Roy S et al (eds) Floods in the Ganga–Brahmaputra–Meghna Delta. Springer, Cham, pp 491–527. https://doi.org/10.1007/978-3-031-21086-0_21
National Informatics Centre (Burdwan District Unit) (2011) Geography. Ministry of Electronics and Information Technology, New Collectorate Building (3rd floor), Burdwan, West Bengal – 713 101, India. http://bardhaman.nic.in/geography.html
Neto F (2003) A new approach to sustainable tourism development: Moving beyond environmental protection. In: Natural resources forum, vol. 27, no. 3. Blackwell, pp 212–222. https://doi.org/10.1111/1477-8947.00056
Paria B, Mishra P, Behera B (2022) Climate change and transition in cropping patterns: district level evidence from West Bengal, India. Environ Challenges 7:100499. https://doi.org/10.1016/j.envc.2022.100499
Paul PK, Mehera C (2016) Impacts of education on sustainable development: a micro study in Burdwan District of West Bengal, India. Am J Educ Res 4(7):551–555
Pearson K (1897) Mathematical contributions to the theory of evolution. On a form of spurious correlation which may arise when indices are used in the measurement of organs. Proc R Soc Lond 60(359–367):489–498. https://doi.org/10.1098/rspl.1896.0076
Pearson KFRS (1901) LIII. On lines and planes of closest fit to systems of points in space. Lond Edinb Dublin Philos Mag J Sci 2(11):559–572. https://doi.org/10.1080/14786440109462720
Pearson K (1914) On certain errors with regard to multiple correlation occasionally made by those who have not adequately studied this subject. Biometrika 10(1):181–187. http://www.jstor.org/stable/2331747
Personnel and Administrative Report Department (2017) The Kolkata Gagette. Government of West Bengal. http://wbpar.gov.in/writereaddata/80%20AR.pdf
Peterson JCK (1910) Bengal District Gazetteers Burdwan 1910. https://archive.org/details/in.ernet.dli.2015.461799
Poli AE, Bisri MS, Lengkong EL (2013) Agropolitan development in East Tomohon, North Sulawesi Indonesia. IOSR J Bus Manage (IOSR-JBM) 13(3):35–40
Praburaj L, Design F, Nadu T (2018) Role of agriculture in the economic development of a country. Shanlax

Int J Commerce 6(3):1–5. https://doi.org/10.5281/zenodo.1323056

Prasetiya A, Suyadi, Bisri M, Soemarno (2014) Analysis of Sendang Agropolitan area development, Tulungagung. Am J Soc Res 4(2):60–66

Qian H, Wong C (2012) Master planning under urban–rural integration: the case of Nanjing, China. Urban Policy Res 30(4):403–421. https://doi.org/10.1080/08111146.2012.737778

Rahaman M (2018) Trend of urbanization in Barddhaman District, West Bengal, India. Int J Curr Adv Res 7(2L):10421–10426

Rahaman S, Kumar P, Chen R, Meadows ME, Singh RB (2020) Remote sensing assessment of the impact of land use and land cover change on the environment of Barddhaman district, West Bengal, India. Front Environ Sci 8:1–127. https://doi.org/10.3389/fenvs.2020.00127

Rama Rao CA, Raju BMK, Subba Rao AVM, Rao KV, Rao VUM, Ramachandran K, Venkateswarlu B, Sikka AK (2013) Atlas on vulnerability of Indian agriculture to climate change. Central Research Institute for Dryland Agriculture, Hyderabad, pp 1–116

Rosdiana H (2014) Evaluation of fiscal policy on agropolitan development to raise sustainable food security (a study case in Bangli Regency, Kuningan Regency and Batu Municipality, Indonesia). Procedia Environ Sci 20:563–572. https://doi.org/10.1016/j.proenv.2014.03.069

Roy J, Bhowmick R, Dolui M (2014) Actions for sustainable development evidence based assessment for the State of West Bengal. Arthaniti 13(1–2):1–18. https://doi.org/10.1177/0976747920140101

Safariah R, Kurniadie D, Widodo T (2016) The development study of Agropolitan Region to optimize natural resources potential in Padang Pariaman. Int J Appl Bus Econ Res 14(2):695–710. https://serialsjournals.com/abstract/98973_695-710.pdf

Sakir M, Tikson DT, Haning T, Susanti G (2017) The effect of Agropolitan program to social economic conditions of the farmers. Scholars J Econ Bus Manage 4(5):357–362

Sarkar S, Ghosh TG (2017) Agricultural development in West Bengal: an inter-temporal analysis. Econ Aff 62(3):483–493. https://doi.org/10.5958/0976-4666.2017.00060.2

Shaffril HAM, Nasir AFA, Idris K, Uli J, D'Silva JL (2010) Agriculture project as an economic development tool to boost socio-economic level of the poor community: the case of Agropolitan project in Malaysia. Afr J Bus Manag 4(11):2354–2361

Singh JS (2006) Sustainable development of the Indian Himalayan region: linking ecological and economic concerns. Curr Sci 90(6):784–788. https://www.jstor.org/stable/24089189

Smith G, Nandwani D, Kankarla V (2017) Facilitating resilient rural-to-urban sustainable agriculture and rural communities. Int J Sust Dev World 24(6):485–501. https://doi.org/10.1080/13504509.2016.1240723

Student (1908) The probable error of a mean. Biometrika 6(1):1–25. https://doi.org/10.2307/2331554

United Nations (2022) The sustainable development goals report 2022. Sustainable Development, Department of Economic and Social Affairs. United Nations, pp. 1–65. https://unstats.un.org/sdgs/report/2022/

Viswanathan PK, Kavya K, Bahinipati CS (2020) Global patterns of climate-resilient agriculture: a review of studies and imperatives for empirical research in India. Rev Dev Change 25(2):169–192. https://doi.org/10.3390/su7078437

Zhong B, Wu S, Sun G, Wu N (2022) Farmers' strategies to climate change and urbanization: potential of ecosystem-based adaptation in rural Chengdu, Southwest China. Int J Environ Res Public Health 19(2):952. https://doi.org/10.3390/ijerph19020952

21

Analysing Sustainable Approaches in MGNREGA Works for Climate Change Adaptation: Case Study of Debra Block, West Bengal, India

Piu Dutta, Debashish Das, and Debasree Bhadra

Abstract

MGNREGA (Mahatma Gandhi National Rural Employment Guarantee Act) is a government scheme in India that guarantees a minimum of 100 days of wage employment per year to every rural household whose adult members volunteer to do unskilled manual work. While the primary objective of MGNREGA is to provide livelihood security to rural households, there is also a focus on incorporating adaptive and sustainable approaches within the program. This scheme aims for Natural Resource Management, Renewable Energy, Sustainable Agriculture, Watershed Development, Skill Development and Capacity Building, and Climate Change Adaptation. The Green Index (G.I.) combines socio-economic and environmental factors to assess overall green performance of a region. By examining the relationship between these indices, the study seeks to shed light on interplay between Socio-Economic Development (S.E.D.), Environmental Improvement (E.I.), and green initiatives. This chapter emphasized on the relation between S.E.D. and E.I. among 14 Gram Panchayat (G.P.) of Debra block, Paschim Medinipur, West Bengal, where 16 major types of works (N) are currently in progress under the MGNREGA. The present chapter explored the regional disparities in S.E.D. by using Socio-Economic Development Index (S.E.D.I.) and E.I. by using Green Index (G.I.). The chapter suggested that if the correlation between the value of S.E.D.I. and G.I. is directly proportional then the development process should move to Loop-1 (as shown in Model 21.1), which represents sustainable development. On the contrary, inverse relation will lead to unsustainable development, as shown in loop 2.

Keywords

Socio-economic development index · Green index · MGNREGA climate change adaptation · Sustainable development

21.1 Introduction

Sustainable Development is a process to develop a local region and at the same time improve the environmental condition and ecosystem services. Environment and sustainable development have been introduced in the world since the year 1960 (Mattei 2015). Sustainable development consists

P. Dutta (✉) · D. Das
Department of Architecture, Jadavpur University, Kolkata, West Bengal, India
e-mail: ddasju.arch@jadavpuruniversity.in

D. Bhadra
Departments of Geography, Lady Brabourne College, Kolkata, West Bengal, India

U. Chatterjee et al. (eds.), *Climate Crisis: Adaptive Approaches and Sustainability*, Sustainable Development Goals Series, https://doi.org/10.1007/978-3-031-44397-8_21

of two principle approach: one is 'needs' and another is 'limitations' or environment's ability (Our common future, chapter: 2 1987). In the year 1983, the term 'sustainable development' was conceived by Indian economist, Nitin Desai, while he was a senior economic consultant to the Brundtland commission formerly known as World Commission on Environment and Development (WCED) that was authenticated by UNEP Governing Council. Sustainable development is essential in modern era, where world population has witnessed the impact of urbanization and modernization status, which only promotes growth, through enormous extraction of natural resources. This is the main cause of climate crises on urban habitat and urban ecosystem. Overall development of any area depends on economic status, educational status, health and nutrition services, magnitude of modernization and urbanization, women status, housing quality, transport and communication services and distribution of essential facilities (Das 1999), as well as environmental condition, climatic condition, ecosystem services, etc.

Being a developing country, the progress of India depends on rural development both qualitatively and quantitatively. According to census 2011, there were 6.4 lakh villages in India and 83.3 crore people (census 2011). Rural people mainly depend on natural resources or primary activities like agriculture, fishing, mining, etc. (Sarkar and Kundu 2016). Enormous dependency of natural resources gives an incredible strain on climate, environment, habitat and ecosystem. MORD has introduced MGNREGA, an innovative programme to improve ecosystem services (water, land, soil, and forest) as well as human systems (employment opportunities, shelter, and communication) (Ravindranath and Murthy 2013; Shah et al. 2012). Implementation of MGNREGA has reduced vulnerability related to climate crisis, water scarcity, and agricultural variability and at the same time it has increased economic status, crop yield, employment opportunities and increased daily wage level. Climatic conditions and ecosystem services as well as rural livelihood both are improved after implementation of MGNREGA. The climate adaptive approaches significant in MGNREGA works focusses on following aspects of sustainable development as given below:

(i) Natural Resource Management: MGNREGA encourages the implementation of works that promote sustainable management of natural resources. This includes activities like water conservation through construction of check dams, water harvesting structures, and recharge wells. Afforestation and soil conservation measures, such as contour stabilizing and tree plantation, are also undertaken to prevent soil erosion and improve environmental sustainability.

(ii) Renewable Energy: To promote sustainable energy practices, MGNREGA works incorporate the use of renewable energy sources. This involves the installation of solar-powered streetlights, solar water pumps for irrigation, and solar panels for electrification of public facilities. By reducing reliance on fossil fuels and promoting clean energy alternatives, these initiatives contribute to both environmental sustainability and rural development.

(iii) Sustainable Agriculture: MGNREGA supports initiatives that focus on sustainable agricultural practices. This includes the promotion of organic farming, vermicomposting, and the use of biofertilizers and biopesticides. The program also encourages the establishment of seed banks to conserve local varieties of crops and promotes agroforestry practices that combine tree planting with agricultural crops, enhancing soil fertility and diversifying income sources.

(iv) Watershed Development: Integrated watershed management is a key component of MGNREGA. Watershed development projects involve the construction of earthen and masonry check dams, percolation tanks, and farm ponds. These structures help in rainwater harvesting, recharging groundwater levels, and improving irrigation facilities. Watershed management also includes soil and moisture conservation measures, afforestation, and water conservation awareness campaigns.

(v) Skill Development and Capacity Building: MGNREGA emphasizes skill development and capacity building of rural workers. By providing training in various trades and vocations, such as construction, masonry, plumbing, and carpentry, the program equips individuals with skills that can lead to sustainable livelihood opportunities beyond the 100 days of guaranteed employment. This helps in reducing dependency on unskilled manual work and enhances the long-term resilience of rural communities.

(vi) Climate Change Adaptation: MGNREGA recognizes the importance of climate change adaptation in rural and urban villages. It promotes climate-resilient infrastructure development, such as construction of flood-resistant roads, bridges, and buildings. The program also encourages the adoption of climate-smart agricultural practices, including drought-tolerant crops and water-efficient irrigation techniques, to mitigate the impacts of climate change on rural livelihoods:

The objectives of this chapter is to develop a progressive index to define the sustainability based on environmental impact of MGNREGA works and the economic, social and cultural development of a society. The study's exploration of the relationship between these indices can provide insights into how rural cities and communities can promote sustainable practices, enhance environmental quality, and improve resilience to climate change. With all these important aspects, this chapter explored the regional disparities in S.E.D. by using Socio-Economic Development Index (S.E.D.I.) and E.I. by using Green Index (G.I.), for which Debra Development Block as an indicator of rural region in West Bengal was selected, where MGNREGA works are actively running since the year 2006. The year of implementation of MGNREGA and the expansion of MGNREGA works were the two key principles for electing this case study. MGNREGA was revealed in this block during the first phase (with effect from second February 2006) of implementation and also the block has high level of work expansion. This block is well connected through rail (South-Eastern Railway), road (NH-6) and water transport (Kangsabati River), and it is located approximately 100 km away from the capital city of West Bengal, Kolkata.

Another fact that we would like to notice that this block is one of places where Rural and Urban livelihood both are presented. The census town Balichak, under Duan -II Gram Panchayat is one and only urban area in Debra block, which is surrounded by rural areas. This is known as 'Urban Village' (Samsurijan et al. 2017). It is also known as Rurban area where urban and rural activities were merged with each other (Mandal 1989; Kolhea and Dhoteb 2016). The study area Balichak is an unplanned urban village, so this region has faced so many climate crises. MGNREGA is an important and very essential programme to rejuvenate the urban village habitat and ecosystem services. Actually, specific study through green index of MGNREGA works on the environment of urban village is very unique and inevitable for sustainable urban environment development.

21.1.1 Information about MGNREGA

MGNREGA was first introduced in second February 2006 in 200 economically disadvantaged districts and after that it was elongated to other 130 districts in 2007–2008 (Ganeriwala 2010; Karthika 2015). Since 2008, India is totally covered under MGNREGA in rural areas, without the 100% urban areas (Ganiee 2014; Yadav 2014). It provides directly 100 days of wage-employment to the desired family. In long run, it has huge potential to enrich environmental protection and ecosystem services, which is linked to water, soil, land and forest. There are so many studies on economic, social, financial and administrative aspect of MGNREGA works (Shah 2007; Kareemulla et al. 2009). Few researchers have focused on the subject of MGNREGA with poverty, gender issues, livelihood security, migration, etc. (Ghosh 2011; Jacob 2008; and Sefiya 2015)), but there have been very few studies conducted to assess the potential of MGNREGA's environmental benefits (Tiwari et al. 2011; Esteaves et al. 2013). In this study, we have emphasized on Green Index (GI), which is the quantitative measurement of ecosystem services of MGNREGA works. As per reports, among 14 Gram Panchayat (G.P.) of Debra block, where 16 major types of works are currently in progress under the MGNREGA. The MGNREGA works were analyzed (Table 21.1) in study area based on their category:

The work under MGNREGA is related to both environment and sustainable development. Environment-related work has the potential to improve environmental quality through the conservation and regeneration of ecosystem services like water, soil or land and forest. Also, the development-related work has the potential to empower rural people and strengthen the livelihood capacity through generating employment opportunity and create durable assets in rural region (Kharkwal and Kuma 2015) Simultaneously, MGNREGA has reduced distress migration from village to urban area (Mistry and Jaswal 2009). Figure 21.1 represents the MGNREGA works according to their impact on environment and socio-economic development.

21.1.2 Impact of MGNREGS on Water Resources

Water-related MGNREGA works are probably to have a direct impact on surface and ground water recharge. Water resource management-related such MGNREGA works (Fig. 21.2) are Farm pond, Check Dam, Earthen Dam, Mini-percolation tank, Earthen bounding, Construction and renovation of canal, Re-excavation of tank or pond, Desilting of irrigation tank or pond, etc. These works also have the potential to control surface water runoff, control soil loss, increase infiltration of water and recharge the ground water, availability of irrigation water, area irrigated by bore wells/open wells, availability of drinking water for humans and livestock (Verma and Shah 2012). The ground water level of Debra block area is 50–115 m during rainy season and 120–195 m in dry season. The number of deep tube wells (DTWs) of water extraction for irrigation were 492 before MGNREGA period (2003–04) and after MGNREGA implementation (2013–2014) the number of DTWs has increased to 2898. This is the sign of ground water availability, which is increased post-MGNREGA period. So, we can accept that MGNREGA works have positive impact on the underground water level in the study region, despite outstretched removal of ground water. Surface water storage

Table 21.1 MGNREGA work category-wise analysis in financial year (2017–18)

Category	Work category name	Sub-work category name
A.	Public works relating to natural resources management	1. Water conservation 2. Watershed management 3. Irrigation 4. Traditional water bodies 5. Afforestation 6. Land development
B.	Individual assets for vulnerable sections	7. Improving productivity of lands 8. Improving livelihoods 9. Development of fallow /waste land 10. Construction of house 11. Promotion of livestock 12. Promotion of fisheries
C.	Rural infrastructure	13. Rural sanitation 14. Road connectivity 15. Play field 16. Disaster preparedness/restoration

Source: Management Information Systems (MIS) data from website

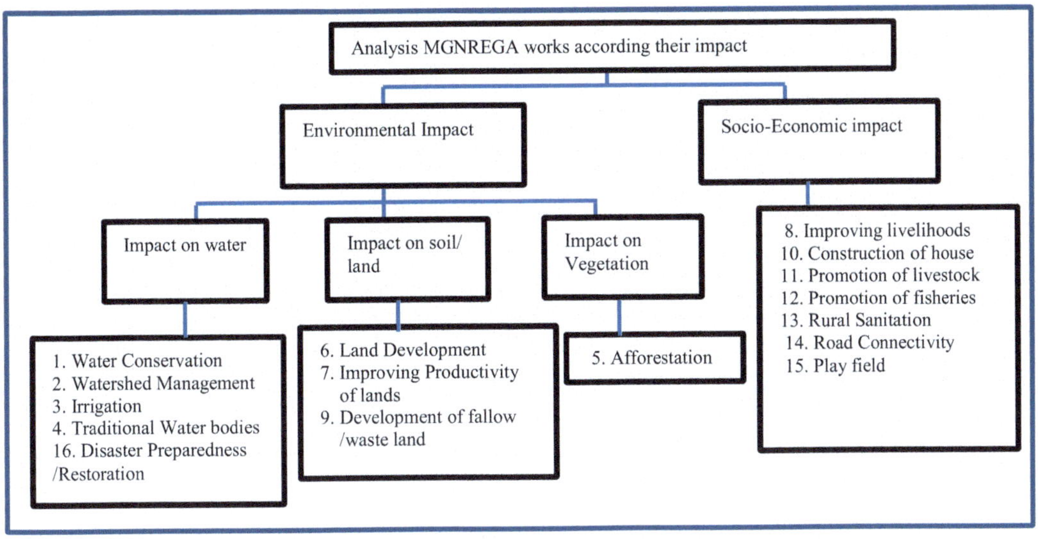

Fig. 21.1 MGNREGA works according their impact. (Source: prepared by author)

Fig. 21.2 (i) Renovation of farm pond, (ii) Earthen Dam, (iii) Renovation of traditional waterbodies with construction of Guard wall

was also increased by the work renovation or desilting of traditional water bodies.

21.1.3 Impact of MGNREGS on Land/ Soil

Land development activities (Fig. 21.3) under MGNREGA involves land levelling, Reclamation of land, Development of waste and fallow land, low land filling and silt removed from desilting of traditional water bodies. Silt was applied to crop land to enhance the land fertility and improve the organic carbon concentration, which is an indicator of nitrogen in soil. Before, the degraded lands or waste land had not been fertile. After land development activities the degraded land are being cultivated and crop yield will be increased. So these activities have not only improved the ecological condition of degraded land but also improved huge financial benefits to local farmers. There has been a significant improvement in crop cultivation and income generation of individual farmers. So, we can accept that MGNREGA works related to land development have positive impact on the soil or land and the work has potential to improve crop productivity and income generation.

Fig. 21.3 (i) Low land filling with silt, (ii) Name plate of land development work, (iii) plantation after land development

Fig. 21.4 (i) Plantation/Social Forestry, (ii) Name plate of plantation work, (iii) Afforestation at the boundary of Playground

21.1.4 Impact of MGNREGS on Vegetation

Drought proofing-related MGNREGA works (Fig. 21.4) such as Plantation, Social Forestry, Social Afforestation at Playground, Plantation beside pond or Khal, Strip plantation at Burial and Cremation Ground, etc. are the works that improve the vegetation density, biomass concentration and carbon sequestration in the rural region (Shah et al. 2012). MGNREGA work increases and improves the forest cover of the country. Plantation work has the potential to improve all environmental quality i.e. air, soil, water, biodiversity. The work covers tree plantation on embankment of tank, wastelands and fallow land, road sides, canal banks, and drainage sides.

21.1.5 Impact of MGNREGS on Socio-economic Development

Several MGNREGA works such as Construction of IAY houses, Houses for S.C/S.T. and low–/middle-class people, repair of village road with moorum, construction of earthen roads, concrete road, toilets, playgrounds, etc. (Fig. 21.5) have large impact on social, economic and cultural improvement, but there is no environmental

Fig. 21.5 (i) concrete road, (ii) House for low or middle class people, (iii) village road with moorum

improvement. So these works have only positive impact on rural livelihood, employment, and rural infrastructure development (Haque 2011). Those MGNREGA works can potentially lead to reduce the migration of landless labourers or unskilled labourers through the massive employment generation at village level (Kumar and Prasanna 2010). Non-environment-friendly materials, such as sand, stones, iron, steel, brick, and cement, etc. are used to construct the rural assets, such as road, house, toilet, etc. Manufacture and transportation of those construction materials have heavy impact on environment. Green House Gas emissions from construction activities are the sum of environment toll of material manufacture and environment toll of transportation of material (Bhaskar et al. 2017).

21.2 Database and Methodology

21.2.1 The Study Area

The study was carried out at Debra Community Development (C.D.) Block of Paschim Medinipur district, West Bengal (Map 21.1). The block covers 342.29 km^2 and is divided into 14 Gram Panchayats and 487 mouzas or villages. Among those, 472 are Inhabitant mouzas and another 15 are depopulated mouzas. The population of the block is 2,84,968 of which 1,44,566 are male and 1,40,402 are female and the population density is 832 per km^2, according to the census 2011.

National Highway (NH-6) covers 15.5 km. in this Block. Maximum population in this region depends on rural sources for survival. The study area is under the hot and humid monsoon region. The average annual rainfall is 1400 mm and a large amount of rainfall (near about 80%) is in monsoon period from June to September. The winter and summer temperature ranges between 10 °C and 40 °C. The region is rich in alluvium soil, which mainly consists of clay particle. So this soil gives high yield in agricultural field. Here, the occupation of most people (48,491 people) is agriculture. In this Block, the total workers are 1,14,400 of which 6681 are employed, 73,160 are semi-employed and 34,559 are unemployed. A vast amount of household 37,188 are Marginal and Small Farmers (1 to 2Ha). The census town Balichak, under Duan-II Gram Panchayat is a large rural settlement or small town, where rural land slowly turnover in a form of urban islands. This census town is lying in a transitional stage between village and town to serve the adjacent areas (Shikalgar and Kamble 2013). This urban region has a population of 13,784. This region is "urban" in nature but are administrated by Rural Local Bodies (Gram Panchayats).

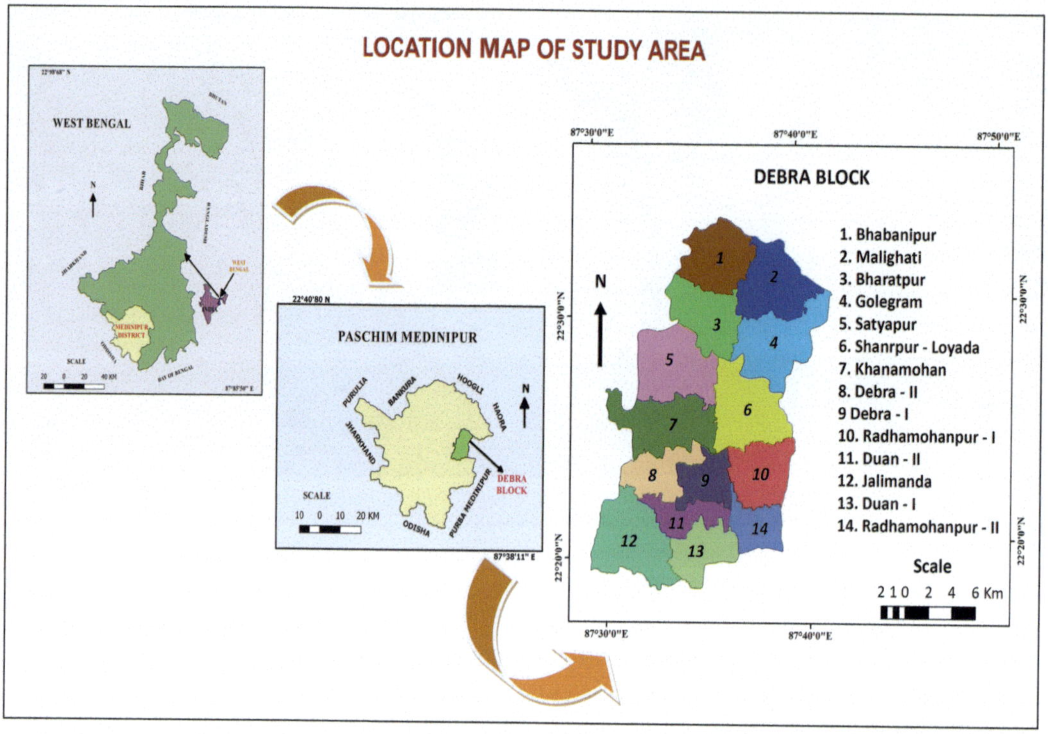

Map 21.1 Location map

21.2.2 Materials and Methods

The concept of sustainability is evaluated by two principle definition. One is the universal definition and another is local definition (Ghaffari and Hashemi 2011). This chapter discusses about second definition where the sustainability is calculated from locals' developmental needs. Based on locals' point of view, the sustainable rural development is determined by these two objectives: (i) assessment of the environmental benefits of MGNREGA works throughout the G.P. level and (ii) measure the disparity and imbalances in the levels of societal development.

In order to examine the societal development and environmental improvement of a particular G.P., we collected two types of data: data related to socio-economic development of all G.P. and another is primary and secondary data of MGNREGA works. The sources of all secondary socio-economic information includes census, district statistical handbook and collected data from related administrative departments and organizational records like MoRD (Ministry of Rural Development), MIS (Management Information Systems) data related to MGNREGA. Summarizing the micro-level or village data we have developed the Panchayat-level data. Through the study area, to interpret the improvement of MGNREGA works, we have collected secondary data from panchayat office and the web site of MGNREGA (www.nrega.nic.in, www.nrega.ap.gov.in). In addition, interview of resource person (highly educated) of different departments were conducted to raise our overall conception about MGNREGA works and assets. Secondary data of MGNREGA work were accumulated from the MIS data of financial year 2017–2018 and the socio-economic data were collected from the last census year 2011.

After accumulation of all secondary data, a field work was conducted at village level for the verification of ground truth. The purpose of this is to verify the asset created by MGNREGA works, which is known as sample survey. We used the process of two-stage cluster sample survey. The sizes of the clusters for this study were large enough to create difficulty to take the information of all the units inside the clusters. At first in each G.P., five villages were randomly selected. Next 10 households (benefited from MGNREGA assets) in each village were selected randomly for the impact assessment of MGNREGA works. So, we selected 700 households for this study. Data were collected through structured questionnaires according to investigation plan. Primary data or field survey was conducted after the interpretation of secondary data or internet-based data. The information of rural socioeconomic condition and environmental condition both were extracted from household interview, such as household livelihoods, characteristics of out migration, household income status, utilization of NREGS earnings, household asset or amenity status, change in forest area, runoff condition during heavy rainy season, condition of soil erosion and siltation after every rainy season, change in area under irrigation, crop diversification, ground water availability throughout the year, utilization of surface water bodies, etc. From this field survey and some literature study, we finalized the indicators of Green Potential and Effectiveness Parameter and analysed the relation between socio-economic condition and environmental condition. Survey data were analysed through some statistical techniques and suitable cartographic presentation.

21.3 Results and Discussion

This study evaluated sustainability of rural development through MRNREGA works. The objectives of this section are (i) to evaluate MGNREGA works according to their environmental aspect for improving the sustainable natural resource management, (ii) exploring inter-G.P. regional imbalances in socio-economic development within 14 G.P. of Debra block, and (iii) to develop a correlation between SEDI and E.I to achieve for sustainable development.

21.3.1 Analysis the Environmental Improvement of MGNREGA Works by Using G.I

21.3.1.1 Concept of Green Index (GI)

The term Green Index is a useful significant index and easy to understand the measurement of green outcomes of MGNREGA works. It is well-known qualitative, quantitative and semi-quantitative type assessment of MGNREGA's environmental potential services such as ecological balance, natural resources management, reclamation of natural ecosystem, etc. Dr. Ravi Chopra proposed the formula $GI_{GP} = \frac{\sum_{i=1}^{N}(GV)_i}{4N}$, where N = No of MGNREGA works undertaken in a G.P. and GV = Green Value, which is the product of Green Potential and Effectiveness parameter. Green Potential and Effectiveness parameter are a set of indicators that were developed on the potential environmental benefit of each and every MGNREGA works. Also the effectiveness value depends on quality of the work and coverage of the work. The range of index value belongs between 0 and 1 where 1 indicates 100% good environmental impact and 0 indicates poor environmental impact.

21.3.1.2 Green Potential

Green Potential is a set of ecological, physical and socio-economic indicators that are related to the structure of sustainable or greening rural development. The structure of Sustainable or greening Rural Development depends on the following environmental effects (Chopra et al. 2012): (i) Improved resource conservation, (ii) Improved resource efficiency, (iii) Strengthened climate resilience of communities, (iv) Reduced negative environmental impacts, (v) Contribution to climate change mitigation. The indicators of Green Potential are divided into four categories:

(i) Indicator P = improving the Productivity (P), which means improving the quality of natural resource or eco system.
(ii) Indicator S = growing the Sustainability (S), which means the sustainable development of natural resource or eco system.
(iii) Indicator I = Decreasing the negative Impact (I), which means decreasing the negative impact of development related work.
(iv) Indicator R = improving Resilience (R), which means enhancing the elasticity to climate change vulnerability.

Table 21.2 refers a set of indicators that were promoted to assess the environmental impact of each MGNREGS works.

21.3.1.3 Effectiveness Parameter

Effectiveness value of any work is the average value of coverage of the work and quality of the work. From the outcomes of MGNREGA works we can measure the adequacy or coverage of all the works. This is the quantitative measurement of effectiveness value. The author distributed the coverage values among three categories namely,

(i) High coverage = (1.0), (ii) Moderate = (0.50), and (ii) Low coverage = (0.25)

On the other hand, the quality or durability of every work is the qualitative measurement of effectiveness value. The standard quality of every work has been described in MGNREGA guideline, which was selected to consider the local development and the environmental services. Also the standard quality has three categories i.e.

(i) High (1.0) = Fulfil all durability features (Good quality).
(ii) Moderate (0.50) = Fulfil average durability features (Neither very good nor very bad) and.
(iii) Low (0.25) = Fulfil partial durability features (bad quality).

The measurement of GI started from a Gram Panchayat level, which is a micro-level study. The formula of GI is applied on each and every G.P. of the study area. At first the Green Value (GV) (product of Green Potential and effectiveness parameters) for every work is computed, after that calculates total Green Value of all works and standardizes by the factor 4 N. Each indicator of green potential value is 1 when MGNREGS work has ability to fulfil the objective. Differently, the value of each indicator is 0, when MGNREGS work does not meet the objectives. The highest green potential score is 4 and

Table 21.2 Indicators of the green potential

Sl. no	Work name	Green Potential (G.P.) Indicator P	Indicator S	Indicator I	Indicator R
1	Water conservation	(i) Accessibility of irrigation water (ii) Increase crop production	(i) Strengthen surface water storage capacity	(i) Decreasing surface runoff (ii) Increasing infiltration for ground water recharge	(i) Zonal flood control during the storms or heavy rainfall
2	Watershed management	(i) Accessibility of irrigation water (ii) Increase crop production	(i) Strengthen surface water storage capacity	(i) Decreasing surface runoff (ii) Increasing infiltration for ground water recharge	(i) Zonal flood control during the storms or heavy rainfall
3	Irrigation	(i) Accessibility of irrigation water (ii) Increased area under irrigation (iii) Increase crop production	(i) Increase the sustainability of ground water reservoir	i) Reducing the arsenic contamination in underground water	(i) Zonal flood control during the storms or heavy rainfall

(continued)

Table 21.2 (continued)

Sl. no	Work name	Green Potential (G.P.)			
		Indicator P	Indicator S	Indicator I	Indicator R
4	Traditional water bodies	(i) Accessibility of irrigation water (ii) Increasing fish farming	(i) Strengthen surface water storage capacity (ii) Improve the soil quality	(i) Decreasing surface runoff (ii) Increasing infiltration for ground water recharge (iii) Conservation of pond eco system	(i) Zonal flood control during the storms or heavy rainfall (ii) Drought proofing
5	Afforestation	(i) Accessibility of (NTFP) non timber Forest product (ii) Increasing infiltration for ground water recharge (iii) Improve soil organic matter	(i) Reclamation & protection of biomass & carbon stock (ii) Biodiversity conservation.	(i) Soil erosion controlled by reducing the velocity of surface runoff (ii) Decreasing siltation rate in water bodies (iii) Improve air quality	(i) Resilience to be a heat island (ii) Controlled ozone depletion & global warming (iii) Halting desertification
6/7/9	Land development/ improving productivity of lands/development of fallow or waste land	(i) Improve soil fertility (ii) Potential to generate increased crop, livestock and forest production.	–	()i) Reduced the risk of water logging during heavy rainy season	(i) Halting land degradation and desertification.
10	Construction of houses	–	–	(i) Increasing the livelihood security (ii) Generate employment opportunities	–
8/11/12	Improving livelihoods/ promotion of livestock/promotion of fisheries			(i) Increasing the livelihood security (ii) Generate employment opportunities	
13	Rural sanitation			(i) Increasing the rural lifestyle (ii) Generate employment opportunities	
14	Rural connectivity	–	–	I) Increase accessibility (ii) Generate employment opportunities	–
15	Play field			(i) Increasing the rural lifestyle (ii) Generate employment opportunities	
16	Disaster preparedness/ restoration	(i) Recharge the ground water by enhancing infiltration	(i) Improve the drainage ecosystem by removing silt and garbage	(i) Controlling runoff velocities, soil erosion and sedimentation in small drainage system	(i) Flood control

Source: Prepared by author from literature survey

the score of effectiveness parameter is 1. So the maximum achieved green value score will be 4x1 = 4. As for example, Table 21.3 represents the Calculation of Green Value and Green Index of a particular G.P.

In the same way we have calculated G.I. value of another 13 G.P. of Debra block. Table 21.4 represents the Green Index (G.I.) of MGNREGA of all Gram Panchayat of Debra Block.

If the G.P. has more environmental conservation related works, such as natural resource (water, soil/land, vegetation) conservation work, those works have the potential to fulfil all the four indicators of Green Potential (Table 21.2), then the Green Index value of G.P. will be increased. Also the coverage and quality, which is the effectiveness parameter of all work, would be maintained properly for increase in G.I. value.

From the above table (Table 21.4) the highest G.I. value of (50%) was found at Snarpur – Loyada G.P., followed by Satyapur and Khanamohan (48%) both, and the lowest value (25%) found at Duan –II G.P. Green Index value of all the 14 G.P. have been prepared in five categories in descending order (Map 21.2).

21.3.2 Analysis of Inter-G.P. Regional Disparity in Socio-Economic Development by Using SEDI

Socio-Economic Development Index (SEDI) is necessary to understand the entire depiction of development of a society. To realize the structure and pattern of inequality in different sectors of society we select some indicators. At first, we have selected four largely accepted components, next 12 indicators were chosen for different sectors (Table 21.5) This societal development index explores the actual variability of inter-G.P. (micro-level) progress and explains the responsibility of indicators to create diversity in development of a society. The value of Socio-Economic Development Index (S.E.D.I.) is the average of all the indicators selected at the meantime forms various sectors. The units of all indicators are not uniform in nature, so they are standardized for easy calculation and understanding. The distribution of socio-economic improvement among fourteen G. P. of Debra block is not homogenous. The simple averages technique is used to calculate and extract the socio-economic development value (Samanta 2015).

Next, we calculated SEDI value of 14 G.P. used this simple index to evaluate the comparative position of every G.P. of Debra block. Table 21.6 represents SEDI value of all G.P. of the study area.

Socio-economic Development Index is a simple statistical index that refers to the levels of development among 14 G.P. of Debra block. The index value is highest in Duan-II (53.66), followed by Radhamohanpur-II (51.65) and lowest in Satyapur (34.41). SEDI value of all the 14 G.P. have been prepared as five categories in descending order (Map 21.3).

Satyapur, Khanamohan, Snarpur - Loyada and Duan -I G.P. belongs to very low to low development stages. It is significant to mention that the performance of these blocks is consistently poor in few sectors of the social development. Economic status of those G.P. is poor because maximum household of this G.P. depends on agricultural sector. Also those G.P. have poor modernization, basically transport and communication system is not good. It can be noticed that few G.P. is in bad state in their development regime due to poor transportation system and job opportunity.

21.3.3 Correlations between S.E.D. and E.I. For Sustainable Development

Analysis of data highlights that the correlation between G.I. and S.E.D.I. of maximum G.P. is inversely proportional. Among 14 G.P., Duan –II G.P., represents strong negative correlation between G.I. and S.E.D.I. Balichak, the Census town of Debra block is under Duan –II G.P. Being the census town, urbanization and modernization rate is highest as well as population density (3739.16/sq. Km.) is also high in Duan –II G.P. and it performs as an administrative, cultural,

Table 21.3 Green value and green index of Radhamohanpur-II G.P

Sl. no.	Work name	Green potential (G. P.)				G. G.P.	Effectiveness value (E. V.)		E.V.	Green value (G. V.) = GP × EV	G.I. = ∑G.V/4 N
		P	S	I	R		A/C	D/Q			
1	Water conservation	1	1	1	1	4	High	High	1	4	$\frac{\sum_{i=1}^{N}(GV)_i}{4N} = \frac{\sum G.V = 16.875}{4*10}$
2	Irrigation	1	1	1	1	4	Moderate	High	0.75	3	
3	Traditional water bodies	1	1	1	1	4	Moderate	High	0.75	3	
4	Afforestation	1	1	1	1	4	Low coverage	Low coverage	0.25	1	
5	Land development	1	0	1	1	3	Moderate	Moderate	0.5	1.5	
6	Improving productivity of lands	1	0	1	1	3	Low coverage	Low coverage	0.25	0.75	
7	Development of fallow /waste land	1	0	1	1	3	Moderate	Low coverage	0.375	1.125	
8	Construction of house	0	0	1	0	1	Moderate	High	0.75	0.75	
9	Road connectivity	0	0	1	0	1	Moderate	High	0.75	0.75	
10	Disaster preparedness/restoration	1	1	1	1	4	Low coverage	Low coverage	0.25	1	
	∑N = 10∑N = 10									∑G.V. = 16.875	G.I. = 0.42

Source: computed by author

Table 21.4 Green Index (G.I.) of MGNREGA at 14 Gram Panchayat of Debra Block(Financial Year 2017–18)

Gram panchayat (G.P.) name	Work type (*N*)	Total works	Estimated outcome (hec.)	Expenditure(in Lakhs)	Green value $\sum$G.V	Green index value(G.I.) = $\sum$G.V/4 N	Green index	Green index In %	Green index	Rank
Bharatpur	9	556	313.75	129.23	15.85	15.85/(4*9)	0.44	44	High	5
Bhabanipur	11	557	184.3	139.88	17.6	17.6/(4*11)	0.4	40	Moderate	7
Debra -I	10	576	503.22	145.06	12.2	12.2/(4*10)	0.31	31	Low	11
Debra -II	10	408	205.19	123.57	18	18/(4*10)	0.45	45	High	4
Duan -I	9	304	373.27	142.86	13.65	13.65/(4*9)	0.38	38	Moderate	8
Duan -II	10	615	126.16	57.01	10.5	10.5/(4*10)	0.25	25	Very low	12
Golgram	12	753	340.17	197.37	16	16/(4*12)	0.33	33	Low	10
Jalimanda	9	1020	505.34	261.13	16	16/(4*9)	0.44	44	High	5
Khanamohan	8	941	136.06	286.22	15.3	15.3/(4*8)	0.48	48	Very high	2
Malighati	13	554	143.45	136.2	18.05	18.05/(4*13)	0.35	35	Low	9
Radhamohanpur -I	10	743	233.11	174.03	18.85	18.85/(4*10)	0.47	47	Very high	3
Radhamohanpur -II	10	777	457.98	159.84	16.875	16.875/(4*10)	0.42	42	High	6
Satyapur	9	892	129.46	254.59	17.15	17.15/(4*9)	0.48	48	Very high	2
Snarpur - Loyada	11	1003	126.13	234.04	22	22/(4*11)	0.5	50	Very high	1

Source: computed by author

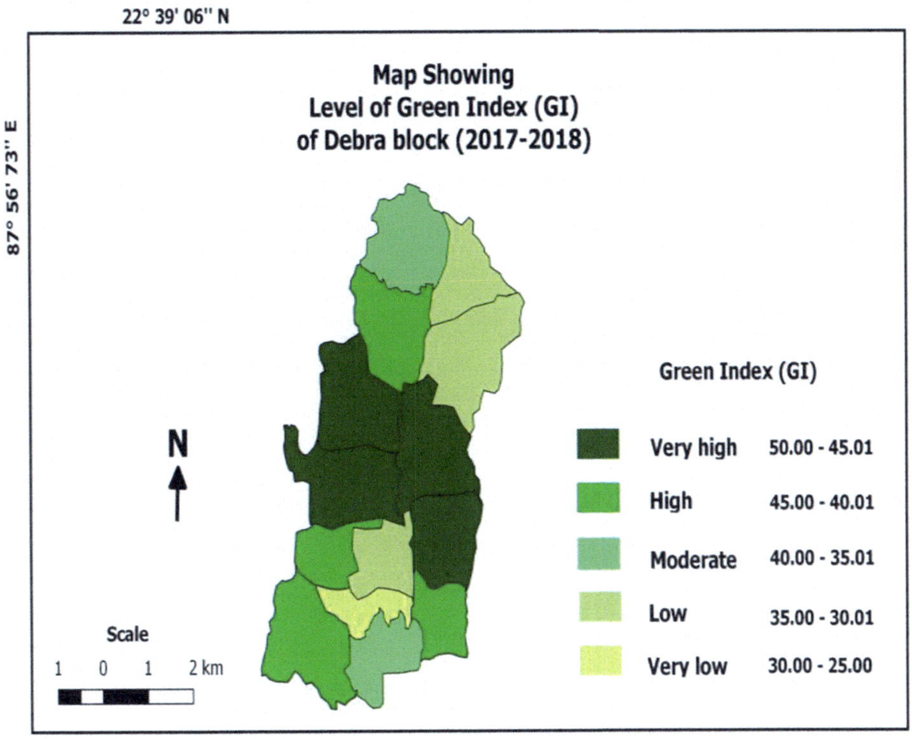

Map 21.2 Level of G.I. of Debra block

Table 21.5 Indicators of Socio-Economic Development Index (S.E.D.I)

Perspective of Socio-Economic Development Index	Indicator	Index (Ij)
Economic growth	Index of worker	(Total worker/ Total Population)* 100
	Index of Households with any member earning more than Rs. 10,000 p.m	(Households earning more than Rs. 10,000 p.m./ Total household)*100
Level of Education	Index of literacy	(Literate people/Total Population)*100
	Index of female literacy	Female literate people/ Total female Population)*100
	Index of school	[Total School (primary, middle, secondary ,higher secondary, Govt. school, Private School)/ total population]*100
Health services	Index of mouza having safe drinking water	(Average no.of hand pump, Tube well, bore whole status/ total mouza)*100
	Index of mouza having nutrition centre	(Average no. of ICDS, Anganwari and other nutrition centre/total mouza)*100
	Index of mouza having health centre	Total no. of Health centre/total mouza)*100
Degree of modernization	Index of urbanization	(Urban population/ Total Population)* 100
	Index of mouza having electrification for all users	(Average power supply for all domestic user/ total mouza)*100
	Index of household with three/ more rooms with concrete wall and roof	(household with three/more rooms with concrete wall and roof/ total household)*100
	Index of mouza having communicated by rail and road	Total no. of National and State highway, major and other district road, Black topped road /total mouza)*100
	$\Sigma n = 12$	S.E.D.I. = Ij = (Σ Iij / n)

Source: prepared by author from literature survey

Table 21.6 Socio-Economic Development Index (S.E.D.I.) of 14 Gram Panchayat of Debra Block (Census data 2011)

	Perspective													Socio-Economic development index (S.E.D.I.) = Ij = (Σ Iij / n)	Socio-Economic development index	Rank
	Economic growth		Level of Education			Health services			Degree of modernization							
Name of the G.P.	Index of worker	Index of Households withany memberearning morethan Rs. 10,000p.m	Index of literacy	Index of female literacy	Index of school	Index of mouza having safe drinking water	index of mouza having nutrition centre	Index of mouza having health centre	Index of Urbanization	Index of mouza having electrification for all users	Index of household with three/ more rooms with concrete wall and roof	Index of mouza having communicated by rail and road				
Bharatpur	46.88	2.24	67.86	60.44	0.33	83.33	89.91	12.1		90.91	3.6	24.24		40.16	Moderate	10
Bhabanipur	45.81	3.27	73.35	67.44	0.27	94.83	93.1	31		77.59	4.15	55.17		45.5	High	4
Debra -I	45.82	8.34	70.28	64.2	0.26	100	85.52	13		78.26	8.28	43.48		43.12	Moderate	5
Debra -II	47.16	8.8	73.06	66.79	0.35	96.67	66.67	6.67		85	7.85	36.67		41.31	Moderate	8
Duan -I	46.72	5.15	77.97	72.61	0.34	58	76	12		80	5.12	16		37.49	Low	11
Duan -II	40.11	18.37	77.95	73.65	0.09	83.33	76.14	42.9	71.43	75	13.53	71.43		53.66	Very high	1
Golgram	46.47	5.65	72.14	65.54	0.37	77.03	76.57	18.9		85.14	7.46	35.15		40.87	Moderate	9
Jalimanda	48.85	4.96	72.47	65.52	0.31	98.84	87.6	9.3		94.19	2.92	16.29		41.77	Moderate	7
Khanamohan	47.15	3.71	74.19	68.91	0.37	65.39	58.97	7.69		77.69	4.41	12.31		35.07	Low	13
Malighati	46.27	4.77	74.45	67.87	0.34	91.43	74.29	28.6		94.29	5.87	77.14		47.11	High	3
Radhamohanpur -I	44.66	3.95	73.52	68.36	0.2	73.08	89.73	19.2		76.92	6.4	53.85		42.49	Moderate	6
Radhamohanpur -II	39.69	8.92	78.41	73.21	0.2	100	96.33	33.3		83.33	6.37	100		51.65	Very high	2
Satyapur	47.13	3.41	68.3	61.15	0.33	76.14	85.61	13.6		64.77	7.73	31.82		34.41	Very low	14
Snarpur - Loyada	44.82	6.77	71.58	65.27	0.4	75.45	64.85	45.5		69.09	4.95	32.73		36.38	Low	12

Source: computed by author

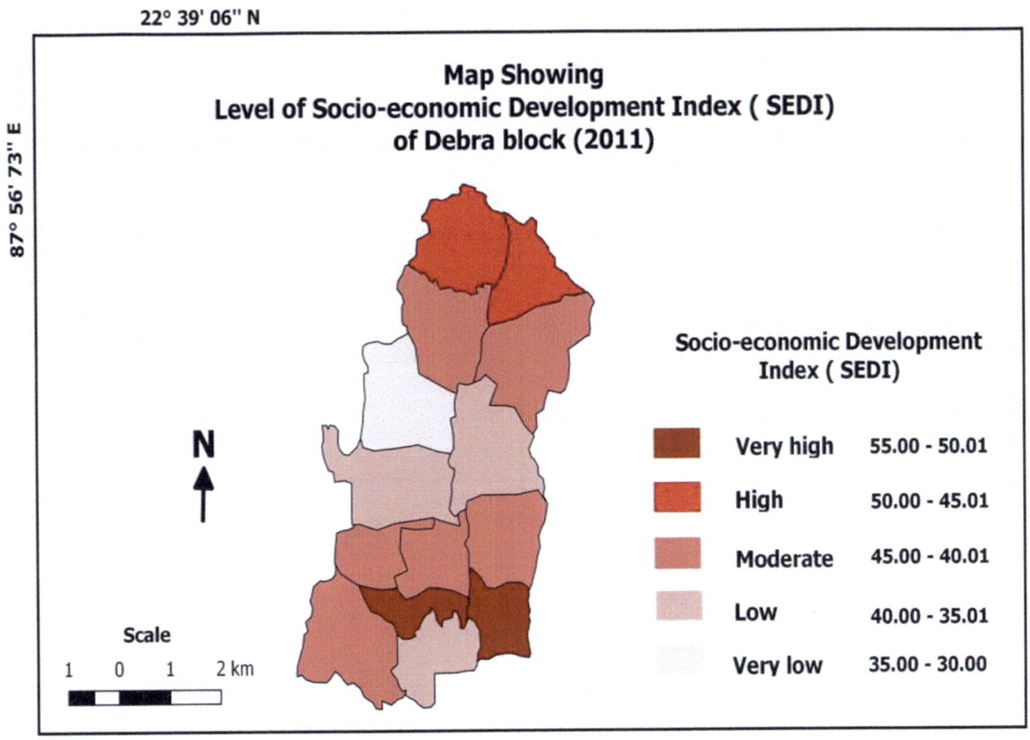

Map 21.3 Level of SEDI of Debra block

economic and health service centre. So, this G.P. has achieved highest SEDI score. Excessive development and unplanned urbanization create climate crisis. According to environmental benefits this G.P. has ranked lowest as it is surrounded by urban environment, which was formed naturally due to urbanization and modernization. The lack of appropriate planning, zonation and unscientific building construction led to unstable growth, haphazard and unhygienic development that destroys the local habitat and ecosystem services. Waterlogged lowlands (Fig. 21.6) that act as a natural sponge during flood and also act as a biodiversity reservoir were converted into land to carry the excessive population growth and modernization and urbanization growth.

In Duan-II G.P. maximum MGNREGA works were related to roads and houses construction, playgrounds, etc. Those works have huge socio-economic benefit, but they do not have any environmental benefits. Those works are constructed with non-environmental-friendly materials such as burnt bricks, cement, steel, sand, stones, iron, etc. and they are not planned in sustainable or green building techniques. So, this type of work has negative impact on the environment and also they decrease the G.I. value of G.P.

The G.P. where SEDI and GI values both are high example Radhamohanpur-II G.P. is a semi-urban area, where both environmental and socio-economic development took place simultaneously. This G.P., achieved maximum development in safe drinking water, transport and communication sector. After Duan-II this G.P. is second highest in socio-economic and environmental development. Among the 14 G.P. of Debra block, we found only 1 G.P. that represents sustainable rural development. Environmental protection and improving rural livelihood promotes village sustainability. Model 21.1 is suggested for achieving sustainable rural development. The relation goes through loop-1; if the value of S.E.D.I. and G.I. is directly proportional then the progress would have positive impact. Among 14 G.P. of Debra block, only G.P. Radhamohanpur-II, where G.I. (42%) and S.E.D.I value (51.65%) both are high, which represent that the development of village has been in a sustainable way. Sustainable rural

Fig. 21.6 (**a**) Small and Shallow waterbodies in 2016. (**b**) After converted into Land in 2019

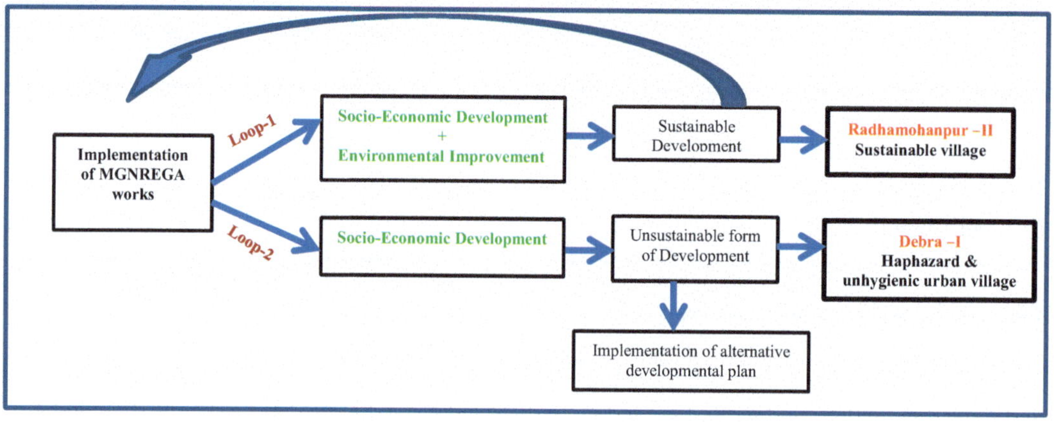

Model 21.1 Rural and urban village sustainability. (Source: Prepared by author)

development has been shown in loop 1, which is the indicator of positive relation between these two developmental processes. Some development guidelines have essential need for these upcoming new centre of urban growth at Radhamohanpur -II G.P., which have huge potential to reach a sustainable urban village in future.

On the contrary, the relation goes through loop 2 if the value of S.E.D.I. and G.I. is inversely proportional and the progress would have negative impact. At Duan –II G.P., where S.E.D.I value (53.66%) is highest but G.I. value (25%) is very poor. Here we found the highest negative correlation among 14 G.P. Unsustainable state of rural development have been shown in loop-2, which is the indicator of negative relation between these two developmental processes.

21.4 Conclusion

This chapter elaborates on the correlation between the Socio-Economic Development Index (S.E.D.I.), and Green Index (G.I.), which provides valuable insights into the interplay between socio-economic development, environmental considerations, and green progress. The findings have significant implications for sustainable development and the achievement of several Sustainable Development Goals.

Through the examination of these indices, it becomes evident that socio-economic development can be achieved while simultaneously promoting environmental sustainability and green practices. The study highlights the importance of integrating green initiatives into socio-economic development strategies, emphasizing that economic growth should not come at the expense of environmental degradation.

The research underscores the need for a holistic approach to sustainable development, where socio-economic progress goes hand in hand with environmental conservation and the promotion of green solutions. It emphasizes the role of cities and communities in adopting sustainable practices, building resilience, and creating inclusive, safe, and sustainable urban environments.

Furthermore, the study highlights the importance of green initiatives in addressing climate change, which is a pressing global challenge. The findings emphasize the role of the Green Index (G.I.) in assessing and promoting effective climate action, enabling policymakers and stakeholders to identify and prioritize green practices that mitigate climate change impacts.

Overall, this study reinforces the notion that sustainable development requires a balanced approach that considers both socio-economic development and environmental sustainability. By integrating green initiatives, fostering inclusive growth, and prioritizing climate action, societies can work towards achieving the SDGs and building a more sustainable and resilient future. The insights from this study can inform policymakers, researchers, and sustainable development practitioners in designing effective strategies and policies for sustainable development at local, regional, and global levels.

References

Bhaskar A, Shah A, Gupta S (2017) 7.5 crore green jobs? Assessing the greenness of MGNREGA work. Indian J Labour Econ 59:441–461

Census of India (2011) District census handbook, Paschim Medinipur, Village and Town Directory

Chopra R, Banerji S, Gera P, Iyer S, Soni P (2012) Greening rural development in India, vol 1. UNDP Publications, India

Das A (1999) Socio-economic development in India: a regional analysis. Develop Soc 28:313–345

Esteaves T, Rao K, Sinha B, Roy S (2013) Environmental benefits and vulnerability reduction through mahatma Gandhi NREGS: a synthesis report. Ministry of Rural Development, Government of India and Deutsche GIZ, New Delhi

Ganeriwala SA (2010) An impact assessment study of the usefulness and sustainability of the assets created under Mahatma Gandhi National Rural Employment Guarantee Act (MGNREGA) in Sikkim. Institute of Rural Management Anand

Ganiee FA (2014) A case studies of rural development Programmes in India. Intl J English Lang Literat Human 1(5):40–47. ISSN 2321-7065

Ghaffari GH, Hashemi N (2011) Rural tourism: a new guideline of sustainable social development. Science in Society Conference, Washington, DC

Ghosh JK (2011) Impact of NREGA on wage rates, food security and rural urban Migration in West Bengal. Agro-Economic Research Centre Visva-Bharati Santiniketan

Haque T (2011) Socio-economic impact of implementation of Mahatma Gandhi National Rural Employment Guarantee Act in India. Social Change 41(3):445–471

Jacob N (2008) The impact of NREGA on rural-urban migration: field survey of Villupuram District, Tamil Nadu, Centre For Public Policy Research

Kareemulla K, Reddy KS, Rao CAR, Kumar S, Venkateswarlu B (2009) Soil and water conservation works through National Rural Employment Guarantee Scheme (NREGS) in Andhra Pradesh - an analysis of livelihood impact. Agric Econ Res Rev 22:443–450

Karthika K (2015) Impact of MGNREGA on socio-economic development and women empowerment. IOSR J Bus Manag 17(7):16–19. ISSN: 2319-7668

Kharkwal S, Kuma A (2015) Socio-economic impact of MGNREGA: evidences from district of Udham Singh Nagar in Uttarakhand, India. Indian J Econ Develop 3(12). ISSN (online): 2320–9836

Kolhea NP, Dhoteb KK (2016) Rurban Centres: The new dimension of urbanism. International conference on emerging trends in engineering. Sci Technol Procedia Technol 24(2016):1699–1705

Kumar R, Prasanna R (2010) Role of NREGA in providing additional employment for tribals and curtailing migration. National Rural Employment Guarantee Act (NREGA): design, process and impact, Delhi: Ministry of Rural Development, Government of India

Mandal RB (1989) System of rural settlements in developing countries, 1st edn. Concept Publishing Company, New Delhi

Mattei L (2015) The Brazilian rural development model in the context of green economy. In: International labour office, global labour university. ILO, No. 33, ISSN: 1866-0541, Geneva

Mistry P, Jaswal A (2009) Study of the implementation of the National Rural Employment Guarantee Scheme (NREGS): focus on migration. Disha, Ahmedabad

Our Common Future, Chapter 2 (1987) Towards sustainable development. Report of the World Commission on Environment and Development. UN Documents

Ravindranath N, Murthy I (2013) Greening of MGNREGS. United Nations Development Programme

Samanta R (2015) Block level disparity in social development: a case study of Paschim Medinipur, West Bengal, India. Intl J Sci Eng Res (IJSER) 3(4):92–95

Samsurijan MS, Awang A, Aiyub K, Hussain MY, Arifin K, Vedevelu K (2017) A review of the "Urban Village" concept: an operational definition in Malaysia. Geografi 5(2):49–58, ISSN 2289-4470/eISSN 2462–2400

Sarkar DN, Kundu K (2016) The economic and social importance of unorganized retailers in rural India. J Rural Stud 43:159–172

Sefiya KM (2015) Mahatma Gandhi National Rural Employment Guarantee Act & Environment. Intl J Manage Soc Sci Res Rev 1(8):9–13

Shah M (2007) Employment guarantee, civil society and Indian Democracy. Econ Polit Weekly 43(45–46):43–51

Shah M, Mann N, Pande V (2012) MGNREGA sameeksha: an anthology of research studies on the mahatma Gandhi national rural employment guarantee (2005) act 2006–2012. Ministry of Rural Development, Government of India, Technical report, ISBN 978 81 250 4725 4

Shikalgar RS, Kamble SM (2013) Geographical Analysis of Rurban Centers in Yerala River Basin. Online Intl Interdisc Res J III(V):152–158. ISSN 2249-9598.

Tiwari R, Somashekhar H, Parama VR, Murthy IK, Kumar MM, Kumar BM, Parate H, Varma M, Malaviya S, Rao AS, Sengupta A, Kattumuri R, Ravindranath NH (2011) MGNREGA for environmental service enhancement and vulnerability reduction: rapid appraisal in Chitradurga district, Karnataka. Economic Political Weekly 46(20):39–47

Verma S, Shah T (2012) Beyond digging and filling holes lessons from case studies of best-performing MGNREGA water assets. Water policy research highlight-42, IWMI-TATA Water Policy Program

Yadav A (2014) Impact of rural employment guarantee act in Kanpur district. Intl J Home Sci Exten Intl Commun Manage 1(2):76–79

Urban Heat: UHI and Heat Stress Threat to Megacities

22

Jagabandhu Panda, Asmita Mukherjee, Animesh Choudhury, and Sreyasi Biswas

Abstract

Urbanization in all its form has influenced heat accretion within the urban extent, making the region somewhat hotter than the neighboring rural areas and consequently leading to the development of an urban heat island (UHI). In the process of urbanization, the megacities could experience temperatures ~2 to 8 °C higher than their surroundings forming an UHI and triggering extreme environmental events. In metropolitan setting, the consequences of heat stress would be more severe with UHI in action. Urban climate, hydrology, soil properties, atmosphere, biological habits, material and energy flux, residents' health and comfort, etc., would be impacted by the rise in land surface temperature (LST) and UHI effect. The increasing urbanization could lead to the growth of impervious surfaces and the loss of green areas, which might induce the observed increase in the overall mean LST. Earlier, air temperature and LST were frequently employed to describe the intensity of UHI, while completely disregarding the human perception of the thermal environment as a crucial outcome in the evaluation of UHI-induced risk. Therefore, it is essential to get insights into the danger that urban heat poses to human health and the environment, focusing on the general inhabitability and environmental sustainability of megacities.

Keywords

Urbanization · UHI · Heat stress · LST · Urban climate

22.1 Introduction

The global urbanization trend experienced a sea change in the past 4–5 decades. This phenomenon became so acute that the global metropolitan population jumped from 10% in 1990 to 55.3% in 2018 (https://population.un.org/wup/Download; Yan et al. 2020). It is further estimated that by 2050, ~68.4% will move into the cities for a better standard of living (Grimm et al. 2008). The urban composition of Indian cities portrays a similar trend of rapid urbanization, registering a 21.05% rise in population from 2011 to 2021 (https://datatopics.worldbank.org/world-development-indicators/). Recent records show that the urban population accounts for ~35.39% (in 2021) of the total population of the country (https://www.macrotrends.net/countries/IND/india/urban-population), which explains the

J. Panda (✉) · A. Mukherjee · A. Choudhury
S. Biswas
Department of Earth and Atmospheric Sciences, National Institute of Technology Rourkela, Rourkela, Odisha, India
e-mail: pandaj@nitrkl.ac.in

U. Chatterjee et al. (eds.), *Climate Crisis: Adaptive Approaches and Sustainability*, Sustainable Development Goals Series, https://doi.org/10.1007/978-3-031-44397-8_22

noticeable expansion of existing urban areas and development of newer regions in the recent years. Over the years, cities have experienced both horizontal as well as vertical expansion. This expansion can be attributed to the unregulated inflow of inhabitants from rural or small towns into the big cities, consequently bringing physical changes to the urban areas (Kuddus et al. 2020).

With uncontrolled population growth, major cities are becoming more and more congested. This alters the existing land use (LU) and gives rise to several concerning issues, including the enhanced contribution of urban heating. It further triggers rising local temperatures, upsetting air circulation, and changes the amount and distribution of precipitation, thunderstorm characteristics, and rainfall occurrences (Baklanov et al. 2016; Li et al. 2013, 2016; Rath and Panda 2020; Rath et al. 2021, 2022). Additionally, slight changes in atmospheric emissions and feedback concerning air pollutants also impart similar effects at various scales. Therefore, the cumulative impacts of overcrowded cities contribute to intensified anthropogenic heat emissions (Crippa et al. 2021) and carbon emissions (Xie et al. 2017), altering surface albedo and roughness (Niyogi et al. 2009), etc. The sustainable development goals report (UN 2019a, b) states that all the major metropolis of the world collectively contributes to approximately 70% of the overall carbon footprints, deteriorating the urban air quality (Crippa et al. 2021). These processes imposed detrimental impacts on the local urban microclimate, modulating the regional weather and climate.

According to the assessment of the United Nations (UN 2014), more than half of the global population already resides in urban centers, which would escalate to ~6 billion by 2041. Urban agglomerations are widely acknowledged to serve varied roles in every society. They are considered the driving force behind technical advancements and economic prosperity. Seto et al. (2011) forecasted that with population increase and consequent urban expansion, significant growth in the gross domestic product (GDP) of the country takes place. However, they can also foster problems growing out of poverty such as inequality, environmental risks, and contagious diseases (McMichael 2000). UN (2014) highlighted that ~67% of the total world population would become a part of the urban economy by 2050, contributing to nearly 60% of the economy worldwide (Crippa et al. 2021) making it extremely essential to study the interconnected dynamics of urbanization, climate, and human health.

One such problem of population explosion and urban expansion includes the effect of urban heat island (UHI) within urban centers (Oke 1982), which corresponds to the temperature disparity between rural–urban set-ups, and this difference is anticipated to widen by ~1 °C every 10 years (Voogt 2002). UHI-related studies sought to better understand its impact by contrasting the temperature fluctuations in rural–urban environments (Schwarz et al. 2012; Almeida et al. 2021). The land surface temperature (LST) products obtained from earth observing satellites has received increased attention to quantify surface UHI (SUHI) for analyzing its spatial and temporal variations. Other approaches in this direction focus on change dynamics (Quan et al. 2014), identification of drivers (Peng et al. 2020), modeling (Wang and Upreti 2019), human health response (Cao et al. 2018), environmental impact assessment (Guan et al. 2019), and adaptation (Gunawardena et al. 2017). It is well acknowledged that urbanization significantly alters regional energy balance and the UHI effect induced by the spatial pattern of cities (Souch and Grimmond 2006). As the UHI mechanism is generally most prominent inside compact urbanized areas, denser urban growth has traditionally been linked with larger UHI intensities. However, several studies have shown that both sprawled (Stone Jr 2012) and denser (Martilli 2014; Schwarz and Manceur 2015) urban layouts may produce stronger UHIs (Debbage and Shepherd 2015). Therefore, a comprehensive review in this direction is required to enrich the existing knowledge regarding the UHI and its effects on megacities.

22.2 UHI Over Mega Cities: Mechanism, Causes, and Heat Stress

With frequent incidences of heat waves, heat stress, and temperature anomalies, the process of UHI intensification has further worsened over the years. It is commonly observed for metropolises to be 3.5–4.5 °C hotter than their adjacent rural counterparts (Zhang et al. 2017). Relatively, warmer city cores have the potential to cause several health issues (Patz et al. 2005). In fact, a significant proportion of climate-related mortalities can be attributed to UHI-induced intense heat episodes (Debbage and Shepherd 2015; Stone et al. 2010).

The UHI phenomenon is intricately connected to several facets of the society and industry. Apparently, UHIs are getting intensified as more and more urban green cover is getting replaced by impermeable surfaces such as roads, buildings, and other man-made structures to benefit the expanding population (Voogt and Oke 2003). Other variables that contribute to UHIs include modifications in the characteristics of urban materials, geometry of cities, anthropogenic heat emissions, climate, geographical location, etc. (EPA 2008). UHIs emerge when the incident solar radiation strikes the urban impervious surfaces and gets absorbed, resulting in an increased land-surface and air temperatures. This happens as the impervious surface has high thermal inertia (Acharya et al. 2021). It continues to absorb heat during the day and slowly releases it at night (Ibrahim et al. 2018). The response of the air temperature is lagged due to air being a poor conductor of heat. When urbanization is dense, the heat released from the impervious surface is trapped within the urban pocket which leads to a hotspot (Sultana and Satyanarayana 2019). Other consequences associated with this occurrence include worsened heat stress (Santamouris 2014), changes in the intensity of rainfall (Han et al. 2014), higher energy requirements, and even possible contributions to global warming (Oke et al. 2017). Air pollution levels are also observed to fluctuate with variations in UHI intensity (Sarrat et al. 2006).

22.2.1 Global Overview and Indian Mega Cities

Over the Asian continent, there has been an increasing temperature trend, leading to a relatively higher number of warmer days than cooler days (Tzavali et al. 2015). Seoul's UHI landscape has been found to have several hotspots that vary diurnally and seasonally in intensity and size, having greater intensity at night under calm and clear conditions (Kim and Baik 2005). The UHI that is concentrated in the heart of the city with high urban density (Boo and Oh 2000) experiences a diurnal shift due to human activities (Kim and Baik 2005). In case of Tokyo, despite a declining population, the city's average temperature is rising at the rate of 3.0 °C every 100 years (Matsumoto et al. 2017). Isfahan has seen a steep increase in UHI intensity due to successive droughts, population expansion, and industrialization, as evident from the UHI Ratio Index (URI) value of 0.16 in 1999 to 0.3 in 2016 (Shirani-bidabadi et al. 2019). New Orleans has been experiencing the worst UHI in the world due to accelerated urbanization and industrialization (https://www.climatecentral.org/climate-matters/urban-heat-islands). Athens has also acquired a strong UHI, with daily mean intensity ranging from 6 to 12 °C in the city center (Tzavali et al. 2015). Brazil, Manaus, Porto Alegre, Belém, and Recife showed the highest values of diurnal UHI, while Rio de Janeiro has a prominent and strong UHI at night (Monteiro et al. 2021). Kolkata's UHI is expanding, with a temperature variation of ~4.72 °C (Halder et al. 2021). The average LST of the urban area in Kolkata is 2.5 times higher than the surrounding regions (Mandal et al. 2022). An increasing LST was noticed in the city center of Lucknow during the period of 2002 to 2014, with the Normalized Difference Vegetation Index (NDVI) on the decline, resulting in a strong UHI phenomenon in the city (Singh et al. 2017). Jaipur showed signatures of UHI due to reflections within walls, intense concreting of the city, sporadic green spaces, automobile emission with little dispersion, and a diversity in the albedo (Mathew et al. 2017). Delhi is also strained by the effects of

UHI, with the average daily maximum intensity (~8.3 °C) recorded in March of 2010 (Mohan et al. 2013). To create a sustainable environment amidst the UHI, Delhi's electricity consumption has accelerated, with the city claiming the position of the highest electricity consumer in India. Regarding UHI intensity, Delhi is at par with cities like Tokyo, Beijing, and London (Mohan et al. 2013). Other studies over Indian regions include cities such as Bengaluru (Ramachandra and Kumar 2010; Siddiqui et al. 2021), Pune (Veena et al. 2020), Hyderabad (Sharma 2016), Bhubaneshwar (Swain et al. 2017), Vishakhapatnam (Devi 2006), Guwahati (Borbora and Das 2014), Kochi (Thomas et al. 2014), and Bhopal (Gupta et al. 2009).

A primary study has been conducted over some major urban agglomerations of the world based on the LST values obtained from Moderate Resolution Imaging Spectroradiometer (MODIS) satellite data. The cities considered for the analysis include London (UK), Beijing (China), Delhi (India), and Kolkata (India). The analysis is carried out for these four cities considering the spatial variability of SUHI (Figs. 22.5, 22.6, 22.7 and 22.8), and both urban and rural LST anomalies for day and night separately (Figs. 22.1, 22.2, 22.3, and 22.4). The rural area of each city has been determined by delineating a 10 km buffer zone from the city boundary. The LST value is derived for a period of 22 years (i.e., from March 2000 to February 2022), retaining only the best-quality pixels. A median LST value (T) for each year, and the average (T_{avg}), maximum (T_{max}), and minimum (T_{min}) values for all considered years are calculated. The LST anomaly is derived as follows:

$$\text{Anomaly} = \left(T - T_{avg}\right) / \left(T_{\max} - T_{\min}\right)$$

Figures 22.1, 22.2, 22.3, and 22.4 illustrate the day and night time LST anomalies of all the considered cities, where the coefficient of determination (R^2) is calculated to define the strength of the linear regression. In the urban region, all the cities show an overall rise in the LST during the day (Fig. 22.1). However, the strength of the linear regression is not so strong in all the cases except for Kolkata ($R^2 = 0.505$), where a strong linear rising trend can be seen. Among the global cities, London shows an increasing linear trend in day time LST. Also, both of these two cities have the highest rate of changing day time LST. At night, London has a decreasing trend (Fig. 22.2) although the strength of the trend is poor ($R^2 = 0.007$). Delhi has a relatively better linear night time LST trend compared to London. The remaining cities exhibit an increasing linear trend in night time LST (Fig. 22.2).

The corresponding rural areas around Delhi show a decreasing trend in day time LST over the years (Fig. 22.3). For the rural areas around Kolkata, the highest rate of increasing "day time LST" is realized among all the cities. All the rural areas around other cities show an increasing trend in day time LST although the R^2 values are positive but not appreciable (Fig. 22.3). A decreasing night time LST is observed for rural areas of Beijing and Kolkata, with the former having a higher rate of decrease (Fig. 22.4). And the considered rural areas around London and Delhi exhibit an increasing trend in the night time LST, with a positive R^2 value (Fig. 22.4).

Furthermore, SUHI is computed to account for the manifestation of higher LST within the urban domain relative to its surrounding rural areas. SUHI intensity values for the four considered cities (i.e., Delhi, Kolkata, London, and Beijing) were derived for the years 2000, 2011, and 2021. A spatial variation of LST, corresponding to the urban region, is estimated for the said years using MODIS data. A median value of the rural LST (T_R) is determined, which is unique to each city and the year. The SUHI is then evaluated for each urban pixel using the following formula:

$$\text{SUHI} = T_{Ui} - T_R$$

where T_{Ui} is the LST value of the ith pixel within the urban region. A severe SUHI has been observed to dominate Beijing in the past years owing to its dense and expanding urbanization (Wang et al. 2020). Figure 22.5a depicts the day time SUHI over Beijing, where it can be seen that there is a spatial expansion from 2000 to 2021; however, the intensity might have decreased over

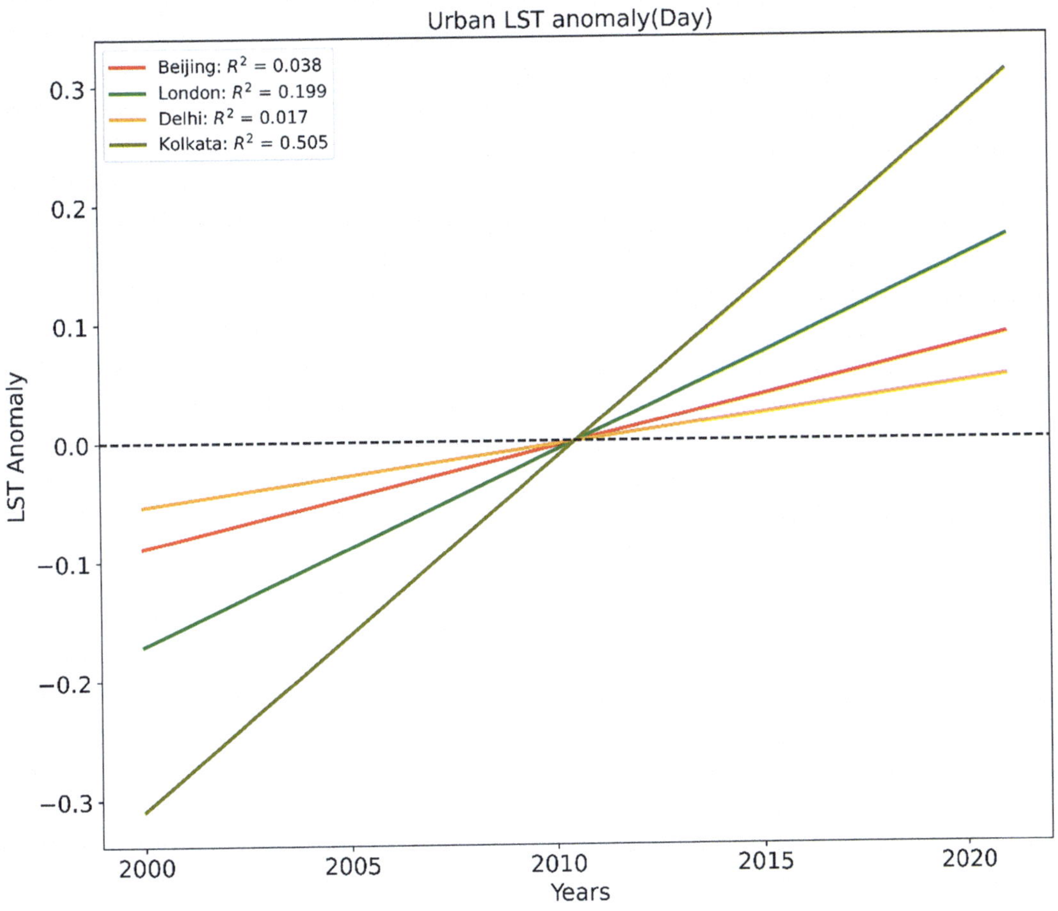

Fig. 22.1 Day time LST anomaly over the urban areas of considered cities

the same time span. Similarly, the night time SUHI over Beijing is illustrated through Fig. 22.5b, where an intense SUHI is seen in 2000, which gradually increased further from 2003 to 2008 and then experienced a decrease from 2009 to 2018.

London shows an SUHI scenario similar to the night time SUHI over Beijing, that is, increase in the intensity from 2000 to 2011 is realized and then a gradual decrease is seen till 2021 (Fig. 22.6a). The night time London SUHI demonstrates the formation of an urban cool island by 2021 (Fig. 22.6b), which agrees to the findings of Kolokotroni and Giridharan (2008). Figure 22.7a and b illustrate the formation of SUHI over Delhi NCR during day and night, respectively. For both day and night, an increased SUHI scenario is realized during the last 20 years or more in general and from 2000 to 2021 in particular, when both the intensity and spatial variation are taken into consideration. In case of Kolkata, the day time SUHI has expanded gradually with upsurge in the intensity (Fig. 22.8a); however, the night time intensity decreased in 2011 and again increased in 2021 (Fig. 22.8b).

22.2.2 Causes of UHI

Initial studies were more concerned about the canopy layer UHI (CLUHI), which was explored based on the observations of air temperature within an urban canopy. However, with the emergence of satellite-based LST measurements that considers both surface temperature variation as well as their temporal variability, the concept of

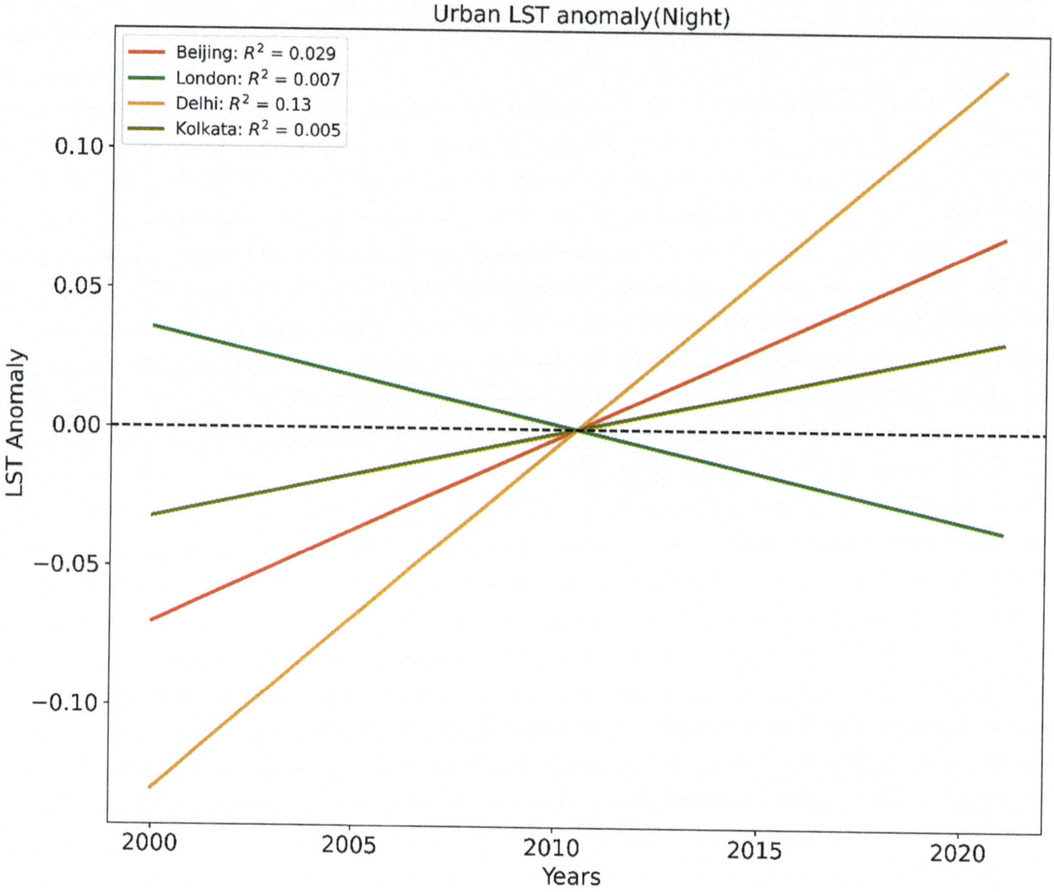

Fig. 22.2 Night time LST anomaly over urban areas of considered cities

SUHI gained more importance (Venter et al. 2021). In fact, researchers became more concerned about the contribution of each of the factors influencing the thermal behavior of urban areas. Therefore, knowledge about the surface energy balance became crucial for comprehending the mechanism behind the urban thermal dynamics (Kuang et al. 2015). The changes in the numerous segments of the surface energy balance theory such as net radiation, sensible heat flux, and latent heat flux perturb UHI structure and intensity. Thus, by analyzing the dynamics in energy level, the impacts imposed by ecoenvironmental elements can be assessed (Oke 1982; Arnfield 2003; Liao et al. 2018). It was further proposed that the changes in radiation balance, evapotranspiration, convection efficiency, heat storage, etc., between urban and rural areas, need to be explored to quantify air/surface temperature anomalies (Hou et al. 2020) so that they could be connected to the UHI formation.

22.2.2.1 Spatio-temporal and Social Factors

Several socio-temporal factors contribute to the emergence of UHI and the variation in its intensity. Impervious surface area (ISA) including paved highways and parking lots, vegetation, water, structures, and bare soil, have impacts on the UHI besides the seasonality, and variation in day and night (Henits et al. 2017). Among various LULC types, several studies have highlighted that vegetation cover and ISA, which accounts for almost 70% (Imhoff et al. 2010), forms the major contributors to UHI variations in and around the cities. While vegetation acts as a cooling agent (Chen et al. 2014), ISA (like asphalt, concrete, etc.) on the other hand reduces the

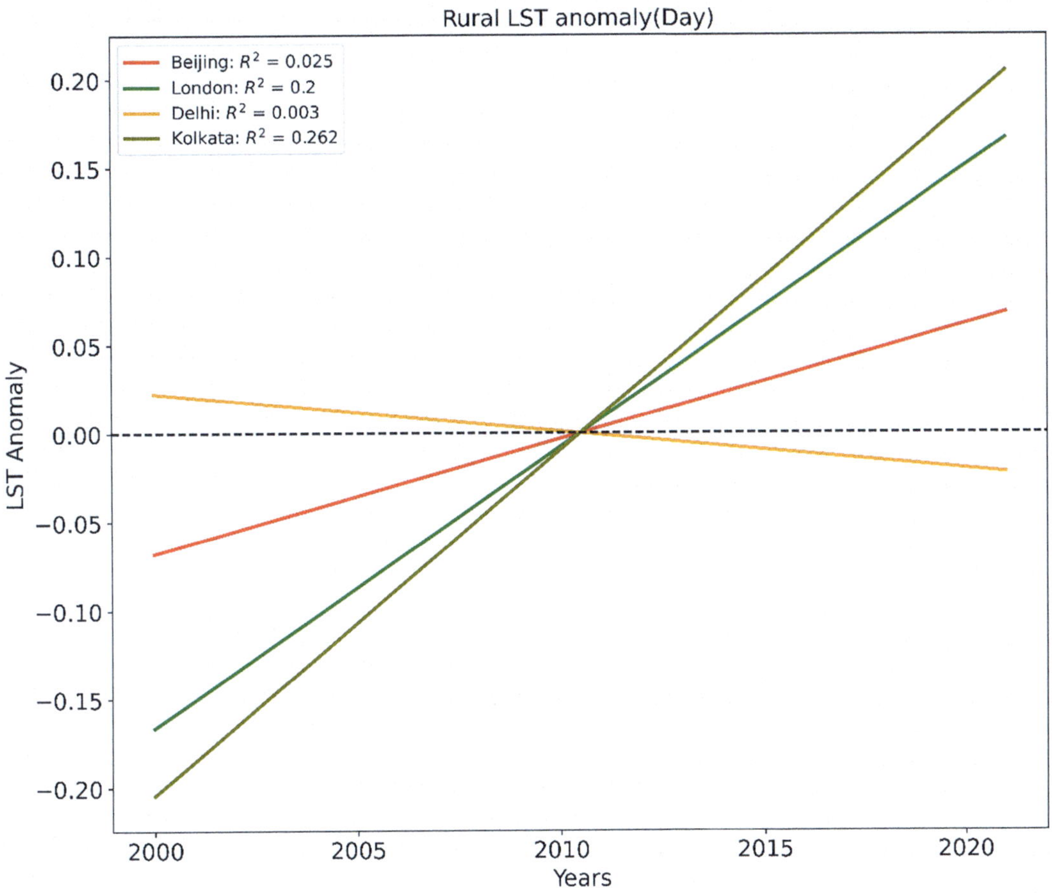

Fig. 22.3 Day time LST anomaly over rural areas considered in this study corresponding to the urban areas

albedo and escalates the LST of the region (Morini et al. 2016). Due to the phenology of various plant species, vegetation has a diverse mitigating effect on UHI (Deilami et al. 2018). Seasonal variations also influence UHI intensity caused by differences in solar insolation and the physiology of plant species. Consequently, UHI intensity is more during summer as compared to winter (Zipper et al. 2016). The geographical location or morphology of an area also plays an important role in determining the UHI intensity. Cities situated within rural outskirts tend to develop heat islands rather than cities located in arid regions as reported in the case of Cairo, Africa (Shahraiyni et al. 2016). Additionally, diurnal variations (UHI intensity is generally higher at night) and city characteristics (namely, construction materials of impervious surfaces, height of buildings (Haashemi et al. 2016), etc.) pose as significant contributors to determining the intensity of UHI (Deilami et al. 2018). Other crucial factors include the extent of the urban area (Li et al. 2017), landscape metrics (Zhou et al. 2011), population composition, economic conditions, etc. On the contrary, urban greenery and inland aquatic bodies significantly contribute in reducing the UHI intensity due to potential cooling through evaporation (Gupta et al. 2019; Li et al. 2013). In some Indian cities like Kolkata (Rath et al. 2022), Hyderabad (Bhanage et al. 2023), Bhopal (Mohammad and Goswami 2022), where there is presence of a large water bodies and adequate greenery, lowering impact of UHI is experienced, and urban cool island (UCI) becomes a possibility at times. Bengaluru city also exhibits similar signatures of UCI owing to adequate greenery prevailing at the core urban areas compared to the peripheral regions

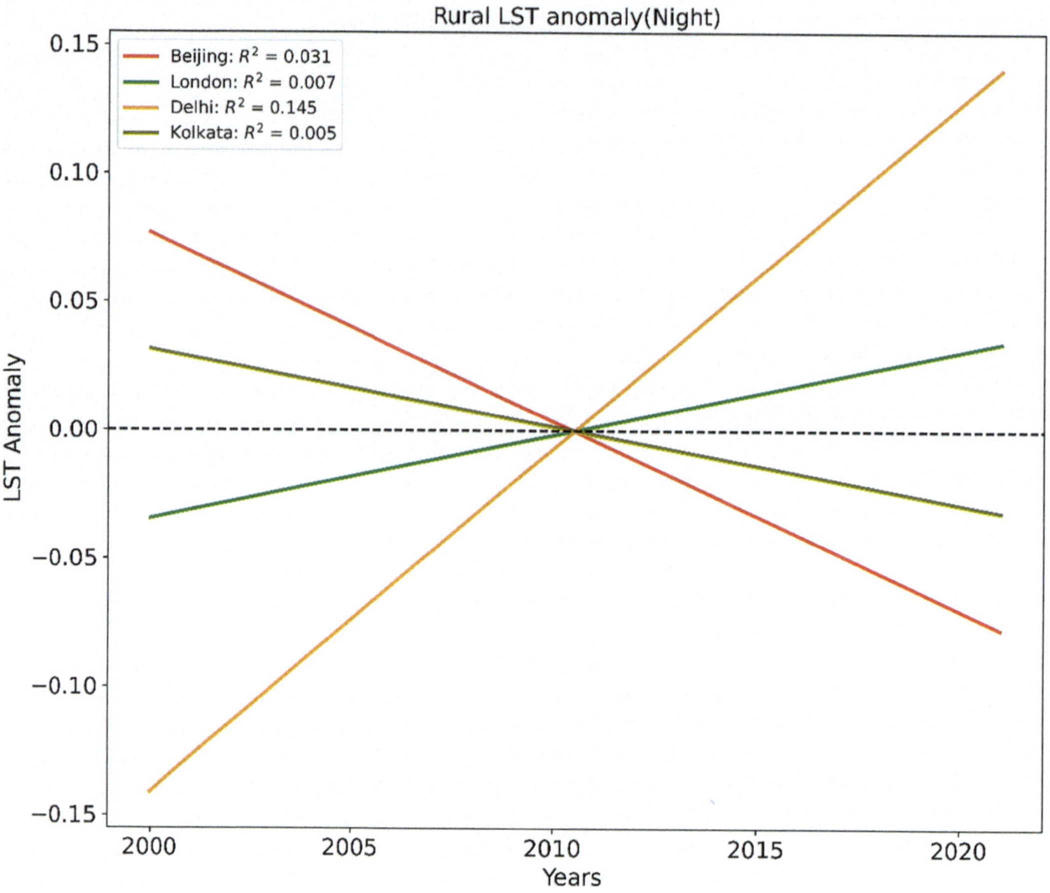

Fig. 22.4 Night time LST anomaly over rural areas considered in this study corresponding to the urban areas

(Ramachandra and Kumar 2010; Siddiqui et al. 2021). However, cities like Ranchi (Mohanta and Sharma 2017) and Ahmedabad (Bhanage et al. 2023) show relatively higher UHI intensity due to enhanced industrial and anthropogenic activities.

22.2.2.2 Meteorological Factors

Apart from anthropogenic drivers, meteorological parameters also impact the UHI occurrence and intensity. For instance, wind speed and cloud cover are two such parameters found to be greatly influencing UHIs (Lee and Baik 2010). Nevertheless, there were inconsistencies regarding the impact of these two on UHIs. For example, the UHI intensity is found to be inversely related to the fourth root of wind speed and cloud cover in Melbourne (Morris et al. 2001), while it is inversely related to the square root of wind speed in Hamburg (Schlünzen et al. 2010). Seoul exhibited a vanishing UHI intensity with increase in the wind speed (Kim and Baik 2002). Some studies also claimed that anticyclones promote the growth of a robust UHI (Ivajnšič and Žiberna 2019). Another study conducted in Hamburg disapproved the effects of atmospheric pressure and water vapor pressure on the UHI intensity and instead placed significant emphasis on the role of relative humidity (Arnds et al. 2017). In the Yangtze River Delta Urban Agglomeration, China (Du et al. 2016) and Debrecen, Hungary (László et al. 2016), the research works have emphasized on the role of precipitation and solar radiation in governing the UHI intensity. Venter et al. (2021) claimed that evapotranspiration and surface roughness are critical parameters that

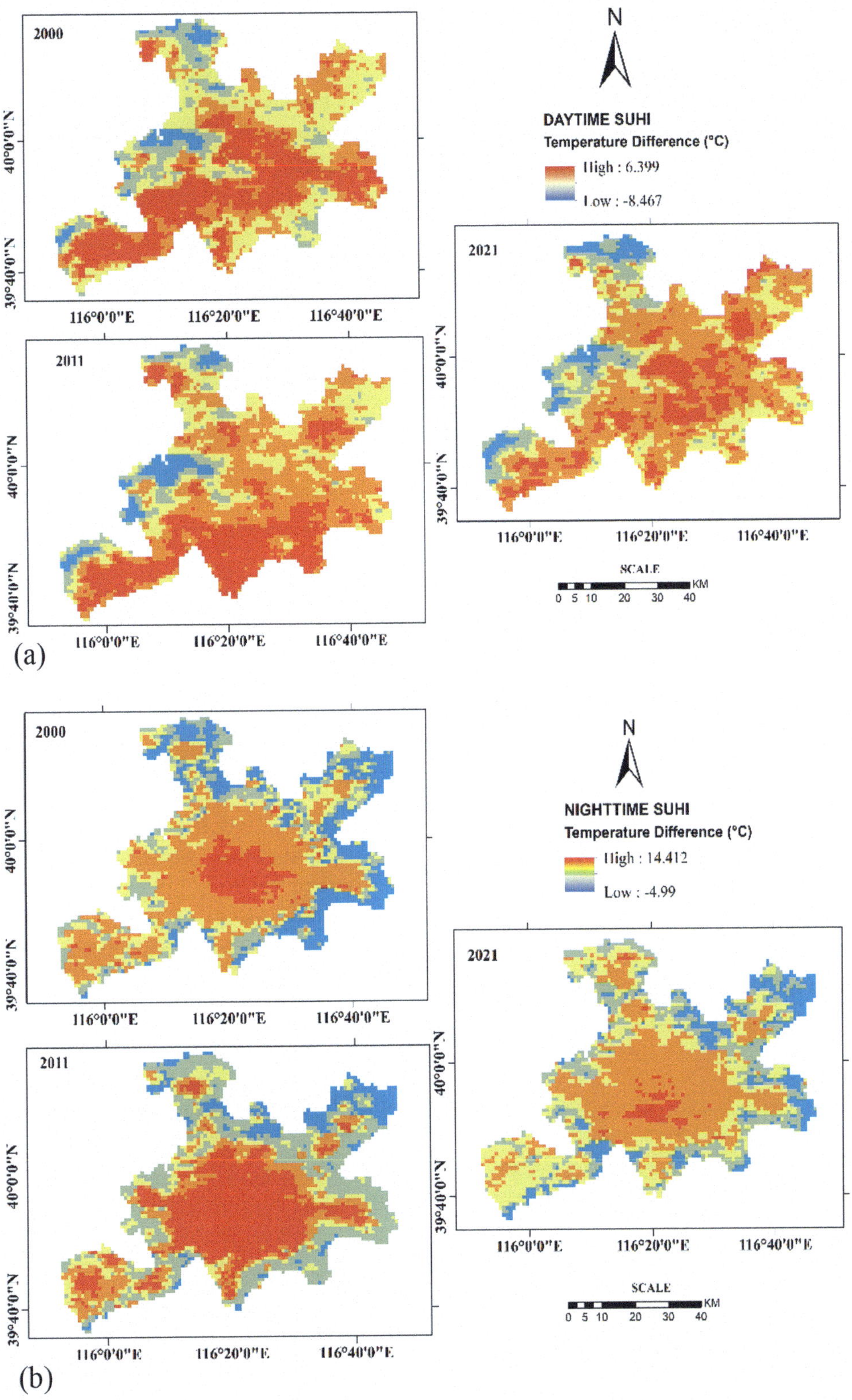

Fig. 22.5 Day and night time SUHI over Beijing

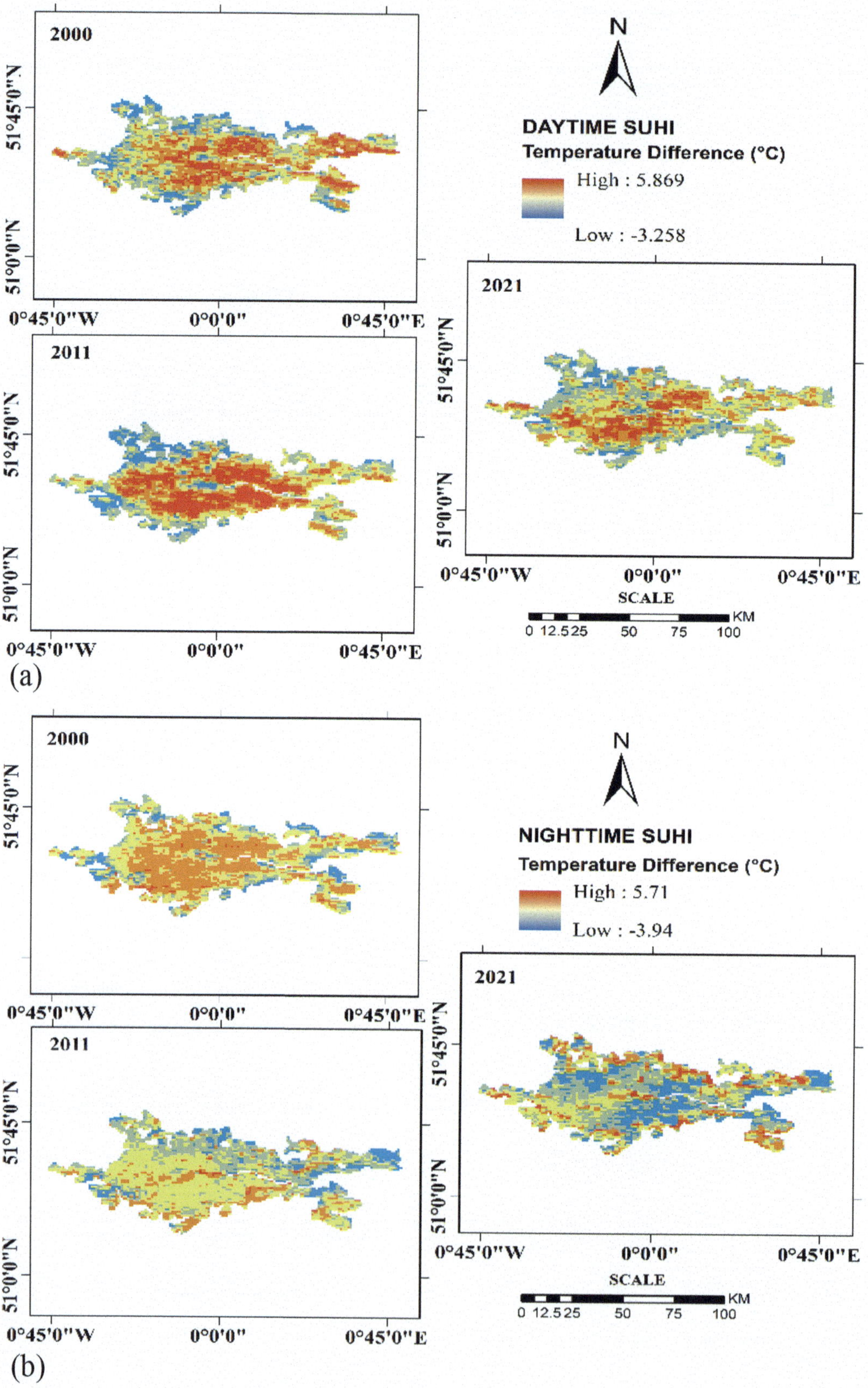

Fig. 22.6 Day and night time SUHI over London

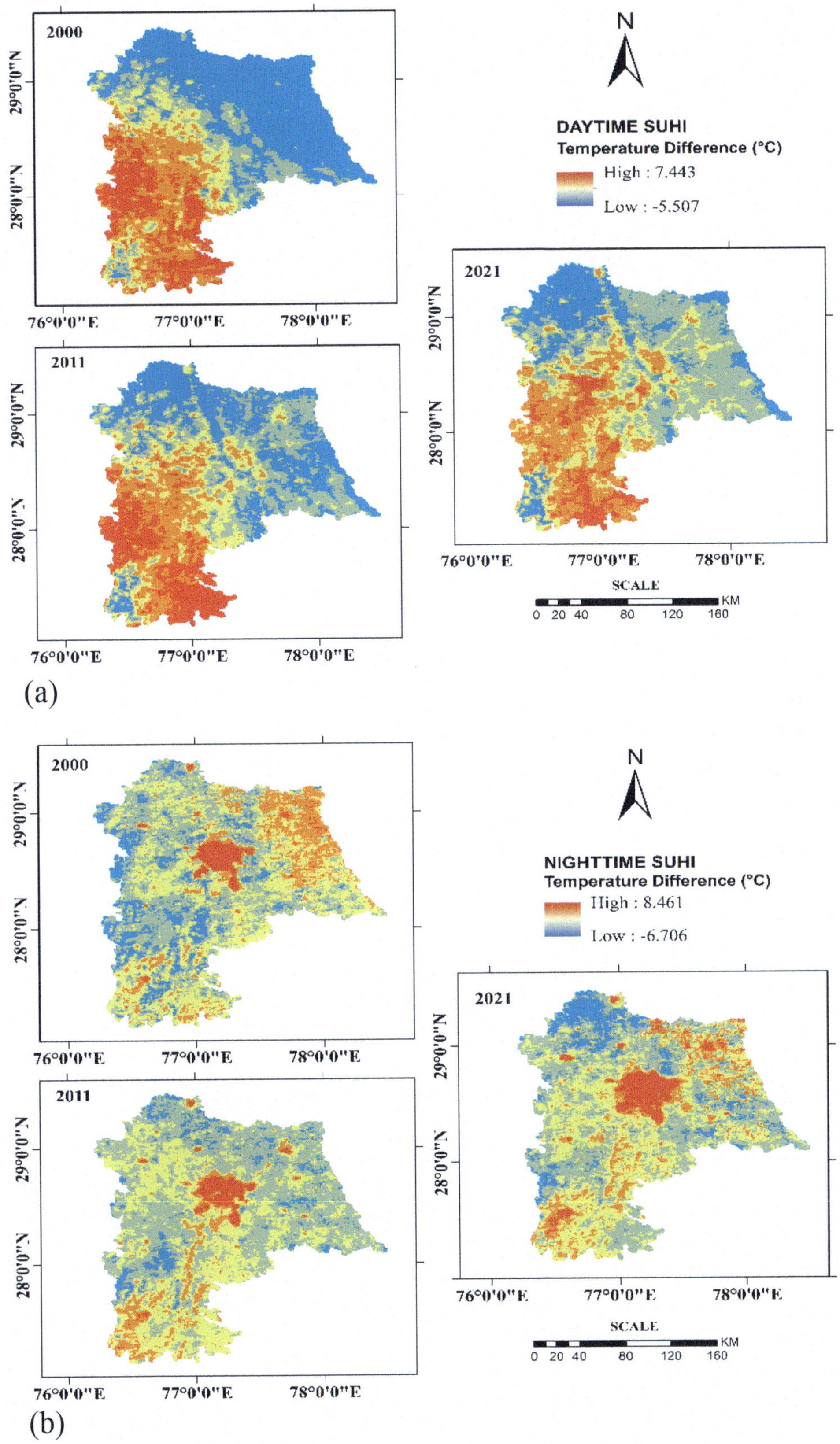

Fig. 22.7 Day and night time SUHI over Delhi-NCR region

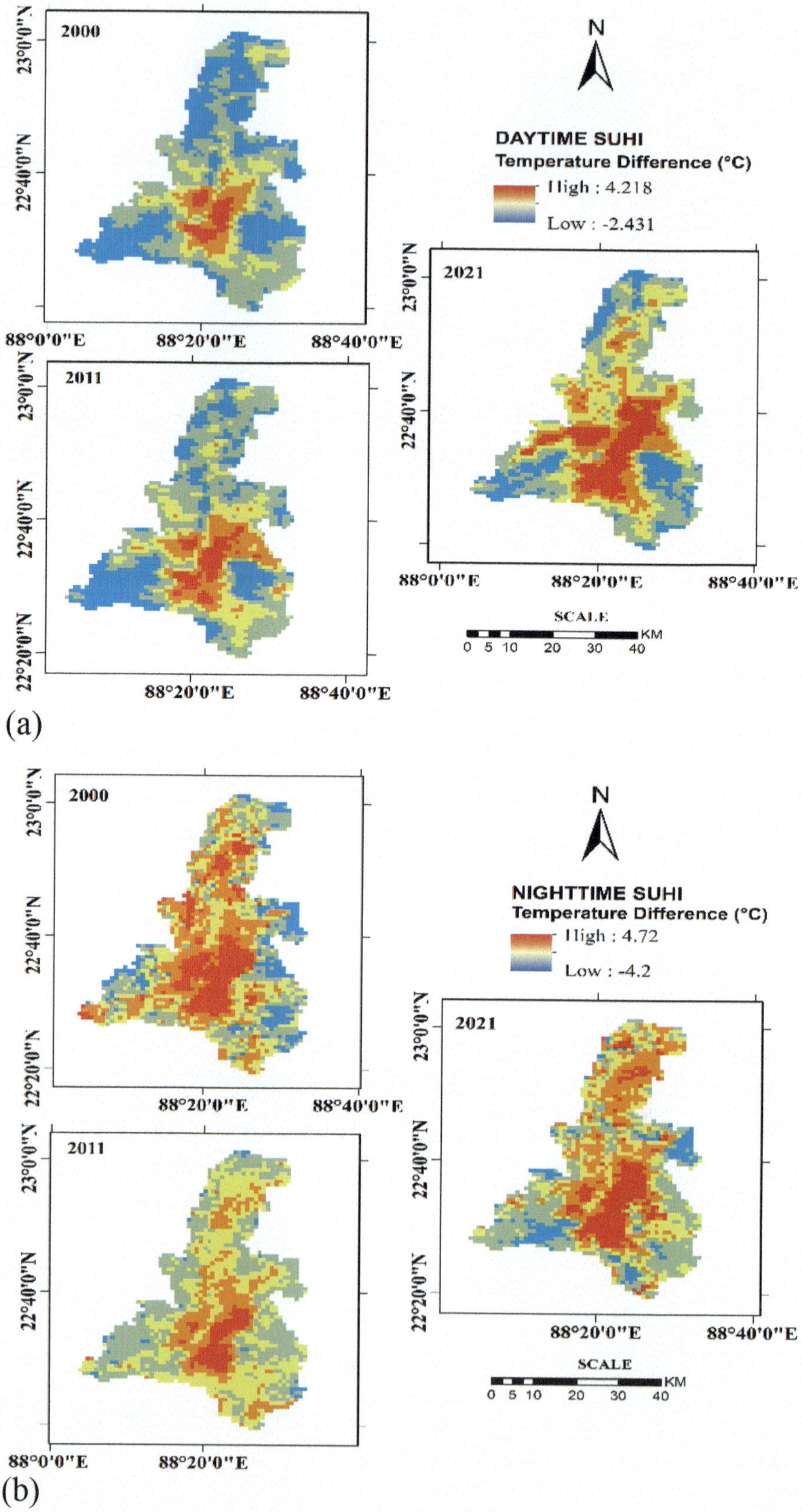

Fig. 22.8 Day and night time SUHI over Kolkata

influence both SUHI and CLUHI intensity during the prevalence of heat waves. Although some researchers argued that meteorology has no role in influencing the UHI intensity (Unger et al. 2001), it is not widely accepted. In light of these dissimilarities, further study is needed to comprehend the impact of environmental conditions on UHI and its variation.

22.3 Threat of UHI and Associated Heat-Related Weather Extremities

Climate forecasts at the global scale indicate that the warming of the atmosphere caused by greenhouse gas emissions is already being experienced and will worsen over this century (IPCC 2013; Lemonsu et al. 2015). Several studies have highlighted that the increase in heat wave frequencies and intensities is one of the most prominent impacts of global warming (Chauvin and Denvil 2007; Beniston et al. 2007). A significant proportion of the world population lives in urban areas, making heat waves and other heat-related weather extremities an issue of great concern (Basara et al. 2010). Therefore, the risks accompanied by heat waves, heat stress, UHI, global warming, etc., cannot be overlooked and should be acted upon with utmost emphasis. Some of the major threats related to UHI and urban heat stress are described in this section.

22.3.1 UHI Effect on Local Weather and Climate

The local microclimate of an urban area is highly influenced by the urban geometry (city size, average height, composition, etc.), thermal characteristics (albedo, energy fluxes, atmospheric moisture content, etc.), and anthropogenic emissions (Santamouris 2007; Niyogi et al. 2009). With the urban expansion and densification, the radiative balance, convective heat transfer between surface and buildings, the motion of air prevailing in the urban region, and heat production inside the city have all been significantly altered (Gaitani et al. 2011). Consequently, air temperature in the urban core is detected to be ~1–6 °C higher than the adjacent rural areas (Santamouris 2007). Heat irregularity experienced in the cities occurs due to hindered airflow and enhanced mechanical (thermal) turbulence (Oke 1995). Since a calm and weak breeze initiates the formation of UHI, the prominent UHI signatures are observed in the afternoon hours (Zhang et al. 2011). Precipitation properties are also impacted due to higher urban temperatures (Li et al. 2013, 2016; Rath and Panda 2020; Panda and Rath 2022; Rath et al. 2021, 2022). Some of the experiments reflected that higher precipitation occurrences (Doan et al. 2021; Sun et al. 2021) were observed within the city limits due to UHI-induced temperature variations (Huff and Vogel 1978), and obstruction offered by the structure of urban canyons (Zhang et al. 2019). However, contradictory results were claimed in the study of Rosenfeld (2000), where it was postulated that due to urban air pollution, the amount of rainfall gets suppressed in cities. Well-developed UHI impacts the location, pattern, and precipitation distribution of thunderstorm events over cities (Lin et al. 2011). In fact, thunderstorms have also been observed to diverge along the circumference of urban areas (Niyogi et al. 2011). Recent studies have highlighted that UHIs can contribute to ~30% of climate change (Huang and Lu 2015) and can even hinder the pattern of jet streams (Zhang et al. 2013). However, the opposite is also true, that is, climate change can impose adverse impacts on the severity of UHI events which is why it is expected to intensify in the coming years (Tuholske et al. 2021). Therefore, it is apparent that the UHI over metropolitan areas can impact the thermodynamic and dynamic processes to consequently alter the local and mesoscale weather (Rath et al. 2021).

22.3.2 Overutilization of Energy Sources

UHI formed over metropolitan areas adversely impact the energy requirements of urban dwellers (Mostafavi et al. 2021). To create a sustainable living environment, the demand for cooling down the interior of buildings is on the rise. Increasing energy consumption and the consequent expenditure are thus inevitable. This is most prominent during summer, which results in the generation of additional energy (Panda et al. 2022). Rising urban temperatures, the subsequent increase in cooling demands, frequency of power failures and blackouts have considerably increased (Maxwell et al. 2018). It is reported that a 0.5°K rise in average UHI intensity can increase the monthly energy usage by ~0.17 kWh/m^2 to 1.84 kWh/m^2 (Su et al. 2021). Studies show that energy used for cooling a building can increase the amount of energy consumption by as much as 19% due to UHI (Li et al. 2018, 2019). However, it is difficult to acquire precise data sets to assess the degree to which UHI affects energy consumption. Therefore, the existing demand for energy will lead to the overutilization of the Earth's energy resources and an upsurge in anthropogenic emissions.

22.3.3 Heat Stress: Impact on Human Health and Well-being

The UHI effect acts as one of the most adverse consequence of urbanization on regional climate which may increase the risk of heat stress (Venter et al. 2021). According to meteorologists, there has been an increase in the frequency of record-breaking high temperatures worldwide. Extreme heat poses a worldwide risk to life (Patz et al. 2005) and due to frequent (intense) heat waves, especially in the summer, the threat is expected to become more severe in the future (Meehl and Tebaldi 2004). From meteorological point of view, heat stress can be defined as the Wet-Bulb globe temperatures (W) (Fischer et al. 2012). Since, W is nonlinear to both temperature and specific humidity (Stull 2011), it can act as an effective measure of the cooling efficiency of perspiration and can thus be useful for analyzing heat stress (Raymond et al. 2017). Environmental (ambient and radiant temperature, humidity, and air movement) and behavioral variables influence thermal comfort in humans (Epstein and Moran 2006). Other factors that influence the heat stress phenomenon include the presence of CO_2, diurnal and nocturnal variations, climate change, etc. (Oleson et al. 2015).

UHI-induced heat stress impacts such as heat fatalities, respiratory difficulties, and blood-pressure changes, have increased over the years (Gamble et al. 2013). High air temperatures in metropolitan areas have further intensified the impact of heat stress (Meehl and Tebaldi 2004). People belonging to economically weaker strata of the society, persons with chronic somatic and mental disorders, older and infant population, pregnant women, etc., are more susceptible to such heat-related discomfort (Błażejczyk and McGregor 2008; Johnson and Wilson 2009). Heat-related fatalities are common in arid, hot, and humid regions of the world (Robinson 2001). However, more deaths are reported every year in temperate regions as people are less acquainted with the high temperatures in those areas (Kalkstein and Davis 1989; Smoyer et al. 2000). The impact of prolonged exposure to heat waves can also prove to be fatal as it can lead to thermoregulatory failure (Błażejczyk and McGregor 2008). Urban inhabitants can face severe symptoms such as heat discomfort, heat stroke, limb swelling, neck rashes, cramps, headache, lethargy, irritability, and weakness. It can lead to severe dehydration and even instigate acute cerebrovascular conditions that cause thrombogenesis (https://www.who.int/health-topics/heatwaves#tab=tab_2). In developing countries like India, where a large proportion of the population is engaged in labor-intense outdoor jobs, heat-related threats can be even more chronic (Murari et al. 2015). Thus, a situation of a public health emergency is inevitable. It also impacts health services like the storage of medicines and increased hospital admissions. Increased heat enhances the potential for disease spread. To some extent, it also impacts the mental health of an individual. It interferes with the day-to-day activities of the inhabitants. Heat-related mortalities are quite frequent across northern India, where 2500 deaths were reported in 2015; Ahmedabad in 2010 saw 800 official deaths (https://harvardpublichealth.org/climate-change/feeling-the-heat/). Increasing temperatures can impose a direct impact on the economy of countries. A report by McKinsey and Company predicted that by 2050 due to rising temperatures, countries like India, Bangladesh, and Pakistan, where the economy heavily depends on outdoor labor, might see an 8–13% decrease in their GDP. Urban centers host agglomerations of a wide range of population composition belonging to different economies. This heterogeneity aggravates the impact of heat stress, especially on the economically underprivileged people residing in unfavorable environmental conditions such as urban slums and other informal settlements (Harlan et al. 2006; Scherer et al. 2013). Additionally, the density of urban areas and the demography of the urban population plays an important role (Harlan et al. 2013; Song et al. 2020) in determining the UHI intensity and consequent heat stress. Therefore, more emphasis should be given to understanding the extent of heat-related threats on Indian cities as this may benefit people belonging to several sectors.

22.3.4 Air Pollution and UHI

The UHI phenomenon and air pollution levels play significant roles in impacting each other (Katsouyanni et al. 1993; Sarrat et al. 2006). Atmospheric thermal and dynamic processes influence the geographical and temporal distribution of UHI and air pollution across urban areas (Wang et al. 2021). Air pollutants have the ability to increase the energy received by the urban land surface by absorbing and reradiating longwave radiation and can therefore induce UHIs (Cao et al. 2016; Liang et al. 2021). As a result, the temperature of the urban surface is amplified by 0.70 ± 0.26 °C (Cao et al. 2016). In Seoul, it was observed that ozone (O_3) concentration was inversely proportional to UHI intensity, while the concentration of carbon mono-oxide (CO), sulfur-dioxide (SO_2), particulate matter (PM), and nitrogen-dioxide (NO_2) are directly proportional to the UHI intensity (Ngarambe et al. 2021). A study on 150 cities of the Beijing-Tianjin-Hebei region of China highlighted that with 10 μg/m^3 increase in annual $PM_{2.5}$ concentrations can elevate the UHI intensity by up to 0.09 °C (Liang et al. 2021). Nevertheless, contradictory claims of the mitigating effect of air pollutants were made by Ding et al. (2022). The impact on human health and mortality was also explored in some studies (Rainham and Smoyer-Tomic 2003). In the case of Japan, it was seen that the collective influence of NO_2 and high temperatures were associated with higher cases of lung cancer (Choi et al. 1997) and heat stroke occurrences (Piver et al. 1999). In fact, persistent exposure to such weather extremities can make young children more vulnerable to the risk of acquiring asthma and other respiratory illnesses later in life (US Climate Change Science Program 2008). Therefore, the threat imposed by pollutants suspended in the air not only has a direct impact on human health but also significantly contributes to UHI and hence the regional climate change.

22.3.5 Effect on Ground and Surface Water

The excessively heated ground surface layer formed as result of UHI can lead to the significant heating of groundwater cradles (Yalcin and Yetemen 2009). When the warm storm water runoff (due to heated pavements and rooftops) drains into sewers, it lifts the temperature of water present in the surface water bodies, impacting millions of aquatic organisms. This can cause rapid temperature spikes in aquatic habitats, making it traumatic and even disastrous to aquatic species. Thermal pollution also aids in the growth of algae. Several studies have confirmed that urban streams are hotter than those flowing through forested regions (Somers et al. 2016).

22.4 Summary and Conclusions

UHI events within cities influence environmental, social, and economic aspects of any metropolitan center. Even though the UHI phenomenon has been extensively studied, it was observed that the impact and future effects of climate change and urban expansion on UHI and the subsequent influence on heat stress has surprisingly received little attention. Therefore, there is a pressing need to investigate the implications of the proliferation of megacities on air pollution, regional weather and climate, and the global climate as more and more people are forced to live in metropolitan areas. Looking at the available literature, it can be said that previous research mostly focused on the understanding of the temporal and spatial variability (LST-based analysis) of UHI developed within the city. Later, the atmospheric (local-scale and micro-scale) and health impacts of UHI and heat stress were also explored.

Despite the fact that developed countries such as the USA, Europe, and several Asian countries have strengthened themselves scientifically and technologically to quantify the intensity of UHI and its correlation with atmospheric parameters and air pollution, not much has been experimented in this direction in India. India is yet to establish a comprehensive mesoscale observational network for urban research. Therefore, to compensate for the lacunae in heat-related studies in India, certain areas need to be explored, namely, (1) developing frameworks to include atmospheric, air pollution, and urban geometry parameters in the UHI study, (2) developing an index to quantify the intensity of UHI, and (3) preparing a well-documented database of heat stress events to expand research on the health impacts of UHI as well. Additionally, numerical modeling may also be useful in UHI research and climate forecasting over the urban areas.

Acknowledgments The authors would like to acknowledge the MODIS portal for land-surface temperature data (https://modis.gsfc.nasa.gov/data/dataprod/mod11.php). Also, the funding from the INSPIRE scheme of the Ministry of Science and Technology (IF190926) and from the Ministry of Earth Sciences through a sponsored project (MoES/16/09/2018-RDEAS-THUMP-2) is highly acknowledged for partly supporting this piece of work.

References

Acharya T, Riehl B, Fuchs A (2021) Effects of albedo and thermal inertia on pavement surface temperatures with convective boundary conditions—a CFD study. PRO 9(11):2078

Almeida CRD, Teodoro AC, Gonçalves A (2021) Study of the urban heat Island (UHI) using remote sensing data/techniques: a systematic review. Environments 8(10):105

Arnds D, Böhner J, Bechtel B (2017) Spatio-temporal variance and meteorological drivers of the urban heat Island in a European city. Theor Appl Climatol 128:43–61

Arnfield AJ (2003) Two decades of urban climate research: a review of turbulence, exchanges of energy and water, and the urban heat Island. Int J Climatol 23(1):1–26

Baklanov A, Molina LT, Gauss M (2016) Megacities, air quality and climate. Atmos Environ 126:235–249

Basara JB, Basara HG, Illston BG, Crawford KC (2010) The impact of the urban heat Island during an intense heat wave in Oklahoma City. Adv Meteorol 2010

Beniston M, Stephenson DB, Christensen OB, Ferro CA, Frei C, Goyette S et al (2007) Future extreme events in European climate: an exploration of regional climate model projections. Clim Chang 81:71–95

Bhanage V, Kulkarni S, Sharma R, Lee HS, Gedam S (2023) Enumerating and modelling the seasonal alterations of surface Urban heat and cool Island: a case study over Indian cities. Urban Sci 7(2):38

Błażejczyk K, McGregor G (2008) Mortality in European cities and its relations to biothermal conditions. Klimat i bioklimat miast, Wydawnictwo Uniwersytetu łódzkiego:313–324

Borbora J, Das AK (2014) Summertime urban heat Island study for Guwahati city, India. Sustain Cities Soc 11:61–66

Boo KO, Oh SN (2000) Characteristics of spatial and temporal distribution of air temperature in Seoul, 1999 (in Korean). J Korean Meteorological Society 36:499–506

Cao C, Lee X, Liu S, Schultz N, Xiao W, Zhang M, Zhao L (2016) Urban heat islands in China enhanced by haze pollution. Nat Commun 7(1):12509

Cao Q, Yu D, Georgescu M, Wu J, Wang W (2018) Impacts of future urban expansion on summer climate and heat-related human health in eastern China. Environ Int 112:134–146

Chauvin F, Denvil S (2007) Changes in severe indices as simulated by two French coupled global climate models. Glob Planet Chang 57(1–2):96–117

Chen A, Yao XA, Sun R, Chen L (2014) Effect of urban green patterns on surface urban cool islands and its seasonal variations. Urban For Urban Green 13(4):646–654

Choi KS, Inoue S, Shinozaki R (1997) Air pollution, temperature, and regional differences in lung cancer mortality in Japan. Arch Environ Health Intl J 52(3):160–168

Crippa M, Guizzardi D, Pisoni E, Solazzo E, Guion A, Muntean M et al (2021) Global anthropogenic emissions in urban areas: patterns, trends, and challenges. Environ Res Lett 16(7):074033

Debbage N, Shepherd JM (2015) The urban heat Island effect and city contiguity. Comput Environ Urban Syst 54:181–194

Deilami K, Kamruzzaman M, Liu Y (2018) Urban heat Island effect: a systematic review of spatio-temporal factors, data, methods, and mitigation measures. Int J Appl Earth Obs Geoinf 67:30–42

Devi SS (2006) Urban heat islands and environmental impact. In: 86th American meteorological society annual meeting. Atlanta, Georgia

Ding Y, Feng H, Zou B, Nie Y (2022) Heterogeneous air pollution controls its correlation to urban heat Island: a satellite perspective. Adv Space Res 69(12):4252–4262

Doan Q-V, Dipankar A, Simón-Moral A, Sanchez C, Prasanna V, Roth M, Huang X-Y (2021) Urban-induced modifications to the diurnal cycle of rainfall over a tropical city. Quart J Roy Meteor Soc 147:1189–1201

Du H, Wang D, Wang Y, Zhao X, Qin F, Jiang H, Cai Y (2016) Influences of land cover types, meteorological conditions, anthropogenic heat and urban area on surface urban heat Island in the Yangtze River Delta Urban agglomeration. Sci Total Environ 571:461–470

EPA (2008) Reducing urban heat islands: compendium of strategies. Heat Island Reduct Activit:1–23

Epstein Y, Moran DS (2006) Thermal comfort and the heat stress indices. Ind Health 44(3):388–398

Fischer EM, Oleson KW, Lawrence DM (2012) Contrasting urban and rural heat stress responses to climate change. Geophys Res Lett 39(3)

Gaitani N, Spanou A, Saliari M, Synnefa A, Vassilakopoulou K, Papadopoulou K et al (2011) Improving the microclimate in urban areas: a case study in the Centre of Athens. Build Serv Eng Res Technol 32(1):53–71

Gamble JL, Hurley BJ, Schultz PA, Jaglom WS, Krishnan N, Harris M (2013) Climate change and older Americans: state of the science. Environ Health Perspect 121(1):15–22

Grimm NB, Faeth SH, Golubiewski NE, Redman CL, Wu J, Bai X, Briggs JM (2008) Global change and the ecology of cities science, 319

Guan X, Shen H, Li X, Gan W, Zhang L (2019) A long-term and comprehensive assessment of the urbanization-induced impacts on vegetation net primary productivity. Sci Total Environ 669:342–352

Gunawardena KR, Wells MJ, Kershaw T (2017) Utilising green and bluespace to mitigate urban heat Island intensity. Sci Total Environ 584:1040–1055

Gupta A, Dey V, Goel A (2009) Mitigation of thermal pollution to enhance urban air quality through remote sensing and GIS. 10th ESRI India user Conf Noida (NCR)

Gupta N, Mathew A, Khandelwal S (2019) Analysis of cooling effect of water bodies on land surface temperature in nearby region: a case study of Ahmedabad and Chandigarh cities in India. Egypt J Remote Sens Space Sci 22(1):81–93

Haashemi S, Weng Q, Darvishi A, Alavipanah SK (2016) Seasonal variations of the surface urban heat Island in a semi-arid city. Remote Sens 8(4):352

Halder B, Bandyopadhyay J, Banik P (2021) Monitoring the effect of Urban development on Urban Heat Island based on remote sensing and geo-spatial approach in Kolkata and adjacent areas, India. Sustain Cities Soc 74(March):103186. https://doi.org/10.1016/j.scs.2021.103186

Han JY, Baik JJ, Lee H (2014) Urban impacts on precipitation. Asia-Pac J Atmos Sci 50:17–30

Harlan SL, Brazel AJ, Prashad L, Stefanov WL, Larsen L (2006) Neighborhood microclimates and vulnerability to heat stress. Soc Sci Med 63(11):2847–2863

Harlan SL, Declet-Barreto JH, Stefanov WL, Petitti DB (2013) Neighborhood effects on heat deaths: social and environmental predictors of vulnerability in Maricopa County, Arizona. Environ Health Perspect 121(2):197–204

Henits L, Mucsi L, Liska CM (2017) Monitoring the changes in impervious surface ratio and urban heat Island intensity between 1987 and 2011 in Szeged, Hungary. Environ Monit Assess 189:1–13

Hou H, Liu K, Li X, Chen S, Wang W, Rong K (2020) Assessing the urban heat Island variations and its influencing mechanism in metropolitan areas of Pearl River Delta, South China. Physics and Chemistry of the Earth, Parts A/B/C 120:102953

Huang Q, Lu Y (2015) The effect of urban heat Island on climate warming in the Yangtze River Delta urban agglomeration in China. Int J Environ Res Public Health 12(8):8773–8789

Huff FA, Vogel JL (1978) Urban, topographic, and diurnal effects on rainfall in the St. Louis Region J Appl Meteor 17:565–577

Ibrahim SH, Ibrahim NIA, Wahid J, Goh NA, Koesmeri DRA, Nawi MNM (2018) The impact of road pavement on urban heat Island (UHI) phenomenon. Civ Eng 9(8)

Imhoff ML, Zhang P, Wolfe RE, Bounoua L (2010) Remote sensing of the urban heat Island effect across biomes in the continental USA. Remote Sens Environ 114(3):504–513

IPCC (2013) The physical science basis, the working group I contribution to the UN IPCC's fifth assessment report (WG1 AR5)

Ivajnšič D, Žiberna I (2019) The effect of weather patterns on winter small city urban heat islands. Meteorol Appl 26(2):195–203

Johnson DP, Wilson JS (2009) The socio-spatial dynamics of extreme urban heat events: the case of heat-related deaths in Philadelphia. Appl Geogr 29(3):419–434

Kalkstein LS, Davis RE (1989) Weather and human mortality: an evaluation of demographic and interregional responses in the United States. Ann Assoc Am Geogr 79(1):44–64

Katsouyanni K, Pantazopoulou A, Touloumi G, Tselepidaki I, Moustris K, Asimakopoulos D, Poulopoulou G, Trichopoulos D (1993) Evidence for interaction between air pollution and high temperature in the causation of excess mortality. Arch Environ Health Intl J 48(4):235–242

Kim YH, Baik JJ (2002) Maximum urban heat Island intensity in Seoul. J Appl Meteorol 41(6):651–659

Kim Y-H, Baik J-J (2005) Spatial and temporal structure of the Urban Heat Island in Seoul. J Appl Meteorol 44(5):591–605. https://journals.ametsoc.org/doi/10.1175/JAM2226.1

Kolokotroni M, Giridharan R (2008) Urban heat Island intensity in London: an investigation of the impact of physical characteristics on changes in outdoor air temperature during summer. Sol Energy 82(11):986–998

Kuang W, Dou Y, Zhang C, Chi W, Liu A, Liu Y, Zhang R, Liu J (2015) Quantifying the heat flux regulation of metropolitan land use/land cover components by coupling remote sensing modeling with in situ measurement. J Geophys Res Atmos 120(1):113–130

Kuddus MA, Tynan E, McBryde E (2020) Urbanization: a problem for the rich and the poor? Public Health Rev 41:1–4

László E, Bottyán Z, Szegedi S (2016) Long-term changes of meteorological conditions of urban heat Island development in the region of Debrecen, Hungary. Theor Appl Climatol 124:365–373

Lee SH, Baik JJ (2010) Statistical and dynamical characteristics of the urban heat Island intensity in Seoul. Theor Appl Climatol 100:227–237

Lemonsu A, Viguie V, Daniel M, Masson V (2015) Vulnerability to heat waves: impact of urban expansion scenarios on urban heat Island and heat stress in Paris (France). Urban Clim 14:586–605

Li XX, Koh TY, Entekhabi D, Roth M, Panda J, Norford LK (2013) A multi-resolution ensemble study of a tropical urban environment and its interactions with the background regional atmosphere. J Geophys Res Atmos 118(17):9804–9818

Li XX, Koh TY, Panda J, Norford LK (2016) Impact of urbanization patterns on the local climate of a tropical city, Singapore: an ensemble study. J Geophys Res Atmos 121(9):4386–4403

Li X, Zhou Y, Asrar GR, Imhoff M, Li X (2017) The surface urban heat Island response to urban expansion: a panel analysis for the conterminous United States. Sci Total Environ 605:426–435

Li X, Zhou Y, Asrar GR, Zhu Z (2018) Developing a 1 km resolution daily air temperature dataset for urban and surrounding areas in the conterminous United States. Remote Sens Environ 215:74–84

Li X, Zhou Y, Yu S, Jia G, Li H, Li W (2019) Urban heat Island impacts on building energy consumption: a review of approaches and findings. Energy 174:407–419

Liang Z, Huang J, Wang Y, Wei F, Wu S, Jiang H et al (2021) The mediating effect of air pollution in the impacts of urban form on nighttime urban heat Island intensity. Sustain Cities Soc 74:102985

Liao W, Rigden AJ, Li D (2018) Attribution of local temperature response to deforestation. J Geophys Res Biogeo 123(5):1572–1587

Lin CY, Chen WC, Chang PL, Sheng YF (2011) Impact of the urban heat Island effect on precipitation over a complex geographic environment in northern Taiwan. J Appl Meteorol Climatol 50(2):339–353

Mandal J, Patel PP, Samanta S (2022) Examining the expansion of Urban Heat Island effect in the Kolkata metropolitan area and its vicinity using multi-temporal MODIS satellite data. Adv Space Res 69(5):1960–1977. https://linkinghub.elsevier.com/retrieve/pii/S0273117721008899

Martilli A (2014) An idealized study of city structure, urban climate, energy consumption, and air quality. Urban Clim 10:430–446

Mathew A, Khandelwal S, Kaul N (2017) Investigating spatial and seasonal variations of urban heat Island effect over Jaipur city and its relationship with vegetation, urbanization and elevation parameters. Sustain Cities Soc 35:157–177

Matsumoto J, Fujibe F, Takahashi H (2017) Urban climate in the Tokyo metropolitan area in Japan. J Environ Sci 59:54–62. https://linkinghub.elsevier.com/retrieve/pii/S1001074216314322

Maxwell K, Grambsch A, Kosmal A, Larson L, Sonti N (2018) Built environment, urban systems, and cities. In: Reidmiller DR, Avery CW, Easterling DR, Kunkel KE, Lewis KLM, Maycock TK, Stewart BC (eds) Impacts, risks, and adaptation in the United States: fourth national climate assessment, volume II. US Global Change Research Program, Washington, DC, pp 438–478

McMichael AJ (2000) The urban environment and health in a world of increasing globalization: issues for developing countries. Bull World Health Organ 78:1117–1126

Meehl GA, Tebaldi C (2004) More intense, more frequent, and longer lasting heat waves in the 21st century. Science 305(5686):994–997

Mohammad P, Goswami A (2022) Exploring different indicators for quantifying surface urban heat and cool Island together: a case study over two metropolitan cities of India. Environ Develop Sustain:1–22

Mohan M, Kikegawa Y, Gurjar BR, Bhati S, Kolli NR (2013) Assessment of urban heat Island effect for different land use–land cover from micrometeorological measurements and remote sensing data for megacity Delhi. Theor Appl Climatol 112:647–658

Mohanta K, Sharma LK (2017) Assessing the impacts of urbanization on the thermal environment of Ranchi City (India) using geospatial technology. Remote Sens Appl Soc Environ 8:54–63

Monteiro FF et al (2021) Assessment of Urban Heat Islands in Brazil based on MODIS remote sensing data. Urban Clim 35:100726. https://linkinghub.elsevier.com/retrieve/pii/S2212095519303554

Morini E, Touchaei AG, Castellani B, Rossi F, Cotana F (2016) The impact of albedo increase to mitigate the urban heat Island in Terni (Italy) using the WRF model. Sustainability 8(10):999

Morris CJG, Simmonds I, Plummer N (2001) Quantification of the influences of wind and cloud on the nocturnal urban heat Island of a large city. J Appl Meteorol Climatol 40(2):169–182

Mostafavi N, Heris MP, Gándara F, Hoque S (2021) The relationship between urban density and building energy consumption. Buildings 11(10):455

Murari KK, Ghosh S, Patwardhan A, Daly E, Salvi K (2015) Intensification of future severe heat waves in India and their effect on heat stress and mortality. Reg Environ Chang 15:569–579

Ngarambe J, Joen SJ, Han CH, Yun GY (2021) Exploring the relationship between particulate matter, CO, SO2, NO2, O3 and urban heat Island in Seoul, Korea. J Hazard Mater 403:123615

Niyogi D, Mahmood R, Adegoke JO (2009) Land-use/land-cover change and its impacts on weather and climate. Boundary Layer Meteorology 133(3):297

Niyogi D, Pyle P, Lei M, Arya SP, Kishtawal CM, Shepherd M et al (2011) Urban modification of thunderstorms: an observational storm climatology and model case study for the Indianapolis urban region. J Appl Meteorol Climatol 50(5):1129–1144

Oke TR (1982) The energetic basis of the urban heat Island. Q J R Meteorol Soc 108(455):1–24

Oke TR (1995) The heat Island of the urban boundary layer: characteristics, causes and effects. Wind Clim Cities:81–107

Oke TR, Mills G, Christen A, Voogt JA (2017) Urban climates. Cambridge University Press

Oleson KW, Monaghan A, Wilhelmi O, Barlage M, Brunsell N, Feddema J et al (2015) Interactions between urbanization, heat stress, and climate change. Clim Chang 129:525–541

Panda J, Rath SS (2022) Observed and simulated characteristics of 2015 Chennai heavy rain event: impact of land-use change, SST, and high-resolution global analyses. Pure Appl Geophys 179(9):3391–3409

Panda J, Mukherjee A, Rath SS (2022) Urban Heat Island and mitigation strategies: international studies and Indian perspectives. (in press). Chapter 2, in book entitled Urban Heat Islands reexamined (Satyaprakash and Anne Edn). ISBN: 979-8-88697-215-3. https://doi.org/10.52305/RKFG7202

Patz JA, Campbell-Lendrum D, Holloway T, Foley JA (2005) Impact of regional climate change on human health. Nature 438(7066):310–317

Peng J, Hu Y, Dong J, Liu Q, Liu Y (2020) Quantifying spatial morphology and connectivity of urban heat islands in a megacity: a radius approach. Sci Total Environ 714:136792

Piver WT, Ando M, Ye F, Portier CJ (1999) Temperature and air pollution as risk factors for heat stroke in Tokyo, July and August 1980–1995. Environ Health Perspect 107(11):911–916

Quan J, Chen Y, Zhan W, Wang J, Voogt J, Wang M (2014) Multi-temporal trajectory of the urban heat Island centroid in Beijing, China based on a Gaussian volume model. Remote Sens Environ 149:33–46

Rainham DG, Smoyer-Tomic KE (2003) The role of air pollution in the relationship between a heat stress index and human mortality in Toronto. Environ Res 93(1):9–19

Ramachandra TV, Kumar U (2010) Greater Bangalore: emerging urban heat Island. GIS Develop 14(1):86–104

Rath SS, Panda J (2020) Urban induced land-use change impact during pre-monsoon thunderstorms over Bhubaneswar-Cuttack urban complex. Urban Clim 32:100628

Rath SS, Panda J, Mohanty S (2021) Urban transition and its impact on mesoscale weather: a review. Methods Appl Geospatial Technol Sustain Urban:187–216

Rath SS, Panda J, Sarkar A (2022) Distinct urban land cover response to meteorology in WRF simulated pre-monsoon thunderstorms over the tropical city of Kolkata. Meteorog Atmos Phys 134(4):76

Raymond C, Singh D, Horton RM (2017) Spatiotemporal patterns and synoptics of extreme wet-bulb temperature in the contiguous United States. J Geophys Res Atmos 122(24):13–108

Robinson PJ (2001) On the definition of a heat wave. J Appl Meteorol Climatol 40(4):762–775

Rosenfeld D (2000) Suppression of rain and snow by urban and industrial air pollution. Science 287(5459):1793–1796

Santamouris M (2007) Heat Island research in Europe: the state of the art. Adv Build Energy Res 1(1):123–150

Santamouris M (2014) On the energy impact of urban heat Island and global warming on buildings. Energ Buildings 82:100–113

Sarrat C, Lemonsu A, Masson V, Guédalia D (2006) Impact of urban heat Island on regional atmospheric pollution. Atmos Environ 40(10):1743–1758

Scherer D, Fehrenbach U, Lakes T, Lauf S, Meier F, Schuster C (2013) Quantification of heat-stress related mortality hazard, vulnerability and risk in Berlin, Germany. DIE ERDE J Geograph Soc Berlin 144(3–4):238–259

Schlünzen KH, Hoffmann P, Rosenhagen G, Riecke W (2010) Long-term changes and regional differences in temperature and precipitation in the metropolitan area of Hamburg. Int J Climatol 30(8):1121–1136

Schwarz N, Manceur AM (2015) Analyzing the influence of urban forms on surface urban heat islands in Europe. J Urban Plann Develop 141(3):A4014003

Schwarz N, Schlink U, Franck U, Großmann K (2012) Relationship of land surface and air temperatures and its implications for quantifying urban heat Island indicators—an application for the city of Leipzig (Germany). Ecol Indic 18:693–704

Seto KC, Fragkias M, Güneralp B, Reilly MK (2011) A meta-analysis of global urban land expansion. PLoS One 6(8):e23777

Sharma V (2016) Green infrastructure measures to mitigate urban heat Island effect: a case of Hyderabad city. Doctoral dissertation, SPA Bhopal

Shirani-bidabadi N et al (2019) Evaluating the spatial distribution and the intensity of Urban Heat Island using remote sensing, case study of Isfahan City in Iran. Sustain Cities Soc 45(May 2018):686–692. https://doi.org/10.1016/j.scs.2018.12.005

Siddiqui A, Kushwaha G, Raoof A, Verma PA, Kant Y (2021) Bangalore: urban heating or urban cooling? Egypt J Remote Sens Space Sci 24(2):265–272

Singh P, Kikon N, Verma P (2017) Impact of land use change and urbanization on Urban Heat Island in Lucknow City, Central India. A remote sensing based estimate. Sustain Cities Soc 32:100–114. https://doi.org/10.1016/j.scs.2017.02.018

Smoyer KE, Rainham DG, Hewko JN (2000) Heat-stress-related mortality in five cities in southern Ontario: 1980–1996. Int J Biometeorol 44:190–197

Somers KA, Bernhardt ES, McGlynn BL, Urban DL (2016) Downstream dissipation of storm flow heat pulses: a case study and its landscape-level implications. JAWRA J Am Water Resour Assoc 52(2):281–297

Song J, Huang B, Kim JS, Wen J, Li R (2020) Fine-scale mapping of an evidence-based heat health risk index for high-density cities: Hong Kong as a case study. Sci Total Environ 718:137226

Souch C, Grimmond S (2006) Applied climatology: urban climate. Prog Phys Geogr 30(2):270–279

Stone B Jr (2012) The city and the coming climate: climate change in the places we live. Cambridge University Press

Stone B, Hess JJ, Frumkin H (2010) Urban form and extreme heat events: are sprawling cities more vulnerable to climate change than compact cities? Environ Health Perspect 118(10):1425–1428

Stull R (2011) Wet-bulb temperature from relative humidity and air temperature. J Appl Meteorol Climatol 50(11):2267–2269

Su MA, Ngarambe J, Santamouris M, Yun GY (2021) Empirical evidence on the impact of urban overheating on building cooling and heating energy consumption. Iscience 24(5):102495

Sultana S, Satyanarayana ANV (2019) Impact of urbanisation on urban heat Island intensity during summer and winter over Indian metropolitan cities. Environ Monit Assess 191(Suppl 3):789

Sun X, Luo Y, Gao X, Wu M, Li M, Huang L et al (2021) On the localized extreme rainfall over the Great Bay Area in South China with complex topography and strong UHI effects. Mon Weather Rev 149(8):2777–2801

Swain D, Roberts GJ, Dash J, Lekshmi K, Vinoj V, Tripathy S (2017) Impact of rapid urbanization on the city of Bhubaneswar, India. Proc Nat Acad Sci India Sect A Phys Sci 87:845–853

Taheri Shahraiyni H, Sodoudi S, El-Zafarany A, Abou El Seoud T, Ashraf H, Krone K (2016) A comprehensive statistical study on daytime surface urban heat Island during summer in urban areas, case study: Cairo and its new towns. Remote Sens 8(8):643

Thomas G, Sherin AP, Ansar S, Zachariah EJ (2014) Analysis of urban heat Island in Kochi, India, using a modified local climate zone classification. Procedia Environ Sci 21:3–13

Tuholske C, Caylor K, Funk C, Verdin A, Sweeney S, Grace K et al (2021) Global urban population exposure to extreme heat. Proc Natl Acad Sci 118(41):e2024792118

Tzavali A, Paravantis JP, Mihalakakou G, Fotiadi A, Stigka E (2015) Urban heat Island intensity: a literature review. Fresenius Environ Bull 24(12b):4537–4554

Unger J, Sümeghy Z, Gulyás Á, Bottyán Z, Mucsi L (2001) Land-use and meteorological aspects of the urban heat Island. Meteorol Appl 8(2):189–194

UN (2014) World urbanization prospects: the 2014 revision, highlights. Department of Economic and Social Affairs

UN 2017 General assembly resolution 71/256, New Urban Agenda A/RES/71/256. Available at: undocs.org/en/A/RES/71/256. Accessed 25 Jan 2017

UN, World urbanization prospects 2018 [EB/OL] (2018). https://population.un.org/wup/Download

UN (2019a) The sustainable development goals report. United Nations, New York

UN (2019b) World urbanization prospects, New York

Veena K, Parammasivam KM, Venkatesh TN (2020) Urban Heat Island studies: current status in India and a comparison with the international studies. J Earth Syst Sci 129:1–15

Venter ZS, Chakraborty T, Lee X (2021) Crowdsourced air temperatures contrast satellite measures of the urban heat Island and its mechanisms. Science. Advances 7(22):eabb9569

Voogt JA (2002) Urban heat Island. In: Douglas I (ed) Encyclopedia of global environmental change. Wiley, New York, pp 660–666

Voogt JA, Oke TR (2003) Thermal remote sensing of urban climates. Remote Sens Environ 86(3):370–384

Wang ZH, Upreti R (2019) A scenario analysis of thermal environmental changes induced by urban growth in Colorado River basin, USA. Landsc Urban Plan 181:125–138

Wang Q et al (2020) Urban Heat Island circulations over the Beijing-Tianjin region under calm and fair conditions. Build Environ 180:107063. https://linkinghub.elsevier.com/retrieve/pii/S0360132320304431

Wang Y, Guo Z, Han J (2021) The relationship between urban heat Island and air pollutants and them with influencing factors in the Yangtze River Delta, China. Ecol Indic 129:107976

Xie R, Fang J, Liu C (2017) The effects of transportation infrastructure on urban carbon emissions. Appl Energy 196:199–207

Yalcin T, Yetemen O (2009) Local warming of groundwaters caused by the urban heat Island effect in Istanbul, Turkey. Hydrogeol J 17(5):1247

Yan C, Guo Q, Li H, Li L, Qiu GY (2020) Quantifying the cooling effect of urban vegetation by mobile traverse method: a local-scale urban heat Island study in a subtropical megacity. Build Environ 169:106541

Zhang DL, Shou YX, Dickerson RR, Chen F (2011) Impact of upstream urbanization on the urban heat Island effects along the Washington–Baltimore corridor. J Appl Meteorol Climatol 50(10):2012–2029

Zhang GJ, Cai M, Hu A (2013) Energy consumption and the unexplained winter warming over northern Asia and North America. Nat Clim Chang 3(5):466–470

Zhang X, Estoque RC, Murayama Y (2017) An urban heat Island study in Nanchang City, China based on land surface temperature and social-ecological variables. Sustain Cities Soc 32:557–568

Zhang D-L, Jin M, Shou Y, Dong C (2019) The influences of urban building complexes on the ambient flows over the Washington–Reston region. J Appl Meteor Climatol 58:1325–1336

Zhou W, Huang G, Cadenasso ML (2011) Does spatial configuration matter? Understanding the effects of land cover pattern on land surface temperature in urban landscapes. Landsc Urban Plan 102(1):54–63

Zipper SC, Schatz J, Singh A, Kucharik CJ, Townsend PA, Loheide SP (2016) Urban heat Island impacts on plant phenology: intra-urban variability and response to land cover. Environ Res Lett 11(5):054023

23 Assessment of LULC Changes and Its Impact on Surface Temperature and Urban Heat Island Conditions in Kolkata During SARS COVID-19 Period

Sohini Sen, Debashish Das, Pankaj Chakroborty, and Raja Ghosh

Abstract

Due to the rising urbanization of cities around the world, significant changes in urban climate can be observed. To ensure sustainable urban development, it is crucial to strike a balance between ecological sustainability and thermal environment. Thus, it is one of the impacts of Urban Heat Island (UHI) on different domains, such as climate change, urban ecology, urban climatology, urban planning, mitigation, and management, urban geography, etc., should be the top research priority. The current work focuses on estimating Land Surface Temperature (LST) in Kolkata Municipal Corporation area, West Bengal, India, with a particular emphasis on Urban Heat Island (UHI) and non-UHI inside the city limit and their interaction with keys like Land Use/Land Cover, biophysical indices such as Normalized Difference Vegetation Index (NDVI), Normalized Difference Water Index (NDWI), Normalized Difference Built-Up Index (NDBI), in the period of pre-COVID (2015–2016) and under SARS COVID-19 (2019–2020) period. For various LULC classes, the highest mean LST increase between 2010 and 2015 was between 30 °C and 32 °C during the pre-COVID-19 period, while the lowest was between 29 °C and 30 °C during the COVID period over the years 2020–2021. LST and the Normal Difference Built-up Index (NDBI) had a positive association, while LST and the Normal Difference Vegetation Index (NDVI) and the Normal Difference Water Index (NDWI) had negative relationships, respectively. The analysis of the gradient between densely populated areas and sparsely populated areas revealed a sharp decrease in mean LST from the city center outward because the NDVI also increased sharply, particularly in the years 2019–2020 in this COVID period, which is a blatant indication of the surface Urban Heat Island effect. The northern, western, and central regions of the city are where the UHI have primarily been constructed. LST in the common UHI ranges from 19.72 °C to 31.69 °C for various years (2010–2020). The built-up area in the northern part of city and port area within Kolkata Municipal Corporation exhibits the highest Land Surface Temperature compared

S. Sen (✉) · D. Das
Department of Architecture, Jadavpur University, Kolkata, West Bengal, India

P. Chakroborty
Department of Geoinformatics, NK College (Mumbai University), Mumbai, Maharashtra, India

R. Ghosh
Synprox Solutions, Kolkata, West Bengal, India

U. Chatterjee et al. (eds.), *Climate Crisis: Adaptive Approaches and Sustainability*, Sustainable Development Goals Series, https://doi.org/10.1007/978-3-031-44397-8_23

to the other region, according to the correlation map of Land Use/Land Cover and Land Surface Temperature for the study years. The findings show that LST and the land use/land cover indices in Kolkata have the strongest association across time. It is observed from the present study that due to pandemic situation when people were homebound, industries were locked, construction sites were deserted and the city had a better ecological status.

Keywords

Urbanisation · Land Surface Temperature · *Urban Heat Island* · LULC Indices · Landsat 5 & 8 OLI (Operational Land Image) · Remote Sensing · SARS COVID-19

23.1 Introduction

Urban regions in emerging nations, particularly in South Asia, have seen a dramatic increase in population and urbanization over the past few decades. According to a survey and UN forecast, Asian cities will host more than half of the world's urban population, and by 2020, there will be 1.5 billion additional people living in urban areas (Hasan et al. 2013). A third of South Asians lived in urban areas in 2000, a figure that was 16% higher than in 1950 and was projected to rise to 33.46% by 2016 (2.6% annual growth) by 2016. Changes in LST and the development of Urban Heat Island (UHI), which is the primary variance of surface temperature between the city core area and the periphery non-urban area, are two of the most immediate effect (Voogt and Oke, 2003).

The shifting Land Use/Land Cover (LULC) pattern in diverse urban areas is investigated using the concepts of Land Surface Temperature (LST) and Urban Heat Island (UHI) (Arnfield 2003; Mirzaei 2015; Rizwan et al. 2008; Rinner and Hussain 2011; Zhao et al. 2016). Urban Heat Island (UHI), which is the primary variance in Surface Temperature (e.g., Beijing, Columbus, Shanghai, Baltimore, Chicago, etc.) to solve many environmental issues. The nature and distribution of LST are influenced by various types of LULC (Chun and Guldmann 2014; Dai et al. 2018; Zhang et al. 2013; Zhou et al. 2011; Coseo and Larsen 2014). There has been a discussion of the linear association analysis between the LST and LULC indexes in several cities throughout the world, including The development of Urban Heat Island (UHI), which is the primary variation of Surface Temperature between the city core area and the peripheral non-urban area, in some recent articles for separate study areas like Brisbane (Deilami and Kamruzzaman 2017), Raipur (Guha et al. 2017), Shanghai (Nie et al. 2016), Addis Ababa (Feyisa et al. 2016), Mexico (Lopez 2017), Philadelphia (Pearsall 2017), Florence and Naples (Guha et al. 2018).

The higher value of NDVI shows the intense vegetation, while lower value shows the water bodies within the Kolkata in this study. NDVI is high at periphery and the low at the heart of the urban agglomeration (Maity and Srivastava 2020).

In urban planning research, determining LULC (Land Use/Land Cover) variations is crucial. However, the variables that should be taken into account in similar studies are LST (Land Surface Temperature) and UHI (Urban Heat Island), which are directly related to LU changes. As a result, methods like Remote Sensing (RS) and Geographic Information Systems (GIS) are frequently employed to gather this kind of data. In this study, the relationship between LULC, NDVI (Normalized Difference Vegetation Index), NDBI (Normalized Difference Built-up Index), NDWI (Normalized Difference Water Index) and LST (Land Surface Temperature) in Kolkata city and its surroundings was studied using Landsat satellite images of 2010, 2015 with 2020 (SARS Covid-19 period) and *UHI* intensity was also demonstrated. The results clearly indicate that though there was some augmentation of urban built-up areas, vegetation areas also increased during the year 2020.

Depending on the building components, such as the existence of green spaces, the city's particular geographic position, and topography, the changes in LST can be observed. Urban built-up areas and barren lands have the greatest LST, and the surface temperature in built-up areas in cities fluctuated throughout time.

Urban built-up areas grew, a positive intensity of UHI was seen, and an Urban Heat Island formation was identified.

Within the Earth's climate system, LST is a crucial factor. It describes processes like the transfer of water and energy between the earth's surface and the atmosphere, and it affects how quickly and when plants grow.

Actually, urban heat islands are hot spots in cities where the temperature is higher than the areas around them. LSE stands for land surface brightness temperature. Although it has a strong relationship with the air temperature, the surface temperature is not the actual temperature. LST may therefore be a sign of UHI.

Sustainable urban planning is required to arrest the rise in LST comprises urban forestry, building water features and fountains, maintaining existing swimming pools and aquaculture ponds, and decreasing construction activity.

The objective of the current study is to find out the relations among the LULC changes with respect to LST, UHI, some spectral indices and also the effects of severe cyclonic disasters in different years of the study period. Also emphasize on the projected LULC change, LST, *UHI* and spectral indices for the year 2030, to maintain the sustainable environment more similar to the year 2020.

Due to the consecutive severe cyclones "Fani" and "Bulbul" in the month of April and November, 2019, respectively, and the devastating super cyclone "Amphun" in the month of May 2020 with heavy to intense rainfall, there were a significant increase of water bodies and vegetation cover which resultant in decrease of the minimum and maximum LST. The same scenario was also observed in the year 2010, as the severe cyclone "Aila" was knocked. Thus, it can infer from the facts that due to the cyclonic effect and heavy rainfall, water bodies and vegetation index had increased and the values of minimum and maximum land surface temperature had decreased.

The observation that the increase of vegetation and water body indices in the year 2010, the land surface temperature of Kolkata Municipal Corporation area had reduced to the tune of 17.25 °C and 28.66 °C as minimum and maximum, respectively. Similarly, it has been observed that the growth of vegetation and water body indices increased in the year 2020 along with the Land Surface Temperature had dropped to the minimum of 17.40 °C and 24.10 °C as maximum. Thus, the changes in Land Surface Temperature helped to control the air pollution as well as the environment within the KMC area.

23.2 Materials and Method

23.2.1 Study Area

The present study area is Kolkata metropolitan, capital of West Bengal and it is the third most populated metropolitan city that runs parallel to the Hooghly River's banks in the eastern section of India. The area of the Kolkata Municipal Corporation is 186.35 km^2. The Kolkata Metropolitan Area is spread over 1886.67 km^2 (728.45 mi^2) and comprises of three municipal corporations (including Kolkata Municipal Corporation) and 39 local municipalities. The city's typical elevation is close to sea level being 17 feet. It was chosen as the study location because it is a large metropolis with a long history, a combination of several Land Use/Land Cover types, and a vulnerability to atmospheric turbulence.

The geographical location of Kolkata is 22.5667° N (Latitude) and 88.3667° E (Longitude) (Fig. 23.1).

23.2.2 Meteorological Status of Kolkata

The climate is tropical, moist, and dry (Koppen climatic classification Aw). The yearly mean temperature is 24.8 °C (80 °F) and the monthly mean temperature ranges from 15 °C to 30 °C (59–86 °F). During dry spells, maximum temperatures

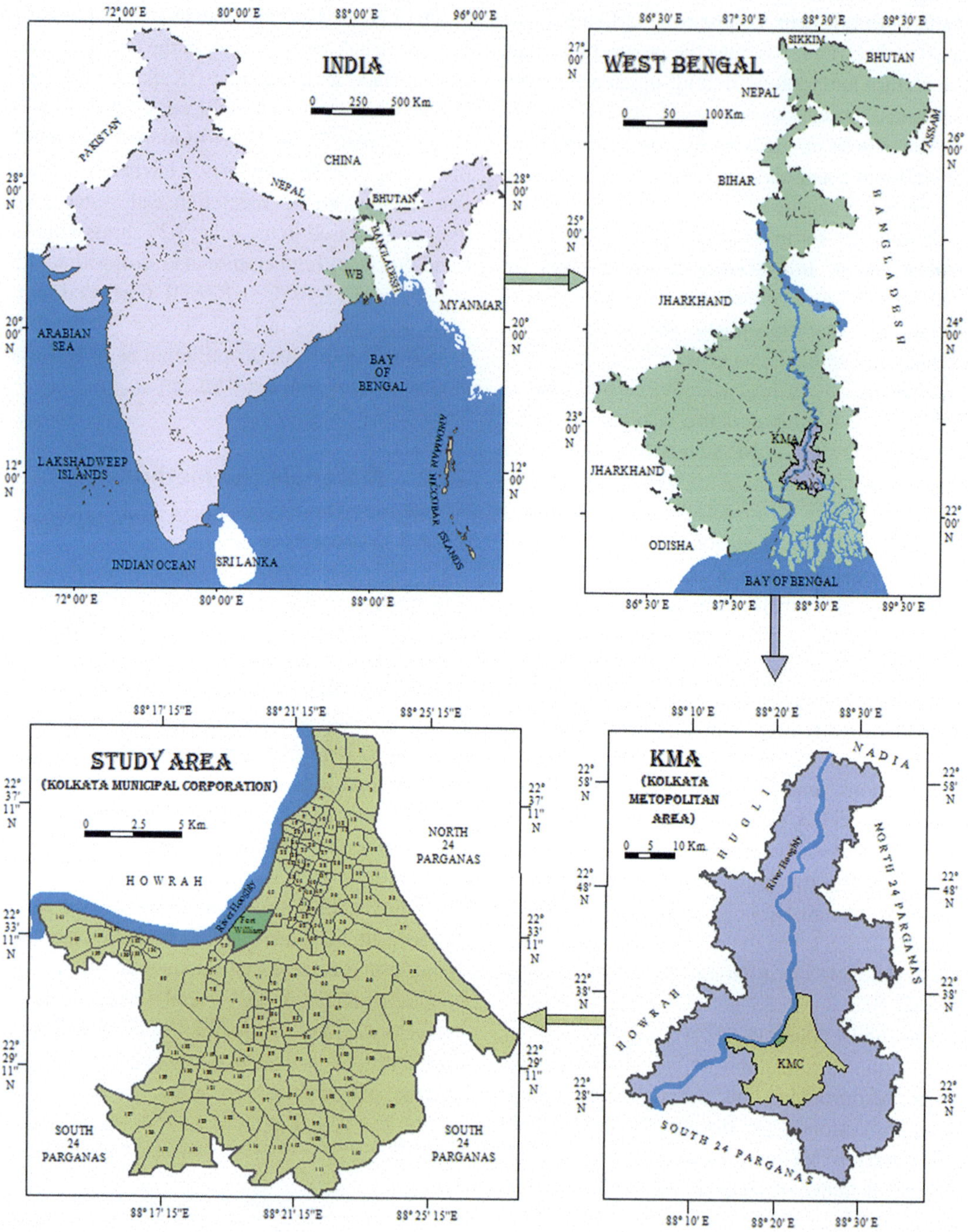

Fig. 23.1 Location map of the study area (KMC)

frequently approach 40 °C (104 °F) in May and June. Summers are hot and humid with low 30 °C temperatures. Winter, typically lasts only around two and a half months, with seasonal temperature lows between December and January falling to 9 °C and 11 °C (48.2 °F and 51.8 °F). Humidity is around 96% in the mornings and 67% in the afternoons. Annual rainfall is 1582 mm (62.3 inches).

23.2.3 Flowchart of Research Work

For the analysis of environmental processes at the local or global scale, remote sensing data are the main resources. These statistics are used to determine the detection of change during the last few decades. Remote sensing data are particularly helpful for mapping, categorization, and analysis of an area. Examples include Landsat data, Sentinel data, Spot images, etc. The resolution, electromagnetic spectrum, energy source, imaging medium, and number of bands can all be used to classify these data. A better degree of classification accuracy will be attained the higher resolution of satellite data (spatial, spectral, radiometric, and temporal resolution) (Fig. 23.2).

23.2.4 Data Used in the Study

The above locations have been preferred due to the availibity of temperature data from the Indian Metereological Department (IMD) (Weather base entry for Kolkata). Government of India and Netaji Subhash International Airport, Kolkata (Table 23.1).

January has been selected for satellite data acquisition month by USGS for the years 2011, 2016 and 2021, so that the effects for the year 2010, 2015, and 2020, respectively, can be used for analysis. Landsat 5 for the year 2010 and Landsat 8 2015 and 2020 have been used to find out the indices, land surface temperature and urban heat islands (Table 23.2).

Usually, Landsat data is utilized for classification, and during classification, there are many bands based on their wavelength (blue, green, red, infrared, thermal, and panchromatic band). Panchromatic band is used to increase the resolution of data. Landsat 5 and 8 Operational Land Image (OLI) with Thermal Infrared Sensor (TIRS) data having 6 and 10 bands, respectively, are used. For analysis of Normalized Difference Vegetation Index (NDVI), Normalized Difference Built-up Index (NDBI), and Normalized Difference Water Index (NDWI), only four bands (Green, Red, NIR, and SWIR) are used.

As the case study area is Kolkata Municipal Corporation, the investigation of *Urban Heat Islands* (*UHI*) and its effects with respect to the Land Use/Land Cover has been initiated by physical verification for collecting the field data in different locations for the year 2010, 2015, and 2020. GIS and Remote Sensing technologies have also been applied for Land Use/Land Cover mapping, the Land Surface Temperature with *Urban Heat Island* and its effects within the case study area. Differential indices like NDVI, NDBI, and NDWI of 2010, 2015, and 2020 are also major key factors for the analysis and observation.

23.2.5 Land Use and Land Cover Map

Maximum likelihood method of supervised classification is used to classify the land use/land cover features from the satellite imageries for the year 2010, 2015, and 2020. Mainly four classes have identified named Water body, Vegetation, Open Space, and Built-up within the study area. The resolution of the satellite imageries became a key factor to identify the more number of classes. After classification ground truth verification and accuracy assessment have been performed.

23.2.6 Normalized Difference Vegetation Index (NDVI)

The Normalized Difference Vegetation Index (NDVI) is a numerical indicator that uses the red and near-infrared spectral bands. NDVI is highly associated with vegetation content. High NDVI values signify regions that reflect more near-infrared light. Higher reflectance in the near-infrared corresponds to denser and healthier vegetation.

Healthy vegetation generally serves as a good electromagnetic spectrum absorber for obvious reasons. Greenery contains chlorophyll, which strongly absorbs blue (0.4–0.5 m) and red (0.6–0.7 m) spectral lines while reflecting green (0.5–0.6 m) spectral lines. Therefore, healthy vegetation

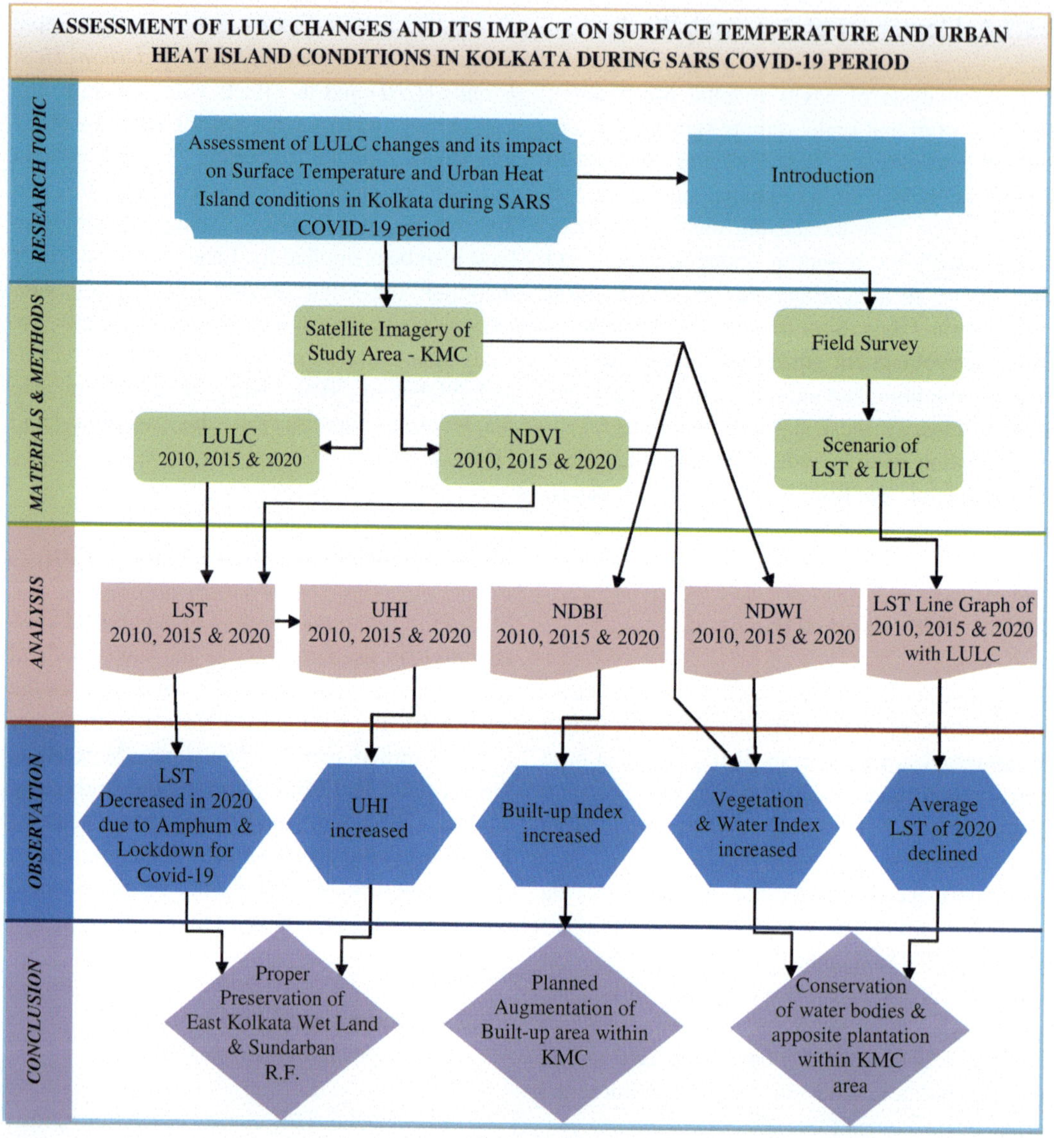

Fig. 23.2 Flowchart showing the steps involved in the research work

appears green to our eye. Healthy plants exhibit a high near infrared (NIR) reflectance between 0.7 and 1.3 m. The interior structure of plant leaves is mostly to blame for this. To compute NDVI, two bands with strong reflectance in the NIR spectrum and high absorption in the red spectrum are utilized. Following formula has been used for the Normalized Difference Vegetation Index.

$$\text{NDVI} = (\text{NIR} - \text{Red}) / (\text{NIR} + \text{Red})$$

$$\text{For Landsat 5 data,}\quad \text{NDVI} = (\text{Band 4} - \text{Band 3}) / (\text{Band 4} + \text{Band 3})$$

$$\text{For Landsat 8 data,}\quad \text{NDVI} = (\text{Band 5} - \text{Band 4}) / (\text{Band 5} + \text{Band 4})$$

Table 23.1 Field survey data used in the study (Primary)

Location	Latitude	Longitude	Temperature (°C) during January (day time)			Land use
			2010	2015	2020	
Mominpur	22° 31′ 37.44″	88° 19′ 19.01″	30	32	29	Built-up
Ultadanga	22° 35′ 43.76″	88° 22′ 57.48″	28	29	27	Dense built-up
Baishnabghata	22° 28′ 14.79″	88° 23′ 30.02″	24	25	25	Mid-rise built-up
Topsia	22° 32′ 22.94″	88° 23′ 14.94″	23	24	25	Built-up
Behala Chowrasta	22° 29′ 11.61″	88° 18′ 49.13″	26	27	28	Dense built-up
Minto Park	22° 32′ 25.41″	88° 21′ 17.86″	25	25	25	Built-up
Shyambazar	22° 36′ 4.58″	88° 22′ 26.22″	28	29	25	Dense built-up
Moulali	22° 33′ 38.20″	88° 21′ 51.90″	28	29	26	Dense built-up
Rabindrabharati	22° 37′ 37.16″	88° 22′ 43.99″	28	28	23	Vegetation with built-up
Hide Road	22° 31′ 31.60″	88° 18′ 20.79″	27	28	25	Built-up
Picnic Garden	22° 31′ 44.44″	88° 22′ 54.31″	25	26	23	Built-up
Victoria Memorial	22° 32′ 29.22″	88° 20′ 29.50″	23	25	21	Vegetation and open space
Paribesh Bhawan	22° 33′ 45″	88° 24′ 30.87″	22	23	22	Vegetation with built-up

Table 23.2 Satellite data used in the study (Secondary)

Sensors	Date acquired	Path/row	Bands	Resolutions (m)	Datum/UTM zone	Source
Landsat 5	24-01-2011	138/44	5, 4, 3, 2 & 6	30 & 120	WGS84/45	UGGS
Landsat 8 OLI (TIRS)	06-01-2016	138/44	6, 5, 4, 3 & 10	30 & 100	WGS84/45	USGS
Landsat 8 OLI (TIRS)	03-01-2021	138/44	6, 5, 4, 3 &10	30 & 100	WGS84/45	USGS

23.2.7 Normalized Difference Built-up Index (NDBI)

There are many indexes available for built-up area analysis. Normalized difference Built-up Index (NDBI) is the most popular indicator used to analyze built-up areas. SWIR has more reflection value than NIR in case of the built-up areas and bare ground. Infrared spectrum reflection is absent from water bodies. NIR reflection on a green surface is greater than SWIR spectrum. NDBI has been calculated by the following formula:

$$\text{NDBI} = (\text{SWIR} - \text{NIR}) / (\text{SWIR} + \text{NIR})$$

$$\text{For Landsat 5 data,} \quad \text{NDBI} = (\text{Band 5} - \text{Band 4}) / (\text{Band 5} + \text{Band 4})$$

$$\text{For Landsat 8 data,} \quad \text{NDBI} = (\text{Band 6} - \text{Band 5}) / (\text{Band 6} + \text{Band 5})$$

The Normalize Distinction The value of the built-up index ranges from −1 to +1. The NDBI has a negative value for water bodies and a positive value for built-up regions. Low NDBI value indicates the vegetation.

23.2.8 Normalized Difference Water Index (NDWI)

Using the near-infrared and short wave infrared spectral regions, optical satellite pictures are used to create the Normalized Difference Water Index (NDWI), a numerical indicator. In most instances, the NDWI can effectively improve water information. It can cause overestimated water bodies and is sensitive to developed terrain. Low reflectance levels are found in water bodies. It only reflects electromagnetic radiation that is visible. Liquid water bodies often reflect more light in the blue (0.4–0.5 m) spectrum than in the green (0.5–0.6 m) or red (0.6–0.7 m) spectrums. The blue region of the visible spectrum is where clear water reflects light the most strongly. Water thus appears blue. Visible spectrum reflectance is increased in turbid water. Beyond near infrared (NIR), there is no reflection. McFeeters (2013) created NDWI to improve the water-related landscape aspects. The near infrared (NIR) and short-wave infrared (SWIR) bands are used in this index. NDWI can be calculated by following formula (according to EOS Data Analytics (https://eos.com/make-an-analysis/ndwi/):

23.2.9 Land Surface Temperature

Thermal band 6 (wavelength 10.40–12.50 μm) of 120 m, has been resample to 30 m of Landsat 5 for the year 2010 (a) and (b) and 10 TIRS 1 (wavelength 10.60–11.19 μm) of 100 m for Landsat 8 OLI TIRS have been used to analyze the LST of KMC area for 2015 (b) and 2020 (c); TIRS, thermal infrared sensor; OLI, operational land imager.

Process of Land Surface Temperature Using the TIR 1 Band (Band 10) of Landsat 8 OLI for 2020

Landsat Level 1 radiometrically calibrated and atmospherically corrected data (LC08_L1TP_138044_20210103_20210308_02_T1_B10) provided by the US Geological Survey (USGS) has been used for the process of Land Surface Temperature evaluation. Thermal band (Band 10 in Landsat-8 OLI) provided as the atmospheric brightness temperature in Kelvin (K). The following factors were taken into account throughout the data selection step to produce high-quality and trustworthy results, including (i), dry-season daytime data, (ii). cloudless pictures, (iii) georectified using the zone 45 north projection of the Universal Transverse Mercator (UTM).

***Step 1**: Conversion to Top of Atmosphere (TOA) Radiance*

Thermal Infra-Red Digital Numbers can be transformed into TOA spectral radiance using the radiance rescaling factor. The following equation

$$\mathrm{NDWI} = (\mathrm{GREEN} - \mathrm{NIR}) / (\mathrm{GREEN} + \mathrm{NIR})$$

$$\text{For Landsat 5 data,}\quad \mathrm{NDWI} = (\mathrm{Band\,3} - \mathrm{Band\,5}) / (\mathrm{Band\,3} + \mathrm{Band\,5})$$

$$\mathrm{NDWI} = (\mathrm{NIR} - \mathrm{SWIR}) / (\mathrm{NIR} + \mathrm{SWIR})$$

$$\text{For Landsat 8 data,}\quad \mathrm{NDWI} = (\mathrm{Band\,5} - \mathrm{Band\,6}) / (\mathrm{Band\,5} + \mathrm{Band\,6})$$

is used to convert a digital number in a Landsat level 1 back to radiance (L).

$$L\lambda = ML^{*} Q_{cal} + AL - O_{i}$$

$$L\lambda = \left(0.0003342^{*}\,\text{Band}10 + 0.10000\right) - 0.29$$

Where,

$L\lambda$ = TOA spectral radiance (Watts/ (m^2 * sr * μm))

ML = Radiance multiplicative Band (3.3420)

AL = Radiance Add Band (0.10000)

Q_{cal} = Quantized and calibrated standard product pixel values (DN)

O_i = correction value for band 10 is 0.29

Output: TOA_{max} = 10.013 & TOA_{min} = 8.44798

Step 2: *Conversion to Top of Atmosphere (TOA) Brightness Temperature (BT)*

Thermal Infra-Red Digital Numbers can be transformed into TOA spectral radiance using the radiance rescaling factor.

Kelvin (K) to Celsius (°C) Degrees

$$\text{BT} = \text{K2} / \ln\left(\text{K1} / L\lambda + 1\right) - 272.15$$

$$\text{BT} = \left(1321.0789 / \ln\left(774.8853 / \text{ToA} + 1\right)\right) - 272.15$$

Where,

BT = Top of atmosphere brightness temperature (°C)

$L\lambda$ = TOA spectral radiance (Watts/(m^2 * sr * μm))

K1 = K1 Constant Band (774.8853)

K2 = K2 Constant Band (1321.0789)

Output: BT_{max} = 30.734 °C & BT_{min} = 19.5027 °C

Step 3: *Normalized Differential Vegetation Index (NDVI)*

A standardized vegetation measure called the Normalized Differential Vegetation measure (NDVI) has been created utilizing near-infrared (Band 5) and red (Band 4) bands.

$$\text{NDVI} = \left(\text{NIR} - \text{RED}\right) / \left(\text{NIR} + \text{RED}\right)$$

$$\text{NDVI} = \left(\text{Band}\,5 - \text{Band}\,4\right) / \left(\text{Band}\,5 + \text{Band}\,4\right)$$

Where,

RED = DN values from the RED band

NIR = DN values from Near-Infrared band

Output: $NDVI_{max}$ = 0.396295 & $NDVI_{min}$ = −0.0889339

Step 4: *Land Surface Emissivity (LSE)*

The formula used to compute the average emissivity of a component of the Earth's surface from NDVI readings is called land surface emissivity (LSE).

$$\text{PV}\left[\left(\text{NDVI} - \text{NDVImin}\right) / \left(\text{NDVImax} + \text{NDVImin}\right)\right]^2$$

Where,

PV = Proportion of Vegetation

NDVI = DN values from NDVI Image

NDVI min = −0.0889339 (Minimum DN values from NDVI Image)

NDVI max = 0.396295 (Maximum DN values from NDVI Image)

Output: PV_{max} = 0.282784 & PV_{min} = 0.00216649

$$\text{LSE} = 0.004^{*}\,\text{PV} + 0.986$$

Where,

0.986 corresponds to a correction value of the equation

Output: LSE_{max} = 0.987131 & LSE_{min} = 0.986009

Step 5: *Land Surface Temperature (LST)*

Using the following formula, the Land Surface Temperature (LST) is the radiative temperature that is determined by the Top of Atmosphere Brightness Temperature, Wavelength of Emitted Radiance, and Land Surface Emissivity (Fig. 23.3)

$$\text{LST} = \left(\text{BT}/1\right) + W^{*}\left(\text{BT}/p\right)^{*} \ln\left(\text{LSE}\right)$$

Where,

BT = Top of atmosphere brightness temperature (°C)

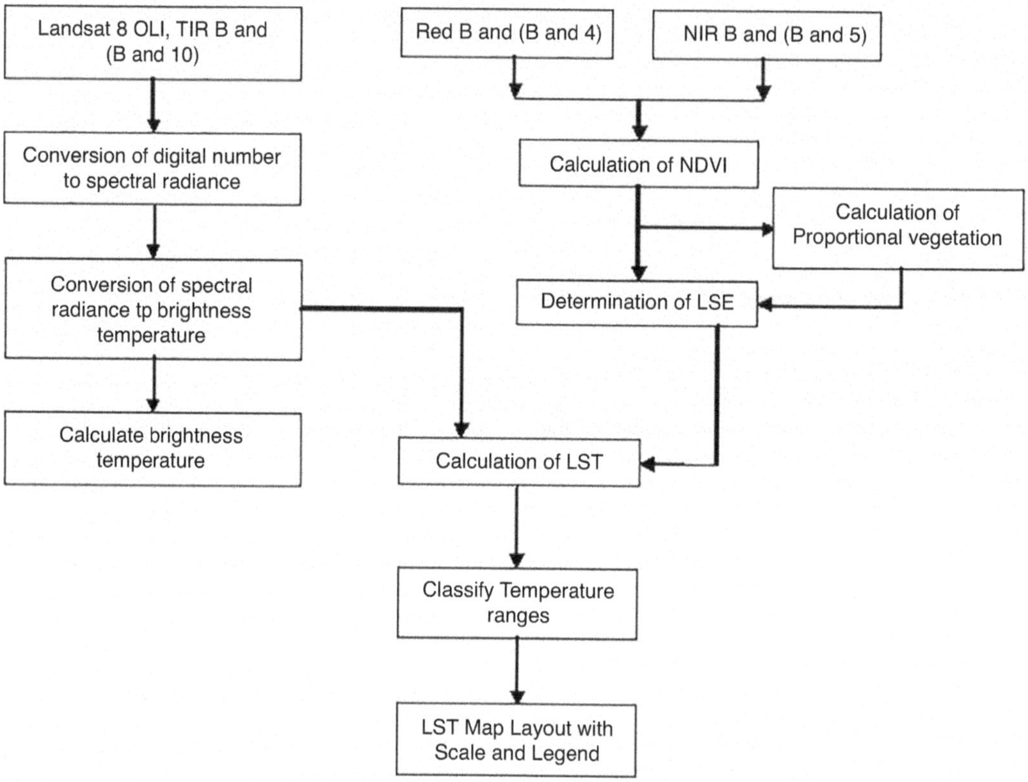

Fig. 23.3 Flowchart of the LST process of 2020

W = Wavelength of emitted radiance (11.5 μm)
p = h * c/s (1.438 * 10^{-2} mk) = 14,388 mk
h = Planck's constant (6.626 * 10^{-34}JS)
s = Boltzmann constant (1.38 * 10^{-23} JK)
c = Velocity of light (2.998 * 10^{8} m/s)
LSE = Land Surface Emissivity

Output: LST_{max} = 29.852 °C & LST_{min} = 19.0264 °C

23.2.10 Urban Heat Island

Urban heat islands has been identified using the formula:

$$\text{UHI}(\text{Urban Heat Islands}) = (T_s - T_m) / \text{SD}.$$

Where,
T_s is the land surface temperature (LST) image.
T_m is the mean of the land surface temperature.
SD is the standard deviation of the LST image.

23.3 Results

23.3.1 LULC Change for the Year 2010, 2015, and 2020

Built-up areas have been converted to Vegetation and Waterbody due to mostly demolition of single dwell units vacant until have converted to multi-storied, which are not natural phenomena. During ground truth verification, it was found true and also satellite imageries of 2015 and 2020 show the same. Surprising information found that in KMC area, after death of parents (owner of single dwell units) successors are interested to stay in separate manner from each other with multistoried flats. Due to Covid-19 pandemic lockdown, the constructions of multistoried dwell units been suspended for a long periods and parallel effects of super cyclones "Bulbul" and "Fani", those vacant areas became filled with water body or shrubs (greeneries), thus in the year 2020, percentage of

Table 23.3 Percentage of LULC distribution within KMC for the year 2010, 2015, and 2020

Sl. No.	LULC	% of Area in 2010	% of Area in 2015	% of Area in 2020
1	Waterbody	5.72	4.75	9.12
2	Vegetation	37.78	28.73	31.21
3	Open space	2.73	1.68	1.21
4	Built-up	53.77	64.84	58.46

Table 23.4 Change of LULC areas from 2010 to 2015

Changed feature (2010–2015)	Changed area (km^2)
Builtup–builtup	87.12
Builtup–vegetation	4.14
Builtup–waterbody	0.43
Builtup–open space	0.89
Open space–builtup	2.06
Open space–vegetation	1.47
Open space–waterbody	0.03
Open space–open space	2.80
Vegetation–builtup	31.37
Vegetation–vegetation	43.45
Vegetation–waterbody	1.35
Vegetation–open space	1.87
Waterbody–builtup	2.59
Waterbody–vegetation	1.50
Waterbody–waterbody	5.23
Waterbody–open space	0.05

water body and vegetation became surprisingly high (Tables 23.3 and 23.4, Figs. 23.4 and 23.5).

Figure 23.6 and Table 23.5 generated by remote sensing and GIS analysis, indicate that the built-up area were 55.77% in 2010, 65.84% in 2015, and 71.46% for the year 2020 which indicates the expansion of 10.07% from 2010 to 2015, and then further 5.62% by the year 2020. Open space area were 2.73% (2010), 0.68% (2015), and further reduced to 0.21% in the year 2020. The vegetation area has been reduced to 29.73% in the year 2015 compare to the year 2010 (37.78%) and further reduced to 19.21% by the year 2020. Waterbody area was 3.72% in the year 2010, 3.75% in the year 2015, and by the year 2020, it has a remarkable increase of 9.12%. This rapid escalation of built-up area with the diminution of vegetation and open space encouraged the intensification of *UHI* in Kolkata Municipal Corporation area and its environment, though in the year 2020, 5.37% increment of water body areas from the year 2015 to 2020, compensate to some extent to decrease the Land Surface Temperature.

The people were homebound due to the primary reason of lockdown and hence the city has broad impact ecological changes. The recent back to back disastrous cyclones in 2019 like "Fani" (April 2019) "Bulbul" (November 2019, very severe cyclone) and "Amphan" (May 2020 and it was also super severe cyclone declared by IMD with heavy rainfall) that we witnessed and also have resulted in a drastic variation of water body changes, like increasing wetland, generating new water bodies, high soil moisture content and uplift of groundwater table, which has also seen a record increase compared to what it was earlier. In this context, we can mention that the severe cyclone "Aila" in the year 2009 (25–27 May) had also left the similar impact on LULC types like water body and vegetation change for the year 2010.

23.3.2 Land Surface Temperature Variation of KMC

LST has been categorized into five zones, namely "very low", "low", "moderate", "high", and "very high". Table 23.6 exemplify that in the year 2010, the minimum and maximum LST values of Kolkata Municipal Corporation area were 17.25 °C and 28.66 °C, the minimum and maximum Land Surface Temperatures were 19.02 °C and 29.85 °C, respectively, of 2015. It seems within the 5 years of time span, the raise of minimum and maximum surface temperatures were 1.77 °C and 1.19 °C. But interestingly in the year 2020, the Land Surface Temperate has been decreased as the minimum 17.40 °C and the maximum is 24.10 °C to the tune of 1.62 °C as minimum and 5.75 °C as maximum which is the sign of improvement in the context of Land Surface

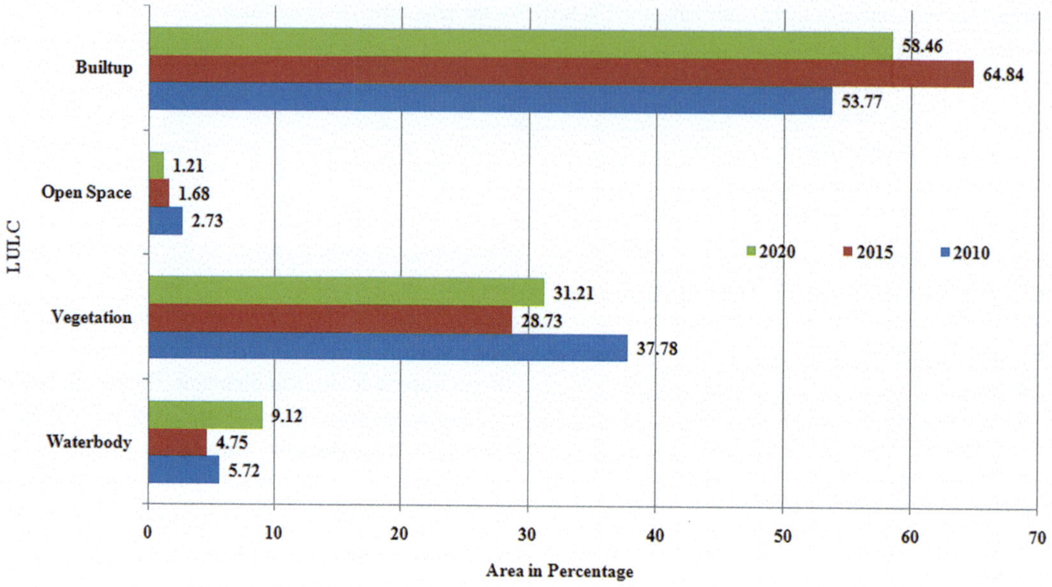

Fig. 23.4 Graphical representation of percentage for the Land Use/Land Cover area change in KMC (2010, 2015 & 2020). KMC Kolkata Municipal Corporation

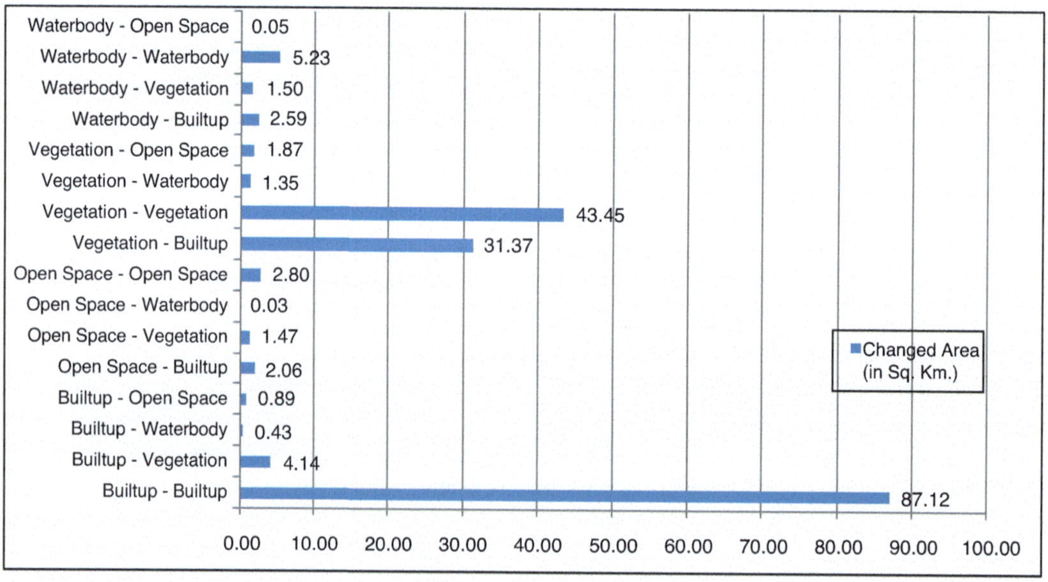

Fig. 23.5 Change of LULC areas from 2010 to 2015

Temperature of Kolkata. The super cyclone “Fani” and “Bulbul” during the month of April and November 2019, respectively, and “Amphun” in the month of May 2020 along with the lockdown (complete and partial) period due to Covid-19 pandemic situation are the major causes to decrease the LST of Kolkata in the year 2020.

Thirteen locations of different land use types within Kolkata Municipal Corporation area have been surveyed using the temperature sensor for the year 2010, 2015, and 2020. The analyzed graph (Fig. 23.7) shows that in the year 2020 (lockdown period of Covid-19 pandemic), Land Surface Temperature was less than the year 2010

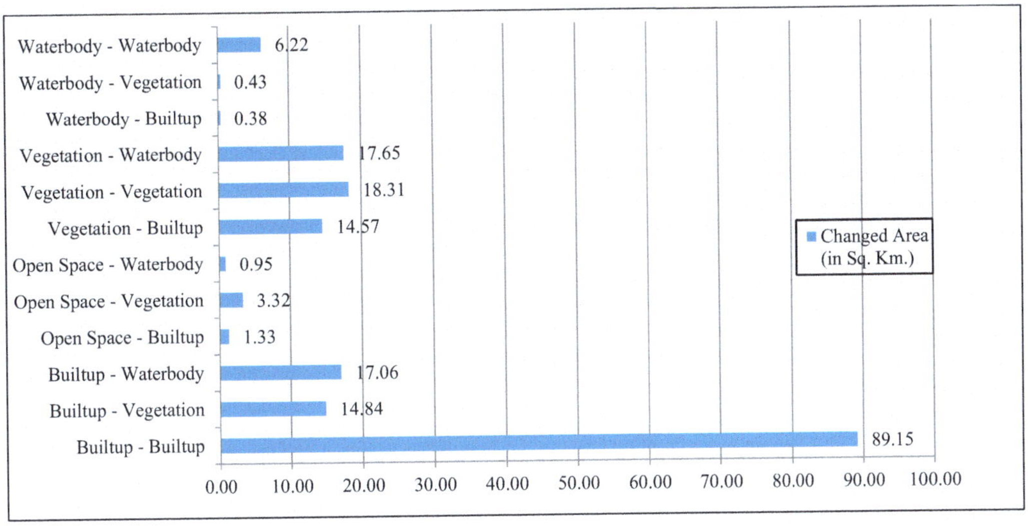

Fig. 23.6 Change of LULC areas from 2015 to 2020

Table 23.5 Change of LULC areas from 2015 to 2020

Changed feature (2015–2020)	Changed area (km²)
Builtup–builtup	89.15
Builtup–vegetation	14.84
Builtup–waterbody	17.06
Open space–builtup	1.33
Open space–vegetation	3.32
Open space–waterbody	0.95
Vegetation–builtup	14.57
Vegetation–vegetation	18.31
Vegetation–waterbody	17.65
Waterbody–builtup	0.38
Waterbody–vegetation	0.43
Waterbody–waterbody	6.22

and 2015 due to the restriction on traffic movements, human activities outside the residential area and closed workshops which caused less fossil fuel emission and environmental hazard in the atmosphere.

23.3.3 Urban Heat Island and Non-urban Heat Islands Areas Within KMC

Table 23.7 shows the *UHI* and non-*UHI* areas for the years 2010, 2015, and 2020. It seems the UHI area in the year 90.68 km^2 out of total area 186.35 km^2 (KMC) for the year 2010. By the year 2015, 27.43 km2 area has been added as *UHI*, whereas 23.63 km^2 has been reduced by the year 2020. Similarly, 95.67 km^2 area was marked as non-*UHI* for the year 2010, reduced to 68.24 km^2 by the year 2015, and again increased to the tune of 23.63 km^2 (91.87) till 2020.

23.3.4 Different Indices Within KMC

Figure 23.8 and Table 23.8 illustrate the Normalized Difference Vegetation Indexes of Kolkata Municipal Corporation for the years 2010, 2015, and 2020. It has been observed that the NDVI value for the year 2010, 0.42 as high and −0.25 as Low and by the year 2015, the high and low values became 0.20 and −0.04, respectively. But in the same context, the high value was 0.40 and −0.09 as low for the year 2020 shown by the indices analysis. Map and table data indicate that during the Covid-19 period of 2020, vegetation index has been improved significantly. The values have been rounded off up to two decimal numbers.

NDBI has been used to identify the built-up index of Kolkata Municipal Corporation area. The resultants show that values of 0.53 (2010), 0.25 (2015), and 0.17 (2020) as high built-up index, whereas −0.47 (2010), −0.32 (2015) and

Table 23.6 Land Surface Temperature variation of KMC for 2010, 2015 and 2020 with covered areas

Land Surface Temperature within Kolkata Municipal Corporation area						
	2010		2015		2020	
Category	Temperature (°C)	Area (km^2)	Temperature (°C)	Area (km^2)	Temperature (°C)	Area (km^2)
Very low	17.25–20.43	06.45	19.02–21.49	8.05	17.40–18.97	27.85
Low	20.43–21.72	25.64	21.49–22.76	40.13	18.97–19.71	61.46
Moderate	21.72–22.62	61.09	22.76–23.70	75.97	19.71–20.36	49.25
High	22.62–23.91	65.00	23.70–25.01	48.18	20.36–21.23	34.97
Very high	23.91–28.66	28.17	25.01–29.85	14.02	21.23–24.10	12.82

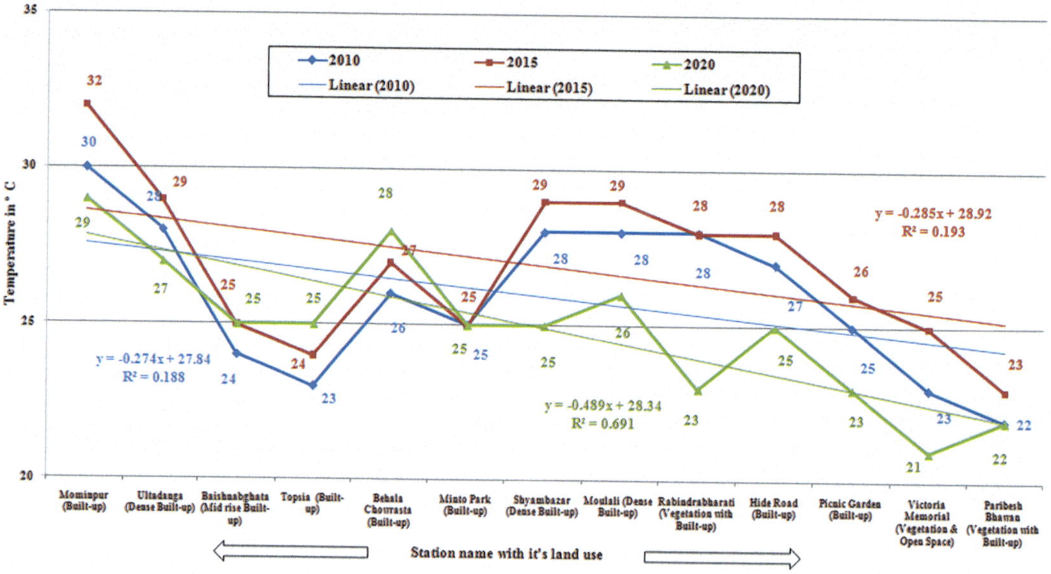

Fig. 23.7 Line graph of surface temperature for sample survey locations within KMC for the year 2010, 2015, and 2020

Table 23.7 *UHI* and Non-*UHI* areas within KMC for the year 2010, 2015, and 2020

Area (under Kolkata Municipal Corporation) in km^2			
	2010	2015	2020
Urban heat island	90.68	118.11	94.48
Non-*urban heat island*	95.67	68.24	91.87

−0.27 (2020) as low due to presence of water bodies, vegetation, and wetland. During the lockdown (COVID-19) period, most of the construction activities, factories were under lock and key. On top of that, consecutive severe cyclonic disastrous with unsystematic rainfall, also have resulted the massive climatic variation and there is a strong possibility, caused a change in the environmental parameters as well. This is the main reason of improvement in water and vegetation index. Table 23.9 shows that the built-up index was improved in the year 2015 compare to 2010, whereas the non-activities of human being due to lockdown caused the huge improvement in the built-up index by the year 2020 (Fig. 23.9).

Table 23.10 and Fig. 23.10 describe that the outcome of normalized difference water index values are 0.56 as high and −0.47 as low for 2010, which had been decreased to 0.27 (high) and −0.25 (low) in 2015, whereas by the year 2020, the values increased as 0.32 (high) and −0.25 (low) within the Kolkata Municipal Corporation area.

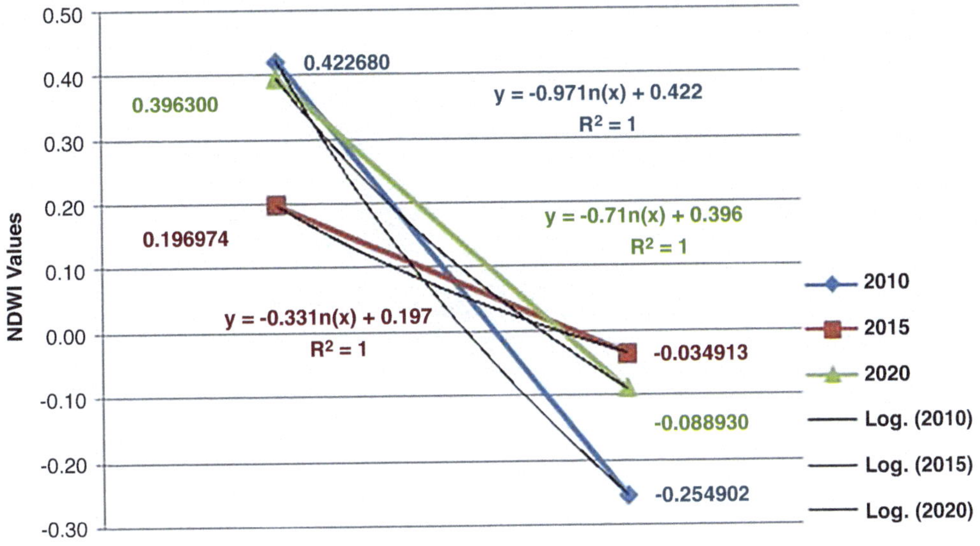

Fig. 23.8 Logarithmic regression analysis of NDVI for 2010, 2015, and 2020

Table 23.8 NDVI values for the years 2010, 2015, and 2020 within KMC area

	Year		
NDVI	2010	2015	2020
High	0.42	0.20	0.40
Low	−0.25	−0.04	−0.09

Table 23.9 NDBI values for the years 2010, 2015, and 2020 within KMC area

	Year		
NDBI	2010	2015	2020
High	0.53	0.17	0.25
Low	−0.47	−0.27	−0.32

23.3.5 Estimated Percentages of LULC, LST, and Spectral Indices for 2030 (Tables 23.11 and 23.12, Fig. 23.11)

The estimated value has been calculated by the formula:

$$\text{Value}(2025) = \{\text{Value}(2015) / \text{Value}(2010)\} * \text{Value}(2020)$$

The estimated value has been calculated by the formula:

$$\text{Value}(2025) = \{\text{Value}(2015) / \text{Value}(2010)\} * \text{Value}(2020)$$

23.4 Discussions

23.4.1 Comparison of Land Use/Land Cover Change of Kolkata Municipal Corporation

The Land Use/Land Cover Change of Kolkata Municipal Corporation area for the year 2010, 2015, and 2020 has shown in Fig. 23.12. The general land use classes of water body, vegetation, open space, and built-up area have been identified by the classification technique of Remote Sensing from the satellite image captured by the sensor of LANDSAT 5 and 8 OLI. The False Color Composite using IR (4), Red (3), Green (2) of Landsat 5 and NIR (5), Red (4) and Green (2) bands of Landsat 8 OLI have been used for the supervised classification (Table 23.13).

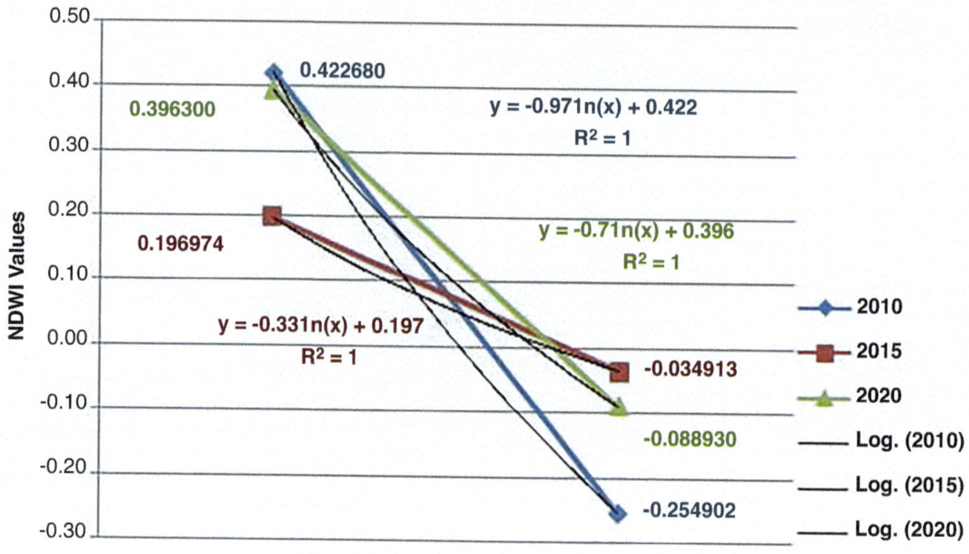

Fig. 23.9 Logarithmic regression analysis of NDBI for 2010, 2015, and 2020

Table 23.10 NDWI values for the years 2010, 2015, and 2020 within KMC area

	Year		
NDWI	2010	2015	2020
High	0.56	0.27	0.32
Low	−0.47	−0.17	−0.25

23.4.2 Comparison of NDVI for Kolkata Municipal Corporation

Due to Covid-19 pandemic lockdown, the constructions of multistoried dwell units been suspended for a long periods and parallel effects of super cyclones "Bulbul" and "Fani", those vacant areas became filled with water body or shrubs (greeneries) (Fig. 23.13).

23.4.3 Comparison Between NDBI and NDWI of Kolkata Municipal Corporation for the Years 2015 and 2020

NDBI analysis of Kolkata Municipal Corporation shown by Fig. 23.14 indicates that in the year 2010, the built-up index value was 0.526882 as high and −0.466667 as low, 2015, the built-up index value was 0.167946 as high and −0.27254 as low for the year 2015, which indicate the downfall of the index values. Again the index values of 2020 signify that, the uplift of built-up index value 0.251219 as high and −0.321822 as low, which clearly indicates that the region covered with mostly concrete, a key factor of hike in the surface temperature.

The Difference in Normalization The built-up index value ranges from −1 to +1. Higher NDBI values imply built-up regions, while lower values represent water bodies. Vegetation has a low NDBI value. Built-up area also has the negative value. The NDWI analysis maps of KMC (Fig. 23.15) signified that the water index value of this region was 0.27254 as high and −0.167946 as low for the year 2015 and the very severe cyclone "Fani" and "Bulbul" in the month of April and November 2019, respectively, affected this region very badly to the extent of 0.321822 as high and −0.251219 as low for the year 2020. Agricultural lands and fisheries are being affected mostly by the consecutive cyclone along with vegetation and built-up area, which extent the hike of surface water table.

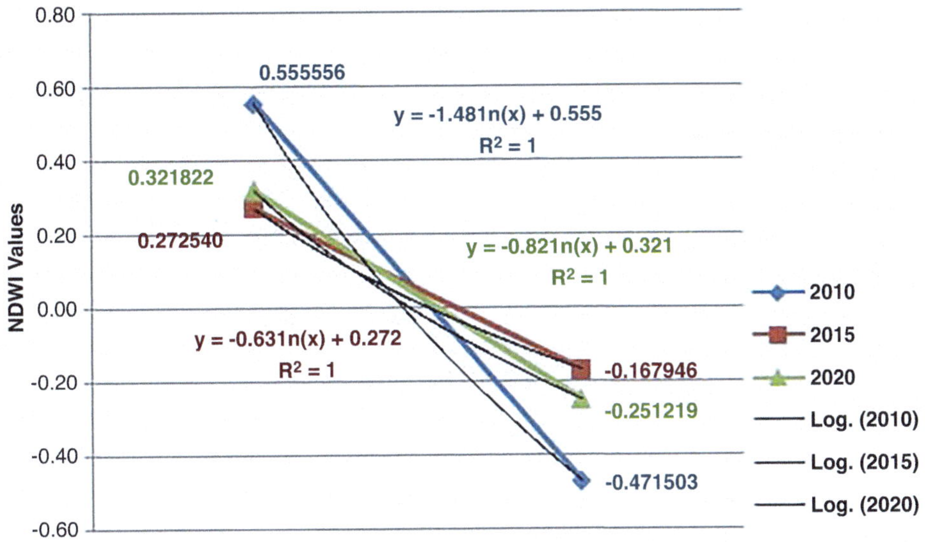

Fig. 23.10 Logarithmic regression analysis of NDWI for 2010, 2015, and 2020

Table 23.11 Different LULC types for 2010, 2015, 2020, and 2030 (estimated) areas

Areas (Kolkata Municipal Corporation) are in percentage (%)				
LULC	2010	2015	2020	2025 (estimated)
Waterbody	5.72	4.75	9.12	7.57
Vegetation	37.78	28.73	31.21	23.73
Open space	2.73	1.68	1.21	0.75
Built-up	53.77	64.84	58.46	70.5

Table 23.12 LST and three spectral indices for the year 2000, 2010, 2015, 2020, and 2025 (estimated) within the Kolkata Municipal Corporation area

	2010	2015	2020	2025 (estimated)
LST_{Max} (°C)	28.65	29.85	24.10	25.11
LST_{Min} (°C)	17.25	19.02	17.40	19.18
$NDVI_{Max}$	0.42	0.20	0.40	0.19
$NDVI_{Min}$	−0.25	−0.03	−0.09	−0.01
$NDWI_{Max}$	0.55	0.27	0.32	0.16
$NDWI_{Min}$	−0.47	−0.17	−0.25	−0.09
$NDBI_{Max}$	0.53	0.25	0.17	0.08
$NDBI_{Min}$	−0.47	−0.32	−0.27	−0.18

23.4.4 Comparison Between the Land Surface Temperatures of Different Years Within KMC

As the acquisition month of satellite imageries was January (winter season), in the year 2010 the minimum and maximum land surface temperature were 17.25 °C and 28.65 °C. By the year 2015, the LST became 19.02 °C and 29.85 °C as minimum and maximum, respectively. Surprisingly in the year 2020, it has been reduced to the tune of 1.60 °C and 5.75 °C as minimum and maximum, respectively. The lockdown for the Covid-19 pandemic and consecutive severe cyclones in the year 2019 and 2020 were the main causes of trim down (Fig. 23.16).

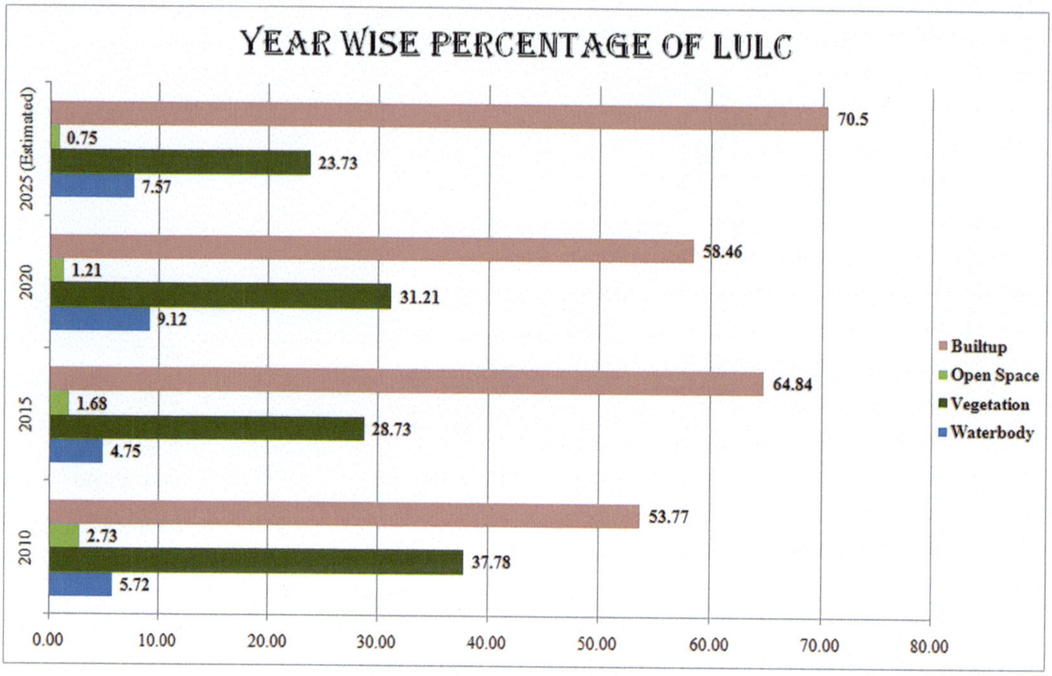

Fig. 23.11 Year wise percentages of LULC within KMC based on Table 23.11

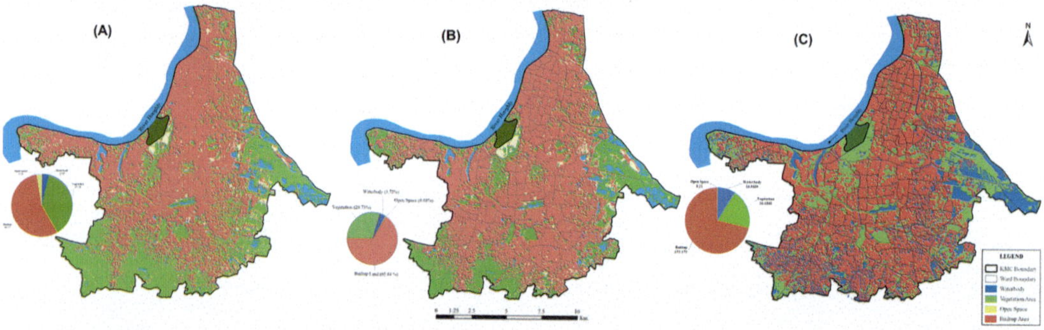

Fig. 23.12 Land Use/Land Cover maps of KMC for the year 2010 (**a**), 2015 (**b**), and 2020 (**c**)

Table 23.13 Accuracy assessment statistics of image-based LULC Classification for the years 2010, 2015, and 2020 in KMC

Sl. No.	Study year	Overall accuracy (%)	Kappa coefficient (%)
1	2010	92	88.03
2	2015	94	91.36
3	2020	96	94.61

23.4.5 LST Profile for the Year 2010, 2015, and 2020 Within KMC

Figure 23.17 illustrates the spatio-temporal distribution of land surface temperature on built-up area. Pixels under the AB profile show the maximum spatio-temporal development of Kolkata, which is mainly due to the density of built-up. The CD profile also shows the medium spatio-temporal development and the EF is the less within the study area.

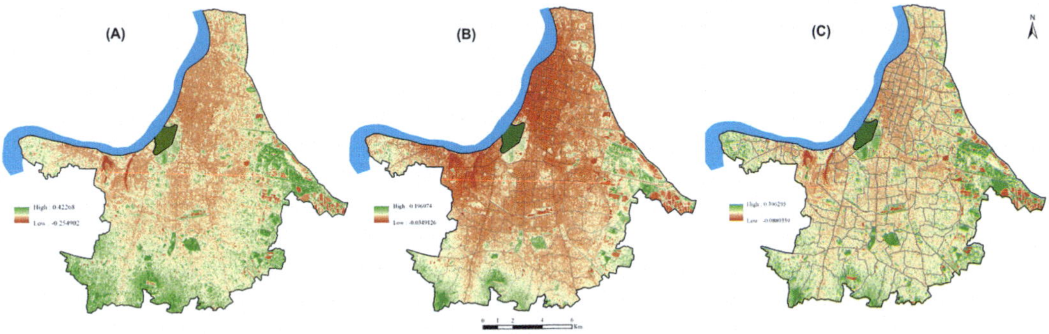

Fig. 23.13 NDVI maps of KMC for the year 2010 (**a**), 2015 (**b**), and 2020 (**c**)

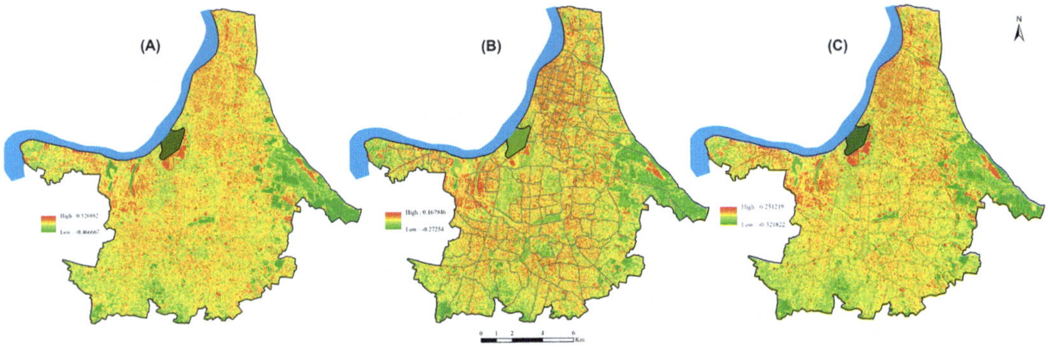

Fig. 23.14 NDBI maps of KMC for the year 2010 (**a**), 2015 (**b**), and 2020 (**c**)

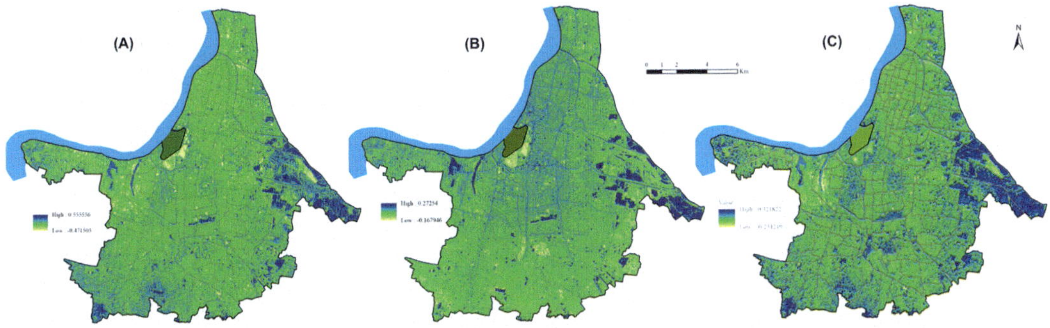

Fig. 23.15 NDWI maps of KMC for the year 2010 (**a**), 2015 (**b**), and 2020 (**c**)

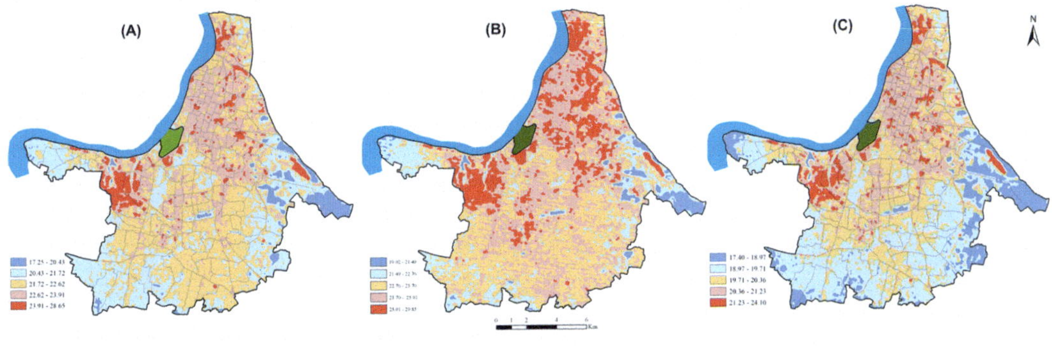

Fig. 23.16 Land surface temperature maps of KMC for the year 2010 (**a**), 2015 (**b**), and 2020 (**c**)

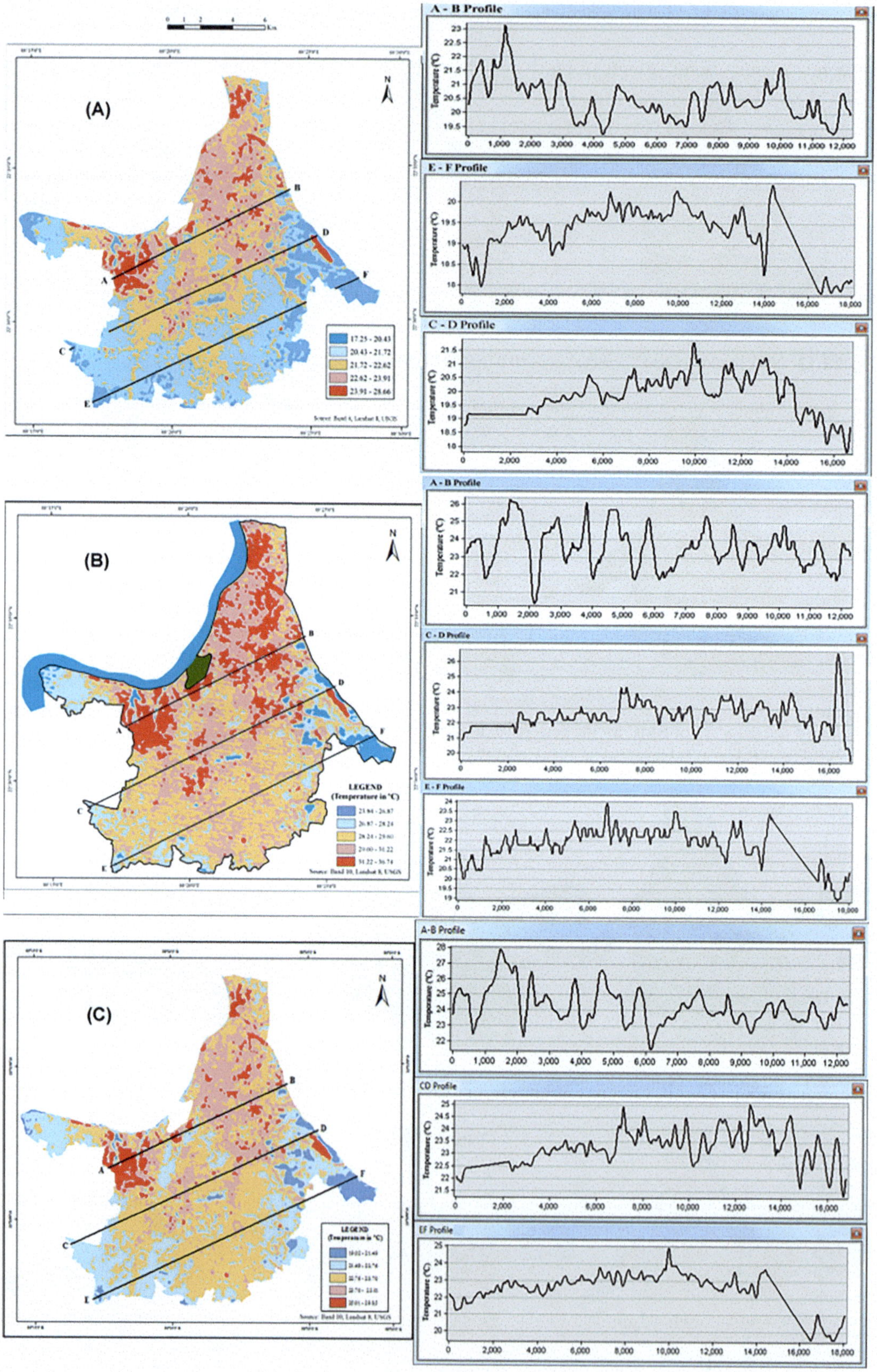

Fig. 23.17 Three LST profiles naming AB, CD and EF across the study area for the years 2010 (**a**), 2015 (**b**), and 2020 (**c**)

23.4.6 Comparison of Urban Heat Island for Kolkata Municipal Corporation (Fig. 23.18)

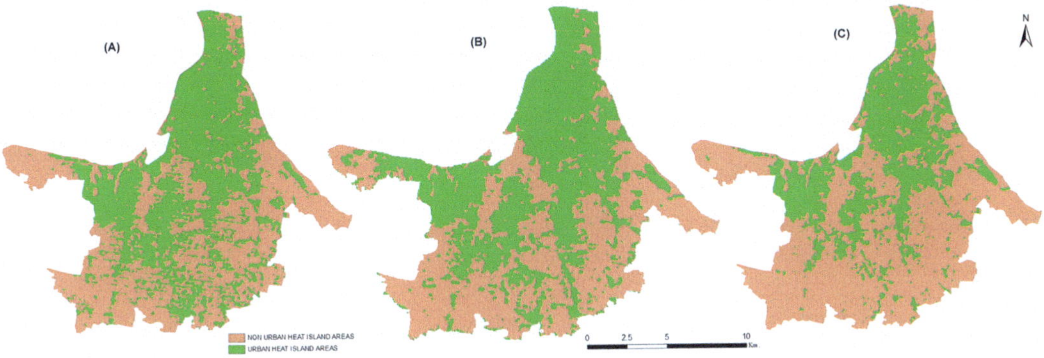

Fig. 23.18 Development of UHI for the year 2010 (**a**), 2015 (**b**), and 2020 (**c**) within the KMC area

23.4.7 Overlay Analysis Between LULC with LST and LULC with UHI in KMC for the Year 2010, 2015, and 2020

The overlay maps (Fig. 23.19) of Land Use/Land Cover and Land Surface Temperature for the year 2015 shows that the built-up of north and port area within Kolkata Municipal Corporation exhibits maximum Land Surface Temperature compare to the other region. The vegetation and water body covered area experienced lowest LST. Different Land Surface Temperature zones overlaid on the Land Use/Land Cover map of Kolkata Municipal Corporation for the year 2010 (a), 2015 (b) and 2020 (c) exemplify that the Garden Reach area and most of the north Kolkata region having land use type built-up comprises the highest surface temperature, whereas the East Kolkata Wetland of water body land use dissipate the lowest surface temperature within the study area. Comparatively, south Kolkata region experiences low surface temperature due to the presence of water bodies and vegetations.

The overlay analysis demonstrates that the Garden Reach area, a portion of central Kolkata and mostly the north Kolkata areas fall under Urban Heat Islands whereas the south, east, and western part of Kolkata are under Non-Urban Islands areas. The correlation maps of LST and UHI on the LULC of Kolkata (Fig. 23.20) for the year 2010 (a), 2015 (b), and 2020 (c) describe that mainly the built-up areas of north Kolkata and Garden Reach areas the very much correlated as the LST of those areas are high and fall under Urban Heat Island zones within the study area. On the other side, south Kolkata and East Kolkata Wetland areas having low surface temperature fall under Non-Urban Heat Islands within the study area.

23.4.8 Area Wise Change in LULC and Average Temperature from 2010 to 2015 and 2015 to 2020

Table 23.14 and Fig. 23.21 show the usual scenario of change in different types of LULC within the study area. Built-up area and average temperature have increased to the tune of 11.07% and 1.48 °C, respectively. On the other side vegetation, water bodies along with open spaces have decreased 9.05%, 0.97%, and 1.05%, respectively.

In the year 2020, the KMC area surprisingly observed the unusual scenario of LULC change by Table 23.15 and Fig. 23.22. Built-up (6.38%), open space (0.47%), and average temperature (3.7 °C) decreased, while the vegetation (2.48%) and water body (4.37%) have increased. Non-activities of human beings due the Covid-19

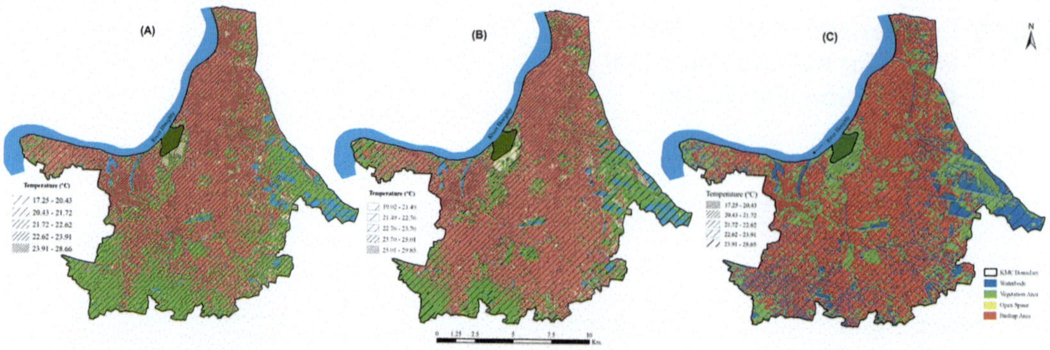

Fig. 23.19 Overlay analysis maps of Land Surface Temperature zones overlaid on the Land Use/Land Cover map of Kolkata Municipal Corporation for the year 2010 (**a**), 2015 (**b**), and 2020 (**c**)

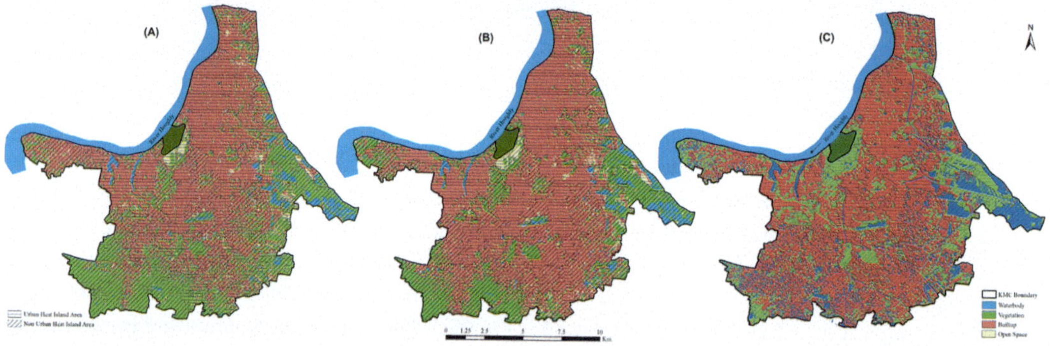

Fig. 23.20 Overlay analysis maps of urban and non-urban heat islands overlaid on the Land Use/Land Cover map of Kolkata Municipal Corporation area for 2010 (**a**), 2015 (**b**), and 2020 (**c**)

Table 23.14 Percentage of area and average temperature change from 2010 to 2015

Built-up	11.07
Vegetation	−9.05
Waterbody	−0.97
Open space	−1.05
Average temperature	1.48

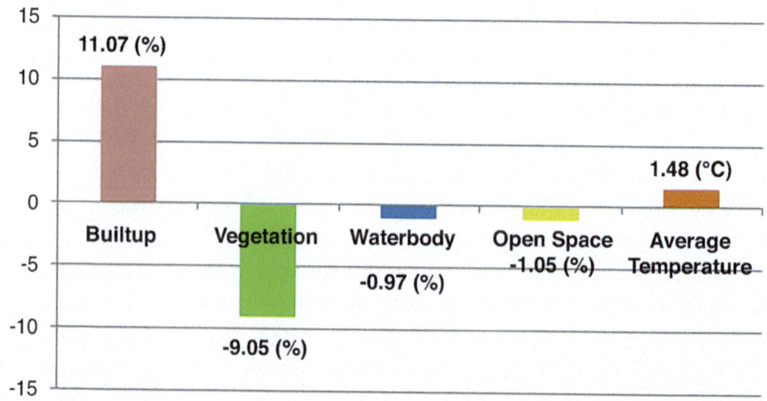

Fig. 23.21 Graphical representation for the change of different LULC areas (in percentage) from 2010 to 2015

Table 23.15 Percentage of area and average temperature change from 2015 to 2020

Built-up	−6.38
Vegetation	2.48
Waterbody	4.37
Open space	−0.47
Average temperature	−3.685

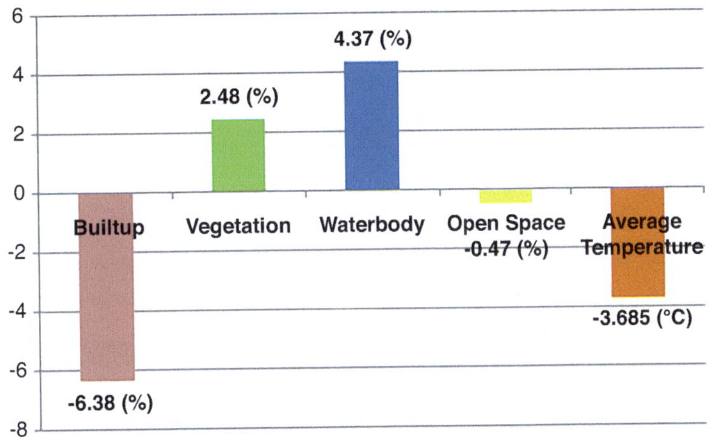

Fig. 23.22 Graphical representation for the change of different LULC areas (in percentage) from 2015 to 2020

pandemic and severe cyclonic disasters discontinue the usual trend of rising scenario of built-up and temperature, resulting into a healthy environment.

Table 23.16 and Fig. 23.23 describe the amount of areas in square kilometer converted into different LULC types within the study area by the year 2020.

23.4.9 Comparison Between NDWI, NDVI, NDBI, LST, UHI, and LULC

The spatial distribution of different indices, LST, UHI, and LULC which have been exaggerated one by other by the Arc Scene tool of ArcGIS software shown by Fig. 23.24.

23.5 Conclusions

Table 23.11 and Fig. 23.11 describe that the status of LULC types will not become comparatively bad by the year 2025, unlike the trend observed during 2010 to 2015. The estimated percentage LULC types by the year 2025 may deteriorate, but to some extent. The LULC of 2020 became the unusual and examine sudden rectification in aspect of the environmental conditions due to Covid-19 pandemic and consecutive severe cyclonic disasters during 2019 and 2020 within the study area (KMC).

Table 23.12 shows that the environmental conditions of unusual and abrupt improvement of atmosphere due to Covid-19 pandemic and consecutive severe cyclonic disasters during 2019 and 2020 within the study area (KMC) sustain, then by the year 2025, the maximum LST will be 25.11 °C and the minimum will be 19.18 °C during the winter season. Similarly, the indices like NDVI, NDBI, and NDWI will also improve and slightly become healthy for the habitants.

According to Fig. 23.18, mainly the northern part of KMC and the port area nominated as the Urban Heat Islands, which were mostly built-up land use type.

Table 23.14 and Fig. 23.21 show the normal scenario of changes from the year 2010 to 2015 within Kolkata Municipal Corporation area. The built-up area increased by diminishing the vege-

Table 23.16 Converted area of LULC type from the year 2015–2020

Changed feature (2015–2020)	Changed area (in km^2)
Built-up–built-up	89.15
Built-up–vegetation	14.84
Built-up–waterbody	17.06
Open space–built-up	1.33
Open space–vegetation	3.32
Open space–waterbody	0.95
Vegetation –built-up	14.57
Vegetation–vegetation	18.31
Vegetation–waterbody	17.65
Waterbody –built-up	0.38
Waterbody–vegetation	0.43
Waterbody–water body	6.22

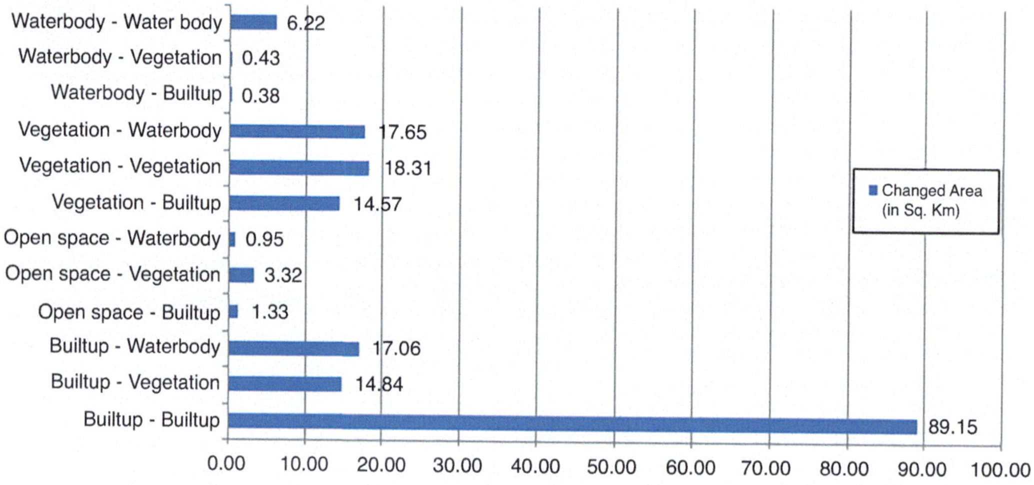

Fig. 23.23 Conversion of LULC features from the year 2015 to 2020 of KMC area

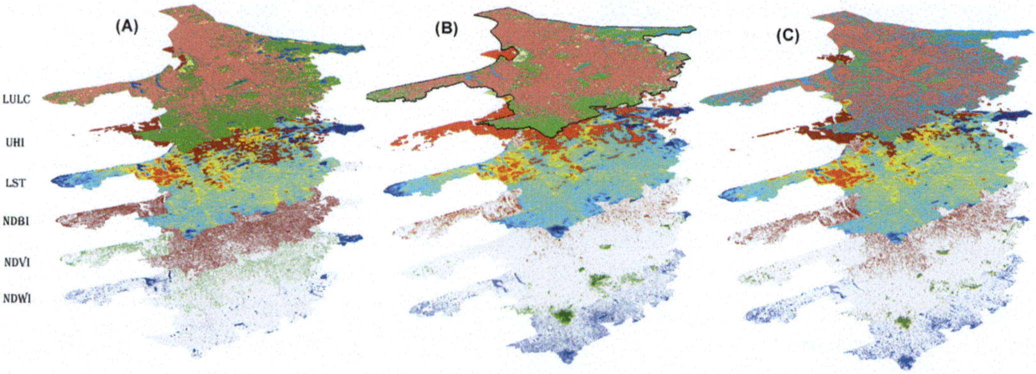

Fig. 23.24 Spatial distribution of NDWI, NDVI, NDBI, LST, UHI, and LULC for the year 2010 (**a**), 2015 (**b**), and 2020 (**c**) within Kolkata Municipal Corporation

tation, water body, and open space areas which resultant the hike of average temperature.

Table 23.15 and Fig. 23.22 show the scenario of unusual changes from 2015 to 2020 within Kolkata Municipal Corporation. Surprisingly, the scenario discontinues for the year 2020 due to the lockdown period of SARS Covid-19 along with the super cyclones "Fani" and "Bulbul" during the month of April and November 2019, respectively. Non-activity of human movement during the lockdown has decreased the built-up and open space areas and increased weeds in the form of vegetation. And also the frequency of super cyclones increased the water bodies which declined the average temperature in 2020.

Table 23.16 and Fig. 23.23 show that between the year 2015 and 2020, the area 1.33, 14.57, and 0.38 km^2. of open space, vegetation, and water body, respectively, have been converted to built-up area though the lockdown due to Coved-19 pandemic starts from March 2019, which is really a matter of concern from the environment aspects within the KMC area.

Figures 23.16 and 23.17 shows that due to the presence of more water bodies and vegetative covers, the southern portion of the study area radiates the least temperature into the environment compare to the middle and northern portion, which is the most heat emittance area for land surface temperature due to intense built-up density and absence of water body with vegetative cover like park and play ground.

Various spectral indices like NDWI, NDVI, NDBI, and LST, UHI have overlay on the LULC maps of KMC for the year 2010 (a), 2015 (b), and 2020 (c) to emphasize the spatial distribution of each (Fig. 23.24). It seems that each theme is correlated with each other. Vegetative areas with water bodies having less built-up generate less UHI zones and radiate less LST. Similarly, visual display of overlay maps confirms that the dense built-up areas radiate high LST and creating more UHI zones within the KMC area.

The East Kolkata Wetlands (an Indian Ramsar site) and the Sundarban Mangrove Forest (an Indian UNESCO World Heritage Site) serve as Kolkata's heat sinks. The radiation from concrete structure and operating waste water treatment plants have been avoided by the city of Kolkata due to the East Kolkata Wetlands. The water flows through fish ponds covering about 4000 hectors that act as solar reactors and complete the majority of their biochemical reactions with the help of solar energy. Through the effective recovery of nutrients, the wetlands create an improved urban facility treating the drained waste water from city and utilize the treated water for horticulture and agriculture. Chemical processes within the wetlands powered by solar energy. So the system is referred to as "one of the rare examples of environmental protection and development management where the local farmers have adopted a complex ecological process for mastering the resource recovery activities." The wetlands area is being reduced due to the construction of multi-story buildings and industries which are using waste water outfall channels without permission, endangering the edible quality of fish and vegetables by adding metals to the canal sludge. In the delta of the rivers Ganges and Brahmaputra on the coast of the Bay of Bengal, the Sundarban is the biggest mangrove forest in the world, encompassing hundreds of islands and a confusing network of rivers, rivulets, and creeks; 90% of the Indian mangrove species are found in the Indian Sundarban, which makes up the majority of the south-western delta and contains more than 60% of the nation's total mangrove forest. The hinterland is shielded by the mangrove forests from storms, cyclones, tidal surges, and the seepage and intrusion of saltwater into inland areas and waterway.

Appendix

Plates 23.1 and 23.2 are the pictorial descriptions of spatial distribution (land use) of Rabindra Sarobar Lake and its surrounding area under the Kolkata Municipal Corporation (South) (Plate 23.3).

Plate 23.1 Aerial view of Rabindra Sarobar, Kolkata (South)

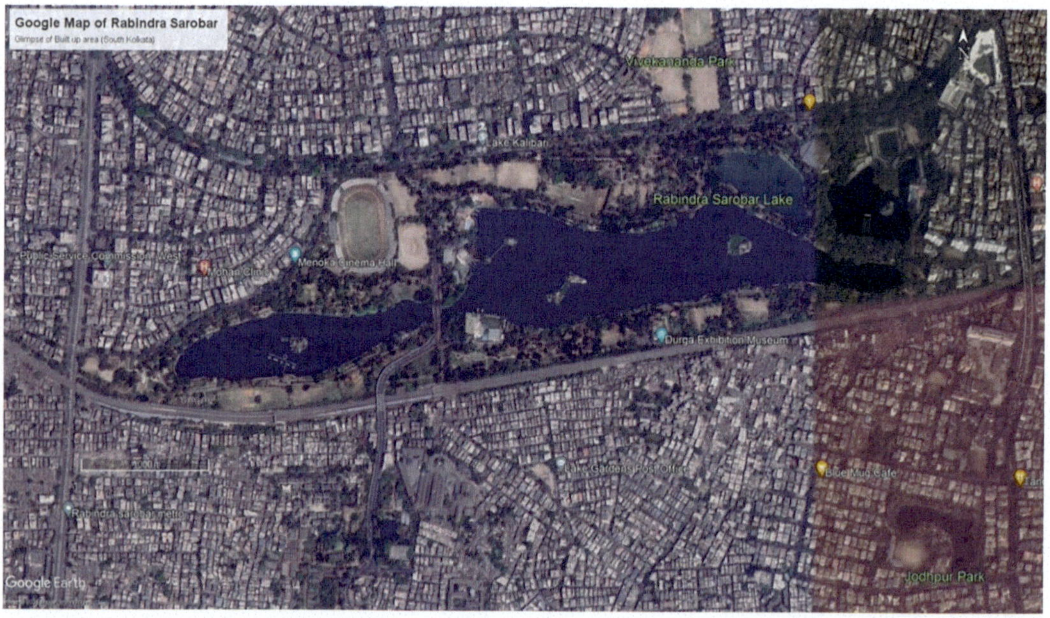

Plate 23.2 Google imagery of Rabindra Sarobar, Kolkata (South)

Plate 23.3 The slum of North Kolkata

References

Arnfield J (2003) Two decades of urban climate research: a review of turbulence, exchanges of energy and water, and the urban Heat Island. Int J Climatol 23:1–26. https://doi.org/10.1002/joc.859

Chun B, Guldmann J-M (2014) Spatial statistical analysis and simulation of the urban Heat Island in high-density central cities. Landsc Urban Plan 125:76–88. https://doi.org/10.1016/j.landurbplan.2014.01.016

Coseo P, Larsen L (2014) How local climate zones predict urban Heat Islands in Eight Chicago Neighborhoods. https://doi.org/10.13140/RG.2.1.2164.1689

Dai Z, Guldmann J-M, Hu Y (2018) Spatial regression models of park and land-use impacts on the urban Heat Island in Central Beijing. Sci Total Environ 626:1136–1147. https://doi.org/10.1016/j.scitotenv.2018.01.165

Deilami K, Kamruzzaman M (2017) Modelling the urban Heat Island effect of smart growth policy scenarios in Brisbane. Land Use Policy 64:38–55. https://doi.org/10.1016/j.landusepol.2017.02.027

Feyisa GL, Meilby H, Jenerette D, Pauleit S (2016) Locally optimized separability enhancement indices for urban land cover mapping: exploring thermal environmental consequences of rapid urbanization in Addis Ababa. Ethiopia Remote Sens Environ 175:14–31. https://doi.org/10.1016/j.rse.2015.12.026

Guha S, Govil H, Mukherjee S (2017) Dynamic analysis and ecological evaluation of urban Heat Islands in Raipur City, India. J Appl Remote Sens 11(3):036020. https://doi.org/10/1117/1.JRS.11.036020

Guha S, Govil H, Dey A, Gill N (2018) Analytical study of land surface temperature with NDVI and NDBI using Landsat 8 OLI and TIRS data in Florence and Naples City, Italy. Eur J Remote Sens 51(1):667–678. https://doi.org/10.1080/22797254.2018.1474494

Hasan S, Zhan X, Ukkusuri S (2013) Understanding urban human activity and mobility patterns using large-scale location-based data from online social media. Proceedings of the ACM SIGKDD International Conference on Knowledge Discovery and Data Mining. https://doi.org/10.1145/2505821.2505823

Li J, Song C, Cao L, Zhu F, Meng X, Wu J (2011) Impacts of landscape structure on surface urban Heat Islands: a case study of Shanghai, China. Remote Sens Environ 115:3249–3263. https://doi.org/10.1016/j.rse.2011.07.008

Lopez JMR (2017) Frontiers of urbanization: identifying and explaining urbanization hot spots in the south of Mexico City using human and remote sensing. Appl Geogr 79:1–10. https://doi.org/10.1016/j.apgeog.2016.12.001

Maity S, Srivastava GL (2020) Assessment of land surface temperature of Kolkata urban agglomeration, West Bengal, India, ISSN 2349-638x, http://www.aiirjournal.com/

McFeeters SK (2013) Using the normalized difference water index (NDWI) within a geographic information system to detect swimming pools for mosquito abatement: a practical approach. Remote Sens 5:3544–3561. https://doi.org/10.3390/rs5073544

Mirzaei PA (2015) Recent challenges in modeling of urban Heat Island. Sustain Cities Soc 19:200–206. https://doi.org/10.1016/j.scs.2015.04.001

Pearsall H (2017) Staying cool in the compact city: vacant land and urban heating in Philadelphia, Pennsylvania. Appl Geogr 79:84–92. https://doi.org/10.1016/j.apgeop.2016.12.010

Rinner C, Hussain M (2011) Toronto's urban heat Island – exploring the relationship between land use and surface temperature. Remote Sens 3:1251–1265. https://doi.org/10.3390/rs3061251

Rizwan AM, Dennis LYC, Liu C (2008) A review on the generation, determination and mitigation of the urban Heat Island. J Environ Sci 20(1):120–128. https://doi.org/10.1002/2017EF000569

Voogt JA, Oke TR (2003) Thermal remote sensing of urban climate. Remote Sens Emviron 86:370–384

Zhang H, Qi Z-f, Ye X-y, Cai Y-b, Ma W-c, Chen M-n (2013) Analysis of land use/land cover change, population shift, and their effects on spatiotemporal patterns of urban heat islands in metropolitan Shanghai, China. Appl Geogr 44:121–133.,ISSN 0143-6228. https://doi.org/10.1016/j.apgeog.2013.07.021

Zhao M, Cai H, Qiao Z, Xu X (2016) Influence of urban expansion on the urban Heat Island effect in Shanghai. Int J Geogr Inf Syst 30(12):2421–2441. https://doi.org/10.1080/13658816.2016.1178389

Zhao ZC, Luo Y, Huang JB (2012) Are there impacts of urban heat island on future climate change? Adv Clim Chang Res 4:1–4

Zhou W, Huang G, Cadenasso M (2011) Does spatial configuration matter? Understanding the effects of land cover pattern on land surface temperature in urban landscapes. Landsc Urban Plan 102:54–63. https://doi.org/10.1016/j.landurbplan.2011.03.009

24 Addressing Climate Change Challenges in South Africa: A Study in KwaZulu Natal Province

Ifeanyi Michael Smarte Anekwe, Helper Zhou, Mphathesithe Mzwandile Mkhize, and Stephen Okiemute Akpasi

Abstract

Climate change has caused significant ecosystem degradation and seriously threatens environmental sustainability. South Africa, including KwaZulu Natal and other provinces, is experiencing climate change challenges that result in adverse effects such as extreme weather conditions, heatwaves, drought and floods. This study aims to critically review KwaZulu Natal's position towards climate action, providing relevant context to the current development and challenges and highlighting government strategies for combating the adverse effects of climate change. By analysing secondary sources, this study reviews key trends and developments in climate action in KwaZulu Natal. It is clear from this study that integrating climate change policies into national sustainable development strategies is crucial for their effectiveness. These policies should be part of broader systems to make national development trajectories more sustainable.

Keywords

Climate change · Adaptation strategies · Mitigation strategies · KwaZulu Natal · South Africa · Sustainable development

I. M. S. Anekwe (✉)
University of the Witwatersrand, Johannesburg, South Africa

H. Zhou
University of KwaZulu Natal, Durban, South Africa

M. M. Mkhize
Mangosuthu University of Technology, Umlazi, South Africa

S. O. Akpasi
Durban University of Technology, Durban, South Africa

24.1 Introduction

The United Nations framework convention on climate change (UNFCCC) describes climate change as the variation in climate triggered by anthropogenic activities which change the constituents of the global atmosphere and the variability in the natural environment monitored within a considerable duration (UNFCCC 1992). Climate change can also be perceived as "any long-term shift in the average weather that a given location experiences caused by the high level of greenhouse gases (GHGs), namely carbon (IV) oxide (CO_2), methane (CH_4), nitrous oxide (N_2O) and chlorofluorocarbon gases (CFCs)'. However, natural forces on Earth release GHGs that shield the Earth from hazardous solar radiation while maintaining a steady temperature and climate. It is widely acknowledged that human activities, including the combustion of fossil fuels, deforestation, and

U. Chatterjee et al. (eds.), *Climate Crisis: Adaptive Approaches and Sustainability*, Sustainable Development Goals Series, https://doi.org/10.1007/978-3-031-44397-8_24

industrial operations, have resulted in a significant concentration of GHGs. These unnecessarily high GHG emissions function like an additional pane of glass in a greenhouse, preventing heat radiation to the space, resulting in higher temperatures and variations in climate that can subsequently lead to floods, droughts, increase in sea level and precarious seasons around the world (Matondo et al. 2004). Most of the current global warming and climate change can be attributed to the burning of coal, oil and gas by first-world countries, including Europe and North America, to enhance industrialisation over the last 150 years. However, people on the ground are currently facing the impacts of climate change, primarily in poorer developing nations. These are not impacts that will occur in 50 years but are occurring presently and will worsen if not remedied. These consequences are eroding the rights of millions of people to life, security, food, water, health, shelter and culture (Bose 2010). Climate change effects are being felt in South Africa just as much as in other developing nations.

During the last five decades, average yearly temperatures have risen by at least 1.5 °C above the reported global mean of 0.65 °C, and extreme rainfall events have become more prevalent. These variations are expected to be more prevalent, as predicted by the 2013 South African Long-Term Adaptation Scenarios and the Fifth Assessment Report of the Intergovernmental Panel on Climate Change (IPCC 2021). The recent report of the IPCC shows that urgent action is needed to mitigate human-induced climate impacts, which have become a permanent condition. The report states that the planet will likely warm by 1.5 °C over the next 20 years (IPCC 2023). This will inevitably lead to regular and intense extreme natural events like floods, heat waves, storms and droughts. Thus, the report further warns of the potential for devastating changes, including the collapse of some species and ecosystems (IPCC 2023). The effects of climate change are already being felt in South Africa, including increasingly distorted seasons, frequent heat waves, wildfires and droughts. The country's coastal cities are also at high risk due to rising sea levels. KwaZulu Natal, the country's second-largest province and home to several coastal towns, has recently experienced devastating weather events such as flooding that can be attributed to climate change. These natural disasters have caused extensive damage to infrastructure, significant economic losses and loss of life. Inevitably, the IPCC report emphasises the need for urgent action to address these climate-related crises by focusing on reducing greenhouse gas emissions, improving resilience to the impacts of climate change and transitioning to sustainable green energy, among others.

These devastating impacts of climate change-related dangers are widely acknowledged to rise (Wei et al. 2020. Following such incidents, municipalities in developing nations have seen severe setbacks in their hard-won economic and social success. Generally, governments cannot anticipate, monitor, reduce, and manage threats and disasters. Typically, long-term development financing is channelled to meet emergency, relief and reconstruction needs in the short term, hindering long-term development objectives (Anderson and Woodrow 2019). Floods, tornadoes and hailstorms are becoming more common in South Africa, as are snow, veld fires, storms, severe rain and winds, and drought (Faling et al. 2012). Where there is a concentration of income inequality, insufficient housing and facilities, population density, capital assets, and commercial and industrial companies, South Africa's urban areas are highly susceptible to climate-related calamities. As a result of the country's rapid urbanisation, people are being compelled to live on unstable hillsides, floodplains and sinkhole zones in more significant numbers (Kreimer et al. 2003). Almost half of South Africa's population is regularly exposed to various dangers and lives in a chronic vulnerability to risks. Despite these negative repercussions, municipalities in South Africa have made only a few efforts to prepare for the likely effects of climate change amid challenges (Joubert 2006). Municipalities tend to consider when, where and how much the local climate may change and how this will impact citizens.

South Africa is already feeling the impacts of climate change, and the effects will only worsen in the foreseeable future. The energy industry in South Africa is the biggest emitter of carbon (IV) oxide and sulphur dioxide due to an ample and low-cost provision of electricity from coal, which generates nearly 70% of the country's energy. Compared to 2000 emissions, South Africa's greenhouse gas emissions increased by 20.5% in 2015. In 2016, South Africa's GHG emissions reached 531 MtCO2e, an increase of 3% compared to the country's emissions in 2015. The transport sector in Egypt was responsible for the continent's highest carbon dioxide (CO_2) emissions. In Egypt, the transport sector released around 65,000 Gg of carbon dioxide. It was followed by South Africa with a total of 53,034 Gg, Nigeria with a total of 35,239 Gg, Libya with 18,200 Gg, Morocco with 17,254 Gg, Kenya with 7287 Gg and Ghana with a total of 6918 Gg (Fig. 24.1) (Ayetor et al. 2021). This emission contributes to global warming. The country accounted for around 1.1% of global emissions in 2005, with sub-Saharan Africa making up the majority of 40% of global emissions (Pegels 2010) and 1.5% in 2019 (Crippa et al. 2020). The per capita emissions rate nearly exceeded the European Union's average per capita emissions of 10.7 tonnes of CO_2e in 2005 (Borello et al. 2018). Energy expenditures and emissions are likely to follow a similar pattern. Emission intensity is predicted to climb as wages increase, and the government of South Africa maintains its efforts to offer global accessibility to power, assuming the current carbon intensity of energy productivity is sustained, at the very least. South Africa's average CO_2/kWh is approximately twice that of developed countries, at over 850 g CO_2/kWh. As a result, CO_2 accounts for over 80% of the country's total greenhouse gas emissions, primarily from electricity generation (Borello et al. 2018). Amidst the high CO_2 emission and after the power outage crisis in 2008, the energy industry (ESKOM) presently focused on reactivating dormant coal-fired power plants and preparing to construct additional coal-fired power plants to cater for the present and subsequent demand. During these crises, the government's priority was to keep the economy running smoothly, irrespective of the high GHG emissions—and, regrettably, minimal emphasis was placed on alternative energy sources (Borello et al. 2018).

Despite these consequences, the initial climate change study in South Africa, like in other regions, was perceived as an environmental issue instead of a developmental one. Similarly, it has taken more than a decade to acknowledge that climate change is among the most severe environmental problems and, more pertinently, that anthropogenic activities contribute directly to climate change and the increased greenhouse effect. The world is concerned that it is taking time to agree on a feasible strategy to cut global GHG emissions, as the first symptoms of a worldwide debate on climate change appeared only in the early 1990s. Therefore, the article reviews the impact and position of KwaZulu Natal province towards climate action by giving relevant context to where the province is, what the challenges are and providing different strategies for combating climate change in South Africa.

24.2 Overview of Climate Change in KwaZulu-Natal (KZN)

24.2.1 KwaZulu-Natal Geography, Economy and Climate

The KwaZulu-Natal province is located on the east coast of South Africa. It has a complex landscape, both in terms of its physical and biological diversity and the different uses and tenures of the landscape (Jewitt et al. 2015). The province's landscape covers an area of 93,307 km (accounting for 7.7% of the nation's landmass) from the mountainous regions of the Drakensberg escarpment with an elevation of over 3000 m in the west to the subtropical areas of the Indian Ocean in the east. However, the slope is not smooth, and many rocky intrusions turn the landscape into uneven steps that rise from 150 m along the coastline to 600 m and then 1200 m in the province's central region, known as the Midlands. The Highveld, or high plateau, lies beyond the Drakensberg (Jewitt et al. 2015). The KZN is characterised by warm to scorching summers, heat waves, forest fires

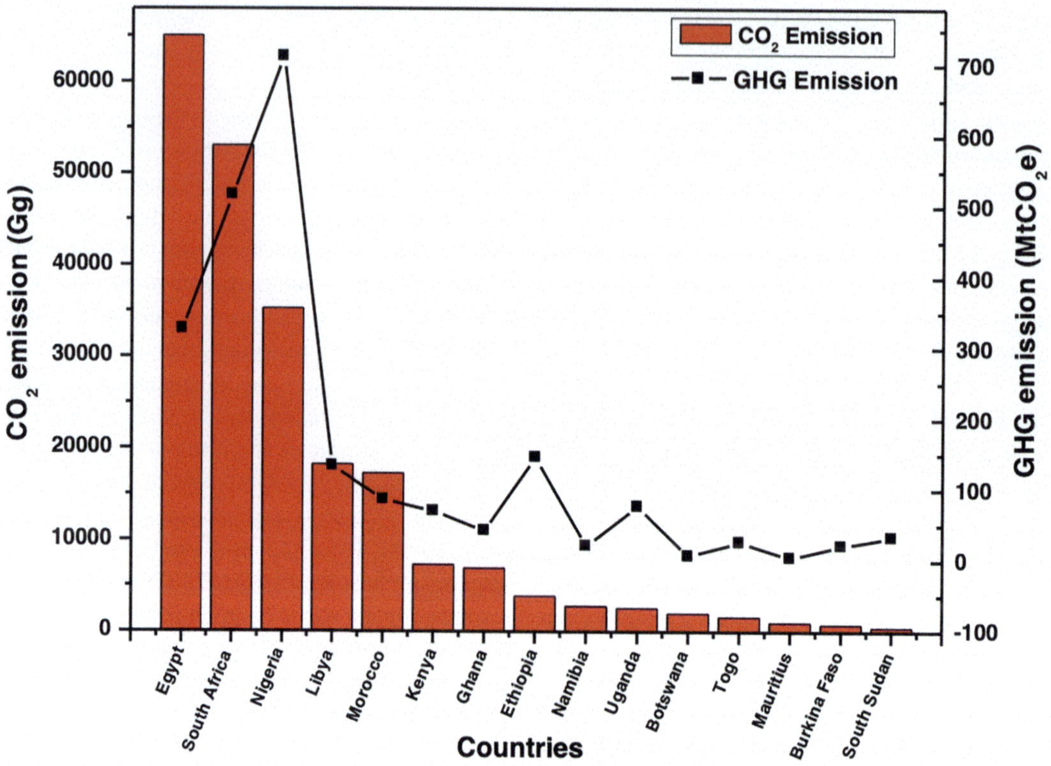

Fig. 24.1 Greenhouse gas emissions and CO_2 emissions from transportation in African countries. (Source: Ayetor et al. 2021)

and temperate winters (Ndlovu et al. 2021). According to Statistics SA (2019), KZN has the second-highest population, estimated at 11.3 million. KZN province comprises 65 municipalities, including one major city (eThekwini municipality), 10 districts and 43 local municipalities.

This province (KZN) has the second-largest economy after Gauteng, contributing R206,8 billion (16.5%) to the country's gross domestic product (GDP) (Mkhabela and Chikandiwa 2016). The manufacturing, trade business and transportation industries are the largest and fastest-growing sectors of the economy. Also, the province is a strong player in agriculture, forestry and fishing, as well as tourism and contributes to the province's GDP. KZN can boast of high export potential and the highest development level in industrialisation, with growth driven by the manufacturing industries (paper and paper products, ferroalloys and chemicals). The unique natural resource endowments in KZN, the exceptional productivity of its manufacturing industry, its well-developed infrastructure and its convenient coastal location likely contributed to the state's growth into a hub of industrial development in sub-Saharan Africa. The central economic regions of the province are Durban, Pietermaritzburg and Richards Bay, with sugar cane plantations, fruit and vegetable cultivation in the coastal belts. KZN is home to two of Africa's major seaports. In contrast, Dube Trade Port—home to King Shaka International Airport—drives economic growth and repositions the country to compete globally (Mkhabela and Chikandiwa 2016).

The climate in KwaZulu-Natal ranges from subtropical to moderate. Along the shore to inland, yearly precipitation ranges from more than 50 inches (1270 mm) to 30 to 40 inches (760 to 1020 mm). Temperatures drop from the frost-free coastal region, although they stay mild. Summers are sometimes hot and humid, with rain, while winters are mild, dry and sunny, making the coast of Southern Africa's primary vacation destination. In the winter, the Drakensberg's higher elevations endure cold temperatures and snowfall. The climate in KwaZulu Natal Province is tourist-friendly all year. Sea temperatures are also reasonably consistent throughout the year, averaging 21 °C, allowing for various water activities year-round, such as diving, fishing, swimming, boating and surfing. However, the largest and most significant metro city in the province, Durban, experiences warmth and sunshine at intervals, coupled with light, consistent rainfall. The air in Durban is dense with moisture, and the city's subtropical latitude brings lengthy, hot summers with rains and extremely moderate winters. Durban has 320 days of sunshine on average every year. In the winter, the average temperature ranges from 16 to 25 °C. During the summer, temperatures range from 23 to 33 °C (between September and April). With an average daily temperature of +/− 32 °C, Durban is the warmest city, with January being the hottest. However, the incidence of climate change in the country has caused variations in temperature and rainfall which have truncated the excellent Durban weather and negatively impacted the city and the province.

24.2.2 Conceptualising Climate Change Impact in KwaZulu-Natal

According to the Department of Environmental Affairs (2017), South Africa is prone to natural disasters such as drought, floods, extreme storms and fires. Recently, this was seen in KwaZulu-Natal, where the province was affected by a short-term drought followed by very high temperatures that affected 256 towns. Various hotspot areas such as Uthukela, Amajuba, Umzinyathi, Uthungulu, Zululand and Umgungundlovu districts of KZN were identified for these negative impacts (South African Weather Service 2020). As a result of these climate impacts, the government has realised that its development is solely influenced by what happens in its regions and the world at large and that success depends only on understanding and responding to these developments (NPC 2011). Due to its dependence on fossil fuels, the country emits excessive carbon (DPME 2019). It is also acknowledged that despite the existence of adaptation strategies, there is still a lack of urgency in implementation due to a lack of funding and capacity. While the government continues to struggle to implement climate adaptation programmes in vulnerable communities in KZN and other parts of South Africa, the industrial and other private sectors in big cities like Durban remain the main contributor to waste and water pollution and the degradation of the ecosystem. Figure 24.2 shows the flow chart of the climate change impact in Kwazulu-Natal.

Although these alterations are directly linked to global warming, historical extreme weather occurrences suggest that KwaZulu Natal is at risk, especially as the frequency of such episodes is expected to rise as the climate changes. A strong storm hit South Africa's east coast in 2007, the outcome of a low-pressure system that has been shut off and a sequence of tropical storms working concurrently. During the Equinox period in March, onshore winds generated swells of up to 10 meters, which also happened to be the highest tides of the year and an 18-year peak in the sun's gravitational pull on the planet (the highest astronomical tide). The destruction was significant throughout South Africa's east coast, totalling almost US$ 12.5 million in the eThekwini Municipality of KwaZulu-Natal province (Mather and Vella 2007). Despite the importance of large-scale catastrophes, a review of lesser incidents affecting at least 100 persons recorded according to the municipal Disaster Management Unit between 2009 and 2012 reveals how often 'small-scale disasters' are in KwaZulu Natal municipality (Roberts and O'Donoghue 2013). Storm damage is the most common event (connected to frequent and severe storms) and fire (due to the requirement to retain houses warm). They approximately correspond to the summer and winter seasons, respectively. Most of these occur in low-income, vulnerable neighbourhoods (Roberts and O'Donoghue 2013).

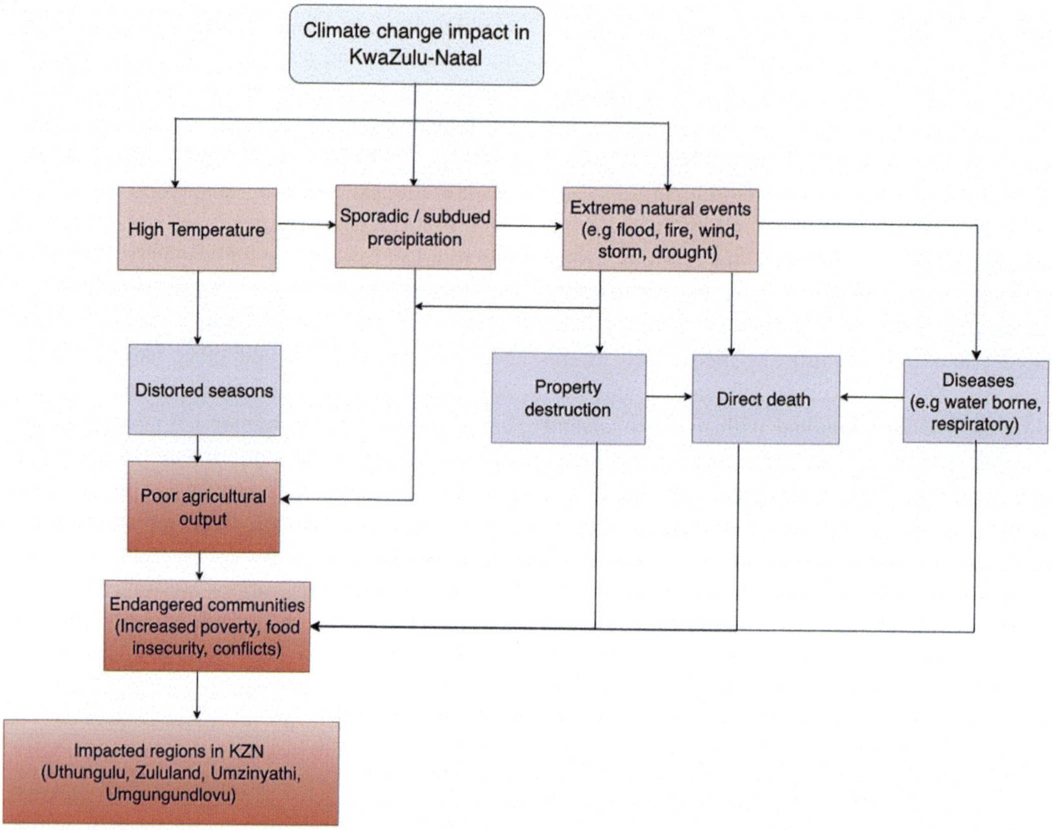

Fig. 24.2 Flow chart of the climate change impact in Kwazulu-Natal

However, even though KZN has made significant progress in terms of development since the year 1994, when democracy was established, considerable human needs and poverty persist while environmental transformation and change continue to accelerate. Consequently, the city's adaptive capacity is insufficient and must be significantly expanded. Although municipalities have been at the forefront of climate adaptation in the province, one of the climate vulnerability studies found that in KZN, 43% of farmers were not warned of the risk that extreme climate conditions could pose, resulting in a 37% decline in sugar production. In the event of climate impacts, especially droughts, 32% of farmers did not receive support from the government to rebuild their agriculture. Of all farmers affected by climate change, 62% recovered, while 33% lost their farms due to extreme weather events in the province (Reid et al. 2005). In addition to its effect on the source of livelihood, Chersich et al. (2018) reported that climate change led to extreme weather conditions, infectious diseases, mental health and specific populations in South Africa. Due to these factors, society cannot address factors that affect climate change instead of focusing on the impacts. However, climate change is not only affecting the livelihood and government. Still, it has also changed how countries look at their economic growth, as countries increasingly have to pay for their carbon emissions and explore other energy sources to move goods through markets (KZNPPC 2016).

As climate change has brought many negative impacts, such as weather changes, heat and drought, KZN province is one of the most affected provinces. For many years, several cases of climate impact have been reported in the province. These include the recent severe storms reported by the Weather Service (2020), which caused significant damage, loss of life and hundreds of homes. KZN province comprises one of the warmest areas, with the coastal regions and the Lowveld of the province (USAID, 2015). Historically, the climate in the province of KwaZulu Natal has been predominantly variable. In the province, most rainfall occurs in the austral summer, between October and March, and total annual precipitation has increased compared to autumnal winter rainfall (Ndlovu et al. 2021). In terms of flooding in South Africa, KZN has been the province with the most flooding for more than 33 years, with the south coast being the most vulnerable area. In the province, there has also been a rise in sea level for some time, the effects of which are already felt through coastal erosion (Hoogendoorn et al. 2016).

Climate change affects the entire planet, not just South Africa. This means that combating climate change requires a joint effort from multiple governments. Climate change is anticipated to raise temperatures in Durban by 1.5 °C to 2.5 °C by 2065 and by 30 °C to 50 °C by 2100, according to current projections (Roberts and O'Donoghue 2013). By 2100, annual rainfall changes are anticipated to increase by up to 500 millimetres, with another 500 millimetres by 2065. This is expected to show itself in the municipal area as an increase in extreme rainfall events, streamflow intensity and extended dry spells between rainfall episodes. Between now and the end of the century, rainfall variability will likely grow by 30% to a potential doubling. Sea level rise is already 2.7 cm per decade along the municipality's coastline in KZN (the eThekwini municipality), and it may accelerate in the future (Roberts and O'Donoghue 2016). The consequences of these modifications are likely to include a rise in both quick and slow-onset catastrophes due to (Roberts and O'Donoghue 2013): (1) A spike in the proportion and severity of cases of severe weather occurrences (floods, hurricanes, etc.), resulting in more significant infrastructure damage, higher insurance costs, as well as human well-being and safety. (2) A reduction in the amount of available water as a result of increasing evaporation and changing patterns of precipitation, which will have the most significant impact on subsistence dryland farmers. (3) A rise in river course, and erosive capability, resulting in more topsoil loss, a drop in agricultural land value and a rise in dam siltation (4) Increased coastline erosion due to elevated sea levels. (5) Increased energy usage due to increased cooling needs in the home. (6) Significant increase in property damage-related financial loss and a decrease in tourism revenue. KZN province contributes to several positive factors in terms of South Africa's development, including mitigating climate change.

24.3 Climate Action Strategies

For this study, climate action will entail stepped-up measures to minimise greenhouse gas emissions for climate change mitigation while also strengthening resilience and adaptive ability to climate-induced effects, which may include climate-related risks in all regions; incorporating climate change initiatives into national policies, strategies and planning; and enhancing education, technology and infrastructural development (Louman et al. 2019). Hence, climate action encompasses mitigation and adaptation strategies. The term mitigation has been defined by the United Nations International Strategy for Disaster Reduction (2002) as 'structural and non-structural measures to limit the adverse effects of natural hazards, environmental degradation and technological hazards', while the approach of adjusting to the existing and prospective impacts of climate change is referred to as adaptation strategies. As a result of this concept, several scientific, technical and administrative climate actions have been explored, but the question is whether the government has ever considered their uniqueness.

The current South African climate action strategies overview shows that most approaches are being controlled and implemented nationally. Hence, the provincial government has contributed little to formulating and implementing climate action strategies. This is evident in the inauguration of the presidential climate change coordinating commission in 2020 which was part of the resolution of the presidential job summit held in

October 2018 to constitute a statutory body that will cater for the transition towards a low-carbon economy in South Africa (Ziervogel and Parnell 2014). In South Africa, there are several mitigation studies have been conducted, such as Ziervogel et al. (2014), Wright et al. (2019) and others. In KwaZulu Natal, only a few mitigation studies were available. They are only related or general studies on climate change, which include many chapters on climate, such as Roberts and Parks (2009), Roberts et al. (2016), Ndlovu et al. (2021), Sithole (2016), Abraha and Savage (2006), Basdew et al. (2017), Strydom and Savage (2017), Mabhaudhi et al. (2019) and Hitayezu et al. (2014) among others. The studies were all conducted in different fields, such as environment, environmental planning, geography, agriculture and water science. The findings of all these studies are different, but their focus is on the status quo of climate change in the province and the country in general.

Several studies have also shown that South Africa must urgently develop appropriate climate action strategies. Even though the government is the leader of society in addressing climate change, stakeholders remain very important in the meantime. According to Ziervogel et al. (2014), countries in the Organisation for Economic Co-operation and Development (OECD) have already established dedicated support agencies internationally to help address climate impacts and implement policies. In contrast to this view, the government argues that as a government, it strives to involve as many stakeholders as possible in its policy-making processes (DEA 2017). Stakeholders such as civil society help shape policy in areas such as biodiversity protection, energy and climate change (The Presidency 2014). In terms of climate change mitigation, South Africa is a signatory to several global strategies internationally, including the 1992 United Nations Framework Convention on Climate Change, the 1997 Kyoto Protocol and the 2016 Paris Agreement, which set out the Intended Nationally Determined Contribution (INDC) (South African Civil Society Working Group 2019). Although South Africa has enacted numerous laws and policy frameworks to address the climate problem, it is acknowledged that the process has been too slow. In its 2019 Voluntary National Review of the SDGs report to the UN, the government recognised the importance of the Paris Climate Agreement by understanding and supporting it through the enactment of legislation and policies that include a carbon tax, renewable energy, clean public transport, energy efficiency, waste management and a land restoration initiative (South African Government 2019).

According to the KwaZulu Natal Department of Cooperative Governance and Traditional Affairs (2019), despite this annual rainfall increase and flooding prevalence, the KwaZulu Natal government has already enacted numerous measures and guidelines to help builders prepare for flood events and better manage storm water runoff in South Africa. The Ethewkini municipality is a leader in climate adaptation among the many existing municipalities in the province. Several activities in the municipality are proposed at the national level aimed at climate adaptation. However, other municipalities have also adopted some climate plans to address climate issues. According to Roberts (2010), Ethekwini Municipality has started planning and implementing climate change activities in the province since 2004. While many frameworks have been enacted at the national level, these have been implemented at the provincial and local levels as part of an attempt to embrace and mitigate climate change. The South African government (2019) reports that the country has so far implemented a comprehensive set of strategies, policies and sectoral plans across key economic sectors to reduce greenhouse gas (GHG) emissions. These strategies or policies include the Integrated Resource Plan (IRP), the Energy Efficiency Strategy, the Industrial Policy Action Plan (IPAP), the Green Transport Strategy (GTS), the Climate Change and Mitigation Plan (CCAMP) for South African Agriculture and Forestry and the National Waste Management Strategy. There have been mixed reactions to these strategies in terms of whether the government is on the right track in tackling climate change. Many feel they are insufficient, but some have made further suggestions to the government on combatting climate change.

In its second Annual Climate Change Report for South Africa, the government outlined numerous efforts to implement strategies to mitigate climate change (DEA, 2017). These include South Africa's regular reporting of its Nationally Determined Contributions (NDCs). It also appears that the country, through its government, is measuring its efforts to implement climate strategies through a monitoring and evaluation system that helps to measure, report and verify. This system goes beyond the objective of assessing the impact of climate change and the effectiveness of policies in South Africa (DEA 2017). While this may be the case, in the meantime, the government remains interested and committed to supporting measures to reduce carbon emissions and limit the resulting climate change (NPC 2015). The Annual Climate Change Report has produced a variety of recommendations for the government to enact climate change legislation. In addition, the South African Low Emission Development Strategy was developed to recognise the contribution of local governments to national efforts to mitigate climate change (Urban LEDS Project II, Local Governments for Sustainability (ICLEI) and UN-Habitat 2020). It is also recommended that government move away from its dependence on fossil fuels and economic reliance on extractive industries such as mining (South African Civil Society Working Group 2020). Thus, much is naturally noted about the government's contribution to climate change; like many developing countries, the government of South Africa is at the forefront of society when it comes to contributing to global problems such as climate change.

24.3.1 KwaZulu Natal Climate Actions

The government of KZN has developed numerous action plans to combat the effect of climate change in the region. These include the Green Economy Strategy (GES) and Durban Climate Change Strategy (DCCS). These action plans are discussed below.

24.3.1.1 KZN Green Economy Strategy (GES)

The transition to a green economy in KwaZulu-Natal is a significant economic growth and job creation opportunity. The aim is not to create a new sector but to transform existing supply chains and production systems to increase resource efficiency, use renewable resources and develop new environmentally friendly products and services (Hlahla et al. 2016). This is necessary to build resilience in the economy and address environmental and social thresholds. The KZN Department of Economic Development and Tourism (DEDT) has taken a participative and capacity-building approach to research and develop a KZN Green Economy Strategy. This included establishing an active 'Community of Innovation' to ensure that the strategy is responsive to the conditions, opportunities and constraints in KwaZulu-Natal. The strategy contains simple goals and objectives (Table 24.1) that the provincial government can work towards, implementing short-term and medium-term interventions. Key Performance Indicators have been developed to measure the success of the strategy's implementation. The KwaZulu-Natal government must champion the implementation of the strategy objectives by developing project and implementation plans that allocate responsibilities, funding requirements and sources (Sutherland et al. 2011). This next step is crucial in transforming the work on unlocking the green economy in KwaZulu-Natal into action.

24.3.1.2 Durban Climate Change Strategy (DCCS)

In 2015, the eThekwini Municipal Council approved the Durban Climate Change Strategy (DCCS) as the city's inaugural cross-sectoral plan to address climate change. The Durban Climate Change Strategy is a comprehensive climate action plan developed by the KwaZulu Natal (KZN) province in South Africa. The strategy aims to tackle the adverse effects of climate change by reducing greenhouse gas emissions and increasing resilience to climate change impacts in the province (EThekwini Municipality 2014; Diga 2017). The strategy is based on four

Table 24.1 Goals and Objectives of KZN Green Economy Strategy (GES) (Sutherland et al. 2011)

Goals	Specific objectives
Goal 1: leverage the green economy by greening government investments	Develop a provincial green infrastructure policy to address the infrastructure development programme
	Developing a greening provincial buildings programme to retrofit energy and water efficiency measures
Goal 2: unlock the green economy through turnkey/pilot projects	Create a green economy fund to facilitate investments in pilot projects
	Establish Centres of excellence in research, development and implementation of GES
Goal 3: create enabling conditions for the development of the green economy	Develop a renewable energy resources map to identify potential renewable resources
	Create a registry of waste resources to link producers and users
	Investigate and develop payment for ecosystem services pilot projects in rural KZN to address community-climate adaptation
	Integrating energy, water and food security into rural developments to get target poverty reduction and climate adaptation

key objectives: Mitigating Greenhouse Gas Emissions, Adaptation to Climate Change, Collaboration and Partnership, and Governance and Implementation. The Durban Climate Change Strategy includes several actions and targets to achieve these objectives. For example, the strategy aims to reduce greenhouse gas emissions from the energy sector by 40% by 2030, increase the use of public transport by 50% by 2030 and restore 10,000 hectares of degraded land by 2030. The strategy also includes targets to increase the resilience of vulnerable communities to climate change impacts, such as providing climate-resilient housing and infrastructure and promoting sustainable agriculture. The strategy recognises the importance of stakeholder engagement and participation in achieving its objectives. The KZN government has committed to engaging with various stakeholders, including communities, civil society organisations, businesses and academic institutions, to develop and implement climate change initiatives. The strategy also includes a plan to raise awareness of climate change among the public through education and outreach campaigns. In terms of governance and implementation, the strategy includes several measures to ensure that this strategy is effectively implemented. The KZN government has established a climate change unit to oversee the implementation of the strategy and ensure that it remains aligned with national and international climate change policies (EThekwini Municipality 2014; O'Donoghue et al. 2022). The unit will also monitor progress towards the strategy's targets and report on progress annually. The strategy also includes provisions for monitoring and evaluating individual initiatives to ensure that they deliver the desired outcomes.

The 2022 Durban Climate Change Strategy (DCCS) report highlights the significant progress in addressing climate change in the city (EThekwini Municipality 2022). The report reinforces the Durban Climate Change Strategy's commitment to reducing greenhouse gas emissions by 2030. Additionally, Durban has increased its climate resilience through adaptation measures such as flood management systems, water conservation and the development of climate-smart infrastructure. The city has also invested in green spaces and urban forests to enhance carbon sequestration and promote biodiversity. The report identifies several ongoing initiatives and future plans to advance further climate action in Durban, such as promoting a green economy, developing sustainable buildings, expanding public transport and integrating climate change considerations into urban planning. The report suggests that Durban remains committed to continued action and progress towards a sustainable and resilient future (EThekwini Municipality 2022).

24.4 Challenges in the Implementation of Climate Action

Challenges have hampered the implementation of climate action in South Africa. Numerous recent efforts have been made from all sectors of society, led by the government, and of course, shortcomings remain in the long run. There is an in-depth critical question on the extent of damage caused by identified challenges, that is, whether the country is on the path to mitigating climate change. The literature has a mixed assessment of the government's response to implementing climate change mitigation strategies. It is argued that there is still a long way to go but that the government may be able to achieve this by 2030 if it changes its attitude and addresses all challenges. Although the government has taken further action in collaboration with various stakeholders, it acknowledges that, like many other developing countries, it still faces some limitations in its climate change strategies due to the effects of climate change (Chersich and Wright 2019).

The factors influencing the successful implementation of climate change mitigation are different; it just depends on which side you look at. Since climate change is a broader area of research, scholars study it differently. In South Africa, studies reveal several institutional barriers to climate change mitigation strategies implementation by the national, provincial and local governments. These factors that hinder climate change implementation are wide-ranging. However, a few were highlighted in the report prepared by Averchenkova et al. (2019), in which they found that (i) lack of policy direction, coherence and coordination, (ii) policy complexity, (iii) inability of human and financial resources, (iv) limited information and (v) lack of public and private sector engagement and consultation are among the key factors that negatively affect the success of climate change implementation, in South Africa. Moreover, these factors lead to progressive climate change and bring serious negative factors that make it impossible to address climate change thoroughly.

24.4.1 Technical Barriers

In literature, technological barriers are one of the major challenges to climate adaptation in South Africa (Kithiia 2011). Only the techniques and adaptability options have been created (Kolikow et al. 2012), and appropriate for local settings can be used by households. Climate adaptations depend on technological advancements, including crop variety development, early warning systems and irrigation techniques. In households in Kwazulu Natal and South Africa, where technology innovation is restricted, accessibility to the entire spectrum of adaptation choices available in the developed world will be restricted. Consequently, the limitation of advanced technologies in South Africa may limit households' (in Kwazulu Natal province) ability to adapt to and use some of the more recent advancements in technology that could enhance food security and related livelihoods by developing early warning systems. It is critical to underline that technical climate adaptation options must be implemented within a legal, economic and institutional framework, which most sub-Saharan African nations lack (Klein et al. 2001). According to a study funded by the UK government on Africa's climate, one of the significant barriers facing climate change research in Africa is a lack of technical knowledge in climate modelling (Washington et al. 2006). A technological gap is also linked to African countries' insufficient financing for agriculture and climate change research (Enete and Amusa 2010).

24.4.2 Scientific Barriers

Inadequate scientific modelling methodologies hinder adaptation assessments in the medium and long run. There is a lack of an effective system that offers significant spatial climate data for South Africa. The latest premium, nationally gridded climate data are from 2000. Current national hydrological data are becoming extremely challenging and expensive to get or develop via modelling. This is a significant impediment to impact modelling in South

Africa. Some sectors, especially biodiversity, agriculture, urban settlements and human health, rely on climate and water as critical resources for impact modelling. In the medium and long terms, inappropriate impact modelling approaches limit adaptation evaluations. In the biodiversity sector, for instance, it is unclear how insects, animals, wildfires and species dispersal rates may adapt to the temperature variation and rainfall patterns or how CO_2 levels may relate to climate-yield reactions such as soil nitrogen and ozone. Increasing knowledge of how the present climate situation affects the human source of living and other sectors has not been represented by an adequate understanding of what climate change may imply for potential well-being. For instance, industrialisation leads to rapid urban densification in cities like Durban, Johannesburg and Cape Town. Nevertheless, the accompanying consequences of increased population densities, infrastructure and destructive weather patterns have not been adequately investigated. There has been little progress in effectively budgeting adaptation strategies and identifying appropriate financing solutions, a significant impediment to implementation.

Many knowledge gaps persist irrespective of almost two decades of climate and impact modelling in South Africa. This affects the prediction chain, which starts with climate scenarios and biophysical repercussions and ends with analysing socio-economic vulnerability and adaptation measures (New and Hulme 2000). The lack of climate scenario products, poor and divergent climate data, insufficient impacts modelling techniques, improper procedure evaluation, impoverished discoverability between consequences evaluations and the climate scenarios, inappropriate socio-economic and vulnerability evaluations and a lack of cross-sectoral incorporation in climate change adaptation partly causes deficiencies. Since some of these shortcomings are irresolvable within realistic timelines, the possible implications of climate change will remain uncharacterised. Hence, the incidence of climate change will aggravate the severe scarcity of water resources, intensify desertification and may affect the volume, timing and distribution of storms that cause flooding evident in some parts of the country like KwaZulu Natal province. Furthermore, the absence of an accurate strategy that can be accepted as 'best practice' makes it impossible to distinguish uncertainties caused by underlying scientific modelling difficulties from those caused by the evaluation process.

In South Africa, integrated scientific assessments have only been carried out in a limited way. These evaluations have the potential to be significant since they combine data from a variety of sources on climate change impacts, vulnerabilities and adaptation objectives. These cross-sectoral methodologies help investigate some of the links between sectors and regional shortcomings. For many adaptation responses, agricultural and water industries, for example, must be evaluated together. One example of research that integrates these two topics is examining the water-stressed Sandveld in the Western Cape region. The study focused on the consequences of climate change on groundwater recharge, which is critical to the agricultural industry (Bourne et al. 2012). Adaptation will necessarily contend with uncertainties regarding climate and its implications and the social, political and economic contexts in which the impacts and solutions will occur (Prosser et al. 2021). This has yet to be represented in South African adaptation research and practice. The development of approaches that support 'robust', 'resilient' or 'adaptive' decision-making frameworks that take into account uncertainty and avoid lock-in to specific scenarios is an essential field for collaborative study (Weaver et al. 2013). The ability to extrapolate case study adaption finding's locations to the national level and provide cross-sector reviews is still limited. This applies to situations, especially emphasised 'climate-resilient development' in South Africa's National Climate Change Response White Paper. The institutional architecture that encourages adaptability and the ability to deal with uncertainty, complicated system feedback and non-stable states is little known.

24.4.3 Administrative Barriers

As previously stated, higher-level contingent barriers might hinder local adaptation. According to various policy-oriented research, political and institutional inefficiencies in South Africa have hampered adaptation opportunities (Magadza 2000; Masters and Duff 2011; Toteng 2011; Wilk et al. 2013). Other scholars argue that climate change and adaptation are frequently overlooked and have low political value to merit politicians' attention and election cycle focus (Kalame et al. 2011; Leck et al. 2011; Toteng 2011).

Due to various South Africa's substantial reliance on fossil fuels, Madzwamuse (2010) claims that the country's sluggish reaction to climate change adaptation is due to persistently dominant mitigation rhetoric. A scarcity of political supporters to stand up for and promote their interests is blamed for the terrible absence of political advocates to speak for and represent their demands and objectives. This claims that to go forward with climate change adaptation, South African governments and the donor community must better grasp the impediments to adaptation. Governments investing in adaptation activities may be hampered by insufficient monitoring of adaptation progress and beneficial effects, according to Berrang-Ford et al. (2011), while Ziervogel and Parnell (2012, 2014) illustrate South Africa's lack of national support for local adaptation. Several commentators have identified hurdles to suitable national policies and frameworks that would promote local-level adaptation in South Africa as a lack of local involvement in policymaking, the omission of social and cultural background, as well as the unintentional undermining of local coping and adaptation techniques (Brockhaus and Djoudi 2008; Dkamela 2011; Eriksen and Marin 2011).

At the local level, government policies, laws and regulations may act as roadblocks to adaption. Short-termism in politics makes it difficult to plan for long-term climate change (Shackleton et al. 2014). According to Conway and Schipper (2011), government discourses and policies that focus on temporary food insecurity and relief and take a disaster-oriented approach to climatic variability can be a drawback to a longer-term strategy that prioritises long-term adaptation, resilience and livelihood security. They also suggest that, as shown in Ethiopia's government, climate change is marginalised because it is perceived as an environmental problem instead of a broad developmental concern. Posing a stumbling block, this is replicated in Botswana (Toteng 2011) and South Africa (Leck et al. 2011; Ziervogel and Parnell 2012, 2014). Indeed, separating adaptation strategy from broader development discourses, programmes and strategies in South Africa conceals the interrelationships between generic and specialised adaptable capability and stifles local-level adaptation (Stringer et al. 2009). However, as noted by Stringer et al. (2009), environmental change narratives that are isolated from the lives and experiences of local people may become discursive obstacles to adaptation support. Factors like bureaucratic inertia can also impede long-term adaptation development. Brockhaus and Djoudi (2008) believe that 'adaptation in the socio-political subsystem lags behind the autonomous reactive adaptation, obstructing the transition from the latter to planned adaptation through reflective and strategic decision-making processes'.

The political environment can also stifle self-adaptation; Stringer et al. (2009) also emphasise how national initiatives typically collide with independent local adaption efforts and how poorly developed mutually supportive links between them can act as a roadblock to adaptation, mainly when cultural, traditional and context-specific factors are ignored (Patt and Schröter 2008). Other types of political barriers specific to South Africa include a failure of government financial support for local planning and research, political rivals to adaptation, inadequate interdepartmental communication and integration, competing priorities (mineral-energy complex), an absence of adaptation mandates and legal constraints (Chevallier 2011).

24.5 Prospects for a Sustainable Climate Action

24.5.1 The Implementation of the Renewable Energy Feed-In Rate (REFIT)

Efforts by the governments of South Africa to stimulate private investment in sustainable technologies already offer potential. However, most countries have a large gap between policy declarations and actual implementation. The implementation of the Renewable Energy Feed-In Rate (REFIT) and recent tariff increases—prompted by financial need rather than environmental concerns—may signal a shift in the trend. Furthermore, as forecasted by the Long-Term Mitigation Scenario (LTMS), these measures will not be sufficient to minimise South Africa's emissions by 30–40% by 2050 (Pegels 2010).

The renewable energy feed-in-tariff, REFIT, is also an option; a well-known renewable energy subsidy is threatened. The higher the consumer's additional fees, the more renewable capacity is created. Given the significant rise in electricity rates, any additional burden on customers may not be acceptable. As a result, the REFIT may only be feasible for modest amounts of electricity generated, a political fact that could render the entire concept ludicrous. Another major stumbling block to the scheme's success is the absence of demand certainty. To overcome this stumbling block, Eskom should be compelled to purchase electricity from renewable energy sources; alternatively, other businesses must be allowed to participate as buyers. The cost–benefit analysis between the need to raise electricity prices and the government's intention to supply low-cost access to electricity, particularly to people experiencing poverty, is unquestionably complex. Such challenges must be carefully addressed in a comprehensive yet detailed manner. During the planning and execution phases, they require substantial human and financial resources. Capacity and knowledge expansion on all levels and the hunt for new financing sources are critical to addressing this setback (Heun et al. 2010).

On a structural level, the LTMS has demonstrated the disparity between science's proposed amount of mitigation and present planning capabilities. Only 76% of the mitigation gap is closed based on concrete action plans in the most extreme scenario. A fraction of the work required should be produced with yet-to-be-released technology and through behavioural change, a somewhat ambiguous term. This fundamental shortcoming must be addressed. Other policy choices, such as standards or regulations, exist in addition to a CO_2 tax that could be considered in future strategic planning. From a different perspective, regulatory authority or policy actions are being taken, which must be consistent with current policies. Several government arms must coordinate their efforts to avoid contradictory methods, including a competitive feed-in tariff and a competitive power procurement process. In the past, this was not always the case (Pegels 2010).

As a result, a solid strategic approach is essential. However, even the best mitigation scenario is meaningless if it does not result in action on the ground. A systematic approach is now required for the feasibility of LTMS mitigation alternatives. A step-by-step implementation approach is necessary for the most demanding practical LTMS mitigation options (Edkins et al. 2009).

24.5.2 Installation of Solar Water Heaters

As part of its demand-side management effort, the energy industry (Eskom) attempted to encourage solar water heaters, a cost-effective incentive that should be widely adopted. However, actual progress towards the modest goal of installing a million solar water heaters in 3 years is slow. One of the key constraints is the testing and installation capacity. The South African Bureau of Standards (SABS) requires all solar water heaters installed under the Eskom programme to pass a test. Testing takes excessive time to fulfil demand because only one test rig is available. Another obstacle is installation capacity, which is limited due to a scarcity of competent workers for expert heater installation. Inadequate local capacity is a barrier that affects the South African renewable energy sector at every educational level. It has a more substantial influence on the development of advanced technologies such as concentrated solar power. Any implementation on a commercial scale will necessitate the development of country-specific technology solutions in South Africa, such as systems for conserving solar water energy centred (Edkins et al. 2009).

In South Africa, most of the barriers to renewable energy adoption, including the installation of solar water heaters, and REFIT increased customers' financial burden, expensive grid extension investment costs, the requirement for further education and research and external financing may be used to tackle issues such as risk cover for early-stage technology. One of the objectives of the international climate change discussion is to reduce greenhouse gas emissions to provide channels for technological and financial support of mitigation actions in the country.

24.5.3 Developing Public Awareness and Acceptance

The broad public understanding of environmental issues is low in many developing countries. This is especially true in the poorer parts of society, where, unsurprisingly, day-to-day difficulties are more pressing than the effects of climate change projected in the decades to come. On the other hand, renewable energies may bring solutions to difficulties that arise in the poor's daily life. They can, for example, be used to provide electricity to isolated rural locations. Off-grid renewable energy technology could be a viable alternative because connecting these regions to the national grid is costly. Governments may also point to additional benefits of renewable energy, including enhanced health due to lower air pollution, job development in expanding markets and increased energy security. In addition to conducting its public awareness efforts, the government may urge civil society organisations to participate in creating public knowledge of environmental problems and the advantages of renewable energy. This could boost public support for programmes encouraging people to transition to alternative energy sources (Edkins et al. 2009).

However, winning public support for costly renewable energy initiatives in developing countries like South Africa is complex. Transformation, particularly poverty reduction, is the principal governmental goal in most countries. Any contradictory goal will be rejected by the people, and politicians who fail to see this will be unlikely to be reflected. The estimated cost to society is thus one of the major impediments to greenhouse gas reduction initiatives. This cost locks up financial resources that cannot be used to help people out of poverty. Furthermore, the costs would be absorbed nationally, while incentives would be distributed at provincial and local levels. South African governments are so discouraged from implementing mitigating strategies. It is consequently critical that environmental goals be linked with development goals or if this is not an alternative, that the detrimental short-term consequences of environmental measures be neutralised. This needs rigorous preparation, in many circumstances, and significant financial resources, both of which may require a systematic approach (Edkins et al. 2009).

24.5.4 Nationally Appropriate Mitigation Actions (NAMAs)

A directory of nationally appropriate mitigation actions (NAMAs) has been proposed under the United Nations Framework Convention on Climate Change (UNFCCC) to minimise the emission of greenhouse gases. NAMAs are designed to be country-driven and flexible. NAMAs include feed-in tariffs for renewable energy, energy efficiency requirements and a national carbon tax. The concept of policies and initiatives for sustainable development may also be included in NAMAs (SD-PAMs). The Sustainable Development Policies and Measures (SD-PAM) methodology acknowledges that economic development is a higher priority for many developing nations than reducing greenhouse gas emissions, as Winkler et al. (2002) reported. Access to power, poverty alleviation and an increasing desire for mobility are all pressing issues in these countries that overshadow climate change mitigation concerns. In the case of South Africa, the government's electrification initiative has improved electricity access while increasing emissions. The SD-PAMs aim to overcome the tension between a country's development goals and the requirement to reduce greenhouse gas emissions. They are 'policies and actions done by a country in pursuit of its domestic policy objectives but tailored to pursue a lower-emission path to those objectives', as noted by the World Resources Institute (Baumert and Winkler 2005). As a result, developing countries like South Africa would begin by setting long-term development goals as their principal objectives. Then, they would determine what policies and actions they could use to make this development route more sustainable. In most circumstances, this would entail a more effective use of renewable energy. As a secondary effect, emission reductions would be highlighted. However, the focus of policy would be on development.

SD-PAMs' unclear environmental consequence is, however, a possible vulnerability. Suppose the 2 °C limit is not to be surpassed. In that case, drastic emission reductions are essential. Current scientific findings indicate that future cooperative efforts will have to be much stronger than in the past, with an early start. SD-PAMs may thus be a feasible component of a multilateral accord, but they will be insufficient without a specific emission reduction or limitation aim. The amount of money needed to switch whole development trajectories to low-carbon growth is significant. Mitigation measures must consequently try to mobilise private financial resources, both domestic and international. Compared to business-as-usual approaches, public financing could be utilised as a source of leverage to cover the increased expenses inflicted on civilisations.

24.6 Future Work Consideration and Recommendations for Effective Climate Action

South Africa is making tremendous advances in understanding the implications of climate change and beginning to adopt and assess adaptation measures at the research and applied levels. Notwithstanding this, much work remains to be done in two distinct parts: (1) Institutional barriers make it more challenging for public and private sector institutions to effectively collaborate and work together to satisfy the country's adaptation needs, and (2) knowledge gaps related to inadequate impact assessments and quantification of the socio-economic costs of climate change.

24.6.1 Addressing Institutional Barriers

The policy framework in which local government operates, which is generally imposed by higher levels of governance, such as provincial and national regulations, results in institutional/regulatory barriers (Burch 2010a; b; Measham et al. 2011). The disparity in responsibility between different levels of government and between government and parastatals presents some issues. Given the potential for climate change to wreak havoc on core constitutional rights, local governments must identify potential climate-related risks and ensure that these hazards are factored into their decision-making. However, the intricacies of this responsibility are not covered by the statute, and it is also shared in complex and ambiguous ways with other areas of government and parastatals. The issues of responsibility, authority and enforcement for environmental management are rampant with ambiguity, as evidenced by various findings, which show that ecosystem management is not adequately integrated into municipalities, which is exacerbated by the fact that respondents do not see environmental measures as a viable solution to climate change adaptation.

Moreover, the legislative framework in South Africa that impacts climate adaptation (e.g., planning, natural resource management and coastal management) does not expressly incorporate climate change and variability, which affects the local government's legal obligation to adopt. Other authors (Burch 2010a; b; Mukheibir et al. 2013) have expressed similar dissatisfaction with inconsistencies in climate change laws at various levels of government. Municipalities are not required to adopt mitigation or adaptation steps, nor are they needed to have a person or a department do so. With little flexibility, local governments are frequently forced to carry out policies established at higher levels of government (Næss et al. 2005). As a result, a lack of national and state attention to climate change indicates a lack of attention to climate adaptation at the local level (Amundsen et al. 2010).

If the government is serious about mitigating the climate problem, it must engage all government and society stakeholders. There are several stakeholders at the local level that have an interest in helping to reduce climate change. If climate action is to be successful, it must be implemented at the local level via provincial administration. Stakeholders at the local level should be trained to play their roles and responsibilities to ensure adequate and effective climate action (South African Civil Society Working Group 2019). However, what seems to be lacking is active leadership in promoting these partnerships. The government is at the centre of this. It must review its existing climate change partnerships to determine their relevance. In addition, formulated indicators to measure such collaborations could help determine whether these multistakeholder engagements are still relevant. Baush and Koziol (2020) believe that the local level is critical to successful climate change mitigation strategies if governments are to mitigate climate change. In the study by Mkhize (2021), multistakeholder partnerships were important in localising global agendas. However, it was recommended that local government institutions and communities must take a solid position to strengthen these partnerships if they are to work to implement the 2030 Agenda. Therefore, the government must work with various key stakeholders to address the factors and impacts of climate change. Although both aspects are essential, the adverse effects of climate change need to be addressed first, with a focus on mitigation.

In the study conducted in KwaZulu-Natal, Sithole (2016) also pointed out that climate change mitigation comes with budgetary and economic complications. Therefore, government institutions like Ethekwini Municipality must develop a cost–benefit model to prioritise climate change mitigation. Sithole (2016) claims that there is an urgent need for support from the international community and the involvement of all stakeholders to improve climate adaptation.

From the international perspective, the South African government can adopt the UK's Climate Change Risk Assessment (CCRA) to facilitate incorporating climate studies into larger development concerns through integrated evaluations (Muller 2007; Ziervogel et al. 2010). The UK's Climate Change Risk Assessment is a worldwide example of such a methodology (Dickson et al. 2012). The CCRA tends to compile available information on climate consequences for over 100 risks (highlighted from a list of roughly 700 entries). The risks will be chosen from several industries depending on the magnitude of the impact and confidence in the evidence foundation. The study focused on the most severe dangers, their relationships and how risks differed by geography. This is currently the evidence base for the UK government's adaptation policy.

24.6.2 Addressing Key Knowledge Gaps

While significant advancements have been made, and a handful of current and practical approaches are addressing some of the highlighted gaps, the knowledge base is insufficient, and much more study is needed. Bridging knowledge gaps and tackling the need to enhance and improve research policy and decision-making procedures is critical. Addressing gaps and eliminating hurdles will necessitate deliberate and collaborative work on the part of all appropriate authorities, departments and multi-sectoral organisations at various governmental levels. There is a need for more research and intervention to close knowledge gaps and resolve the key challenges or barriers through a multi-stakeholder procedure that is organised and participatory, as well as to enhance the understanding and create an enabling environment at different levels to develop climate adaptation further. Apart from tackling the gaps and demands, it is critical to recognise that climate change adaptation is a multifaceted, long-term and ongoing cycle that will necessitate changes to methods, resources, programmes and strategies to meet best the new conditions (Rizvi and van Riel 2018). As a result, all stakeholders involved must be flexible and committed over the long term.

The concept of managing knowledge (i.e. creating, preserving and sharing it) as well as performing research can serve as an effective tool for bridging the knowledge gap. Action learning is critical in sustaining adaptation strategies throughout the adaptation process, and it must be strengthened and maintained to successfully direct initiatives, influence legislation, promote cooperation and integration among organisations, the private and public sectors, and countries, deepen our understanding, and raise awareness. The available (stored) knowledge will offer insightful information on present gaps linked to ecosystem-based and climate change adaptation, which will be used to enlighten key stakeholders and other professional groups involved in the adaptation process (from planning to actual implementation and integration). Furthermore, it may be valuable in guiding and strengthening adaptation programmes. These approaches also entail the collaboration of researchers, stakeholders and policymakers in formulating research requirements. As a result, policymakers and decision-makers can make the best adaptation recommendations based on prevailing uncertainty.

To inform policy/decision-making, improving and establishing information-sharing mechanisms that allow for greater incorporation, uptake and exchange of data and knowledge is essential. Information-sharing mechanisms and services (i.e. Internet global/national knowledge systems, portals and forums) must collate and adequately handle a large amount of available information, data and knowledge products surrounding climate change adaptation. This will increase information exchange, cross-fertilisation and learning by making existing knowledge immediately accessible to multiple user organisations and the public. In terms of data and knowledge products, such a network will also facilitate identifying and evaluating what is currently known, the effectiveness of knowledge products, and what is still necessary for effective impact modelling. Researchers and policymakers should compose the obtained information and knowledge products in an easily accessible style suited to diverse target groups to reach a broader audience (Rizvi and van Riel 2018).

Furthermore, in bridging the knowledge gap associated with climate change adaptation measures, capacity building should be a part of any adaptation plan for efficient, multi-stakeholder, collaborative networks that integrate scientific and local, traditional knowledge, to increase public awareness, to generate and disseminate knowledge continuously, to improve learning and to ensure that required resources and funding are allocated despite insecurity. Uncertainty will continue to play a role in adaptation decision-making. As a result, they should be recognised and adequately considered during the adaptation process to aid decision-making. Researchers should share their concerns with practitioners, policymakers and decision-makers and offer advice on coping with them. It is also necessary to have a high degree of cooperation and engagement among scholars, practitioners, policymakers and decision-makers. A robust adaption solution will be identified through an iterative procedure of action learning, assessing and revising plans as needed (Groot et al. 2014).

Given the ever-changing environment, improving the Monitoring and Evaluation (M&E) system is critical and ensuring their continuation in justifying adaption efforts, such as projects, policies and programmes. This is critical for making decisions and implementing effective adaptation strategies. Moreover, conducting M&E will aid in creating an evidence base for climate change adaptation. M&E also offers a valuable way to improve 'learning by doing' and the ability to adapt remedial measures to the dynamic environment. As a result, it should be included in developing any adaptation policy or strategy. M&E actions might occur during the adaptation process and after implementation (Tschakert and Dietrich 2010; Andrade Pérez et al. 2010).

24.7 Conclusion

Climate change has contributed to environmental degradation, and the problem is gradually taking centre stage as it threatens the sustainability of the ecosystem and the existence of life on Earth. The present study revealed that KwaZulu-Natal had been tremendously affected by climate change. To this end, the government of KZN has implemented diverse climate action strategies. The Green Economy Strategy (GES) focuses on facilitating a transition to a sustainable economy, while the Durban Climate Change Strategy (DCCS) aims to reduce greenhouse gas emissions and enhance the resilience of Durban city to climate change impacts. Thus, both of these strategies involve setting goals and objectives to guide government actions and engaging stakeholders to garner support and enhance capacity for climate action. Given the current state of the climate crisis, it is vital to establish more viable and strengthen the existing mitigation and adaptation strategies as soon as possible. It is critical to emphasise that there is no ultimate remedy to climate change and that all the approaches mentioned in this study should be implemented if they are economically practical. Mitigative and adaptative solutions that lessen susceptibility to current and future climate variability and change are crucial in the context of KZN and South Africa's urgent socio-economic development requirements and threatened ecological services. Minimising the so-called 'mitigation deficiency', or one's exposure and susceptibility to climate variability and observable change, is essential to long-term adaptation measures. National responses should focus more extensively on integrated development pathways as climate change is redefined as a development challenge in South Africa's new National Development Plan 2030. Several stakeholders and government sectors from all three levels of government—national, provincial and local should be included in the formulation and implementation of climate change policies and plans. This chapter highlights the importance of research to address knowledge gaps, plan adaptation measures and formulate effective climate change policies in South Africa.

Acknowledgement We want to thank the Climate and Environmental Action Group (CEAG), South Africa for their invaluable support in providing the resources necessary for this study. Their contribution has allowed us to conduct this research, and we sincerely appreciate their assistance. Thank you, CEAG, for your commitment to promoting environmental awareness and sustainability.

References

Abraha MG, Savage MJ (2006) Potential impacts of climate change on the grain yield of maise for the midlands of KwaZulu-Natal, South Africa. Agric Ecosyst Environ 115:150–160

Amundsen H, Berglund F, Westskog H (2010) Overcoming barriers to climate change adaptation—a question of multilevel governance? Environ Plan C Govern Policy 28(2):276–289

Anderson MB, Woodrow PJ (2019) Rising from the ashes: development strategies in times of disaster. Routledge, New York

Andrade Pérez A, Herrera Fernandez B, Cazzolla Gatti R (2010) Building resilience to climate change: Ecosystem-based adaptation and lessons from the field. IUCN, Gland, p 164

Averchenkova A, Gannon KE, Curran P (2019) Governance of climate change policy: a case study of South Africa. Grantham Research Institute on Climate Change and the Environment and Centre for Climate Change Economics and Policy, London

Ayetor GK, Mbonigaba I, Ampofo J, Sunnu A (2021) Investigating the state of road vehicle emissions in Africa: a case study of Ghana and Rwanda. Transport Res Interdiscip Perspect 11:100409

Basdew M, Jiri O, Mafongoya PL (2017) Integration of indigenous and scientific knowledge in climate adaptation in KwaZulu-Natal, South Africa. Change and Adaptation in Socio-Ecological Systems 3(1):56–67

Baumert K, Winkler H (2005) SD-PAMs and international climate agreements. In: Bradley R, Baumert KA (eds) Chapter 2 in growing in the greenhouse: protecting the climate by putting development first. World Resources Institute, Washington, DC, pp 15–23

Baush T, Koziol K (2020) New policy approaches for increasing response to climate change in small rural municipalities. Sustainability 12(1894):1–17

Berrang-Ford L, Ford JD, Paterson J (2011) Are we adapting to climate change? Glob Environ Chang 21(1):25–33

Borello A, Kissoon S, Trois C (2018) Feasibility of landfill gas upgrade for use as a fuel source for refuse trucks: a case study in South Africa. Detritus 1(1):69

Bose BK (2010) Global warming: energy, environmental pollution, and the impact of power electronics. IEEE Ind Electron Mag 4(1):6–17

Bourne A, Donatti C, Holness S, Midgley G (2012) Climate change vulnerability assessment for the Namakwa District municipality: full technical report. Cape Town, South Africa, Conservation South Africa

Brockhaus, M., & Djoudi, H. (2008). Adaptation at the interface of forest ecosystem goods and services and livestock production systems in northern Mali. JSTOR

Burch S (2010a) In pursuit of resilient, low carbon communities: an examination of barriers to action in three Canadian cities. Energy Policy 38(12):7575–7585

Burch S (2010b) Transforming barriers into enablers of action on climate change: insights from three municipal case studies in British Columbia, Canada. Glob Environ Chang 20(2):287–297

Chersich FM, Wright CY (2019) Climate change adaptation in South Africa: a case study on the role of the health sector. Glob Health 15(22):1–16

Chersich MF, Wright CY, Venter F, Rees H, Scorgie F, Erasmus B (2018) Impacts of climate change on health and wellbeing in South Africa. Int J Environ Res Public Health 15(1884):1–14

Chevallier R (2011) Political barriers to climate change adaptation implementation in SADC. Overcoming Barriers to Climate Change Adaptation Implementation in Southern Africa. Africa Institute of South Africa, Pretoria, pp 1–19

Conway D, Schipper ELF (2011) Adaptation to climate change in Africa: challenges and opportunities dentified from Ethiopia. Glob Environ Chang 21(1):227–237

Crippa M, Guizzardi D, Muntean M, Schaaf E, Solazzo E, Monforti-Ferrario F, Olivier J, Vignati E (2020) Fossil CO2 emissions of all world countries. Luxembourg, European Commission, pp 1–244

Department of Environmental Affairs (2017) South Africa's 2nd annual climate change report. Department of Environmental Affairs, Pretoria

Department of Planning Monitoring and Evaluation (DPME) (2019) Medium term strategic framework 2019–2024. Retrieved from Pretoria, South Africa: https://www.dpme.gov.za/keyfocusareas/outcomesSite/MTSF_2019_2024/2019-2024%20MTSF%20Comprehensive%20Document.pdf

Dickson E, Baker JL, Hoornweg D, Asmita T (2012) Urban risk assessments: understanding disaster and climate risk in cities. World Bank Publications, Washington, DC

Diga K (2017) A review of poverty reduction within local climate change initiatives: a case of eThekwini municipality

Dkamela GP (2011) The context of REDD+ in Cameroon: drivers, agents and institutions, vol 57. CIFOR

Edkins M, Winkler H, Marquard A (2009) Large-scale rollout of concentrating solar power in South Africa

Enete AA, Amusa TA (2010) Challenges of agricultural adaptation to climate change in Nigeria: a synthesis from the literature. Field Actions Sci Rep J Field Actions 4

Eriksen SH, Marin A (2011) Pastoral pathways: climate change adaptation lessons from Ethiopia

EThekwini Municipality (2014). Durban climate change strategy. eThekwini Municipality

EThekwini Municipality (2022). Durban climate change strategy. eThekwini Municipality

Faling W, Tempelhoff JW, Van Niekerk D (2012) Rhetoric or action: are south African municipalities planning for climate change? Dev South Afr 29(2):241–257

Groot A, Hollaender K, Swart R (2014) Productive science-practice interactions in climate change adaptation: lessons from practice. A CIRCLE-2 research policy brief. Foundation of the Faculty of Sciences, Lisbon, Portugal, p 32

Heun M, Van Niekerk J, Swilling M, Meyer A, Brent A, Fluri T (2010) Learnable lessons on sustainability from the provision of electricity in South Africa. Energy Sustain

Hitayezu P, Zegeye EW, Ortmann GF (2014) Some aspects of agricultural vulnerability to climate change in the KwaZulu-Natal midlands, South Africa: a systematic review. J Hum Ecol 48(3):347–356

Hlahla S, Goebel A, Hill TR (2016) Green economy: a strategy to alleviate urban poverty and safeguard the environment? KwaZulu-Natal, South Africa. In Urban forum, vol 27, pp 113–127. Springer, Cham

Intergovernmental Panel on Climate Change. (IPCC) (2021) Summary for policymakers. In: Masson-Delmotte V, Zhai P, Pirani A, Connors SL, Pean C, Berger S, Cau N, Chen Y, Goldfarb L, Gomis MI, Haung M, Leitzell K, Lonnoy E, Matthews JBR, Maycock TK, Waterfield T, Yelekci O, Yu R, Zhou B (eds) Climate change 2021: the physical science basis. Contribution of working group I to the sixth assessment report of the Intergovernmental Panel on Climate Change. Cambridge University Press, Cambridge

IPCC (2023) Climate Change 2023: Synthesis Report. Contribution of Working Groups I, II and III to the Sixth Assessment Report of the Intergovernmental Panel on Climate Change: https://www.ipcc.ch/report/ar6/syr/downloads/report/IPCC_AR6_SYR_SPM.pdf

Hoogendoorn G, Grant B, Fitchett JM (2016) Disjunct perceptions? Climate change threats in two-low lying south African coastal towns. Bulletin of Geography Socio-economic Series 31(31):59–71

Jewitt D, Goodman PS, Erasmus BFN, O'Connor TG, Witkowski ETF (2015) Systematic land-cover change in KwaZulu-Natal, South Africa: implications for biodiversity. South Afr J Sci 111(9/10):1–9

Joubert LS (2006) Scorched: South Africa's changing climate

Kalame FB, Kudejira D, Nkem J (2011) Assessing the process and options for implementing National Adaptation Programmes of Action (NAPA): a case study from Burkina Faso. Mitig Adapt Strateg Glob Chang 16(5):535–553

Kithiia J (2011) Climate change risk responses in east African cities: need, barriers and opportunities. Curr Opin Environ Sustain 3(3):176–180

Klein RJ, Nicholls RJ, Ragoonaden S, Capobianco M, Aston J, Buckley EN (2001) Technological options for adaptation to climate change in coastal zones. J Coast Res:531–543

Kolikow S, Kragt ME, Mugera AW (2012) An interdisciplinary framework of limits and barriers to climate change adaptation in agriculture. Working paper, School of Agricultural and Resource Economics, University of Western Australia, p 1202

Kreimer A, Arnold M, Carlin A (2003) Building safer cities: the future of disaster risk. World Bank Publications

KwaZulu Natal Provincial Planning Commission (KNPPC) (2016) KwaZulu Natal provincial planning commission provincial growth and development strategy. Draft 4 for comments. KwaZulu Natal. South Africa

Leck H, Sutherland C, Scott D, Oelofse G (2011) Social and cultural barriers to adaptation implementation. Overcoming Barriers to Climate Change Adaptation Implementation in Southern Africa, pp 61–88

Louman B, Keenan RJ, Kleinschmit D, Atmadja S, Sitoe AA, Nhantumbo I, de Camino Velozo R, Morales JP (2019) SDG 13: climate action—impacts on forests and people. In: Sustainable development goals: their impacts on forests and people, pp 419–444

Madzwamuse M (2010) Drowning voices: the climate change discourse in South Africa. Policy Brief 5(8):1–8

Magadza CH (2000) Climate change impacts and human settlements in Africa: prospects for adaptation. Environ Monit Assess 61(1):193–205

Masters L, Duff L (2011) Overcoming barriers to climate change adaptation implementation in southern Africa. African Books Collective

Mather A, Vella G (2007) Report on the march 2007 coastal erosion event for the KwaZulu-Natal minister of agricultural and environmental affairs. Durban, South Africa

Matondo JI, Peter G, Msibi KM (2004) Evaluation of the impact of climate change on hydrology and water resources in Swaziland: part I. Phys Chem Earth Parts A/B/C 29(15–18):1181–1191

Measham TG, Preston BL, Smith TF, Brooke C, Gorddard R, Withycombe G, Morrison C (2011) Adapting to climate change through local municipal planning: barriers and challenges. Mitig Adapt Strateg Glob Chang 16(8):889–909

Mkhabela P, Chikandiwa CT (2016) Insights and challenges of strategy and its financial component execution in the public sector

Mkhize MM (2021) Localising the sustainable development goals in rural municipalities through multistakeholder partnerships: a case study of Nkandla local municipality. MSocSc dissertation, University of KwaZulu Natal

Mukheibir P, Kuruppu N, Gero A, Herriman E (2013) Cross-scale barriers to climate change adaptation in local government, Australia. National climate change adaptation research facility

Muller M (2007) Adapting to climate change: water management for urban resilience. Environm Urbanisat 19(1):99–113

Næss LO, Bang G, Eriksen S, Vevatne J (2005) Institutional adaptation to climate change: flood responses at the municipal level in Norway. Glob Environ Chang 15(2):125–138

National Planning Commission (NPC) (2011) National development plan 2030 our future-make it work. Government Printers, Pretoria

National Planning Commission (NPC) (2015) National planning commission handover report. National Planning Commission of South Africa, Pretoria

Ndlovu M, Clulow AD, Savage MJ, Nhamo L, Magidi J, Mabhandhi T (2021) An assessment of the impacts of climate change variability and change in KwaZulu-Natal Province, South Africa. Atmosphere 12(427):1–22

New M, Hulme M (2000) Representing uncertainty in climate change scenarios: a Monte-Carlo approach. Integr Assess 1(3):203–213

O'Donoghue S, Morgan D, Leck H, Haydvogl K (2022) The Durban climate change strategy: lessons learnt from the 2021 strategy review and implementation plan. Town Reg Plan 81:84–96

Patt AG, Schröter D (2008) Perceptions of climate risk in Mozambique: implications for the success of adaptation strategies. Glob Environ Chang 18(3):458–467

Pegels A (2010) Renewable energy in South Africa: potentials, barriers and options for support. Energy Policy 38(9):4945–4954

Mabhaudhi T, Chimonyo VGP, Hlahla S, Massawe F, Mayes S, Nhamo L, Modi AT (2019) Prospects of orphan crops in climate change. Planta 250:695–708

Prosser IP, Chiew FH, Stafford Smith M (2021) Adapting water management to climate change in the Murray–Darling basin, Australia. Water 13(18):2504

Reid P, Massey R, Vogel C (2005) Climate and development: experiences of farmers in KwaZulu-Natal, South Africa. In: Schulze RE (ed) Climate change and water 61 resources in Southern Africa: studies on scenarios, impacts, vulnerabilities and adaption. ISBN 1943-8168. Water Research Commission, Pretoria, South Africa

Rizvi A, van Riel K (2018) EbA nature based solutions for climate change adaptation–knowledge gaps. IUCN Knowledge Series–Working Paper

Roberts D, O'Donoghue S (2013) Urban environmental challenges and climate change action in Durban, South Africa. Environment and Urbanisation 25(2):299–319

Roberts D (2010) Prioritizing climate change adaptation and local level resilience in Durban. South Africa Environment and Urbanization 22(2):397–413
Roberts D, O'Donoghue S (2016) CITY VIEW: Durban, South Africa. In: State of the world. Springer, pp 337–342
Roberts JT, Parks BC (2009) Ecologically unequal exchange, ecological debt, and climate justice: the history and implications of three related ideas for a new social movement. Int J Comp Sociol 50(3–4):385–409
Roberts P, Henshilwood CS, van Niekerk KL, Keene P, Gledhill A, Reynard J, Lee-Thorp J (2016) Climate, environment and early human innovation: stable isotope and faunal proxy evidence from archaeological sites (98-59ka) in the southern cape. South Africa PloS one 11(7):e0157408
Shackleton S, Cobban L, Cundill G (2014) A gendered perspective of vulnerability to multiple stressors, including climate change, in the rural eastern cape. South Africa Agenda 28(3):73–89
Sithole B (2016) Investigating possible impact of climate change on sugarcane production in KwaZulu-Natal, South Africa. MSc dissertation, University of KwaZulu-Natal
South African Civil Society Working Group (2019) Citizen's report South Africa: ensuring an inclusive reporting process for HLPF 2019. Africa Monitor
South African Government (2019) Department of Energy Annual Report 2019. Retrieved from Pretoria, South Africa: https://www.gov.za/sites/default/files/gcis_document/202101/doe-annual-report-2019-20.pdf
South African Civil Society Working Group (2020) Open SDG Club South Africa 2020 symposium report: 'inclusive and transformative recovery: delivering on the SDGs decade of action. Africa Monitor, Johannesburg
South African Weather Service (2020) Annual state of the climate of South Africa 2020. South Africa, Pretoria
Statistics SA (2019) Mid-year population estimates. Statistics SA, Pretoria
Stringer LC, Dyer JC, Reed MS, Dougill AJ, Twyman C, Mkwambisi D (2009) Adaptations to climate change, drought and desertification: local insights to enhance policy in southern Africa. Environ Sci Pol 12(7):748–765
Strydom S, Savage M (2017) Potential impacts of climate change on wildfire dynamics in the midlands of KwaZulu-Natal, South Africa. Clim Chang 143:385–397
Sutherland C, Scott D, Diederichs N, Mander M, van Niekerk M, Sim V (2011) Unlocking the green economy in KwaZulu-Natal. School of Development Studies and Futureworks! for the KwaZulu-Natal Department of Economic Development and Tourism, Durban
The Presidency (2014) Annual Report 2014/2015. Retrieved from Pretoria, South Africa: https://www.thepresidency.gov.za/sites/default/files/The%20Presidency%20Annual%20Report%202013-2014.pdf
Toteng E (2011) Political and state system barriers to the implementation of climate change adaptation in Botswana. Overcoming barriers to climate change adaptation implementation in Southern Africa, p 21
Tschakert P, Dietrich KA (2010) Anticipatory learning for climate change adaptation and resilience. EcologySoc 15(2):11. Available through: http://www.ecologyandsociety.org/vol15/iss2/art11/
UNFCCC (1992) United nations framework convention on climate change: text. UNEP/WMO Information Unit on Climate Change, Geneva
United States Agency of International Development (USAID) (2015) Climate change information fact sheet South Africa. United States Agency of International Development, USA
Urban-LEDS Project II, Local Governments for Sustainability (ICLEI) and UN-Habitat (2020) Multilevel Governance for climate change in South Africa (Online), Available: https://unhabitat.org/sites/default/files/2021/01/urban-leds-country-factsheet-south-africa.pdf. Accessed 9 Sept 2021
Washington R, Harrison M, Conway D, Black E, Challinor A, Grimes D, Jones R, Morse A, Kay G, Todd M (2006) African climate change: taking the shorter route. Bull Am Meteorol Soc 87(10):1355–1366
Weaver CP, Lempert RJ, Brown C, Hall J, Revell D, Sarewitz D (2013) Improving the contribution of climate model information to decision making: the value and demands of robust decision frameworks. Wires Clim Change 4:39–60
Wei K, Ouyang C, Duan H, Li Y, Chen M, Ma J et al (2020) Reflections on the catastrophic 2020 Yangtze River basin flooding in southern China. The Innovation 1(2)
Wilk J, Andersson L, Warburton M (2013) Adaptation to climate change and other stressors among commercial and small-scale South African farmers. Reg Environ Chang 13(2):273–286
Winkler H, Spalding-Fecher R, Mwakasonda S, Davidson O (2002) Sustainable development policies and measures. In: Options for protecting the climate; world resource institute: Washington. USA, DC
Wright CY, Norval M, Kapwata T, Du Preez DJ, Wernecke B, Tod BM, Visser WI (2019) The incidence of skin cancer in relation to climate change in South Africa. Atmos 10(10):634
Ziervogel G, Parnell S (2012) South African coastal cities: governance responses to climate change adaptation: Gina Ziervogel and Susan Parnell. In Climate change at the city scale. Routledge, pp 237–257
Ziervogel G, Parnell S (2014) Tackling barriers to climate change adaptation in South African coastal cities. In Adapting to climate change. Springer, pp. 57–73
Ziervogel G, Shale M, Du M (2010) Climate change adaptation in a developing country context: the case of urban water supply in Cape Town. Clim Dev 2(2):94–110
Ziervogel G, New M, Garderen EA, Midgley G, Taylor A, Hamann R, Stuart-Hill S, Myers J, Warburton M (2014) Climate change impacts and adaptation in South Africa. WIREs climate change, pp 1–17

Part V

Climate Crisis and Land Water and Forest Sustainability

Evaluating the Potential Impact of Climate Change on Glacier Dynamics in Western Himalayas, India

25

Suhail A. Lone and Gh. Jeelani

Abstract

Glaciers are the main source of freshwater supply to millions of people living in different Himalayan basins. A reliable amount of water generated from these glaciers is highly significant for the social and economic development of the region. However, due to recent climate change, these potential water resources are under serious threat of retreat. In the present study, we assessed the response of some major glaciers of the western Himalayas using time series satellite data in association with changing meteorological variables (temperature and precipitation). The temperature and precipitation time series data from 1901 to 2022 suggested that average temperature has increased and average precipitation has decreased across the western Himalayas. The results suggested an average increase of 0.2 °C decade^{-1} from 1901 to 2022 with a significant increase of 0.09 °C year^{-1} from 1991 to 2022 and a decrease of annual precipitation by 0.032 mm year^{-1}. From the time series satellite dataset, it was observed that the glaciers in the Kashmir Himalayas (Kolahai, Hoksar, and Sheshram glaciers) have lost 21% of the glacier extent in the last 40 years from 1980 to 2020. The glaciers in Ladakh Himalayas are also showing a significant decreasing trend. The study indicated that continuous loss of glaciers under changing climatic scenarios can have a profound impact on regional hydrology, which will negatively impact the economic stability of the region. The study further suggests that the continuous recession of glaciers will increase the probability of Glacier Lake Outburst floods (GLOF) leading to disasters across the western Himalayas.

Keywords

Glaciers · Hydrology · Himalaya · Climate · Landsat · Time-series

25.1 Introduction

The Himalayas host the world's richest cryospheric cover outside polar regions. A reliable water supply generated from this cryospheric cover is highly significant in governing the social and economic stability of the region (Immerzeel et al. 2010; Lone et al. 2016, 2022b, 2023; Scott et al. 2019; Lone and Jeelani 2020; Jeelani et al. 2020, 2021a). Apart from the economic significance, Himalayan cryospheric cover plays an essential role in supporting the atmospheric phenomenon of southeast Asia (Bolch et al. 2012; Somers 2019; Lone et al. 2020). The health of Himalayan cryospheric cover is highly dependent

S. A. Lone (✉) · G. Jeelani
Department of Earth Sciences, University of Kashmir, Srinagar, India

U. Chatterjee et al. (eds.), *Climate Crisis: Adaptive Approaches and Sustainability*, Sustainable Development Goals Series, https://doi.org/10.1007/978-3-031-44397-8_25

upon the movement and sustenance of the two largest weather systems viz., Western Disturbance and Indian Summer Monsoons (Immerzeel et al. 2015; Jeelani et al. 2017, 2020; Jeelani and Lone 2019; Qazi et al. 2020; Lone et al. 2023). However, the variability in the moisture circulation pattern and modification in the climatic variables (Temperature, Precipitation) have negatively affected the health of cryospheric cover across the world, particularly at higher altitudes (Beniston 2003; Wilson et al. 2017; Lone et al. 2017, 2021a, b; Jeelani et al. 2021b). In the Himalayas, glaciers are shrinking and receding at a faster rate than the world's other glaciers (Dobhal et al. 2013; Lone et al. 2017). Recent studies in the Himalayas have postulated that glaciers are retreating at a rate of 0.4% year^{-1} due to variability in temperature and precipitation (Bolch et al. 2012). Based on time series temperature and precipitation data, the studies have suggested that temperature has increased and precipitation has decreased during the last 100 years (Lone et al. 2019, 2021a, 2022a, b). The surface and subsurface hydrology of the western Himalayas is highly dependent upon the meltwater generated from snow and glacier cover. The studies have postulated that ~70% of annual stream water discharge and groundwater recharge is sustained by water derived from snow and glacier melt (Lone et al. 2021b, 2022a). Therefore, a decrease in precipitation, particularly during the winter season, can have a significant effect on the health of glaciers in general and regional hydrology in particular (Vuille et al. 2003; Jeelani et al. 2012, 2013). In the last 2–3 decades' various studies have been carried out in the Himalayas to assess the response of high-altitude glaciers to recent climate change (Bahuguna et al. 2014; Chand and Sharma 2015; Swain et al. 2017; Patel et al. 2018 and references therein). Most of these studies have assessed that the topography and climate of the region are highly responsible for the recession of glaciers (King et al. 2017; Debnath et al. 2019). Different remote sensing-based methods including on-screen delineation, supervised and unsupervised classification image rationing, etc. have been widely used to understand the response of these glaciers (Villaba et al. 2003; Nathawat et al. 2008; Shukla et al. 2009; Kamp et al. 2011; Pandey et al. 2011; Brahmbhatt et al. 2012; Ghosh et al. 2014). However, most of these studies are laboratory-based without much field validation and uncertainty analysis.

Keeping in view the dynamic nature and response of mountainous glaciers to different climatic variables, the regular and detailed monitoring of glaciers is critically important for economic stability and water resource management policies at local and regional scales. Therefore, the main aim of the present study is to evaluate the current receding rate of a few important glaciers (Kolahoi, Hoksar, and Shesram from Kashmir and Parachik, Drung-Drung from Ladakh) in the western Himalayas using geospatial technology and field validations. The record of monitoring the glacier recession can help us to better understand the response of glaciers to changing climate on a long-term basis. The study can be useful for constructing past climatic conditions and equilibrium-line altitude (ELA), and will add more insights into water resource management and ecosystem restoration policies across the Himalayas.

25.2 Study Area

The study presents the climatologically two different basins: the Liddar Basin from the Kashmir Himalayas and the Suru Basin from the Ladakh Himalayas. The Liddar Basin lies in the UT of Jammu and Kashmir and the Suru Basin lies in the UT of Ladakh. The basins lie within geographical coordinates of Latitude 33°50′ to 34°00′N, 33°00′ to 34°00′N and longitudes 75°00′ to 75°30′E, 75°00′ to 78°00′E, respectively. Both the basins lie within the greater Himalayan range but due to the highly complex physiography of the Himalayas, the basins exhibit distinct climatic conditions viz.; temperate in Liddar and dry-cold arid in Suru basin (Fig. 25.1).

In the Himalaya, though the Western Disturbances and Indian Summer Monsoons are two major moisture source contributors, the

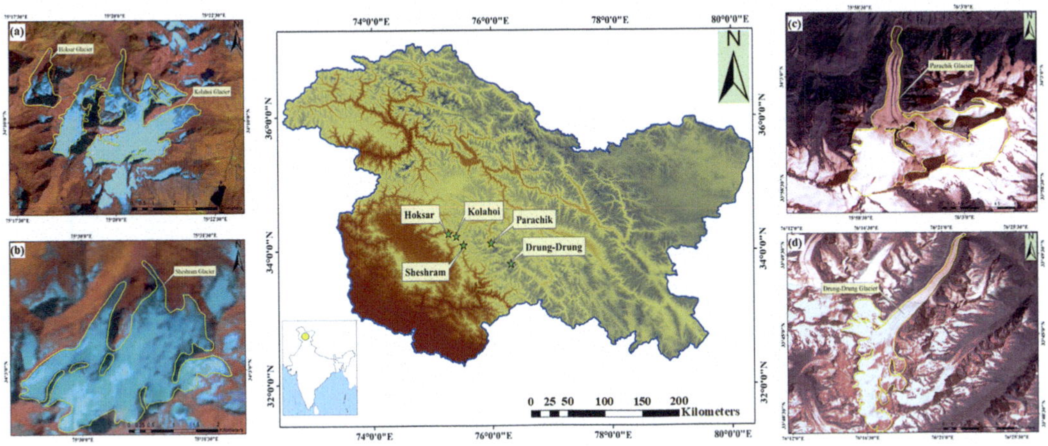

Fig. 25.1 Study area map showing the location of glaciers with respect to India and the Union territory of Jammu and Kashmir and Ladakh. The inset maps **a–d** show the zoomed-in view of these glaciers

complex physiographic setup significantly affects the spatio-temporal distribution of precipitation in these basins. Liddar basin receives an average precipitation of ~1200 mm recorded at Pahalgam weather station, out of which about ~60% amount is received during the winter season while, the Suru basin receives an average precipitation of ~115 mm recorded at Leh weather station, out of which more than ~80% is received during the winter season (Fig. 25.2, Lone et al. 2017, 2019). The average temperature varies from 1.2 °C to 26 °C and −19 °C to 6 °C in the Liddar and Suru basins, respectively (Jeelani et al. 2017; Lone et al. 2017, 2019). It is important to mention here that due to the unavailability of local weather stations in the Suru basin, no records are available about the amount of precipitation received at an altitude <5000 m, asl.

25.3 Materials and Methods

In the present study, time-series data from 1901 to 2022 of different climatic variables (Temperature and Precipitation) along with remote sensing satellite data was used to understand the response of a few important glaciers to changing temperature and precipitation variables in the Himalayas. The study mainly focused on evaluating the changes in the snout position and glacier area in the last few decades. To fulfill our requirements, the Landsat data with different sensor resolutions from 1980 to 2022 for the Kashmir Himalayas from 1992 to 2022 for Ladakh Himalayas (https://earthexplorer.usgs.gov/) corroborated by Google Earth and Field validation data. Image selection for glacier mapping was done keeping in view the ablation period, snow cover, and cloud conditions for better results. The snout position of glaciers on satellite data was ascertained by locating the terminal moraines (Bolch et al. 2010; Lone 2016; Lone et al. 2017). The on-screen digitization of the snout position was done at an average scale of 1:150,000 in Arc-GIS 10.7.1. Image processing methods including filtering, FCC, and band rationing were applied to discriminate glacial features for precise delineation of present snot position and area (Paul et al. 2002; Jaan et al. 2015; Lone et al. 2017). The long-term meteorological parameters (temperature and precipitation) were acquired from the National Ocean and Atmospheric Administration (NOAA) and weather stations located Liddar and Suru basins.

25.3.1 Uncertainty Estimations

The varying snow cover, cloud cover, shadows, and image resolution lead to different levels of accuracy. The errors in length estimation and glacier area delineation are due to image

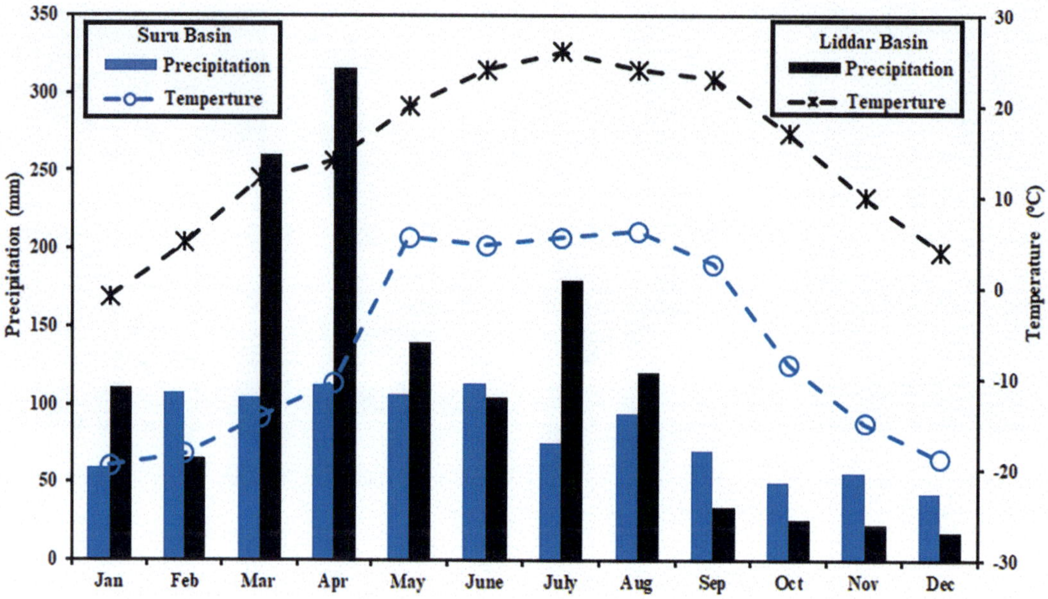

Fig. 25.2 Plot depicting the overall temporal variability of temperature and precipitation in Suru and Liddar basins. It is evident from the plot that the Suru basin receives less precipitation than the Liddar basin

co-registration. Hence, error estimation becomes crucial to identify the accuracy level and importance of results. In the present study, the snout change was evaluated using the formula:

$$U = \sqrt{a^2 + b^2} + \sigma$$

a and b denotes the resolution of images and σ denotes uncertainty in image georeferencing. The image georeferencing error was evaluated visually as 14 m, 8 m, 7 m, and 0 m, for the 1980 Landsat MSS (80 m resolution), 1992 TM, 2000 ETM+, and 2022 OLI images when georeferencing with the base image of Landsat OLI 2022. The uncertainties linked with glacier area estimations were calculated using the buffer method following (Wang et al. 2009; Bolch et al. 2010). The region of the buffer was set as double the digitization error around each glacier. The uncertainty measurements of each glacier (U_{area}) were also attained using following formula.

$$U_{area} = 2UV$$

U is glacier area uncertainty and V denotes the pixel resolution of an image. Therefore, the uncertainty in the area of the glaciers was assessed to be ±0.0098 km^2 for the MSS 1980 image, ± 0.0028 km^2 for the 1992 TM image, 0.0027 km^2 for the 2000 ETM+ image, and 0.0025 km^2 for the 2015 TM image.

25.4 Results and Discussions

25.4.1 Long-Term Temperature and Precipitation

Long-term statistical information on atmospheric variables (temperature and precipitation) indicated an increase and decrease in temperature and precipitation in the western Himalayas. Reanalysis data from the National Ocean and Atmospheric Administration (NOAA) was used as primary data sources to analyze different meteorological parameters in this region. The long-term temperature data exhibited an average rise of about 0.166 °C from the 1901 to 2022 period (Fig. 25.3). However, the temperature and precipitation showed a substantial temporal trend. The temperature first exhibited a rise of 0.02 °C year^{-1} from 1901 to 1930 and 1974 to 1990, then a decline of 0.0002 °C year^{-1} from

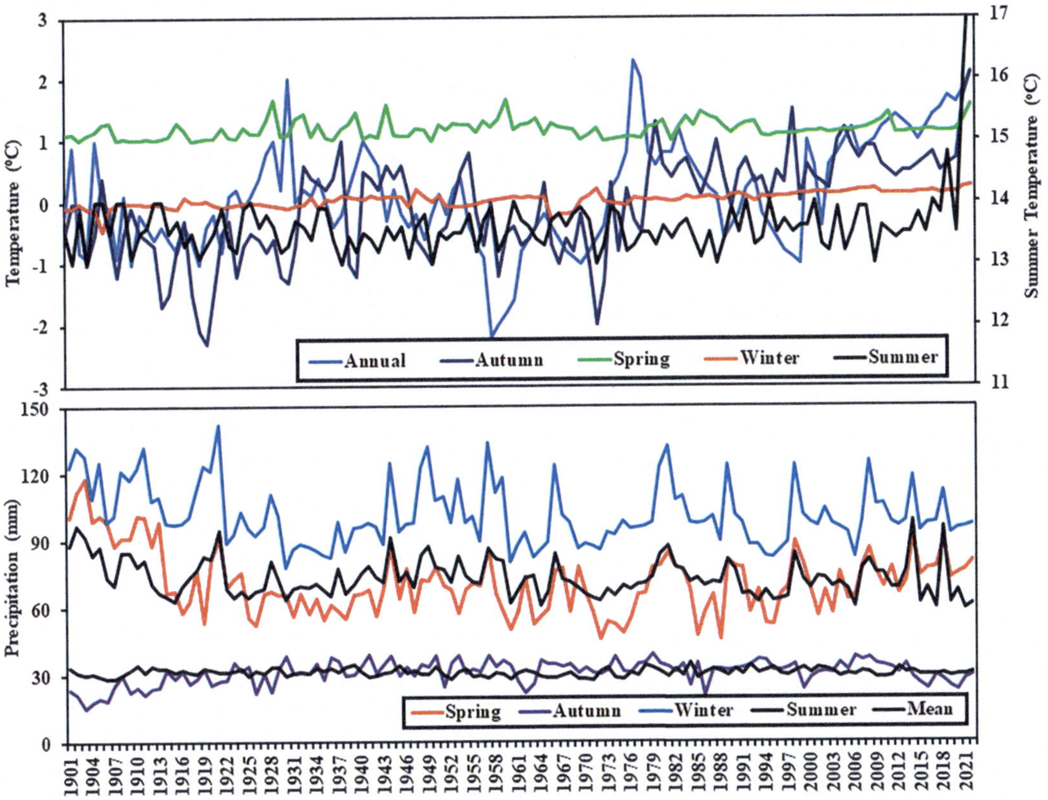

Fig. 25.3 Time-series plot showing the variability of mean annual temperature and precipitation from 1901 to 2022. The plots depicted that the mean annual temperature has increased by 1.5 °C decade^{-1} and mean annual precipitation has decreased by 0.032 mm year^{-1}

1931 to 1973, followed by a major increase by 0.10 °C year^{-1} from 1991 to 2022. Similarly, the precipitation also showed a variable trend. Although, the average annual precipitation has reduced by about 0.032 mm year^{-1}. However, an increase in average precipitation amount by 0.8 mm year^{-1} was observed from 1910 to 1930 (Fig. 25.3). Apart from the interannual variability the seasonal trends in temperature and precipitation are also quite evident from the meteorological data of the western Himalayas. The time-series temperature trend showed a little bit increase in summer (0.006 °C) but a significant increase in the winter season (0.08 °C). But precipitation amount decreased in winter (0.18 mm year^{-1}) and increased in other seasons viz.; spring (0.21 mm year^{-1}), summer (0.09 mm year^{-1}), and autumn (0.18 mm year^{-1}).

Similar conclusions have been put forward by studies carried out in different parts of the Himalayas and outside (Auer and Boehm 1994; Diaz and Bradley 1997; Sharma et al. 2000; Jamelli et al. 2004; Bhutiyani et al. 2007; Shekhar et al. 2010; Jeelani et al. 2012, 2013; Kohler et al. 2014; Rao et al. 2016; Bari et al. 2016; Ferreira et al. 2018; Shrestha et al. 2019; Lone et al. 2019).

25.4.2 Glacier Recession

To understand the time-series response we have mapped five important glaciers which include Kolahoi, Hoksar, and Sheshram of Liddar basin from the Kashmir Himalayas and Parachik and Drung-Drung of Suru basin from Ladakh

Himalayas using optical remote sensing data from last 43 years (Fig. 25.4).

Kolahoi glacier (Fig. 25.4) is the major glacier in the Liddar basin followed by Sheshram and Hoksar glaciers. From 1980 to 2022 the area of Kolahoi glacier reduced from 13.55 to 10.45 km^2 followed by Sheshram (5.90–4.78 km^2) and Hoksar glaciers (1.38–0.78 km^2), respectively. Similarly, the Drung-Drung glaciers from the Suru basin reduced from 50.81 km^2 in 1990 to 48.85 km^2 in 2022 while, the area of Parachik glacier has declined from 49.38 km^2 in 1990 to 47.89 km^2 in 2022 with an areal change of 1.96 and 1.49 km^2, respectively (Fig. 25.5).

The results indicated that the overall area of all the studied glaciers has declined substantially during the observation period. The total glaciated extent of three major glaciers from the Liddar basin in 1980 was 20.83 km^2 which declined to ~16.01 km^2 in 2022, the total area of Parachik and Drung-Drung glaciers was about ~100.19 km^2 in 1992 which declined to ~98.11 km^2 in 2022 (Table 25.1).

Therefore, the Liddar basin has witnessed a deglaciation of 3.53 km^2 in 43 years and the Suru basin has witnessed a decline of ~3.45 km^2 in 30 years (Table 25.1). However, the rate of glacier retreat is highly variable due to variations in size, elevation, slope, aspect, and physiography of two climatologically different basins. It has been postulated by different studies that glacier response to receding with time is highly dependent upon the thickness of glaciers which in turn depends upon the glaciated area (Chaohai and

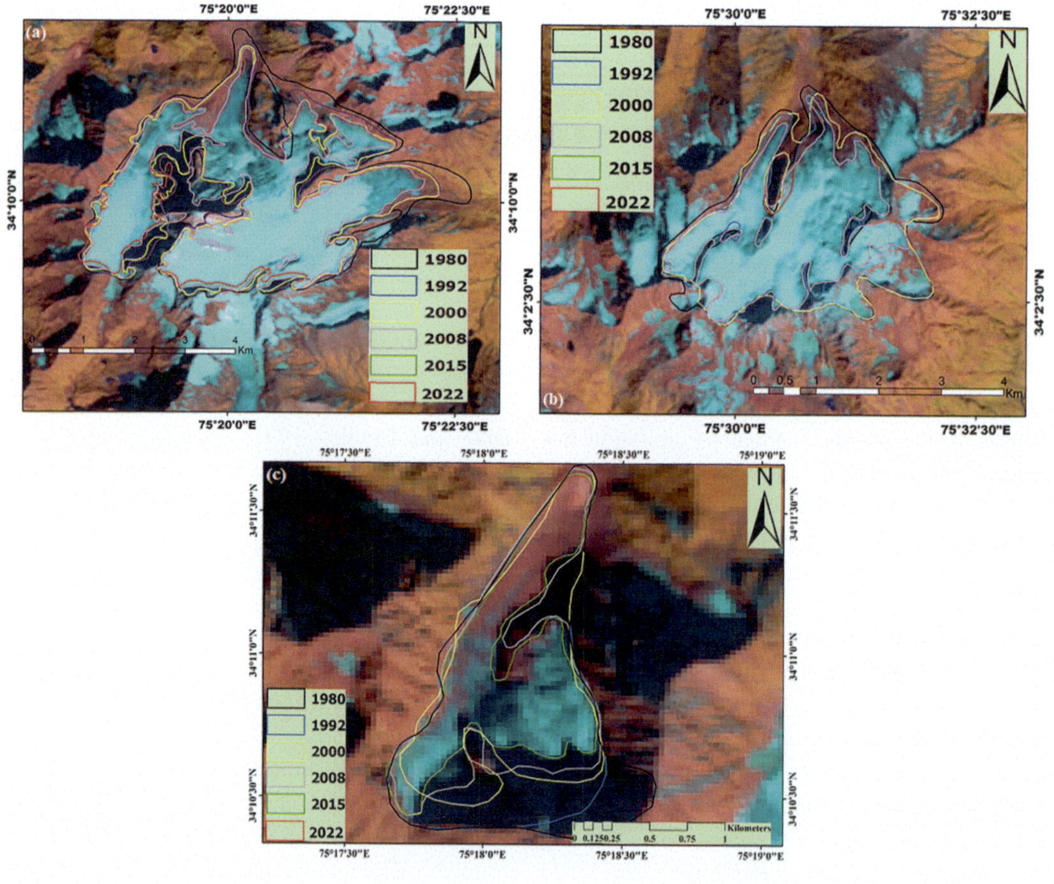

Fig. 25.4 The plots show the time-series areal change of some major glaciers (**a**) Kolahoi (**b**) Shesram and (**c**) Hoksar in the Liddar basin. It can be observed from these layouts that the glacier area has decreased from 1980 to 2022

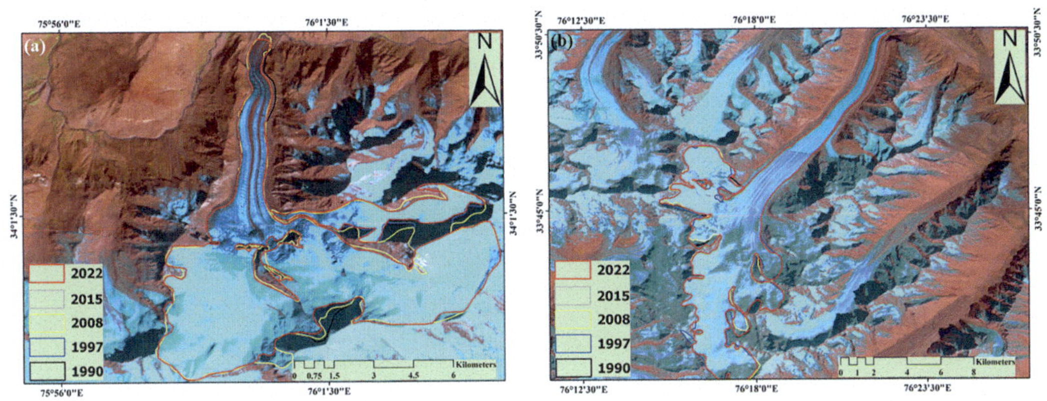

Fig. 25.5 The plots are showing the time-series areal change of two major glaciers (**a**) Parachik and (**b**) Drung-Drung in the Suru basin. It can be observed from these layouts that the glacier area has decreased from 1990 to 2022

Sharma 1988; Jóhannesson et al. 1989). It has been reported that small-sized glaciers are retreating at a faster rate than large-sized glaciers.

Recent studies have also suggested that changing climate effects are more prominent in small glaciers than large glaciers (Kulkarni et al. 2007; Armstrong et al. 2009) and continuous temperature rise can further impact the melting rate of small glaciers. Another important variable that impacts the glacial recession is aspect and snout elevation. To evaluate the aspect effect, the sub-basins were divided into various sun-facing classes. It was perceived that south and northeast-facing glaciers are retreating more than the north-facing. Also, to understand the elevation effect on glacier recession. It was observed that Liddar basin glaciers which lie at an altitude of 3500–5000 m, asl have receded more than Suru basin glaciers which lie at an altitude of 4200–6000 m, asl (Fig. 25.6).

25.5 Climate Change and Future Hydrological Regimes

The climate, hydrology, and ecosystem sustainability of southeast Asia are critically dependent upon the meltwaters generated from the cryospheric nectar of the Himalayas. The snow and glaciers are important components of the Himalayan hydrological cycle (Jeelani et al. 2012; Gruber et al. 2017; Lone et al. 2022a, b, 2023). These water resources play an incredible part in sustaining the livelihood of the major Himalayan basins. But the negative impacts of a warming climate have drastically affected the normal cycle of cryospheric advance and retreat which in turn have disturbed the aquatic and terrestrial ecosystems (Immerzeel et al. 2010; Lone et al. 2021a, b). The variability in the long-term meteorological parameters suggested that observed modification in hydrological regimes of Himalayan river basins is directly associated with past and present climatic trends and it is expected that these regimes may change more due to continuous increase in winter temperature and decrease in precipitation particularly in western Himalayas were surface and sub-surface local and regional hydrology is dominantly controlled by meltwaters (Miller et al. 2012; Nepal et al. 2021; Shah and Lone 2019; Lone et al. 2021a, 2022a; Lone and Jeelani 2022). The observed and predicted variability in meteorological parameters specifies that the cryosphere of the Himalayas is vulnerable to negative consequences of climate change (Barnett et al. 2005; Sakai and Fujita 2017; Li et al. 2019; You et al. 2020). The World's largest irrigated agriculture system of the Indus River Basin (IRB) is highly dependent upon cryospheric waters (Laghari et al. 2012). Therefore, these unexceptional externalities will make the public water supply and agro-economy of this region highly susceptible leading to instability, particularly in mountainous communities.

Table 25.1 Statistical summary of time-series receding rate of some major glaciers of Greater Himalayas

Glacier name	Lat. (N)	Long.(E)	Elevation (m, asl)	Year						Total area change (Km²)
				1980	1992	2000	2008	2015	2022	
Kolahoi	34°12′	75° 17′	3300	13.55	13.01	12.45	12.24	11.18	10.4	**3.1**
Hoksar	34°13′	75° 19′	3500	1.38	1.34	1.24	1.14	1.01	0.78	**0.6**
Sheshram	34°05′	75°30′	3600	5.90	5.81	5.68	5.51	5.20	4.78	**1.12**
Parachik	34°51′	34°51′	3890	**1990**		**1997**	**2008**	**2015**	**2022**	**1.49**
				49.38		49.01	48.80	48.36	47.89	
Drung-Durung	33°50′	76 °22′	4500	50.81		50.29	49.91	49.75	48.85	**1.96**

Fig. 25.6 Field photographs showing the present snout position and condition of glaciers. (**a**) Kolahoi (**b**) Sheshram (**c**) Parachik (**d**) Drung-Drung. It can be observed from these photographs that glaciers are receding at an enormous rate

25.6 Conclusion

In the present study, a time series of optical remote sensing and meteorological datasets were used to assess and understand the response of some major glaciers of the western Himalayas to recent climate change. The time series temperature and precipitation data of the Himalayas have suggested that an increase of temperate by 0.2 °C decade^{-1} and a decrease in precipitation by 0.032 mm year^{-1} was observed from 1901 to 2022. The results showed that the areal extent of all the glaciers from both basins declined significantly during the observation period. The study further suggested total glaciated area of three major glaciers from the Liddar basin has declined from 20.83 km^2 in 1980 to ~16.01 km^2 in 2022 and the area of Parachik and Drung-Drung glaciers from the Suru basin has declined from ~100.19 km^2 in 1992 to ~98.11 km^2 in 2022. Therefore, the Liddar basin and Suru basins have witnessed a deglaciation of ~3.53 km^2 and ~3.45 km^2 and 43 and 30 years, respectively. It was also observed that Liddar basin glaciers show more recession than Suru basin glaciers due to aspect and altitudinal variability. The result of this study would be significant for water resource management policies and will serve as primary information to different water-related organizations to sustainably manage the water resources in the Himalayas.

Acknowledgments The authors are highly thankful to the United States Geological Survey (USGS) and the National Oceanic and Atmospheric Administration (NOAA) for open access to Satellite and meteorological datasets.

References

Armstrong R, Alford D, Racovitneau A (2009) Glaciers as indicators of climate change: the special case of the high elevation glaciers of the Nepal Himalayas. Water storage, a strategy for climate change adaptation (No. 56). Report Kathmandu ICIMOD, pp 14–16

Auer I, Böhm R (1994) Combined temperature-precipitation variations in Austria during the instrumental period. Theor Appl Climatol 49(3):161–174

Bahuguna IM, Rathore BP, Bhambhatt R, Sharma M, Dhar S, Randhawa SS, Kumar K et al (2014) Are the Himalayan glaciers retreating? Curr Sci 106(7):1008–1013

Bari SH, Rahman MTU, Hoque MA, Hussain MM (2016) Analysis of seasonal and annual rainfall trends in the northern region of Bangladesh. Atmos Res 176:148–158

Barnett TP, Adam JC, Lettenmaier DP (2005) Potential impacts of a warming climate on water availability in snow-dominated regions. Nature 438(7066):303–309

Beniston M (2003) Climatic change in mountain regions: a review of possible impacts. Clim Chang 59(1–2):5–31

Bhutiyani MR, Kale VS, Pawar NJ (2007) Long-term trends in maximum, minimum and mean annual air temperatures across the Northwestern Himalaya during the twentieth century. Clim Chang 85(1–2):159–177

Bolch T, Yao T, Kang S, Buchroithner MF, Scherer D, Maussion F, Huintjes E, Schneider C (2010) A glacier inventory for the western Nyainqentanglha Range and Nam Co Basin Tibet and glacier changes 1976–2009. Cryosphere 4:419–433

Bolch T, Kulkarni A, Kääb A, Huggel C, Paul F, Cogley JG et al (2012) The state and fate of Himalayan Glaciers. Science 336:310–314

Brahmbhatt RM, Bahuguna I, Rathore BP, Kulkarni AV, Shah RD, Nainwal HC (2012) Variation of snowline and mass balance of glaciers of Warwan and Bhut Basins of Western Himalaya using remote sensing technique. J Indian Soc Remote Sens 40(4):629–637

Chand P, Sharma MC (2015) Glacier changes in the Ravi basin, North-Western Himalaya (India) during the last four decades (1971–2010/13). Glob Planet Change 135:133–147

Chaohai L, Sharma CK (1988) Report on first expedition to glaciers in the Pumqu (Arun) and Poiqu (Bhote-Sun Kosi) river basins Xizang (Tibet) China. Beijing Science Press, p 192

Debnath M, Sharma MC, Syiemlieh HJ (2019) Glacier dynamics in Changme Khangpu basin, Sikkim Himalaya, India, between 1975 and 2016. Geosciences 109(6):259

Diaz HF, Bradley RS (1997) Temperature variations during the last century at high elevation sites. Clim Chang 36:253–279

Dobhal DP, Mehta M, Srivastava D (2013) Influence of debris cover on terminus retreat and mass changes of Chorabari Glacier, Garhwal region, central Himalaya, India. J Glaciol 59(217):961–971

Ferreira RN, Nissenbaum MR, Rickenbach TM (2018) Climate change effects on summertime precipitation organization in the Southeast United States. Atmos Res 214:348–363

Ghosh S, Pandey AC, Nathawat MS, Bahuguna IM (2014) Contrasting signals of glacier changes in Zanskar valley, Jammu & Kashmir, India using remote sensing and GIS. J Indian Soc Remote Sens 42:817–827

Gruber S, Fleiner R, Guegan E, Panday P, Schmid MO, Stumm D, Wester P, Zhang Y, Zhao L (2017) Inferring permafrost and permafrost thaw in the mountains of the Hindu Kush Himalaya region. Cryosphere 11(1):81–99

Immerzeel WW, Van Beek LP, Bierkens MF (2010) Climate change will affect the Asian water towers. Science 328(5984):1382–1385

Immerzeel WW, Wanders N, Lutz AF, Shea JM, Bierkens MFP (2015) Reconciling high-altitude precipitation in the upper Indus basin with glacier mass balances and runoff. Hydrol Earth Syst Sci 19(11):4673–4687

Jaan O, Lone SA, Malik R, Lone A, Mir W, Nissa A (2015) Morphotectonic and morphometric analysis of Vishav Basin left bank tributary of Jhelum River SW Kashmir Valley India. Int J Econ Environ Geol 6(2):17–26

Jamelli V, Pech VP, Chochillan C, Brunstein D (2004) Geomorphic variability of debris flows and recent climate changes in the French Alps. Clim Chang 64:77–102

Jeelani G, Lone SA (2019) Establishing the effect of aridity on the stable isotopes ($\delta^{18}O$ and δD) of precipitation in cold desert, Ladakh. Int J Water Resour Arid Environ 8(2):161–168

Jeelani G, Feddema JJ, Veen CJ, Stearns L (2012) Role of snow and glacier melt in controlling river hydrology in Liddar watershed (western Himalaya) under current and future climate. Water Resour Res 48(12)

Jeelani G, Kumar US, Kumar B (2013) Variation of $\delta^{18}O$ and δD in precipitation and stream waters across the Kashmir Himalaya (India) to distinguish and estimate the seasonal sources of streamflow. J Hydrol 481:157–165

Jeelani G, Shah RA, Fryar AE, Deshpande RD, Mukherjee A, Jerome P (2017) Hydrological processes in glacierized high altitude basins of western Himalaya. Hydrogeol J 26(2)

Jeelani G, Lone SA, Nisa AU, Mukherjee A, Deshpande RD (2020) Sources and processes of groundwater arsenic mobilization in upper Jhelum basin, western Himalayas. J Hydrol 591:125292

Jeelani G, Lone SA, Nisa AU, Deshpande RD, Padhya V (2021a) Use of stable water isotopes to identify and estimate the sources of groundwater recharge in an alluvial aquifer of Upper Jhelum Basin (UJB), western Himalayas. Hydrol Sci J 66(16):2330–2339

Jeelani G, Lone SA, Lone A, Deshpande RD (2021b) Groundwater resource protection and spring restoration in Upper Jhelum Basin (UJB), western Himalayas. Groundw Sustain Dev 15:100685

Jóhannesson T, Raymond C, Waddington ED (1989) Time-scale for adjustment of glaciers to changes in mass balance. J Glaciol 35:355–369

Kamp U, Byrne M, Bolch T (2011) Mapping glacier fluctuations between 1975 and 2008 in the greater Himalaya range of Ladakh, North-Western India. Int J Mt Sci 8:374–389

King O, Quincey DJ, Carrivick JL, Rowan AV (2017) Spatial variability in mass loss of glaciers in the Everest region, Central Himalayas, between 2000 and 2015. Cryosphere 11(1):407–426

Kohler T, Wehrli A, Jurek M (2014) Mountains and climate change: a global concern. Sustainable Mountain Development Series Centre for Development and Environment (CDE), Swiss Agency for Development and Cooperation (SDC) and Geographica Bernensi, Bern, p 136

Kulkarni AV, Bahuguna IM, Rathore B, Singh SK, Randhawa SS, Sood RK, Dhar S (2007) Glacial retreat in Himalaya using Indian remote sensing satellite data. Curr Sci 92(1):69–74

Laghari AN, Vanham D, Rauch W (2012) The Indus basin in the framework of current and future water resources management. Hydrol Earth Syst Sci 16(4):1063–1083

Li H, Liu L, Shan B, Xu Z, Niu Q, Cheng L, Liu X, Xu Z (2019) Spatio-temporal variation of drought and associated multi-scale response to climate change over the Yarlung Zangbo River Basin of Qinghai–Tibet Plateau China. Remote Sens 11(13):1596

Lone S (2016) Mapping the glacial-geomorphological landforms in East Liddar Valley, NW Himalaya Kashmir India. J Geogr Nat Disast S6:4. https://doi.org/10.4172/2167-0587.S6-004

Lone SA, Jeelani G (2020) Stable water isotopic evidence for the moisture source and composition of surface runoff in Ladakh, upper Indus river basin (UIRB). In: Proceedings of the 5th International electronic conference on water sciences 16–30 November 2020 MDPI Basel. https://doi.org/10.3390/ECWS-5-07903

Lone SA, Jeelani G (2022) Appraising the groundwater potential of Liddar Sub-Basin (Western Himalayas) using geospatial techniques. In: Climate change impact on groundwater resources. Springer, Cham, pp 329–344

Lone SA, Lone AA, Jeelani G (2016) Characterization of groundwater potential of Sindh watershed Western Himalayas. J Res Dev 16. ISSN 0972-5407

Lone SA, Jeelani G, Deshpande RD (2017) Evaluating the sensitivity of glaciers to climate by using stable water isotopes and remote sensing environ. Earth Sci 76:598

Lone SA, Jeelani G, Deshpande RD, Mukherjee A (2019) Stable isotope ($\delta^{18}O$ and δD) dynamics of precipitation in a high altitude Himalayan cold desert and its surroundings in Indus river basin, Ladakh. Atmos Res 221:46–57

Lone SA, Jeelani G, Mukherjee A, Coomar P (2020) Geogenic groundwater arsenic in high altitude bedrock aquifers of upper Indus river basin (UIRB) Higher Himalayas. Appl Geochem 113:104497

Lone SA, Jeelani G, Mukherjee A, Coomar P (2021a) Arsenic fate in upper Indus river basin (UIRB) aquifers: controls of hydrochemical processes, provenances, and water-aquifer matrix interaction. Sci Total Environ 795:148734

Lone SA, Jeelani G, Deshpande RD, Mukherjee A, Jasechko S, Lone A (2021b) Meltwaters dominate groundwater recharge in the cold arid desert of Upper Indus River Basin (UIRB), western Himalayas. Sci Total Environ 786:147514

Lone SA, Jeelani G, Padhya V, Deshpande R (2022a) Identifying and estimating the sources of river flow in the cold arid desert environment of Upper Indus River Basin (UIRB), western Himalayas. Sci Total Environ 795:154964

Lone SA, Jeelani G, Alam A, Bhat MS, Farooq H (2022b) Effects of changing climate on the water resources of upper Jhelum basin (UJB), India. In: Mukherjee A (ed) Riverine systems: understanding the hydrological, hydrosocial and hydroheritage connections. Springer, Cham, pp 139–154

Lone SA, Jeelani G, Deshpande RD, Sultan Bhat M, Virendra P (2023) Assessing the hydrological controls on spatio-temporal patterns of streamwater in glacierized mountainous Upper Indus River Basin (UIRB) western Himalayas. J Hydrol 619:129310

Miller JD, Immerzeel WW, Rees G (2012) Climate change impacts on glacier hydrology and river discharge in the Hindu Kush–Himalayas. Mt Res Dev 32(4):461–467

Nathawat MS, Pandey AC, Rai PK, Bahuguna IM (2008) Spatio-temporal dynamics of glaciers in Doda valley, Zanskar Range, Jammu & Kashmir, India. In: Proceedings of the international workshop on snow, ice, glacier and avalanches. IIT Bombay, pp 256–264

Nepal S, Khatiwada KR, Pradhananga S, Kralisch S, Samyn D, Bromand MT, Jamal N, Dildar M, Durrani F, Rassouly F, Azizi F (2021) Future snow projections in a small basin of the Western Himalaya. Sci Total Environ 795:148587

Pandey AC, Ghosh S, Nathawat MS (2011) Evaluating patterns of temporal glacier changes in Greater Himalayan range, Jammu & Kashmir, India. Geocarto Int 26:321–338. https://doi.org/10.1080/10106049.2011.554611

Patel LK, Sharma P, Fathima TN, Thamban M (2018) Geospatial observations of topographical control over the glacier retreat, Miyar basin, Western Himalaya, India. Environ Earth Sci 77(5):190

Paul F, Kääb A, Maisch M, Kellenberger T, Haeberli W (2002) The new remote-sensing-derived Swiss glacier inventory: I methods. Ann Glaciol 34(1):355–361

Qazi NQ, Jain SK, Thayyen RJ, Patil PR, Singh MK (2020) Hydrology of the Himalayas. In: Himalayan weather and climate and their impact on the environment. Springer, Cham, pp 419–450

Rao VB, Franchito SH, Gerólamo ROP, Giarolla E, Ramakrishna SSVS, Rao BRS, Naidu CV (2016) Himalayan warming and climate change in India. Am J Clim Change 5(4):558

Sakai A, Fujita K (2017) Contrasting glacier responses to recent climate change in high-mountain Asia. Sci Rep 7(1):1–8

Scott CA, Zhang F, Mukherji A, Immerzeel W, Mustafa D, Bharati L (2019) Water in the Hindu Kush Himalaya. In: The Hindu Kush Himalaya assessment. Springer, Cham, pp 257–299

Shah RA, Lone SA (2019) Hydrogeomorphological mapping using geospatial techniques for assessing the groundwater potential of Rambiara river basin, western Himalayas. Appl Water Sci 9:64

Sharma KP, Moore B III, Vorosmarty CJ (2000) Anthropogenic, climatic and hydrologic trends in the Kosi Basin, Himalaya. Clim Chang 47:141–165

Shekhar MS, Chand H, Kumar S, Srinivasan K, Ganju A (2010) Climate-change studies in the western Himalaya. Ann Glaciol 51(54):105–112

Shrestha S, Yao T, Adhikari TR (2019) Analysis of rainfall trends of two complex mountain river basins on the southern slopes of the Central Himalayas. Atmos Res 215:99–115

Shukla A, Gupta RP, Arora MK (2009) Estimation of debris cover and its temporal variation using optical satellite sensor data – a case study in Chenab basin Himalayas. Ann Glaciol 55:444–452

Somers LD (2019) Mountain water resources, groundwater and climate change in the Peruvian Andes. McGill University, Montreal

Swain AK, Mukhtar MA, Majeed Z, Shukla SP (2017) Depth profiling and recessional history of the Hamtah and Parang glaciers in Lahaul and Spiti, Himachal Pradesh, Indian Himalaya. Geol Soc London Spec Publ 462:35–49

Villaba R, Lara A, Boninsegna JA, Masiokas M, Delgado S, Aravena JC, Roig FA, Schmelter A, Wolodarsky A, Ripalta A (2003) Large-scale temporal changes across the southern Andes: 20th century variations in the context of the past 400 years. Clim Chang 59:177–232

Vuille M, Bradley RS, Werner M, Keimig F (2003) 20th-century climate change in the tropical Andes: observations and model results. Clim Chang 59:75–99

Wang Y, Hou S, Liu Y (2009) Glacier changes in the Karlik Shan, eastern Tien Shan, during 1971/72–2001/02. Ann Glaciol 50:39–45

Wilson AM, Gladfelter S, Williams MW, Shahi S, Baral P, Armstrong R, Racoviteanu A (2017) High Asia: the international dynamics of climate change and water security. J Asian Stud 76(2):457–480

You Q, Chen D, Wu F, Pepin N, Cai Z, Ahrens B, Jiang Z, Wu Z, Kang S, AghaKouchak A (2020) Elevation dependent warming over the Tibetan plateau: patterns, mechanisms and perspectives. Earth Sci Rev:103349

A Tale of Crab Collectors and Fatteners: Negotiating Climate Change in Indian Sundarbans

26

Anindya Basu and Adrija Bhattacharjee

Abstract

The aquaculture sector is termed as the 'Sunrise Sector' which apart from providing nutrition and food security to the people provides employment opportunities. India currently ranks second in aquaculture production in the world (Ministry of Fisheries, Animal Husbandry and Dairying, Year ender review 2021 on highlight key initiatives and achievements pertains to Department of Fisheries, Ministry of Fisheries, animal husbandry and dairying for the year 2021, PIB, New Delhi, 2021). Inland and marine aquaculture are two important components of the sector. Brackish water aquaculture has been in practice from since 1970s in the *bheris* (anthropogenic impoundments along coastal wetlands) and with frequent coastal inundation this trend has been increasing. Frequent occurrence of tropical cyclones over the Bay of Bengal and its resultant sea-level rise, embankment breaching, and increasing salinity in the agricultural fields are major factors behind the struggle of coastal communities. But coastal people of Indian Sundarban have found wild crab harvesting and crab fattening as viable alternative economic activity due to its greater survival capacity and high demand in foreign as well as local markets. As this livelihood avenue is comparatively new to the Indian context, not much in-depth study has been done about how far crab culture has been effective in tackling the adversities of climate change. This chapter attempts to understand the latent marginalization related to mud crab cultivation and how far it has been successful in providing a viable occupational alternative. The chapter has carried out a perception survey of the mud crab farmers and also tried to provide a pen picture regarding women's participation in mud crab rearing.

Keywords

Aquaculture · Mud crabs · Crab harvesting · Crab fattening

26.1 Introduction

India, predominantly an agriculture-dependent country recognizes the importance of fisheries as an important sector. The aquacultural sector is termed as the 'Sunrise Sector' which apart from providing nutrition and food security to the people ensures employment. Fisheries provides job opportunities to around 28 million people (Meena 2022) and as per an estimate by the National Fisheries Development Board contributes 1.07%

A. Basu (✉) · A. Bhattacharjee
Department of Geography, Diamond Harbour Women's University, Sarisha, West Bengal, India

U. Chatterjee et al. (eds.), *Climate Crisis: Adaptive Approaches and Sustainability*, Sustainable Development Goals Series, https://doi.org/10.1007/978-3-031-44397-8_26

of the country's GDP (Financial Express 2020). India currently ranks second in aquaculture production in the world (Ministry of Fisheries, Animal Husbandry and Dairying 2021). Over the last six decades (1951 to 2012) the sector saw a spectacular 11-fold increase from 0.75 million tons to 9.6 million tons (FAO 2021) and since 2014 it has maintained a double-digit average annual growth and accounts for 7.56% of global production (Ministry of Fisheries, Animal Husbandry and Dairying 2021). The sector has not only helped to ensure the country's livelihood security but has also helped to earn a reasonable amount of foreign exchange; the export earnings reached Rs.46,662.85 crores during 2019–20 (Ministry of Fisheries, Animal Husbandry and Dairying 2021).

Inland and marine aquaculture are two important components of the sector. Brackish water aquaculture has been in practice since 1970s in the *bheris* (anthropogenic impoundments along coastal wetlands) and with frequent coastal inundation this trend has been increasing. The first experimental brackish aquaculture farm at Kakdwip, West Bengal was monitored by the Central Inland Fisheries Research Institute under the Indian Council of Agricultural Research in 1973 (FAO, 2021). Even long back in the 1990s, this kind of semi-intensive farming had a yield of 4–6 tons/ha (Surendran et al. 1991) and micro-credit facilities were available to the small land-holders (owning less than 2 ha), who were the majority, incentivizing the sector, especially the growing shrimp farming. The tidal water is impounded within the *bheris* with the help of *bunds* (embankments) and often there are sluice gates which allows passage of tidal water and assorted fishes. Due to the presence of natural nutrients, the cost of external input is pretty less, and at a low-cost good yield of marketable-sized fish and shrimp is procured.

Sea-level rise is one of the major threats to entire South and Southeast Asia (IPCC 2001) and it is a threat to the community living in Sundarban too who are already vulnerable in terms of socio-economic and ecological aspects (Mazumdar et al. 2010). To promote resilience and capacity to adapt to stress and change, successful adaptation strategies especially in terms of livelihood choices have to be implemented (Parry et al. 2007). Frequent storm surges, and increasing salinity did lead to reduced agricultural yield, and thrust on tiger prawn (*bagda*) culture increased. The profitability of shrimp farming faced a dip due to the increasing salinity and surface temperature rise leading to the frequent onset of viral diseases. With dwindling yield rate and frequent disease outbreaks the fish farmers started looking for viable alternative avenues and keeping in tune with climate change, the culture of mud crab (*Scylla serrata*) colloquially known as *jatikankra* or *habakankra*, one of the major crustacean decapods under Portunidae family (Piatek 1981) became a major hit; it could deal with the climatic adversities, high salinity (Rahman 2016) and fetched high export prices making the crab farming highly remunerative (Ghosh 2019).

26.2 Literature Review

The initial demand for mud crabs was seen in Malaysia and Singapore, which apart from their home-grown crabs depended on imports from Indonesia, India, Sri Lanka, and the Philippines since the 1980s and Bangladesh joined the league a bit later (Angell 1992). Recently, countries like China, Taiwan, Singapore, Thailand, Hong Kong, Japan, the USA, and some parts of Europe have also become important importing nations (Azra and Ikhwanuddin 2016; Chakrabarty 2017). Kathirvel and Srinivasagam (1992a, b) traced the growth of Indian crab culture practice; compared to the 1960s an eight-fold increase in mud crab production was witnessed in the 1980s and in recent years it emerged as an export-oriented trade with potential. After the sectoral contribution is analyzed, it is seen that the brackish water crab has the lion's share (more than 65%), while the marine sector lags far behind (Kathirvel and Srinivasagam 1992a, b). Out of the 15 edible crabs available in India, the mud crab, Scylla serrata (Forskal), is 4.5% of the crab landings (Banerji 1969). It was highlighted that 167,193 ha of brackish water area where shrimp farming thrived can be suitable for crab farming too.

Overton and Macintosh (1997) noticed that mud crab culture which was previously practiced traditionally only on a small scale is now emerging as a demanding economic activity that is replacing farmers' interest from commercial shrimp cultivation to crab farming in a commercial way. Wang et al. (2016) have observed that traditional Chinese major carp polyculture practices inflict negative impacts on the environment and its sustainability whereas Chinese mitten crab cultivation as a traditional as well as sustainable farming practice has experienced a major spring in response to a search for a tenable economic activity. The authors also established a positive linear correlation between crab yields and stocking density where a negative correlation between their yields and pond size and capture size. Rahman et al. (2018) have given soft-shell mud crab farming the distinctive dress of the most resilient economic opportunity for natural disaster-prone coastal areas of Bangladesh which creates significant employment opportunities for marginal households as well as women. Rahman et al. (2020) calculated the Human Development Index that is linked to mud crab fishery and have done a SWOT analysis which indicates positive factors predominate over negative factors revealing that mud crab farming can appear as an alternative livelihood opportunity for climatically vulnerable coastal communities. Washim et al. (2022) commented that mud crab has become the most popular farming practice of the south-west coast of Bangladesh and tried to find out the suitable stocking density of juvenile mud crab that should be economically viable. Nandi et al. (2016) have spotted that uncertainty of species-level identity and wild harvesting of under-sized crabs are major constraints of capture fishery and crab fattening practice in Indian Sundarbans.

Mud crab fishing grounds are common in remote areas of Indian Sundarban too, prominent crab landing areas are Basanti, Sonakhali, Canning, Kakdwip, and Namkhana (Raj 1992). Though crab culture is a year-round activity, an abundance of production varied seasonally along the southwest coast and east coast of India (Kathirvel and Srinivasagam 1991). Chandrasekaran and Natarajan (1994) from their studies identified the annual fluctuations and stated that from April to September the concentration was highest. Ali et al. (2020) while conducting a study about the nature of breeding in the Sundarban area of Bangladesh, identified March–April as the peak breeding season, i.e., it is the peak time of spawning; while the second peak was observed in August–September. Mia et al. (2007) through a study in Bangladesh tried to correlate stocking density and production; it was seen that higher stocking density ensured greater total production while lower stocking density (2500 carbs/hectares) ensured a healthier size of the crabs. On the other hand, according to Rahman et al. (2017), a stocking density of around 5000–15,000/ha might yield the highest productivity depending on culture intensity. Nirmale (2012) while enquiring about the effect of the lunar cycle and tides on the behavior and availability of mud crabs in the Ratnagiri coastal area, India came to know that the amount and quality of catch are formidable during night hours, high tide and on new moon days.

The export items primarily involve live crab meat and frozen crab meat, canned crab meat export has not made much progress in India (Kathirvel and Srinivasagam 1992a, 1992b). The quantity of exported live mud crab meat increased steadily– from 36 mt in 1987–88 to 654 mt in 1990–91 (Raj 1992). Twenty years down the line, the production increased sharply. Live mud crab export from India was higher in 2011–2012 (3054 mt) compared to 2010–2011 (2963 mt). Frozen and other forms of crab export declined in 2011–2012 (2848 mt) compared to 2010–2011 (3251 mt) (Thampi et al. 2015). The growing demand for mud crab in Asia, Europe, and America has stimulated an increase in mud crab production in the Indian subcontinent (Bir et al. 2020). In India, the present rate of molted crabs sold by grassroots-level crab farmers to crab pond operators is Rs. 120–180 per kg. After pond fattening, the market price of crabs increases to at least Rs. 250–300 per kg (Kulasekarapandias and Panigrahi 2009).

Sahu (2013) have found that though women play a significant role in crab farming in the Gosaba block of South 24 Parganas their participation is mostly confined to feed application and is remarkably less in the marketing sector. They have also mentioned about lack of seed availability, the presence of middlemen, the lack of proper transport facilities, and proper market information as major constraints of crab farming.

26.3 The Pioneers and Peers

As per Shelley (2008), the pioneer in the production of farmed mud crab culture has been China, followed by Vietnam for quite a long. At present, Bangladesh has also stepped up in mud crab production. Fisheries have always been an integral part of Bangladesh's economy – both inland and marine. Shrimp culture also became very popular and profitable but a gradual shift to crab farming took place of its lesser vulnerability to climate change and disease outbreaks (Paul and Vogl 2011; Islam et al. 2015a, b; Akber et al. 2017). Initially, conventional pond crab culture was practiced with a higher mortality rate, due to a lack of quality seeds, required stock density, rudimentary management practices (Rahman et al. 2020). A new lease of crab culture was devised, by the Bangladesh Fisheries Research Institute through artificial breeding and hatchery conditions. Bangladesh has handled mud crab aquaculture in commercial scale in a very promising manner, with crab fattening making its way in mid-1990s (Azam et al. 1998; Azra and Ikhwanuddin 2016; Hungria et al. 2017) which has led to high export earnings and ensuring employment generation for the marginalized coastal population (Jahan and Islam 2016; Rahman et al. 2020). There are three distinct farming strategies- a) grow-out culture involving rearing of crabs in ponds; b) crab fattening where soft-shell crabs are grown and c) soft-shell crab fattening which is done through box or cage culture (Rahman et al. 2018).

26.4 Perceived Climate Change Impacts on Mud Crabs and Related Local Livelihoods

A multi-agency regional fisheries program, Bay of Bengal Program (BOBP) involving seven countries around Bay of Bengal – Bangladesh, India. Indonesia, Malaysia, Maldives, Sri Lanka, and Thailand provided attention to modernization of pisciculture in the Indo-Pacific region. As the global demand for mud crab (Scylla sp.) increased sharply in the 1990s for its size and meat content, BOBP encouraged the growth of artisanal fisherfolk engaged in crab framing. In Central Java, growout crab culture was more rampant but elsewhere due to a high mortality rate primarily due to climate change vagaries leading to seed shortages. Due to fast turnover rate and good survival crab fattening became more popular amongst small-scale crab farmers through a variety of floating cages and pens. Often the seed stock for crab fattening is procured from local markets and dealers while in Malaysia, stocking material for fattening operations was predominantly imported from Thailand, Sri Lanka, and Indonesia (Angell 1992).

The natural growout system, which came into use earlier was mostly pond-based in mangrove-covered areas. The initial form of aquaculture was extensive, involving large mangrove silviculture ponds where the stocking rate was very low and no supplemental feeding was required; but while the commercial demand increased the yield rate faced a decline due to climatic uncertainties. To keep with the demand, the growout method adopted an intensive approach, with higher stocking rates and supplemental feeding arrangements (Keenan 1999).

As lean crabs failed to claim a high price from local or export markets, fattening was done in a controlled environment leading to monoculture of crabs, leading to biodiversity loss, but keeping commercial viability in mind. The ideal temperature (20–30 degree Celsius), salinity (5–25 ppt), dissolved oxygen (4-8 mg/l), and salinity (7.5–8.5) are maintained in the artificial set-up, be it a pond or pen so that the stocking density is not affected by the impact of climate change (Huq et al. 2015). Mohapatra et al. (2007) conducted a study in Chilika lagoon and derived a positive relation between salinity and crab landing, concluding that crab farming will be able to maintain its viability through the phase of global warming and climate change to certain extent.

When ponds are used for crab fattening purpose, appropriate bunding and trenching are done so that the crabs cannot escape from the water body and also create burrows for shelter during several phases of their life cycle (Table 26.1). Water exchange in such ponds is done either by tides or through simple concrete sluice gates. But due to cyclonic events, this entire system often gets disrupted; the embankments are breached, captive crabs either flee freely or due to intrusion of saline water the water pH balance for crab culture is affected (Angell 1992). Taking into commercial viability into account, rearing juvenile crabs for a period of 4 months in a medium-sized pond generates about 50,000 rupees profit while with similar capital and time investment crab fattening realizes a profit margin of around 87,000 rupees (Kathirvel et al. 2004).

In most cases, farming or fattening of mud crab is totally dependent on capture of wild crablets. Due to the changing climate the supply of crablets has dwindled for which indiscriminate catching in terms of area and season is putting more stress on the wild stock. Gradually, there has been a noticeable decline in the mud crab population in the natural habitat throughout the Indian coastal area due to this over-exploitation of wild seeds (Thampi et al. 2015).

Table 26.1 Typical life cycle of mud crab

Stage	Characteristics
Zoea	Initial larva stage after hatching from an egg
Megalopa	A zoea grows by molting four times during a period of 12 to 15 days As it molts for the fifth time, it transforms into a megalopa, which has functional claws
Juvenile crab	Adulting stage is achieved about a month after hatching
Young adult	Reaching sexual maturity at 18 to 24 months. Often grading is done at this stage
Mating	Adult crabs generally mate in warmer months
Spawning and hatching	Female crabs migrate offshore to spawn and restart the cycle

Source: Modified from Bir et al. (2020)

26.5 Study Area

Coastal ecosystem are complex and dynamic areas that are continuously affected by biophysical changes, demographic pressure, pollution and other factors (Campbell et al. 2006); these extend favorable conditions for the development of fisher community. There is a high market demand of mud crab (both live crabs and frozen crab meat), in domestic as well as in export arena, especially in south-east Asian region (Pripanapong and Tongdee 1998) for its taste and fetches high price. The predominance of mud crab culture is seen along the estuarine, swampy eastern coast of India and mangrove areas of deltaic Indian Sundarban is a natural habitat for them (Overton and Macintosh 1997).

Sundarban, a deltaic ecosystem formed at the confluence of the Ganges, the Brahmaputra, and Meghna Rivers, spread over India and Bangladesh is famous as the largest mangrove forest globally and is designated as UNESCO World Heritage Site for its rich biodiversity. The Indian section of Sundarban comprises of 102 islands, out of which 54 are inhabited, housing over 4.5 million people. This the mangrove dominated deltaic area is guarded by mangroves and surrounded by earthen embankment to absorb storm surge impacts during extreme weather events and to regulate the ingression of saline water (Sandilyan and Kathiresan 2015). Residents, who have settled in here are primarily dependent on agriculture and aquaculture (Raha et al. 2013). With increasing cyclonic frequency, embankment breaching and rising salinity affect the ecosystem and the productivity of traditional agriculture and aquaculture has dwindled (Ghosh and Roy 2022). To deal with the marginalization of the ecological refugees due to the climatic vagaries, gradual change in aquacultural practices have taken place in tune with that (Basu et al. 2021).

It was observed that between 1980–2007 the decadal rate of global sea surface temperature warming was around 0.06 °C, while for the Sundarban area, it was as high as 0.5 °C per decade (Ghosh 2012). This severe global warming, resultant climate change experienced in terms of increasing cyclone intensity and frequency put the sensitive mangrove ecosystem into danger (Ghimire and Vikas, 2012). The 2011 State Action Plan on Climate Change (SAPCC) accredited that the devastating cyclone Aila reaffirmed that the southern coast of West Bengal is extremely susceptible to storm surges and cyclones (GoWB Government of West Bengal (GoWB) 2010), affecting inland fisheries and agriculture.

Total population of the entire South 24 Parganas district is 81,61,961 out of which in Gosaba and Basanti Community Development blocks it is 2,44,721 and 3,31,973 respectively. The population density is almost similar in both – 825 persons/sq. km for Gosaba and 821 persons/sq. km for Basant. But there is a discrepancy in percentage of literate population and percentage of people belonging to below poverty line (BPL). Interestingly, Gosaba has a literacy rate of 69.67% and it is only 58.12% in Basanti while BPL population percentage is 38.02 and 64.89 respectively which depicts an inversely related scenario (Census of India 2011). Since, both the blocks are vulnerable in terms of socio-economic and climatic criteria, there is an urgent need for the residents to pursue livelihood activities that will be in tune with their expertise and changing climatic regime.

Inhabitants of Sundarban – the 'climate change hotspot' who are highly vulnerable to climate-change have found out ways to counter sea level rise, coastal erosion, tidal surge and saline water inundation by (Islam and Wahab 2005; Uddin et al. 2013) introducing brackish and saline water aquacultural practices involving shrimps and crabs. Mud crabs are extremely common in mangrove dominated areas as their habitat conditions and dietary requirements are well fulfilled (Hill 1976).

Initially, it was the marginal farmers who faced the brunt of climate change, which affected their age-old agriculture, fresh water fisheries and livestock rearing who looked for alternative vocation and collection of crab seeds though extremely strenuous, became highly remunerative for them (Zafar and Siddiqui 2000). The small-scale crab farmers with limited capital started to rear small, lean mud crabs in saline ponds until they reached a marketable price and earned a slim profit margin. Lack of access to modern technology mainly due to financial crunch deterred the marginal and small-scale farmers from thinking big in terms of production scale and managerial dimension.

In West Bengal, due to presence of brackish water, mud crab culture is being popularized in districts of South 24 Parganas, North 24 Parganas, and Purba Medinipur; the highest concentration is in the estuarine areas of South 24 Parganas where traditional crab cultivation and crab fattening is going on side-by-side (Roy et al. 2021). It was roughly estimated that over 30,000 residents are involved in crab culturing and fattening in Indian Sundarban (Fig. 26.1).

Due to presence of higher number of crab farmers Basanti and Gosaba blocks of the said district have been selected for intensive field survey during 2022. (Basanti viz. Gadkhali, Majidbari, Radhaballavpur, Ramkrishnapur; Gosabaviz. Arampur, Bali, Bijoynagar, Bipradaspur, Pakhirlalay,Satjelia,).The prominent mud crab landing areas are Kakdwip, Namkhana, Gosaba, Basantiand Satjelia. The main wholesale market and break of bulk point is at Canning (Pramanikand Nandi 2012).

26.6 Methodology

The study involves detailed literature review to understand the rapidly changing aquacultural scenario in Sundarban area. In depth primary survey, mostly involving qualitative research techniques like Focus Group Discussion, case-studies, was done to understand the struggle for existence of the climate-change ridden people of Indian Sundarban specifically, most of whom have opted for aquaculture to survive. Survey comprised of three segments of crab farmers involved with carbling collection, traditional pond-based rearing, and crab fattening.

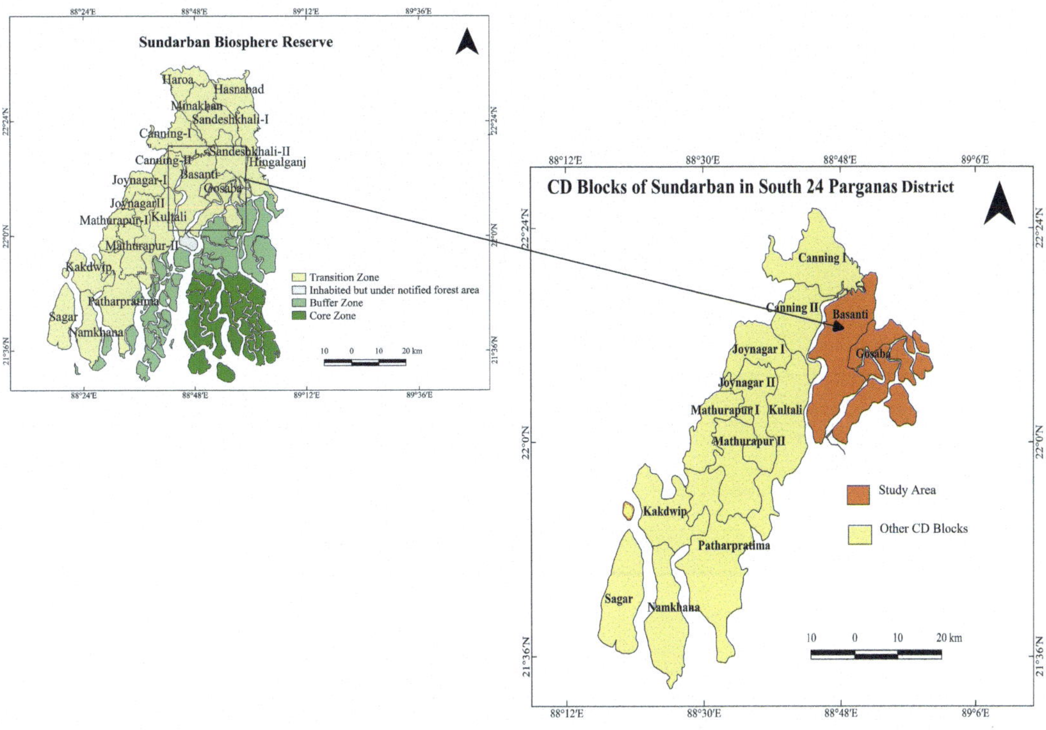

Fig. 26.1 Introducing the study area, prominent mud crab culture zones of South 24 Parganas, West Bengal, India

A SWOT analysis has been done to identify the ways forward to ensure sustainable crab farming (Fig. 26.2).

26.7 Objectives

The main objectives are to understand the nuances of mud crab culture exposing the latent marginalization related with it; to conduct perception survey of the mud crab farmers bringing out the ground reality and to highlight the role of women in the crab rearing process.

26.8 Result and Discussion

The rising frequency and intensity of tropical cyclones and resultant embankment breaching, flooding have hampered cultivation, rising temperature and salinity even affected the alternative vocation of shrimp aquaculture too. Initially, crab cultivation was started to deal with the climate change and global warming-related hazards. So, the livelihood avenue which started as an economic necessity has now showed a new way for higher profits and a better standard of living.

26.8.1 Crab Rearing: Gradual Evolution

Crab rearing or fattening ponds or *gher* are created with great care as the crabs have a tendency to escape by burrowing through the embankments. Normally the crab ponds are small tidal ones with comparatively high embankment walls and bamboo stilt support of about 60 cm depth, to retain the crabs, and the enclosures have a strong nylon net fencing supported by bamboo stilts. While preparing the pond every season, first the water is drained out, the bottom sludge is removed, bottom soil is completely dried up and treated with lime for around a week. Then the

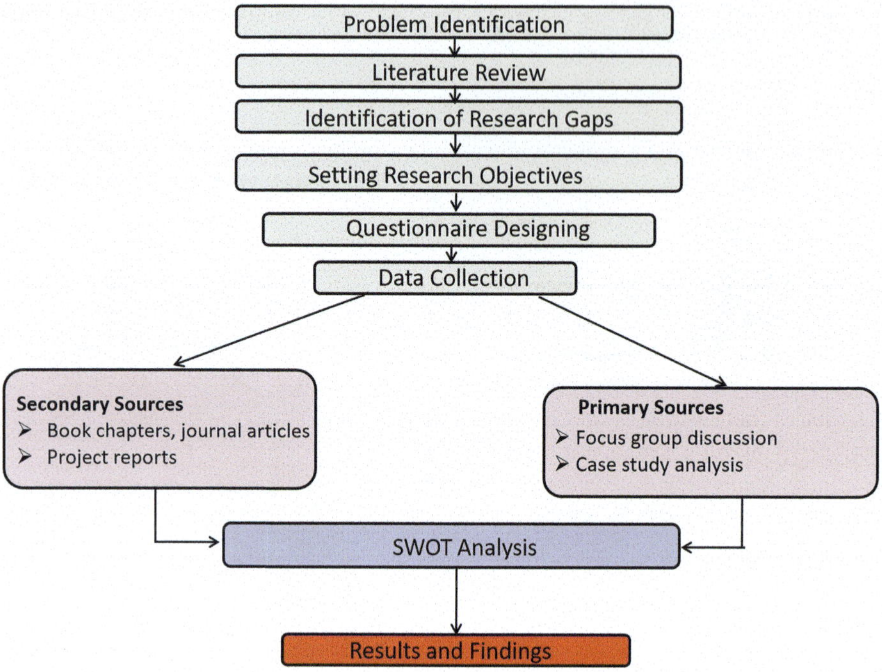

Fig. 26.2 Methodology flowchart to understand the nuances of crab collection and fattening

gher is filled up with tidal water at least up to 3 feet and then mud crabs are stocked (Rahman 2016). To maintain the water quality of the pond or fattening chambers for healthy growth of crabs, dolomite and zeolite are applied at an interval of 15–20 days, and tidal water is also changed through the inlet-outlet system (Rahman 2016). All through the culture phase, a pond water depth of 1 m is maintained.

There is a seasonal element involved in crab farming; though the preparation goes on year-round – during April–July culturing takes a back-seat as there is lesser availability of wild seeds. Before every season the pond water is drained out, dried out and then is liming is done. Fresh water is again let in through sluice gates or pumped in to maintain the required water level (around 1 to 1.5 m). Then stocking is done with the crab seeds or soft-shelled mud crabs (juvenile crabs within 200 gm of weight). After the selection of appropriate crab seed, they are acclimatized with the pond water by introducing them into plastic bowls and then are let in the prepared ponds in the wee hours of the morning to avoid physico-chemical stress (Wedjatmiko 1994).

Since there is a tendency of the larger crabs to consume the smaller ones i.e. cannibalism, similar smaller-sized crabs are reared at one go; if the rearing pond size is extensive, different compartments are created – where crabs of different sizes, at different stages of culture cycle are reared simultaneously. As cannibalism is sharply reduced with similar sized fattened crabs at the same pond (Say and Ikhwanuddin 1999), those can be stocked at higher densities (15 crabs/m^3) compared to grow-out systems (1 crab/ m^3). Depending on the crab size the harvest is completed within 3–8 months. Crab feed is predominantly bivalve meat or trash fish (like Puntius sp., Johnius sp., *Setipinnaphasa, Coiliadussumieri, Tilapia nilotica*, soft-shelled snails, and mussel meat) and the amount daily fed is 5–10% of the body weight to ensure healthy molting and growth (Kerala Agricultural University 2021). To ensure high price proper monitoring and sampling of the produce is done and the abnormal ones are weeded out. The length and width are measured so that they can be classified accordingly for proper selling price bracket. Besides, water quality monitoring is also done periodically to resist disease outbreak.

Apart from natural crab culture and rearing, popularity of non-conventional culture system -crab fattening is catching up fast. It involves stocking of 'soft shell' crabs within small impoundments for a period of approximately 21 days until the shell hardens leading to further 'fleshing' and higher economic return (Sahu 2013). Currently, native crab species from Sundarban and species from Gujarat are predominantly fattened for one or two cycles depending on the growth and desired weight with the present market rate for fattened females being around Rs. 900/kg.

As per the documentations, in West Bengal, the crab fattening technology was introduced by Sri Dibakar Majumdar of East Ganeshnagar village of South 24 Parganas, who currently maintains over 7000 boxes (Ghosh 2019). The technique closely followed the Vietnamese concept using perforated, submerged boxes made of durable plastic, in small to medium-sized ponds, for stocking male and female crabs individually in each box compartment for molting. As his venture bore success, manufacturing of fattening boxes started due to high market demand was initiated under the brand of Bengal Mud Crab Aquaculture with a length of 33.5 cm, height of around 15 cm and weight of 700 gm approximately and was sold at a rate of Rs. 200 per box (Ghosh 2019).

For stocking the crabs, lean male species are chosen which are around less than 400 gm while the female counterparts with under-developed ovaries weighing around 150–200 gm are selected. The crabs are harvested when the shell harden, before the next molting when the desired size is reached. Mud crab is fattening is highly appropriate for small-scale aquaculture as the turnover is very quick, with the gap between investment and return remaining very brief. Besides, short production time ensures lesser disease outbreak ensuring a healthy survival rate of around 90%, while that of the traditional grow-out systems with raising of young crab seeds hovers around 50%. In West Bengal, the large crabs with a width of over 14 cm and of stipulated weights (over 500 gm for male crabs and over 300 gm for female crabs) are mostly earmarked for export purpose predominantly in China, Singapore and Malaysia through same-day air-lifting (Sahu 2013). The entire mud crab export is handled by nine exporters located in Kolkata city (Pramanik and Nandi 2012).

26.8.2 Knowing Who Matters: Demographic Profile

The crabling collectors belong to the lowest rung and are mostly landless laborers or fringe cultivators who have very little resource backing. A crabling collector in search of healthy stock ventures deep into the forested area for a trip of 7–10 days, risking their lives and in lieu of that earns a handsome amount of around Rs. 15,000 which is not available to them for any other livelihood options. Mustaq Ahmed of Satjelia shared his experience "Often we need to go to the densely forested estuaries, ignoring the dictate of the forest department authority, as in the remote areas the crab seed catch is higher in amount and the quality is also good which fetches high market value. We fight with tigers and crocodiles every day to earn our livelihood, taking extra risk do not bother us anymore. We all here will die anyway either in abject poverty or in tiger attack."

Almost 37% of the crab farmers and fatteners possess small sized water bodies (less than 5 kathas), majority (61%) of the crab rearers possess medium sized water bodies (5–10 kathas). [1 Katha = 126.44 sq. m] Only 2% of the respondents have ownership of large water bodies (more than 10 kathas) – this is the section which is comparatively well off and are getting interested to pursue the more profit generating crab fattening business (Fig. 26.3).

As in the study area, traditional joint family system is predominant (82% of the surveyed households) the vocation pursued in most cases is done collectively, involving a common hearth. Though presently, the younger generation (73%)

Fig. 26.3 Slices of reality – Aspects of crab rearing in South 24 Parganas, West Bengal, India a. Crab fattening going on in ponds; b. Embankment construction around '*gher*'; c. Preparing Pond for the next cycle; d. Filling up pond with brackish water; e. Continuing grow-out culture; f. Setting aside the quality produce; g. Getting ready for market; h. Traditional aquacultural pond; i. Modernized soft-shell crab boxes. (Source: Field photographs by the authors, 2022)

is moving out in search of higher education, secured job and a better standard of living, the older members are still trying find out ways to battle with the nature. The level of education of the respondents are mostly (65%) limited to primary and secondary while in case of women respondents' illiteracy or bare alphabet familiarity is more common. But lack of exposure to institutional education has not been an impediment for the crab cultivators to adapt to the new techniques of crab culture effectively. Since systematic mud crab culture in this area does not have a long history, maximum length of experience very few respondents (4%) stated was of around 7 years, majority (72%) claimed an experience of around 5 years. Rest 14%, who are predominantly associated with crab fattening, a more modern technique has been engaged in the vocation for less than 3 years.

Md. Mustakim of Ramkrishnapur rued that "adaptation is the only way out for us. To cope with the climatic vagaries, we have changed our vocation, learnt new techniques and have toiled hard only to ensure that we are alive, profit is a mirage for marginal farmers like us."

26.8.3 The Ground Reality: Perceptions of the Stakeholders

When asked about the income, the respondents highlighted that there is a seasonal element – in the winter season, the large or moderately-sized produce fetched high returns of approximately Rs. 1000/kg. If profit generation is in question, then it depends on many factors as disclosed by the crab farmers like growth cycle, survivability, individual crab size and negotiating capability with the middlemen.

Hannan Alam of Jharkhali said "Very little amount is retained by as surplus income. The profit margin varies greatly from Rs. 100-300 per kg depending on the demand-supply scenario." Biplab Naskar from Pakhirlalay expressed his plight "During the recent COVID lockdown, the entire system almost collapsed. The local demand was low, transportation system broke down and our mainstay export was just stalled." Manas Purkait a crab cultivator from Bijoynagar, Gosaba admitted "God willing, if everything falls into place then from an investment of rupees one lakh in a year six lakhs can be earned from mud crab culture; in a year we can have up to nine proper yields."Dibakar Jana of Majidbarigave an insight to the profit generation mechanism. He divulged "The full-grown female mud crabs are gems in the international market, during winter season the consignments for South Asia fetches price as high as Rs. 2000/kg. But it is not possible to have that kind of mono-culture continuously and there is a mix of lean female and male counterparts which caters to the domestic market and fetches lesser profit."

As the crab farmers are mostly economically marginalized and from rural interiors, their direct contact with the city is limited. Either the small aquaculturists hand over the produce to the local aggregators or at best take them to the nearest landing areas and hand over to the middlemen to transfer the produce to the city or to the licensed exporters. But this complex chain robs off the just amount of profit from the actual producer. The respondents seemed to be pretty aware of this fact but they confessed that there was no other viable way-out until there is some governmental intervention that ensures a Minimum Support Price and pick-up of produce from neighborhood. Samrat Dhali of Bali pointed out the loopholes clearly "The healthy crabs fetch the maximum price. As proper farm management is not there, the harvested crabs have to be tied with nylon ropes with great care so that they are not injured during the long transportation journey and in that process, we do lose out few from internal injury raising the mortality rate."

26.8.4 The Gendered View: Involvement of Women

The aquacultural activities have been gender neutral as since inception women have been proactive in the production process (FAO 1987) and crab cultivation is not an exception. For women, it is also an avenue for contributing economically and a welcome break from household chores. The activities involves both male and female of a household at various work stages (Kevane and Wydick 1999). Women in Sundarban are part of the marginal workforce and many of them are involved as unskilled labor in risky and strenuous vocation like crab catching (Ghosh et al. 2018). Tedious jobs like collecting crablings, feeding the crabs are often carried out by women while monetary aspects like purchase of materials, post-harvest management and marketing are taken care of by men (Shaleesha and Stanley 2000) – as both are involved in decision-making at various stages, a gender equality in the entire process can be noticed (Sharma 2003). As crab fattening is gradually being promoted, institutional loan facilities are being extended to the Self-Help Groups of women.

Two Focused Group Discussions (FGDs) were carried out with the women involved in crab farming, one in Arampur, Gosaba and the other in Jharkhali, Basanti. The interaction with the first group occurred in the retail market area of Gosaba; most of the women were accompanying their husbands who were bargaining and settling deals. Several issues cropped up while having an informal discussion with them – few, who were associated with collection of crab seeds and subject to abject poverty, complained of long hours in the waist-deep water which caused skin diseases and affected their intimate health too. Two of them also rued that they lost their male family member to tiger attack, as for quality seed collection they ventured into the prohibited dense mangrove zone. Most of them complained that though there is provision of micro-credit system they faced a lot of difficulties in setting up their own Self Help Group carrying out crab farming. The other group at Jharkhali also echoed the same problem of institutional financial assistance and also highlighted the high handedness of the middle men and exporters who try to 'rob off their legitimate profit'. Another underlying emotion that was palpable was that even though women worked as hard as their male counterparts they were not given due importance in decision-making like selection of techniques, determining marketable size etc. but were treated more as laborers even by their own family members.

26.9 Hindrances Faced

The issues faced involves lack of seed availability; dearth of institutional loan availability; exploitation by the middleman; lack of proper transport facility; large-scale cold storages for post-harvest storage, accurate market information, suitable insurance framework and regular training facilities curbing their full potential to achieve a high return from their investment.

At times, very young crablets or under-sized crabs are harvested which needs to be controlled through legal provision like that in Australia, as such rampant practices will affect the diversity (Shelley 2008). In the storm-stricken, saline Sundarban crab culture is probably the only option as of now. I have been carrying out brackish water-based crab fattening for last 3 years and

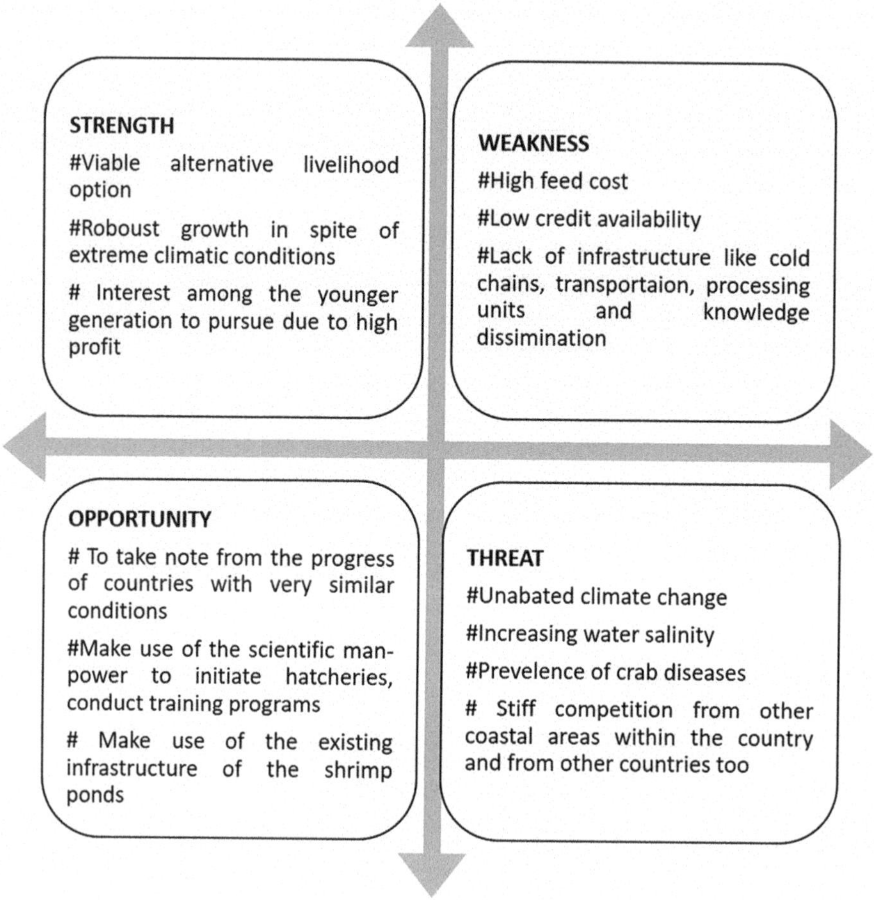

Fig. 26.4 SWOT Analysis to gauge the prospect of crab farming in in South 24 Parganas, West Bengal, India

the return has been handsome. But gathering suitable wild seeds is being an increasing issue as due to profitability many are opting for crab culture and huge stress in being exerted on the natural stock. A government supported hatchery in Sundarban is the need of the hour." But it has to borne in mind that hatched domesticated stock is not free from all problems, they are not as robust as the wild ones and are more prone to diseases (Shelley 2008).

The main concerning is the marketing chain, which is dominated by brokers, wholesalers and distributers whose intervention forces the crab cultivators to sell their produce at a compromised rate, in spite of generating a hefty revenue from the end users i.e., consumers (Fig. 26.4). There are many scientific studies, gatherings and exchanges but that is mainly limited to the academicians and there has to be continuous efforts to disseminate that to fish farmers and primary producers.

26.10 Concluding Observations

The mud crab fishery supports the livelihood of a huge number of crab collectors, fatteners, growers, depot-holders, traders, brokers, transporters and exporters (Molla et al. 2009). According to a recent information of Department of Fisheries of Bangladesh, some 3,00,000 to 4,00,000 people are directly or indirectly involved with mud crab aquaculture for their sustenance (Chakrabarty 2017). Since the world market for live mud crab is expanding it is possible for India to emerge as the largest exporter in the world with a turnover of 100,000 tons worth of about Rs. 2000 crores every year (Kulasekarapandias and Panigrahi 2009).

Still the mud crab trade is subject to wild capture mostly from coastal mangrove and there is no absolute procedure for hatchery production (Bir et al. 2020). Rahman et al. (2017) pointed out that in Bangladesh, most farmers purchase seed crabs from depots, while few collect directly from crabbers exploiting wild stocks. While, Quinitio et al. (2011) dealing with cases in Philippines noted that producing good-quality captive brood stock is a precondition for productive domestication. Problems identified in crab hatcheries by Quinitio et al. (2002) are egg loss in females due to fungal infection and epibiotic fouling; luminescent bacterial attack; molt death syndrome (MDS) during in seed production; and severe cannibalism among megalopa (Quinitio et al. 2015). Another perspective has been pitched in by Sarower et al. (2012) that the high mortality rate of the embryo and larvae might be due to low dose of disinfectant usage and improper transportation of eggs.

The common procedures followed to keep away pests and predators from affecting the earthen ponds are application of lime, usage of tobacco dust, ammonium sulfate; besides usage of fence, nets are made to enclose the ponds to avoid escape of crabs. The use of antibiotics for treatment has proven to improve larval survival but excessive use of antibiotics might lead to the development of resistant strains of bacteria and abnormalities that become apparent at the juvenile stage. The inconsistent survival from later zoea to megalopa stage due to molt death syndrome (MDS) remains the major problem in mud crab hatchery. It has been suggested that poor nutrition, low water temperature and prophylaxis application in the zoeal stage are some of the causes of MDS. In the nursery phase when the crabs molt, cannibalism becomes a serious concern. Adequate feed supply along with use of bio-materials like leaves, twigs can be effectively used to reduce it (Rahman et al. 2017). Another labor intensive method, trimming of claws have also been implemented in some cases to avoid this cannibalism (Quinitio and Estepa 2011). Proper training regarding handling crab aquaculture catering issues like stocking, feed, cage or pen maintenance, handling cannibalism and harvesting are required so that the farmers are well-equipped to maximize profit (Ladra 1992). Many scholars have spoken in favor of imposing ban, of about 2 months, during peak breeding season ensuring safe migration spawning of berried females.

From the study it can concluded that in the Indian Sundarban area apart from natural crab culture and rearing, the popularity of non-conventional crab fattening is increasing. The crab farmers, who primarily deal with small to medium-sized ponds, from their field experience now emphasize on appropriate selection of crab seeds, water quality monitoring, systematic nurturing and scientific grading to ensure that farming is carried forward effectively. The severe impacts of climate change affected the livelihood avenues of the study area, resulting in a mass-migration of the male youth for which the role of women in this vocation has been very prominent. The women participate in the socio-economic mobilization in crab culture as they collect crablings, lime the pons, prepare the crab feed by mincing trash fishes or feeding the crabs.

26.11 Ways Forward

To realize the potential of 'blue economy' infrastructural and organizational overhaul is needed. This sector apart from providing livelihood opportunities, export earnings – is an affordable and rich source of animal protein for the common people. So, focused attention is needed through policy framework and financial support to ensure sustainable, inclusive and responsible growth. To sustain this promising, alternative livelihood generous funding from governmental and non-governmental agencies are required and exposure to modern aquacultural practices through training programs are required. Besides, excessive dependency on wild mud crab seeds might lead to over-exploitation and hamper the natural stock; so, establishment of hatchery is required to flourish commercially (Rahman 2016).

Since there has been a sharp rise in aquacultural practices especially shrimp and crab farms, due to the apparent high monetary return, certain legislative irregularities are being witnessed. There are several age-old acts like Fisheries Act (1897), Environment (Protection) Act (1986), Water (Prevention and Control of Pollution) Act (1974) etc. which needs to be followed for ecological sustainability. To deal with the environmental crisis created by the aquacultural practices, Supreme Court of India in 1996 passed a historic judgment prohibiting setting up of aquacultural farms in coastal areas within the Coastal Regulation Zone.

From Ministry for Fisheries and Animal husbandry a separate Ministry of Fisheries, Animal Husbandry and Dairying headed by a Minister of Cabinet rank was formed which had a special allocation budget of Rs. 3737 crores to tap the huge potential of this sector involving both inland and marine fisheries (John 2021). Under Pradhan Mantri MatsyaSampada Yojana (PMMSY) with a financial package of Rs. 20,000 crores, aquacultural practices have been safeguarded including that of prawn and crab pens through provisions of livelihood and nutritional support for the aquaculture practitioners during the lean phases, ensuring higher quality breeding stock, provision of ice plants and cold storages, setting up of fish retail market and kiosks alongside concentrating on aquatic health by setting up testing labs and disease diagnostic centers. Imparting skill to the marginalized fishermen is important so that they can step up the production and for that *Matsya Vigyan Kendras* in collaboration with the Central Fisheries Research Institute have been set up (Financial Express 2020). Besides, Fisheries and Aquaculture Infrastructure Development Fund (FIDF) is also there to provide concessional finance for improving infrastructural facilities through banking institutions. A very interesting self-learning app has been launched by the Fisheries ministry and funded by National Fisheries Development Board (NFDB), Hyderabad called "MatsyaSetu" where self-explanatory course modules have been uploaded to make the aqua farmers well conversant with modern technologies and management practices (Ministry of Fisheries, Animal Husbandry and Dairying 2021).

To tap the potential of the country in aquaculture, a Conference on Fisheries and Aquaculture was organized by Confederation of Indian Industry (CII) with the aim of "Showcasing India as a hub for Aquaculture and fisheries Investment" on January, 2022 in collaboration with the Ministry of Fisheries, Animal Husbandry and Dairying, National Fisheries Development Board, and the Marine Products Export Development Authority (Meena 2022).The target set by the government is to reach a production level of 22 million metric tons by 2024–25. It was also reported that 12.89 lakh metric tons of fisheries products were exported for Rs.46,662 crores during 2019–20 (Anand 2022). However, to achieve this attainable target – quality seed, smart farming technology and adherence to food safety standards are essential.

References

Akber MA, Islam MA, Ahmed M, Rahman MM, Rahman MR (2017) Changes of shrimp farming in southwest coastal Bangladesh. Aquac Int 25:1883–1899

Ali MY, Hossain MB, Sana S, Rouf MA, Yasmin S, Sarower MG (2020) Identifying peak breeding season and estimating size at first maturity of mud crab (Scylla olivacea) from a coastal region of Bangladesh. Heliyon 6(6):e04318

Anand S (2022) Indian fisheries sectors should deploy more scientific methods of production. Retrieved from https://economictimes.indiatimes.com/news/economy/agriculture/indian-fisheries-sectors-should-deploy-more-scientific-methods-of-production-parshottam-rupala/articleshow/89039210.cms?from=mdr. Accessed on 30th Dec 2022

Angell CA (1992) The mud crab. A report of the seminar on the mud crab culture and trade, swat Thani, Thailand. Bay of Bengal Programme, Madras, India

Azam K, Kamal D, Mostofa M (1998) Status and potential of mud crab (Scylla serrata) in Bangladesh. Khulna University, Bangladesh, Integrated management of Ganges floodplains and Sundarbans ecosystem

Azra MN, Ikhwanuddin M (2016) A review of maturation diets for mud crab genus Scylla broodstock: present research, problems and future perspective. Saudi J Biol Sci 23(2):257–267

Banerji SK (1969) Crustacean production in India. Bull CMFRI, No 14:259–272

Basu A, Nandy G, Kar NS (2021) Evaluating the viability of shrimp aquaculture to impede climate change in the Sundarban: experiences from Bangladesh. In: South Asia climate change. Routledge, India, pp 156–188

Bir J, Islam SS, Sabbir W, Islam R, Huq KA (2020) Ecology and reproductive biology of mud crab Scylla spp: A study of commercial mud crab in Bangladesh. Intl J Acad Res Develop 5(2):01–07

Campbell J, Whittingham E, Townsley P (2006) Responding to coastal poverty: should we be doing things differently or doing different things?. Environment and livelihoods in tropical coastal zones: managing agriculture-fishery-aquaculture conflicts:274–292

Census of India (2011) District census handbook, North and South Twenty Four Parganas, West Bengal, Series-20 Part XII-A & B, Directorate Of Census Operations, West Bengal. Registrar General of India, Ministry of Home Affairs, Government of India

Chakrabarty BK (2017) Present status of mud crab and potentials of crablets culture in Bangladesh. National Fish Week 2017 Compendium, Department of Fisheries, Ministry of Fisheries and Livestock, Bangladesh

Chandrasekaran VS, Natarajan RG (1994) Seasonal abundance and distribution of seeds of mud crab Scylla serrata in Pichavaram mangrove, Southeast India. J Aquacult Trop, pp 343–349

Financial Express (2020) India's blue economy net getting bigger! Country ranks third in fisheries and second in aquaculture. Retrieved from https://www.financialexpress.com/opinion/indias-blue-economy-net-getting-bigger-country-ranks-third-in-fisheries-and-second-in-aquaculture/1867607/. Accessed on 30th Dec 2022

Food and Agricultural Organization (FAO) (2021) National Aquaculture Sector Overview, India. Retrieved from https://www.fao.org/fishery/en/countrysector/in/en. Accessed on 30th Dec 2022

Food and Agriculture Organization (FAO) (1987) ADCP/REP/87/28-women in aquaculture. In: Proceedings of the ADCP/NORAD workshop on women in aquaculture, 13–16 April, 1987, Rome

Ghosh A (2012) Living with changing climate–impact, vulnerability and adaptation challenges in Indian Sundarbans. Centre for Science and Environment, New Delhi, p 91

Ghosh S (2019) Fattening of mud crab Scylla serrata in estuarine region of South-Eastern West Bengal, India. Aquacult Asia Magaz 23(1):11–19

Ghosh S, Roy S (2022) Climate change, ecological stress and livelihood choices in Indian Sundarban. In: Climate change and community resilience. Springer, Singapore, pp 399–413

Ghosh U, Bose S, Bramhachari R (2018) Living on the edge: climate change and uncertainty in the Indian Sundarbans. In: STEPS working paper 101. STEPS Centre, Brighton

Government of West Bengal (GoWB) (2010) West Bengal State Action on Climate Change, Kolkata

Hill BJ (1976) Natural food, foregut clearancerate and activity of the crab Scyllaserrata. Mar Biol 34:109–116

Hungria DB, dos Santos Tavares CP, Pereira LÂ, de Assis Teixeira da Silva U, Ostrensky A (2017) Global status of production and commercialization of soft-shell crabs. Aquac Int 25:2213–2226

Huq KA, Rahaman SB, Hasanuzzaman AFM (2015) Mud crab culture as an adaptive measure for the climatically stressed coastal fisher-folks of Bangladesh. Environ Manage Govern 2015:175–198

IPCC (Intergovernmental Panel on Climate Change) (2001) Technical summary. In: McCarthy JJ, Canziani OF, Leary NA, Dokken DJ, White KS (eds) Climate change 2001: impacts, adaptation, and vulnerability. Contribution of working group II to the third assessment report of the intergovernmental panel on climate change. Cambridge University Press, Cambridge/New York

Islam MS, Wahab MA (2005) A review on the present status and Management of Mangrove Wetland Habitat Resources in Bangladesh with emphasis on mangrove fisheries and aquaculture. Aquatic Biodiver II:165–190. Dordrecht

Islam MS, Aleem NA, Rahman MM (2015a) Mud crab aquaculture, present status, prospect and sustainability in Bangladesh. World Aquacult 46(3):58–60

Islam GT, Islam AS, Shopan AA, Rahman MM, Lázár AN, Mukhopadhyay A (2015b) Implications of agricultural land use change to ecosystem services in the Ganges delta. J Environ Manag 161:443–452

Jahan H, Islam MS (2016) Economic performance of live crab (Scylla Serrata) business in the southwest coastal region of Bangladesh. Intl J Fisher Aquat Stud 4(1):453–457

John SB (2021) Indian aquaculture: poised for a quantum leap. Retrieved from https://www.dailypioneer.com/2021/state-editions/indian-aquaculture%2D%2Dpoised-for-a-quantum-leap.html. Accessed on 30th Dec 2022

Kathirvel M, Srinivasagam S (1992a) Taxonomy of the mud crab, Scylla serrata (Forskal), from India. In: Angell CA (ed) The mud crab. A report of the seminar on the mud crab culture and trade, Swat Thani, Thailand. Bay of Bengal Programme, Madras, India

Kathirvel M, Srinivasagam S (1992b) Resource and exploitation of mud crab Scylla serrata (Forskal) in India. In: Angell CA (ed) The mud crab. A report of the seminar on the mud crab culture and trade, swat Thani, Thailand. Bay of Bengal Programme, Madras, India

Kathirvel M, Kulasekarapandian S, Balasubramanian CP (2004) Mud crab culture in India. Bull Central Inst Brackishwater Aquacult, p 17

Keenan CP (1999) Aquaculture of the mud crab, genus Scylla-past, present and future. In Aciar Proceedings, Australian Centre for International Agricultural

Kerala Agricultural University (2021) Crab Fattening-Infotech Portal. Retrieved from http://www.celkau.in/Fisheries/CultureFisheries/Crabs/crabfattening.aspx. Accessed on 30 Dec 2022

Kevane M, Wydick B (1999) Social norms and the time allocation of women's labor in Burkina Faso. Review of Development Economics Manuscript:8076

Kulasekarapandias S, Panigrahi A (2009) Training manual on mud crab breeding and culture. Central institute of Brackishwater Acquaculture, Chennai, India

Ladra DF (1992) Mud crab fattening practices in The Philippines. BOBP, Madras, India, pp 151–154

Mahadevia Ghimire K, Vikas M (2012) Climate change–impact on the Sundarbans, a case study. Int Sci J Environ Sci 2(1):7–15

Mazumdar PG, Kanjilal B, Mondal S, Mandal A (2010) Dynamics of resilience in the Sunderbans delta, India: a focus on health and climate shocks. Institute of Health Management Research

Meena S (2022) India continues to maintain its position as the world's second-largest fish producer. Retrieved from https://krishijagran.com/animal-husbandry/india-continues-to-maintain-its-position-as-the-worlds-second-largest-fish-producer/. Accessed on 30 Dec 2022

Mia YM, Shah MR, Alam JM, Begum M (2007) Optimization of stocking density of mud crab. Scylla serrata culture in brackishwater earthen ponds in Bangladesh, Sindh University Research Journal 39(1):25–32

Ministry of Fisheries, Animal Husbandry and Dairying (2021) Year ender review 2021 on highlight key initiatives and achievements pertains to Department of Fisheries, Ministry of Fisheries, animal husbandry and dairying for the year 2021, PIB, New Delhi

Mohapatra A, Mohanty RK, Mohanty SK, Bhatta KS, Das NR (2007) Fisheries enhancement and biodiversity assessment of fish, prawn and mud crab in Chilika lagoon through hydrological intervention. Wetl Ecol Manag 15:229–251

Molla M, Islam M, Islam S, Salam M (2009) Socio-economic status of crab collectors and fatteners in the southwest region of Bangladesh. J Bangladesh Agricult Univ 7:411–419

Nandi NC, Pramanik SK, DevRoy MK (2016) Mud crab culture: relevance of species identity in production economics with reference to Sundarban coast. J Fisher Sci 10(4):84–89

Overton JL, Macintosh DJ (1997) Mud crab culture: prospects for the small-scale Asian farmer. Infofish Intl 5:26–32

Parry ML, Canziani O, Palutikof J, Van der Linden P, Hanson C, (Eds.). (2007) Climate change 2007-impacts, adaptation and vulnerability: working group II contribution to the fourth assessment report of the IPCC, vol 4. Cambridge University Press, Cambridge

Paul BG, Vogl CR (2011) Impacts of shrimp farming in Bangladesh: challenges and alternatives. Ocean Coastal Manage 54(3):201–211

Piatek MA (1981) Mangrove crab, Scylla serrata and its utilization with particular reference to Mozambique, FAO, Rome, p 100

Pramanik SK, Nandi NC (2012) Crab fattening (chamber Chas) – a promising enterprise in Indian part of Sundarban. J Environ Sociobiol 9:78

Pripanapong S, Tongdee NA (1998) Review of the mud crab (Scylla sp.) fisheries and culture in Thailand. Newsletter of Danish-SE Asian collaboration in tropical coastal ecosystems research and training project, Denmark, Thailand and Malaysia, 2(2), 7–10

Quinitio ET, Estepa FDP (2011) Survival and growth of mud crab, Scylla serrata, juveniles subjected to removal or trimming of chelipeds. Aquaculture 318:229–234

Quinitio ET, Parado-Estepa FD, Rodriguez E (2002) Seed production of mud crab Scylla spp. Aquacult Asia 7(3):29–31

Quinitio ET, de la Cruz JJ, Eguia MRR, Parado-Estepa FD, Pates G, Lavilla-Pitogo CR (2011) Domestication of the mud crab Scylla serrata. Aquac Int 19:237–250

Quinitio ET, Parado-Estepa FD, Huervana JJ, Burlas MR (2015) Updates on the seed production of mud crab. In: Resource enhancement and sustainable aquaculture practices in southeast Asia: challenges in responsible production of aquatic species: Proceedings of the international workshop on resource enhancement and sustainable aquaculture practices in southeast Asia 2014 (RESA) (207–212). Aquaculture Department, Southeast Asian Fisheries Development Center

Raha AK, Zaman S, Sengupta K, Bhattacharyya SB, Raha S, Banerjee K, Mitra A (2013) Climate change and sustainable livelihood programme: a case study from Indian Sundarban. J Ecol 107(335348):64

Rahman MM (2016) Small-scale mud crab Scylla sp. fattening in Shyamnagar, Satkhira. Bangladesh. J Zool 44(1):163–166

Rahman MM, Islam MA, Haque SM, Wahab A (2017) Mud crab aquaculture and fisheries in coastal Bangladesh. World Aquacult 48(2):47–52

Rahman MM, Haque SM, Wahab A, Egna H, Brown C (2018) Soft-shell crab production in coastal Bangladesh: prospects, challenges and sustainability. World Aquacult 49(3):43–47

Rahman MM, Haque SM, Galib SM, Islam MA, Parvez M, Hoque M, Brown C (2020) Mud crab fishery in climate vulnerable coastal Bangladesh: an analysis towards sustainable development. Aquac Int 28(3):1243–1268

Raj MM (1992) A review of the mud crab (Scylla serrata) fishery on the east coast of India and in Kerala state. In Angell CA (ed) The mud crab. A report of the seminar on the mud crab culture and trade, swat Thani, Thailand. Bay of Bengal Programme, Madras, India

Roy M, Dey A, Chatterjee D (2021) Mud crabs farming: an alternative sustainable livelihood approach in Sundarban. West Bengal, India

Sahu S (2013) Livelihood generation of women by mud crab (Scylla sp.) fattening in deltaic Sundarban of West Bengal. Intl J Sci Res 10(4):2134–2136

Sandilyan S, Kathiresan K (2015) Mangroves as bioshield: an undisputable fact. Ocean Coastal Manage 103:94–96

Sarower MG, Aktar N, Mostafa M, Sabbir W, Islam MS (2012) Some aspects of captive breeding biology of mud crab, Scylla serrata in Bangladesh. J Innovat Develop Strat 6:1–6

Say WC, Ikhwanuddin AM (1999) Pen culture of mud crabs, genus Scylla, in the mangrove ecosystems of Sarawak, East Malaysia. In Aciar proceedings. Australian Centre for International Agricultural, pp 83–88

Shaleesha A, Stanley VA (2000) Involvement of rural women in aquaculture: an innovative approach. Naga The ICLARM Quarterly:23(3)

Sharma C (2003) The impact of fisheries development and globalization processes on women of fishing communities in the Asian region. Asia Pacific Research Network, APRN J:8

Shelley C (2008) Capture-based aquaculture of mud crabs (Scylla spp.). In Lovatelliand A, Holthus PF (eds) Capture-based aquaculture. Global overview. FAO FisheriesTechnical paper. no. 508, pp 255–269, Rome

Surendran V, Madhusudhan Reddy K, Subba Rao V (1991) Semi-intensive shrimp farming-TASPARC's experience at Nellore. Fishing Chimes, February 1991:23–29

Thampi SR, Yohannan C, Mandal A, Kumaran G, Sethuramalingam A, Srinivasan P, Kumar J (2015) Status of mud crab industry in India. In: Proceedings of the international seminar-workshop on mud crab aquaculture and fisheries management, 10–12 April 2013, Rajiv Gandhi Centre for Aquaculture (MPEDA), Tamil Nadu, India

Uddin MS, van Steveninck EDR, Stuip M, Shah MAR (2013) Economic valuation of provisioning and cultural services of a protected mangrove ecosystem: A case study on Sundarbans reserve Forest, Bangladesh. Ecosyst Serv 5:88–93

Wang Q, Liu J, Zhang S, Lian Y, Ding H, Du X, De Silva SS (2016) Sustainable farming practices of the Chinese mitten crab (Eriocheir sinensis) around Hongze Lake, lower Yangtze River basin. China Ambio 45(3):361–373

Washim MR, Rubel ASA, Ahmmed S, Rahman SL, Islam L (2022) Stocking density optimization of juvenile mud crab (Scylla olivacea) cultivation in bamboo fetched earthen ponds. Aquacult Aquarium Conserv Legislation 15(1):384–394

Wedjatmiko D (1994) The effect of feeding frequency on mud crab (Scylla serrata) growth. Coastal Aquacult Newsl 6(3):37–39

Zafar M, Siddiqui MZH (2000) Occurrence and abundance of four brachyuran crabs in the Chakaria Sundarban of Bangladesh. Chittagong Univ J Sci 24(2):105–110

Climate Crisis and Wetland Ecosystem Sustainability

27

Suchetana Karmakar, Sk Saruk Islam, Krishnendu Sen, Sayani Ghosh, and Sujoy Midya

Abstract

Wetlands and its native ecology are being seriously threatened by extreme global climate change pattern. This chapter summarizes recent research on climate vulnerability of wetlands, as well as their contribution to climate change mitigation. Hopefully, the carbon sequestered by vegetated coastal wetlands and peatlands is among the highest on the planet, equaling that of the world's forest ecosystems. Unfortunately, due to relative sea level rise (RSLR) flood-plain wetlands are deteriorating and gradually declining which projected 3.4 Pg blue carbon loss by the year of 2100. Globe has experienced a significant change in climate, with an upward trend in temperature (0.60 °C over the past 112 years), as well as changes in the pattern and frequency of rainfall. The protection and restoration of ecosystems as well as the facilitation of fluctuations in ecosystem distributions due to climate change are necessary to realize the full potential of wetland carbon sequestration and adaptation benefits. The sustainable development goal (SDG) addresses worldwide wetland and its ecosystem management services. We have analyzed recent research updates provided by Global Wetland Ecohydrological Network (GWEN), Global Wetland (CIFOR) Dataset and Global Lakes and Wetlands Database (GLWD). The study included the development of a model that estimates the risk of losing ecosystem services under various climatic crises under the SDGs.

Keywords

Global Wetland Ecohydrological Network (GWEN) · Global Wetland (CIFOR) Dataset · Global Lakes and Wetlands Database (GLWD) · Wetland ecosystem · Global warming · Carbon sequestration

Abbreviations

CBD	The Convention on Biological Diversity
GHGs	Green House Gases
GLWD	Global Lakes and Wetlands Database

S. Karmakar · S. S. Islam · S. Midya (✉)
Department of Zoology, Raja Narendra Lal Khan Women's College, Midnapore, West Bengal, India

K. Sen
Department of Plant Pathology, Bidhan Chandra Krishi Viswavidyalaya, Mohanpur, West Bengal, India

Department of Microbiology, Vidyasagar University, Midnapore, West Bengal, India

S. Ghosh
Department of Sociology, The University of Burdwan, Burdwan, West Bengal, India

U. Chatterjee et al. (eds.), *Climate Crisis: Adaptive Approaches and Sustainability*, Sustainable Development Goals Series, https://doi.org/10.1007/978-3-031-44397-8_27

GWEN	Global Wetland Ecohydrological Network
HAT	Highest Astronomical Tide
IHA	The Indicators for Hydrologic Alteration
INWCP	Indian National Wetland Conservation Programme
IPCC	The Intergovernmental Panel and Climate Change
LAT	Lowest Astronomical Tide
NbS	Nature-based Solution
NEE	Net Ecosystem Exchange
Pct	Precipitation
SDGs	Sustainable Development Goals
Temp	Temperature

27.1 Introduction

Wetlands are found in all over the world in every biome. Wetlands are aquatic ecosystem or an area with water table that are either permanently or seasonally covered by landscape throughout the year and located in the transition zone between aquatic and terrestrial ecosystem (McCauley et al. 2015). Some wetlands, such as tidal marshes, fens, peatlands, swamps, wet grasslands, flood plain, rivers, lakes, saltpans, underground aquifers, oases, rice paddies and a wide variety of inland and coastal wetlands are promoting an abundance of water (Schuerch et al. 2018). Wetlands are among the most productive ecosystem which supply a large number of ecosystem services by a distinctive biodiversity that is characterized by species diversity and their complexity (Neogi et al. 2016). Various wetland ecosystems services such as inland fisheries, groundwater recharge, flow regulation, waste assimilation, sediment transport, erosion control, deriving natural products etc., can benefit the economy and environment and therefore can be important for NbS and multiple SDG (Campbell 2019). In ecology wetland play a vital role over natural cycle by supporting the wide range of multiple biotas like decomposers, well managed rice paddy systems, estuaries and coastal capture fisheries, pest and disease regulation etc., (Ronco et al. 2020). Almost all of the world's migratory waterbirds feeding and breeding habitat depends on wetlands. Wetlands are hydrologically connected to other landscapes through ground water, rivers and lakes and water, nitrogen, and sulfur cycles throughout the world's ecosystems. These freshwater ecosystems provide a medley of primary productivity, nutrient abundance, and shallow water and helps to aggregate impacts on catchment scales (Cunillera-Montcus et al. 2022). Although wetlands are natural source of blue carbon in which plant communities and anaerobic soils can both store CO_2 as carbon sinks then play a role in climate system through releasing it into the atmosphere as carbon dioxide which energize the hydroclimatic changes in their catchments (Gallego-Sala et al. 2018). And with warmer condition, nitrous oxide emissions, methane emissions from wetlands affect carbon sequestration and greenhouse gas emissions and leads to higher microbial activity, higher nitrification, denitrification rate (Rogers et al. 2019; Li et al. 2004). Due to global warming, the temperature of wetlands gradually increased and reduced the flow level which leads to decrease the quantity of water and diversity of aquatic species and increase water stress. The climate change (rapidly changes between dry and wet soil) accelerates higher rate of decomposition than photosynthesis and release the excess nutrients to the water bodies causing eutrophication, brownification, acidification, oligotrophic water bodies, oxygen starvation. Wetlands excite the intensity of extreme natural events like precipitation, tropical cyclones, floods, droughts etc., (Campion 2012). The changes of total annual runoff due to shifted the timing of floods and fresh water pulses have the significant impact on fresh water ecosystem (Gibson et al. 2002). This alternation of runoff generally leading to increase sediments, contamination of water and overflow of water. Due to surface water warming, evaporation increasing from the surface of water. Sublimation of seasonal snowpack and glaciers appears to be increasing that leads to reserve the water in glaciers and meltdown the snowpack, resulting decreases in summer and autumn flows (Akhtar et al. 2021). The warmer condition increases the timing of snowpack melting which leads to shifts from snow to rainfall and rising the sea levels for which mortality of saline intolerant organisms increased (Rogers et al. 2019). Climate change can lead to increase the intensity of precipitation at high latitude and reduce it at low latitude, resulting changes the structure of habitats, increases saltwater intrusion, increases temperature, seasonality, shifts hydrology-based life cycle stages etc., (Dadhwal et al. 2020; Mander et al. 2015). In winter and early spring, water discharge and sediments increased due to climate change which could be harmful for development of young organisms. Many landscapes have moved to dry and cold wet landscapes due to a variability in hydro-climatic variables. Climate change poses a significant threat to wetland sustainability, affecting their hydrological, ecological and biogeochemical processes (Adade Williams et al. 2020; Rahel 2002). Due to climate change, the function of wetland ecosystem services might be turned into disservices. Wetland ecosystems play a vital role in maintaining the Earth's biodiversity and ecological balance. Climate change poses a significant threat to wetland sustainability, affecting their hydrological, ecological and biogeochemical processes. The purpose of this chapter is to explain how climate change may affect over wetland ecosystems and its long-term viability.

27.2 Climate Change and Wetland Hydrology: Understanding the Impacts on Water Levels and Flow Regimes

Wetlands are very vulnerable due to effects of climate change in various way such as changing hydrological regimes, precipitation pattern, temperature changes, frequent evapotranspiration pattern, rainfall pattern etc. (Mulholland et al. 1997). Because wetlands are sharply depending upon hydrological cycle. The climate change causes a potential alternation in biogeochemical and physiochemical properties along with carbon dynamics, soil structure, wetland biota, primary and secondary productions, species biodiversity, rainfall, temperature, water level etc. The main impact from climate change for wetlands are alternation of hydrological regimes which afford the pattern of wetting and drying of wetlands and rivers, stream affected ground water level etc. because ecological processes cannot be fostered by low water flow, there need a high flow variation (Dasgupta et al. 2020; Cizkova et al. 2013). The increase in temperature due to global warming leads to glaciers melting, cyclones, storms etc. In conformity with IPCC, the temperature of earth became warmer up to (1.8–4 °C) which is the major cause of increase in sea level of 18–59 cm. (Sorokin 2015). This enhancement of water temperature causes increase in frequency and time duration of hypo toxic episodes which is responsible for drastic reduction of reproductive success and reduce the adaptability of various species (Diaz and Rosenberg 1995). This water temperature and hydrological regimes of freshwater ecosystems are regulated by the regional and local climate with topology and vegetation area (Gresswell 1999). Due to global warming, the enhancement of temperature leads to increase or decrease of precipitation regimes. And for the oscillation of water flow, the seasonal cycles are responsible which is controlled by the wind and rainfall, helps to stimulate the fluctuation of water level. The enhancement of temperature leads to evaporation from surface area (water) and transpiration by plants execute the retrenchment in runoff. This rising evapotranspiration degree melts snow cover and change precipitation pattern. In higher latitudes, there is a lot of water vapor being transported. As a result, these regions experience high levels of rainfall. On the other hand, in lower latitudes, precipitation is typically low. This changing in precipitation, temperature and evaporation responsible for changes the intensity of drought and floods (Trenberth 2011). This random fluctuation of natural variability affects the shallow seasonal wetlands with aquatic species. The way of river flow with the environmental flow components and wetlands that is filled with rainfall and ground water by rate of change in pressure with time is termed as water regimes. Streamflow regimes are related with substrate particle size, organic matter, sediments, dissolved oxygen concentration to environmental channel geomorphology. In riverine ecosystems, the flow regimes depend on some indicators like magnitude, frequency, timing, elevation, rate of change which are approached by IHA and also depend on volume and frequency of winter and summer discharge. River flow fluctuation stimulate the abiotic interactions which affect the fresh water organisms, water quality, water temperature and sediment transport. Thus, the hydrological cycle or the water level is in the way to increasing with intensity of rainfall.

27.3 The Role of Wetlands in Carbon Cycling and Climate Change Mitigation

Wetland's ecosystems can store 20–30% of earth soil which is fully saturated (Bridgham et al. 2006). This creates large amount of organic matter and stock it in semi decomposed form which can store carbon by absorbing from atmosphere, through photosynthesis, aggregation of organic matter in anoxic soil, plant biomass and release methane by decreasing the decomposition rate, this process of capturing and stabilizing carbon as carbon sink or carbon pool known as carbon sequestration (Rogers et al. 2019; Kayranli et al. 2010). This carbon sequestration includes a set of function like respiration of aerobic zone, methanogenesis (formation of methane by microbes), fermentation of dissolved organic carbon, sedimentation, retrenchment of sulfate, biodegradation, volatilization and nitrate in anoxic soil. Carbon exuberant wetlands store higher amount of carbon as per unit area, nitrogen, methane by microbial activity of dissolved organic matter in soil and leads to emit a significant amount of GHGs (Bharati et al. 2015; Kayranli et al. 2010). These gaseous products can be formed under both anaerobic condition (formation of carbon dioxide and methane by the decomposition of organic matter) and also aerobic condition (formation of carbon dioxide). This aerobic and anaerobic situation in wetlands regulate the process of methane production in dried wetlands by absorbing atmospheric methane by methanophiles and methane oxidizing bacteria in unsaturated soil (Kayranli et al. 2010). Due to this microbial oxidation, the liquefied organic carbon alters to inorganic compounds and mineralize it. In the aerobic zone, the rate of respiration covert carbohydrate to carbon dioxide whereas the rate of fermentation convert carbohydrate to chemical compounds like carbon dioxide and ethanol because this high productivity, low decomposition rate, high water table leads to store more carbon in the soil, sediments from catchments and detritus and triggered to horizontal transport of carbon within ecosystems (Dasgupta et al. 2020). This aerobic and anaerobic condition are depending upon water table of wetlands because water logging of wetland soils bounds oxygen diffusion into sediment contours for generating anaerobic condition (Dasgupta et al. 2020; Kayranli et al. 2010). This anaerobic soil condition produces nitrous oxide and induce methane emission and increase the organic carbon deposition by decrease the rate of decomposition. Due to this carbon rich wetlands, wetland surrounding plants can grow speedily than the rate of production of decompose. Increasing the quantity of carbon di oxide and GHG gases in the atmosphere leads to climate change mitigation, frequency of stress, sensitivity of ecosystem. The organic carbon that is captured by ocean and coastal ecosystem (mainly intertidal and shallow water) including tidal salt marshes, mangrove forest, seagrass meadows in sediment and sea inclined soil refer as blue carbon (BC) which despite an important role to mitigate of climate change like GHG emissions, seaweed contributes, carbonate invention for the high carbon stowage and organic carbon sequestration from vegetated coastal wetland ecosystem (Gallego-Sala et al. 2018). BC ecosystem increased BC storage with increasing vertical acceleration of soil and sediments due to degradation of coral reefs, sea level rise ranks, marine heat waves, ocean acidification, rainfall variability, storms, land sinking, raised CO_2, hydrological changes, occurrence of sea walls and migrate up to landward area undesirably regulate BC stocks through enhance C stock/sequestration by failure of remineralization, restoration, loss of ecosystem regulation in ocean and coastal ecosystem resulting emission of GHGs to atmosphere (Rogers et al. 2019; Woodroffe et al. 2016; Schuerch et al. 2018). This trigger to loss of regulation of BC in BC ecosystem. BC provide an opportunity for adaptation strategies and climate change mitigation by following the 2006 IPCC guidelines for National Greenhouse Gas Inventories. In Littoral wetlands, production rate of methane is low than freshwater wetlands due to salinity because salinity inhibit methane production and also nitrous oxide and also drainage oxidative wetland losses net sources of methane from sediment and therefore can switch carbon sinks. Ocean acidification is the crucial stressors for coastal area, shallow marine seawaters, and estuaries because the atmospheric CO_2 deposit to ocean and reacts with salt water to make carbonic acid and decrease the pH and carbonate concentration in ocean bed which is used by corals, oysters, planktons to build consolidated skeletons or shells. On the other hand, in Northern peatlands, common cushion plants can trap the adjacent air temperature and rise vegetation canopy humidity by dropping wind shear and evapotranspiration (Cavieres et al. 2007) and in drying warming condition, water level affect CO_2 fluxes by increasing respiration from lower water level and NEE from higher water levels (Gallego-Sala et al. 2018) then it might have significances for carbon cycling (Salvador et al. 2014). Vegetation type control the water level and carbon cycling in peatlands because a degenerating water table rise the amount of anoxic soil and rise CO_2 production and at the same time minimize CH_4 (Fig. 27.1) (Silvola et al. 1996).

27.4 The Effects of Sea Level Rise on Coastal Wetland Ecosystems

The mean sea level means a particular average point over a period of month or year at surface level of ocean. According to global scale, sea level rise among (14–80 cm) or near about 75% over the end of the twenty first century (IPCC 2019). The intergovernmental panel on climate changes decides that sea levels rising occurs due to thermal expansion (global warming-GHGs) in ocean water through melting polar ice, ice sheets and mountain glaciers (CSIRO and Bureau of Meteorology 2015). Because of the global warming, 90% of earth's ice-covered water is at risk to melting which is relatively immutable at Antarctic ice sheet, Greenland ice sheet and high latitude mountain glaciers while global water are contained by ocean, so tempering ice-covered earth surface increase the global sea level (expand ocean water volume) (Fig. 27.2) (IPCC 2019). Alternation of sea level can be due to interactions with land, intensity of precipitation, surface runoff variation, ground water flow pattern, snow to rainfall emigrate, seasonality, air sea fluxes and ocean dynamics, land elevation, vertical movement etc. (CSIRO and Bureau of Meteorology 2015). Earth's gravitational field trigger to shift ordination of water on earth surface and uplift-down vertical land movement for regional and local scale. This plate movements are imposed due to tectonic cause, consolidation, glacial isostatic adjustment-GIA2, neo-tectonic cause, anthropogenic cause etc. (Emery and Aubrey 1991). Revocation of inadequate ground water, oil accelerate low lying coastal area to decrement, peatland annihilation by oxidation and erosion, this enhances sea level raise locally. Buteven though this is extremely precarious, increasing ground water mining causes sea level elevation while increasing terrestrial water storage causes sea level to drop. Due to variations in the global thermal expansion's magnitude across space, it alters ocean circulation like gulf stream, wind fields, tropical-stroms which are indirectly responsible for sea level uplift (Warrick et al. 2000). Moreover, combine effects of currents, wind flow, waves pace leads to raise sea level. Sea level rise is related with tide (HAT and LAT), storm depending upon gravitational fascination of moon and sun. Storm surges increase extreme sea levels (a combination of tides, wind waves, storm pace) at the coast (CSIRO and Bureau of Meteorology 2015). This sea level rising for the climate change would be responsible for vulnerability of world's coastal areas and has impacts on ecological services (IPCC 2019). coastal areas

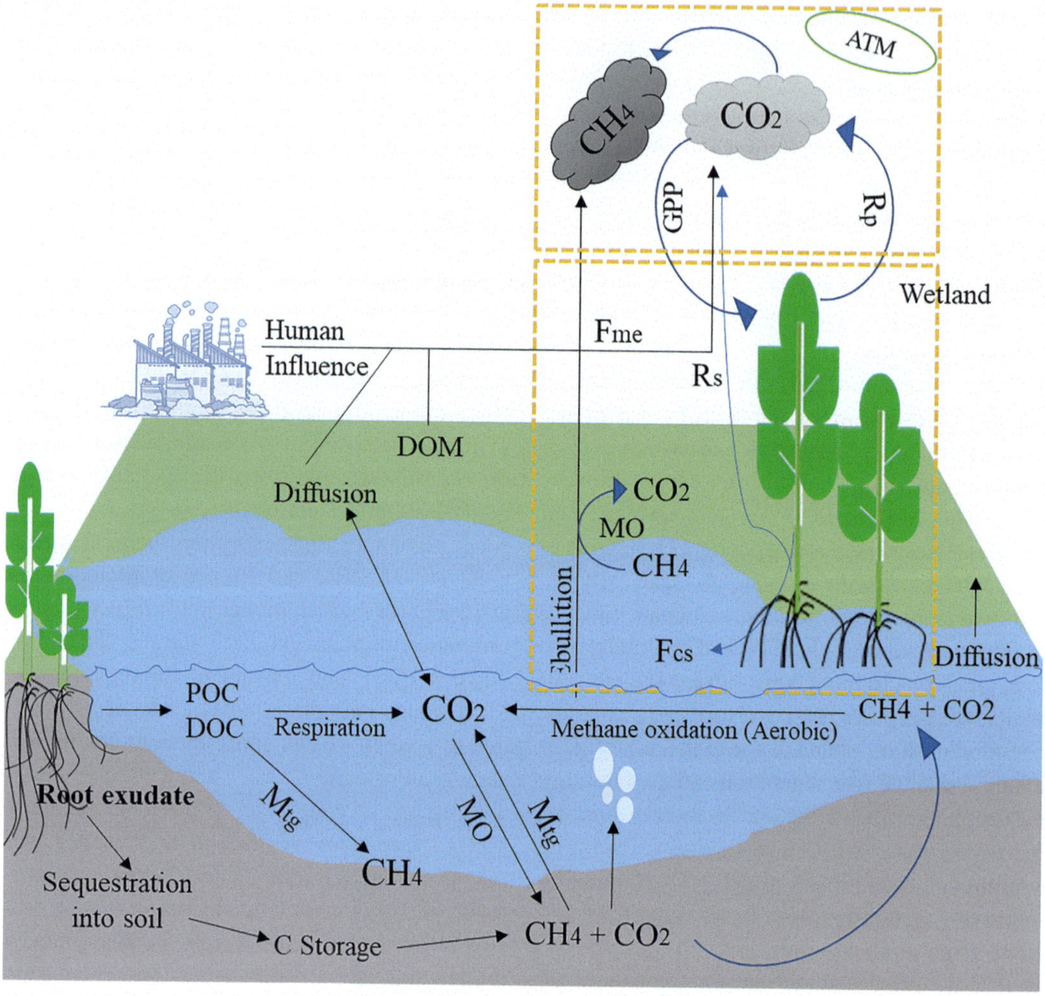

Fig. 27.1 Conceptual model of wetland carbon sequestration and exchanges with atmosphere. Where, particulate organic carbon (POC), carbon dioxide (CO_2), dissolved organic carbon (DOC), methane (CH_4), carbon sequestration (F_{cs}), methane emissions (F_{me}), gross primary productivity (GPP), plant respiration (R_P), soil respiration (R_s), decomposition of organic matter (DOM), methane oxidation (MO) and atmosphere (ATM)

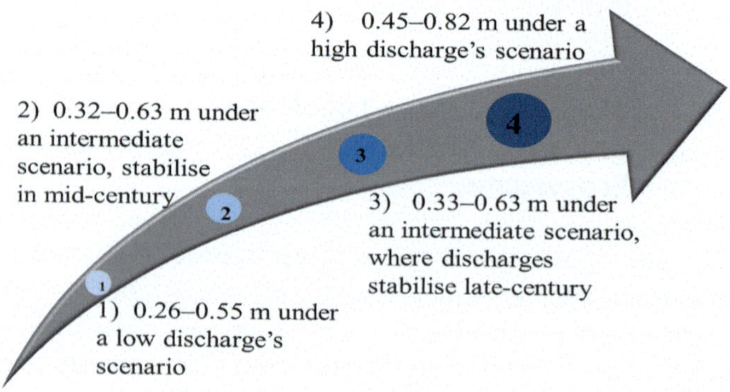

Fig. 27.2 The scenario of global sea levels is expected to rise by IIASA, on the basis of period 1986–2005

are very essential for human habituation, ecological services, socio-economic demand (Sachs et al. 2001). In near coastal zone (in the direction of seawards-close to the coastline), the population density is widely growing by net coastal migration for urbanization.

Rising sea level affects coastal areas such as increased erosion of beaches, shorelines, bay shores, tidal flats, river deltas, marshes, wetlands by wave action, surges, storms, rocky beaches, sediment budgets, rocky cliffs etc. Increased flooding and submergence in estuaries, salt marshes, deltaic plains and lagoon shores. Irregular storm surges in low lying coastal area and tidal flooding waves penetrate into inland. Increased salt concentration in estuaries, coastal aquifers, salt marshes, coastal rivers (Rogers et al. 2019). Drainage problem in coastal population areas and water stress in some area, changes of thermal stratification, periodic inundation. Loss of coastal habitat, wetlands and infrastructures including commercial activities, airports, transportation, agriculture, aquaculture, power plants, warehouses, salt ponds, highways, marines etc. (Nicholls 2002). Coastal ecosystems are way to destruction in coral reefs, estuaries, tidal flat area, associated wetlands for the lack of sediment replenishment. But coastal area of North and West are less vulnerable for sea level raise affect because there has less erodible rock, glacial, riverine, natural land removal.

27.5 The Impact of Global Warming on Wetland Biodiversity and Species Distribution

CBD is quite concerned about how climate change could affect biodiversity. One of the biggest threats toward the region's biodiversity and ecosystem services is climate change (Sonwa et al. 2017). The research's general conclusion is that many species have expanded their geographic ranges in relation to abrupt changes in temp. and pct. Patterns, usually poleward also upward. (Doney et al. 2012). Climate change (mainly changes in temp., rainfall and pct.) has an impact on species diversity in a number of ways, including shifts in distribution, higher extinction rates, altered reproductive schedules, and lengthened plant growth seasons. As rainfall and snowmelt patterns change, various rivers and lakes will see changes in their flow regimes. Several species' spawning and eating patterns will be impacted by this. The duration of the seasons for flora and fauna would be affected by raising global temperatures by 0.798 °C and the level of CO_2 from 280 to 379 ppm similar on pre-industrial concentrations (Linderholm 2006). The effects of climate change are especially dangerous to several species that are currently in danger. The majority of the surface of the Earth is covered by oceans, which make up the planet's largest habitat. The species is widely distributed over 3000 meter above the sea level, which reaches up to 3700 m above the ocean's surface and is mostly limited to the highest peaks, and its distribution moves upward higher altitudes (Gottelli and Sillero-Zubiri 1992). Coastal regions are habitat of the maximum productive and diversified ecosystems on the world, such as mangroves, coral reefs, as well as seagrass beds. Fresh water sustains at least 100,000 species (approximately 6% of all identified species), although making up only 0.01% of the surface water and roughly 0.8% of the Earth's area. 19 Upwards of 20% of the freshwater fish species found around the world are now extinct, endangered. More biodiversity reductions are occurring in freshwater species than in the majority of terrestrial environments. According to research by (Groffman et al. 2014), In response to climate change, terrestrial plants and animals moved in the direction of higher elevations at a pace of 0.011 km per decennary and toward higher latitudes at a pace of 16.9 km per decennary. Migration decorations (timings) of species which use seasonal wetlands (such as migratory birds) and trace seasonal variations in vegetation (such as herbivores) could alter due to climate change, which may potentially increase interactions with people, especially in areas with limited rainfall (Thirgood et al. 2004). The detrimental effects on many wetlands could be made worse by anthropogenic responses to climate change. The need for fresh water to serve urban and agricultural demands, for instance, is anticipated to rise as a result of human reactions to a warming environment. Reduced river and stream discharge and a reduction of ecosystem services are likely outcomes.

27.6 The Relationship Between Climate Change and Wetland Invasive Species

Invasive species are those whose populations tend to increase and spread, harming the environment, the economy, or public health. They are typically non-native, which means that they are species that were previously found somewhere and have relocated into a new area where they are freed from predators or competition that would have limited their spread in their native area. Invasive species have the ability to change ecological processes including the frequency of fires, the cycling of water and nutrients, and others. High spreading capacities, fast development with quick generation times, and great environmental tolerance are typical characteristics of invasive species. The ability of a species to adapt in to the new environments and endure rapid changes in their environment, such as those brought on by climate fluctuation and change, is considerably improved by these features. In response to the population's fluctuation and changes in temp., moisture and biotic interactions, invasive species' distribution will almost likely vary as an outcome of climate change. As projected by climate change scenarios, invasive organisms are inherently equipped to flourish in regions with an abundance of resources.

Species distributions will very probably change as a consequence of climate change (Parmesan 2006). Invasive insects' movements, phenology, growth rates, dispersal and survival can all be influenced favorably by rising temperatures. The capability of species gets into and effectively colonize renewed areas will depend on the subtraction of temp., or moisture limits on dispersal and survival (Hellmann et al. 2008). Invasive plants can develop dramatically and favorably as a result of elevated CO_2, whereas invasive insects and pathogens, which are farther impacted by temperature and pct variations, may experience little to no direct impact (Gallego-Sala et al. 2018). Due to their effects on tree and plant development, mortality, decomposition rates and ecosystem processes, invasive insects, diseases, and plants have the potential to interfere with the sequestration of carbon in forests and their sequestration rates. Our knowledge of the hazards to the significant atmospheric GHG sink that those forests illustrate will improve as we get more insight into invasive species outbreaks. In many cases, climate change will result in future increases in CO_2accessibility, temp., and nitrogen deposition off agriculture also vehicles (Bradley et al. 2010). Invasive red brome within the Sonoaran Desert is one example of how plants can invade arid ecosystems like western shrublands and in places with limited resources like serpentine grasslands due to this increase in resource availability (Weltzin et al. 2003). For instance, the winter pine processionary moth, a significant pest of pine trees, has increased flight activity as a result of warmer night time temperatures; Rainfall might be expanding Argentine ants more widely across California; and kudzu, one invasive plant species is likely to multiply by several hundred miles to the north and west (Walther et al. 2009; Bradley et al. 2010). Abiotic filters that control the prosperity of invasive species in aquatic habitats will change as an effect of climate change. Elements of climate change, such as altered heat regimes, decreased lake ice cover, rehabilitated streamflow regimes, amplified salinity, and augmented water-development operations such as canal and reservoir construction, may have an impact on aquatic invasive species (Poff et al. 2002). The pool of prospective colonists may shift as a result, which may affect the likelihood that non-native species will launch, change the impact of invasive species already present, and necessitate starting or increasing prevention and control activities. Water temperature rise with the respect of air temperature for climate change so when, Warmer weather and less winter hypoxia would increase the range of exotic species to include warm-water or hypoxia-intolerant species while decreasing the success of invasive cold-water species (Rahel 2002). If non-native salmonids have limited native fish to warm headwater reaches in Patagonian streams, such a condition might be advantageous to native species (Pascual et al. 2007). Trophic interactions will likely remain impacted through climate change. It is anticipated that climate change will have an effect on several fungus-borne insect diseases that are significant in controlling insect populations. By influences on spore formation, the success of host infection, or selection pressures, changing climate can actually influence invasive pathogens. Others are more sensitive to variations in host stress, whereas other invasive pathogens are susceptible to fluctuations in temperature, humidity, and duration and quantity of precipitation. Drivers of climate changes include invasive species and climate change, which are most likely to interact beyond ecological processes and have long-lasting effects on eco-evolution (Cunillera-Montcusi et al. 2022).

27.7 The Economic Value of Wetlands in the Face of Climate Change

Twelve percent of all species of animals and far over 40% of the biosphere's species can be found in freshwater wetlands. Wetlands are accountable for being extremely productive ecosystems that offer several significant advantages. The advantages relate to "goods and services," their purposes (such as ground flood prevention and ground water recharge), their by-products (such as a place for wood collection or study), and their characteristics (such as a landscape feature with aesthetic value or religious significance). By supporting a large concentration of aquatic fauna, mammals, native and migratory bird varieties, permanently residing amphibians, reptiles, and numerous species of insects, etc., wetlands are protecting the native ecosystem. Moreover, they serve as important flora and water reservoirs.

Wetlands offer room for activities like human habitation, agriculture, alternative energy, and animal habitat. They also serve as a nursery environment for numerous freshwater fish that are fished beyond of the wetland but are of great economic importance (Nayak and Bhushan 2022). Environmentally beneficial ecological processes are connected to regulation action, reprocessing of nutrients and human excrement, safeguarding watersheds, controlling floods naturally, and regulating flow. Purification of water with carbon sequestration, climate regulation, water holding capacity etc., carriers' function for activities including human habitation, agriculture, energy generation, wildlife habitats, wildlife harvesting and resource gathering, tourism, and recreation, as well as habitat and nascent habitat for species of plants and animals, outturn of vegetable, cotton, wheat etc., (Gallego-Sala et al. 2018). The wetland services with the largest average economic values per hectare per year, at $492 and $464 respectively, include flood management and storm attenuation, as well as the availability of recreation and facilities. When it comes to supporting large populations of endemic species, individual wetlands might be crucial. For instance, Lake Tanganyika in Central Africa is home to 632 native animal species (Ronco et al. 2020). For many local communities, wetlands have religious and historical significance. Wetland management does not seem to be a real alternative because decision-makers frequently lack basic awareness of the benefits of wetlands, especially their economic worth, which is the primary cause of that much wetland deterioration.

27.8 Traditional Knowledge and Community-Based Approaches to Wetland Adaptation to Climate Change

Throughout ceremonies and traditional ceremonies, indigenous people pass on generational understanding of the environment, including includes knowledge of how to adapt to climate change (Gopal 2013). Despite the abundance of research on local adaptation to climate change, there are few synthesizes of this data. There are evaluations of how developing economies are coping with climate change, yet very few of them have taken into account indigenous knowledge systems. Indigenous or traditional knowledge, specifically in the developing world, has gained recognition in past few decades as being essential for the management of natural resources and sustainable lives. Wetlands are now widely acknowledged to play a significant eco-hydrological function in sustainable development in addition to offering a wide range of ecological as well as socio-economic advantages to the surrounding community (Silvius et al. 2000; Stuip et al. 2002). Likewise, the international community's 2015–2030 Agenda For sustainable development expressly recognized that indigenous people's indigenous knowledge and ancestral lands must be protected in order to achieve true sustainable development (Campbell 2019). The decolonization of climate change learning includes IKS-based climate change adaption techniques with scientific knowledge already developed in the traditional educational systems of especially in the developing contexts. This is a reference to adaptation approaches that incorporate cognitive, behavioural, and cultural elements, such as agricultural production, watershed management as well as irrigation techniques, soil improvement techniques, native cropping patterns, native animal breeds, control of pests and diseases food preservation, as well as livelihood diversification. Research has found that in these situations, wetland utilization is frequently premised on community management techniques that have developed over time and through improvement of "indigenous" or "local" knowledge, via the transmission of ancient knowledge and/or the procedure of innovation propagation (Adade Williams et al. 2020). In the last three decades, the reputation of wetlands has substantially changed from being a largely ineffective wastelands to being one of important international refuges for species and essential natural resource to indigenous peoples. Participants reported that practical example and close observation are used to transmit ancestry knowledge from one generation the next. Participants mentioned that the parents likely passing on their knowledge of, among several other things, cropping rhythms, productivity variations, medicinal herbs, and soil fertility. Yet, participants acknowledged noted their personal interactions or direct observation of many other farmers served as a valuable source of information about wetlands (Bataille et al. 2021). This serves as the foundation for early recommendations for wetland runoff and agriculture in certain wetlands, while in others it serves as a catalyst for creativity or narrow adaptations. The wetland committee provides crucial for advancing sustainable management practices at the local level, providing a platform to wetland stakeholders to potentially share data on their perspectives, improvements, and observations. The Hazard Impact Assessment (HIA) advises monitoring environmental change rates, like as sea level rise as well as erosion, to lower the cost of maintaining and updating infrastructure but also to handle the urgent need for knowledge to more accurately estimate percentages for climate-induced environmental issues. on the other hands, would be incorrect to draw a distinction among EBA (Ecosystem Based adaptation) and CBA (Climate Based adaptation) based on whether or not the former acknowledges the importance of wetlands and the regulation and adaptation of those ecosystems in sustainable people-centered adaptation. It is simple to locate projects, for instance, that enhance vegetation management; utilize a comprehensive approach to groundwater recharge; utilize "natural remedies" to reduce hazards (for instance, rising vegetation on slopes to decrease the risk of landslides); improving agro - forestry; raise urban vegetation as well as green spaces (Akhtar et al. 2021).

They located protracted water sources without the need of expensive technology, discovered and separated fertile soils from bad ones, then cultivated land using simple techniques that may have had minimal effects on wetlands and forests. The idea of collecting rainwater and creating rain came from terms like "musena" or "dambos," the placement of freshwater resources, and thus the ritualization of rainwater through the usage of holy sites or shrines. Due to their connections to biodiversity and process that sustains life—water—rain shrines include perennial wetlands that are revered as holy places, divine, and sacred natural designs (Makondo and Thomas 2018). A life-sustaining system including both biodiversity and human existence, rainmaking, particularly through shrines, seemed to be an aspect of significant importance. In this respect, doing damaging operations like felling trees or moving into or close to perennial wetlands were discouraged. In addition, limits on tree removal near wetlands and waterways may also have simply been an old-fashioned conservation tactic. This management approach preserved the river systems, wetlands, and related resources (Garbisch 2019).

27.9 The Role of Wetlands in Protecting Against Natural Disasters in a Changing Climate

In recent years, there has been a significant increase in the frequency and intensity of natural disasters worldwide. Many countries have experienced devastating floods, droughts, and storms, resulting in loss of life, damage to infrastructure, and economic losses. In the United States, for example, the annual cost of damages caused by natural disasters has averaged around $100 billion over the past decade (FEMA 2021). In India, over 11,000 people died due to floods and landslides in 2021 alone (India Today 2021, October 2). Wetlands play a critical role in protecting against natural disasters such as floods, droughts, and storms. They provide a range of ecosystem services that help to reduce the impacts of these events on people and the environment. In the face of a fluctuating climate, the preface of wetlands in disaster risk reduction has become even more important.

Wetlands act as natural buffers against floods by absorbing and storing excess water. They can also slow down floodwaters, reducing the speed and volume of water flowing downstream. This aid to alleviate the risk of downstream flooding and prevent decay of riverbanks. For example, in Bangladesh, wetlands called Haors have been found to reduce flood damage by up to 90% (UNDP Bangladesh 2021). Similarly, in the US, at the time of Hurricane Sandy in 2012, wetlands along the coastline of New York and New Jersey reduced flood levels by up to 30% (Temmerman et al. 2013). Wetlands can also help to mitigate the impacts of droughts by acting as natural water storage systems. During periods of low rainfall, wetlands can release stored water into rivers and streams, helping to maintain water levels and support ecosystems downstream. In South Africa, wetlands have been found to provide up to 25% of the country's water resources (Mander et al. 2015). Wetlands also help to maintain groundwater levels, which can be critical during times of drought. In Australia, the Ramsar-listed Gwydir Wetlands provided water during the millennium drought, which helped to sustain local communities, agriculture, and ecosystems (Kingsford et al. 2010).

In addition, wetlands can help to protect against the impacts of storms by reducing the intensity of storm surges and waves. Coastal wetlands, such as mangroves and salt marshes, act as natural barriers against storm surges by absorbing and dissipating wave energy. In Louisiana, coastal wetlands have helped to reduce the impacts of storm surges and hurricanes on coastal communities (Vogelmann et al. 2020). The mangrove forests in Thailand and Indonesia protected coastal communities from the devastating impacts of the 2004 Indian Ocean Tsunami (Barbier et al. 2011).

The role of wetlands in protecting against natural disasters varies depending on the type of wetland and the location. In China, wetlands such as floodplain wetlands and rice paddies have been shown to reduce flood risk and damage. For example, a study in the Yangtze River Basin found that wetlands reduced peak flood levels by 10–20% (Wang et al. 2020). In India, wetlands such as the Dal Lake in Kashmir have helped to reduce the impacts of floods caused by heavy rainfall. However, wetlands in the country are also under threat from urbanization and pollution, which can reduce their ability to provide these important ecosystem services (Government of India 2021). Moreover, wetlands can also help to reduce the impacts of other natural disasters, such as droughts and wildfires. For example, in California, wetlands play a significant role in reducing the risk of wildfires by acting as natural firebreaks, slowing down or stopping the spread of fires (Wetlands International 2021). Wetlands can lessen the effects of droughts by preserving water levels and serving as a supply of water with both people and wildlife (Mander et al. 2015).

Despite the many benefits of wetlands, these ecosystems continue to face multiple threats. Wetlands are often drained or filled in for agricultural or urban development, which can have serious consequences for the environment and human communities (UNDP Bangladesh 2021). Climate change also poses a threat to wetlands, as rising sea levels and changing precipitation patterns can alter the hydrology and functioning of these ecosystems (Temmerman et al. 2013). In order to protect wetlands and the many ecosystem services they provided, it is essential that we take action to conserve and restore these ecosystems. This can involve measures such as protecting wetlands through legislation, implementing sustainable land use practices, and restoring degraded wetland habitats (Wang et al. 2020).

27.10 The Potential of Wetlands to Serve as Carbon Sinks Under Future Climate Scenarios

Wetlands are important ecosystems that provide manifold ecosystem services, flood regulation, inclusive water purification and carbon sequestration. Since that wetlands have the ability to store significant quantities of carbon in their soils and vegetation, there has been an increase in curiosity in the potential of wetlands to act as carbon sinks in recent years (Bridgham et al. 2006; Jacob et al. 2021). However, wetlands are under threat due to land-use change, drainage, and climate change.

27.10.1 Carbon Sequestration Potential of Wetlands

Wetlands are important carbon sinks due to their ability to store carbon in both above-ground and below-ground biomass, as well as in the soil. The carbon stored in wetland soils is called blue carbon, and it can remain stored for hundreds to thousands of years (Mcleod et al. 2011). Recent studies suggest that wetlands can sequester carbon at rates of up to 3000 g CO_2 per square meter per year, making them among the most efficient carbon sinks in the world (Lovelock et al. 2017). Furthermore, in the Okavango Delta in Botswana found that inundated wetlands sequestered more carbon than dry lands, highlighting the importance of water management in wetland conservation for carbon storage (Musiwa et al. 2019). Wetlands are found in every continent, and their carbon sequestration potential varies depending on their type, location, and management practices. In Indonesia, for example, peatlands are important carbon sinks, storing up to 60 billion tons of carbon (Gallego-Sala et al. 2018,Hooijer et al. 2010). Nonetheless, peatland degradation and fires have led to significant carbon emissions, making Indonesia the third-largest producer of greenhouse gases in the world. In the United States, the restoration of wetlands in the Mississippi River Basin has been found to sequester up to 5.5 million metric tons of carbon per year (Windham-Myers et al. 2018). This means that the protection and restoration of wetlands have a significant role to play in reducing climate change. Wetlands have the potential to sequester significant amounts of carbon, with estimates suggesting that they may store up to 30% of the world's terrestrial carbon (IPCC 2019). In India, wetlands cover an estimated area of 15.26 million hectares, accounting for about 4.63% of the country's landmass (Dadhwal et al. 2020). A study conducted in India found that the carbon stocks in the wetlands of Keoladeo National Park in Rajasthan were equivalent to the annual carbon emissions of a small town (Bharati et al. 2015). Similarly, the wetlands in the Sundarbans delta in West Bengal were found to sequester over 4.5 million metric tons of carbon annually, equivalent to the carbon emissions of over 1 million passenger cars in a year (Dasgupta et al. 2020). However, wetland degradation and loss continue to occur in India, threatening their carbon storage capacity. Wetland area in India has declined by over 30% since the 1970s, largely due to anthropogenic activities such as land-use change, infrastructure development, and pollution (Dadhwal et al. 2020). This trend is particularly concerning given that IPCC has identified wetland conservation and restoration as a crucial method of reducing climate change (IPCC 2019). Despite the challenges, there are examples of successful wetland conservation and restoration efforts in India. The Chilika Lake in Odisha, for instance, has been designated as a Ramsar Wetland of International Importance and has seen successful conservation efforts that have resulted in the recovery of its ecological health and carbon sequestration potential (Chilika Development Authority 2021). Similarly, the restoration of the Vellayani wetland in Kerala has resulted in increased carbon sequestration and improved biodiversity (Jacob et al. 2021).

27.10.2 Impact of Wetland Loss on Global Carbon Budgets

Despite the potential of wetlands to sequester carbon, wetland loss is a major contributor to carbon emissions (Bridgham et al. 2006). It is estimated that wetland degradation and loss result in the forsake of 0.2–1.5 Pg CO_2 per year, equivalent to approximately 6–18% of annual anthropogenic emissions (Roulet et al. 2007). Wetland loss can occur due to land-use change, such as conversion to agriculture or urbanization, as well as drainage and water management practices (Davidson et al. 2015). Climate change is also a threat to wetlands, as rising temperatures and changing precipitation patterns can alter wetland hydrology and vegetation composition (IPCC 2019). Wetlands have the potential to serve as important carbon sinks under future climate scenarios, but their loss and degradation must be addressed to avoid significant carbon emissions. The protection and restoration of wetlands should be considered in climate change mitigation strategies, and policies should be put in place to promote wetland conservation and sustainable use.

27.11 The Importance of Wetlands for the Preservation of Indigenous Cultures and Ways of Life in the Context of Climate Change

Wetlands are critical ecosystems that provide numerous benefits to both humans and the environment, including water storage and purification, carbon sequestration, and habitat for a variety of plant and animal species. In addition to their ecological significance, wetlands also have cultural importance, particularly for indigenous communities around the world. With regard to climate change, the preservation of wetlands and the cultures that depend on them has become increasingly important.

Indigenous cultures around the world have strong connections to wetlands and the resources they provide. For example, the Baka people of Cameroon rely on wetlands for hunting and gathering, as well as for traditional medicine (Sunderland et al. 2014). Similarly, the Penan people of Malaysia use wetlands for fishing and gathering forest products (Colfer et al. 2015). These communities have developed deep knowledge and understanding of their local wetlands, including the complex relationships between water, vegetation, and wildlife. This knowledge is passed down through generations, and forms the basis for their cultural identities and ways of life. However, the preservation of wetlands and indigenous cultures is threatened by a range of socio-economic and anthropogenic factors (Jones et al. 2021). Land-use change, such as the conversion of wetlands to agriculture or urbanization, is a major contributor to wetland loss. This can have significant impacts on indigenous communities, who may lose access to the resources they rely on for their livelihoods. Additionally, climate change is altering wetland hydrology and vegetation, further threatening the traditional practices and knowledge of indigenous peoples (Jones et al. 2021).

The loss of wetlands and associated indigenous cultures can have significant socio-economic impacts. For example, the degradation of wetlands in the Pantanal region of Brazil has impacted the livelihoods of the Guato people, who rely on wetlands for fishing and hunting. Similarly, the loss of wetlands in the Mekong Delta in Vietnam has led to food and water insecurity for local communities. These impacts are particularly pronounced for indigenous peoples, who often have limited access to alternative resources and livelihoods.

Wetland restoration has been identified as an important strategy for preserving both the ecological and cultural values of wetlands. For example, the restoration of wetlands in the Murray-Darling Basin in Australia has provided employment opportunities for indigenous communities, while also improving water quality and restoring important cultural sites (Kingsford 2013). Similarly, the restoration of wetlands in the Everglades in the United States has improved habitat for endangered species, while also preserving traditional fishing and hunting practices of the Miccosukee Tribe (National Park Service 2015).

In India, wetlands play a vital role in supporting indigenous communities and their traditional practices. For example, the Loktak Lake in Manipur, India is an important wetland for the indigenous fishing communities, who rely on the lake for their livelihoods (Khuraijam et al. 2013). However, the lake has been threatened by pollution and encroachment, leading to the loss of fish populations and traditional fishing practices. Similarly, the Sundarbans mangrove forest in West Bengal, India is a critical ecosystem for the indigenous groups that depend on the forest for fishing and collecting forest products (Dasgupta et al. 2017). However, the forest is threatened by rising sea levels and climate change, which could impact the livelihoods and cultural practices of these communities.

In Canada, the First Nations have a deep connection to wetlands, which have sustained their cultures and ways of life for thousands of years (Environment and Climate Change Canada 2019). Wetlands provide resources such as wild rice and fish, which are essential to their diets and traditional practices. Similarly, in India, wetlands are important for the preservation of indigenous cultures and ways of life. The Bhojpuri-speaking people of Bihar have a long history of living in and around wetlands, which provide them with resources such as fish, wild vegetables, and medicinal plants (Bhatnagar and Sengupta 2018). Wetlands also have cultural and spiritual significance for the Bhojpuri people, serving as important sites for festivals and other traditional practices. However, wetlands in both Canada and India are facing increasing pressures from development, land-use change, and climate change. The loss and degradation of wetlands can have significant socio-economic and socio-anthropogenic effects, particularly on indigenous communities that rely on these ecosystems for their livelihoods and cultural traditions.

Therefore, wetland restoration and conservation efforts are crucial for preserving indigenous cultures and ways of life, as well as for mitigating and adjusting to the impacts of climate change. Initiatives like the Ramsar Agreement, an international agreement for the preservation and wise use of wetlands, and the INWCP, which aims to conserve and manage wetlands in India, are important steps towards achieving these goals (Ramsar Convention 2022).

27.12 Impacts on Human Livelihoods and Adaptation Strategies

Livelihoods There are numerous definitions, but generally speaking, livelihoods are seen as the capacities, resources (with both economic and social resources), and activities requiredas a way to provide assistance. A sustainable way of life can bear pressure, bounce back from shocks, and maintain or increase its capabilities and resources both now and in the next without jeopardizing the foundation of natural resources (Carney et al. 1999). Wetlands offer a wide variety of ecological services that support the daily lives of many coastal people. Wetlands are thought to offer a variety of "ecosystem services" that are advantageous to human livelihoods and communities. It is generally accepted that the loss in ecosystem services results in fewer advantages for human livelihoods and, as a result, reduced net human well-being. Wetlands' contributions to people's livelihoods result from how their ecological roles interact with those of human society. Many people who live in and around these wetlands rely solely on natural resources for daily livelihoods, making them significant hubs of social, cultural, including recreational activity, crop production, natural resource extraction, grazing pastures (Gopal 2013; Chong 2005). On the other hand, wetlands play a role ecologically like ground water storage for drinking, filtration, toxic retention etc. which is very beneficial for wetland surrounding peoples. The Millennium Ecosystem Assessment (MEA) from 2005 showed the close ties among ecosystem services and human well-being, as well as how changes in human well-being have been brought about by the reduction of ecosystem services throughout biomes. Wetlands are essential to the survival of more than 80% of tribal and reserved caste populations as well as it offer livelihoods to hundreds of millions of individuals. But now degradation of wetlands is a significant issue that causes severe livelihood crises in rural, urban, but also suburban settings. Wetlands used to be considered unproductive which providing few livelihoods and the cause of illness in many regions. It is well known that wetlands are particularly important for the security of food, water, and livelihoods for the people who live there and nearby (Ricaurte et al. 2014). Since 1900, there has been an estimated global loss of wetlands of about 50%, which has had a significant impact on ecosystem services, species diversity, and human livelihoods (ICSU-UNESCO-UNU 2008). Wetlands have significantly decreased as a result of anthropogenic pressures like urbanization, agricultural expansion, industrial emissions, especially changes in the climate (Bassi et al. 2014). Community members who live in wetlands depend on the ecosystem services provided by such wetlands. In terms of means of subsistence, ecosystem services are "natural capital." Many processes convert this natural capital into other resources for sustaining livelihoods, such as human capital which includes health. Communities who depend on wetlands for their livelihoods may suffer if ecosystem services continue to deteriorate since they have few other choices for diversifying their sources of income. The depletion of resources, including land, grass, timber, and even historically trained labor, is displacing tradition on the one hand and creating a crisis of livelihoods on the other. Peoples are largely reliant on natural forests for both revenue and subsistence because to severe livelihood restrictions such widespread poverty, land encroachment, and an absence of development options. In contrast, degraded wetlands serve a variety of purposes that might be detrimental to human health and way of life, such as providing a breeding ground for parasites that spread diseases like schistosomiasis and malaria. Every sort of environmental degradation contributes to a variety of human issues, including global warming, the greenhouse effect, health risks, problems with livelihood, etc. Natural catastrophes like floods, droughts, and earthquakes interrupt livelihoods, posing substantial health risks to people and, in the worst situations, causing fatalities (Fig. 27.3).

Wetlands provide important ecosystem services such as water provision, flood control, and carbon sequestration. Climate change has led to changes in the functioning of wetlands, affecting the livelihoods of people who depend on them. Wetland conservation and restoration can help to enhance the resilience of wetlands to climate change. Integrated management approaches that involve local communities and stakeholders can help to ensure the sustainable use of wetland resources (Table 27.1).

27.13 Conclusion

Wetlands are among the most productive ecosystems on the planet, providing a wide range of important services, including water purification,

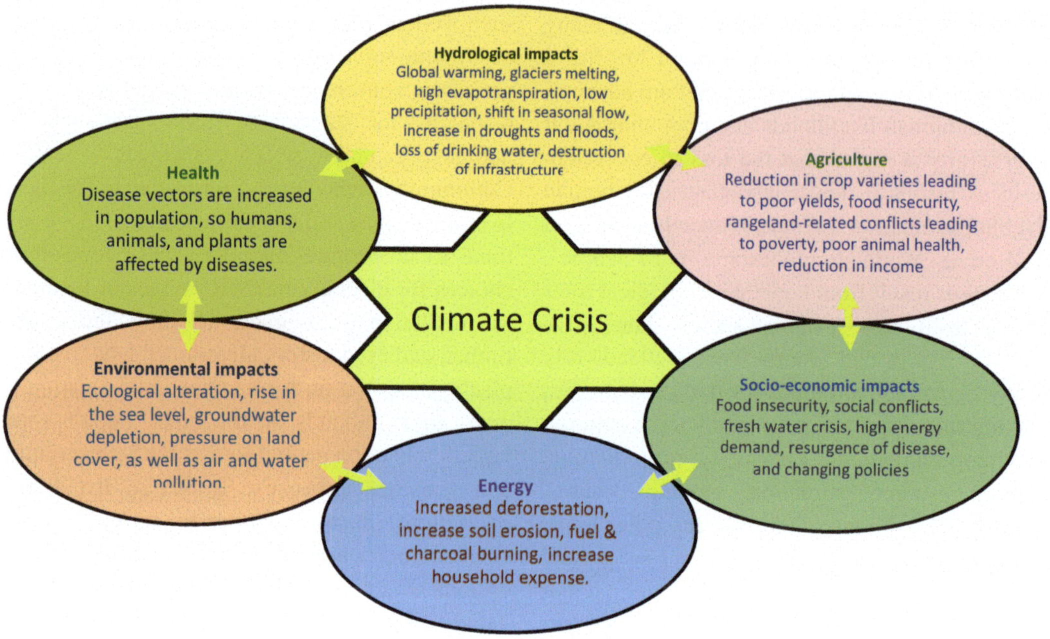

Fig. 27.3 Impact of climate changes on different wings of the environment

Table 27.1 List of various strategy adaptation approaches

Strategies	Approaches	Example of adaptation tactics	References
Maintain and enhance hydrologic processes and water quantity	In wetlands, neighboring uplands, and groundwater recharge zones, maintain and promote infiltration and water storage	A wetland must have buffers maintained or built around it for at minimum the first 100 m (328 feet)	Faber-Langendoen et al. (2016)
	Sustain and reclaim the natural hydrologic regime	Reclaim resident time and hydrologic processes by modifying or removing compacted soils in damaged wetlands	Sax et al. (2017)
	Reestablish the hydrologic operation of waterways that are connected to wetlands and reestablish the processes of stream channels	Increase channel sinuosity-related functions and modify levees, dams, and also other hardened infrastructure to reestablish the natural hydrologic regime	Yochum (2017)
Maintain and enhance water quality of wetland habitats	Moderate surface water temperature increases	Maintains headwater fed by groundwater to encourage summer flows that are colder to wetlands downstream	Erwin (2009)
	Reduce soil erosion and sediment deposition	Limit the effects of sedimentation on riparian wetlands by using permitted mechanisms for sediment movement in the dam	Kondolf et al. (2014)
	Decrease the export and loading of pollutants and other nutrients	Eliminate legacy phosphorus from declining streams and headwater lakes	Sharpley et al. (2013)
Maintain and restore wetland vegetation	Sustain and reclaim wetland structure	Use earth-moving equipment to produce dips and to roughen the surface soil in order to restore microtopographic variation	Caners et al. (2019)

(continued)

Table 27.1 (continued)

Strategies	Approaches	Example of adaptation tactics	References
	Sustain and expand diversity of plant species, as well as their morphologies, phenologies, life history and ecological niches	By burning, mowing, or planting in erratic patterns and by constructing rotating management units, increase the complexity and variability of microsite	Larkin et al. (2016)
	Promote proper fire in fire-adapted wetlands	It can be essential to consider when using fire in particular situations where peatlands are transitioning	Grau-Andres et al. (2017)
	Avoid the establishment of invasive species and lessen their effects where they already exist	Maintaining surveillance on invasion-prone areas and quickly eradicating any new infestations	Boos et al. (2010)
Facilitate transformation of wetland communities by adjusting species composition	Prefer native species and genotypes that are anticipated to be suited to future conditions and restore them to life	Using GIS and SDM software to assist with seed provenancing decision-making	Ramalho et al. (2017)
	In the suitable seed transfer zones, increase the genetic variation of seed mixtures	Collect seed from various wild resources within a designated "regional admixture provenancing" seed transfer zone	Bucharova et al. (2019)
	Transfer endangered species to areas that are anticipated to offer better habitat	Create a plan for assisted migration that explains all the variables that affect a species' or population's susceptibility	McLachlan et al. (2007)
Adjust wetland systems to cope with altered hydrology	Control systems to manage declining water levels and a scarcity of water	By strategically removing invasive shrubs, regulate the transition from open wetlands to wetlands that are dominated by shrubs	National Park Service (2015)
	Manage systems to cope with increased water abundance and higher water levels	Install carp barriers to anticipate increased invasion with flooding and surpassing water levels	Johnson and Havranek (2013)
	Create new and improved wetlands and manage them to accommodate shifts in hydrologic variability	Move riparian areas upslope or downslope to match existing or anticipated circumstances, and/or make stream channels more sinuous, to adapt the location and amount of wetland regions to fresh or fluctuating water levels.	Perry et al. (2015)
Sketch and alter infrastructure to accommodate future situations	Reinforce infrastructure to encounter predictable situations	Change undersized culverts to bottomless culverts by using the bank full width as a design guide and a USFS stream simulation design that enables sediment and debris to avoid a collision during greater flows or floods	Barnard et al. (2015)
	Reroute or relocate infrastructure, or use temporary structures	To restore the deteriorated condition of wetlands and riparian zones, relocate field roads	Daigle (2010)
	Use ecological or low-impact development in the designs	To decrease runoff, deposit sediments, and lessen the likelihood of erosion, improve drainage, consolidate slopes, and replant surrounding ground cover with vegetation	Strauch et al. (2015)
	Withdraw infrastructure and readapt system	To rehabilitate the riparian ecology and re-connect the channel and floodplains, remove levees that raise flood stage and flow velocity	Acreman et al. (2003)

flood control, and carbon sequestration. However, the sustainability of these critical ecosystems is threatened by the ongoing climate crisis. As temperatures continue to rise, precipitation patterns shift, and sea levels rise, wetlands are facing significant challenges that threaten their long-term health and viability. Climate change poses a significant threat to wetland ecosystem sustainability. Wetlands play a crucial role in carbon sequestration and are important for biodiversity and human livelihoods. Integrated management approaches that involve local communities and stakeholders are needed to ensure the sustainable use of wetland resources and adapt to the impacts of climate change. To know the better understanding of changes in ecosystems service and biodiversity, we urge researchers and scientist to work together across the disciplines to explore the importance of changing the climate. By this improvement of multidisciplinary understanding at varied tempo-spatial scales, to support the adaptive governance in gradually increasing present scenario of sustainable ecosystems.

References

Acreman M, Riddington R, Booker D (2003) Hydrological impacts of floodplain restoration: a case study of the river Cherwell, UK. Hydrol Earth Syst Sci Discuss 7(1):75–85

Adade Williams P, Sikutshwa L, Shackleton S (2020) Acknowledging indigenous and local knowledge to facilitate collaboration in landscape approaches – lessons from a systematic review. Land 9(9):331

Akhtar N, Syakir Ishak MI, Bhawani SA, Umar K (2021) Various natural and anthropogenic factors responsible for water quality degradation: a review. Water 13(19):2660

Barbier EB, Hacker SD, Kennedy C, Koch EW, Stier AC, Silliman BR (2011) The value of estuarine and coastal ecosystem services. Ecol Monogr 81(2):169–193

Barnard RJ, Yokers S, Nagygyor A, Quinn T (2015) An evaluation of the stream simulation culvert design method in Washington state. River Res Appl 31(10):1376–1387

Bassi N, Kumar MD, Sharma A, Pardha-Saradhi P (2014) Status of wetlands in India: a review of extent, ecosystem benefits, threats and management strategies. J Hydrol Regional Stud 2:1–19

Bataille CY, Malinen SK, Yletyinen J, Scott N, Lyver POB (2021) Relational values provide common ground and expose multi-level constraints to cross-cultural wetland management. People Nature 3(4):941–960

Bharati L, Bhattacharjee J, Nath B, Bhattacharya S (2015) Carbon stock and sequestration potential of Keoladeo National Park, India. Carbon Manag 6(3–4):81–90

Bhatnagar P, Sengupta M (2018) Wetlands and indigenous communities in India. In: Handbook of wetland management. Springer, Cham, pp 601–622

Boos T, Kearns K, LeClair B, Panke B, Scriver B, Williams B (2010) A field guide to terrestrial invasive plants of Wisconsin. Wisconsin Department of Natural Resources, Madison. DNR PUB-FR 436-2010

Bradley BA, Blumenthal DM, Wilcove DS, Ziska LH (2010) Predicting plant invasions in an era of global change. Trends Ecol Evol 25(5):310–318

Bridgham SD, Megonigal JP, Keller JK, Bliss NB, Trettin C (2006) The carbon balance of North American wetlands. Wetlands 26(4):889–916

Bucharova A, Bossdorf O, Holzel N, Kollmann J, Prasse R, Durka W (2019) Mix and match: regional admixture provenancing strikes a balance among different seed-sourcing strategies for ecological restoration. Conserv Genet 20(1):7–17

Campbell JY (2019) No sustainable development without indigenous peoples. International Institute for Sustainable Development, pp 1–3

Campion BB (2012) Urban wetland ecology and floods in Kumasi, Ghana. Doctoral dissertation, Universität Bremen

Caners R, Crisfield V, Lieffers V (2019) Habitat heterogeneity stimulates regeneration of bryophytes and vascular plants on disturbed minerotrophic peatlands. Can J For Res 49:281–295

Carney D, Drinkwater M, Rusinow T, Wanmali S, Singh N (1999) Livelihood approaches compared. DFID

Cavieres LA, Badano EI, Sierra-Almeida A, Molina-Montenegro MA (2007) Microclimatic modifications of cushion plants and their consequences for seedling survival of native and non-native herbaceous species in the high Andes of Central Chile. Arct Antarct Alp Res 39(2):229–236

Chilika Development Authority (2021) Chilika Lake. Retrieved from http://www.chilika.com/

Chong J (2005) Protective values of mangroves and coral ecosystem: a review of methods and evidence. IUCN, Gland

Cizkova H, Kvet J, Comin FA, Laiho R, Pokorný J, Pithart D (2013) Actual state of European wetlands and their possible future in the context of global climate change. Aquat Sci 75:3–26

Colfer CJP, Awang SA, Basuki I, Kusters K, Sheil D (eds) (2015) Human dimensions of tropical peatlands and their multi-functionality: a special issue from the International Peat Congress, Kuching, Sarawak, Malaysia. Int J Sustain Dev World Ecol 22(6):481–622

CSIRO and Bureau of Meteorology (2015) Climate change in Australia information for Australia's natural resource management regions: technical report. CSIRO and Bureau of Meteorology. Available online at http://www.climatechangeinaustralia.gov.au/en/publications-library/technical-report/. Accessed 18 Apr 2016

Cunillera-Montcus D, Beklioglu M, Canedo-Arguelles M, Jeppesen E, Ptacnik R, Amorim CA et al (2022) Freshwater salinisation: a research agenda for a saltier world. Trends Ecol Evol

Dadhwal VK, Rao KV, Saxena R, Menon M (2020) Wetlands of India: status, conservation and management. Curr Sci 118(4):514–526

Daigle P (2010) A summary of the environmental impacts of roads, management responses, and research gaps: a literature review. BC J Ecosyst Manag 10(3):65–89

Dasgupta R, Shaw R, Islam A, Rahman MM (2017) Climate adaptation, coastal zone management and community resilience in Bangladesh: policies, institutions and livelihoods. Springer

Dasgupta S, Bose P, Ghosh T, Mitra A, Mukhopadhyay A (2020) Carbon sequestration potential of wetlands in the Sundarbans delta, India. Environ Monit Assess 192(7):458

Davidson NC, Fluet-Chouinard E, Finlayson CM (2015) Global wetland outlook: state of the world's wetlands and their services to people. Ramsar Convention Secretariat, Gland

Diaz RJ, Rosenberg R (1995) Marine benthic hypoxia: a review of its ecological effects and the behavioural responses of benthic macrofauna. Oceanogr Mar Biol Annual Rev 33(245):3

Doney SC, Ruckelshaus M, Duffy JE, Barry P, Chan F, English CA, Galindo HM et al (2012) Climate change impacts on marine ecosystems. Annu Rev Mar Sci 4:11–37

Emery KO, Aubre DG (1991) Sea levels, land levels and tide gauges. Springer, New York

Environment and Climate Change Canada (2019) Wetlands and indigenous peoples. https://www.canada.ca/en/environment-climate-change/services/wetlands-indigenous-peoples.html

Erwin KL (2009) Wetlands and global climate change: the role of wetland restoration in a changing world. Wetl Ecol Manag 17(1):71–84

Faber-Langendoen D, Nichols B, Walz K, Rocchio J, Lemly J, Gilligan L (2016) Nature serve ecological integrity assessment: protocols for rapid field assessment of wetlands. Version 2. NatureServe, Arlington. Available upon request

FEMA (2021) FEMA's budget and costs of disasters. Retrieved from https://www.fema.gov/sites/default/files/documents/FEMA%20Budget%20and%20Costs%20of%20Disasters_0.pdf

Gallego-Sala AV, Charman DJ, Brewer S, Page SE, Prentice IC, Friedlingstein P et al (2018) Latitudinal limits to the predicted increase of the peatland carbon sink with warming. Nat Clim Chang 8(10):907–913

Garbisch E (2019) Wetland restoration: changing techniques, changing goals, changing climate. In: Right W (ed) Discovering the nature of wetlands, p 191

Gibson RN, Barnes M, Atkinson RJA (2002) Impact of changes in flow of freshwater on estuarine and open coastal habitats and the associated organisms. Oceanogr Mar Biol Annu Rev 40:233

Gopal B (2013) Future of wetlands in tropical and subtropical Asia, especially in the face of climate change. Aquat Sci 75:39–61

Gottelli D, Sillero-Zubiri C (1992) The Ethiopian wolf – an endangered endemic canid. Oryx 26(4):205–214

Government of India (2021) Damage to property and crops due to heavy rainfall/floods. Retrieved from https://pib.gov.in/PressReleasePage.aspx?PRID=1741395

Grau-Andres R, Gray A, Davies G (2017) Sphagnum abundance and photosynthetic capacity show rapid short-term recovery following managed burning. Plant Ecol Divers 10(4):353–359

Gresswell RE (1999) Fire and aquatic ecosystems in forested biomes of North America. Trans Am Fish Soc 128(2):193–221

Groffman PM, Kareiva P, Carter S, Grimm NB, Lawler J, Mack M, Matzek V, Tallis H (2014) Ecosystems, biodiversity, and ecosystem services. climate change impacts in the United States: the third national climate assessment. In: Melillo JM, Richmond T, Yohe GW (eds) U.S. Global Change Research Program 2000–2010

Hellmann JJ, Byers JE, Bierwagen BG, Dukes JS (2008) Five potential consequences of climate change for invasive species. Conserv Biol 22(3):534–543

Hooijer A, Page S, Canadell JG, Silvius M, Kwadijk J, Wösten H, Jauhiainen J (2010) Current and future CO_2 emissions from drained peatlands in Southeast Asia. Biogeosciences 7(5):1505–1514

ICSU-UNESCO-UNU (2008) Ecosystem change and human well-being: research and monitoring priorities based on the Millennium Ecosystem Assessment. International Council for Science, Paris

India Today (2021, October 2) Floods, landslides killed over 11,000 in India in 2021, majority in Himachal, Uttarakhand: Report. Retrieved from https://www.indiatoday.in/india/story/floods-landslides-killed-over-11-000-in-india-in-2021-majority-in-himachal-uttarakhand-report-1865545-2021-10-02

IPCC (2019) IPCC special report on climate change and land. Retrieved from https://www.ipcc.ch/srccl/

Jacob S, Mohan M, Menon R, Pillai SM (2021) Carbon sequestration and biodiversity enhancement through wetland restoration: a case study of Vellayani wetland in Kerala, India. J Environ Manag 282:111969

Johnson JA, Havranek AJ (2013) Effectiveness of temporary carp barriers for restoring wild rice growth in Upper Clam Lake: 2010 To 2013. Report to St. Croix Tribal Environmental Services. Natural Resources Department/Freshwater Scientific Services LLC, Webster/Maple Grove. http://freshwatersci.com/Downloads/UpperClamSouthBayWR2013.pdf

Jones B, Chalise SR, Subedi R, Bhandari RK (2021) Wetland conservation and indigenous communities: a review of challenges and opportunities. Environ Manag 67(4):649–663

Kayranli B, Scholz M, Mustafa A, Hedmark A (2010) Carbon storage and fluxes within freshwater wetlands: a critical review. Wetlands 30:111–124

Khuraijam J, Bhagat RM, Arunkumar L (2013) Fishery and biodiversity conservation of Loktak Lake in Northeast India. Proc Natl Acad Sci B Biol Sci 83(2):177–186

Kingsford RT (2013) Ecological impacts of dams, water diversions and river management on floodplain wetlands in Australia. Austral Ecol 38(4):356–375

Kingsford RT, Porter JL, Thomas RF (2010) Protecting wetlands in a changing landscape in Australia. Aquat Conserv Mar Freshwat Ecosyst 20(3):243–251

Kondolf GM, Gao Y, Annandale GW, Morris GL, Jiang E, Zhang J, Cao Y, Carling P, Fu K, Guo Q (2014) Sustainable sediment management in reservoirs and regulated rivers: experiences from five continents. Earth's Future 2(5):256–280

Larkin DJ, Bruland GL, Zedler JB (2016) Heterogeneity theory and ecological restoration. In: Palmer MA, Zedler JB, Falk DA (eds) Foundations of restoration ecology. Island Press, Washington, DC, pp 271–300

Li C, Cui J, Sun G, Trettin C (2004) Modeling impacts of management on carbon sequestration and trace gas emissions in forested wetland ecosystems. Environ Manag 33:S176–S186

Linderholm HW (2006) Growing season changes in the last century. Agric For Meteorol 137(1–2):1–14

Lovelock CE, Atwood TB, Baldock J, Duarte CM, Hickey S, Lavery PS, Masque P, Macreadie PI, Ricart AM, Serrano O, Steven A (2017) Assessing the risk of carbon dioxide emissions from blue carbon ecosystems. Front Ecol Environ 15(5):257–265

Makondo CC, Thomas DS (2018) Climate change adaptation: linking indigenous knowledge with western science for effective adaptation. Environ Sci Pol 88:83–91

Mander M, Wolski P, Nkuna K, Warchol J, Trollope W (2015) A review of the hydrology of the South African wetland biome, WRC report no. TT 640/15. Water Research Commission, Pretoria

McCauley LA, Anteau MJ, van der Burg MP, Wiltermuth MT (2015) Land use and wetland drainage affect water levels and dynamics of remaining wetlands. Ecosphere 6(6):1–22

McLachlan JS, Hellmann JJ, Schwartz MW (2007) A framework for debate of assisted migration in an era of climate change. Conserv Biol 21(2):297–302

Mcleod E, Chmura GL, Bouillon S, Salm R, Björk M, Duarte CM, Lovelock CE, Schlesinger WH, Silliman BR (2011) A blueprint for blue carbon: toward an improved understanding of the role of vegetated coastal habitats in sequestering CO_2. Front Ecol Environ 9(10):552–560

Mulholland PJ, Best GR, Coutant CC, Hornberger GM, Meyer JL, Robinson PJ et al (1997) Effects of climate change on freshwater ecosystems of the South-Eastern United States and the Gulf Coast of Mexico. Hydrol Process 11(8):949–970

Musiwa K, Segosebe E, Garegae K, Totolo O (2019) Carbon storage in wetlands of the Okavango Delta, Botswana. Afr J Environ Sci Technol 13(4):105–118

National Park Service (2015) The everglades: restoring the river of grass. U.S. Department of the Interior. https://www.nps.gov/ever/learn/nature/restoration.htm

Nayak A, Bhushan B (2022) Wetland ecosystems and their relevance to the environment: importance of wetlands. In: Handbook of research on monitoring and evaluating the ecological health of wetlands. IGI Global, pp 1–16

Neogi SB, Dey M, Lutful Kabir SM, Masum SJH, Kopprio GA, Yamasaki S, Lara RJ (2016) Sundarban mangroves: diversity, ecosystem services and climate change impacts

Nicholls RJ (2002) Analysis of global impacts of sea-level rise: a case study of flooding. Phys Chem Earth A/B/C 27(32–34):1455–1466

Parmesan C (2006) Ecological and evolutionary responses to recent climate change. Annu Rev Ecol Evol Syst 37:637–669

Pascual MA, Cussac V, Dyer B, Soto D, Vigliano P, Ortubay S, Macchi P (2007) Freshwater fishes of Patagonia in the 21st century after a hundred years of human settlement, species introductions, and environmental change. Aquat Ecosyst Health Manage 10(2):212–227

Perry LG, Reynolds LV, Beechie TJ, Collins MJ, Shafroth PB (2015) Incorporating climate change projections into riparian restoration planning and design. Ecohydrology 8(5):863–879

Poff NL, Brinson MM, Day JW Jr (2002) Aquatic ecosystems & global climate change: potential impacts on inland freshwater and coastal wetland ecosystems in the United States. Pew Center on Global Climate Change, Arlington

Rahel FJ (2002) Using current biogeographic limits to predict fish distributions following climate change. In: American Fisheries Society symposium. American Fisheries Society, pp 99–112

Ramalho CE, Byrne M, Yates CJ (2017) A climate-oriented approach to support decision-making for seed provenance in ecological restoration. Front Ecol Evol 5:Article 95

Ramsar Convention (2022) What are wetlands? Retrieved from https://www.ramsar.org/wetland/wetlands

Ricaurte LF, Wantzen KM, Agudelo E, Betancourt B, Jokela J (2014) Participatory rural appraisal of ecosystem services of wetlands in the Amazonian Piedmont of Colombia: elements for a sustainable management concept. Wetl Ecol Manag 22:343–361

Rogers K, Kelleway JJ, Saintilan N, Megonigal JP, Adams JB, Holmquist JR et al (2019) Wetland carbon storage controlled by millennial-scale variation in relative sea-level rise. Nature 567(7746):91–95

Ronco F, Büscher HH, Indermaur A, Salzburger W (2020) The taxonomic diversity of the cichlid fish fauna of ancient Lake Tanganyika, East Africa. J Great Lakes Res 46(5):1067–1078

Roulet NT, Moore TR, Bubier J, Lafleur PM, Juutinen S, Tuittila ES (2007) Northern peatlands in global climatic change: a synthesis and review.

Ecosystems 10(4):585–610. https://doi.org/10.1007/s10021-007-9041-5
Sachs JD, Mellinger AD, Gallup JL (2001) The geography of poverty and wealth. Sci Am 284(3):70–75
Salvador F, Monerris J, Rochefort L (2014) Peatlands of the Peruvian Puna ecoregion: types, characteristics and disturbance. Mires Peat 15(3):1–17
Sax MS, Bassuk N, van Es H, Rakow D (2017) Long-term remediation of compacted urban soils by physical fracturing and incorporation of compost. Urban For Urban Green 24:149–156
Schuerch M, Spencer T, Temmerman S, Kirwan ML, Wolff C, Lincke D et al (2018) Future response of global coastal wetlands to sea-level rise. Nature 561(7722):231–234
Sharpley A, Jarvie HP, Buda A, May L, Spears B, Kleinman P (2013) Phosphorus legacy: overcoming the effects of past management practices to mitigate future water quality impairment. J Environ Qual 42(5):1308–1326
Silvius MJ, Oneka M, Verhagen A (2000) Wetlands: lifeline for people at the edge. Phys Chem Earth B Hydrol Oceans Atmos 25(7–8):645–652
Silvola J, Alm J, Ahlholm U, Nykanen H, Martikainen PJ (1996) CO_2 fluxes from peat in boreal mires under varying temperature and moisture conditions. J Ecol:219–228
Sonwa DJ, Dieye A, El Mzouri EH, Majule A, Mugabe FT, Omolo N et al (2017) Drivers of climate risk in African agriculture. Clim Dev 9(5):383–398
Sorokin LV (2015) Simplified Sea level estimation based on logistic equation and paleoclimate data. Discrete Contin Models Appl Comput Sci 4:18–29
Strauch RL, Raymond CL, Rochefort RM, Hamlet AF, Lauver C (2015) Adapting transportation to climate change on federal lands in Washington state, U.S.A. Clim Chang 130(2):185–199
Stuip MAM, Baker CJ, Oosterberg W (2002) The socio-economics of wetlands. Wetlands International and RIZA
Sunderland T, Achdiawan R, Angelsen A, Babigumira R, Ickowitz A, Paumgarten F, Reyes-García V (2014) Challenging perceptions about men, women, and forest product use: a global comparative study. World Dev 64:S56–S66
Temmerman S, Meire P, Bouma TJ, Herman PMJ, Ysebaert T, De Vriend HJ (2013) Ecosystem-based coastal defence in the face of global change. Nature 504(7478):79–83
Thirgood S, Mosser A, Tham S, Hopcraft G, Mwangomo E, Mlengeya T et al (2004) Can parks protect migratory ungulates? The case of the Serengeti wildebeest. Anim Conserv 7(2):113–120
Trenberth KE (2011) Changes in precipitation with climate change. Clim Res 47(1–2):123–138
UNDP Bangladesh (2021) How wetlands protect Bangladesh from floods. Retrieved from https://www.bd.undp.org/content/bangladesh/en/home/stories/how-wetlands-protect-bangladesh-from-floods.html
Vogelmann JE, Howard ME, Yang Z, McAfee SA (2020) Louisiana coastal wetland assessment: 2010–2016. US Geol Surv Sci Investig Rep:2020–5073
Walther GR, Roques A, Hulme PE, Sykes MT, Pyšek P, Kühn I et al (2009) Alien species in a warmer world: risks and opportunities. Trends Ecol Evol 24(12):686–693
Wang S, Liu H, Zuo Q, Cai X, Luan S, Zhang W, Wang X (2020) The role of natural wetlands in reducing flood risk in the Yangtze River basin, China. J Hydrol 589:125168
Warrick RA, MAccines KL, Pittock AB, Kench PS (2000) Climate change, severe storms and sea level. In: Floods. Parker Taylor and Francis Group, London, p 139
Weltzin JF, Belote RT, Sanders NJ (2003) Biological invaders in a greenhouse world: will elevated CO_2 fuel plant invasions? Front Ecol Environ 1(3):146–153
Wetlands International (2021) Wetlands and wildfires. Retrieved from https://www.wetlands.org/wetlands-and-wildfires/
Windham-Myers L, Crooks S, Emmett-Mattox S, Byrd K, Drexler JZ, Howard J, Pidgeon E, Schile LM, Stagg CL, Troxler T, Whigham DF, Woodrey M, Zielinski C (2018) Restoring US federal wetlands: values, successes, and challenges. Front Ecol Environ 16(7):365–372. https://doi.org/10.1002/fee.1819
Woodroffe CD, Rogers K, McKee KL, Lovelock CE, Mendelssohn IA, Saintilan N (2016) Mangrove sedimentation and response to relative sea-level rise. Annu Rev Mar Sci 8:243–266
Yochum SE (2017) Guidance for stream restoration, Technical note TN-102.3. U.S. Department of Agriculture, Forest Service, National Stream & Aquatic Ecology Center, Fort Collins.

28 Land Suitability Assessment for Mulberry-Based Agroforestry Using AHP and GIS Technique in the Northwestern Himalayan Region of Kashmir Valley, India to Achieve Sustainable Agriculture

Ruyida Mushtaq, Rajesh Kumar Yadav, Abida Fayaz, Pervez Ahmed, Harmeet Singh, and Jaipreet Singh

Abstract

Land suitability assessments are necessary to reduce human impact on the environment and determine the best use of land. The Baramulla district of the Kashmir valley has been searched for in this study as a potential location for mulberry agroforestry. We assessed the suitability of the land for mulberry agroforestry in the Baramulla district of India using basic parameters such as slope, temperature, rainfall, LULC, and physical and chemical properties of the soil and applied the weighted multicriteria evaluation (MCE) technique in a geographic information system (GIS) environment. Utilising the accessible literature and interviews with regional experts, the AHP approach was used to score the criteria. Through weighted overlay analysis, a map of the suitability of mulberry agroforestry for the research area was created, incorporating numerous criteria. The Mulberry agroforestry suitability map for the study region reveals that 16.99% of the land was extremely suitable, 28.96% was moderately suitable, 38.18% was just marginally suitable, and 15.87% was unsuitable. The results have also been thoroughly examined using Google Earth images, in comparison with Sentinel-2b LULC output and on field visits utilising GPS in order to increase the reliability of the data. Farmers, regional planners, and the government can use the research study's findings to guide their decisions regarding prospective mulberry agroforestry areas in the region and the methods that will provide a sustainable agroforestry management strategy and sustainable land use planning.

Keywords

Land suitability · Analytical hierarchy process · GIS · Mulberry agroforestry · Baramulla district · Kashmir valley

R. Mushtaq (✉) · A. Fayaz
Department of Geography and Disaster Management, University of Kashmir, Srinagar, India

R. K. Yadav
Mohanlal Sukhadia University, Udaipur, Rajasthan, India

P. Ahmed · H. Singh
Department of Geography, University of Kashmir, Srinagar, India

J. Singh
Department of Geography, Cluster University of Kashmir, Srinagar, Jammu and Kashmir, India

U. Chatterjee et al. (eds.), *Climate Crisis: Adaptive Approaches and Sustainability*, Sustainable Development Goals Series, https://doi.org/10.1007/978-3-031-44397-8_28

28.1 Introduction

A land suitability analysis is necessary to make the greatest possible use of the land resources that are now at our disposal. It aids in creating strategies for increasing agricultural productivity as well as plans for sustainable land usage (FAO 1985). The main objective of land suitability management is to increase the use of sustainable agriculture (FAO 1993; Hopkins 1977; Hilali 2018; Habibie et al. 2019; Herzberg et al. 2019; Erener and Uzgeun 2008; Mazahreh and Bsoul 2018; Patil et al. 2012). Land suitability is based on a combination of social and economic factors, including soil adaptability. Land potentiality describes a region's innate ability to harvest common crops based on the soil and geographical characteristics of the land (USDA 1961; FAO 1976). Therefore, due to changes in soil characteristics, terrain features, land use, and water availability, land suitability in each region differs depending on the type of farm. The classification and administration of land units must take into account the most effective environmental considerations (FAO 2007). Due to human activity, land resources in mountainous regions have gradually decreased as a result of increasing urbanisation and development activities, which reduce the viability of current agricultural systems (Bandooni 2004; Alam et al. 2020). Making the best use of the available natural resources is crucial today to achieving sustainable agriculture. The degradation of the land threatens natural resources. Inappropriate land use and management are the main causes of degradation. Site selection is a crucial component of sustainable land use management systems. Agroforestry practises that are sustainable are required because of land deterioration. One of the best strategies to reduce soil degradation is to evaluate the land's suitability. Mulberry agroforestry offers a variety of benefits for the environment and economy, and these benefits have greater regional and local significance (FAO 1983; Torralba et al. 2016; Shahid et al. 2021).

The mulberry (Morus spp.) is a deciduous woody tree with rapid growth that is native to China and the foothills of the Himalayas in India (Wang et al. 2013). By feeding leaves to silkworm (Bombyx mori) larvae, mulberry agroforestry is widely acknowledged to serve a vital economic role in the production of mori silk (Zhang et al. 2018). Throughout millennia, mulberry silk has significantly boosted local economies, and it has also played a vital role in the development of global trade networks and globalisation. Currently, UNESCO has designated mulberry silk as an intangible cultural heritage. If supply chain management is done carefully, it helps promote a circular economy that is sustainable and may even aid in reducing carbon emissions Holland et al. 2019). The regulation of environmental protection, agroforestry, carbon storage, and soil management are some of the ways that mulberry can help achieve the SDGs, according to a recent study (Li et al. 2020; Zhang et al. 2018; Yuan and Zhao 2017). For decades, the Food and Agricultural Organisation has worked to develop effective mulberry agroforestry techniques to address issues with food and nutrition security and poverty reduction (FAO 2007). By 2050, the Indian government aims to double the amount of land used for agroforestry (Nath et al. 2021).

Fallow areas, pasture land, and wasteland are all potentially suitable alternatives for mulberry agroforestry techniques. Agroforestry practises, food access, and access to healthcare can all benefit from thoughtful implementation (Nair and Toth 2016; Rosenstock et al. 2016). A careful selection of the land is essential for the effective implementation of agroforestry practises and the achievement of optimal yields. Agroforestry mainly enhances regions of land with little rainfall and nutrient-poor soil. Researchers employed geospatial technology, which combines remote sensing (RS) and geographic information systems (GIS), to discover suitable acreage for agroforestry activities (Ahmad et al. 2021; Ahmad and Goparaju 2017). When looking for good agroforestry land, factors like climate, topography, land use, vegetation, and soil are crucial (Ahmad et al. 2019; Zomer et al. 2019; Herzberg et al. 2019).

Baramulla Mulberry Agroforestry Suitability Mapping revealed several villages and sites that would be appropriate for planting trees but require policy attention and investment to achieve the SDGs. When trees are grown within an agroforestry framework, poverty will be reduced (SDG 1), food security to end hunger (SDG 2), nutrition security (SDG 3), gender equality (SDG 5), clean water and sanitation (SDG 6), affordable energy (SDG 7), responsible agricultural production (SDG 12), climate change mitigation and adaptation (SDG 13), degraded land and wasteland restoration (SDG 15), and biodiversity enhancement (SDG 15) will be achieved among poverty-ridden societies and the environment (Zomer et al. 2019).

Mulberry agroforestry land has a lot of potentials to be expanded in the Baramulla district due to its favourable environmental and climatic characteristics. When choosing locations that are suited for agricultural land use, geophysical factors, various topological circumstances, and the climatic environment must all be taken into consideration (Malczewski 2000; Kuria et al. 2011; Kuta et al. 2016; Feizizadeh and Blaschke 2012). Therefore, current and accurate land use and land cover (LULC), as well as other environmental and topographical data, should be taken into account to establish the most suitable mulberry agroforestry zone. LULC, slope, elevation, and soil characteristics are significant factors that are frequently used to assess a land's quality and its potential for agricultural output. When deciding whether a site is suited for agricultural growth, physiographic information, meteorological conditions (rainfall and temperature), and soil quality are all taken into consideration (Pramanik 2016; Ahmad et al. 2021).

Saaty was the first to create AHP, which established a hierarchical approach for solving complicated problems of land management with the best options (Everest et al. 2021; Cengiz and Akbulak 2009; Chen et al. 2010). The analytical hierarchy process (AHP) is often used to make multi-criteria decisions on the suitability of a land area for a certain field. Based on pairwise comparisons of numerous factors in accordance with their relative relevance, it establishes the weight of importance for various land uses (Cengiz and Akbulak 2009). For the site suitability assessment of agricultural land use, the weighted overlay method (WOM) and the analytical hierarchy process (AHP) have been utilised Akinci et al. (2013), Garcia et al. (2014) So, several researchers have applied the AHP method to identify suitable sites for agricultural land use (Sajjad and Nasreen 2016; Feizizadeh and Blaschke 2012; Chen et al. 2010; Shahram 2010; Barakat et al. 2017; Mokarram and Aminzadeh 2010; Joerin et al. 2001; Chandio et al. 2014; Mesgaran et al. 2017; Kazemi and Akinci 2018; Barakat et al. 2017; Habibie et al. 2019) GIS techniques are also utilised to construct several criteria maps that are used to develop the site suitability model for agricultural development (Rasheed and Venugopal 2009).

The research region also offers land that can be used for mulberry agroforestry because of the great elevational variation and accessibility to natural resources. Therefore, evaluating mulberry agroforestry land's suitability is a trustworthy and acceptable strategy for establishing sustainable agriculture. The study's present objective was to assess the land's suitability for mulberry agroforestry using the AHP method, compare it to the area's current land use in 2020, and generate concrete data that landowners and decision-makers could use to evaluate the land in the Baramulla district of the Kashmir valley.

28.2 Study Area

Biodiversity has greatly increased in the Baramulla region due to the wide variety of flora and animals. This district, which is in the union territory of Jammu and Kashmir, is located on the Jhelum River's banks between 34.1980° N Longitude and 74.3636° E Latitude. It is a thriving agricultural and horticultural region (Hussain 2018; Romshoo et al. 2020; Raza et al. 1978). The Baramulla region is known for its production of apples, walnuts, rice, maize, mulberries, and other fruits. Khan (2008) and Lone and Sen (2014) Most of the district is covered with brown forest soil. Both a silty clay loam and a sandy clay loam make up its granular texture (NBSS 2020). The area's unique geography contributes to the abundance of both the flora and the animals. 1,008,039 people reside in the Baramulla

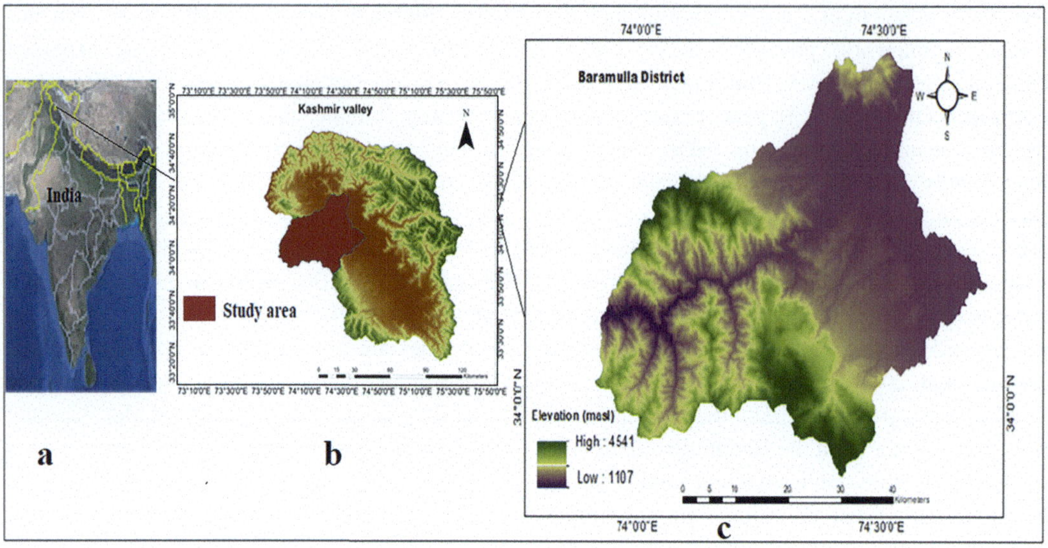

Fig. 28.1 Location map of study area (**a**) India, (**b**) Kashmir valley, (**c**) study area. (Source (CARTOSAT-DEM2.5 m))

district as of 2011 (Census of India 2011). The average annual high and low temperatures in Baramulla are 19.8 °C and 7.5 °C, respectively, and 1400 mm of precipitation falls here each year (Gujree et al. 2022). Baramulla is the primary silk cocoon producer and has active mulberry nurseries in the Kashmir valley (Mushtaq et al. 2022). The production of organic honey and silk in this area has great potential. Jams, pulp, and juice are value-added products created from mulberry fruits (Fig. 28.1).

28.3 Material and Methods

The current study's goal is to identify potential and ideal mulberry agroforestry sites based on several restrictions and criteria using a site suitability analysis. This study makes use of primary data from on-site field surveys. Also necessary at this point was expert advice from a crop specialist. Using a desk search of the data, interviews with local agronomists and researchers at SKAUST Mirgund, and a review of several academic references, the criteria for the optimum locations to develop mulberry agroforestry were established. Using Arc GIS 10.4, climate, soil, and satellite data were gathered, and criteria maps were produced. Next, the acquired data is transformed into a shape file format for weighted overlay processing, as depicted in Tables 28.3 and 28.4. ArcGIS 10.4 is being used to process the data for the study, which is being carried out in the Kashmir Valley's Baramulla region. The FAO land suitability classification was then used to reclassify these criteria maps. The weights assigned to each criterion were determined using the AHP multi-criteria evaluation process. Using a multi-criteria evaluation (MCE) model, the research area's mulberry agroforestry suitability map was finally created. In order to determine which factors were most crucial for mulberry agroforestry suitability, they were graded and given relative weights (FAO 2007; Bera 2002; Uniyall et al. 2014; Handique et al. 2016) (Fig. 28.2).

28.4 Generation of Different Thematic Layers for Suitable Land Assessment of Mulberry Agroforestry

Thematic maps were made and rasterized in a GIS environment for each pertinent criterion. The slope layer was produced using Bhuvan Cartosat-1 digital elevation model (DEM) data

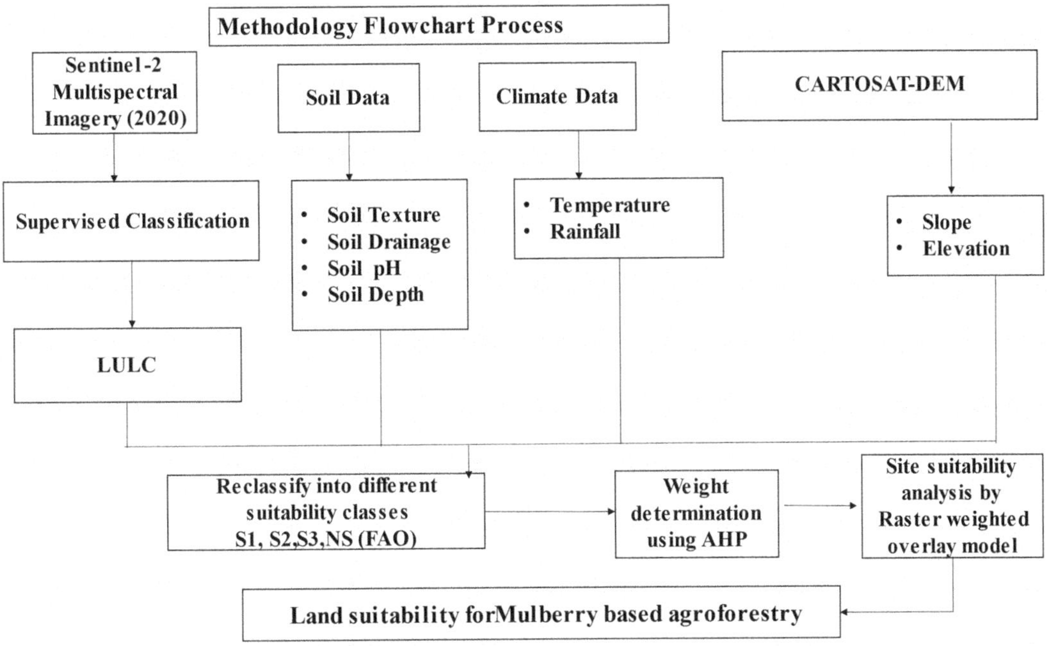

Fig. 28.2 Methodology flowchart process used in the study

with a spatial resolution of 2.5 m (Bhuvan, the India Geo-Platform of the Indian Space Research Organisation, ISRO 2016). Much depends on the characteristics of the soil to determine whether a land area is suitable for farming (Bhandari et al. 2014; Dominati et al. 2016; Ebrahimi et al. 2019). The depth, texture, drainage, and pH of the soil were taken into account while determining if mulberry agroforestry was appropriate.

To identify the depth, texture, drainage, pH, and other physical and chemical features of the soil, the National Bureau of Soil Survey and Land Use Planning mapped the soil at a scale of 1:250000. (NBSS & LUP 2020). The maps of different soil characteristics were also created using information from soil testing facilities. (Directorate of Agriculture 2022). The primary climatic variables that have a substantial impact on mulberry agroforestry are temperature and rainfall. Three weather stations that cover the study area have had their theme layers of temperature and precipitation acquired from the IMD department, Srinagar, and power access climatic data. (Nasa Data Access Viewer, 2022 https://power.larc.nasa.gov/data-access-viewer/). Based on the average temperature of the past 20 years (2000–2020) at the various stations, a temperature map has been constructed using the IDW technique in the GIS environment (Fig. 28.9). The ArcGIS software's IDW tool has been utilised to construct the thematic layer of rainfall using data from multiple rainfall stations. Land use and land cover (LULC) maps are produced using Sentinel-2b satellite images with a 10 m resolution. The eight primary components stated in land use and land cover classifications are agriculture, forests, dense forests, wasteland, scrubland, water, built-up area, and snow or glacier land (Anderson et al. 1976) (Figs. 28.3, 28.4, 28.5, 28.6, 28.7, 28.8, 28.9, and 28.10).

28.5 Standardisation of Criteria Layers for Mulberry Agroforestry Suitability

Before doing a weighted overlay analysis for land suitability, the criterion layers must be standardised. As a consequence, we used a GIS programme to convert the required vector layers to raster format. The final suitability map for mulberry agroforestry was then created by reclassifying all

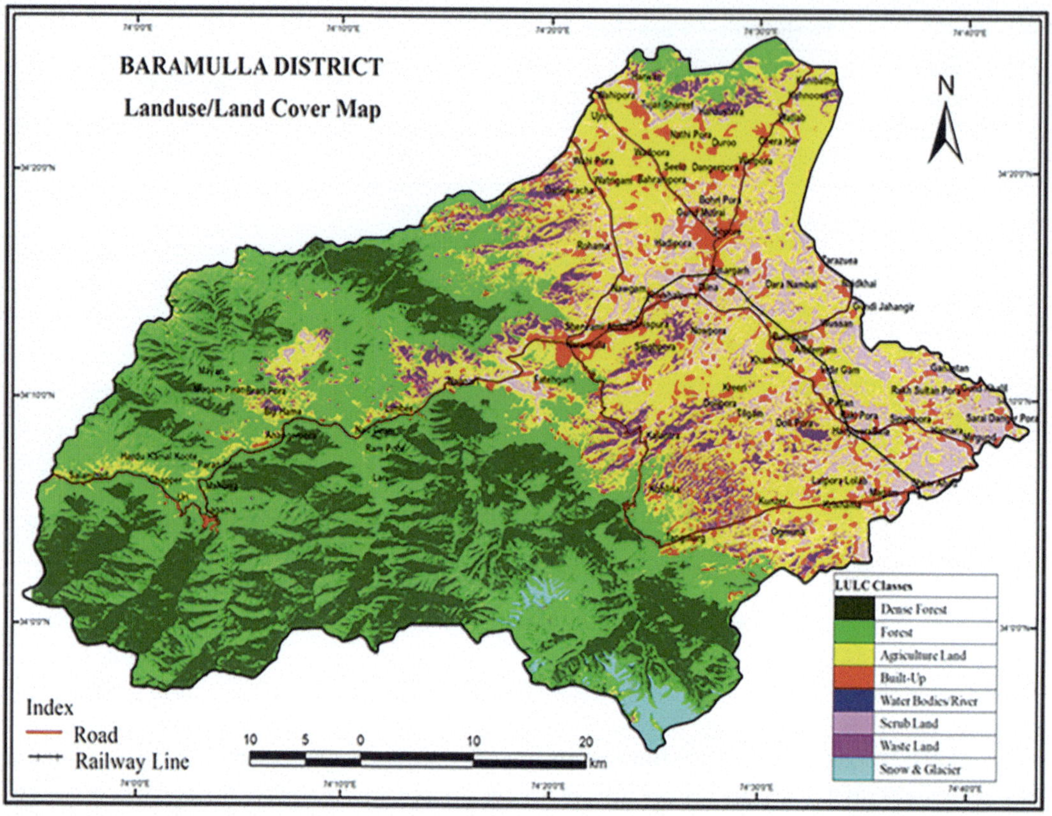

Fig. 28.3 LULC of Baramulla district 2020

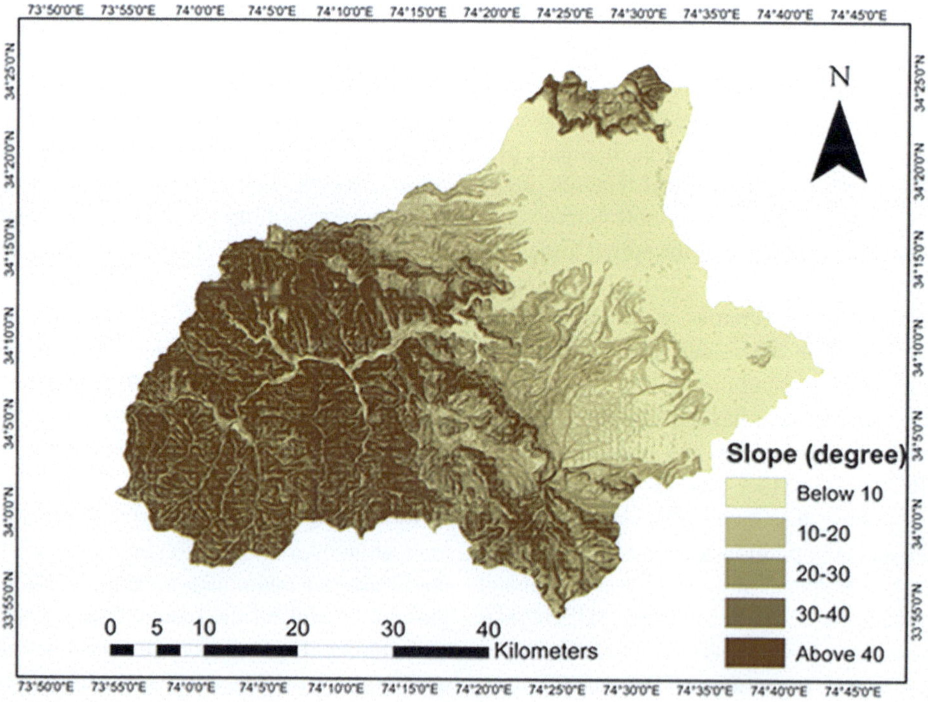

Fig. 28.4 Slope map of Baramulla. (Source: CARTOSAT-DEM)

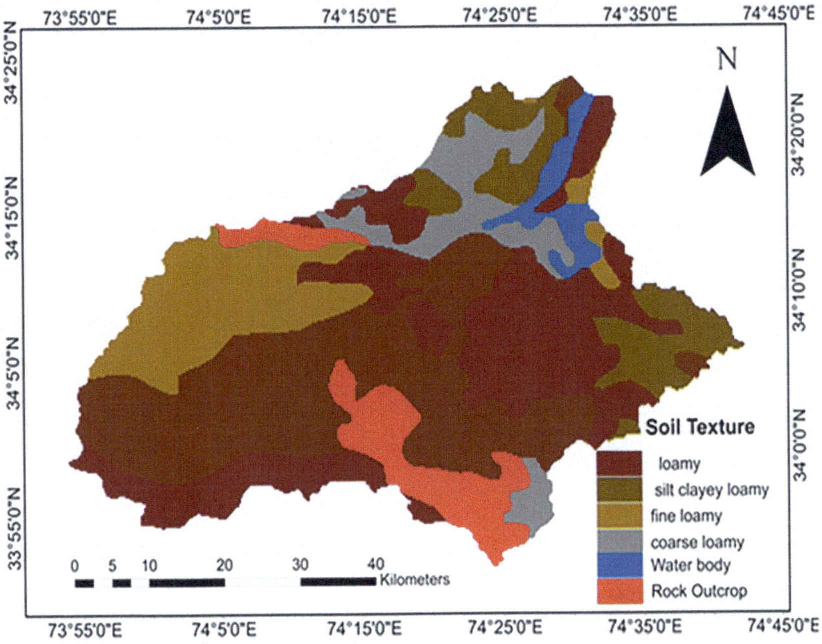

Fig. 28.5 Soil texture of Baramulla (NBSS and LUP)

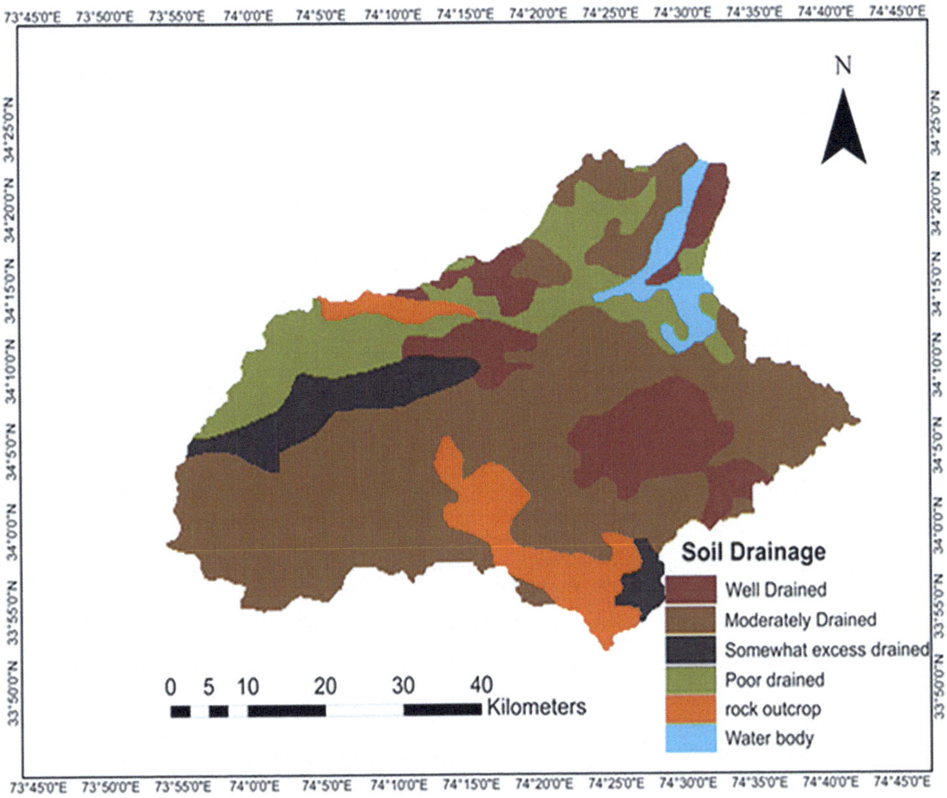

Fig. 28.6 Soil Drainage of Baramulla (NBSS & LUP)

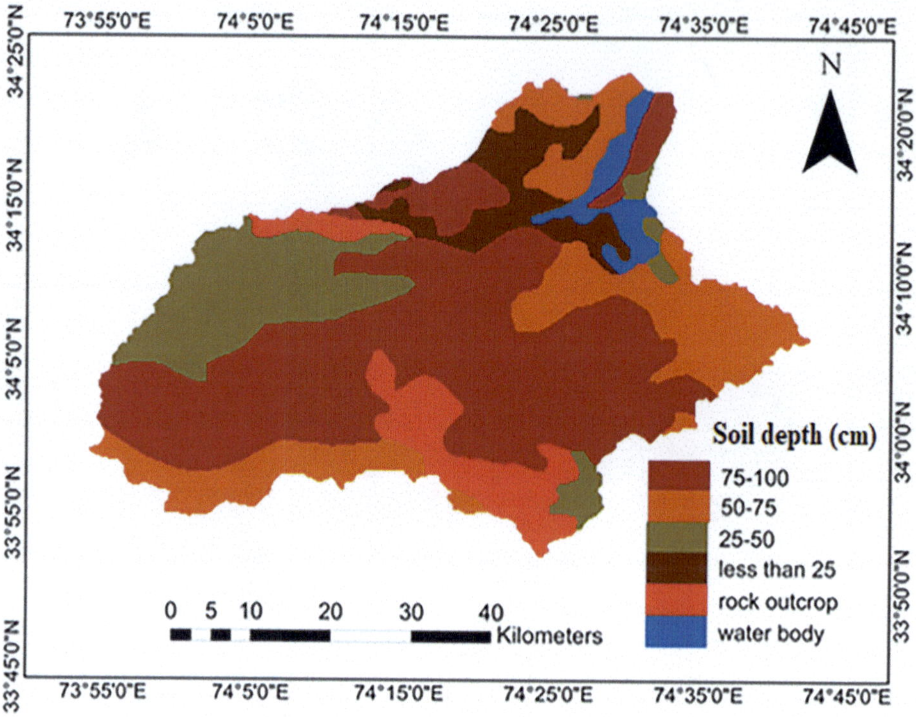

Fig. 28.7 Soil Depth of Baramulla (NBSS & LUP)

Fig. 28.8 Soil pH of Baramulla (NBSS & LUP)

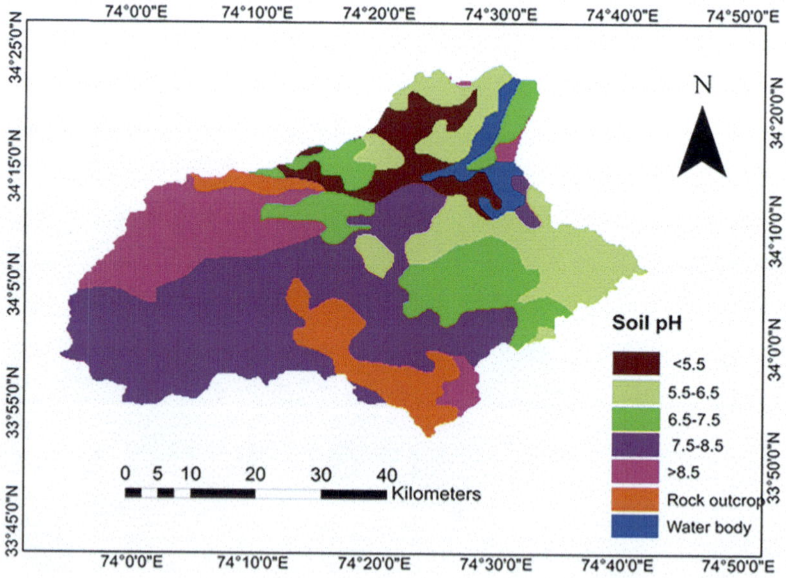

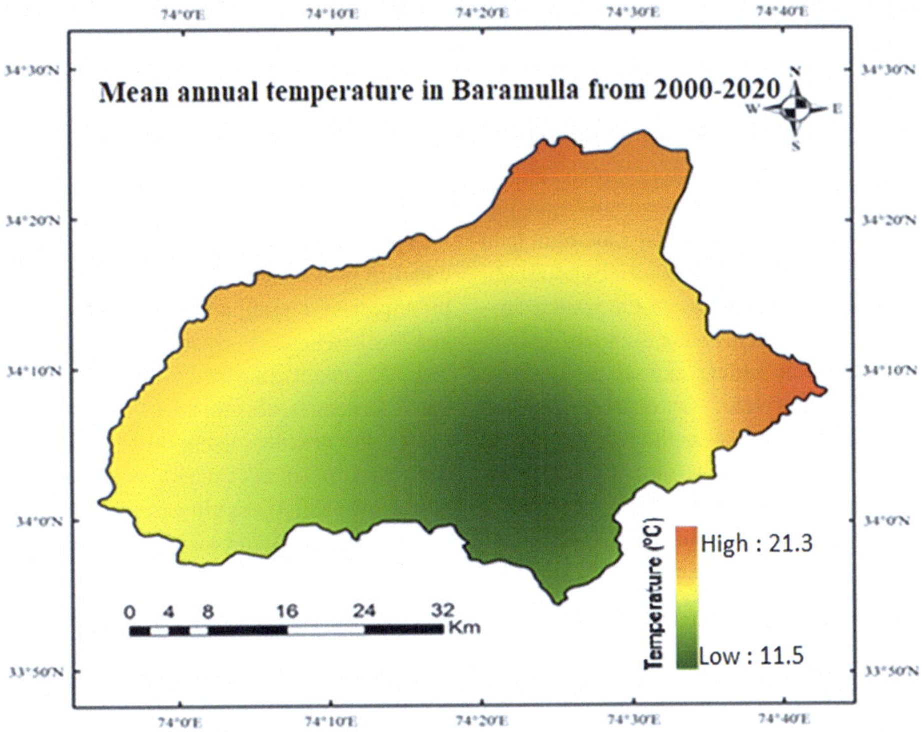

Fig. 28.9 Mean annual average growing temperature from 2000–2020 in Baramulla

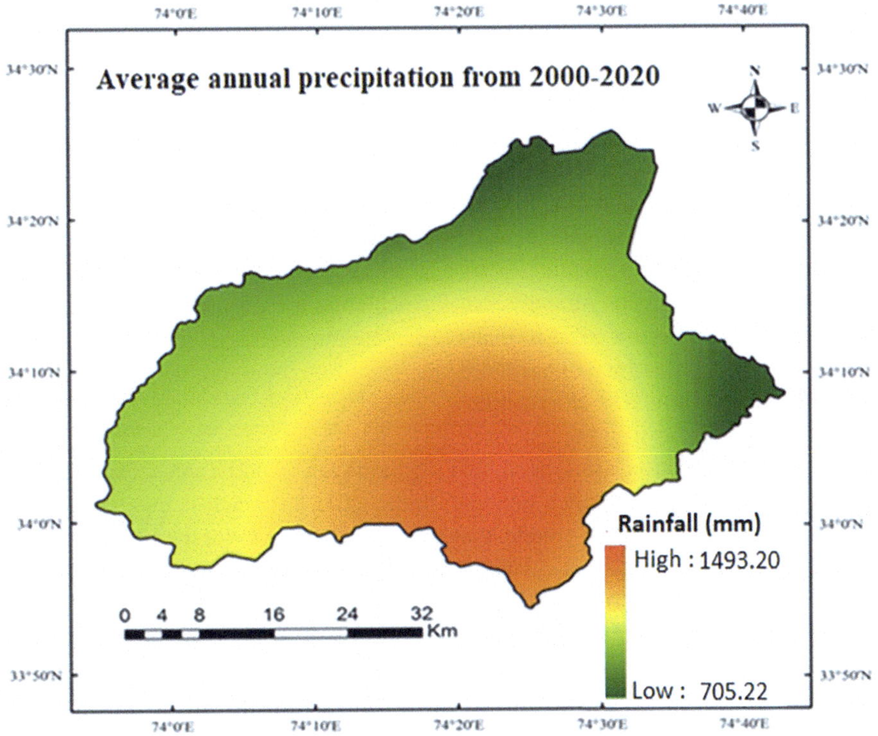

Fig. 28.10 Average annual precipitation from 2000–2020 in Baramulla

Table 28.1 Land Suitability Classification (FAO)

Symbol	Suitability	Description
S1	High	Land is more favourable for mulberry productivity
S2	Moderate	Land has some restrictions for mulberry productivity
S3	Marginal	Land with major restrictions for sustained mulberry productivity
S4	Not suitable	Land with extreme restrictions for sustained mulberry productivity

raster layers, which were then used as input data for the weighted overlay method (Bera 2002). The reclassification technique in the spatial analyst inbox of Arc-GIS software standardises the value of all selected criteria for the analysis of comparative significance. Table 28.4 lists the standardisation standards applied to mulberry agroforestry suitability. According to the FAO's guidelines, there are four categories of appropriateness for land: highly suitable (S1), moderately suitable (S2), marginally suitable (S3), and not suitable (S4). All of these factors were taken into account when analysing the potential mulberry agroforestry zone (Table 28.1).

28.6 Analytical Hierarchy Process for Determining Land Suitability

AHP, which is based on the multiple criterion approach, is one of Saaty's most successful methods for making decisions (Saaty 2012). The method can be applied to a group of parameters to create a hierarchical structure that simplifies decision-making by assigning weights to each criterion (Kiker et al. 2005; Pramanik 2016). AHP is a systematic and reliable strategy that integrates objective and subjective ideas to make decisions by assessing a wide range of variables based on expert opinions in a GIS environment (Ying et al. 2007). The proper weighting of each parameter and its normalisation are crucial for getting appropriate results because they are the only factors that affect the final conclusion (Zarkesh et al. 2010). AHP is regarded as one of the most promising techniques for determining suitability for agroforestry since it provides quick, extremely accurate, and affordable findings.

The comparison matrix, which theoretically equals n (n-1)/2 for the n number of components in a pairwise comparison matrix (Patil et al. 2012), is where the reciprocity conditions are most frequently applied. After computing the pairwise matrix, the relative weights and eigenvectors are obtained using Saaty's method (Chen et al. 2010; (Tables 28.3 and 28.4). The analytical hierarchy method's ability to identify and quantify the inconsistencies of decision-makers is another one of its key characteristics (Park et al. 2011). The efficiency criteria of the analytic hierarchy technique are estimated using the consistency relationship (CR), which is evaluated by Eq. (28.1).

$$CR = CI / RI \tag{28.1}$$

The CR is represented by Eq. (28.1), where CI stands for consistency index and RI stands for random index.

The consistency relationship assists in identifying potential outcomes and assesses the logical consistency of the decision-makers/conclusions judgements (Chen et al. 2010). It represents the likelihood that the matrix verdicts were generated at random (Tercan and Dereli 2020; Ennaji et al. 2018). The two most crucial variables in calculating the CR are the consistency index and the random index.

$$CI = (\lambda_{max} - 1) / (n - 1) \tag{28.2}$$

Equation (28.2) represents the Consistency Index (CI), in which k max is the principle or highest eigenvector of the computed matrix and n is the matrix order. According to Saaty, the random index (RI) is the mean consistency index value based on the computed matrix order. The weight values in the matrix exhibit irregularities if the CR value is [0.10], and the approach (AHP) may fail to yield useful results (Saaty 2012). The computed weight values are precise, and the calculated CR in this experiment was 0.039, which is within acceptable bounds. In order to do a weighted overlay analysis in GIS, the resulting weight values are then converted to percentages, as shown in Table 28.4.

Based on a review of pertinent literature and in-person interviews with the District Agriculture Officer, agriculturists, farm managers, and local farmers in 2021–22, the classes and value ranges of the chosen evaluation criteria for the land suitability of mulberry agroforestry (Table 28.3) were established. The local experts in the Baramulla district who are well-versed in mulberry cultivation were asked to share their opinions on the relative weight of each criterion. The suitability levels were: highly suitable (S1), moderately suitable (S2), marginally suitable (S3), and not suitable (N) based on the FAO land suitability categorisation's framework.

28.7 Weighted Overlay Analysis

The land suitability for mulberry agroforestry in Baramulla was determined using a weighted overlay analysis employing the weight values of specific elements generated from the Analytic Hierarchy Process and defined scores of sub-criteria. (Tables 28.2, 28.3, and 28.4). Land suitability for mulberry is classified into four levels, according to the Food and Agricultural Organisation (FAO): highly suitable mulberry land, moderately suitable mulberry land, marginally suitable mulberry land, and not suitable for mulberry. In this study, environmental factor maps were employed to evaluate the mapping of mulberry agroforestry suitability. To incorporate these into the study, each cell from each map layer must be classified on a similar preference scale, such as 1–9, with 1 representing low suitability and 9 representing great suitability. The formula for the weighted approach method is as follows:

$$S = (wisij) / (wi)$$

Where *wi* is the weight factor map,

S_{ij} is the *i*th spatial class weight of the *j*th factor map,

S is the spatial unit value in the output map.

These weighted factor maps were integrated into a GIS platform using WOM analysis to prepare a mulberry agroforestry suitability map. Since the weighted overlay in Arc Map was used to construct the final suitability maps based on the influence percent for each layer, each layer contributes to the influence in accordance with the kind of land use (Malczewski 2000; Cengiz and Akbulak 2009; Pramanik 2016). Once the composite layers and their weights were gathered, the map of the relevant areas was made using the MCE method in Arc GIS 10.4. With the help of ArcGIS 10.4's spatial analysis capabilities and a weighted overlay, the suitability map for mulberry agroforestry was found (Fig. 28.11). In the study area, highly appropriate sites exhibit the most favourable or best conditions for mulberry agroforestry growth. The next priority areas, which require careful consideration of all relevant criteria and decisions regarding the viability of investment over alternative opportunities,

Table 28.2 Pairwise comparison matrix of eight thematic layers

Criteria	Slope	Temp	LULC	Rainfall	Soil texture	Soil drainage	Soil depth	Soil pH
Slope	1	2	3	4	5	6	7	8
Temp	1/2	1	2	3	4	5	6	7
LULC	1/3	1/2	1	2	3	4	5	6
Rainfall	1/4	1/3	1/2	1	2	3	4	5
Soil texture	1/5	1/4	1/3	1/2	1	2	3	4
Soil drainage	1/6	1/5	1/4	1/3	1/2	1	2	3
Soil depth	1/7	1/6	1/5	1/4	1/3	1/2	1	2
Soil pH	1/8	1/7	1/6	1/5	1/4	1/3	1/2	1

Table 28.3 Normalised comparison matrix of thematic layers

Criteria	Slope	Temperature	LULC	Rainfall	Soil texture	Soil drainage	Soil depth	Soil pH	Normalised weight
Slope	0.3189	0.4165	0.3254	0.2461	0.2343	0.2530	0.2069	0.1976	0.2710
Temperature	0.1980	0.2176	0.2658	0.2655	0.2480	0.2250	0.2155	0.1944	0.2287
LULC	0.1225	0.1088	0.1432	0.1774	0.1856	0.1823	0.1745	0.1676	0.1790
Rainfall	0.0922	0.0736	0.0681	0.0878	0.1344	0.1378	0.1414	0.1398	0.1498
Soil texture	0.0746	0.0644	0.0457	0.0445	0.0625	0.0926	0.1054	0.1122	0.0590
Soil drainage	0.0614	0.0436	0.0345	0.0297	0.0312	0.0485	0.0712	0.0843	0.0485
Soil depth	0.0562	0.0373	0.0286	0.0229	0.0208	0.0234	0.0356	0.0567	0.0350
Soil pH	0.0471	0.0321	0.0232	0.0176	0.0157	0.0156	0.0176	0.0279	0.0290

Maximum eigen value ((ƛ)_max) =8.39, $n = 8$, C.R = 0.039, $R = 1.41$

Table 28.4 Normalised weight and area (%) for eight selected parameters based on AHP and rank assigned to their sub-classes based on mulberry suitability

Criteria	Classes	Area	Rank	AHP-normalised weight	Influence (%)
Slope (°)	Below 10	20.38	9	0.271	27.1
	10–20	36.45	7		
	20–30	14.63	5		
	30–40	19.65	1		
	Above 40	8.89	1		
Temperature°c	20–24	16.05	7	0.2287	22.87
	24–28	21.06	6		
	28–32	62.22	3		
	<15 and > 32	0.67	1		
LULC	Agriculture	25.8	8	0.179	17.9
	Scrubland	10.54	5		
	Wasteland	4.97	3		
	Water	1.26	1		
	Built up area	4.47	1	0.1498	14.98
	Snow/ glacier	1.05	1		
	Dense forest	25.25	1		
	Forest	27.33	2		
Rainfall	500–750	6.28	8		
	750–2000	17.75	5		
	2000–3400	52.4	2		
	<500 and > 3400	23.57	1		
Soil texture	Loamy	12.75	8	0.059	5.9
	Silt clayey loamy	18.61	5		
	Fine loamy	50.75	3		
	Coarse loamy	17.89	1		
Soil drainage	Well drained	13.45	7	0.0485	4.85
	Moderate drained	19.8	5		
	Excessively drained	48.5	2		
	Poorly drained	18.25	1		
Soil depth	75–100	14.4	7	0.035	3.5
	50–75	20.9	6		
	25–50	45.8	3		
	Less than 25	18.9	1		
Soil p H	6.5–7.5	14.4	6	0.029	2.9
	5.5–6.5, 7.5–8.5	20.9	3		
	8.5–9.5	45.8	3		
	>4.5,<9.5	18.9	1		

are those that are moderately and marginally suited. The sites that are restricted by frost, flash floods, ice cover, unsuitable elevation, and other LULC kinds are represented by the unsuitable area. Finally, to corroborate the findings, the GPS-collected occurrence locations of mulberry agroforestry from the field study were superimposed over the final suitability map shown in Fig. 28.11 (Fig. 28.12, Table 28.2).

28.8 Result and Discussions

The main goal of an applied AHP-based land suitability study is to identify potential locations that are suitable for mulberry agroforestry and that will increase productivity and sustainability in the Baramulla region. The investigation of the appropriateness of the land for agroforestry is an essential source of knowledge for planning and

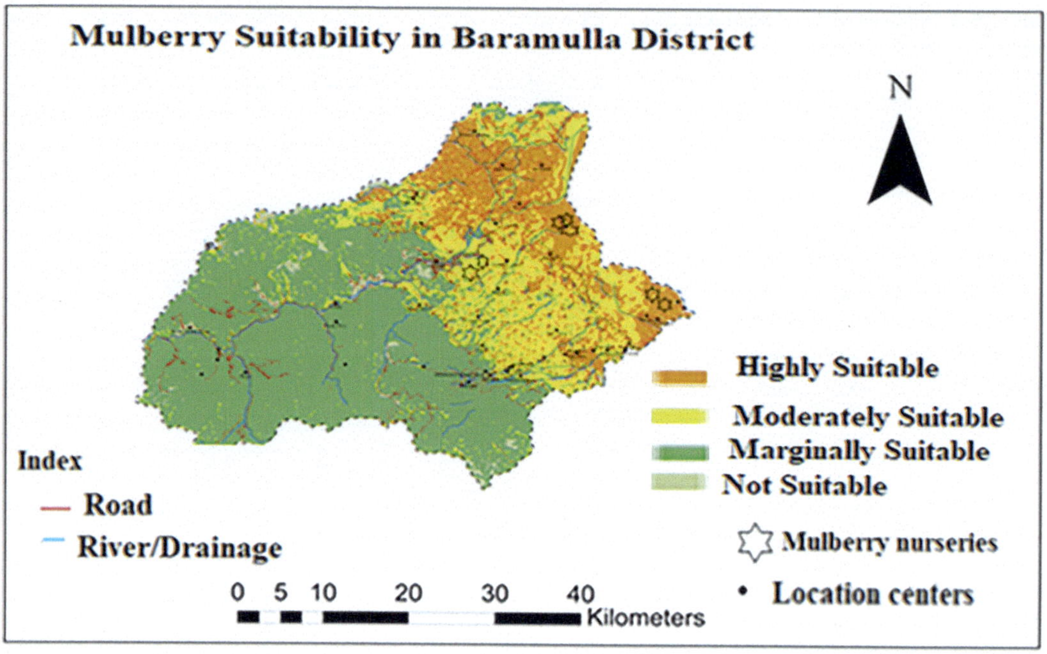

Fig. 28.11 Potential sites for mulberry agroforestry cultivation along with GPS locations for verification in Baramulla district

Fig. 28.12 Field observations of mulberry suitability along with GPS locations in Baramulla

development. Using the multi-criteria decision-making approach and chosen geographic datasets, including temperature, precipitation, slope, LULC, soil texture, depth, pH, and soil drainage, the best area of land for mulberry production was found. Each data layer has been given a weight using the multi-criteria decision-making method with AHP. Then, using ArcGIS software, all thematic layers were combined to provide a map indicating the suitability of the area for mulberry growing. Four classifications of mulberry suitability have been made: "highly suitable," "moderately suitable," "marginally suitable," and "not suitable." (FAO 1976). The physical and chemical characteristics of the soil have an impact on mulberry productivity.

28.9 Soil Texture

Soil texture is a significant component since it affects most of the physical characteristics of the soil (Haider 2013; Mustafa et al. 2011). The relative ratio of clay, silt, and sand in loamy soil establishes the textural class when combined and compared to texture triangles. Mulberry agroforestry, according to Sánchez (2000), performs poorly in heavy clay soils but thrives in rich loams and is well adapted to silty, clayey, loamy soils. The very high clay content accounts for the marginal applicability. As a result, site sustainability for mulberry agroforestry is influenced by soil texture. According to the reclassified texture map, 12.75% of the Baramulla district is fairly favourable for mulberry agroforestry (Table 28.4), with 18.61 percent and 50.75 percent of the district being moderately and marginally suitable, respectively. The very high clay content accounts for the marginal applicability. Area-based land suitability maps were created using the soil quality indicators of texture, drainage, pH, and depth.

28.10 Soil Depth

Soil depth is considered extremely appropriate for soil to be between 75 and 100 cm deep and have a pH between 6.5 and 7.5 (S1). Soil depth refers to the estimated depth in cm at which root growth is unrestricted by any physical or chemical impediment, such as an impenetrable or toxic layer. Figure 28.7 shows that 19.35% of the district is moderately acceptable for mulberry production with soil depths less than 30 cm, while 14.4% of the district is exceptionally favourable for mulberry cultivation with soil depths greater than 50 cm. Root development is hindered in shallow soils, which reduces the plant's capacity to absorb water and nutrients (Barbour et al. 2000). Furthermore, shallow soils don't have enough room to store water, which causes the plant to experience water stress during extended dry spells. This may clarify the importance of depth in the production of mulberry agroforestry.

28.11 Soil pH

Mulberry agroforestry can tolerate high soil alkalinity and acidity (Bera 2002). It grows in soils that are somewhat alkaline and acidic. A soil pH of 6.5 to 8.5, on the other hand, is regarded as optimal for mulberry growth. Most of the areas under mulberry cultivation (14.4%) showed pH within range and was found highly suitable; 20.9% was moderately suitable; and 45.8% was marginally suitable.

28.12 Soil Drainage

The soil drainage reclassified map shows that 18.25% of the Baramulla district is not suitable for mulberry production, as shown in Table 28.4. The remainder of the district is suitable in high (13.45%), moderate (19.8%), and marginal (48.5%) proportions, as depicted in Table 28.4. Mulberry health is dependent on soil drainage, which indicates the rate at which free moisture escapes from the soil. Poorly drained soils are more likely to support root rot, infections, and fungal development. Mulberry agroforestry requires deep, well-drained soils (Bera 2002).

28.13 Slope

Slope plays a significant role in mulberry agroforestry. The five groups used to categorise the slope map produced from CARTO-DEM were: level to nearly level (below 10 degrees), gently sloping (10–20 degrees), moderately sloping (20–30 degrees), sharply sloping (30–40 degrees), and very steeply sloping (> 40 degrees). According to the slope of the reclassification map, 20.38% of the Baramulla district is extremely suited for mulberry agroforestry (Table 28.4). The remainder of the land has moderate (36.45%)

and marginal (14.63%) suitability for mulberry production, while 19.65% is unsuitable, as shown in Table 28.4.

28.14 Climate

The parameters for mulberry agroforestry cultivation are temperature and rainfall. Climate factors such as precipitation and temperature influence the yield and growth of crops and plants (Chen 2016; Hijmans et al. 2015). Agroforestry climates are defined by mean annual precipitation greater than 1200 mm and mean annual temperatures between 20 and 30 °C. (Ahmad et al. 2019) A mulberry crop's climate adaptability can be expressed in terms of the mean growing season temperature (°C), which was divided into four groups: (20–24) highly suitable, (24–28), moderately suitable (28–32), marginally suitable, 20, and > 32 (not suitable). The classification of rainfall maps was split into four categories: highly suitable (500–750 mm), moderately suitable (750–2000 mm), marginally suitable (2000–3400 mm), and not suitable (500 and > 3400 mm). According to the analysis of the results, the average annual precipitation of Anantnag district varied significantly from 705.22 mm to 1493.20 mm from 2000 to 2020. As far as the mean annual growing season temperature (°C) is concerned, it ranges from 21.3 °C and 11.5 °C, respectively from 2000–2020, as mentioned in Fig. 28.9. Out of the total area, 16.69% was highly suitable for mulberry cultivation.

The findings collectively show that there is enormous potential for mulberry agroforestry in the research area. The bottom part of this district is most suited for the growth of mulberry agroforestry due to the presence of more recent alluvium soil, a lower slope, and the presence of soil micronutrients. (Wani and Shaista 2016) Therefore, it is plainly clear from the results that the lower part of the Baramulla district has more fertile and productive soil than the higher part.

In the Baramulla district, there are different levels of topographical suitability for mulberry farming. The main issues preventing the ground from being particularly suitable include high acidity, poor drainage, and steep hills. Areas designated as "extremely favourable" provide the optimal texture, depth, pH, drainage, and slope criteria for high mulberry agroforestry growth." Farmers in these regions should feel confident investing in mulberry agroforestry due to the favourable growth circumstances. The results might be used by the administration of the union region of Jammu and Kashmir to suggest to potential investors where they could make significant returns on their mulberry agroforestry investments. However, value chain participants may plant and engage in mulberry agroforestry in areas that are only slightly or moderately suitable. Liming and the use of farming strategies that increase drainage, such as excavating fallows and trenches, may be beneficial in areas with poor pH and drainage. Slopes greater than 40% should ideally not be favourable.

By ensuring a sustainable mulberry agroforestry system, applying resilient agricultural practises, and increasing the region's agricultural output and income, the study's findings have implications for achieving the Sustainable Development Goals (SDG) of the United Nations. (Agroforestry Module sustainable forest management 2021) It is suggested that in order to choose the best crop for the land, this study be carried out with more precise and accurate spatial data on many characteristics of the land and environment incorporated into the methodological framework used in the present study. This will result in a more manageable conclusion for the region's sustainable growth. Our method of evaluating the potential of a location for mulberry agroforestry allows us to rehabilitate damaged or abandoned agricultural fields. The scientific land evaluation approach can also be used to determine the best combination of crops and mulberry tree species to increase production and ensure the sustainable growth of the local community. (Rao 2022).

28.15 Conclusion

Using the FAO framework, GIS, and a multi-criteria evaluation approach, we assessed the suitability of the land in the Baramulla area for mulberry agroforestry. This study illustrated how the weights of physical factors and land suitability may be determined using an integrated AHP raster-weighted overlay GIS. The results of the research region indicated that, respectively, 16.99%, 28.96%, 38.18%, and 15.87% of the land were very, moderately, marginally, and not properly suitable for mulberry agroforestry. Using a multi-criteria method and raster weighted overlay analysis, mulberry agroforestry suitability mapping is created, and it is an effective tool for decision-makers to create a sustainable environmental management system and livelihood. As a result, this promotes farmer wellbeing and guarantees the study area's sustainability. An action plan needs to be created in order to increase land efficiency in the Baramulla district. Enhancing compatibility may be possible by using organic matter, pH-specific fertilizers, and better irrigation systems. Mulberry agroforestry can be a practical approach for tackling land degradation and climate change mitigation and adaptation while maintaining farmers' ecological and economic sustainability, given the significant potential for mulberry agroforestry expansion in the region as revealed in this study.

The results of the study could support more sustainable regional growth, better mulberry agroforestry management, and district-wide land use planning. The numerous field trips to this region revealed that the farmers still work their fields with conventional agricultural inputs and procedures. The framework was created by renowned international organisations that study land management after reviewing several studies. Given the region's limited amount of land and the fact that some of it are in critical condition, it is essential to act quickly and carefully to enhance the land's quality and discover alternative sources of food through local resource management in order to reduce the strain on agricultural land.

Mulberry agroforestry not only preserves soil fertility and supports biodiversity but also offers a sizable source of income for the local farmers. It must be done in marginal but currently appropriate locations. Making a plan to support the small and medium-sized agro-based businesses there will aid in the recovery of the region's economy and sources of income because the majority of the mulberry agroforestry produced there is organic. In order to reduce population pressure and support the long-term growth of the mulberry agroforestry sector, it is essential to create alternate livelihood policies for this area. The aforementioned comments highlight the fact that this study will assist decision-makers in creating mulberry agroforestry programmes to boost land production.

References

Agroforestry Module sustainable forest management (2021). http://www.fao.org/sustainable-forest-management/toolbox/modules/agroforestry/basic-knowledge/en/

Ahmad F, Goparaju L (2017) Geospatial approach for agroforestry suitability mapping to enhance livelihood and reduce poverty, FAO based documented procedure (Case study of Dumka District, Jharkhand, India). Biosci Biotechnol Res Asia 14(2):651–665. https://doi.org/10.13005//2491

Ahmad F, Uddin MM, Goparaju L (2019) Agroforestry suitability mapping of India geospatial approach based on FAO guidelines. Agrofor Syst 93(4):1319–1336

Ahmad F, Uddin MM, Goparaju L (2021) Tree suitability modelling and mapping in Nepal, a geospatial approach to scaling agroforestry. Model Earth Syst Env 7(1):169–179

Akinci H, Ozalp AY, Turgut B Agriculture (2013) Land use suitability analysis using Gis and Ahp technique. Comput Electron Inagricult 97:71–82

Alam A, Bhat MS, Maheen M (2020) Using landsat satellite data for assessing the land use and land cover change in Kashmir Valley. GeoJournal 85:1529–1543. https://doi.org/10.1007/s10708-019-10037-x

Anderson JR, Hardy EE, Roach JT, Witmer RE (1976) A land-use and land-cover classification system for use with remote sensor data. Us geological survey professional paper 964, Washington, DC

Bandooni SK (2004) Land resource management and development in hill areas. Research India Press

Barakat A, Ennaji W, El Jazouli A, Amediaz R, Touhami F (2017) Multivariate analysis and Gis-based soil suitability diagnosis for sustainable intensive agriculture in Beni-Moussa irrigated sub perimeter (Tadla plain, Morocco). Model Earth Syst Environ 3(1):3

Barbour JR, Read RA Barnes RL (2000) Moraceae-Mulberry Family Morus L. Mulberry. Woody Plant Seed Manual:728–732

Bera AK (2002) Land suitability evaluation for mulberry plantation: a remote sensing and Gis based approach, Govt. of India, (India). Cazri Campus. Rrssc, Isro/Department of Space

Bhandari S, Jhadav ST, Kumar S (2014) Land capability classification and crop suitability assessment in a watershed using Rs and Gis—a case study of watershed in Dehradun, Uttarakhand. Ssarsc Intl J Geosci Geoinform 1(1):1–15

Bhuvan, India Geo-Platform of the Indian Space Research Organization (ISRO) (2016). http://bhuvan.nrsc.gov.in/. Accessed 21 Aug 2016

Cengiz T, Akbulak C (2009) Application of analytical hierarchy process and geographic information systems in land-use suitability evaluation: a case study of Du¨Mrek Village (C¸ Anakkale, Turkey). Int J Sustain Develop World Ecol 16(4):286–294,2009

Census Of India (2011) District census handbook, Baramulla, Village And Town Directory. Directorate Of Census Operations, Baramulla 10(12b)

Chandio IA, Matori AN, Yusof K, Talpur MH, Aminu M (2014) Gis-based land suitability analysis of sustainable hillside development. Procedia Eng 77:87–94

Chen XPS (2016) An analysis of climate impact on landscape design. Atmos Clim Sci 6:475–481

Chen H, Liu G, Yang Y, Ye X, Shi Z (2010) Comprehensive evaluation of tobacco ecological suitability of Henan Province based on Gs. Agric Sci China 9(4):583–592

Directorate of Agriculture (2022) Soil maps in Kashmir. http://diragrikmr.nic.in/soil_maps.html

Dominati E, Mackay A, Bouma J, Green S (2016) An ecosystems approach to quantify soil performance for multiple outcomes: the future of land evaluation? Soil Sci Soc Am J 80(2):438–449

Ebrahimi MH, Nejadsoleymani MR, Mansouri D (2019) Land suitability map and ecological carrying capacity for the recognition of touristic zones in the Kalat region, Iran: a multi-criteria analysis based on AHP And GIS Asia-Pac. J Reg Sci 33:1–22

Ennaji, W., Barakat, A., El Baghdadi, M., Oumenskou, H., Aadraoui, M., Karroum, LA., Hilali, T (2018) A GIS-based multi-criteria land suitability analysis for sustainable agriculture in The Northeast Area of Tadla Plain (Morocco). J Earth Syst Sci 127(6):79

Erener A, Uzgeun HS (2008) Analysis on landslide hazard mapping methods: regression models versus weight rating. Intl Arch Photogram Remote Sens Spatial Sci 37:(B8) Beijing

Everest, T., Sungur, A., Ozcan, H. (2021) Determination of agricultural land suitability with a multiple-criteria decision-making method in North-Western Turkey. Int J Environ Sci 18(5):1073–1088

FAO (1976) A framework for land evaluation, Soil Bulletin 32. Food and Agriculture Organization of the United Nations, Rome

FAO (1983) Guidelines: land evaluation for rain fed agriculture. Fao Soils Bulletin 52. FAO, Rome

FAO (1985) Guidelines: land evaluation for irrigated agriculture. Soils Bulletin 55. Food and Agriculture Organization of the United Nations (Fao), Rome, Italy

FAO (1993) Guidelines for land use planning. Develop ment series 1. Food and Agriculture Organization of the United Nations, Rome

FAO (2007) Land evaluation towards a revised framework. Food and Agriculture Organization of the United Nations Italy, Rome

Feizizadeh B, Blaschke T (2012) Land suitability analysis for Tabriz County, Iran: a multi-criteria evaluation approach using Gis. J Environ Plan Manag 56:1–23

Garcia, J.I., Alvarado, A., Blanco J, Jimenez E, Maldonado AA, Cortés G (2014) Multi-attribute evaluation and selection of sites for agricultural product warehouses based on an analytic hierarchy process. Comput Electron Agricult:60–69

Gujree I, Ahmad I, Zhang F, Arshad A (2022) Innovative trend analysis of high-altitude climatology of Kashmir Valley, North-West Himalayas. Atmos 13(5):764. https://doi.org/10.3390/Atmos13050764

Habibie MI, Noguchi R, Shusuke M, Ahamed T (2019) Land suitability analysis for maize production in indonesia using satellite remote sensing and Gis-based multicriteria decision support system. Geojournal:1–31

Haider J (2013) Land suitability assessment for crop cultivation by using remote sensing and GIS. J Geograph Geol 5:65–67

Handique BK, Das PT, Goswami J, Goswami C, Singh PS, Prabhakar CJ, Bajpeyi CM, Raju PL (2016) Expansion of sericulture in India using geospatial tools and web technology. Curr Sci 111(8):1312–1318

Herzberg R, Pham TG, Kappas M, Wyss D, Tran CT (2019) Multi-criteria decision analysis for the land evaluation of potential agricultural land use types in a hilly area of Central Vietnam. Land 8:90

Hijmans RJ, Cameron SE, Parra JL, Jones PG, Jarvis A (2015) Very high resolution interpolated climate surfaces for global land areas. Int J Climatol 25

Hilali A (2018) GIS-based multi-criteria land suitability analysis for sustainable agriculture in the northeast area of Tadla plain (Morocco). J Earth Syst Sci 127(6):79

Holland C, Numata K, Rnjak-Kovacina J, Seib FP (2019) The biomedical use of silk: past, present. Future Adv Healthc Mater 8(1)

Hopkins LD (1977) Methods for generating land suitability maps: a comparative evaluation. J Am Inst Plann 43(4):386–400

Hussain M (2018) Systematic geography of Jammu and Kashmir. Rawat Publications, New Delhi

Joerin F, Theriault M, Musy A (2001) Using GIS and outranking multicriteria analysis for land-use suitability assessment. Int J Geogr Inf Sci 15(2):153–174

Kazemi H, Akinci H (2018) A land use suitability model for Rainfed farming by multi-criteria decision-making analysis (Mcda) And geographic information system (GIS). Ecol Eng 116:1

Khan AR (2008) Geography of Jammu and Kashmir. Gulshan Books, Srinagar

Kiker, G.A., Bridges T.S., Varghese, A., Seager T.P., Linkov ,V. (2005) Application of multicriteria decision analysis in environmental decision making. Integr Environ Asses Manag 2:95–108

Kuria D, Ngari D, Waithaka E (2011) Using geographic information system (GIS) to determine land suitability for Rice crop growing in the Tana Delta. J Geogr Region Plan 4:525–532

Kuta AA, Opaluwa YD, Zitta N, Ojatah E, Ugolo EM (2016) Application of GIS sieve mapping and overlay techniques for building site suitability analysis in part of Fut, Gidan Kwano, Minna, Nigeria. Indian J Sci Technol 46(9):1–6

Li Y, Wang Y, Qing H, Yang Y (2020) Calculation and evaluation of carbon footprint In Mulberry production: a case of Haining in China. Int J Environ Res Public Health 17(4):1339. https://doi.org/10.3390/ijerph17041339

Lone AL, Sen V (2014) Horticulture sector in Jammu and Kashmir. Economy Eur Acad J 11(2)

Malczewski J (2000) On the use of weighted linear combination method in GIS: common and best practice approaches. Trans GIS 4(1):5–22

Mazahreh SM, Bsoul HD (2018) Gis approach for assessment of land suitability for different land use alternatives in semi-arid environment in Jordan: case study (Al Gadeer Alabyad-Mafraq). Inform Process Agricult 2018:1–18

Mesgaran MB, Madani K, Hashemi H, Azadi P (2017) Iran's land suitability for agriculture. Sci Rep 7(1):7670

Mokarram M, Aminzadeh F (2010) Gis-based multicriteria land suitability evaluation using ordered weight averaging with fuzzy quantifier: a case study in Shavur plain, Iran. Int Arch Photogram Remote Sens Spat Inf Sci 38(2):508–512

Mushtaq R, Fayaz A, Singh A, Raja T, Singh H, Ahmed P (2022) Spatio temporal distribution of sericulture concentration in North Western Himalayan Region of Kashmir Valley, J & K, India. Sustainab Agri Food Environ Res:12. https://doi.org/10.7770/Safer-V12n1-Art2682

Mustafa AA, Singh M, Sahoo RN, Ahmed N, Khanna M (2011) Land suitability analysis for different crops: a multi criteria decision making approach using remote sensing and Gis. Researcher 3:61–84

Nair PKR, Toth GG (2016) Measuring agricultural sustainability in agroforestry systems. In: Lal R, Kraybill D, Hansen DO (eds) Climate change and multidimensional sustainability in African agriculture. Springer, Cham, pp 365–394

Nasa Data Access Viewer (2022). https://power.larc.nasa.gov/data-access-viewer/

Nath AJ, Kumar R, Devi NB (2021) Agroforestry land suitability analysis in the eastern Indian Himalayan region. Environment 4:100199. https://doi.org/10.1016/J.Envc.2021.100199

NBSS & Department of Agriculture (2020) Soils of India series: soils of Jammu and Kashmir for optimizing land use, NBSS Publication no 78. Government of J & K

Park S, Jeon S, Kim S, Choi C (2011) (2011) prediction and comparison of urban growth by land suitability index mapping using GIS And RS in South Korea. Landsc Urban Plan 99(2):104–114

Patil VD, Sankhua RN, Jain RK (2012) Analytic hierarchy process for evaluation of environmental factors for residential land use suitability. Int J Comput Eng Res 2(7):182–189

Pramanik MK (2016) Site suitability analysis for agricultural land use of Darjeeling District using AHP and GIS techniques. Model Earth Syst Environ. https://doi.org/10.1007/S40808-016-0116-8

Rao U (2022) Andhra Pradesh farmers reap carbon credits from agriculture. https://article.wn.com/view/2022/03/24/

Rasheed S, Venugopal K (2009) Land suitability assessment for selected crops in Vellore District based on agro ecological characterisation. J Indian Soc Remote Sens 37(4):615–629

Raza M, Din A, Ali M (1978) The valley of Kashmir, a geographical interpretation, land, vol I. Vikas Publication, New Delhi, pp 1–32

Romshoo SA, Rashid I, Altaf S, Dar GH (2020) Jammu and Kashmir state : an overview. Biodiversity of the Himalaya, Jammu and Kashmir State, pp 129–166

Rosenstock TS, Lamanna C, Chesterman S (2016) The scientific basis of climate-smart agriculture: a systematic review protocol. In: Ccafs (Climate Change, Agriculture and Food Security) Working Paper No. 136. Cgiar, Copenhagen

Saaty, TI (2012) Decision making for leaders: the analytic hierarchy process for decisions. In: A complex world (3 Rev edn). Rws Publications, Pittsburgh

Sajjad H, Nasreen I (2016) Assessing farm-level agricultural sustainability using site-specific indicators and sustainable livelihood security index: evidence from Vaishali District, India. Community Dev 47(5):602–619

Sánchez MD (2000) Mulberry: an exceptional forage available almost worldwide. http://www.fao/mulberry/papers/html/mulbwar2.htm

Shahid M, Rahman KU, Haider S, Gabriel HF, Khan AJ, Pham QB, Pande CB, Linh NTT, Anh DT (2021) Quantitative assessment of regional land use and climate change impact on runoff across Gilgit Watershed Environ. Earth Sci 80(2021):743

Shahram A (2010) Land suitability analysis for wheat using multicriteria evaluation and Gis method. Res J Biol Sci 5(9):601–605

Tercan E, Dereli MA (2020) Development of a land suitability model for citrus cultivation using Gis qnd multi-criteria assessment techniques in antalya Province of Turkey. Ecol Indic 117:106549

Torralba M, Fagerholm N, Burgess PJ, Moreno G, Plieninger TD (2016) European agroforestry systems enhance biodiversity and ecosystem services? a meta-analysis. Agric Ecosyst Environ 230(2016):150–161

Uniyall D, Kandwall D, Kimothi MM, Dhaundial VK(2014) Geospatial technology for identification of suitability sites for mulberry plantation: a case study of Uttarakhand state. In: Proceedings of ISPRS Tc viii international symposium on operational remote sensing application: opportunities, progress and challenges, December 9–12, Hyderabad, India, pp 1–4

USDA (1961) The agriculture handbook no. 210: land-capability classification. United States Department of Agriculture. United States Government Printing Office, Washington, DC

Wang Y, Xiang L, Wang C, Tang X (2013) Antidiabetic and antioxidant effects and phytochemicals of mulberry fruit (Morus Alba L.) polyphenol enhanced extract. Plos One 8(7)

Wani MA, Shaista N (2016) Available nitrogen (N) response through organic carbon And fertility index of Baramulla District of lesser Himalayas. Res J Agricult Sci 7(4/5):732–734

Ying X, Guang-Ming Z, Gui-Qiu C, Lin T, Ke-Lin W, Dao-You H (2007) Combining AHP with GIS in synthetic evaluation of eco-environment quality—a case study of Hunan Province. China Ecol Model 209(2–4):97–109

Yuan Q, Zhao L (2017) The Mulberry (Morus Alba L.) fruit – a review of characteristic components and health benefits. J Agric Food Chem 65:10383–10394. https://doi.org/10.1021/Acs.Jafc.7b03614

Zarkesh MK, Ghoddusi J, Zaredar N, Soltani MJ, Jafari S, Ghadirpour A (2010) Application of spatial analytical hierarchy process model in land use planning. J Food Agric Environ 8(2):970–975,2010

Zhang H, Ma F, Luo X, Li X (2018) Effects of mulberry fruit (Morus Alba L.) consumption on health outcomes: a mini-review antioxid. Basel Switz 7(5):69

Zomer RJ, Neufeldt H, Xu J, Ahrends A, Bossio D, Trabucco A, Van Noordwijk M, Wang M (2019) Global tree cover and biomass carbon on agricultural land: the contribution of agroforestry to global and national carbon budgets scientific reports 6:29–987

Climate Crisis and Coastal Risk Management

29

N. P. P. S. Nugawela, A. S. Mahaliyana, and G. Abhiram

Abstract

The global human population reached 8 billion recently and is expected to surge to 9.7 billion in 2050. Demographics and infrastructure on earth have expanded extensively in the last few centuries due to the advancement of science and technology and people seeking for better life conditions. Anthropogenic activities including urbanization, industrialization, and deforestation have made drastic changes in global climate causing issues in life and geospatial characteristics. Approximately, 25% of metropolitans are located within 100 km of shoreline and 0.6 billion people live in coastal urban centers which are below 10 m from the sea level. The majority of the world's busiest cities are in coastal areas to gain the advantage of accessibility and international transport and trade. The coastal zone, being the interface between the terrestrial and the oceanic environment, is predominantly and critically affected by the climate crisis. In addition to urbanized populations, a number of sensitive ecosystems such as coral reefs, mangroves, seagrass beds, salt marshes, etc. are also part of coastal regions. Climate changes including an increase in atmospheric concentrations of greenhouse gases, the rise of air and ocean temperatures, and glazier-melting and sea-level rise are risk multipliers. These issues are even threatening the peace and security of mankind due to conflicts emerging at national, regional, and international levels. Hence, strategic management of the climate crisis and its risk to the coastal zone is vital for the health and well-being of the global bio-geosphere. The Holistic approach in risk assessment and management is more promising rather focusing on single-factor instruments since the latter is well documented as a failure approach to address the crisis.

Keywords

Climate crisis · Coastal regions · Strategic management · Holistic approach · Risk assessment

29.1 Introduction to the Climate Crisis and Its Impact on Coastlines

The coastal region represents a highly dynamic ecological system where various geospheres interact, making it one of the most vibrant eco-

N. P. P. S. Nugawela · A. S. Mahaliyana
G. Abhiram (✉)
Faculty of Animal Science and Export Agriculture, Department of Animal Science, Uva Wellassa University of Sri Lanka, Badulla, Sri Lanka
e-mail: aqt17032@std.uwu.ac.lk; sachintha@uwu.ac.lk; abhiram@uwu.ac.lk

U. Chatterjee et al. (eds.), *Climate Crisis: Adaptive Approaches and Sustainability*, Sustainable Development Goals Series, https://doi.org/10.1007/978-3-031-44397-8_29

systems on the planet (Łabuz 2015). Studying the coastline is of utmost importance due to the significant concentration of major cities in low-lying areas. Approximately 40% of the global population resides within a 100-kilometer proximity to coastal regions (Muafiah 2019). On a global scale, over 600 million individuals currently inhabit coastal areas situated at an elevation of 10 meters or lower. According to existing migration trends, this figure is projected to surpass 1 billion by the year 2050 (Barnard et al. 2019; Manda and Klein 2019). All of them are at a high-risk state due to the climate crisis. The coastal zone is extremely dynamic and sensitive since this serves as a place where the land and the sea interact (Antunes do Carmo Antunes do Carmo 2018). The climate crisis is a major environmental issue that must be addressed internationally and has regional implications that can only be managed regionally with appropriate strategies (Kiguchi et al. 2021). The acceleration of climate change contributes to the rise in sea levels (SLR), as well as increases in sea surface temperature (SST), storm intensity, and storm surges. These factors play a significant role in coastal erosion, coastal flooding, sediment transport, and other consequences that impact the coastal ecosystem (Nicholls et al. 2007; Barnett and Adger 2014; Anarde et al. 2018; Bruno et al. 2020; Nicholls et al. 2021). Global climate change is expected to affect the frequency, intensity, timing, and distribution of hurricanes and tropical storms, along with the oceanic and atmospheric circulation patterns (Michener et al. 1997).

Coastal ecosystems play a crucial role in providing essential ecosystem services, such as regulating disturbances and nutrient cycling. They act as vital intermediaries between the land and the ocean, particularly in safeguarding against coastal erosion and flooding (Harley et al. 2006; Seabloom et al. 2013; Schuit et al. 2021). Coastal habitats, such as estuaries, coral reefs, and wetlands, are at risk of being impacted by the climate crisis. Even moderate levels of climate change have the potential to increase their vulnerability (Paw and Thia-Eng 1991). The population's concentration in coastal areas negatively influences the ecosystem as a result of changing land use patterns, overuse of natural resources, and the rising susceptibility to coastal dangers (Bruno et al. 2021). Overall, natural ecosystems are anticipated to severely impacted by climate change than societal systems (Smith 1997). According to the Intergovernmental Panel on Climate Change (IPCC) report, they published that climate change has been a key contributor to the loss of biodiversity across many ocean and coastal environments (very high confidence) (IPCC 2022). For instance, the Millennium Ecosystem Assessment investigated the protective functions of various coastal features such as coastal wetlands, mangroves, dunes, and broad beaches. This assessment encompassed multiple countries, including India, Bangladesh, and New Zealand, to understand their role in mitigating coastal flooding (van Slobbe et al. 2013). Mangroves and wetlands are likely to experience various effects such as alterations in their ecosystem, an increase in algal blooms, and a decline in water quality and oxygen levels. These effects are a result of the overall "High" impact of climate change variables on these habitats. (Amuzu et al. 2018). Since some coastal wetland regions are anticipated to vanish due to the effects of climate change in the twenty-first century, the loss of wetlands should be of great concern (Santoro et al. 2013).

In recent decades, the dynamics of traditional livelihoods have undergone rapid transformations due to various factors. These include the escalating frequency of climate change disasters, expanding urbanization, unplanned road development, industrialization, and population growth (Ataur Rahman and Rahman 2015). Before the introduction of typical Western practices, coastal zone communities around the world relied on their traditional knowledge to minimize disaster risks (Ataur Rahman and Rahman 2015). Especially, there has been an increase in the vulnerabilities and dangers in coastal areas since the middle of the previous century, and a more noticeable increase is expected after the middle of the present century (Antunes do Carmo Antunes do Carmo 2018). Through the twentieth century, the average global SLR was at a rate of 0.5–1.7 mm a year, while the average global SST has increased by roughly 0.6 °C since 1950, causing an increase in atmospheric warming near the coasts (Nicholls et al. 2007). Effective management of the coast is therefore essential from both a social and an ecological viewpoint (Barzehkar et al. 2021).

Since climate change was identified as a risk in the late 1980s, the primary emphasis has been on mitigation efforts, such as the reduction of atmospheric greenhouse gas emissions, rather than adaptation measures (Islam and Ghosh 2019). The past coastal defense initiatives have frequently been undertaken without taking into account the negative effects that certain measures could create on the stretches of the coast close to the zone of interest and without considering the hardening of the latter (Foti et al. 2020). In 1972, the United States implemented its first systematic coastal zone management strategy with the approval of the U.S. Congress of the Coastal Zone Management Act to manage coastal zones holistically several other countries began some sort of coastal management initiative in the late 1970s and early 1980s (Post and Lundin 1996; Kittinger and Ayers 2010). In the past, several shore protection installations, including "hard" (seawalls, groins, breakwaters, and revetments) and "soft" (dune construction and beach replenishment) constructions, have been built to reduce coastal erosion and floods (De Ruig 1998; Mimura and Nunn 1998; Parkinson and Ogurcak 2018).

Since any alteration in climatic processes would subsequently have an impact on the coastal zone, anticipating how coastal zones will adapt to short- and long-term climate fluctuations is one requirement for effective coastal management (Gornitz et al. 1994). However, rather than focusing on a single factor instrument, it is essential to dive into the multiple factors to identify the true picture of the climate crisis effect on coastal risk management. Therefore, the approach of this book chapter is to examine the actual vulnerability behind coastal risk management due to the climate crisis and discover the different management strategies to promote the sustainable development of the coastal area in a global context. This book chapter provides evidence for the significance of integrating knowledge and governance systems as well as including adaptive processes in policy action on climate change adaptation by analyzing available comprehensive coastal management and risk assessment frameworks. Hence, the main objectives of this book chapter were;

(a) To assess the potential economic, social, and environmental impacts in the coastal zone due to the climate crisis.
(b) To identify successful and innovative strategies to promote sustainable development of the coastal zone.
(c) To identify the community and the stakeholder involvement to minimize coastal risk and,
(d) To recognise the potential to strengthen the partnership between stakeholders and the community.

Ultimately it gives broader knowledge regarding climate crisis impacts on the coastal environment and encourages the community to promote sustainable development by emphasizing mitigation techniques of climate change impacts and future vulnerability.

29.2 Understanding the Risk

Coastal areas are currently seriously threatened by erosion and flooding concerns, and this tendency will worsen in the future as a result of climate change and expanding urbanization processes (Bruno et al. 2021). Typically, the coastal risk is a specific phenomenon, whether of natural or manmade origin resulting in the loss of lives, property, and productive capacity (Foti et al. 2020). Besides the climate crisis, other factors including geological, hydrodynamic, biological, and anthropogenic activities influence changes in the coastal zone (Łabuz 2015). However, the pace of coastal erosion/accretion along the world's sandy coastlines is already changing as a result of the effects of the climate crisis on hydrodynamic forces, including SLR, changes in wave conditions, storm surge, and changes in the sediment supply from rivers

(Cai et al. 2009; Dastgheib et al. 2018; Dastgheib et al. 2022). Further adaptation to the climate crisis has become a challenging issue globally (Dastgheib et al. 2022).

Coastal areas are particularly vulnerable to risks brought on by extreme weather in terms of their physical, ecological, and economical surroundings (Barzehkar et al. 2021). As per the report IPCC, there are 127 significant risks associated with different levels of climate change, covering various regions and sectors. These risks are further consolidated into eight overarching categories, which include low-lying coastal systems, terrestrial and ocean ecosystems, critical physical infrastructure, networks and services, living standards and equity, human health, food security, water security, and peace and migration (IPCC 2022). Especially, increased flood severity may have a massive impact on land use and cover changes, substantial damage to the infrastructure of economic, social, or cultural significance, and other effects including serious health problems from accidents or fatalities (Amuzu et al. 2018). Researchers must first determine which places are most susceptible to erosion, and measuring the extent of the erosion is crucial for coastal zone management and intervention measures (Mohanty et al. 2017).

29.2.1 Consequences of Climate Change

29.2.1.1 Sea Level Rise (SLR)

Sea level rise is a long-term threat, as it will take decades for the impacts of climate change to result in outcomes and it is mainly caused by global warming (Walsh et al. 2004; Klemas 2009). The expected SLR and increased storminess accelerate coastal processes, which may in turn cause massive ecological change and economic harm to coastal zones (Orvikut et al. 2003). The predictions for the twenty-first century are significantly more frightening and disturbing, calling for spectacular and unexpected SLR at rates up to eight times higher than the twentieth century (Djouder and Boutiba 2017). Sea level rise has been identified as one of the primary drivers behind historical changes in shorelines (Mohanty et al. 2017). The retreat of shorelines as a result of coastal erosion or persistent passive submersion is one of the expected effects of SLR (Le Cozannet et al. 2014). The progressive rise in sea levels has the potential to worsen erosion and gradually submerge coastal areas that are situated at lower elevations (Mohanty et al. 2017). According to prior research, the United States, Malaysia, Germany, Thailand, India, Sri Lanka, and other nations are particularly vulnerable to the effects of rising sea levels (Paw and Thia-Eng 1991; Teh and Voon 1992; Weerakkody 1997; Sterr 2008; Rao et al. 2009; Sallenger et al. 2012; Elsharouny 2016; Wdowinski et al. 2016) (Fig. 29.1a). The consequences of this rise in sea level include an increased risk of erosion, saltwater intrusion into rivers and underground aquifers, coastal flooding, changes in sediment deposition patterns, and the risk of inundation (Mimura, 1999). For example, coastal flooding and coastal erosion-like incidences reported in several countries due to the sea level rise (Heberger et al. 2011; Pramanik et al. 2016). Since 1988, the IPCC has taken proactive steps to address this type of problem at the interface of science and real-world decision-making (Hinkel et al. 2015). Therefore, SLR could potentially be one of the most detrimental impacts of the climate crisis, as it poses a significant risk to coastal regions.

29.2.1.2 Storm Surge

Storm surges are caused by the force of a hurricane's wind pushing water toward the shore (Klemas 2009). Storm surges are long gravity waves, the same kind of wave as tides and tsunamis (Murty et al. 1986). Storm surge heights are influenced by numerous factors, including the width and slope of the coastal shelf, the depth of the sea, the severity of the storm, and the speed and direction of the storm's forward motion. Tropical and extratropical storm activity may alter as a result of global warming driven by an increase in greenhouse gas concentration in the atmosphere (Zhang et al. 2000). Storm intensity and wind strength increase due to ocean warming, which also increases storm surge (Zhang 2019). Low-lying regions across the globe including China, Bangladesh, India, Sri Lanka,

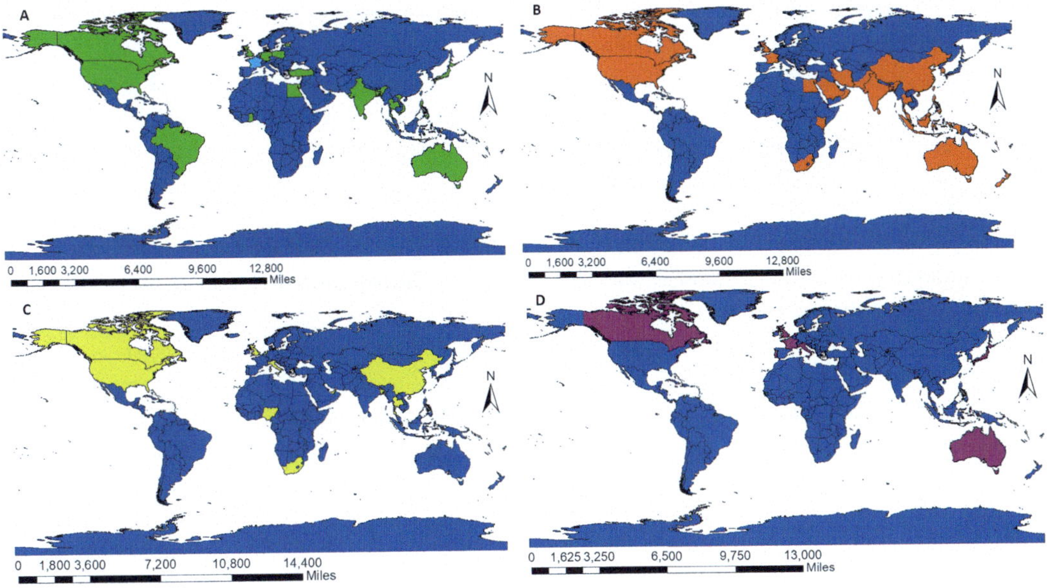

Fig. 29.1 Highly affected countries by climate crisis effects based on previously conducted studies; (**a**) Sea level rise (green), (**b**) Storm surge (Orange), (**c**) Sea surfaces temperature (Yellow), (**d**) Storm intensity (Purple)

and others, are highly susceptible to the devastating effects of storm surges (Neumann et al. 2015; Zhang et al. 2019) (Fig. 29.1b). These surges can create catastrophic flooding conditions and increases cyclone intensity causing significant damage and disruption to affected communities (McInnes et al. 2003; Wheeler 2009). However, studying the spatial extension of storm surges, also known as the "spatial footprint," which is the simultaneous exceeding of critical thresholds over a predetermined area as a result of a single storm surge occurrence, has received less attention (Enríquez 2020). The impact of climate change can lead to the occurrence of storm surges, which pose a severe threat to both coastal communities and ecosystems. Understanding the risk behind the storm surge is essential for minimizing the risk and managing the coastal environment.

29.2.1.3 Increased Sea Surface Temperature (SST)

Sea surface temperature serves as a gauge of dynamic processes taking place in the ocean (Schumann et al. 1995). Since the upper ocean stores a large portion of the residual heat that drives air anomalies, monitoring variations in SST is crucial to understand global climate change (Schumann et al. 1995). In addition to raising air temperature, greenhouse warming also enhances SST (Singh et al. 2001). Normally, tropical storms develop into cyclones when the sea surface temperature exceeds 26 °C (Singh et al. 2001). Increase of SST could result in coral bleaching and massive mortality even if coral does not possess any thermal adaptation or acclimatization (Nicholls et al. 2007). China, the United States and South Africa are a few regions affected by SST (Schumann et al. 1995; Jones et al. 2009; Zhang et al. 2010) (Fig. 29.1c). Tropical storms become cyclones, summer upwelling, seasonal variations and precipitation in coastal area are some influences of increasing SST (Schumann et al. 1995; Singh et al. 2001; Lenderink et al. 2009). Given that the rise in SST poses a significant threat to coastal ecosystems, it is crucial to accurately identify and assess the risks associated with it in order to effectively mitigate or reduce their impact. However, monitoring SST becomes a laborious task due to the complicated nature of measurements at sea and the size of the area to be covered (Gómez-Gesteira et al. 2008).

29.2.1.4 Storm Intensity

A storm event is a high-energy phenomenon that can be described by several factors, such as wave height, wind speed, and storm surge (Gervais et al. 2012). It is obvious that, at time horizons of a few decades, storm intensity may represent a more significant and immediate threat than sea level rise (Seabloom et al. 2013). Storms can also be thought of in terms of damage and costs since they contribute to a variety of risks, such as beach and dune erosion, dune overtopping and related flooding, and sporadically breaching the coastal system (Gervais et al. 2012). Italy, Canada, Netherlands and Portugal are a few countries affected by extreme storm events (Zhang et al. 2000; Ford et al. 2018; Armaroli et al. 2012) (Fig. 29.1d). Increasing probability of bridge failure, seasonal changes are some influences of storm intensity (Zhang et al. 2000; Anarde et al. 2018) Most of the coastal ecosystem resilience deteriorates due to erosion, flooding and acute disturbance associated with extreme storm events (Schuit et al. 2021). Extreme storm events can have a profound impact not only on coastal ecosystems but also on the well-being of coastal communities. As such, it is imperative to develop effective management strategies that can mitigate the severe consequences of these events.

29.3 Case Studies of Successful and Unsuccessful Management Strategies

Coastal risk management is designed to avoid and mitigate unacceptable risks in the coastal environment (Idier et al. 2013). Coastal planning measures are divided into three conceptual categories shoreline armoring, accommodation to stabilize erosion, and shoreline change and retreat from the shoreline (Dyckman et al. 2014). However, coastal managers and engineers have several options for addressing erosion, including beach nourishment, the construction of sturdy sea walls, or the simple moving or abandonment of coastal property as defence mechanisms (Smith et al. 2009). Depending on the severity of climate conditions, success and failure could decide on different management strategies. In this section, both successful and unsuccessful management strategies are discussed in order to achieve the second objective.

29.3.1 Integrated Coastal Management (ICM)

Integrated coastal management (ICM) is a set of concepts that offer a coherent, balanced approach in managing coastal resources and preserving sustainable development (Huang 1997). Integrated coastal management is being more recognized as a critical policy solution to address numerous issues that coastal zones are currently confronting (e.g., climate change) (Ojwang et al. 2017). It provides a framework for the discussion that follows about the benefits and drawbacks of using ICM as the foundation for coastal planning (Norman 2009). Excellent examples of ICM approaches that are significantly advancing coastal zone maintenance can be found on a global scale including Ireland (O'Hagan and Ballinger 2010), Malaysia (Mokhtar and Ghani Aziz 2003), Antigua, Barbuda (Ramsey et al. 2015), Brazil and Indonesia (Wever et al. 2012). Some risk managers advocate for more proactive and integrated planning and management of coastal zones as a powerful tool for enhancing sustainable development (Klein et al. 2001). The concept and application of ICM still have some limitations as the lack of long-term coastal protection planning procedures, the ongoing divergence between local action and national, regional, and organizational policy, the failure to address the environmental effects of coastal urbanization, better integrating regional and urban planning and integrated coastal management, are needed to address the increased challenge of expected climate change and lack of political commitment to implement ICM into practice in the face of serious development demands (Norman 2009). The development of integrated coastal management (ICM) has coincided with the recognition of the severity of the possible effects of climate change (Celliers et al. 2013).

29.3.1.1 Integrated Coastal Zone Management -Odisha

The Odisha Integrated Coastal Zone Management (ICZM) project employs a comprehensive and interactive framework, drawing upon multiple disciplines, to promote the sustainable management of coastal zones. It adopts an inclusive and consultative approach, involving all stakeholders in the decision-making process. The project follows an adaptive approach due to the dynamic nature of coastal ecosystems and aims to gather input from stakeholders through grassroots-level workshops, as their participation is a key element of the project. In order to enhance accessibility for stakeholders, the State Project Management Unit (SPMU) has introduced a toll-free hotline and established a dedicated website that features an interactive grievance redressal platform. The project aims to coordinate various coastal economic sectors to achieve optimal socio-economic outcomes, including resolving sectoral conflicts and mediating beneficial trade-offs, while promoting ICZM-compliant coastal infrastructures. Adopting a multi-sectoral approach, the project will guide key economic sectors, ensuring an effective coastal planning and management system is in place to oversee their activities.

29.3.2 Soft Engineering Defences

29.3.2.1 Beach Nourishments

Beach nourishment is the process of restoring or creating, and afterward maintaining, a sufficient protective or desirable recreational beach by manually or hydraulically depositing sand directly on an eroding shore (Speybroeck 2006) (Fig. 29.2a). This was first implemented in the US in the 1920s, and it is currently the main coastal management strategy for sandy beaches all across the US Atlantic and Gulf Coasts (Qiu and Gopalakrishnan 2018). Despite the fact that recreational benefits are also taken into account, beach nourishments are primarily motivated from a storm damage-reduction benefit perspective (Hanson et al. 2002). Especially, state resource agencies prioritize beach nourishment as a means to prevent the adverse impacts of hardened structures on the recreational and ecological values of coastal beaches (Peterson and Bishop 2005). Numerous types of research have been conducted to assess the efficacy of beach nourishment as a long-term and cost-effective management strategy to minimize climate change consequences (Parkinson and Ogurcak 2018). Yao et al. (Yao et al. 2022) confirmed that in the years between 2013 and 2019, when hurricanes were inactive, the Louisiana CPRA's beach nourishment project (which included sand replenishment, sand barrier installation, and dune vegetation planting) was successful in stopping coastal erosion and stabilizing the shoreline at Bay Champagne. However, they further indicated that Hurricanes Zeta and Ida made landfall on this beach for two consecutive years, but beach nourishment was not adequate to withstand them.

Beach nourishment occasionally has the explicit purpose of enhancing the comfort of its users, either by expanding the space accessible for beach activities or by modifying the grain size of its sediments (Pinto et al. 2020). Monitoring the sediment size composition is crucial throughout the project as restoring the original granulometry of beach sands becomes exceedingly difficult once they have been replaced with incompatible sediments. Additionally, maintaining natural sediments possesses great biological significance (Peterson and Bishop 2005). However, any assumption that nourishment projects are ecologically benign is derived from an insufficient and flawed body of science (Peterson and Bishop 2005). A high volume of sand may hurt the beach's quality if it is implemented there (Hanson et al. 2002). Besides, Yao et al. (2022) confirmed that beach nourishment is not strong enough to withstand intense hurricanes. According to Parkinson & Ogurcak, (2018), the findings conclusively reveal that beach nutrient application is not a viable method for reducing the consequences of climate change along the Florida panhandle.

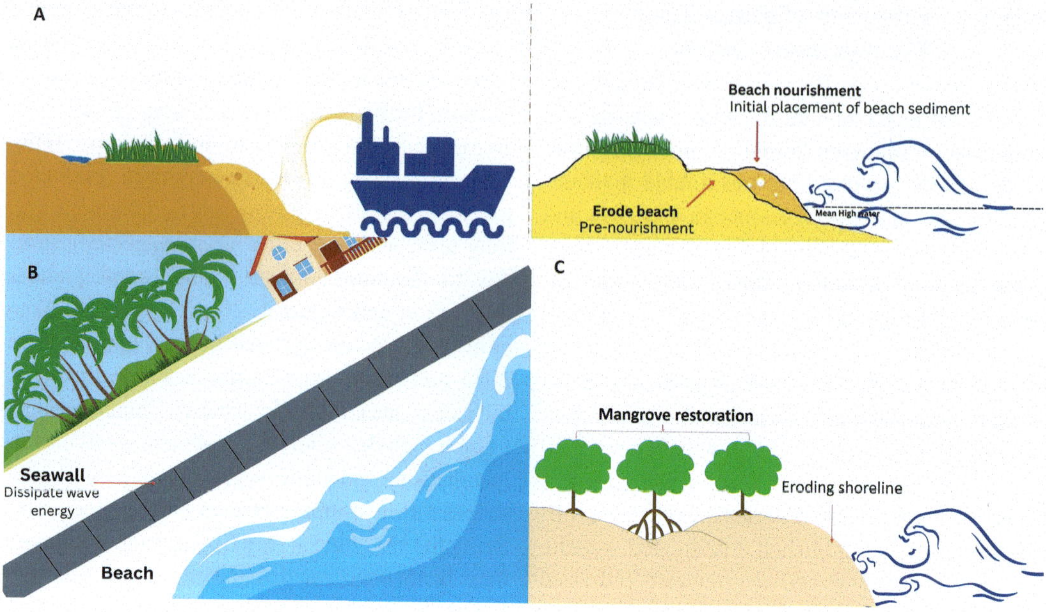

Fig. 29.2 Management strategies to avoid coastal erosion and coastal flooding: (**a**) Beach nourishment, (**b**) Seawall, and (**c**) Mangrove restoration

29.3.2.2 Duxbury Beach Dune Restoration Project, Massachusetts

The Duxbury Beach Dune Restoration Project in Massachusetts was named a Best Restored Beach for its resiliency, ecological, and recreational benefits. Duxbury Beach Reservation (DBR) completed a large dune restoration project in the winter of 2018–2019, which was the third step in the project's conception, design, and implementation. The project was designed by the Woods Hole Group and performed by SumCo Eco-Contracting, with oversight from DBR. The project covered the dune structure between the first and second over-sand vehicle crossovers, and it's part of the efforts to strengthen and preserve Duxbury Beach's resilience. The project, funded by a coastal resiliency grant and private donations, restored 3500 ft. of the dune, improving the resilience of 15 miles of shoreline while also providing nesting habitat for state and federally listed beach-nesting species. The project was a comprehensive effort involving community planning, permitting, and restoration of one of the narrowest and most at-risk sections of the barrier beach system. The restored dune is now less vulnerable to coastal erosion and provides continued access for recreationists, landowners, and emergency personnel.

29.3.3 Hard Engineering Defences

29.3.3.1 Seawalls

The trapezoidal shape of a seawall structure, where the base must be extended proportionally with its height, has been recognized as the reason construction costs increase geometrically with height (Ng and Mendelsohn 2005) (Fig. 29.2b). The fundamental benefit of a seawall is that it offers excellent defence against erosion and coastal floods (Savo et al. 2017; Alves et al. 2020). When constructed and maintained properly, it serves as a protective barrier, mitigating erosion between the land and the sea. This approach occupies less space, particularly when utilizing vertical seawalls, thereby reducing construction costs (Alves et al. 2020). According to Mase et al. (2015), the differences in wave height change for a seawall at a target location have a greater impact on overtopping rates than the differences in SLR trends (rapid, medium, or low).

According to a recent analysis, several sections of seawalls have crumbled due to being overly brittle, applied incorrectly, and unable to stop large waves from breaching (Alves et al. 2020). Therefore, retrofitting existing seawalls has the potential to improve coastal resilience by enabling them to adjust and react to changes in climatic conditions (Dong et al. 2020). Smooth and vertical seawalls are less effective as they dissipate wave energy instead of reflecting wave energy out to the sea (Shieh et al. 2010). Besides, hard protection can affect the degree to which a coastal ecosystem performs overall, deteriorate the quality of ecosystem services, and cause habitat loss or reduction in species diversity (Bongarts Lebbe et al. 2021).

29.3.3.2 The Gold Coast Seawall Project- Australia

The Gold Coast Seawall serves as an example of a long-term coastal protection project that clearly defines the roles and responsibilities of both public and private land tenure holders. The project was initiated by the local government to address the issue of uncoordinated coastal protection by establishing design standards, alignment, and allocating responsibilities, thus setting clear rules. While private property owners still need to be involved, this approach reduces their costs of procurement and risks associated with ad hoc actions by adjacent property owners (Ware and Banhalmi-Zakar 2017).

29.3.4 Ecosystem-Based Solutions

29.3.4.1 Mangrove Restoration

Mangroves are salt-tolerant woody plants that are frequently found in or near intertidal zones in tropical and subtropical areas (Arifanti 2020; Virni Budi Arifanti et al. 2022). Mangroves are known as an equilibrium system, therefore their power to resist change can be used to evaluate their protective role (Powell et al. 2007). They have evolved adaptations in their morphology, physiology, and reproductive techniques to exist in the harsh intertidal zones (Arifanti 2020). Especially, mangrove ecosystem is especially highly susceptible to the reduction of greenhouse emissions though can be used as a coastal risk management strategy (Virni Budi Arifanti et al. 2022). Tri et al. (1998) indicated that there is a powerful argument for mangrove rehabilitation as a crucial element of a sustainable coastal management approach (Fig. 29.2c). Pham et al. (2018) showed that locals are aware of the importance of mangrove forests in their livelihoods because they are currently at risk from tropical storms, which seem to be occurring more frequently as a result of climate change. However, mangrove planting failures are frequently the result of a lack of baseline knowledge regarding the factors that contribute to mangrove degradation or deforestation, problems with property ownership, weak law enforcement, and complex governance around mangrove replanting and conservation (Virni Budi Arifanti et al. 2022).

29.3.4.2 Mangrove Action Project-CBEMR Mangrove Restoration (around Globe)

The Community-Based Ecological Mangrove Restoration (CBEMR) technique, promoted by the Mangrove Action Project. This process involves engaging local stakeholders from the beginning and focuses on mitigating mangrove stressors while promoting natural regeneration (CBEMR 2021). Unlike many other restoration projects, CBEMR strives to work in harmony with nature by taking into account the unique ecology and biology of mangroves, thereby replicating natural processes to rehabilitate degraded areas. Utilizing natural regeneration not only results in a more diverse and resilient mangrove ecosystem, better equipped to withstand the effects of climate change, but also provides a potentially more cost-effective solution as it eliminates the need for costly nurseries and planting efforts.

29.3.5 Decision Supportive Systems (DSS)

Decision Support Systems (DSS) are regarded as advanced tools for addressing climate change-related challenges. They assist decision-makers in the sustainable management of natural resources, aiding in the implementation of ICM strategies and adaptation plans (Santoro et al. 2013). Typically, a DSS will include applicable environmental models, databases, and evaluation tools such as Graphic User Interface (GUI) and a Geographic Information System (GIS) which are necessary for spatial issues including erosion and flood risk (Zanuttigh et al. 2014). In the context of climate change in coastal zones, Decision Support Systems (DSS) represent a cutting-edge solution that bridges the gap between scientific information on climate change and the specific needs of regional and local coastal stakeholders. These innovative tools are designed to facilitate effective communication and provide the necessary information required for informed decision-making (Santoro et al. 2013). Santoro et al. (2013) indicated that this DSS may become a location to promote a deliberative climate through thoughtful and scientifically supported cognitive pathway design. The DSS enables the creation of flooding maps and hydraulic vulnerability maps starting from environmental data and scenarios, Wave transmission function, erosion, and flood risk models (Kane et al. 2014).

29.3.6 GIS-Based Tools

29.3.6.1 Satellite Remote Sensing (SRS)

Remote sensing extremely useful approach for shoreline change studies with a Geographic Information System (GIS) which provides high resolution, multi-spectral database, synoptic and repetitive data coverage and it is cost-effective in comparison to the conventional techniques (Chand and Acharya 2010). The significance of remote sensing technologies for tracking storms and their consequences on coastal populations has already been established (Klemas 2009). SRS is one of the greatest ways to look into how shorelines change over time, as it is affordable and has access to temporal data in the same areas on the ground (Warnasuriya et al. 2018). Applications of the SRS are the coastal topography and bathymetry, wind and wave regime, absolute SLR and land use information (Klein et al. 2001). Remote sensing satellites are one of the most effective and successful methods which enable monitoring of the earth's surface to identify the actual vulnerability of coastline changes in the present scenario.

29.4 Innovative Risk Management Strategies

Innovation encourages the development of novel concepts with novel approaches to address both current and emerging demands (De La Vega-Leinert and Nicholls 2008). The necessity of innovative strategies to mitigate the corresponding storm surge, erosion, and floods is accelerating due to climate change-induced coastal consequences, which are dramatically changing the discipline of coastal management planning (Dyckman et al. 2014). The potential of floating structures has emerged as a consequence of the quest toward novel methods for reducing especially flood risk in coastal areas, particularly deltas and estuaries of large rivers (Burcharth et al. 2014). The different productive innovative risk management strategies which apply in coastal risk management in a global context are shown in Table 29.1.

Stakeholder involvement in environmental management may contribute to the pursuit of innovative solutions to complicated problems involving coastal risk management (Eaton et al. 2021). Following a common vision of current and upcoming developments and their possible effects, stakeholder involvement can affect settlement transformation (Ciampa et al. 2021). Stakeholder engagement is achieved through educating important stakeholders about the risks of climate change to the fulfillment of national development priorities. Additionally, stakeholders are informed of the potential costs and benefits that can be gained through the implementation of adaptation strategies in both the short and long term (Elsharouny 2016). Making long-term decisions about the coast currently necessitates considering the complexities and uncertainties of

Table 29.1 Innovative management strategies and their preliminary objectives

Innovative management strategies	Objectives	References
Floating breakwaters (FBs) and wave energy converters (F-WECs) (Fig. 29.3a, b)	Flood risk mitigation	Burcharth et al. (2014)
Decision-supportive system (DSS)	Communicating risk	Santoro et al. (2013)
Beach drainage systems (BDS) (Fig. 29.3c)	Reduce coastal erosion	Archetti et al. (2019)
Sand engines (Fig. 29.3d)	Saving coastal areas from SLR The development of spaces that are both environmentally friendly and appealing for recreational purposes.	Stive et al. (2013)

climate change as well as the extent of stakeholder support (Tompkins et al. 2008). The Sendai Framework for Disaster Risk Reduction 2015–2030 also underlines the significance of exchanging ideas, experiences, and pragmatic guidance in the formulation of community-based, Nationalistic, regional, and universal plans and strategies with stakeholders (Bruno et al. 2021). However, competition between different stakeholders for land and marine uses frequently leads to violent conflicts and the collapse of the functional integrity of resource systems (Thia-Eng 1993). At various scales, actionable information and data are required for coastal risk managers, which also indicates that the relationship between policymakers and researchers must improve (Bongarts Lebbe et al. 2021).

Understanding how people interpret climate change in their daily lives, particularly in small-scale communities that depend on the environment for their survival requires a subjective assessment of community experiences as well as perceptions of climate change (Escudero and Mendoza 2021). The majority of community-based or co-management systems run into financial and enforcement issues, which keeps local consultation and engagement at a minimum (Wever et al. 2012). The community's involvement enables interactive consultations and debates between decision-makers and stakeholders, launching a collaboration between stakeholders as an intricate network of diverse representation and viewpoints (Ciampa et al. 2021). As an example, communities' active participation and empowerment in coastal management are still developing and fragmented, especially in Brazil and Indonesia (Wever et al. 2012). The involvement of local communities was essential to completing restoration tasks like dredging tidal channels, dispersing mangrove seedlings, elevating beds of soil, and monitoring (Escudero and Mendoza 2021).

29.5 Looking to the Future

Coastal areas are currently seriously threatened by erosion and flooding concerns, and this tendency will indeed worsen in the future as a result of climate change and expanding urbanization processes (Bruno et al. 2021). In the foreseeable future, there will inevitably be an increase in coastal risk as a result of the combined impacts of the coastal climate changes and its effects on the constantly expanding human use of the coastal zone (Ranasinghe and Jongejan 2018). Those who live in those coastal and low-lying areas are not well equipped to handle the issues that climate unpredictability may face shortly (Islam 2022). However, a more holistic, integrated, and, adaptive approach with iterative feedback and revision has been encouraged by the development of the broader philosophy of ICM, flood and coastal risk management approaches, and the policy framework through Shoreline Management Plans (SMP) (Creed et al. 2018). Climate change is now developing at a rate that is comparable to or greater than the expected lifespans of urban growth patterns and some capital assets, opposite to what planners have historically believed to be the case (Land et al. 2016). However, we continue to face new challenges in the areas of obser-

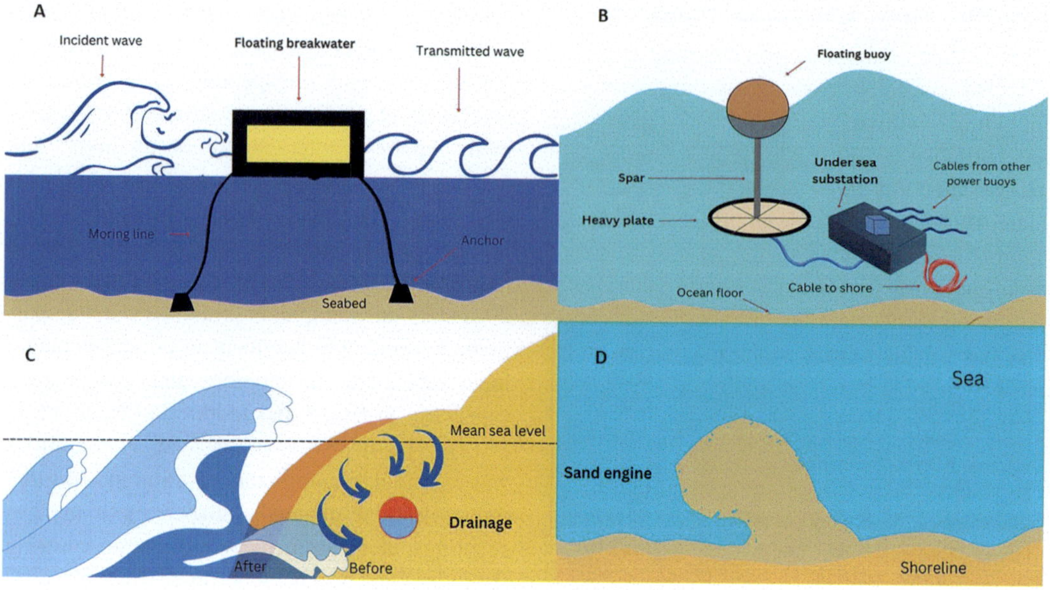

Fig. 29.3 Innovative risk management strategies: (**a**) Floating breakwaters (FBs), (**b**) Wave energy converters (F-WECs), (**c**) Beach Drainage Systems (BDS), and, (**d**) Sand engine Stakeholder engagement and community involvement

vations, modeling, and impact studies as we attempt to understand the causes of observed changes that are related to the climate (Cazenave and Cozannet 2014). Despite climate models having numerous flaws, the improvement of new mathematical tools and a greater understanding of atmospheric processes will make them faultless in tackling the challenges of complicated climatic processes in the future (Dastagir 2015).

It is crucial to take a holistic approach that considers both actual and perceived risks associated with potential response options to effectively manage risks and make informed decisions regarding climate change impacts (Simpson et al. 2021). The Holistic approach in risk assessment and management is more promising rather focusing on single-factor instruments since the latter is well documented as a failure approach to address the crisis. While adaptation to climate change and disaster risk reduction are distinct concepts, they share a common focus on reducing vulnerability and implementing sustainable and adaptable long-term strategies to mitigate the potential negative impacts (Schipper 2009). Currently, The IPCC is taking a holistic approach by producing extensive Assessment Reports that cover the current scientific, technical, and socio-economic understanding of climate change, as well as the potential impacts and future risks associated with it. These reports also explore potential strategies for mitigating the rate at which climate change is occurring. It is essential to adopt an integrated approach that involves regional, national, and local government agencies to effectively manage coastal zones. This requires a holistic consideration of various sectoral policies and their integration into coastal zone management issues (Nianthi and Shaw 2015).

29.6 Conclusion

The severity of the climate crisis and its impact on coastal environments and communities cannot be overstated. This book chapter examines various climate crisis impacts, including SLR, increasing SST, storm surges, and intensifying storms. Taking a holistic perspective, the chapter focuses on the effects of multiple climate crises effects and identifies the vulnerability level of each crisis effect. It is crucial to understand the current state of the climate crisis and assess its risks to coastal regions. Low-lying countries including Maldives, Vietnam, and the Netherlands are particularly vulnerable to SLR impacts. As global attention turns toward this issue, it is increasingly important to adopt a comprehensive approach to risk management that considers multiple factors. It introduces different strategies to mitigate the risks, building on existing approaches that have made strides in identifying risks and mitigation techniques. In addition to traditional methods, it is also critical to explore innovative strategies including sand engines, DSS, BDS, FB-s, and FWECs to address the unique challenges faced by coastal communities. By implementing these risk management strategies, coastal regions can better prepare for the potential impacts of climate change and build resilience in their livelihoods. To ensure a sustainable and resilient future for all, it is essential to continue researching and developing effective risk management strategies. Strong partnerships between governments, stakeholders, and local communities are necessary to foster cooperation and collaboration in addressing the challenges posed by the climate crisis.

References

"CBEMR" Mangrove Restoration - Mangrove Action Project (2021, October 22). Mangrove action project. https://mangroveactionproject.org/mangrove-restoration/

Alves B, Angnuureng DB, Morand P, Almar R (2020) A review on coastal erosion and flooding risks and best management practices in West Africa: what has been done and should be done. J Coast Conserv 24(3). https://doi.org/10.1007/s11852-020-00755-7

Amuzu J, Jallow BP, Kabo-Bah AT, Yaffa S (2018) The climate change vulnerability and risk management matrix for the coastal zone of The Gambia. Hydrology 5(1). https://doi.org/10.3390/hydrology5010014

Anarde KA, Kameshwar S, Irza JN, Nittrouer JA, Lorenzo-Trueba J, Padgett JE, Sebastian A, Bedient PB (2018) Impacts of hurricane storm surge on infrastructure vulnerability for an evolving coastal landscape. Nat Hazards Rev 19(1). https://doi.org/10.1061/(asce)nh.1527-6996.0000265

Antunes do Carmo JS (2018) Climate change, adaptation measures, and integrated coastal zone management: the New protection paradigm for the Portuguese coastal zone. J Coast Res 34(3):687–703. https://doi.org/10.2112/JCOASTRES-D-16-00165.1

Archetti R, Damiani L, Bianchini A, Romagnoli C, Abbiati M, Addona F, Airoldi L, Cantelli L, Gaeta MG, Guerrero M, Pellegrini M, Saccani C, Barbanente A, Saponieri A, Simeone V, Tarantino E, Bruno MF, Doglioni A, Motta Zanin G, … Molfetta MG (2019) Innovative strategies, monitoring and analysis of the coastal erosion risk: the stimare project. Proc Intl Off Polar Eng Conf 3:3836–3841

Arifanti VB (2020) Mangrove management and climate change: a review in Indonesia. IOP Confer Ser Earth Environ Sci 487(1). https://doi.org/10.1088/1755-1315/487/1/012022

Arifanti VB, Kauffman JB, Subarno JB, Ilman M, Tosiani A, Novita N (2022) Contributions of mangrove conservation and restoration to climate change mitigation in Indonesia. Glob Chang Biol 28(15):4523–4538. https://doi.org/10.1111/gcb.16216

Armaroli C, Ciavola P, Perini L, Calabrese L, Lorito S, Valentini A, Masina M (2012) Critical storm thresholds for significant morphological changes and damage along the Emilia-Romagna coastline, Italy. Geomorphology 143–144:34–51. https://doi.org/10.1016/j.geomorph.2011.09.006

Ataur Rahman M, Rahman S (2015) Natural and traditional defense mechanisms to reduce climate risks in coastal zones of Bangladesh. Weather Clim Extrem 7:84–95. https://doi.org/10.1016/j.wace.2014.12.004

Barnard PL, Erikson LH, Foxgrover AC, Hart JAF, Limber P, O'Neill AC, van Ormondt M, Vitousek S, Wood N, Hayden MK, Jones JM (2019) Dynamic flood modeling essential to assess the coastal impacts of climate change. Sci Rep 9(1):1–13. https://doi.org/10.1038/s41598-019-40742-z

Barnett J, Adger WN (2014) "Climate dangers and atoll countries": climate change (2003). Cities Global South Reader:221–226

Barzehkar M, Parnell KE, Soomere T, Dragovich D, Engström J (2021) Decision support tools, systems and indices for sustainable coastal planning and management: a review. Ocean Coast Manag 212(May). https://doi.org/10.1016/j.ocecoaman.2021.105813

Bongarts Lebbe T, Rey-Valette H, Chaumillon É, Camus G, Almar R, Cazenave A, Claudet J, Rocle N, Meur-Férec C, Viard F, Mercier D, Dupuy C, Ménard F, Rossel BA, Mullineaux L, Sicre MA, Zivian A, Gaill F, Euzen A (2021) Designing coastal adaptation strategies to tackle sea level rise. Front Mar Sci 8(November):1–13. https://doi.org/10.3389/fmars.2021.740602

Bruno MF, Saponieri A, Molfetta MG, Damiani L (2020) The DPSIR approach for coastal risk assessment under climate change at regional scale: the case of apulian coast (Italy). J Marine Sci Eng 8(7):1–18. https://doi.org/10.3390/JMSE8070531

Bruno MF, Motta Zanin G, Barbanente A, Damiani L (2021) Understanding the cognitive components of coastal risk assessment. J Marine Sci Eng 9(7). https://doi.org/10.3390/jmse9070780

Burcharth HF, Zanuttigh B, Andersen TL, Lara JL, Steendam GJ, Ruol P, Sergent P, Ostrowski R, Silva R, Martinelli L, Nørgaard JQH, Mendoza E, Simmonds D, Ohle N, Kappenberg J, Pan S, Nguyen DK, Toorman EA, Prinos P et al (2014) Innovative engineering solutions and best practices to mitigate coastal risk. Coast Risk Manage Changing Climate (Issue December). https://doi.org/10.1016/B978-0-12-397310-8.00003-8

Cai F, Su X, Liu J, Li B, Lei G (2009) Coastal erosion in China under the condition of global climate change and measures for its prevention. Prog Nat Sci 19(4):415–426. https://doi.org/10.1016/j.pnsc.2008.05.034

Cazenave A, Cozannet GL (2014) Sea level rise and its coastal impacts. Earth's Future 2(2):15–34. https://doi.org/10.1002/2013ef000188

Celliers L, Rosendo S, Coetzee I, Daniels G (2013) Pathways of integrated coastal management from national policy to local implementation: enabling climate change adaptation. Mar Policy 39(1):72–86. https://doi.org/10.1016/j.marpol.2012.10.005

Chand, P., & Acharya, P. (2010). Shoreline change and sea level rise along coast of Bhitarkanika wildlife sanctuary , Orissa : An analytical approach of remote sensing and statistical techniques. Intl J Geomat Geosci, 1(3), 436–455

Ciampa F, Seifollahi-aghmiuni S, Kalantari, Z., Sofia, C., & Ferreira, S. (2021). Flood mitigation in Mediterranean coastal regions: problems , Solut Stakeholder Involvement:1–22

Creed R, Baily B, Potts J, Bray M, Austin R (2018) Moving towards sustainable coasts: a critical evaluation of a stakeholder engagement group in successfully delivering the mechanism of adaptive management. Mar Policy 90:184–193. https://doi.org/10.1016/j.marpol.2017.12.009

Dastagir MR (2015) Modeling recent climate change induced extreme events in Bangladesh: a review. Weather and Climate Extremes 7:49–60. https://doi.org/10.1016/j.wace.2014.10.003

Dastgheib A, Jongejan R, Wickramanayake M, Ranasinghe R (2018) Regional scale risk-informed land-use planning using probabilistic coastline recession modelling and economical optimisation: East Coast of Sri Lanka. J Mar Sci Eng 6(4):1–32. https://doi.org/10.3390/jmse6040120

Dastgheib A, Martinez C, Udo K, Ranasinghe R (2022) Climate change driven shoreline change at Hasaki Beach Japan: a novel application of the probabilistic coastline recession (PCR) model. Coast Eng 172(December 2021):104079. https://doi.org/10.1016/j.coastaleng.2021.104079

De La Vega-Leinert AC, Nicholls RJ (2008) Potential implications of sea-level rise for Great Britain. J Coast Res 24(2):342–357. https://doi.org/10.2112/07A-0008.1

De Ruig JHM (1998) Coastline management in The Netherlands: human use versus natural dynamics. J Coast Conserv 4(2):127–134. https://doi.org/10.1007/BF02806504

Djouder F, Boutiba M (2017) Vulnerability assessment of coastal areas to sea level rise from the physical and socioeconomic parameters: case of the Gulf Coast of Bejaia. Algeria. https://doi.org/10.1007/s12517-017-3062-5

Dong, S., Abolfathi, S., Salauddin, M., Tan, Z. H., & Pearson, J. M. (2020). Enhancing climate resilience of vertical seawall with retrofitting – a physical modelling study. Appl Ocean Res, 103, 0–47. https://doi.org/10.1016/j.apor.2020.102331

Dyckman CS, St. John C, London JB (2014) Realizing managed retreat and innovation in state-level coastal management planning. Ocean and Coastal Management 102(PA):212–223. https://doi.org/10.1016/j.ocecoaman.2014.09.010

Eaton WM, Brasier KJ, Burbach ME, Whitmer W, Engle EW, Burnham M, Quimby B, Kumar Chaudhary A, Whitley H, Delozier J, Fowler LB, Wutich A, Bausch JC, Beresford M, Hinrichs CC, Burkhart-Kriesel C, Preisendanz HE, Williams C, Watson J, Weigle J (2021) A conceptual framework for social, behavioral, and environmental change through stakeholder engagement in water resource management. Soc Nat Resour 34(8):1111–1132. https://doi.org/10.1080/08941920.2021.1936717

Elsharouny MRMM (2016) Planning coastal areas and waterfronts for adaptation to climate change in developing countries. Procedia Environ Sci 34:348–359. https://doi.org/10.1016/j.proenv.2016.04.031

Enríquez AR (2020) Spatial footprints of storm surges along the global coastlines. J Geophys Res Oceans:1–15. https://doi.org/10.1029/2020JC016367

Escudero M, Mendoza E (2021) Community perception and adaptation to climate change in coastal areas of Mexico. Water (Switzerland) 13(18). https://doi.org/10.3390/w13182483

Ford JD, Couture N, Bell T, Clark DG (2018) Climate change and Canada's north coast: research trends, progress, and future directions. Environ Rev 26(1):82–92. https://doi.org/10.1139/er-2017-0027

Foti E, Ester R, Martina M (2020) Coastal defence techniques and climate change: a review. Rendiconti Lincei Scienze Fisiche e Naturali:October 2018. https://doi.org/10.1007/s12210-020-00877-y

Gervais, M., Balouin, Y., & Belon, R. (2012). Geomorphology morphological response and coastal dynamics associated with major storm events along the Gulf of Lions Coastline, France. Geomorphology, 143–144:69–80. https://doi.org/10.1016/j.geomorph.2011.07.035

Gómez-Gesteira M, deCastro M, Alvarez I, Gómez-Gesteira JL (2008) Coastal Sea surface temperature warming trend along the continental part of the Atlantic arc (1985–2005). J Geophys Res Oceans 113(4):1–9. https://doi.org/10.1029/2007JC004315

Gornitz VM, Daniels RC, White TW, Birdwell KR (1994) All use subject to JSTOR terms and conditions development assessment database: rise the of a coastal risk to vulnerability in the U. S. Southeast. Coastal Estuarine Res Federat 12:327–338

Hanson H, Brampton A, Capobianco M, Dette HH, Hamm L, Laustrup C, Lechuga A, Spanhoff R (2002) Beach nourishment projects, practices, and objectives – a European overview. Coast Eng 47(2):81–111. https://doi.org/10.1016/S0378-3839(02)00122-9

Harley CDG, Hughes AR, Hultgren KM, Miner BG, Sorte CJB, Thornber CS, Rodriguez LF, Tomanek L, Williams SL (2006) The impacts of climate change in coastal marine systems. Ecol Lett 9(2):228–241. https://doi.org/10.1111/j.1461-0248.2005.00871.x

Heberger M, Cooley H, Herrera P, Gleick PH, Moore E (2011) Potential impacts of increased coastal flooding in California due to sea-level rise. Clim Chang 109(SUPPL. 1):229–249. https://doi.org/10.1007/s10584-011-0308-1

Hinkel J, Jaeger C, Nicholls RJ, Lowe J, Renn O, Peijun S (2015) Sea-level rise scenarios and coastal risk management. Nat Clim Chang 5(3):188–190. https://doi.org/10.1038/nclimate2505

Huang JCK (1997) Climate change and integrated coastal management: a challenge for small Island nations. Ocean Coast Manag 37(1):95–107. https://doi.org/10.1016/S0964-5691(97)00042-2

Idier D, Rohmer J, Bulteau T, Delvallée E (2013) Development of an inverse method for coastal risk management. Nat Hazards Earth Syst Sci 13(4):999–1013. https://doi.org/10.5194/nhess-13-999-2013

IPCC (2022) Fact Sheets | Climate Change 2022: Impacts, Adaptation and Vulnerability. In Fact Sheet. https://www.ipcc.ch/report/ar6/wg2/about/factsheets/%0A

Islam M (2022) Is climate change migration an adjustment to extreme weather events? A study on the coastal areas of Bangladesh. Weather Climate Soc 14(4):1247–1260. https://doi.org/10.1175/WCAS-D-21-0124.1

Islam A, Ghosh RC (2019) Assessment on climate change adaptation: a study on coastal area of Khulna District in Bangladesh. Aust J Eng Innovat Technol 1(6):14–20. https://doi.org/10.34104/ajeit.019.14020

Jones BM, Arp CD, Jorgenson MT, Hinkel KM, Schmutz JA, Flint PL (2009) Increase in the rate and uniformity of coastline erosion in Arctic Alaska. Geophys Res Lett 36(3):1–5. https://doi.org/10.1029/2008GL036205

Kane IO, Vanderlinden JP, Baztan J, Touili N, Claus S (2014) Communicating risk through a DSS: a coastal risk centred empirical analysis. Coast Eng 87:240–248. https://doi.org/10.1016/j.coastaleng.2014.01.007

Kiguchi M, Takata K, Hanasaki N, Archevarahuprok B, Champathong A, Ikoma E, Jaikaeo C, Kaewrueng S, Kanae S, Kazama S, Kuraji K, Matsumoto K, Nakamura S, Nguyen-Le D, Noda K, Piamsa-Nga N, Raksapatcharawong M, Rangsiwanichpong P, Ritphring S et al (2021) A review of climate-change impact and adaptation studies for the water sector in Thailand. Environ Res Lett 16(2). https://doi.org/10.1088/1748-9326/abce80

Kittinger JN, Ayers AL (2010) Shoreline armoring, risk management, and coastal resilience under rising seas. Coast Manag 38(6):634–653. https://doi.org/10.1080/08920753.2010.529038

Klein RJT, Nicholls RJ, Ragoonaden S, Capobianco M, Aston J, Buckley EN (2001) Technological options for adaptation to climate change in coastal zones. J Coast Res 17(3):531–543

Klemas VV (2009) The role of remote sensing in predicting and determining coastal storm impacts. J Coast Res 25(6):1264–1275. https://doi.org/10.2112/08-1146.1

Łabuz TA (2015) Environmental impacts—coastal erosion and coastline changes:381–396. https://doi.org/10.1007/978-3-319-16006-1_20

Land LI, Butler WH, Deyle RE, Mutnansky C (2016) Accelerating sea level rise in Florida's coastal communities

Le Cozannet G, Garcin M, Yates M, Idier D, Meyssignac B (2014) Approaches to evaluate the recent impacts of sea-level rise on shoreline changes. Earth Sci Rev 138:47–60. https://doi.org/10.1016/j.earscirev.2014.08.005

Lenderink G, Meijgaard E, Selten F (2009) Intense coastal rainfall in The Netherlands in response to high sea surface temperatures: analysis of the event of august 2006 from the perspective of a changing climate. Clim Dyn 32(1):19–33. https://doi.org/10.1007/s00382-008-0366-x

Manda AK, Klein WA (2019) Adaptation strategies to address rising water tables in coastal environments under future climate and sea-level rise scenarios. Coastal zone management: global perspectives, regional processes, Local Issues. Elsevier Inc. https://doi.org/10.1016/B978-0-12-814350-6.00017-3

Mase H, Tamada T, Yasuda T, Karunarathna H, Reeve DE (2015) Analysis of climate change effects on seawall reliability. Coast Eng J 57(3):1–18. https://doi.org/10.1142/S0578563415500102

McInnes KL, Walsh KJE, Hubbert GD, Beer T (2003) Impact of sea-level rise and storm surges in a coastal community. Nat Hazards 30(2):187–207. https://doi.org/10.1023/A:1026118417752

Michener WK, Blood ER, Bildstein KL, Brinson MM, Gardner LR (1997) Climate change, hurricanes and tropical storms, and rising sea level in coastal wetlands. Ecol Appl 7(3):770. https://doi.org/10.2307/2269434

Mimura N, Nunn PD (1998) Trends of beach erosion and shoreline protection in rural Fiji. J Coast Res 14(1):37–46

Mohanty PC, Mahendra RS, Nayak RK, Kumar TS (2017) Impact of sea level rise and coastal slope on shoreline change along the Indian coast. Nat Hazards 89(3):1227–1238. https://doi.org/10.1007/s11069-017-3018-9

Mokhtar, M. B., & Ghani Aziz, S. A. B. A. (2003). Integrated coastal zone management using the ecosystems approach, some perspectives in Malaysia. Ocean Coast Manag, 46(5), 407–419. https://doi.org/10.1016/S0964-5691(03)00015-2

Muafiah AF (2019) No TitleΕΛΕΝΗ. Αγαη. 8(5):55

Murty TS, Flather RA, Henry RF (1986) The storm surge problem in the bay of Bengal. Prog Oceanogr 16(4):195–233. https://doi.org/10.1016/0079-6611(86)90039-X

National Beach Nourishment Database (n.d.) National beach nourishment database. https://asbpa.org/national-beach-nourishment-database/

Neumann B, Vafeidis AT, Zimmermann J, Nicholls RJ (2015) Future coastal population growth and exposure to sea-level rise and coastal flooding – a global assessment. PLoS One 10(3). https://doi.org/10.1371/journal.pone.0118571

Ng W, Mendelsohn R (2005) The impact of sea level rise on Singapore the impact of sea level rise on Singapore. July 2014. https://doi.org/10.1017/S1355770X04001706

Nianthi KWGR, Shaw R (2015) Climate change and its impact on coastal economy of Sri Lanka. Global Challenge 2015:1–21. http://citeseerx.ist.psu.edu/viewdoc/download?doi=10.1.1.455.6242&rep=rep1&type=pdf

Nicholls RJ, Wong PP, Burket VR, Codignotto J, Hay JE, McLean RF, Ragoonaden S, Woodroffe CD (2007) Coastal systems and low-lying areas. Climate change 2007: impacts. Adapt Vulnerab:315–356

Nicholls RJ, Lincke D, Hinkel J, Brown S, Vafeidis AT, Meyssignac B, Hanson SE, Merkens JL, Fang J (2021) A global analysis of subsidence, relative sea-level change and coastal flood exposure. Nature Clim Change 11(4). https://doi.org/10.1038/s41558-021-00993-z

Norman BJ (2009) Integrated coastal management to sustainable coastal planning thesis submission for doctor of philosophy October 2009. October, p 223

O'Hagan AM, Ballinger RC (2010) Implementing integrated coastal zone management in a national policy vacuum: local case studies from Ireland. Ocean Coast Manag 53(12):750–759. https://doi.org/10.1016/j.ocecoaman.2010.10.014

Ojwang L, Rosendo S, Celliers L, Obura D, Muiti A, Kamula J, Mwangi M (2017) Assessment of coastal governance for climate change adaptation in Kenya. Earth's Future 5(11):1119–1132. https://doi.org/10.1002/2017EF000595

Orvikut K, Jaagus J, Kont A, Ratas U, Rivis R (2003) Increasing activity of coastal processes associated with climate change in Estonia. J Coast Res 19(2):364–375

Parkinson RW, Ogurcak DE (2018) Beach nourishment is not a sustainable strategy to mitigate climate change. Estuar Coast Shelf Sci 212(February):203–209. https://doi.org/10.1016/j.ecss.2018.07.011

Paw JN, Thia-Eng C (1991) Climate changes and sea level rise: implications on coastal area utilization and management in south-East Asia. Ocean Shoreline Manage 15(3):205–232. https://doi.org/10.1016/0951-8312(91)90043-2

Peterson CH, Bishop MJ (2005) Assessing the environmental impacts of beach nourishment. Bioscience 55(10):887–896. https://doi.org/10.1641/0006-3568(2005)055[0887:ATEIOB]2.0.CO;2

Pham TD, Kaida N, Yoshino K, Nguyen XH, Nguyen HT, Bui DT (2018) Willingness to pay for mangrove restoration in the context of climate change in the cat Ba biosphere reserve, Vietnam. Ocean Coast Manag 163(June):269–277. https://doi.org/10.1016/j.ocecoaman.2018.07.005

Pinto CA, Silveira TM, Teixeira SB (2020) Beach nourishment practice in mainland Portugal (1950–2017): overview and retrospective. Ocean Coast Manag 192(May). https://doi.org/10.1016/j.ocecoaman.2020.105211

Post JC, Lundin CG (1996) Guidelines for integrated coastal zone management. Guidelines for integrated coastal zone management

Powell N, Osbeck M, Bach Tan S, Vu Canh T (2007) Mangrove restoration and rehabilitation for climate change adaptation in Vietnam. World Resour Report 22. http://www.worldresourcesreport.org

Pramanik MK, Biswas SS, Mondal B, Pal R (2016) Coastal vulnerability assessment of the predicted sea level rise in the coastal zone of Krishna–Godavari delta region, Andhra Pradesh, east coast of India. Environ Dev Sustain 18(6):1635–1655. https://doi.org/10.1007/s10668-015-9708-0

Project Description (n.d.). https://www.iczmpodisha.org/project-description.php

Qiu Y, Gopalakrishnan S (2018) Shoreline defense against climate change and capitalized impact of beach nourishment. J Environ Econ Manag 92:134–147. https://doi.org/10.1016/j.jeem.2018.08.013

Ramsey V, Cooper JAG, Yates KL (2015) Integrated coastal zone management and its potential application to Antigua and Barbuda. Ocean Coast Manag 118:259–274. https://doi.org/10.1016/j.ocecoaman.2015.04.017

Ranasinghe R, Jongejan R (2018) Climate change, coasts and coastal risk. J Marine Sci Eng 6(4):6–9. https://doi.org/10.3390/jmse6040141

Rao KN, Subraelu P, Rao TV, Malini BH, Ratheesh R, Bhattacharya S, Rajawat AS, Ajai (2009) Sea-level rise and coastal vulnerability: an assessment of Andhra Pradesh coast, India through remote sensing and GIS. J Coast Conserv 12(4):195–207. https://doi.org/10.1007/s11852-009-0042-2

Sallenger AH, Doran KS, Howd PA (2012) Hotspot of accelerated sea-level rise on the Atlantic coast of North America. Nat Clim Chang 2(12):884–888. https://doi.org/10.1038/nclimate1597

Santoro F, Tonino M, Torresan S, Critto A, Marcomini A (2013) Involve to improve: a participatory approach for a decision support system for coastal climate change impacts assessment. The north Adriatic case. Ocean Coast Manag 78:101–111. https://doi.org/10.1016/j.ocecoaman.2013.03.008

Savo V, Morton C, Lepofsky D (2017) Impacts of climate change for coastal fishers and implications for fisheries. Fish Fish 18(5):877–889. https://doi.org/10.1111/faf.12212

Schipper ELF (2009) Meeting at the crossroads?: exploring the linkages between climate change adaptation and disaster risk reduction. Clim Dev 1(1):16–30. https://doi.org/10.3763/cdev.2009.0004

Schuit M, Biryukov J, Beck K, Yolitz J, Bohannon J, Weaver W, Miller D, Holland B, Krause M, Freeburger D, Williams G, Abbott E, Wahl V, Ratnesar-shumate S, Dabisch P (2021). Isolates Ce Pt E Us Cr Ip T Ce Pt E D Us Cr. 1–32

Schumann EH, Cohen AL, Jury MR (1995) Coastal Sea surface temperature variability along the south coast of South Africa and the relationship to regional and global climate. J Mar Res 53(2):231–248. https://doi.org/10.1357/0022240953213205

Seabloom EW, Ruggiero P, Hacker SD, Mull J, Zarnetske P (2013) Invasive grasses, climate change, and exposure to storm-wave overtopping in coastal dune ecosystems. Glob Chang Biol 19(3):824–832. https://doi.org/10.1111/gcb.12078

Shieh A, Kandula S, Greenberg A, Kim C (2010) Seawall: performance isolation for cloud datacenter networks. 2nd USENIX workshop on hot topics in cloud computing, HotCloud 2010

Simpson NP, Mach KJ, Constable A, Hess J, Hogarth R, Howden M, Lawrence J, Lempert RJ, Muccione V, Mackey B, New MG, O'Neill B, Otto F, Pörtner HO, Reisinger A, Roberts D, Schmidt DN, Seneviratne S, Strongin S et al (2021) A framework for complex climate change risk assessment. One Earth 4(4):489–501. https://doi.org/10.1016/j.oneear.2021.03.005

Singh OP, Khan TMA, Aktar F, Sarker MA (2001) Recent sea level and sea surface temperature changes along the Maldives coast. Mar Geod 24(4):209–218. https://doi.org/10.1080/01490410175322785l

Smith JB (1997) Setting priorities for adapting to climate change. Glob Environ Chang 7(3):251–264. https://doi.org/10.1016/S0959-3780(97)00001-0

Smith MD, Slott JM, McNamara D, Brad Murray A (2009) Beach nourishment as a dynamic capital accumulation problem. J Environ Econ Manag 58(1):58–71. https://doi.org/10.1016/j.jeem.2008.07.011

Speybroeck J (2006) Defence alternative? A Review 435(April):419–435

Sterr H (2008) Assessment of vulnerability and adaptation to sea-level rise for the coastal zone of Germany. J Coast Res 24(2):380–393. https://doi.org/10.2112/07A-0011.1

Stive MJF, De Schipper MA, Luijendijk AP, Aarninkhof SGJ, Van Gelder-Maas C, Van Thiel De Vries JSM, De Vries S, Henriquez M, Marx S, Ranasinghe R (2013) A new alternative to saving our beaches from sea-level rise: the sand engine. J Coast Res 29(5):1001–1008. https://doi.org/10.2112/JCOASTRES-D-13-00070.1

Teh TS, Voon PK (1992) Impact of sea level rise on the coastal zone of Malaysia. Background paper, Malaysian National Conservation Strategy. Economic Planning Unit, Kuala Lumpur

Thia-Eng C (1993) Essential elements of integrated coastal zone management. Ocean Coast Manag 21(1–3):81–108. https://doi.org/10.1016/0964-5691(93)90021-P

Tompkins EL, Few R, Brown K (2008) Scenario-based stakeholder engagement: incorporating stakeholders preferences into coastal planning for climate change. J Environ Manag 88(4):1580–1592. https://doi.org/10.1016/j.jenvman.2007.07.025

Tri NH, Adger WN, Kelly PM (1998) Natural resource management in mitigating climate impacts: the example of mangrove restoration in Vietnam. Glob Environ Chang 8(1):49–61. https://doi.org/10.1016/S0959-3780(97)00023-X

van Slobbe E, de Vriend HJ, Aarninkhof S, Lulofs K, de Vries M, Dircke P (2013) Building with nature: in search of resilient storm surge protection strategies. Nat Hazards 65(1):947–966. https://doi.org/10.1007/s11069-012-0342-y

Walsh KJE, Betts H, Church J, Pittock AB, McInnes KL, Jacket DR, McDougall TJ (2004) Using sea level rise projections for urban planning in Australia. J Coast Res 20(2):586–598. https://doi.org/10.2112/1551-5036(2004)020[0586:USLRPF]2.0.CO;2

Ware D, Banhalmi-Zakar Z (2017) Funding coastal protection in a changing climate: lessons from three projects in Australia. ACCARNSI discussion paper, October, 14

Warnasuriya TWS, Gunaalan K, Gunasekara SS (2018) Google earth: a new resource for shoreline change estimation—case study from Jaffna peninsula, Sri Lanka. Mar Geod 41(6):546–580. https://doi.org/10.1080/01490419.2018.1509160

Wdowinski S, Bray R, Kirtman BP, Wu Z (2016) Increasing flooding hazard in coastal communities due to rising sea level: case study of Miami Beach, Florida. Ocean Coast Manag 126:1–8. https://doi.org/10.1016/j.ocecoaman.2016.03.002

Weerakkody U (1997) Potential impact of Accelerated Sea-level rise on beaches in Sri Lanka. J Coast Res 24:225–242

Wever L, Glaser M, Gorris P, Ferrol-Schulte D (2012) Decentralization and participation in integrated coastal management: policy lessons from Brazil and Indonesia. Ocean Coast Manag 66:63–72. https://doi.org/10.1016/j.ocecoaman.2012.05.001

Wheeler D (2009) Sea-level rise and storm surges a comparative analysis of impacts in developing countries. April

Yao Q, Cohen MCL, Liu KB, de Souza AV, Rodrigues E (2022) Nature versus humans in coastal environmental change: assessing the impacts of hurricanes zeta and Ida in the context of beach nourishment projects in the Mississippi River Delta. Remote Sens 14(11). https://doi.org/10.3390/rs14112598

Zanuttigh B, Simcic D, Bagli S, Bozzeda F, Pietrantoni L, Zagonari F, Hoggart S, Nicholls RJ (2014) THESEUS decision support system for coastal risk management. Coast Eng 87:218–239. https://doi.org/10.1016/j.coastaleng.2013.11.013

Zhang F (2019) Impacts of ocean warming, sea level rise, and coastline management on storm surge in a semienclosed bay. J Geophys Res Oceans. https://doi.org/10.1029/2019JC015445

Zhang K, Douglas BC, Leatherman SP (2000) Twentieth-century storm activity along the U.S. East Coast. J Clim 13(10):1748–1761. https://doi.org/10.1175/1520-0442(2000)013<1748:TCSAAT>2.0.CO;2

Zhang L, Wu L, Lin X, Wu D (2010) Modes and mechanisms of sea surface temperature low-frequency variations over the coastal China seas. J Geophys Res Oceans 115(8):1–13. https://doi.org/10.1029/2009JC006025

Zhang Y, Li G, Guo T (2019) Spatiotemporal patterns and evolution of storm surge threats along the southeastern coastline of China. Atmos 10(2). https://doi.org/10.3390/atmos10020061

The manufacturer's authorised representative in the EU is Springer Nature Customer Service Centre GmbH, Europaplatz 3, 69115 Heidelberg, Germany. If you have any concerns regarding our products, please contact ProductSafety@springernature.com

Printed and bound by CPI Group (UK) Ltd, Croydon, CR0 4YY
01/10/2026
02225855-0020